Advances in Oil and Gas Exploration & Production

Series Editors

Zhi Geng, Institute of Geology and Geophysics, Chinese Academy of Sciences, Beijing, China

Rudy Swennen, Department of Earth and Environmental Sciences, K.U. Leuven, Heverlee, Belgium

The book series *Advances in Oil and Gas Exploration & Production* is dedicated to publishing scientific monographs that cover a broad range of topics in geophysical and geological research related to both conventional and unconventional oil and gas systems. These topics are explored from both exploration and production perspectives.

The series aims to build a comprehensive and diverse library of reference works that reflect the current state of research on selected themes—ranging from specific techniques employed in petroleum geoscience to regionally focused studies. All volumes are authored and edited by leading experts actively working in their respective fields.

The *Advances in Oil and Gas Exploration & Production* series includes single- and multi-authored books, edited volumes, and conference proceedings. A Book Proposal Form is available on our website, or you may request one directly from the Publishing Editor, Qiao Shu, at **qiao.shu@ springernature.com**.

Yongsheng Ma · Maowen Li ·
Hairuo Qing

Editors

Petroleum Geology and Exploration of Deep Marine Strata in China

Editors
Yongsheng Ma
Sinopec Group Company
Beijing, China

Maowen Li
Sinopec Petroleum Exploration
and Production Research Institute
Beijing, China

Hairuo Qing
Department of Geology
University of Regina
Regina, SK, Canada

ISSN 2509-372X   ISSN 2509-3738 (electronic)
Advances in Oil and Gas Exploration & Production
ISBN 978-3-032-02495-4   ISBN 978-3-032-02496-1 (eBook)
https://doi.org/10.1007/978-3-032-02496-1

This Springer imprint is published by the registered company Springer Nature Switzerland AG
The registered company address is: Gewerbestrasse 11, 6330 Cham, Switzerland

If disposing of this product, please recycle the paper.

Preface

Deep-earth, deep-sea, and unconventional resources represent three major frontiers for the development of the oil and gas industry. Deep petroleum exploration typically refers to drilling operations targeting hydrocarbon reserves at depths exceeding 4500 m, often extending beyond 8000 m. The significance of venturing into these depths is underscored by the substantial portion of the world's remaining hydrocarbon resources believed to be located there. It is estimated that 60% of the world's new oil and gas discoveries are in deepwater formations; for China, resources in deep and ultradeep strata account for 34% of the country's total oil and gas resources.

Ultradeep exploration is a global endeavor, with significant activity in several key regions. Since 2016 China has drilled more than 90 ultradeep exploration wells, of which 25 have drilling depths greater than 8000 m. China's onshore Tarim Basin has emerged as a global leader in onshore ultra-deep exploration. The Tarim Basin is home to the groundbreaking "Deep Earth No. 1" project (the Shunbei oil and gas field). A landmark achievement was the Ta Ke 1 Well, which reached a depth of 10,910 m in 2025, and the discovered oil validates a new theoretical model that predicted hydrocarbon generation potential beyond 10,000 m. Sinopec also drilled the Yuejin 3-3XC Well to a design depth of 9472 m, making it Asia's deepest well at the time. By May 2025, 454 wells in the Tarim Basin exceed 8000 m in depth.

The offshore South Atlantic Margin is a premier global region for offshore ultradeep oil and gas exploration. The Guyana–Suriname Basin has seen a series of massive discoveries by a consortium led by ExxonMobil, with recoverable resources exceeding 17 billion barrels of oil equivalent. Recent projects such as the Whiptail project continue to develop these resources. Brazil's presalt fields, located beneath thick salt layers in very deep water, are another world-class hub, holding an estimated 20–30 billion barrels of recoverable resources. Active exploration and development are also ongoing in the U.S. portion of the Gulf of Mexico, off the coast of West Africa and East Africa, the Eastern Mediterranean, and Australia's Northwest Shelf.

According to available statistics, deep hydrocarbon reservoirs have been discovered in nearly one-quarter of the world's petroliferous basins, with passive marginal and foreland basins being the most enriched types. In terms of the geological age of deep reservoirs, the top five most productive sequences are in Paleogene, Upper Paleozoic, Cretaceous, Neogene, and Jurassic formations. However, currently known global oil and gas reservoirs buried greater than 6000 m are predominantly concentrated in Mesozoic and Cenozoic clastic formations. In contrast, the deep to ultradeep hydrocarbon reservoirs in China are distributed mainly in Paleozoic marine carbonate sequences within three major cratonic basins: Tarim, Sichuan, and Ordos.

China's deep to ultradeep marine strata in the Tarim, Sichuan, and Ordos Basins represent critical frontiers for hydrocarbon exploration. The exploration of these basins has been made possible by advancements in technology and geological understanding. The cratonic area of the Tarim Basin hosts Cambrian marine source rocks and Ordovician-Carboniferous transitional facies, with hydrocarbon generation centers shaped by multiple tectonic orogenies and paleouplifts. Hydrocarbon accumulation is concentrated in Carboniferous sandstone, Ordovician karstic weathered crust, and fractured carbonate reservoirs, often near regional caprocks or unconformities. Ultradeep strata (> 8000 m) exhibit extremely hard and complex fault systems, as seen in the Ta Ke 1 Well, which encountered formations that were more than 500 million years old and whose records were established in drilling engineering. Strike-slip faults in deep carbonate reservoirs (e.g., the Sinian Dengying Formation) control reservoir quality and hydrocarbon migration, with hard-linked fault zones offering the most productive reservoirs.

On the other hand, the Sinian Dengying Formation in central Sichuan is characterized by layered strike-slip fault systems that segment reservoirs into zones of varying quality. These faults act as primary migration pathways for hydrocarbons. Intersalt dolostones in the Triassic Feixianguan Formation (NE Sichuan) formed in evaporative lagoons, influenced by hypersaline fluids and contemporaneous seawater seepage, creating low-porosity reservoirs with complex diagenetic histories.

In contrast, the Ordovician strata in the Ordos Basin experienced multiple tectonic phases, including initial deposition, uplift, rapid subsidence, and differential erosion. These processes shaped hydrocarbon generation centers, with source rocks in the Ordovician Majiagou Formation entering high to postmature thermal evolution stages, dominated by gas generation. The basin's western margin features Ordovician under-salt carbonate reservoirs and organic-rich marine shale, contributing to large-scale gas accumulations.

Unlike deep to ultradeep reservoirs in other countries, deep to ultradeep reservoirs of ancient marine strata in China are characterized by multiphase tectonic modifications. The complexity of enrichment mechanisms and distribution patterns in these deep marine strata is governed by (1) the divergent tectonic evolution of the three major cratonic basins, (2) the unique sedimentological characteristics of Neoproterozoic–Paleozoic hydrocarbon-generating organic materials and diverse deposits such as dolomite and gypsum-salt rocks, and (3) the multiphase accumulation processes controlled by the long-term evolution of key geological elements and tectonic activity. Therefore, it is essential to conduct comparative studies across these basins, focusing on the specific characteristics of various types of deep to ultradeep marine reservoirs. Research should focus on the sedimentary-tectonic and thermal evolution history of each basin; investigate the formation and preservation mechanisms of large-scale reservoirs; dynamically reconstruct hydrocarbon generation, expulsion, retention, and multiphase accumulation processes; and ultimately reveal the enrichment mechanisms and distribution patterns of deep to ultradeep hydrocarbons to facilitate large-scale discoveries.

The prediction of diverse ultradeep reservoir types, along with corresponding drilling and stimulation technologies, remains underdeveloped, which poses significant economic risks in ultradeep exploration. The complexity of deep to ultradeep reservoir targets makes accurate identification and detailed characterization challenging. Seismic imaging technology involving broadband excitation, wide azimuth observations and high-density three-dimensional acquisition increases subsurface imaging precision, with applications in fault mapping and reservoir characterization across all three basins. Artificial intelligence-driven platforms streamline seismic data processing, fault identification, and reservoir modeling, reducing exploration timelines.

High-angle strike-slip faults complicate reservoir connectivity, whereas ultradeep drilling faces tool failures and formation instability, as seen in ultradeep projects in the Tarim Basin. Technological advancements such as the 12,000 m automated drilling rig and geotechnical engineering integration enable access to > 10,000 m of strata, as demonstrated in the Tarim Basin. This monograph is a synthesis of key geological features, exploration strategies, and challenges across these basins, and it integrates recent research and discoveries funded by China's Deep Earth Project (which has the aim of enabling exploration below 10,000 m by 2030), which targets unconventional resources and conducts crust–mantle interface studies. The use of integrated approaches combining seismic, geochemical, and microbiological data will refine predictions of hydrocarbon migration in composite basins. Extreme pressures, high temperatures, tool failures, and ecological impacts necessitate sustainable practices and advanced materials.

Chapter 1 briefly introduces the global trends in petroleum exploration, the current status of petroleum exploration in China, the scientific problems faced in understanding deep marine strata, and the technological challenges in drilling deep petroleum targets. It covers the recent progress that has been made in improved geological understanding and the main advances in related geophysical and engineering techniques, and it ends with a brief discussion of potential areas of future development.

Chapter 2 discusses the evolution of deep marine sedimentary filling and the development of high-quality hydrocarbon source rocks in cratonic basins in China. It describes the sedimentary filling processes in marine basins and their differential evolution paths in different geographic locations; describes methods and new results of paleoclimate, paleoenvironmental and paleoecological reconstruction; and interprets the results in terms of multilayered interactions and their relationships with global geological events. The highlights of this chapter are the accumulation mechanisms and models of organic-rich shales in the context of tectonostratigraphic frameworks.

Chapter 3 describes new theories and evidence for Sinian–Cambrian tectonic sedimentary differentiation and large-scale reservoir development in the Sichuan, Tarim and Ordos Basins, with an emphasis on the impact of intracontinental rifts on reef beach facies reservoirs, uplift and depression in cratons on granular beach reservoirs, and paleo-uplift on karst reservoirs.

Chapter 4 introduces the temporal and spatial framework of structural deformation in the three cratonic basins in response to the far-field influence of the supercontinent convergence events. Structural deformation types and mechanisms in cratons during Meso-Cenozoic periods are discussed, with an emphasis on intracratonic strike-slip faults at depth and their control on hydrocarbon accumulation. Finally, the influence of deformation type on petroleum occurrence in deep marine strata is discussed.

Chapter 5 focuses on geothermal regimes, differential thermal maturation, and hydrocarbon fluid distribution, reporting new data compilation from the three cratonic basins. Factors affecting the current and historic geothermal regimes are discussed, the thermal maturation patterns found in typical tectonic units are described, and maps illustrating how different types of petroleum fluids occur are presented.

Chapter 6 addresses isotopic and molecular tracers and the determination of effective source contributions to oil and gas accumulations in deep reservoirs. It discusses regionally significant deep petroleum source rocks, with a focus on their geochemical characteristics, and describes geological processes affecting the isotopic and molecular tracers commonly used for petroleum source correlation. Examples are introduced to demonstrate how to reconstruct petroleum source models for ultradeep reservoirs.

Chapter 7 introduces three major types of deep carbonate reservoirs, discusses fault–fluid coupling and its impact on deep carbonate reservoirs, describes the origin and distribution of the subsalt deep dolostone reservoirs, and delineates geological conditions for microbiolite buildups and their impact on hydrocarbon reservoir potential in deep buried ancient strata.

Chapter 8 discusses the dynamic mechanism and prediction methods of deep marine hydrocarbon accumulation in China. It begins with the introduction of the medium conditions and the temperature–pressure environment of deep hydrocarbon accumulations, basic geological features of deep hydrocarbon accumulations in China, and methods for the evaluation of hydrocarbon generation and expulsion dynamics and accumulation contributions of deep source rocks. Dynamic boundaries of deep hydrocarbon accumulations and their geological significance are discussed, as are dynamic models and methods for evaluating deep oil and gas accumulations.

Chapter 9 presents detailed case studies of deep oil and gas accumulations in the Tarim and Sichuan Basins, using Tahe and Shunbei as examples of liquid oil accumulation in ultradeep reservoirs and Puguang and Yuanba as representatives of deep gas fields. Following the consideration of key geological elements of deep petroleum systems, potential areas of oil and gas exploration in deep carbonate reservoirs are discussed.

Chapter 10 introduces the petroleum geology and development history of Puguang, a giant sour gas field in the Sichuan Basin, where liquid and solid sulfur precipitation has the potential to become a major problem following a decade of production. The results of the sulfur solubility experiments are discussed in light of the determining conditions for sulfur precipitation in sour gas reservoirs. Permeability results are also presented in the form of flow experiments involving two-phase systems (with water–liquid sulfur and gas–liquid sulfur) and three-phase systems (gas–water–liquid sulfur).

Chapter 11 focuses on the fluid flow mechanism and numerical simulation method in deep fracture-cavity reservoirs in the Tarim Basin. The geological models of deep fracture-cavity reservoir systems are discussed, the stress sensitivity characteristics and fluid flow law of fractured-vuggy media are introduced, the multifield coupled numerical simulation methods for fractured-cavity reservoirs are described, and the problem areas and future work for the commercial production of fault-controlled deep carbonate reservoirs are discussed.

Chapter 12 addresses deep shale gas accumulation, flow mechanisms and development methods in the Sichuan Basin. It starts with a brief introduction to hydrocarbon expulsion and retention, followed by a discussion of the formation and preservation of effective porosity values in deep shales, as well as the dynamic evolution of a source–reservoir system and in situ hydrocarbon accumulation. Fluid phases and flow mechanisms in deep shale reservoirs are reviewed, together with the dynamic evaluation and numerical modeling of deep shale gas production.

Chapter 13 focuses on safe and efficient drilling and completion control methods for high-temperature–high-pressure oil and gas wells, methods for predicting formation pressures in deep marine shale reservoirs, methods for accelerating axial–torsional coupled impact rock breaking in hard formations, methods for controlling shale gas well casing–cement sheath integrity, methods for controlling wellbore stability in high-stress fractured formations, and methods for stimulating volume fracturing in tight shale/carbonate reservoirs.

Ultradeep petroleum exploration is a high-stakes endeavor fueled by the need to secure future energy supplies. It is a field defined by extraordinary technological ambition and equally formidable challenges related to extreme environments, economic constraints, and safety imperatives. China's actions in the Tarim, Sichuan, and Ordos Basins exemplify the country's leadership in unlocking deep to ultradeep marine hydrocarbons. Breakthroughs in seismic imaging, ultradeep drilling, and AI-driven reservoir modeling have revolutionized exploration, while tectonic complexity and reservoir heterogeneity remain key challenges. Future success will depend on a relentless focus on technological innovation, unwavering commitment to safety and environmental stewardship, and collaborative efforts to close critical knowledge gaps. The journey into the deep earth is as much about advancing human engineering and geoscience as it is about managing profound risk.

Beijing, China   Yongsheng Ma
Beijing, China   Maowen Li
Regina, Canada   Hairuo Qing

Acknowledgments

The coeditors and authors of this monograph sincerely thank the following colleagues and peers for their critical reading and helpful suggestions for an earlier version of the individual chapters included here: Dr. Baojun Bai, Missouri University of Science and Technology; Dr. Zhuoheng Chen, Geological Survey of Canada; Dr. Yuan Di, Peking University; Dr. Gregor P. Eberli, University of Miami; Dr. Martin Fowler, Geological Survey of Canada; Dr. Haiping Huang, University of Calgary; Dr. Zhiyang Li, Texas A&M International University; Dr. Hairuo Qing, University of Regina; Dr. Liang Xue, China University of Petroleum; Dr. Mengwei Zhao, Sinopec Canada Energy Limited; Dr. Donggao Zhao, University of Texas at Austin; Drs. Fang Wan, Ping Xiong, Kebin Zhao, Xiaoxiao Ma, and Enze Wang made great efforts in providing administrative support and in contacting contributing authors and potential reviewers, without which everything is almost impossible.

We sincerely thank the Natural Science Foundation of China, for generous financial support (Grants U24B6001 and U19B6003), and Sinopec Group Company, for match funding and permission to publish the work.

Contents

1 Status and Outlook of Deep Marine Petroleum Exploration and Development in China 1
Yongsheng Ma, Maowen Li, Xiaoxiao Ma, and Enze Wang

2 Evolution of Deep Marine Sedimentary Filling and Development of High-Quality Hydrocarbon Source Rocks in Cratonic Basins, China 35
Yifan Li, Tailiang Fan, Juye Shi, Ming Jiang, Tan Zhang, Mingzhi Kuang, and Wangwei Liu

3 Tectonic–Sedimentary Differentiation of Cratonic Basins in China and Its Controlling Effects on the Development of Large-Scale Carbonate Reservoirs 57
Shuangjian Li, Yingqiang Li, Shoutao Peng, Yunqing Hao, Juntao Zhang, Jingbin Wang, and Xuhui Xu

4 Structural Transformation in Cratonic Basins and Influence on Oil and Gas Accumulation in Deep Marine Strata 89
Xuhui Xu, Jianlin Lu, Chenyu Zhang, Mingzhe Deng, Shang Deng, Peng Deng, Shuaiqiang Shan, and Zongxin Zuo

5 Geothermal and Geopressure Regimes in Ultra-Deep Sedimentary Strata, Tarim and Sichuan Basins 121
Nansheng Qiu, Jian Chang, Anlai Ma, Yifeng Liu, Qianqian Feng, and Dan Li

6 Diversity and Effectiveness of Deep Marine Source Rocks in the Sichuan and Tarim Basins 157
Maowen Li, Enze Wang, Xiaoxiao Ma, Qianglu Chen, Liyu Zhang, and Kun Yuan

7 Multiple Types of Deep Carbonate Reservoirs in China: Consensuses and New Developments 177
Dongya Zhu, Jingbin Wang, Donghua You, Kaibo Shi, Juntao Zhang, Chongyang Wu, Qian Ding, Huili Li, Xunyu Cai, Quanyou Liu, Zhiliang He, and Bo Liu

8   **Dynamic Mechanism and Prediction Methods for Deep Marine Hydrocarbon Accumulation** 213
Xiongqi Pang, Hong Pang, Tao Hu, and Junqing Chen

9   **Typical Examples and Exploration Potentials of Large-Scale Petroleum Accumulations in Deep and Ultradeep Carbonate Reservoirs in China** 261
Xunyu Cai, Huili Li, Zicheng Cao, Bisong Li, Jun Han, Xiuxiang Zhu, Qingfeng Zhang, Zuxin Xu, and Chongyang Wu

10 **Geological Characteristics and Development Techniques of Ultra-deep Carbonate Sour Gas Reservoirs: A Case Study of the Puguang Gas Field** 297
Daqian Zeng, Tong Li, Yuchun You, Rui Zhang, Qian Li, Zhaojie Song, Song Peng, and Cheng Zhang

11 **Fluid Flow Mechanism and Numerical Simulation Method in Deep Fracture-Cavity Reservoirs, Tarim Basin** 331
Zhijiang Kang, Yun Zhang, Hongkai Li, Dawei Wu, Ke Sun, and Ziyan Deng

12 **Deep Shale Gas Accumulation, Flow Mechanisms and Development Methods in the Sichuan Basin** 357
Dongfeng Hu, Xianming Xiao, Weihong Wang, Ruobing Liu, Baojian Shen, Jing Wang, and Tao Yuan

13 **Advances in Drilling and Completion for Deep Marine Reservoirs** 385
Yijin Zeng, Gensheng Li, Bing Hou, Huaizhong Shi, Mao Sheng, Daqi Li, Junhai Chen, Rengguang Liu, Yang Xia, Shiming Wei, Wenhao He, and Yayun Zhang

Contributors

Xunyu Cai China Petroleum and Chemical Corporation (Sinopec), Beijing, China

Zicheng Cao Northwest Oilfield Company, SINOPEC, Urumqi, China

Jian Chang China University of Petroleum, Beijing, China

Junhai Chen Sinopec, Beijing, China

Junqing Chen State Key Laboratory of Petroleum Resources and Engineering, China University of Petroleum (Beijing), Beijing, China;
College of Geosciences, China University of Petroleum (Beijing), Beijing, China

Qianglu Chen State Key Laboratory of Shale Oil and Gas Enrichment Mechanisms and Efficient Development, SINOPEC, Beijing, China;
Petroleum Exploration and Production Research Institute, SINOPEC, Beijing, China

Mingzhe Deng Sinopec Petroleum Exploration and Production Research Institute, Beijing, China

Peng Deng Sinopec Petroleum Exploration and Production Research Institute, Beijing, China

Shang Deng Sinopec Petroleum Exploration and Production Research Institute, Beijing, China

Ziyan Deng China University of Geosciences, Beijing, China

Qian Ding Petroleum Exploration and Production Research Institute of SINOPEC, Beijing, China;
Key Laboratory of Geology and Resources in Deep Stratum of SINOPEC, Beijing, China

Tailiang Fan School of Energy Resources, China University of Geosciences (Beijing), Beijing, China

Qianqian Feng China University of Petroleum, Beijing, China

Jun Han Northwest Oilfield Company, SINOPEC, Urumqi, China

Yunqing Hao Key Laboratory of Geology and Resources in Deep Strata, SINOPEC, Beijing, China;

Petroleum Exploration and Production Research Institute, SINOPEC, Beijing, China

Wenhao He China University of Petroleum-Beijing, Beijing, China

Zhiliang He Petroleum Exploration and Production Research Institute of SINOPEC, Beijing, China

Bing Hou Sinopec, Beijing, China;
China University of Petroleum-Beijing, Beijing, China;
China University of Petroleum-Beijing atKaramay, Karamay, China

Dongfeng Hu State Key Laboratory of Shale Oil and Gas Enrichment Mechanisms and Efficient Development, SINOPEC, Beijing, China;
Sinopec Exploration Company, Chengdu, Sichuan, China

Tao Hu State Key Laboratory of Petroleum Resources and Engineering, China University of Petroleum (Beijing), Beijing, China;
College of Geosciences, China University of Petroleum (Beijing), Beijing, China

Ming Jiang School of Energy Resources, China University of Geosciences (Beijing), Beijing, China

Zhijiang Kang Sinopec Petroleum Exploration and Production Research Institute, Beijing, China

Mingzhi Kuang School of Energy Resources, Chengdu University of Technology, Chengdu, China

Bisong Li Exploration Company, SINOPEC, Chengdu, China

Dan Li China University of Petroleum, Beijing, China

Daqi Li Sinopec, Beijing, China

Gensheng Li China University of Petroleum-Beijing, Beijing, China

Hongkai Li Sinopec Petroleum Exploration and Production Research Institute, Beijing, China

Huili Li Petroleum Exploration and Production Research Institute of SINOPEC, Beijing, China;
Key Laboratory of Geology and Resources in Deep Stratum of SINOPEC, Beijing, China

Maowen Li State Key Laboratory of Shale Oil and Gas Accumulation Mechanisms and Effective Development, SINOPEC, Beijing, China;
Sinopec Petroleum Exploration and Production Research Institute, Beijing, China

Qian Li State Key Laboratory of Shale Oil and Gas Accumulation Mechanisms and Effective Development, Beijing, China;
Sinopec Petroleum Exploration and Production Research Institute, Beijing, China

Shuangjian Li Key Laboratory of Geology and Resources in Deep Strata, SINOPEC, Beijing, China;
Petroleum Exploration and Production Research Institute, SINOPEC, Beijing, China

Tong Li State Key Laboratory of Shale Oil and Gas Accumulation Mechanisms and Effective Development, Beijing, China;
Sinopec Petroleum Exploration and Production Research Institute, Beijing, China

Yifan Li School of Energy Resources, China University of Geosciences (Beijing), Beijing, China

Yingqiang Li Key Laboratory of Geology and Resources in Deep Strata, SINOPEC, Beijing, China;
Petroleum Exploration and Production Research Institute, SINOPEC, Beijing, China

Bo Liu Peking University, Beijing, China

Quanyou Liu Peking University, Beijing, China

Rengguang Liu Sinopec, Beijing, China

Ruobing Liu State Key Laboratory of Shale Oil and Gas Enrichment Mechanisms and Efficient Development, SINOPEC, Beijing, China;
Sinopec Exploration Company, Chengdu, Sichuan, China

Wangwei Liu Wuxi Institute of Petroleum Geology, Sinopec Petroleum Exploration and Development Research Institute, Wuxi, China

Yifeng Liu Petroleum Exploration and Production Research Institute, SINOPEC, Beijing, China

Jianlin Lu Sinopec Petroleum Exploration and Production Research Institute, Beijing, China;
Sinopec Petroleum Exploration and Production Research Institute, Beijing, China

Anlai Ma Zhejiang University, Zhoushan, China

Xiaoxiao Ma State Key Laboratory of Shale Oil and Gas Accumulation Mechanisms and Effective Development, Beijing, China;
Sinopec Petroleum Exploration and Production Research Institute, Beijing, China;
State Key Laboratory of Shale Oil and Gas Enrichment Mechanisms and Efficient Development, SINOPEC, Beijing, China

Yongsheng Ma State Key Laboratory of Shale Oil and Gas Accumulation Mechanisms and Effective Development, Beijing, China;
Sinopec Group Company, Beijing, China

Hong Pang State Key Laboratory of Petroleum Resources and Engineering, China University of Petroleum (Beijing), Beijing, China;

College of Geosciences, China University of Petroleum (Beijing), Beijing, China

Xiongqi Pang State Key Laboratory of Petroleum Resources and Engineering, China University of Petroleum (Beijing), Beijing, China;
College of Geosciences, China University of Petroleum (Beijing), Beijing, China

Shoutao Peng Key Laboratory of Geology and Resources in Deep Strata, SINOPEC, Beijing, China

Song Peng Exploration and Development Research Institute of SINOPEC Zhongyuan Oilfield, Puyang, Henan, China

Nansheng Qiu China University of Petroleum, Beijing, China

Shuaiqiang Shan Sinopec Petroleum Exploration and Production Research Institute, Beijing, China

Baojian Shen State Key Laboratory of Shale Oil and Gas Enrichment Mechanisms and Efficient Development, SINOPEC, Beijing, China;
Exploration and Production Reservoir Institute, Key Laboratory for Marine Oil and Gas Exploitation, Sinopec, Beijing, China

Mao Sheng China University of Petroleum-Beijing, Beijing, China

Huaizhong Shi China University of Petroleum-Beijing, Beijing, China

Juye Shi School of Energy Resources, China University of Geosciences (Beijing), Beijing, China

Kaibo Shi Peking University, Beijing, China

Zhaojie Song China University of Petroleum—Beijing, Beijing, China

Ke Sun China University of Petroleum, Beijing, China

Enze Wang State Key Laboratory of Shale Oil and Gas Enrichment Mechanisms and Efficient Development, SINOPEC, Beijing, China;
Sinopec Petroleum Exploration and Production Research Institute, Beijing, China

Jing Wang State Key Laboratory of Petroleum Resources and Prospecting, China University of Petroleum (Beijing), Beijing, China

Jingbin Wang Petroleum Exploration and Production Research Institute of SINOPEC, Beijing, China;
Key Laboratory of Geology and Resources in Deep Stratum of SINOPEC, Beijing, China

Weihong Wang State Key Laboratory of Shale Oil and Gas Enrichment Mechanisms and Efficient Development, SINOPEC, Beijing, China;
Exploration and Production Reservoir Institute, Key Laboratory for Marine Oil and Gas Exploitation, Sinopec, Beijing, China

Shiming Wei China University of Petroleum-Beijing, Beijing, China

Chongyang Wu Petroleum Exploration and Production Research Institute of SINOPEC, Beijing, China;
Key Laboratory of Geology and Resources in Deep Stratum of SINOPEC, Beijing, China

Dawei Wu Sinopec Petroleum Exploration and Production Research Institute, Beijing, China

Yang Xia China University of Petroleum-Beijing, Beijing, China

Xianming Xiao School of Energy Resources, China University of Geosciences (Beijing), Beijing, China

Xuhui Xu Geophysical Exploration Technology Research Institute, SINOPEC, Nanjing, China;
Sinopec Geophysical Research Institution, Nanjing, China

Zuxin Xu Exploration Company, SINOPEC, Chengdu, China

Donghua You Petroleum Exploration and Production Research Institute of SINOPEC, Beijing, China

Yuchun You State Key Laboratory of Shale Oil and Gas Accumulation Mechanisms and Effective Development, Beijing, China;
Sinopec Petroleum Exploration and Production Research Institute, Beijing, China

Kun Yuan State Key Laboratory of Shale Oil and Gas Enrichment Mechanisms and Efficient Development, SINOPEC, Beijing, China;
Petroleum Exploration and Production Research Institute, SINOPEC, Beijing, China

Tao Yuan State Key Laboratory of Shale Oil and Gas Enrichment Mechanisms and Efficient Development, SINOPEC, Beijing, China;
Sinopec Exploration Company, Chengdu, Sichuan, China

Daqian Zeng State Key Laboratory of Shale Oil and Gas Accumulation Mechanisms and Effective Development, Beijing, China;
Sinopec Petroleum Exploration and Production Research Institute, Beijing, China

Yijin Zeng Sinopec, Beijing, China

Cheng Zhang Research Institute of Production Engineering, Zhongyuan Oilfield, Puyang, Henan, China

Chenyu Zhang Sinopec Petroleum Exploration and Production Research Institute, Beijing, China

Juntao Zhang Petroleum Exploration and Production Research Institute of SINOPEC, Beijing, China;
Key Laboratory of Geology and Resources in Deep Stratum of SINOPEC, Beijing, China

Liyu Zhang State Key Laboratory of Shale Oil and Gas Enrichment Mechanisms and Efficient Development, SINOPEC, Beijing, China;
Petroleum Exploration and Production Research Institute, SINOPEC, Beijing, China

Qingfeng Zhang Exploration Company, SINOPEC, Chengdu, China

Rui Zhang State Key Laboratory of Shale Oil and Gas Accumulation Mechanisms and Effective Development, Beijing, China;
Sinopec Petroleum Exploration and Production Research Institute, Beijing, China

Tan Zhang Institute of Sedimentary Geology, Chengdu University of Technology, Chengdu, China

Yayun Zhang Sinopec, Beijing, China

Yun Zhang Sinopec Petroleum Exploration and Production Research Institute, Beijing, China

Dongya Zhu Petroleum Exploration and Production Research Institute of SINOPEC, Beijing, China;
Key Laboratory of Geology and Resources in Deep Stratum of SINOPEC, Beijing, China

Xiuxiang Zhu Northwest Oilfield Company, SINOPEC, Urumqi, China

Zongxin Zuo Sinopec Petroleum Exploration and Production Research Institute, Beijing, China

Abbreviations

α_{Fi}	Volume fraction of fracture i
$[C_{Fi}]$	Compliance matrix of fracture i
$[T_i]$	Coordinate transformation matrix of fracture i
$k_{r\beta}$	Relative permeability to phase
$\vec{F}$	Mass or heat flux term
ϕ	Porosity
ρ_R	Density of the rock grains
ρ_β	Density of β
("S$_1$ + S$_2$"/TOC)	Hydrocarbon generation potential index
"A"	Soluble organic matter extracted from rock samples by chloroform solvents
ΔV_{f3}	The fluid volume related to F$_3$
ASDL	The maximum burial depth of active source rocks
ASDLg	The ASDLs for natural gas
ASDLo	The ASDLs for natural oil
bcm	Billion cubic meters
B$_g$	The gas formation volume factor at current reservoir pressure
B$_{gi}$	The initial gas formation volume factor
BHAD	The buoyancy-driven hydrocarbon accumulation depth
B$_k$	The factor of light hydrocarbon compensation, which is the ratio of lost-to-residual light hydrocarbons quantitatively related to the major factors established before
bnb	Billion barrels
bt	Billion tons
CFVR	Carbonate Fractured-Vuggy Rocks
C$_p$	The compressibility coefficient of reservoir rock, MPa^{-1}
CPD	Capillary Pressure Displacement
C_R	Specific heat of the rock
C$_r$	Sulfur content in gas sample, g/Nm^3

C_w	The compressibility coefficient of formation water, MPa^{-1}
D	Density
D	The diffusion coefficient
dc/dz	Hydrocarbon concentration gradient
DK	Permeability Decline Rate
E_{hc}	The hydrocarbon expulsion intensity, t/km^2
ESCP	Effective Stress Coefficient for Permeability
F_{9g}	The capillary pressures for gas in the rocks
F_{9o}	The capillary pressures for oil in the rocks
FIVE-INJECTION METHOD	Five-Step Water Injection Strategy
g/cm^3	Gram per cubic centimetre
GT	The variation rates of the geothermal gradient
h	The reservoir thickness, m
H	The source rock thickness
H/C	The atomic ratios of hydrogen to carbon
HADL	The hydrocarbon accumulation depth limit
HC	Hydrocarbon
HCI_o	The original hydrocarbon generation potential index
HCI_p	The residual hydrocarbon generation potential index
HET	The hydrocarbon expulsion threshold
HFU	Heat flow unit
HI	The hydrogen index
hm	Hundred meters
h_β	Specific enthalpy in phase β
k	Absolute permeability
K	The effective permeability of the water-producing formation, D
K	Thermal conductivity
K_0	The initial permeability of the reservoir, mD
Ke	The cumulative amount ratio of the hydrocarbons expelled from the source rocks to the cumulative amounts of generated hydrocarbons
km^2	Square kilometers
K_s	The permeability under sulfur saturation S_S, mD
Ksr	The expansion coefficient of the skeleton content
KTI	Kerogen-type index
LHED	Liquid hydrocarbon expulsion depth
M	Mass or heat accumulation term

m	Meters
m^3	Square kilometer
m^3/d	Cubic meters per day
m^3/km^2	Cubic meters per square kilometer
Ma	Mega annum
mD	MilliDarcy
MPa	Megapascal
mt	Million tons
N_f	Number of fractures in the considered element
O/C	The atomic ratios of oxygen to carbon
OM	The hydrocarbons generated from organic matter
p	Internal stress
p	Reservoir pressure, MPa
p_c	Confining pressure
P_c	The capillary pressure
Pc	The capillary pressure in the source rock
P_e	The hydrocarbon expansion pressure
P_e and P_w	The formation pressure and wellbore injection pressure, MPa
P_i	The initial and current reservoir pressures
p_p	Pore pressure
P_w	The overlying hydrostatic pressure
q	Source term
Q	The volume of formation water injected, m^3/d
q_e	The hydrocarbon expulsion rate: the amount of hydrocarbon expelled per unit organic carbon after the source rock reaches the HET
Q_e	The total hydrocarbon amount expelled, kg/m^3 for oil or m^3/m^3 for gas
$q_e(Z)$	Hydrocarbon expulsion rate per unit TOC, mg/g
Qe'	The amount of hydrocarbon expelled in early expulsion
Q_{e1}	The hydrocarbon amount of Q_{ed} was contributed by F_1
Qed	The amount of hydrocarbon diffused out of the source rock
Q_{ed}	The hydrocarbon amount expelled as a diffusion phase, kg/km^2 or m^3/km^2
Qed'	The amount of hydrocarbon diffused out of the source rock in early expulsion
Q_{ei}	The oil/gas amount expelled by each of the other eight driving forces

Q_{eog}	The gas amount expelled from the source rocks as an oil-solution phase, kg/km^2 or m^3/km^2
Qes	The amount of hydrocarbon expelled in a separate phase
Q_{es}	The hydrocarbon expulsion amount as a free phase, kg/km^2 or m^3/km^2
Qew	The amount of hydrocarbon dissolved in water
Q_{ew}	The hydrocarbon amount expelled as water solution, kg/km^2 or m^3/km^2
Q_{ew}'	The amount of hydrocarbon dissolved in water in early expulsion
Q_p	The total hydrocarbon amount generated in kg/m^3 for oil or m^3/m^3 for gas
Qr	The amount of residual hydrocarbon in the source rock
Q_r	The amounts of residual hydrocarbon
Q_{rg}	The total hydrocarbon amount remaining, kg/m^3 or m^3/km^3
Q_{rgb}	The amount of adsorbed gas
Q_{rgo}	The amount of oil-dissolved gas
Q_{rgw}	The amount of water-dissolved gas
Qrm	The minimal amount of hydrocarbon needed to saturate the source rock porosity
Q_{rm}	The residual hydrocarbon amount
Q_{ro}	The total hydrocarbon amount remaining, kg/m^3 or m^3/km^3
R	The throat radius of reservoirs
r	The throat radius of the surrounding rocks
R_e	Efficiency of hydrocarbon expulsion
R_e	The hydrocarbon amount expelled, kg/T toc or m^3/Tc
R_e and R_w	The control radius and well radius, m
Ro	Thermal evolution degree
R_p	The hydrocarbon amount generated in kg/t_{toc} or m^3/t_c
RSM	Response Surface Method
S	Diffusion area
S (n)	The source rock area, m^2
S_1	The amount of free hydrocarbons already present in the rock sample
S_2	The hydrocarbon content of the source rock from the pyrolysis of kerogen
S_e	Rate of hydrocarbon expulsion
SRI	Source rock index
S_S	The saturation of the sulfur deposition in the reservoir

S_{wc}	The initial water saturation in the reservoir, %
S_β	Saturation of β
T	Current temperature
t	Diffusion period
T	Reservoir temperature, K
t	Tons
T_0	Initial temperature
tcf	Trillion cubic feet
tcm	Trillion cubic meters
THM	Thermal-Hydro-Mechanical
TOC	The organic matter content
tt	Trillion tons
u	The viscosity of the formation water, mPa·s
u_β	Specific internal energy in phase β
Ve	The hydrocarbon expulsion rate
V_e	The hydrocarbon expulsion velocity or the hydrocarbon amount expelled per square meter of the source rock
V_e	Velocity of hydrocarbon expulsion
We	The water influx from natural aquifer, m^3
W_i	The volume of natural aquifer, m^3
Z	Buried depth
Z	Gas deviation factor
α	Biot coefficient
β_T	Coefficient of thermal expansion
γ_g	Ratio of specific heat
ΔP	The reservoir pressure depletion, MPa
$\Delta \Phi$	The extra porosity of the source rock with abnormal compaction, %
θ	The wetting angle of hydrocarbon/water
μi	Fluid viscosity
μm^2	Square micrometer
μ_β	Viscosity of phase β
$\rho(Z)$	The source rock density
ρ_o	The crude oil density in t/m^3
σ	External stress
$\sigma_{eff}{}^B$	Effective stress (Biot 1941)
$\sigma_{eff}{}^K$	Effective stress, relationship between permeability (K)
$\sigma_{eff}{}^T$	Effective stress (Terzaghi 1923)
σ_{net}	Net stress
φ	Source rock porosity
Φn	The source rock porosity with normal compaction in %

Status and Outlook of Deep Marine Petroleum Exploration and Development in China

Yongsheng Ma, Maowen Li, Xiaoxiao Ma, and Enze Wang

Abstract

A significant proportion of the world's ultra-deep wells have been drilled in China to search for oil and gas in the deeper parts of the petroliferous Tarim, Sichuan, Ordos and Bohai Bay Basins. These basins are often referred to using the term "superimposed basin" because such a basin is in fact composed of several basins that were formed at different geological times and vertically stacked because of complex tectonics. Based on a brief review of the general trends in global oil and gas exploration and development and the history of major deep oil and gas discoveries in China, the geological and engineering characteristics of deep petroleum systems in these basins are analyzed. Long-term exploration practices and theoretical research have led to gains of significant knowledge of the deep petroleum habitats in "superimposed basins" and, thus, have provided an impetus for the proposal of several valuable exploration paradigms. An overview of the technological advances and challenges is provided in areas such as deep target evaluation, field development design and reservoir engineering. Deep resource potential, exploration realms and play distribution, and key geoscientific and technological research directions in China are discussed, with reference to the needs for geological, geophysical and engineering integration and exploration and development cross vergence along the core value chain of the crusade into the deeper earth.

Keywords

Deep marine strata · Deep petroleum resource · Exploration and development · Engineering technology · Development direction

Abbreviations

Billion tons	Bt
Million tons	Mt
Billion cubic meters	Bcm
Trillion cubic meters	Tcm
Billion barrels	Bnb
Trillion cubic feet	Tcf
Trillion tons	Tt

Y. Ma · M. Li (✉) · X. Ma · E. Wang
State Key Laboratory of Shale Oil
and Gas Accumulation Mechanisms
and Effective Development,
Beijing, China
e-mail: limw.syky@sinopec.com

Y. Ma
Sinopec Group Company,
Beijing, China

M. Li · X. Ma · E. Wang
Sinopec Petroleum Exploration
and Production Research Institute,
Beijing, China

© The Author(s) 2026
Y. Ma et al. (eds.), *Petroleum Geology and Exploration of Deep Marine Strata in China*, Advances in Oil and Gas Exploration & Production, https://doi.org/10.1007/978-3-032-02496-1_1

1.1 Global and Domestic Settings

1.1.1 Global Trends in Petroleum Exploration

Deep sedimentary strata, deep-water offshore areas and unconventional oil and gas reservoirs are the three main domains that have increasingly gained prominence in global oil and gas exploration and development. Therefore, it is inevitable for the sustainable development of the oil and gas industry to insist on efficient exploration and effective development.

Deep oil and gas resources have been the main source of the increase in the global proven reserves in the past decade (Fig. 1.1), constantly breaking through the burial depth limit for effective resource preservation. From 2008 to 2018, the world's oil and gas reserve addition from reservoirs whose burial depth was greater than 4000 m was 23.4 billion tons (bt oil equivalent),

accounting for more than 60% of the world's total oil and gas reserve addition during the same period (Zhao 2019 and references therein). The maximum drilling depth of the oil and gas wells reached 12,869 m.

Deep-water offshore areas have become another focal point of oil and gas reserve addition in recent years, and major discoveries are concentrated mainly in passive continental margin basins on both sides of the Atlantic Ocean, East Africa and the Mediterranean Sea. From 2008 to 2018, new proven oil and gas reserves in the world's offshore regions were 21.7 bt oil equivalent, accounting for 68% of the world's total new proven oil and gas reserves, with deep water and ultradeep water accounting for more than 40% of the world's total new recoverable reserves.

As oil and gas exploration in many sedimentary basins has advanced, the deterioration in the quality of oil and gas resources has become marked, and unconventional oil and gas resources have become important for global oil and gas

Fig. 1.1 Annual new additions of global proven oil (**a**) and gas reserves (**b**) from 2008 to 2018, and relative percentages of these reserves in different reservoir depths. Modified from Zhao (2019)

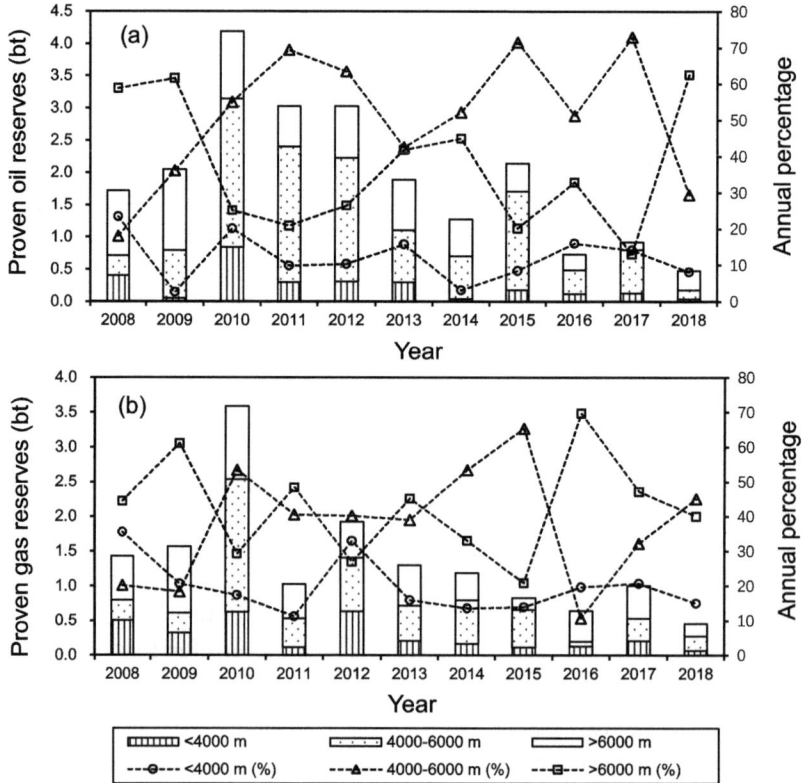

development in the future. With respect to the total global recoverable oil and gas resources, which is approximately 5 trillion tons (tt, oil equivalent), the ratio of conventional to unconventional resources is approximately 1:4. Global annual oil and gas production is expected to exceed 5 bt and 5 trillion cubic meters (tcm), respectively, in 2040, with unconventional oil and gas production accounting for 15.2% and 32.6% of oil and gas production, respectively.

Natural gas has attracted increasing attention as the main transitional resource in the global transition to green energy. Therefore, deep sedimentary strata and deepwater offshore and unconventional oil and gas resources have become the focus of oil and gas exploration worldwide. Many types of carbonate rocks (deep water, deep burial, subsalt, and tight lithology), marine turbidite in passive continental margins, complex piedmont zones, and deep shale oil and gas will become the major breakthrough fields of global oil and gas exploration in the future.

Since 2014, international oil prices have fluctuated between low and medium levels. The COVID-19 pandemic and the oil price slump have combined to affect the production and operation of global oil and gas enterprises in an unprecedented manner. The integrated exploration and development model, which is based on traditional oil and gas exploration plus new business operation modes, has achieved great success in the past and is also the inevitable choice for oil and gas companies to survive and develop in the future.

To respond to low oil prices, international oil and gas companies must take the following strategic routes: (1) optimize investment, giving more prominence to risk exploration in strategic areas, and enhancing oil and gas discoveries in deep-water and deep-seated plays; (2) give greater prominence to the role of technology in reducing discovery costs and improving competitiveness, vigorously developing low-cost technologies, and fully integrating information, intelligence, and big data into oil and gas operations; (3) give greater prominence to the development of companies' core competitiveness, focusing technology and capital on oil and gas-rich basins and focusing on large-scale production capacity-building projects;

and (4) pay equal attention to oil and gas and accelerate the development of the natural gas industry.

1.1.2 Characteristics of Petroleum Exploration in China

Tables 1.1 and 1.2 list the major oil and gas discoveries made in China since 2000. Oil exploration in China is currently in the middle stage of development, and natural gas exploration is in the early and middle stages. Since 2000, China's oil and gas exploration has made significant progress, with six distinct characteristics.

The first is the development of deep sedimentary strata in the sedimentary basins in central and western China. Breakthroughs have been made in the deep marine carbonates of the Tarim, Sichuan and Ordos Basins and in the clastic rocks of the Junggar Basin, where more than a dozen oil and gas fields have been found. Notable examples include the Tahe oil and gas field and Kalasu gas field in the Tarim Basin and the Puguang, Yuanba and Anyue gas fields in the Sichuan Basin.

The second is the shift in the focus of petroleum exploration in the superimposed basins to address the complex piedmont zone. Explorations of the Ten-zan Piedmont of the Tarim Basin and the southern, northern, and eastern Piedmonts of the Junggar Basin, Longmen Mountains and Micang–Daba Mountains Piedmonts of the Sichuan Basin have revealed several major gas discoveries in deep tight sandstone reservoirs. Typical examples include those in the Kuqa foreland thrust belt and the Xujiahe Formation in the western Sichuan Basin.

The third is the strengthening of oil and gas exploration in strike-slip fault zones. Important oil and gas discoveries have been made successively in the Shunnan, Shuntuo and Shunbei fields along the strike-slip fault zones in the low uplift and slope areas between the Tabei and Tazhong paleo-uplifts of the Tarim Basin. In the central Junggar, Bohai Bay and Ordos Basins, such targets along the strike-slip fault zones could also be potential exploration domains for new discoveries and reserve additions.

Table 1.1 Major oil discoveries in China since 2000, with technical recoverable reserves > 22.5 mt oil equivalent

No.	Oil field	Cumulative proven reserve		Technical recoverable reserve		Exploration domain
		mt	bnb	mt	bnb	
1	Red River	184.4	1.2	22.8	0.1	Tight sandstones
2	Jiyuan	1317.7	8.3	247.0	1.6	
3	Huaqing	521.7	3.3	100.0	0.6	
4	Xifeng	333.7	2.1	57.2	0.4	
5	Nanpu	360.9	2.3	76.6	0.5	Offshore
6	Bozhong 25-1 South	116.9	0.7	24.9	0.2	
7	Caofeidian 11-1	171.3	1.1	25.7	0.2	
8	Penglai 19-3	501.7	3.2	118.7	0.7	
9	Penglai 9-1	271.2	1.7	38.1	0.2	
10	Fanyu 5-1	54.8	0.3	24.8	0.2	
11	Harahatang	225.8	1.4	42.4	0.3	Carbonates
12	Tahe	1381.8	8.7	196.7	1.2	
13	Chunfeng-Chunguang	128.9	0.8	44.5	0.3	Maturely explored basins
14	Beier	161.4	1.0	30.8	0.2	
15	Gulong	106.5	0.7	23.8	0.1	
16	Daqingzi Well	159.1	1.0	32.4	0.2	
17	Daluhu	105.8	0.7	15.6	0.1	
18	North Kun	112.7	0.7	22.4	0.1	Piedmont belts

The fourth area is characterized by the rapid expansion of deep shale gas exploration. With the establishment of national shale gas demonstration zones in the Fuling, Changning–Weiyuan and Zhaotong fields, Sinopec recently made a breakthrough in deep shale gas exploration in the Weirong region. Important progress has also been made in the exploration of deep shale gas in new strata, new types and new basins in China.

The fifth is new breakthroughs in nonmarine (mostly lacustrine) shale oil exploration. With the application of long horizontal interval horizontal well subsection fracturing technology, local breakthroughs have been made in the Junggar, Bohai Bay, Songliao, Sichuan, Santanghu and Jianghan Basins. Moreover, the production technology route for shale oils with medium to low maturity and much larger resources has begun to be explored.

Finally, new discoveries have been made in the fine exploration of mature basins in eastern China. Many breakthroughs have been made in recent years in the exploration of "buried hill"

reservoirs in the Bohai Bay Basin both on land and in sea areas, gradually deepening the geological understanding of buried hill reservoirs. The deep sag areas and the internal reservoir–caprock assemblage of the buried hills in faulted basins are important for the effective replacement of oil and gas resources in eastern China.

After nearly 20 years of practice, China's oil exploration and development has achieved stable development, with natural gas exploration and development increasing markedly. The development of deep strata, deep water and unconventional oil and gas has become an important field for the future. In addition, the complexity of exploration objects and the poor quality of resources have become the new characteristics of China's oil and gas exploration and development and present many new challenges to the development of the oil and gas industry.

Oil and gas exploration and development in China should pay equal attention to both "resource-oriented" and "technology-oriented" strategies, taking the new areas, new realms

Table 1.2 Major gas discoveries in China since 2000, with technical recoverable reserves > 22.5 mt oil equivalent

No.	Gas field	Cumulative proven reserve		Technical recoverable reserve		Exploration domain
		bcm	tcf	bcm	tcf	
1	Xushen	221.8	7.8	104.6	3.7	Volcanics
2	Changling No. 1	70.6	2.5	38.9	1.4	
3	Kelameili	105.3	3.7	58.6	2.1	
4	Sulige	1644.8	58.1	866.4	30.6	Tight sandstones
5	Daniudi	454.6	16.1	206.7	7.3	
6	Yangliupu	55.0	1.9	25.1	0.9	
7	Eastern Hubei	64.0	2.3	30.8	1.1	
8	Yanan	241.8	8.5	118.2	4.2	
9	Zizhou	115.2	4.1	68.0	2.4	
10	Shenmu	333.4	11.8	167.3	5.9	
11	Chengdu	221.0	7.8	92.7	3.3	
12	Tieshanpo	37.4	1.3	28.0	1.0	
13	Hechuan	229.9	8.1	103.5	3.7	
14	Puguang	412.2	14.6	291.6	10.3	Deep carbonates
15	Longgang	72.0	2.5	46.0	1.6	
16	Luojiazhai	83.6	3.0	62.5	2.2	
17	Anyue	1018.4	36.0	634.3	22.4	
18	Yuanba	230.3	8.1	117.7	4.2	
19	Tazhong 1	353.5	12.5	216.4	7.6	
20	Dongping	51.9	1.8	29.8	1.1	Piedmont belts
21	Kelasu	437.9	15.5	255.1	9.0	
22	Dina 2	175.2	6.2	113.9	4.0	
23	Akemomu	44.6	1.6	31.3	1.1	
24	Dabei	109.3	3.9	69.2	2.4	
25	Kela 2	284.0	10.0	212.9	7.5	Offshore
26	Ningbo 22-1	102.0	3.6	52.4	1.8	
27	Ningbo 17-1	106.7	3.8	64.6	2.3	
28	Liwan 3-1	47.6	1.7	34.4	1.2	
29	Lingshui 17-2	102.0	3.6	59.3	2.1	
30	Dongfang 13-2	68.6	2.4	44.9	1.6	

and medium-deep strata in central and western China as the main battlefields of conventional oil and gas exploration, the "three new" fields (new areas, new realms and new strata) and low-grade resources in eastern China as the potential direction of increasing reserves, and unconventional oil and gas as the strategic replacement fields of increasing reserves and production. The innovation of geological theory and the research and development of engineering technology and equipment constitute the core of efficient exploration and effective development.

1.1.3 Deep Oil and Gas Exploration in China

The threshold value for classifying petroleum reservoirs based on the present-day burial depth varies in the literature. However, the petroleum industry generally adopts 4500 m and 6000 m as the thresholds for "deep" and "ultradeep" carbonate reservoirs, respectively (Dyman et al. 2002; Standardization Administration of the People's Republic of China 2012; Bai and Cao 2014). The corresponding values for "deep" and "ultradeep" clastic reservoirs and shale gas reservoirs are 3500 m and 4500 m, respectively (Ministry of Land and Resources of China 2014). As exploration for petroleum plays at shallow–medium depths in most Chinese onshore basins has matured, deep and ultradeep targets have gradually gained prominence (Mou and Cai 2007; Cai 2011; Sun et al. 2013; Feng et al. 2016a, b; Jin et al. 2018), with the deepest exploration wells in the Sichuan and Tarim Basins approaching a depth of 9000 m.

The fluid phase in deep reservoirs varies widely under different paleo- and present-day high-temperature and high-pressure regimes (Pang et al. 2014). In general, increased tectonic stresses in deep-seated plays tend to have a high probability for the development of overpressured reservoir segments. The mechanical strength of deep buried rocks often changes as a function of burial depth and fluid pressure. Rock porosity decreases with increasing burial impact on the phase state and dynamics of oil, gas, and water in porous media and ultimately on hydrocarbon migration, accumulation, and preservation in deep reservoirs (Jin and Cai 2007; Jin 2005, 2010; Ma et al. 2017). Therefore, increased burial depth for deep reservoirs generally means more stringent requirements for exploration geologists and geophysists to reduce the increased uncertainty in predicting oil and gas distribution and for drilling and completion engineers to minimize the associated risk and cost.

Sedimentary basins in China have developed on the tops or near the peripherals of several small Proterozoic cratonic blocks. Deep sedimentary strata in these basins were deposited within multiple tectonic successions, with long histories of evolution. The frameworks of each prototype basin dictate the deposition and spatial relationships of potential petroleum source beds, reservoirs and caprocks. Subsequent transformation

and reformation during the evolution of superimposed basins control source rock maturation, carbonate dissolution and dolomitization, trap formation, and ultimately hydrocarbon accumulation, alteration, and redistribution (Ma et al. 2010a, b, 2011; He et al. 2016). Unconformities, fracture zones and trap assemblies significantly control oil and gas habitats, and the alteration processes involved in high-temperature and high-pressure ultradeep reservoirs are clearly complex (Luo et al. 2016; Pang et al. 2015; Li 2016). Understanding these processes is key for overcoming the limitations of deep oil and gas exploration.

In the past decade, a significant proportion of ultradeep petroleum wells worldwide have been drilled in China. As shown in Tables 1.1 and 1.2, discoveries made in marine carbonate reservoirs include the Tahe and Shunbei oil and gas fields in the Tarim Basin; the Anyue, Puguang, Yuanba and Pengzhou gas fields in the Sichuan Basin; and the Jingbian West gas field in the Ordos Basin (Ma et al. 2010a, b, 2017; Qi 2014), as well as several giant shale gas fields (Guo et al. 2016a, b). Giant gas discoveries have also been made from siliceous clastic reservoirs in the Mahu Depression of the Junggar Basin and the Kuqa Depression of the Tarim Basin and volcaniclastic rocks from the Carboniferous in the Junggar Basin, Cretaceous in the Songliao Basin, and Permian in the Sichuan Basin, demonstrating not only the large resource potential but also the great complexity in the petroleum systems of sedimentary basins in China (Tuo 2002; Ma et al. 2020, 2021).

Deep sedimentary strata in China represent a domain of great opportunities with significant challenges. Several generations of geologists and engineers have worked and explored tirelessly to solve key geological problems and have made a series of innovative engineering breakthroughs and substantial discoveries. It remains a phenomenal task for Chinese petroleum geologists and engineers to develop new theories and technologies for oil and gas exploration in deep strata. This work is important not only for sustainable economic development in China but also for deep petroleum explorers elsewhere who need geological analogs.

1.2 Status of Deep Petroleum Exploration and Development

1.2.1 Three Development Stages

Deep petroleum exploration and development in China began in the 1960s and has experienced three different stages of development (Li et al. 2020):

(1) Early Exploration Stage (1966–1980)

Starting with the Songji-6 well in the Songliao Basin, with a total drilling depth of 4770 m, three other deep wells with drilling depths greater than 5000 m were drilled from 1966 to 1976 in the Huanghua and Jiyang depressions of the Bohai Bay Basin and the Qianjiang Depression of the Jianghan Basin. Nvji-1 in the Sichuan Basin, the first ultradeep well in China, was completed in 1976, with a drilling depth of 6011 m. During this period, more than 100 deep wells were drilled, with 10 being ultradeep wells. The Guanji-1 well in the Sichuan Basin was the deepest (7175 m in drilling depth), followed by the Gu-2 well (7002 m) in the Junggar Basin. No substantial discoveries were made at this stage because of the lack of good knowledge of deep sedimentary strata and the limited access to advanced drilling technologies.

(2) Breakthrough and Discovery Stage (1980–1990s)

The first major breakthrough in deep marine strata was made in 1984 by the drilling of the Shacan-2 well in the Tarim Basin, where a significant amount of oil was encountered from the Ordovician carbonate reservoirs. High oil flow was obtained in 1988 from the Lunnan-2 well, also in the Tarim Basin. High gas yields were produced in 1989 from the Tiandong-1 well in the Sichuan Basin, with payzones in the Carboniferous reservoirs. The Donghetang oilfield was subsequently confirmed in 1990 in the Tarim Basin. As a result, great discoveries have been made from multiple stratigraphic horizons and different play types in the Tarim and Sichuan Basins.

(3) Large-Scale Development Stage (Since 1986)

Marked by the commercial development of the Tahe oilfield in the Tarim Basin in 1998, deep petroleum exploration and development in China entered the large-scale economic development phase. Major oil discoveries in this period include the Halahatang and Shunbei fields in the Tarim Basin, which were made in 1998 and 2015, respectively. Major deep gas discoveries in the Tarim Basin include the Kela-2 (1998), Shunnan (2013) and Kelasu (2018) fields, whereas those in the Sichuan Basin include the Longgang and Puguang (2000–2003), Yuanba (2007), Chuanxi (2010), and Anyue (2011) gasfields. Construction of two gas pipelines that link the gas production bases in western China to consumers in the major urban centers in the east has fundamentally changed the landscape of the Chinese energy supply markets.

1.2.2 Major Oil and Gas Discoveries

In the past decade, deep-seated oil and gas plays have become the dominant sources for increasing reserves and stabilizing production in China, with major discoveries being made successively in the Tarim, Sichuan, Bohai Bay, Ordos and Songliao Basins (He et al. 2019; Li et al. 2020; Ma 2020). According to the Ministry of Natural Resources of China (2018), more than 20 deep oilfields had been found in China by the end of 2018, with total cumulative geological reserves of approximately 4 bt, of which 0.57 bt had been produced. The total cumulative reserves from more than 10 deep gas fields exceeded 5 trillion cubic meters (tcm), 0.43 tcm of which had been produced.

1.2.2.1 Tarim Basin

The first encounter of crude oils in the Paleozoic marine sequence in the Shacan-2 well in 1984 was the turning point for the petroleum industry in China, which kickstarted the decisive battle in oil and gas exploration in the Tarim Basin.

To date, Sinopec has discovered the Tahe, Shunbei, Yuejin and Shunnan deep oil and gas fields, with total proven oil and gas reserves of 1.665 billion tons (bt) and 94.583 billion cubic meters (bcm), respectively. Sinopec's cumulative production from these fields has exceeded 140 million tons (mt, oil equivalent). The discovery of Tahe, the first giant marine oilfield with more than one billion tons of oil resources, was signaled by the oil testing results of the Sha-46, Sha-47 and Sha-48 wells in 1997. The bulk of the Tahe oilfield was confirmed from 1997 to 2001. Stereoscopic exploration has been carried out since 2002, with significant reserve additions at the periphery of the Tahe field and the underlying subsalt plays to the south. By the end of 2018, the cumulative proven oil reserves of the Tahe field had reached 1.35 bt; an annual production capacity of 7 mt was constructed, and cumulative oil production exceeded 100 mt by August 2019.

Testing of high oil flows from the Shunbei 1-1H well in 2015 along strike-slip fault zones signaled the discovery of the Shunbei oil and gas field, with an estimated total petroleum reserve of 1.7 bt (oil equivalent), including 1.2 bt of oil and 500 bcm of gas. This ultradeep fracture–cavity-controlled field represents a new and important type of petroleum accumulation in the Tarim Basin. The construction of an annual production capacity of one mt was completed in the Shunbei field in 2020, with cumulative oil and gas production levels of 3.33 mt and 1.497 bcm, respectively, by the end of 2021.

PetroChina has discovered and developed the deep Halahatang, Tazhong, Lungu, Hadexun, and Donghetang oilfields and the deep Kelasu, Tazhong-1, Kela-2, Dila-2 and Dabei gasfields in the Tarim Basin. By the end of 2018, PetroChina's cumulative proven oil and gas reserves were 790 mt and 1.6 tcm, respectively. In recent decades, several giant deep and ultradeep gas fields in Mesozoic nonmarine clastic reservoirs have been discovered in the thrust belts of the Kuqa foreland depression, thus confirming the first giant ultradeep gas province with gas reserves greater than 1 tcm in China (Du et al. 2016). An annual gas production capacity of 15 bcm was constructed.

1.2.2.2 Sichuan Basin

Since 2000, Sinopec has discovered 4 major deep gas fields in the Sichuan Basin with single-field proven gas reserves of multiple hundred bcm, including the Puguang and Yuanba sour gas fields in the Upper Permian–Lower Triassic reef–shoal reservoirs in the NE Sichuan Basin, the Chuanxi gas field along the erosional surface in the Middle Triassic Leikoupo Formation in the western Sichuan Basin, and the Weirong deep shale gas field in the Upper Ordovician–Lower Silurian Wufeng–Longmaxi formations in the southern Sichuan Basin. PetroChina has discovered and developed the Anyue gas field in Sinian–Cambrian strata in the Gaoshiti–Moxi region, with proven gas reserves greater than 1 tcm.

The discovery of the Puguang gas field was marked by a production test of the Puguang-1 well drilled in 2003, with a daily gas yield of 423,700 m^3. By the end of 2021, this field had a cumulative gas reserve of 457 bcm; an annual production capacity of 10.5 bcm was completed in 2013, with cumulative gas production breaking the hallmark of 100 bcm by the end of 2021.

The Yuanba gas field was discovered from reservoirs with depths of 6240–6950 m, following the test of the Yuanba-1-c1 well with a daily gas flow of 503,000 m^3. This field had a cumulative gas reserve of 230.3 bcm by the end of 2018, and the cumulative gas production had exceeded 24 bcm by January 2022.

Gas exploration along the thrust belts of the Longmen Mountains and Chuanxi Depression in the western Sichuan Basin has also produced excellent results. In the marine sequence of the Middle Triassic Leikoupo Formation, Sinopec began in 2007 to implement the scientific drilling well Chuanke-1, with a daily gas flow of 868,000 m^3 obtained in 2010. Subsequent drillings in the Xinchang and Pengzhou areas led to great breakthroughs in the T_2l_4 member of the Leikoupo Formation and hence the discovery of the Chuanxi gas field. By the end of 2018, the Chuanxi field had controlled and predicted gas reserves of 216.61 bcm and 44.16 bcm, respectively, with more than 100 bcm of proven gas reserves by June 2022. An annual production capacity of 3 bcm is in construction.

In the nonmarine sequence, the X851 well drilled in 2000 produced an open daily gas flow of 3,260,000 m^3 from the deep buried tight sandstone reservoirs of the Upper Triassic Xujiahe Formation in the Xinchang area, western Sichuan Basin, indicating significant gas potential in the deep buried nonmarine clastic sequences in the Sichuan Basin. This suggestion has been confirmed by subsequent findings in the Xujiahe Formation in the central Sichuan Basin as well as other deeply buried nonmarine strata in the northeastern Sichuan Basin.

PetroChina drilled the Gaoshi-1 well in 2011 in the Gaoshiti–Moxi region, and a high gas flow of 1,020,000 m^3/day was encountered from the second member of the Sinian Dengying Formation (Z_3d_2), leading to the discovery of the Anyue gasfield. The proven gas reserves of this field exceeded 1 tcm by the end of 2018, and an annual gas production of 7.3 bcm was established, with a cumulative production of 72.5 bcm by the end of June 2021.

1.2.2.3 Bohai Bay Basin

Drilling of the Ren 4 well in the Jizhong depression in 1975 resulted in a daily oil flow of 1014 tons from the old basement rocks of the Mesozoic–Cenozoic Bohai Bay Basin, marking the beginning of petroleum exploration from the "buried hill" reservoirs. Following the initial stage of unaka exploration (1960–1995) and a later stage of multitype reservoir exploration (1996–2013), the industry has entered the stereoscopic exploration stage for buried hills, with many new discoveries. For example, Sinopec discovered the Chengdao, Gaoqing, Dawangzhuang, and Chezhengbei buried hill oil plays, with proven oil reserves exceeding 236 mt in the Jiyang Depression by the end of 2018. Recent findings of the giant Bozhong-19-6 gas condensate field by China National Offshore Oil Corporation, with proven gas and condensate reserves of approximately 100 bcm and 100 mt, respectively, have broken new ground for deep petroleum exploration in the Bohai Bay Basin.

1.2.2.4 Deep Shale Gas Discoveries

Following the breakthrough in the Jiaoye-1 well, marine shale gas exploration and development in the Sichuan Basin experienced a period of rapid growth. Three national demonstration projects for shale gas development were established in the Fuling, Changning–Weiyuan and Zhaotong fields, with significant technological advances being achieved for the development of gas shales with shallow to medium burial depths (< 3500 m). The results of basin-wide resource assessment indicate that more than 60% of the in-place shale gas resources occur in deep shale reservoirs at depths of 3500–4500 m. The cumulative proven shale gas reserves in the Upper Ordovician–Silurian Wufeng–Longmaxi formations in the Sichuan Basin have exceeded 1.8 tcm thus far.

As exploration activities increase and geological understanding deepens, industry efforts on deep shale gas targets have been made, with significant findings. In March 2015, high commercial gas flow was encountered in Sinopec's Weiye-1HF well, marking the discovery of Weirong, the first deep shale gas field in China. Shale reservoirs in the Weirong field range from 3500 to 4000 m, with an average of 3750 m. The cumulative proven shale gas reserves of the Weirong field were 124.7 bcm in 2018, and the construction of an annual production capacity of 3 bcm was completed in 2022.

As Sinopec and PetroChina put more exploration efforts into the deep gas shales in the Sichuan Basin, many significant breakthroughs have been made, revealing the great resource potential of deep gas shales. In Sinopec's Yongchuan block, six wells drilled in the Yongye-1 area produce gas from the Wufeng–Longmaxi shales at 3700–4300 m in depth, with single-well daily gas yields ranging from 80,000 to 140,000 m^3. The proven gas reserves in the Yongchuan block have reached 21.4 bcm. In Sinopec's Dingshan–Dongxi block, the tests of 4 deep shale gas wells with depths ranging from 3800 to 4500 m produced single-well daily gas flows of 50,000–310,000 m^3, and Dongyeshen-1, with a reservoir depth of 4219 m, was the first deep shale gas well in China with a reservoir depth exceeding 4200 m. In PetroChina's Yuxi block, a daily gas yield of 456,700 m^3 was obtained after fracking from the Zu-202H1 well, with a vertical reservoir depth of 3925.5 m. A daily gas yield of 223,700 m^3 was obtained after fracking from the Huang-202 well on the western flank of the Cucumber Mountain structure in PetroChina's Bishan–Hejiang block, with a vertical reservoir depth of 4086.1 m. The record for the initial daily gas flow (1,379,000 m^3 after fracking) was set by the Lu-203 well in PetroChina's Luxian-Changning block in March 2019.

1.2.3 Geological and Engineering Characteristics

Table 1.3 lists the stratigraphy, type, depth, reserve, temperature and pressure of the selected deep oil and gas fields in the Tarim and Sichuan Basins in China. Deep petroleum reservoirs exhibit several distinct geological and engineering characteristics.

1.2.3.1 High Resource Volumes

Deep petroleum resources in China are concentrated in the Tarim and Sichuan Basins, where several deep and ultradeep oil and gas fields have been discovered. As shown in Fig. 1.2, more than 80% of the proven oil and gas reserves in the Tarim Basin occur in reservoirs with burial depths greater than 5000 m. Tahe is the largest producing oilfield in the Tarim Basin, with a reservoir depth of 5350–6200 m. The reservoir depths of other producing fields in the Tarim Basin are greater than 6000 m (Table 1.3). Of the 20 giant gas fields (with proven gas reserves greater than 30 bcm or 1 trillion cubic feet) in the Sichuan Basin, six occur in reservoirs with present-day burial depths of 3500–5000 m, accounting for 24.9% of the total proven reserves, and five in reservoirs with burial depths greater than 5000 m, accounting for 37.9% of the total proven reserves (Wei et al. 2019). Yuanba, with a reservoir depth of 6240–7300 m, is the first giant gas field with ultradeep reef reservoirs in the world and remains the deepest giant gas field with marine carbonate reservoirs in

Table 1.3 Key reservoir parameters of major deep and ultra-deep oil and gas fields in China

Exploration domain/field	Reservoir age	Reservoir depth (m)	Resource (reserve)		Max temperature (°C)	Max pressure (°C)	Reservoir type[a]
Tarim Basin							
Tahe	O	5350–6200	(1.35 bt)	(9.5 bnb)	165	90	A
Shunbei	O	6500–9000	1.7 bt	14.5 bnb	190	104	A
Halahatang	O	5900–7100	(0.25 bt)	(2.1 bnb)	170	104	A
Shuntuoguole	O	6600–8300	> 2 tcm	> 70.6 tcf	220	180	A
Kuchekelasu-Dabei	J, T	6000–8000	3.5 tcm	123.6 tcf	175	128	B
Sichuan Basin							
Sinian-Cambrian (NE)	Z, $\in$	8000–10,000	457–541.4 bcm	16.1–19.1 tcf	180 (CS-1, 8420 m)	About 150	B
Leikoupo Fm. (W)	T_3l	5000–8000	(216.6 bcm)	(7.6 tcf)	166 (Ck-1, 7560 m)	About 130	C, B
Yuanba	P_2ch-T_1f	6420–7300	(219.9 bcm)	(7.8 tcf)	160	147	B
Anyue	Z_2d-$\in_1l$	4500–6000	(1 tcm)	(35.3 tcf)	161	78	B
Puguang	P_2ch-T_1f	4800–5500	(412.1 bcm)	(14.6 tcf)	135	57	C
Deep shale gas	O_3w-S_1l	3500–4500	> 12.5 tcm	> 441.4 tcf	155	97	C

Data from Qi (2016), Wei et al. (2017a, b) and Li et al. (2018, 2019)

[a] Reservoir type: A. fracture-cavity; B. fracture-pore; C. pore

Fig. 1.2 Proven oil and gas reserves in the Tarim Basin as a function of reservoir depth. *Data source* Ministry of Natural Resources of the People's Republic of China (2018)

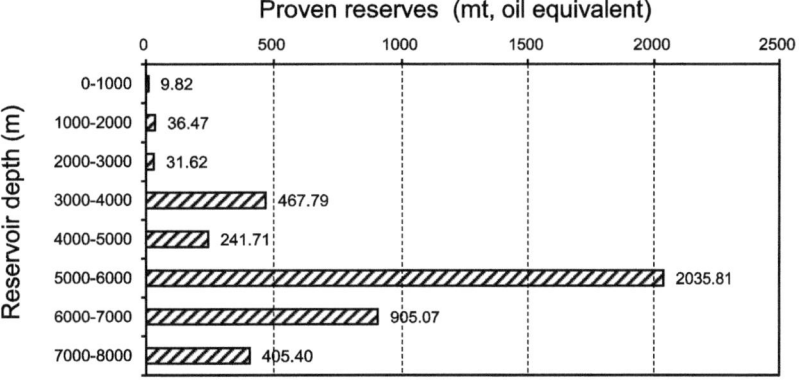

China. Tests of ultradeep wells (Xinshen-1, reservoir depth: 6421 m; Pengzhou-1, reservoir depth: 6050 m) in the Chuanxi gasfield revealed a daily gas flow of more than 1 million m^3, confirming the great potential of the ultradeep carbonate reservoirs of the Middle Triassic Leikoupo Formation in the Sichuan Basin.

1.2.3.2 Multiple Production Horizons

Statistical analysis of global oil and gas occurrence in sedimentary basins revealed five major deep oil- and gas-bearing sequences, which decreased in importance in the order of Neogene, Upper Paleozoic, Cretaceous, Paleogene and Jurassic (Zhao 2019). Compared with global basins, sedimentary basins in China are characterized by superimposed basins with multicyclic development, and deep petroleum resources are present in Sinian, Cambrian, Ordovician, Silurian, Upper Paleozoic, Mesozoic, and Cenozoic strata, with wide stratigraphic coverages (Fig. 1.3). Deep marine carbonate reservoirs in China are distributed in the infrastructural layers of superimposed basins, most with old geological ages (Zhao et al. 2014). In contrast, the distributions of deep clastic reservoirs display large regional

Fig. 1.3 Proven gas reserves and total gas resources in the Sichuan Basin as a function of reservoir stratigraphy. Only conventional gas is included. *Data source* Ministry of Natural Resources of the People's Republic of China (2018)

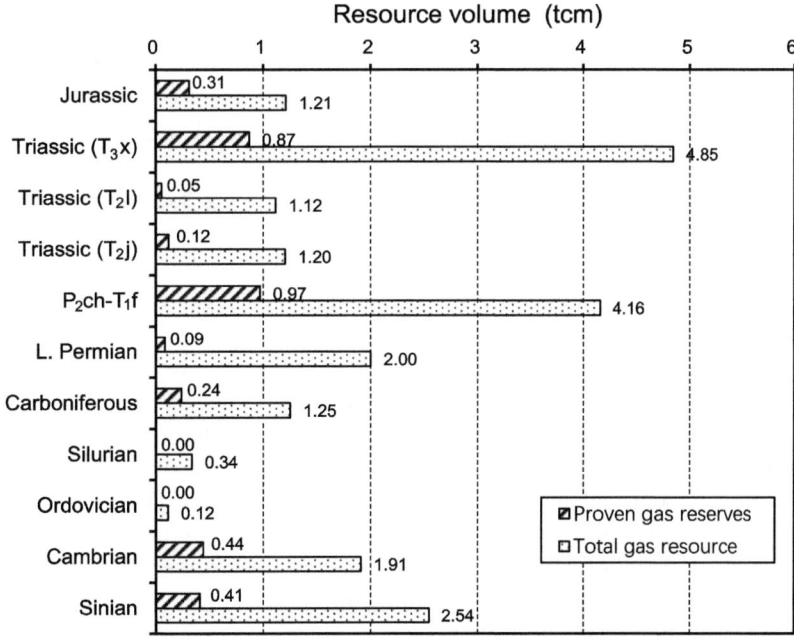

differences, with Paleozoic and Mesozoic dominance in western China and Paleogene dominance in eastern China (Feng et al. 2016a, b). Deep volcanic reservoirs are present mainly in Permian–Triassic strata in western China and in Jurassic, Cretaceous and Paleogene strata in eastern China.

1.2.3.3 Variable Reservoir Types

Deep oil and gas reservoirs in China include carbonate, clastic, and volcanic rocks. Deep clastic reservoirs occur widely in many sedimentary basins in China, with large variations in porosity types and physical properties. Porosities in clastic reservoirs are dominated by residual intergranular pores, with those in western China displaying higher porosity and permeability than their counterparts in eastern China due to differences in burial history, geothermal regime, pressure evolution, and diagenesis (Feng et al. 2016a, b).

Deep carbonate reservoirs can be classified genetically into four basic types: reef–shoal, karst, dolomite, and fractured reservoirs. Tectonic activity, stratigraphic sequence, lithofacies, geofluids, and geological age are important factors that control the formation and evolution of high-quality petroleum reservoirs in deep carbonate strata.

Hydrocarbon storage in the marine carbonate reservoirs of the Tarim Basin is dominated by fractures and cavities but varies significantly at different locations. The Ordovician carbonate reservoirs in the Tahe oilfield consist of large karst caves and fractures, and each cave unit acts as a separate petroleum trap, with complex three-dimensional morphological assemblages that vary in size up to 6–7 orders of magnitude. In contrast, ultradeep fractures and cavity oil and gas pools in the Shunbei region are controlled by intracratonic strike-slip faults. The Ordovician reservoirs originated from successive reformation processes related to fracturing, burial dissolution, and sequence boundaries. The resulting reservoir storage is dominated by caves, fractures, and voids. Oil and gas in the Shunbei region occur along fracture zones, with relatively high formation fracture pressures, large oil columns (> 510 m), and high reservoir temperatures (> 150 °C). The complexity of reservoir attributes

leads to significant challenges in reservoir prediction, target assessment, development mode selection and engineering technology optimization (Jiao 2018; Qi 2019).

The deep marine carbonate reservoirs in the Sichuan Basin can be divided into three main genetic types: fracture–pore, pore, and cavity (Table 1.2). The Puguang gas field is produced from the reef–shoal reservoirs of the upper Permian Changxing Formation–Lower Triassic Feixianguan Formation, with hydrocarbon storage being dominated by dissolution and intergranular pores. The gas distribution in this field is characterized by multistage accumulation, oil-to-gas transformation, and late-stage emplacement. The ultradeep reef reservoirs in the Yuanba gasfield consist of reef and shoals in the Changxing Formation and oolitic bank deposits in the Feixianguan Formation, with biomediated dissolution cavity and dolomite intercrystalline dissolution pores being the dominant storage spaces. The gas distribution in the Yuanba field is characterized by individual reefs and shoals forming separate gas pools. Although the reef–shoal reservoirs in the Puguang and Yuanba gas fields formed along the platform margins of the ancient Kaijiang–Liangping Trough, the tectonostratigraphic settings in the two areas differed. As a result, almost 94% of the proven gas reserves in the Puguan gasfield came from the thick, high-quality reservoirs in the Lower Triassic Feixianguan Formation, whereas the dominant reservoirs in the Yuanba field are in the Upper Permian Changxing Formation.

1.2.3.4 Complex Pressure Regimes

Deep petroleum accumulations in China generally display complex pressure systems. For example, large vertical variations in reservoir pressure can be observed in the Changxing and Feixianguan Formations in the Yuanba gasfield in the Sichuan Basin. The reef–shoal reservoirs in the Changxing Formation are normally pressured with pressure coefficients within 1.01–1.12, where H_2S-rich (approximately 5.22%) dry gas is produced from fracture–pore-type reservoirs. In contrast, the oolitic reservoirs in the Feixianguan Formation are overpressured (with pressure coefficients of 1.95–1.96) and produce gas with only moderate amounts of H_2S (Ma et al. 2014). In the Anyue gasfield, deep reservoirs in the Z_1d_2 and Z_1d_4 members of the Lower Sinian Dengying Formation are characterized by high temperature and normal pressure (with pressure coefficients of 1.06–1.13), but those in the Lower Cambrian Longwangmiao Formation are in high-temperature and overpressured zones (with pressure coefficients of 1.50–1.67) (Wang et al. 2016).

1.2.3.5 Difficulties in Well Construction

Deep petroleum systems are commonly associated with compounded complexity in surface topography, formation lithology, pressure systems, reservoir fluids, and engineering mechanics. Deep petroleum drillings are often characterized by a slow penetration rate, long drilling circles, high cost and high risk, and difficulties in the optimization of deep well design and the quality control of engineering operations. Thus, more stringent requirements are needed for drilling speed, drilling fluids and well cementing. Deep and ultradeep well drilling in the Sichuan Basin faces many challenges, including but not limited to unknown factors in deep strata, high temperature and high pressure, and extrahard and abrasive formations. The current burial depths of the Sinian–Cambrian strata in the northeastern Sichuan Basin range from 8000 to 10,000 m, with bottom-hole temperatures and formation pressures reaching 180 °C and 150 MPa, respectively. The marine reservoirs in the Middle Triassic Leikoupo Formation in the western Sichuan Basin are now at similar burial depths, with formation temperatures and pressures reaching 166 °C and 130 MPa, respectively. The drilling rate through a portion of the nonmarine Upper Triassic Xujiahe Formation in the western Sichuan Basin was lower than 1 m/ h. Mud loss during the penetration of the Middle Triassic Leikoupo Formation reached 53,653 m^3 in the Chunsheng-1 well; high mud loss was also recorded during the penetration of the Xujiahe and Leikoupo formations in the Yongfu-1 well (up to 10,270 m^3). Drilling of the Sinopec Mashen-1 well (with a footage of 8418 m) over 549 days clearly revealed the high degree of difficulty involved in the construction of ultradeep wells.

1.2.3.6 Necessity in Reservoir Stimulation

Deep and ultradeep reservoirs are characterized by deep burial, high temperature, high pressure, tight matrices, complex fabrics, and strong heterogeneity. This is particularly true for fracture- and cavity-controlled marine carbonates and deep gas shales, where superimposition and complex interplay of multiple geological factors lead to great challenges for deep petroleum exploration and development. Artificial stimulation is often necessary to increase the per-well productivity and economic efficiency of deep tight reservoirs.

Deep carbonate reservoirs contain multiple media (cavities, pores, and fractures) with strong heterogeneities. The results of laboratory and field studies reveal that it is necessary to perform acid fracturing, acidification, or high-pressure water injection to realize the effective development of fracture–cavity reservoirs. In the Tahe oilfield, for example, 2277 acid fracturing operations were carried out from 1999 to 2015 on 350 mt of producing reserves, with a cumulative oil production of 24.9 mt.

Deep shale gas reservoirs exhibit high fracture pressure and high horizontal stress differential and thus tend to produce simple fractures with small rock volumes upon hydraulic fracking. In the Weirong deep shale gas field, the temperature and pressure of the Wufeng–Longmaxi shales range from 127.4 to 135 °C and 68.7 to 77.5 MPa, respectively. These shales have deep burial depths (3550–3880 m), high in situ stress (86.0–97.7 MPa), high horizontal in situ stress differential (7–17 MPa), and low rock brittleness indices (< 0.50), with few natural fractures. Hydraulic fracturing in this field often encounters difficulties, including high operation pressures, narrow pressure windows, low sensitive sand concentrations, and high difficulty in sand pumps. Therefore, small fracking volumes and rapid production declines were common phenomena at the early stages of the pilot production test in the Weirong field.

1.3 Exploration Paradigms and Technological Innovation

Recent industrial practices indicate that significant petroleum resource potential occurs in deep and ultradeep sedimentary strata, which are among the most important domains for sustained resource development in China. Major discovery became possible only after the industry had committed serious investment in deep exploration. A shift in exploration paradigms due to improved geological understanding often holds the key for breakthroughs in deep and ultradeep exploration ventures. Advances in engineering technology and equipment integration provide safeguards for the large-scale commercial development of deep oil and gas fields.

1.3.1 Exploration Paradigms and Target Evaluation

Sedimentary basins in China vary in origin, with complex histories and different mechanisms of oil and gas accumulation. Following many decades of tireless exploration practices and research efforts, different exploration paradigms have been developed for and target evaluation techniques tailored to deep marine carbonate, tight sandstone, and shale reservoirs.

1.3.1.1 Deep Marine Carbonate Exploration in the Tarim Basin

Significant deep and ultradeep oil and gas discoveries in marine carbonate strata in China have been made primarily from reef–shoal and karst reservoirs.

Earlier deep petroleum exploration in the Tarim Basin followed a paradigm that focuses on paleo-uplifts (e.g., Tabei and Tazhong) and paleo-slopes that are near source kitchens, major fault belts, and unconformities, with large-scale primary oil and gas pools as drilling targets. This led to the discovery of the giant Tahe oilfield. The fractured–

vuggy reservoirs in the Tahe field formed because of early Paleozoic (Caledonian) and late Paleozoic (Hercynian) karstification. Fracture–cavity bodies occur in vertical zonation and are largely determined by the orientation of paleofaults, the geometry of paleofractures, and the directions of paleocurrents (Li and Fan 2011).

In the past few years, the petroleum exploration paradigm in the Tarim Basin has shifted away from paleo-uplifts to strike-slip fault-controlled fracture–cavity reservoirs around paleo-uplifts and paleo-slopes and near source kitchens, with late-stage, medium- to large-sized, primary oil and gas pools as exploration targets. A ternary model was proposed to explain the origin of the strike-slip fault-controlled fracture–cavity reservoirs, highlighting the important role of fault activity and dissolution processes in the reformation of the intraplatform shoal reservoirs, where dissolution occurs pene-contemporaneously during faulting and deep burial. A conceptual model was formulated in Sinopec to emphasize "Cambrian source rocks, fracture-controlled reservoirs, vertical fluid migration, and late-stage hydrocarbon accumulation". Adoption of this concept in exploration deployment led to the discovery of the Shunbei oil and gas field in the Shuntuoguole lower uplift, which is located between the Tabei and Tazhong uplifts (Jiao 2017, 2018; Deng et al. 2018, 2019).

As seismic acquisition, imaging and combined seismic-logging prediction techniques have improved, detailed descriptions of the external configurations of strike-slip faults, the internal structures of fault-controlled cavities and fractures, and the content and saturation of reservoir fluids have become possible. It is also possible to use seismic data in the optimization of ultradeep carbonate reservoir targets, with the signal-to-noise ratio doubled and the frequency band widened by 5–10 Hz. Integration of geology and engineering, from well trajectory optimization, high-efficiency drilling, and completion to reservoir stimulation, has provided important support for the exploration and development of the Shunbei field.

1.3.1.2 Deep Marine Carbonate Exploration in the Sichuan Basin

High levels of gas accumulation in the Sichuan Basin often occur near major source kitchens, and their spatial distributions are commonly affected by dynamic adjustments in response to regional orogenic events. Therefore, preservation conditions are vital for giant oil and gas pools to survive successive tectonic movements. During the early stage of deep gas exploration in the Sichuan Basin, Sinopec carried out a detailed sequence stratigraphy and facies analysis and established sedimentary models for large-scale effective reservoirs of the Permian–Triassic reef–flat facies along the Kaijiang–Liangping platform margins. The resulting oil and gas accumulation model, and hence the exploration paradigm, was summarized as "multiplex hydrocarbon sources, near-source enrichment, ternary controls on reservoirs, superimposed constraints on accumulation, and sustained preservation". Here, *multiplex hydrocarbon sources* refer to hydrocarbons derived from the thermal maturation of source rocks, as well as those originating from in-reservoir oil to gas cracking; *near-source enrichment* places a special emphasis on oil and gas pools in close proximity to the effective source kitchen; *ternary controls on reservoir* pay attention to the important role of the original lithofacies, fracture development and fluid–rock interactions in determining reservoir porosity and permeability; *superimposed constraints on accumulation* highlight the fact that petroleum accumulation is more complex in superimposed basins than in monocyclic basins; and *sustained preservation* is most relevant for oil and gas accumulations in old reservoirs with long residence times where diffusion and leakage become pervasive and in reservoirs with super-high temperatures and high sulfate contents where hydrocarbon thermal cracking and thermochemical sulfate reduction occur. The implementation of this exploration paradigm led to the discovery of the giant Puguang and Yuanba gas fields (Ma et al. 2006, 2010a, b).

Moreover, a three-step comprehensive seismic technique for ultradeep carbonate reservoir prediction involving sedimentary facies analysis, forward modeling and seismic facies analysis, and facies-controlled multiparameter reservoir inversion was established. The application and optimization of this technique in the Puguang and Yuanba fields have generated excellent results, with the drilling coincidence rate of the predicted reservoir thickness reaching 93% in the Yuanba field. As a result, several wells in the Yuanba field were tested with more than 1 million cubic meters of daily gas flow, with a predicted high-productivity enrichment zone close to 100 km^2 (Guo 2011; Guo et al. 2018).

PetroChina developed and improved oil and gas exploration theories in Sinian–Lower Paleozoic carbonate strata during the discovery process of the giant Anyue gasfield. The proposed petroleum accumulation model, and hence related exploration paradigm, was centered on "old intracratonic troughs, old inherited uplifts, old reef–shoal bodies, old source kitchens, and sustained closed petroleum systems" (Zou et al. 2014; Wei et al. 2018). High-precision seismic techniques have been developed for the characterization and prediction of high-energy shoal carbonate reservoirs in the Lower Cambrian Longwangmiao Formation and karst fracture–cavity reservoirs in the Sinian Dengying Formation, thus facilitating the efficient exploration and development of the Anyue field (Liu et al. 2013, 2016; Yang et al. 2019).

1.3.1.3 Deep Tight Gas Exploration in the Tarim and Sichuan Basins

Deep tight gas exploration in China is exemplified by accumulations in the Kuqa foreland thrust belts of the Tarim Basin and in the Xujiahe Formation of the western Sichuan Basin. Tight gas accumulations in the Mesozoic sandstone reservoirs along the Kuqa foreland thrust belts are characterized by folded blanket-like source beds, ultradeep tight sandstone reservoirs, thick evaporates, and imbricated structures. A methodology framework was established for evaluating the basin/tectonic unit, play and target levels in foreland thrust belts, with highly advanced techniques for seismic prestack depth domain imaging, modeling, and comprehensive reservoir evaluation (Yu et al. 2019).

These techniques have expedited the discovery of an ultradeep gas province with more than 1 tcm of proven reserves in the Kuqa Depression.

1.3.1.4 Deep Shale Gas Exploration in the Sichuan Basin

Successful deep shale gas exploration in China has occurred mostly in the southern and northeastern areas of the Sichuan Basin, targeting the Upper Ordovician–Lower Silurian Wufeng–Longmaxi shales. The Weirong shale gas field has been discovered, together with significant findings in the Yongchuan and Luzhou blocks in southern Sichuan, as well as several blocks in northeastern Sichuan. The key geological controls for deep shale gas accumulation in the Sichuan Basin include the high hydrocarbon source potential, good shale reservoir quality, and favorable preservation conditions. Two types of shale gas accumulation models have been established based on the structural style of the gas shales. One is structurally controlled, with the Fuling shale gas field being a typical example, where the Wufeng–Longmaxi gas shales occur in a faulted anticline structure (Jiaoshiba). Another is "continuous" shale gas accumulation over a large area with little structural disturbance, represented by the Weirong deep shale gas field. Structurally controlled shale gas accumulations often occur in small but "focused" geographic areas with relatively high gas intensity per volume, whereas continuous accumulations tend to have much greater amounts of gas resources because of their areal extent.

A framework of methodologies, techniques and workflows has been established for the integrated evaluation of geological, engineering, and economic "sweet spots" of deep shale reservoirs in the Sichuan Basin (Ma et al. 2018). A range of seismic techniques have been developed for sweet spot prediction, considering key geological and engineering parameters such as the shale thickness, total organic carbon content, porosity and permeability, brittleness, depth, pressure gradient, gas in place, basement faulting and structural complexity, horizontal pressure differential, degree of natural fracturing, lateral landing target and orientation, and capacity for hydrofracturing.

1.3.2 Development Theories and Techniques

Several field development and reservoir engineering techniques and methods have been tailored to meet the specific needs of deep and ultradeep oil and gas development in China, with significant advancements in deep oil and gas field development theories (Li et al. 2016, 2017, 2020; Ma et al. 2020, 2021, 2022).

1.3.2.1 Development in Deep Marine Carbonate Reservoirs

Key technological progress in deep carbonate oil and gas development includes (1) precise reservoir description and geological modeling of fracture–cavity and reef reservoirs; (2) improved mechanistic understanding and numerical modeling of fluid flows in fracture–cavity reservoirs; (3) optimization of well spacing design and injection–production parameters for fracture–cavity reservoirs; and (4) theories and methods of water flooding development and gas-injection-enhanced oil recovery.

Tahe oilfield development in the Tarim Basin occurred during the early discovery–production test stage (1997–2000), rapid production increment stage (2001–2010), and stable production stage (2011 onward). Water injection experiments began in 2006 in some blocks, and the water and gas injection development mode was later found to be more conducive to achieving effective development with steady production. A highly innovative technology and management mode for the volumetric development of fracture–cavity reservoirs was established, including multiple shot arrays, multiple-process potential tapping, and volumetric water and gas injection development. As a result, the precision of the imaging of the fracture–cavity bodies decreased from the initial error margins of 30 m to 10 m, with a 92.7% coincidence rate achieved by drilling for fracture–cavity bodies with diameters greater than 5 m. The oil recovery ratio reached 20% in the tested blocks.

The Yuanba gas field is the world's deepest sour gas field produced from reef reservoirs.

Early production tests from vertical wells yielded subeconomic results. Through a period of painstaking practices, suitable technologies for precise reef reservoir characterization and horizontal well production have been established, with a substantial increase in per-well production.

1.3.2.2 Development in Deep Tight Gas Reservoirs

Deep tight gas production from nonmarine clastic reservoirs in the Kuqa Depression of the Tarim Basin is problematic due to the complexity of gas–water relationships and the strong heterogeneity in water invasion patterns. Extensive studies have investigated static gas–water distributions, microscopic water invasion mechanisms, and dynamic water invasion. Thus, significant progress has been made in the improved description of gas–water distribution, better interpretation of water invasion patterns, and integrated static and dynamic evaluation of water–control development (Jia et al. 2019).

1.3.2.3 Development of Deep Shale Gas Reservoirs

With respect to deep shale gas development in the southern Sichuan Basin, key progress is shown in three different aspects. An improved understanding of gas flow mechanisms in deep shales was achieved through gas adsorption/desorption–low flux flow–stress sensitivity experiments. Integrated physical and mathematical modeling was realized through physical simulation experiments, incorporating mathematical models for gas flows in coupled multiphase and multiscale fractured porous media, embedded discrete fracture models and fluid flow simulation methods, and microseismic-based postfracturing modeling methods for horizontal wells. Learning curves have been shortened through exploration and development practices in the optimization of shale gas "sweet spot" evaluation, well spacing design, volumetric fracturing parameters for horizontal wells, production system design, and platform factory style management modes.

1.3.3 Advances in Engineering Technology

The successes in deep oil and gas exploration and development have been highly dependent on the ceaseless advances made by petroleum engineers in well construction and reservoir simulation. Significant progress has been made in safe and rapid drilling, well completion tests, reservoir stimulation techniques, and logging and drilling equipment for ultradeep wells (Lu 2018).

1.3.3.1 Deep Well Construction Capacity

The deep well construction capacity in China has exceeded 10,000 m, and several ultradeep wells with greater than 13,000 m of drilling depth are now in the planning stages. By the end of 2019, Sinopec had drilled 272 wells that were more than 7000 m deep, of which 33 wells were over 8000 m deep. By May 2022, PetroChina had completed 41 wells with drilling depths greater than 8000 m. By May 2025, PetroChina and Sinopec had drilled a total of 454 ultradeep wells with drilling depths greater than 8000 m in the Tarim Basin.

The first deep well in China with a drilling depth greater than 6000 m, Nvji-1, was drilled in 1976 in the Sichuan Basin, with a drilling depth of 6011 m. The first deep well in China with a drilling depth greater than 7000 m, Guanji-1, was drilled in 1978 in the Sichuan Basin, with a drilling depth of 7175 m. Since 2000, the record depth and number of ultradeep wells in China have been broken many times with the continuous improvement in drilling technologies. The Tashen-1 well drilled in the Tarim Basin in 2006 reached a depth of 8408 m. The Mashen-1 well drilled in the Sichuan Basin in 2008 reached a depth of 8418 m. The Shunbei P2H well drilled in the Tarim Basin in 2017 reached a depth of 8433 m. The Luntan-1 well drilled in the Tarim Basin in 2019 reached a depth of 8882 m. The SY001-H6 well, which was completed in June 2022 in the Sichuan Basin, reached a depth of 9010 m, with a bottom-hole temperature of 180 °C, a formation pressure of 130 MPa and a H_2S concentration of 6 g/m^3, and remains the deepest well in Asia.

1.3.3.2 Safe and Rapid Drilling Technology

Advances in safe and rapid technology for drilling ultradeep wells have been made from the efficient drilling of hard rock formations, safety control for drilling fracture–cavity reservoirs, and drilling fluids for deep–ultradeep wells to well cementing under multiple pressure regimes. In particular, the density of the super high-density drilling mud has reached a density of 3.00 g/cm^3 and is temperature resistant up to 240 °C; the high-temperature and high-density drilling mud has reached a density of 2.30 g/cm^3 and is temperature resistant up to 220 °C; the high-temperature resistant, environmentally friendly and low-friction mud is temperature resistant up to 180 °C, with a lubrication factor of 0.12; and the high-temperature anti-gas channeling cement slurry system has a density of 1.90 g/cm^3 at a circulation temperature of 220 °C.

1.3.3.3 Ultradeep Well Logging Tools

Ultradeep well logging equipment has been developed, with temperatures and pressures of up to 200 °C and 175 MPa, respectively. High-pressure leakage well logging tools, construction safety techniques, and online detection systems for ultradeep well logging are also established. Wireline logging data were acquired 19 times from 9 ultradeep wells in the Shunbei oil and gas field, with the deepest data in the Peng-1 well being logged at a depth of 8450 m.

1.3.3.4 Completion Test and Reservoir Stimulation

Various acid fracturing techniques have been developed to meet the specific demands for stimulating deep carbonate reservoirs, e.g., deep penetration and distance communication, long section intervals and large fracture heights, complex fracturing in fractured zones, and acidizing fluid systems with temperatures up to 160 °C. Technologies developed domestically have met the basic needs for the completion test of deep to ultradeep wells with reservoir depths up to 8000 m and for acid fracturing treatment of reservoirs with temperatures up to 160 °C. Oil and gas develop-ment practices in the Puguang gas field and Tahe oil and gas field indicate that these technologies are particularly useful for carbonate reservoirs with deep burial, high temperature, high pressure, and high closure stress.

1.3.3.5 Deep Shale Gas-Related Technologies

Drilling speed, well quality, stimulation volume and per-well gas productivity are important measures for the engineering quality of a deep shale gas well. Extensive studies have been conducted on sweet spot evaluation, integrated geological and engineering design, precise formation evaluation, new oil-based drilling fluids, new elastic to tough and foamed cement slurries, and high-efficiency fracture fluids (Lu and Ding 2018). As a result, significant progress has been made in the development of customized solutions in the drilling, cementing, and fracturing engineering of deep shale gas reservoirs. Substantial progress has been made in deep-speed drilling technology, with noticeable breakthroughs in pilot tests of water-based drilling fluids. Cementing technology can be used to construct deep shale gas wells with high-quality wellbore integrity at depths less than 4000 m, with integrated foamed cement slurry cementing equipment being manufactured. The optimization of hydraulic fracturing technologies in the Fuling, Weirong and Dingshan–Dongxi gas fields has generated excellent results.

1.4 Key Development Realms and Future Research Directions

As indicated in the previous sections, deep petroleum exploration and development in China has made enormous progress, making it a significant and integral part of the national strategy for securing energy supply. However, owing to the complexity of petroleum systems in deep sedimentary strata, many geological uncertainties remain to be resolved, and new technological advances need to be made to make increasingly greater breakthroughs in deep petroleum resource exploitation.

1.4.1 Deep Resource Distribution

The results of the 2015 China National Oil and Gas Resources Assessment (Strategic Research Center of Oil and Gas Resources 2017a) indicate that the total petroleum resource in the deep and ultradeep strata in China is 76.3 bt of oil equivalent, accounting for 35% of the total petroleum resources in China. The total oil resources in deep and ultradeep strata in China amount to 26.6 bt, accounting for 21% of the total oil resources in China (Fig. 1.4). The gas resources in deep and ultradeep strata in China total 49.7 tcm, accounting for 55% of the total gas resources in China (Fig. 1.5). The proven oil and gas reserves in the deep and ultradeep strata in China are relatively low, accounting for only 13% and 10% of the national oil and gas resources, respectively.

Undiscovered geological oil and gas resources in China are 23.1 bt and 44.8 tcm, corresponding to 26% and 56% of the total oil and gas resources in China, respectively.

Geographically, the undiscovered oil resources in eastern and western China are 5.8 bt and 13.6 bt, or 25% and 59% of the undiscovered total oil resources in China, respectively (Fig. 1.4). Natural gas occurs mainly in central and western China, with 15.6 tcm and 14.7 tcm of undiscovered gas resources accounting for 35% and 33% of the undiscovered total deep gas resources in China, respectively (Fig. 1.5). Deep petroleum resources in China occur mainly in six sedimentary basins (Tarim, Sichuan, Bohai Bay, Ordos, Songliao and Junggar).

According to the results of the 2015 China National Dynamic Evaluation of Shale Gas Resources (Strategic Research Center of Oil and Gas Resources 2017b), the shale gas resources in sedimentary strata with burial depths shallower than 4500 m were estimated to total 121.9 tcm,

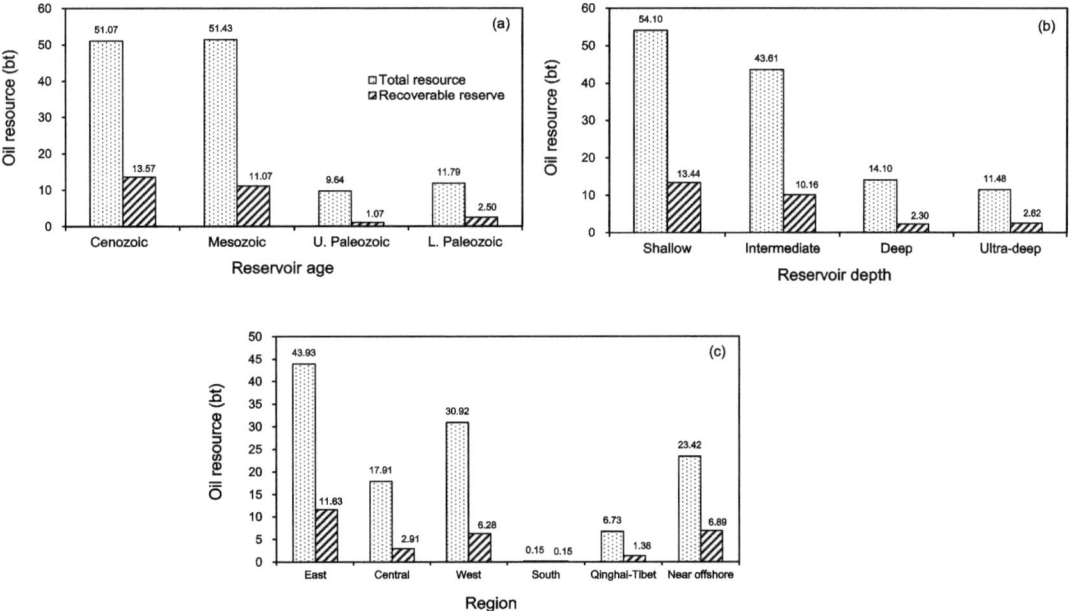

Fig. 1.4 Estimated oil resources and recoverable reserves in China as a function of reservoir age, burial depth and geographic region. *Data source* Ministry of Natural Resources of the People's Republic of China (2018)

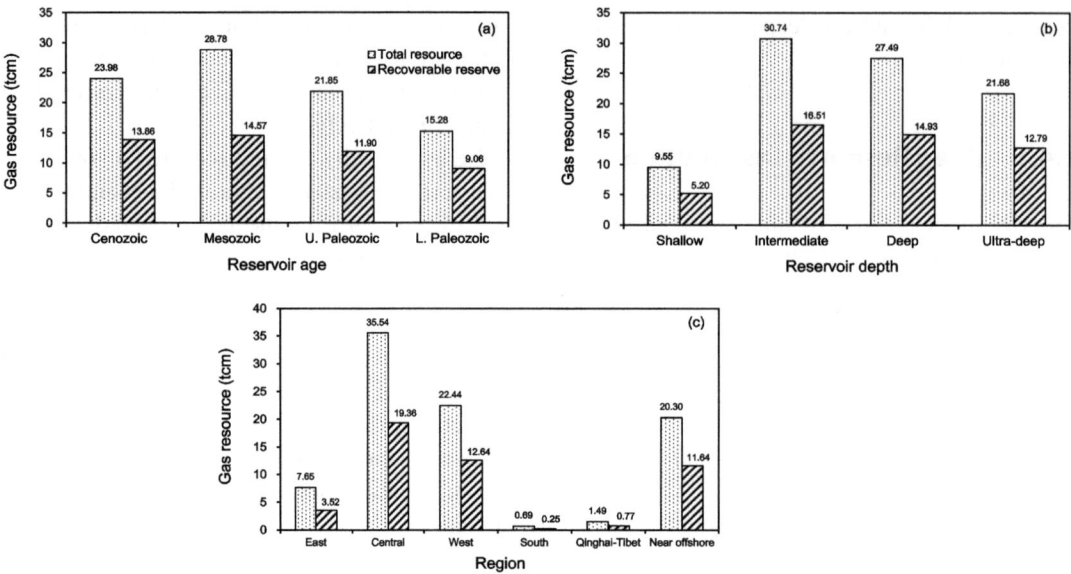

Fig. 1.5 Estimated natural gas resources and recoverable reserves in China as a function of reservoir age, burial depth and geographic region. *Data source* Ministry of Natural Resources of the People's Republic of China (2018)

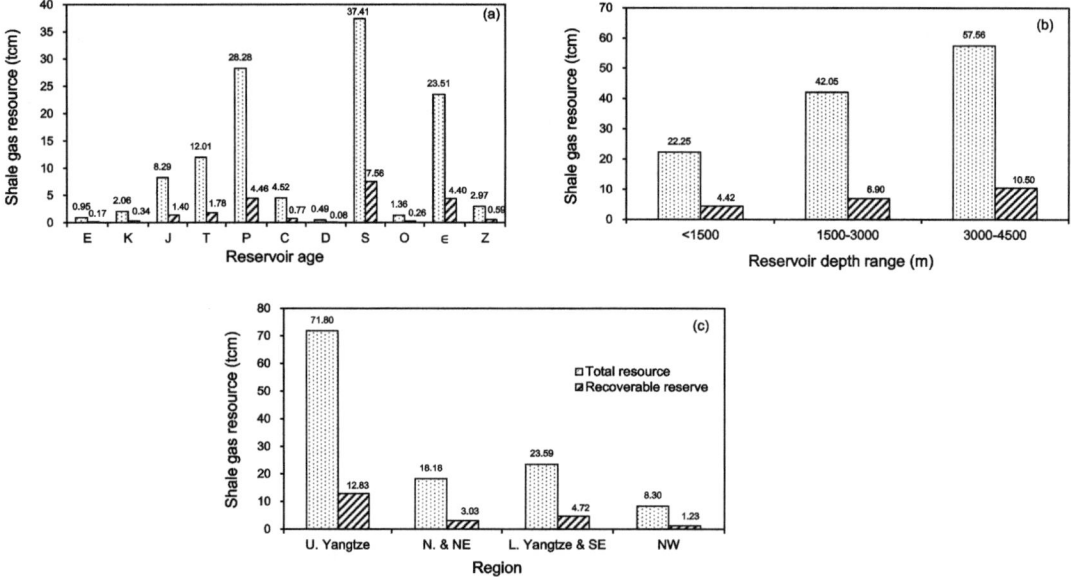

Fig. 1.6 Estimated shale gas resources and recoverable reserves in China as a function of reservoir age, burial depth and geographic region. *Data source* Ministry of Natural Resources of the People's Republic of China (2018)

of which 21.8 tcm is recoverable (Fig. 1.6). No national assessment was performed during the last round for deep shale gas resources in reservoirs with burial depths greater than 4500 m. Slightly more than half of the national total shale gas resources and recoverable reserves exist in the Sichuan Basin and adjacent areas, with values of 63.8 tcm and 11.3 tcm, respectively. The total shale gas resources in deep shale gas reservoirs with burial depths ranging from 3500 to 4500 m were estimated to be 12.5 tcm (Ma et al. 2018; Ma 2017).

1.4.2 Key Development Realms

Sedimentary basins in western China are characterized by superimposed basins, with multiple source beds and multiple vertically stacked plays (Jin 2014). "Buried hill" traps occur widely in the basement rocks underlying the Mesozoic–Cenozoic rift basins in eastern China. These basins have excellent deep petroleum potential, with a high probability of finding more giant-, large- and medium-sized oil and gas fields. Deep sedimentary strata in the Tarim and Sichuan Basins, buried hill traps in the Bohai Bay Basin and deep shale gas represent four major realms for deep petroleum exploration and development in China.

1.4.2.1 Tarim Basin

The Tarim Basin, with a geographical area of $560,000 \, km^2$, consists of an old Paleozoic cratonic basin and superimposed younger Mesozoic–Cenozoic foreland basins (Fig. 1.7). Hydrocarbon source rocks occur in the Cambrian to Mesozoic

strata, with multiple petroleum plays. Multistage tectonic activities led to multiple unconformities, which act as excellent conduits for hydrocarbon migration and thus exert significant control on multiphase hydrocarbon accumulation, destruction and redistribution. Petroleum accumulation in the Tarim Basin occurs in different types of reservoirs, with variable trapping styles and gas-to-oil ratios. Most of the oil and gas reservoirs are in deep and ultradeep strata. The total oil and gas resources in the Tarim Basin are estimated to be 12.07 bt and 14.7 tcm, respectively (Kang 2018).

The main petroleum exploration and development realms in the Tarim Basin include the cratonic marine sequence in the platform-basin region, the nonmarine sequence in the Kuqa Depression, and the cratonic marine to foreland nonmarine sequence in the Southwestern Depression. The Cambrian–Ordovician marine carbonates and Silurian and Mesozoic–Cenozoic sandstone reservoirs are the main exploration targets in the platform basin region. The Cretaceous–

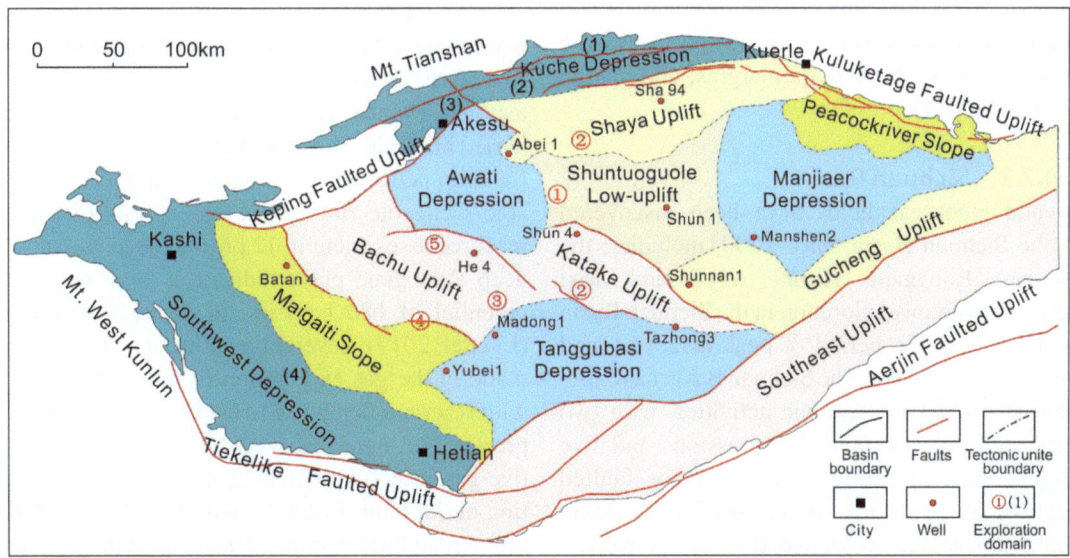

Fig. 1.7 Potential deep oil and gas exploration domains in the Tarim Basin: deep marine plays: ① the fracture-cavity and interlayer karst reservoirs in the Yingshan and Penglaiba formations in the circum-Awati to Manjiaer transition zone; ② the epidiagenetic paleokarst reservoirs in the Yijianfang, Yingshan and Penglaiba formations in the Tazhong and Tabei uplifts; ③ the reef flat karst reservoirs in the Liangditage Formation in the Tazhong uplift; ④

the pre-salt reservoirs in the Cambrian strata; and ⑤ the marine clastic reservoirs in the platform-basin region. Deep non-marine petroleum plays: (1) the pre-salt ultra-deep Cretaceous strata in the Kuqa foreland thrust belts; (2) the Jurassic-Triassic strata in the northern Kuqa compressional structural zone; (3) the pre-Mesozoic strata in the southern Kuqa margins; and (4) the Southwestern Depression

Paleogene nonmarine clastic sequences are the focus in the Kuqa Depression. In contrast, Cretaceous–Paleogene carbonates and Neogene clastic reservoirs are the targets in the Southwestern Depression (Qi 2019).

As shown in Fig. 1.7, deep marine petroleum plays in the Tarim Basin include (1) the fracture–cavity and interlayer karst reservoirs in the Lower–Middle Ordovician Yingshan ($O_{1-2}y$) and Lower Ordovician Penglaiba formations (O_1p) in the circum-Awati to Manjiaer transition zone; (2) the epidiagenetic paleokarst reservoirs in the Middle Ordovician Yijianfang (O_2yj), Yingshan ($O_{1-2}y$) and Penglaiba formations (O_1p) in the Tazhong (Katake) and Tabei (Shaya) uplifts; (3) the reef flat karst reservoirs in the Upper Ordovician Liangditage Formation (O_3l) in the Tazhong uplift; (4) the presalt reservoirs in the Cambrian strata; and (5) the marine clastic reservoirs in the platform-basin region.

Deep nonmarine petroleum plays in the Tarim Basin include (1) presalt ultradeep Cretaceous strata in the Kuqa foreland thrust belts; (2) Jurassic–Triassic strata in the northern Kuqa compressional structural zone; (3) pre-Mesozoic strata in the southern Kuqa margins; and (4) the Southwestern Depression.

1.4.2.2 Sichuan Basin

Several giant gas fields have been discovered in the Sichuan Basin since 2000, including the Puguang, Longgang, Yuanba, Anyue and Moxi fields. Petroleum exploration in the Sichuan Basin is still in a boom period of peak discovery and rapid reserve growth. The Sichuan Basin has experienced multicyclic tectonic activities, with two large intracratonic faulted depressions and five regional unconformities. Large rifts or faulted depressions controlled the deposition of the main petroleum source rocks and thus the petroleum source kitchens. Large platform margins and regional unconformities significantly influence the distribution of petroleum reservoirs. Ample evidence from the exploration results indicates that the conditions for giant gas accumulations are definitely present in the deep and ultradeep strata of the Sichuan Basin, as high-quality reservoirs occur widely in deep and ultradeep strata.

Five main realms for deep petroleum exploration and development have been identified in the Sichuan Basin (Fig. 1.8): (1) paleouplifts and paleoslopes (e.g., the Central Sichuan and Dazhou–Kaijiang paleouplifts); (2) high-energy facies along the platform margins or trough margins (e.g., the Deyang–Anyue intracratonic faulted depression, the Kaijiang–Liangping trough margins, the shoal dolomite along the western Sichuan platform margins, the mound–shoal complexes in the Sinian Doushantuo and Dengying formations deposited along the eastern Sichuan paleorift margins, and those along the Chengkou–western Hubei trough margins); (3) major unconformities (e.g., Middle Triassic Leikoupo Formation in the western and northeastern Sichuan Basin); (4) the nonmarine reservoirs in the Upper Triassic Xujiahe Formation; and (5) new exploration frontiers in the Ordovician strata in northern and central Sichuan Basin.

1.4.2.3 Deep Strata and Buried Hills in the Rift Basins of Eastern China

Sedimentary basins in eastern China consist of several rift basins formed as the result of different tectonic events. Deep sedimentary strata and buried hills in the basement are important habitats for deep petroleum accumulation in these Mesozoic–Cenozoic rift basins. The results of a gas resource assessment in 12 prolific deep and ultradeep Paleogene plays indicate a large resource potential of 680 bcm in the Bohai Bay Basin (Li et al. 2019). Several breakthroughs have been made in the gas exploration of buried hill plays with reservoir depths up to 6000 m, indicating the presence of high-quality and thus highly productive gas reservoirs (Zhao et al. 2011). Reactivation of old faults inside buried hills often leads to segmented fluid systems. Thus, both subtle buried hills (such as hillsides, broken walls, and mountainsides) and the negative segments of faulted blocks between the buried hills become potential hydrocarbon traps when the necessary source, reservoir and seal conditions are present. It is also possible to find giant gas fields in the oil-rich Bohai Bay Basin because the deep sedimentary strata are currently within the gas–condensate window.

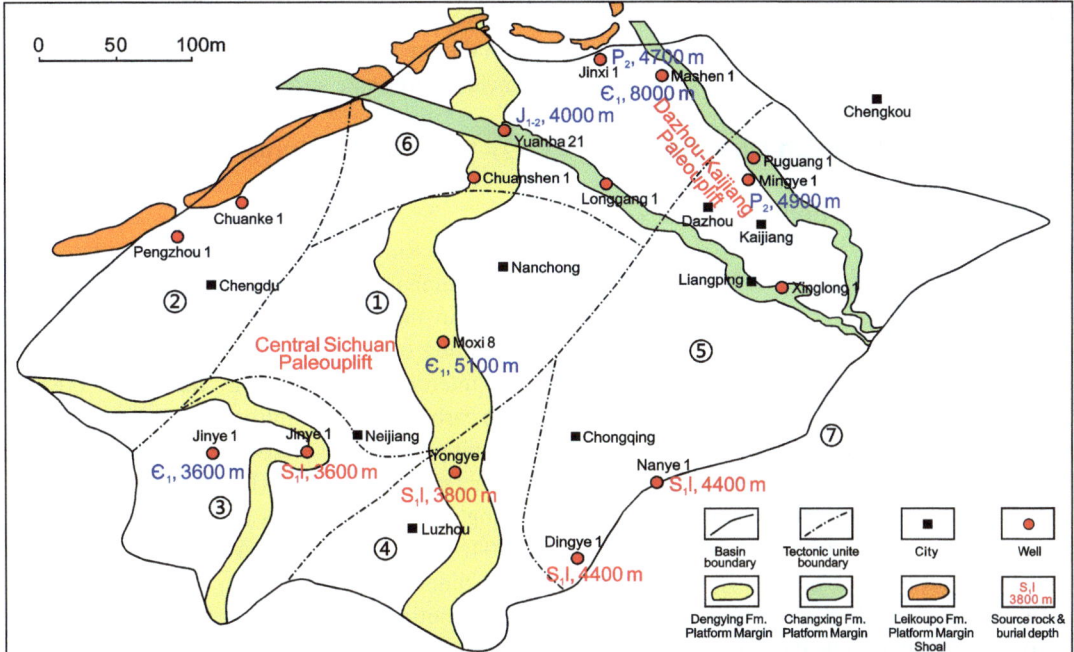

Fig. 1.8 Potential deep oil and gas exploration domains in the Sichuan Basin. Tectonic subdivisions: ① Central Sichuan gentle structure zone; ② West Sichuan low amplitude structure zone; ③ Southwest Sichuan low amplitude structure zone; ④ South Sichuan low amplitude structure zone; ⑤ East Sichuan high amplitude structure zone; ⑥ North Sichuan low amplitude structure zone; and ⑦ Qiyueshan Fault belt

1.4.2.4 Deep Shale Gas

Although sedimentary basins in China are rich in shale gas resources in several stratigraphic intervals, commercial development has been conducted mainly from the Upper Ordovician–Lower Silurian Wufeng–Longmaxi Formations in the Sichuan Basin and peripheral areas. Future development realms for deep shale gas resources in China are distributed in the Sichuan Basin and adjacent regions, the Tarim Basin, the Great Huabei Basin, and the Ordos Basin. The Wufeng–Longmaxi shales in the Sichuan Basin constitute the most realistic development area for deep shale gas resources in China, as the base of the Lower Silurian black shales with burial depths greater than 3500 m in the Sichuan Basin and peripherals are distributed in an area of 128,000 km^2. This area is twice as large as that for the Silurian shales with burial depths less than 3500 m.

The main deep shale gas plays in the Sichuan Basin and peripherals include (1) Silurian black shales in the southern, southeastern, and north-eastern Sichuan Basin; (2) Cambrian black shales in the central and northeastern Sichuan Basin; (3) Permian marine to nonmarine shales in the northeastern Sichuan Basin; and (4) Jurassic nonmarine shales in the northern Sichuan Basin.

The main deep shale gas plays in the Tarim Basin include (1) the Cambrian marine black shales in the Yuli and Tadong regions; (2) the Ordovician marine shales in the Yuli and Tadong regions; (3) the Carboniferous shales in the Bachu–Maigaiti region; and (4) the nonmarine Jurassic shales in the Baicheng–Yangxia and Caohu–Mandong regions.

In the Great Huabei Basin, the main deep shale gas plays include Carboniferous–Permian marine to nonmarine shales in southern Huabei and in the Bohai Bay Basin.

In the Ordos Basin, the main target for deep shale gas exploration lies in the Lower Paleozoic marine sequence, as indicated by the results of the Zhong 4 well on the western margin of the Ordos Basin, where commercial gas flow was obtained from the black shales in the Middle Ordovician Wulalike Formation (He et al. 2019).

1.4.3 Future Research Directions

1.4.3.1 Petroleum Geoscience Issues

(1) Formation and Evolution of Marine Prototype Basins

Numerous studies have focused on improving our basic understanding of the basement properties and tectonic frameworks of sedimentary basins in China, with the dynamic processes of basin evolution and superimposition being reconstructed (Zhang 1997; Xu et al. 2005; Ingersoll 2012; Baitsch-Ghirardello et al. 2014; Koptev et al. 2016; Huang et al. 2008). Several types of superimposed marine basins have been recognized, with detailed reconstructions of basin prototypes at different geological periods for the Sichuan and Tarim Basins (Wang et al. 1990; Zhou and Lin 1995; Ma 2009). Depositional models for sedimentary fillings in southern China (Ma 2009) and the Ordos Basin have been established. Studies of the Sinian–Early Cambrian aulacogen in the central Sichuan Basin (Liu et al. 2013; Wang et al. 2017), the Middle–Late Proterozoic rift trough in the Ordos Basin (Sun and Wang 2016), the Nanhuaa–Sinian rift trough and the Cambrian–Ordovician "Manjiaer aulacogen" in the Tarim Basin have provided circumstantial evidence for the tectonic differentiation of marine cratonic basins in China (Aydin 2000; Budd et al. 1995; Chen et al. 2009; Han et al. 2017; He et al. 2010, 2011, 2019; Jia et al. 2005; Lin 2002; Luo et al. 1994; Wu et al. 2018; Zhang 2008; Zhang et al. 2011; Zhou et al. 2015). However, the late-stage deformation and destruction along the orogenic belt at basin margins increased the difficulties in the reconstruction of basin prototypes and sedimentary facies for small residual cratonic blocks. Therefore, detailed correlation studies at various scales are necessary for determining the origin and controlling factors of early rifting processes, the mechanisms underlying tectonic and sedimentary differentiation across different tectonic blocks, and the preservation conditions for oil and gas in ancient carbonate strata.

(2) Source Rock Distribution and Hydrocarbon Generation

Deep marine petroleum source rocks occur predominantly in Paleozoic deep petroleum plays in the Tarim, Sichuan and Ordos Basins, including organic-rich black shales deposited in the intercraton depressions, laminated organic-rich carbonates in the intracraton depressions, and swampy shales and mudstones deposited in broad intracraton depressions (Liang et al. 2000, 2009; Rong et al. 2006; Tenger et al. 2016; Zhong et al. 2004; Fu and Jia 1984; Wang and Han 2011). In Chinese cratonic basins, the stable Precambrian continental crust background and associated low thermal flux and low geothermal gradients tend to delay the oil-to-gas conversion process in the Middle–Lower Paleozoic to Sinian strata. In contrast, thick Mesozoic–Cenozoic sedimentary covers provide favorable conditions for late-stage gas generation and preservation in Paleozoic to Sinian strata. The hydrocarbon generation centers for the main petroleum source beds in the Tarim, Sichuan and Ordos Basins have different vertical stacking patterns from those of the overlying sediments, leading to significant differences in the petroleum enrichment zones and hence in exploration plays (Ma et al. 2008). Hydrocarbons in these cratonic basins are potentially derived from multiple source beds, generated during multiple periods, and charged from multiple source kitchens (Dai 1992; Xu 1994; Liu et al. 2005, 2009; Zhao et al. 2001, 2005). Therefore, more research efforts should be made on source rocks to identify high-quality source facies, develop unbiased oil and gas source tracers, select reliable maturity indicators, and establish criteria for effective source rock evaluation. In particular, the Lower Cambrian Yiertusi Formation, Cambrian–Ordovician, and Lower Cambrian Qiongzhusi Formation need to be tested for their effectiveness as the dominant source rocks in the Tarim, Ordos and Sichuan Basins, respectively.

(3) Carbonate Deposition and Diagenesis

Deep marine carbonates in China occur mainly in the Middle–Late Proterozoic to Triassic strata in the cratonic Sichuan, Tarimim and Ordos Basins and in the rift basins and adjacent orogenic belts of the Huabei region (Fan et al. 2007; Ma et al. 2007; Zhao et al. 2010; Zou et al. 2011; Guo 2011; Huang et al. 2014). Controlled by the difference in basement activity, the sedimentary filling sequences in these basins vary widely, with several carbonate platform margins being formed along the small craton blocks. As more breakthroughs have been made in deep marine petroleum exploration, the importance of primary interparticle pores and near-surface secondary dissolution pores in carbonate reservoirs has been gradually realized (Li et al. 2016; Hu et al. 2018). The main task of petroleum exploration in carbonate rocks is to identify favorable source facies for primary porosity formation and ideal diagenetic settings for porosity generation and preservation. Therefore, it is highly desirable to use seismic and geophysical logs and map the facies distributions of deep carbonate rocks within a sequence stratigraphic framework. Moreover, changes in the overall environmental conditions can be used to explain the rock–fluid interactions from the initial deposition to the present-day conditions and their influence on the transformation of hydrocarbon storage in deep reservoirs. Large efforts should be directed to three different types of carbonate reservoirs to: (1) further investigate the tectonic and sedimentary differentiations in different carbonate platforms and explain their origin, distribution and influence on reef shoal development; (2) address old dolomite problems with new approaches to understand the genesis, key geological controls and quantitative models for spatial prediction; and (3) study ancient reefs and microbiolites, with a clear focus on geological modeling for predicting large reservoir bodies.

(4) Reservoir Characterization and Distribution Prediction

A ternary model was proposed to explain the genesis of high-quality carbonate reservoirs char-acteristic of small ancient cratons in China (Ma et al. 2010a, b). Platform margin reef, tidal flat shoal, and intertidal dolomitic flat facies play a dominant role in the development of primary porosity in carbonate reservoirs. Tectonic and geostress coupling leads to fracture development and creates opportunities for potential carbonate dissolution. Fluid–rock interactions control dissolution processes and porosity preservation in deep carbonate strata (Jin et al. 2006; Liu et al. 2008; Chen 2008). In light of the large variations in reservoir lithofacies and dissolution characteristics, it is important to strengthen the studies in the following areas: (1) tectonic–lithofacies paleogeography and carbonate platform characterization; (2) key geological controls and quantitative models for reservoir deposition and development; (3) genetic models for ultradeep dolomite reservoirs; (4) geological models assisting geophysical techniques for reservoir description and prediction; (5) temporal and spatial relationships between reservoir porosity formation and petroleum fluid charge; and (6) the roles of evaporate rocks in reservoir formation and petroleum accumulation.

(5) Petroleum Accumulation and Distribution Models

Case studies of the major oil and gas fields in the deep marine carbonate strata of the Tarim, Sichuan and Ordos Basins indicate several models of oil and gas accumulation, with the reef shoal reservoirs, karst reservoirs in the buried hills and along the paleo-uplifts and paleo-slopes favoring oil and gas accumulation over a large area (Zhao et al. 2012; Pang et al. 2012; Jia and Pang 2015; Luo et al. 2016). The oil and gas distributions in these basins clearly differ. In the Sichuan Basin, the dominant oil and gas habitats include intracratonic sags, paleo-uplifts and basin-range structures (Ma et al. 2006; Zou et al. 2014; Liu et al. 2016; Guo et al. 2018). In the Tarim Basin, oil and gas are distributed mainly in fracture–cavity reservoirs along uplifts and slopes (He et al. 2010). In contrast, source–reservoir relationships appear to play a more important role in the Ordos Basin (Wang et al. 2005). Further studies need to address the uncertainty in the source rock distribution in deep strata and develop effective methods for probing the oil (gas)-source relationships. In light of multistage hydrocarbon accumulation and redistribution, the validity of many molecular and isotopic tracers must be contested before they can be used to reconstruct the dynamic processes of hydrocarbon generation, migration, accumulation, and dispersion in deep strata. Given the strong heterogeneity in deep strata, taking a systematic approach to characterize deep petroleum system elements and linking these elements to global geological events is highly recommended.

(6) Deep Marine Shale Gas Accumulation and Distribution

Many studies have revealed the basic geological conditions for marine shale gas accumulation in the Sichuan Basin (Guo and Liu 2013; Hu et al. 2014; Liu 2015; Guo 2014; Guo et al. 2016a, b; He et al. 2017; Wei et al. 2017a, b; Zou et al. 2016). Whether rapid growth in shale gas production in the Sichuan Basin can be realized in the next few years depends on whether more commercial shale gas production can be sustained outside of the established gas fields in the Upper Ordovician–Lower Silurian Wufeng–Longmaxi shales with burial depths greater than 3500 m or whether new shale gas plays can be found in the Cambrian and Permian strata. More studies should focus on detailed shale bed correlation, fine structural interpretation, quantitative fracture prediction, and integrated characterization of the shale gas content, mode of occurrence, and frackability.

1.4.3.2 Deep Oil and Gas Exploration Theories

Deep marine petroleum exploration in China faces several difficult theoretical and technological problems. The key to solving these problems is to prioritize the following fundamental questions related to deep petroleum exploration theories.

(1) Deep Marine Petroleum Resource Potential and Distribution

Most deep marine oil and gas discoveries in China have been made in Paleozoic strata in the Tarim, Sichuan and Ordos Basins, with a consensus on the main source rocks in these basins. However, as more data become available from shale gas exploration, geological models for the creation, deposition, and preservation of the organic-rich Wufeng–Longmaxi shales have evolved over the arguments concerning whether black shales were deposited in deep or shallow water. An increased understanding of these shales indicates that the hydrocarbon source rocks and thermal evolution and accumulation processes in Paleozoic marine strata need to be understood further. Reassessment of deep marine petroleum resources and potential play areas has thus become critical for future deep petroleum exploration. Additionally, the exploration prospects and resource potential in the underlying carbonate reservoirs of the Nanhuaan–Sinian deposit and those of older ages are worth investigating.

(2) Diversity in Marine Carbonate Reservoirs and Complexity in Traps

Marine carbonate reservoirs in the Tarim, Sichuan and Ordos Basins can be divided into four major genetic types: (a) reef–shoal reservoirs; (b) karst fracture–cavity reservoirs; (c) dolomite reservoirs; and (d) fractured reservoirs. However, carbonate reservoirs in different basins and different tectonic units display large variations in their characteristics. Deeper insights are necessary to understand the reasons behind these differences and the implications for oil and gas accumulation. Hydrothermal dolomites associated with tectonic faulting and magmatic activity have been constantly encountered in recent exploration. For practical reasons, however, many explorers are still puzzled by the following questions: (a) How can the key trap factors for such carbonate reservoirs be more accurately defined? (b) How can the boundaries of a trap that is associated with an unconformity, a fault, a karst, a group of fractures, or a hydrothermal dolomite body be determined? (c) How can the top and bottom seals for a trap associated with the Ordovician karst fracture–cavity reservoirs in the Tarim Basin be defined? (d) How significant are the widespread tight carbonate rocks for "continuous" oil and gas accumulation?

(3) Multiphase Petroleum Charges and Preservation Effectiveness

In the superimposed basins, Paleozoic carbonate strata have experienced multiple orogenic events, leading to the redistribution and even destruction of the early accumulated oil and gas. Excellent conditions for oil or gas preservation clearly exist, at least for a considerable period of geological history, when the giant Tahe oil field or Puguang gas field formed. In contrast, the Majiang area beyond the southern margin of the Sichuan Basin represents the other end of the spectrum, where two phases of petroleum accumulation and destruction occurred. According to Gao et al. (2012), the first phase of small-scale oil accumulation and destruction in the Majiang area occurred during the Late Caledonian; the second phase of oil charging led to the formation of a giant oil field during the Hercynian, which was converted to a gas field during the Indosinian and eventually destroyed during the Yanshanian. Some of these processes were related to regional tectonic activity and thus controlled the distribution of oil and gas throughout the entire basin. In contrast, some are related to local factors, with limited impact at the local scale. Therefore, it is crucial for petroleum explorers to study the impact of tectonism on oil and gas destruction or preservation, determine the timing of key tectonic events, and understand the key controls of oil and gas occurrence.

(4) Ultradeep Petroleum Exploration Realms and Theoretical Guiding Principles

As new discoveries are constantly being made in deep petroleum exploration, new types of carbonate plays have emerged, and many traditional concepts in petroleum geology have to be redefined. To date, the burial depth of effective oil and gas reservoirs has approached 9000 m. There is an urgent need to determine (a) what the reservoir depth limit is for petroleum exploration; (b) whether there is a depth limit for oil and gas preservation; and (c) whether nanoscale pores are effective for hydrocarbon storage in carbonate reservoirs. Early studies have reached a consensus that effective petroleum source rocks determine the oil and gas distribution, whereas paleo-environment, paleo-uplifts and paleo-slopes dictate reservoir development. For deep exploration target selection, one must consider the traditional geological evaluation box by incorporating resource assessment, engineering feasibility and economic viability organically into a single evaluation scheme.

(5) Effectiveness of Deep Marine Shale Gas Reservoirs

Hydrocarbon source potential, reservoir quality and preservation conditions are three pillars of deep shale gas accumulation, with two contrasting models (structural and continuous) being established from the study of the Fuling and Weirong fields (Guo 2014; Ma et al. 2018). Further work is necessary to recognize the differences in the mechanical properties, organic pore preservation, gas occurrence mode and multiphase flow mechanism between shallow and deep burial shales to fill the void in the criteria for delineating deep shale gas enrichment zones and high-productivity zones. The gas content, frackability, and economic efficiency must be considered. More attention should be given to "sweet spot" selection, low-cost technology optimization, and the adoption of cost reduction measures.

1.4.3.3 Exploration and Development Technologies

Despite the significant progress over the past decades, deep petroleum exploration and development in China will continue to face great challenges associated with deep targets, including but not limited to (a) high temperature and high pressure; (b) complexity in deep structures and deep seismic wave fields; and (c) low-grade reserves and low single-well productivity. Therefore, an integrated system for exploration and development must be established with strong interdisciplinary crossing and integration involving geologists, geophysicists, and reservoir engineers to achieve more efficient exploration and effective development in deep strata.

(1) Drilling Target Optimization and Risk Evaluation and Decision-Making System

During long-term exploration practices in the Tarim, Sichuan and Ordos Basins, a series of oil and gas distribution models that are suitable for the specific geological conditions in the area of interest have been gradually developed and established, which have not only guided regional exploration and discovery but also enriched geological theories and developed target evaluation technologies for deep marine oil and gas exploration in China. Based on these works, a decision-making system for deep exploration target optimization and risk evaluation should be established, which has three characteristics: (a) integrating geology, engineering and economy; (b) spanning the exploration, development, and engineering value chain; and (c) serving both reserve exploration and production capacity construction.

(2) Geophysical Technology and Reservoir Geological Modeling

Geophysical exploration of deep carbonate, clastic, and volcanic reservoirs often faces problems such as complex terrenes, low signal-to-noise ratios, low resolution, low imaging accuracy, and low amplitude preservation. An earlier study (Sun et al. 2015) revealed that wide frequency, amplitude preservation, high accuracy, and information integration are key for imaging deep complicated structures and predicting deep complicated reservoirs. Technical countermeasures, such as wide-frequency seismic acquisition, rock physical modeling of complex reservoirs, high-accuracy prestack amplitude-preservation imaging, comprehensive evaluation of complex reservoirs, nonseismic techniques, and drilling

steering with seismic data, need to be developed for petroleum explorers. Geological modeling is powerful for reservoir evaluation, management, and 3D numerical simulation. Existing modeling methods and software can meet the general need for 3D graphical visualizations of elementary interwell reservoir prediction, but the precision is insufficient to reflect the geological reality of deep petroleum reservoirs. Large efforts are still necessary to improve ultradeep 3D high-resolution seismic techniques using controlled sources in complex terrenes, including design, acquisition, processing, interpretation, and application in integrated reservoir geological modeling.

(3) Deep Marine Petroleum Development

Deep marine petroleum development in China often occurs in rather hostile subsurface environments, where old rocks, deep burial, complex reservoir types, high-temperature and high-pressure, and multiscaled complex fluids are present (Yao et al. 2018). High-priority deep marine petroleum development studies should focus on reservoir porosity and permeability, petrophysics, flow characteristics and optimized development schemes. Key technologies need to be developed in areas such as (a) multiscaled reservoir descriptions with applications in both exploration and development based on the strong integration of geology, engineering and reservoir management; (b) geological and numerical modeling of coupled deep thermal–hydrological–mechanical processes; (c) reservoir engineering optimization; (d) stimulation for substantially improving gas production in deep and ultradeep reservoirs; and (e) innovative low-cost technology for deep shale gas development and enhanced recovery.

(4) Optimized Ultra-Deep Well Fast Drilling and Completion Techniques

Drilling ultradeep wells often faces challenges associated with high-amplitude structures, high temperature and high pressure, fractured beds, and the cooccurrence of blowout and leakage, leading to great difficulties in the design of well casing programs. Given that the ultradeep well drilling capacity in China approaches 9000–10,000 m, strengthening studies on techniques for predicting deep geology, engineering and environmental factors and on optimized drilling, completion, and fracturing designs and supporting technologies is necessary. The focus should be on (a) optimized fast drilling techniques for ultradeep directional and horizontal wells; (b) highly effective cementing techniques for high-temperature and high-pressure formations with lost circulation and channeling; (c) well logging and downhole drill-stem testing techniques for high-temperature and high-pressure strata; (d) acid fracturing techniques for high-temperature and high-pressure carbonate reservoirs; and (e) optimized fast drilling, completion and fracturing techniques for deep shale gas horizontal wells with long horizontal sections.

(5) Information Technology, Artificial Intelligence and Big Data

To better meet the business needs of integrated comprehensive study, production organization and decision-making, a core value chain analysis of deep oil and gas exploration and development should be conducted. An intelligent exploration system based on big data, an exploration and development expert database based on big data cloud computing, and an intelligent cloud platform should be established, with the following basic functions: (a) systematic integration of big data from deep exploration, reservoir evaluation, field development, and production; (b) large data analysis and intelligent reservoir prediction of seismic data volumes; (c) large data analysis and intelligent interpretation of logging interpretation data; and (d) research and application of large data analysis in the optimization of favorable target areas for further exploration.

1.5 Prospects for Future Development

The global energy industry is now working toward a clean and low-carbon transition. However, oil and gas will remain the main sources of global energy consumption for a long time to come. The next few years constitute a critical period for China's economic transformation to high-quality development. New drivers of economic growth continue to grow along with domestic oil and gas consumption. Under the baseline scenario, China's oil and gas demand in 2030 will reach 700 mt and 550–600 bcm, respectively. At present, China's dependencies on imported oil and gas are more than 70% and 40%, respectively. Ensuring both national energy security and sustained green development is important for domestic oil and gas exploration and development.

Affected by the COVID-19 pandemic, global economic growth has slowed, and international oil prices will maintain a high probability of wide fluctuations. Moreover, the long-term complexity of international trade friction is further revealed, as it faces the risk of containment with some key technologies, equipment, and software. During this period of great change, the high dependence on imported oil and gas will seriously restrict China's national energy security. It is necessary for China to overcome these difficulties in the long term, break through the bottleneck of domestic oil and gas supply and expand the supply scale to slow down the overdependence on imported oil and gas.

Restricted by geological conditions, geological knowledge, engineering technology, the policy environment and mechanism systems, China's domestic annual oil and gas output should ideally be kept at the level of the 400 mt oil equivalent for a long time. It is not easy to add at least 2 bt oil equivalent of new oil and gas reserves annually. This is the main contradiction and strategic choice that China's upstream oil and gas industry must face in the next few years.

To achieve the sustainable development of oil and gas exploration and development in China, the key is to adopt an appropriate development strategy relying on science and technology and paying equal attention to both oil and gas, from both conventional and unconventional resources, and utilizing sites both in sea and on land. Exploration and development should be strengthened based on key basins. A new national strategy for energy security should be adopted by promoting revolutions in energy consumption and curbing irrational energy consumption, in energy supply and establishing a diversified supply system, in energy technology and industrial upgrading, and in the energy system, opening up the fast track of energy development and strengthening international cooperation on all fronts to achieve energy security in an open environment.

China is rich in deep marine oil and gas resources, with various types of oil and gas reservoirs, which is the key to searching for large oil and gas fields and realizing sustainable development. The geological and engineering conditions of deep marine oil and gas reservoirs are complex; efficient exploration and effective development are difficult, and opportunities and challenges coexist. It is necessary to further strengthen basic research with multidisciplinary cross-integration, deepen cross-company and cross-industry collaboration to develop common engineering technology, and make new progress in deep oil and gas exploration and development. Moreover, exploration and development integration, geological and engineering integration, and management and operation mode innovation should be strengthened, and the efficiency and effectiveness of exploration and development should be constantly enhanced.

Acknowledgements The authors acknowledge the financial support of Natural Science Foundation of China (Projects U24B6001 and U19B6003) and Sinopec S&T project (P19032), and Sinopec management for permission to publish this work.

References

Aydin A (2000) Fractures, faults, and hydrocarbon entrapment, migration and flow. Mar Petrol Geol 17(7):797–814

Bai GP, Cao BF (2014) Characteristics and distribution patterns of deep petroleum accumulations in the world. Oil Gas Geol 35(1):19–25

Baitsch-Ghirardello B, Gerya TV, Burg JP (2014) Geodynamic regimes of intra-oceanic subduction: implications for arc extension vs shortening processes. Gondwana Res 25(2):546–560

Budd DA, Saller AH, Harris PM (1995) Unconformities and porosity in carbonate strata. AAPG Mem 63:1–63

Cai XY (2011) Recent advances in Sinopec on oil and gas exploration and technology. In: Proceedings of the fourth annual convention of China petroleum geology, Beijing, 8–10 June 2011. Petroleum Industry Press

Chen DZ (2008) Structural-hydrothermal dolomitization and dolomite reservoir. Oil Gas Geol 29(5):614–622

Chen HL, Luo JC, Guo QY (2009) Deformation history and tectonic evolution of southeastern Tarim Basin in Mesozoic and Cenozoic. Geotecton Metallog 33(1):38–45

Dai JX (1992) Identification of various hydrocarbon gases. Sci China (Ser B Chem) 22(2):183–193

Deng S, Li HL, Zhang ZP et al (2018) Characteristics of differential activities in major strike-slip fault zones and their control on hydrocarbon enrichment in Shunbei area and its surroundings, Tarim Basin. Oil Gas Geol 39(5):38–48

Deng S, Li H, Zhang Z et al (2019) Structural characterization of intracratonic strike-slip faults in the central Tarim Basin. AAPG Bull 103(1):109–137

Du JH, Yang T, Li X (2016) Oil and gas exploration and discovery of PetroChina Company Limited during the 12th five-year plan and the prospect during the 13th five-year plan. China Petrol Explor 21(2):1–15

Dyman TS, Crovelli RA, Bartberger CE et al (2002) Worldwide estimates of deep natural gas resources based on the US Geological Survey World Petroleum Assessment 2000. Nat Resour Res 11(6):207–218

Fan TL, Yu BS, Gao ZQ (2007) Characteristics of carbonate sequence stratigraphy and its control on oil-gas in Tarim Basin. Mod Geol 21(1):57–65

Feng JH, Cai XY, Mou ZH et al (2016a) Oil and gas exploration of China Petroleum and Chemical Corporation during the 12th five-year plan and the prospect for the 13th five-year plan. China Petrol Explor 21(3):1–13

Feng JR, Gao ZY, Cui JG et al (2016b) The exploration status and research advances of deep and ultra-deep clastic reservoirs. Adv Earth Sci 31(7):718–736

Fu JM, Jia RF (1984) Dispersed organic matter in carbonate: the state of occurrence, evolution and petroleum potential. Geochim 13(3):1–9

Gao B, Wo YJ, Zhou Y et al (2012) Hydrocarbon accumulation phases of Majiang paleo-oil reservoir in Guizhou. Oil Gas Geol 33(3):417–423

Guo TL (2011) Sequence strata of the platform edge in the Changxing and Feixianguan formations in the Yuanba area, northeastern Sichuan Basin and their control on reservoirs. Acta Petrol Sin 32(3):387–394

Guo XS (2014) Shale gas enrichment mechanisms and exploration techniques in the Jiaoshiba block, Fuling field. Science Press, Beijing

Guo TL, Liu RB (2013) Implications from marine shale gas exploration breakthrough in complicated structural area at high thermal stage: taking Longmaxi Formation in Well JY1 as an example. Nat Gas Geosci 24(4):643–651

Guo XS, Hu DF, Li YP et al (2016a) Analyses and thoughts on accumulation mechanisms of marine and lacustrine shale gas: a case study in shales of Longmaxi Formation and Da'anzhai Section of Ziliujing Formation in Sichuan Basin. Geosci Front 23(2):18–28

Guo XS, Hu DF, Wei ZH et al (2016b) Discovery and exploration of Fuling shale gas field. China Petrol Explor 21(3):24–37

Guo XS, Hu DF, Li YP et al (2018) Discovery and theoretical and technical innovations of Yuanba gas field in Sichuan Basin, SW China. Petrol Explor Dev 45(1):14–26

Han X, Deng S, Tang L et al (2017) Geometry, kinematics and displacement characteristics of strike-slip faults in the northern slope of Tazhong uplift in Tarim Basin: a study based on 3D seismic data. Mar Petrol Geol 88:410–427

He ZL, Peng ST, Zhang T (2010) Controlling factors and origin of Ordovician reservoirs in the Tahe region, Tarim Basin. Oil Gas Geol 31(6):743–752

He DF, Li DS, Zhang GW et al (2011) The formation and evolution of the multicyclic basins in the Sichuan. Chin J Geol 46(3):589–606

He ZL, Jin XH, Wo YJ et al (2016) Hydrocarbon accumulation characteristics and exploration domains of ultra-deep marine carbonates in China. China Petrol Explor 21(1):3–14

He ZL, Hu ZQ, Nie HK et al (2017) Characterization of shale gas enrichment in the Wufeng-Longmaxi Formation in the Sichuan Basin and its evaluation of geological construction-transformation evolution sequence. Nat Gas Geosci 28(5):724–733

He DF, Ma YS, Liu B et al (2019) Main advances and key issues for deep-seated exploration in proliferous basins in China. Earth Sci Front 26(1):1–12

Hu DF, Zhang HR, Ni K et al (2014) Preservation conditions and main controlling factors for marine shale gases in the southeastern edge of the Sichuan Basin. Nat Gas Ind 34(6):17–23

Hu AP, Pan LY, Hao Y et al (2018) Characteristics, origin and distribution of dolomite reservoirs in the Permian Qixia-Maokou formations, Sichuan Basin. Mar Petrol Geol 23(2):39–52

Huang BC, Zhou YX, Zhu RX (2008) Formation and evolution of Chinese continents based on paleomagnetic study. Geosci Front 15(3):348–359

Huang SJ, Lan YF, Huang KK et al (2014) Vug fillings and records of hydrothermal activity in the Middle Permian Qixia Formation, western Sichuan Basin. Acta Petrol Sin 30(3):687–698

Ingersoll RV (2012) Tectonics of sedimentary basins, with revised nomenclature. In: Busby C, Perez AA (eds) Tectonics of sedimentary basins: recent advances. Wiley-Blackwell, Hoboken

Jia CZ, Pang XQ (2015) Research progresses and main development directions of deep hydrocarbon geological theory. Acta Petrol Sin 36(12):1457–1469

Jia CZ, Wei GQ, Li BL (2005) Superimposed-composite characteristics of micro-craton basins and its bearing petroleum systems, central-western China. Geol J China Univ 11(4):479–482

Jia AL, Tang HF, Han YX et al (2019) The distribution of gas and water and development strategy for deep-buried gas field in Kuqa Depression, Tarim Basin. Nat Gas Geosci 30(6):908–918

Jiao FZ (2017) Significance of oil and gas exploration in NE strike-slip fault belts in Shuntuoguole area of Tarim Basin. Oil Gas Geol 38(5):831–839

Jiao FZ (2018) Significance and prospect of ultra-deep carbonate fault-karst reservoirs in Shunbei area, Tarim Basin. Oil Gas Geol 39(2):207–216

Jin ZJ (2005) New progresses in research of China's typical superimposed basins and reservoiring of hydrocarbons (Part I): classification and research methods of superimposed basins. Oil Gas Geol 26(5):553–562

Jin ZJ (2014) A study on the distribution of oil and gas reservoirs controlled by source-cap rock assemblage in unmodified foreland region of Tarim Basin. Oil Gas Geol 35(6):763–770

Jin ZJ, Cai LG (2007) Inheritance and innovation of marine petroleum geological theory in China. Acta Geol Sin 81(8):1017–1024

Jin ZJ, Zhu DY, Hu WX et al (2006) Geological and geochemical signatures of hydrothermal activity and their influence on carbonate reservoir beds in the Tarim Basin. Acta Geol Sin 80(2):245–253

Jin ZJ, Zhou Y, Yun JB et al (2010) Distribution of gypsum-salt cap rocks and near-term hydrocarbon exploration targets in the marine sequences of China. Oil Gas Geol 31(6):715–724

Jin ZJ, Cai XY, Liu JL et al (2018) The recent exploration progress and resource development strategy of China. Petroleum and Chemical Corporation. China Petrol Explor 23(1):14–25

Kang YZ (2018) The resource potential and exploration for oil and gas in the Tarim Basin. Petrol Sci Bull 3(4):369–375

Koptev A, Burov E, Calais E et al (2016) Contrasted continental rifting via plume-craton interaction: applications to Central East African Rift. Geosci Front 7(2):221–236

Li Z (2016) Research frontiers of fluid-rock interaction and oil-gas formation in deep-buried basins. Bull Mineral Petrol Geochem 35(5):807–816, 805

Li Y, Fan ZH (2011) Developmental pattern and distribution rule of the fracture-cavity system of Ordovician carbonate reservoirs in the Tahe Oilfield. Acta Petrol Sin 32(1):101–106

Li Z, Li JW, Zhang PT et al (2016) Key structural-fluid evolution and reservoir diagenesis of deep-buried carbonates: An example from the Ordovician Yingshan Formation in Tazhong, Tarim Basin. Bull Mineral Petrol Geochem 35(5):827–838

Li Y, Wu SH, Hou JG et al (2017) Progress and prospects of reservoir development geology. Petrol Explor Dev 44(4):569–579

Li XZ, Guo ZH, Hu Y et al (2018) Efficient development strategies for large ultra-deep structural gas fields in China. Petrol Explor Dev 45(1):111–118

Li J, She YQ, Gao Y et al (2019) Onshore deep and ultra-deep natural gas exploration fields and potentials in China. China Petrol Explor 24(4):403–417

Li Y, Xue ZJ, Cheng Z et al (2020) Progress and development directions of deep oil and gas exploration and development in China. China Petrol Explor 25(1):45–57

Liang DG, Zhang SC, Zhang BM et al (2000) Understanding on marine oil generation in China based on Tarim Basin. Earth Sci Front 7(4):534–547

Liang DG, Guo TL, Chen JP et al (2009) Some progresses on studies of hydrocarbon generation and accumulation in marine sedimentary regions, southern China (Part 3): controlling factors on the sedimentary facies and development of Palaeozoic marine source rocks. Mar Petrol Geol 14(2):1–19

Lin ZM (2002) Geological conditions for the formation of giant Ordovician oilfield in the Tahe, Tarim Basin. Geol Rev 48(4):372–376

Liu RB (2015) Analyses of influences on shale reservoirs of Wufeng-Longmaxi Formation by overpressure in the south-eastern part of Sichuan Basin. Acta Sedimentol Sin 33(4):817–827

Liu WH, Zhang DW, Gao B et al (2005) Multiple gas sources and significance. Oil Gas Geol 26(4):393–401

Liu SG, Huang WM, Chen CH et al (2008) Hydrothermal activity and effect on petroleum mineralization and accumulation in the Sinian-Paleozoic sequence, Sichuan Basin. Acta Petrol Mineral 28(3):41–50

Liu WH, Chen MJ, Guan P et al (2009) Ternary geochemical tracing system in natural gas generation and accumulation and its application. Science Press, Beijing

Liu SG, Sun W, Luo ZL (2013) Xinkai geofracturing orogeny and petroleum exploration in the Lower Paleozoic and older strata in Sichuan Basin. J Chengdu Univ Technol (Sci Technol Ed) 40(5):511–520

Liu SG, Wang YG, Sun W et al (2016) Control of intracratonic sags on the hydrocarbon accumulation in the marine strata across the Sichuan Basin, China. J Chengdu Univ Technol (Sci Technol Ed) 43(1):1–23

Lu BP (2018) Key technologies and equipment for ultra-deep oil and gas exploration and development engineering. China Petroleum and Chemical Corporation, Beijing

Lu BP, Ding SD (2018) New progress and development prospect in shale gas engineering technologies of Sinopec. Petrol Drill Tech 46(1):1–9

Luo ZL, Liu SG, Zhao XK et al (1994) C-type subduction zone and its role in the orogenic belts in mid-western China. In: Luo Z (ed) The rising of the Longmen orogenic belts and the formation and evolution of the Sichuan Basin. Chengdu University of Technology Press, Chengdu

Luo XR, Zhang LK, Fu XF et al (2016) Advances in dynamics of petroleum migration and accumulation in deep basins. Bull Mineral Petrol Geochem 35(5):876–890

Ma YS (2009) Sequence stratigraphy and paleogeography in southern China. Science Press, Beijing

Ma XH (2017) A golden era for natural gas development in the Sichuan Basin. Nat Gas Ind 37(2):1–10

Ma YS (2020) Marine oil and gas exploration in China. Geological Publishing House and Springer-Verlag GmbH Germany, Beijing

Ma YS, Mou CL, Tan QY et al (2006) On the Kaijiang-Liangping Trough. Oil Gas Geol 27(3):326–331

Ma YS, Guo XS, Guo TL et al (2007) The Puguang gas field: new giant discovery in the mature Sichuan Basin, southwest China. AAPG Bull 91(5):627–643

Ma YS, Zhang SC, Guo TL et al (2008) Petroleum geology of the Puguang sour gas field in the Sichuan Basin, SW China. Mar Petrol Geol 25(4–5):357–370

Ma YS, Cai XY, Zhao PR et al (2010a) Distribution and further exploration of the large-medium sized gas fields in Sichuan Basin. Acta Petrol Sin 31(3):347–354

Ma YS, Cai XY, Zhao PR (2010b) Formation mechanism of deep-buried carbonate reservoir and its model of three-element controlling reservoir: a case study from the Puguang oilfield in Sichuan. Acta Geol Sin 84(8):1087–1095

Ma YS, Cai XY, Zhao PR (2011) The research status and advances in porosity evolution and diagenesis of deep carbonate reservoir. Earth Sci Front 18(4):181–192

Ma YS, Cai XY, Zhao PR (2014) Characteristics and formation mechanisms of reef-shoal carbonate reservoirs of Changxing-Feixianguan formations, Yuanba gas field. Acta Petrol Sin 35(6):1001–1011

Ma YS, He DF, Cai XY et al (2017) Distribution and fundamental science questions for petroleum geology of marine carbonate in China. Acta Petrol Sin 33(4):1007–1020

Ma YS, Cai XY, Zhao PR (2018) China's shale gas exploration and development: understanding and practice. Petrol Explor Dev 45(4):561–574

Ma YS, Li MW, Cai XY et al (2020) Mechanisms and exploitation of deep marine petroleum accumulations in China: advances, technological bottlenecks and basic scientific problems. Oil Gas Geol 41(4):655–672, 683

Ma YS, Li MW, Cai XY et al (2021) Advances in basic research on the mechanisms of deep marine hydrocarbon enrichment and key exploration technologies. Petrol Geol Exp 43(5):737–748

Ma YS, Cai XY, Yun L et al (2022) Practice and theoretical and technical progress in exploration and development of Shunbei ultra-deep carbonate oil and gas field, Tarim Basin, NW China. Petrol Explor Dev 49(1):1–17

Ministry of Land and Resources of the People's Republic of China (2014) Regulation of shale gas resources/reserves estimation: DZ/T 0254-2014. China Standard Publishing House, Beijing

Ministry of Natural Resources of the People's Republic of China (2018) The bulletin of national oil and gas reserves in 2018. Geological Publishing House, Beijing

Mou SL, Cai XY (2007) Never ending. Petroleum Industry Press, Beijing

Pang XQ, Zhou XY, Jiang ZXQ et al (2012) Hydrocarbon reservoirs formation, evolution, prediction and evaluation in the superimposed basins. Acta Geol Sin 86(1):1–10

Pang XQ, Jiang ZX, Huang HD et al (2014) Formation mechanisms, distribution models, and prediction of superimposed, continuous hydrocarbon reservoirs. Acta Petrol Sin 35(5):795–828

Pang XQ, Zhu WL, Lv XX et al (2015) A study on hydrocarbon thresholds controlling reservoir accumulation and predictive evaluation of favorable accumulations areas in eastern Bohai Bay Basin. Acta Petrol Sin 36(10):1167–1187

Qi L (2014) Exploration practice and prospects of giant carbonate field in the Lower Paleozoic of Tarim Basin. Oil Gas Geol 35(6):771–779

Qi LX (2016) Oil and gas breakthrough in ultra-deep Ordovician carbonate formations in Shuntuoguole uplift, Tarim Basin. China Petrol Explor 21(3):38–51

Qi LX (2019) Characteristics and enlightenment of ultra-deep carbonate fault-karst reservoirs in Shunbei area, Tarim Basin. In: Abstracts of the 8th annual meeting of China association of petroleum geologists, Beijing, 1–3 June 2019

Rong JY, Fang ZJ, Zhou ZH et al (2006) Originations, radiations and biodiversity changes: evidence from the Chinese fossil record. Science Press, Beijing

Standardization Administration of the People's Republic of China (2012) Vocabulary of drilling engineering for the petroleum and natural gas: GB/T 28911-2012. China Standard Publishing House, Beijing

Strategic Research Center of Oil and Gas Resources, Ministry of Land and Resources (2017a) National petroleum resource assessment. China Land Press, Beijing

Strategic Research Center of Oil and Gas Resources, Ministry of Land and Resources (2017b) National shale gas resource assessment. Geological Publishing House, Beijing

Sun S, Wang TG (2016) Middle Neoproterozoic geology and petroleum resource in eastern China. Science Press, Beijing

Sun LD, Zou CN, Zhu RK et al (2013) Formation, distribution and potential of deep hydrocarbon resources in China. Petrol Explor Dev 40(6):641–649

Sun LD, Fang CL, Sa LM et al (2015) Innovation and prospect of geophysical technology in the exploration of deep oil and gas. Petrol Explor Dev 42(4):414–424

Tenger B, Liu WH, Xu YC et al (2016) Comprehensive geochemical identification of highly evolved marine carbonate rocks as hydrocarbon-source rocks as exemplified by the Ordos Basin. Sci China (Ser D Earth Sci) 49(4):384–396

Tuo JC (2002) Research status and advances in deep oil and gas exploration. Adv Earth Sci 17(4):565–571

Wang TG, Han KY (2011) On Meso-Neoproterozoic primary petroleum resources. Acta Petrol Sin 32(1):1–7

Wang HZ, Yang SN, Liu BP (1990) Tectono-paleogeography and biological paleogeography in China and adjacent regions. China University of Geosciences Press, Wuhan

Wang SC, Fu ST, Li XZ et al (2005) The influence to the accumulation of oil and gas, of the development of the arcuate structure zone in the transition zone from the plate to the geosyncline in the west Ordos Basin. Nat Gas Geosci 16(4):421–427

Wang ZC, Wang TS, Wen L et al (2016) Basic geological characteristics and accumulation conditions of Anyue giant gas field, Sichuan basin. China Offshore Oil Gas 28(2):45–52

Wang ZC, Zhao WZ, Hu SY et al (2017) Control of tectonic differentiation on the formation of large oil and gas fields in craton basins: a case study of Sinian-Triassic of the Sichuan Basin. Nat Gas Ind 37(1):9–23

Wei GQ, Wang ZH, Li J et al (2017a) Characteristics of source rocks, resource potential and exploration direction of Sinian and Cambrian in Sichuan Basin. Nat Gas Geosci 28(1):1–13

Wei XF, Li YP, Wei ZH et al (2017b) Effects of preservation conditions on enrichment and high yield of shale gas in Sichuan Basin and its periphery. Petrol Geol Exp 39(2):147–153

Wei GQ, Li J, Yang W et al (2018) Major progress of the Chinese natural gas geological theory and new exploration discovery since the eleventh five-year plan. Nat Gas Geosci 29(12):1691–1705

Wei GQ, Yang W, Liu MC et al (2019) Distribution rules, main controlling factors and exploration directions of giant gas fields in the Sichuan Basin. Nat Gas Ind 39(6):1–12

Wu LL, Mei LF, Paton DA et al (2018) Deciphering the origin of the Cenozoic intracontinental rifting and volcanism in eastern China using integrated evidence from the Jianghan Basin. Gondwana Res 64:67–83

Xu YC (1994) Natural gas genetic theory and application. Science Press, Beijing

Xu XH, Gao CL, Huang ZG et al (2005) Three major active tectonic stages in the origin of sedimentary basins in China. Oil Gas Geol 26(2):119–127

Yang YM, Yang Y, Yang G et al (2019) Gas accumulation conditions and key exploration & development technologies of Sinian and Cambrian gas reservoirs in Anyue gas field. Acta Petrol Sin 40(4):493–508

Yao J, Huang ZQ, Liu WZ et al (2018) Key mechanical problems in the development of deep oil and gas reservoirs. Sci Sin (Phys Mech Astron) 48(4):044701

Yu YJ, Yang T, Guo BC et al (2019) Major advances and outlook for oil and gas exploration, theory and technology of foreland thrust belts in China. Acta Geol Sin 93(3):545–564

Zhang YC (1997) Prototype analysis of petroliferous basins in China. Nanjing University Press, Nanjing

Zhang CH (2008) Intracratonic deformation and dynamic mechanisms. Geosci Front 15(3):140–149

Zhang GW, Guo AL, Dong YO et al (2011) Continental geology, tectonics and dynamics. Geosci Front 18(3):1–12

Zhao Z (2019) Analysis of global oil and gas exploration characteristics during the past decade. Oil Forum 38(3):58–64

Zhao MJ, Zeng FG, Qin SF et al (2001) The discovery and confirmation of two types of natural gases derived from thermal cracking. Nat Gas Ind 21(1):35–39

Zhao WZ, Wang ZY, He HQ et al (2005) Gas formation mechanism of marine carbonate source rocks in China. Sci China (Ser D Earth Sci) 35(7):638–648

Zhao ZJ, Zhang YB, Pan M et al (2010) Cambrian sequence stratigraphic framework in Tarim Basin. Geol Rev 56(5):609–620

Zhao XZ, Jin FM, Wang Q et al (2011) Niudong 1 ultra-deep and ultra-high temperature subtle buried hill field in Bohai Bay Basin: discover and significance. Acta Petrol Sin 32(6):915–927

Zhao WZ, Wang ZC, Hu SY et al (2012) Large-scale hydrocarbon accumulation factors and characteristics of marine carbonate reservoirs in three large onshore cratonic basins in China. Acta Petrol Sin 33(S2):1–10

Zhao WZ, Hu SY, Liu W et al (2014) Petroleum geological features and exploration prospect in deep marine carbonate strata onshore China: a further discussion. Nat Gas Ind 34(4):1–9

Zhong NN, Lu SF, Huang ZL et al (2004) TOC changes in the process of thermal evolution of source rock and its controls. Sci China (Ser D Earth Sci) 47(S2):141–149

Zhou ZY, Lin HL (1995) Stratigraphy, paleogeography and plate tectonics in northwestern China. Nanjing University Press, Nanjing

Zhou XB, Li JH, Wang HH (2015) Nanhuaan-Sinian basin types and tectonic setting for early basin initiation. Geosci Front 22(3):290–298

Zou C, Xu C, Wang Z et al (2011) Geological characteristics and forming conditions of the platform margin large reef-shoal gas province in the Sichuan Basin. Petrol Explor Dev 38(6):641–651

Zou CN, Du JH, Xu CC et al (2014) Formation, distribution, resource potential and discovery of the Sinian-Cambrian giant gas field, Sichuan Basin, SW China. Petrol Explor Dev 41(3):278–293

Zou CN, Dong DZ, Wang YM et al (2016) Shale gas in China: characteristics, challenges and prospects (II). Petrol Explor Dev 43(2):166–178

Evolution of Deep Marine Sedimentary Filling and Development of High-Quality Hydrocarbon Source Rocks in Cratonic Basins, China

2

Yifan Li, Tailiang Fan, Juye Shi, Ming Jiang, Tan Zhang, Mingzhi Kuang, and Wangwei Liu

Abstract

In the three cratonic basins of China, ancient source rocks are widely distributed, and understanding the mechanisms of hydrocarbon enrichment in Mesoproterozoic to early Cambrian source rocks has significant implications for ultradeep oil and gas exploration in China. This article provides a systematic summary of the source rocks in the Mesoproterozoic Chuanlinggou Formation of the North China Craton, the Neoproterozoic Doushantuo Formation of the Yangtze Craton, and the early Cambrian Yurtus Formation of the Tarim Craton. The formation processes of organic-rich shales are analyzed from the perspectives of sedimentary filling, paleoclimate, paleoenvironment, and specific geological events. It is concluded that the source rocks of the Chuanlinggou Formation developed mainly in Chuanlinggou Member 3, which experienced inner shelf sedimentation. A warm and humid climate, upwelling currents, and volcanic activity promoted nutrient input, triggered high primary productivity, and led to organic matter enrichment. Ferruginous seawater facilitated organic matter preservation. In the Doushantuo Formation of the Chuanbei depression, organic-rich intervals occur in Doushantuo Member 3. Under a warm climate, intense weathering resulting from the amalgamation of the Gondwana supercontinent promoted nutrient input and bottom-water anoxia, leading to organic matter enrichment. In the Yuertusi Formation, the enrichment of organic matter is attributed primarily to the extremely high productivity caused by intense upwelling currents and the oxygen-depleted environment resulting from transgression.

Y. Li · T. Fan (✉) · J. Shi · M. Jiang
School of Energy Resources, China University of Geosciences (Beijing), Beijing, China
e-mail: fantl@cugb.edu.cn

T. Zhang
Institute of Sedimentary Geology, Chengdu University of Technology, Chengdu, China

M. Kuang
School of Energy Resources, Chengdu University of Technology, Chengdu, China

W. Liu
Wuxi Institute of Petroleum Geology, Sinopec Petroleum Exploration and Development Research Institute, Wuxi, China

Keywords

Shale · Paleoclimate · Paleoenvironment · Yangtze Block · North China Block · Tarim Basin · Precambrian · Organic matter

2.1 Introduction

Early studies suggested that hydrocarbon source rocks were limited prior to the Phanerozoic, with significant hydrocarbon source rocks distributed only in Cambrian and younger strata. It was

Y. Ma et al. (eds.), *Petroleum Geology and Exploration of Deep Marine Strata in China*, Advances in Oil and Gas Exploration & Production, https://doi.org/10.1007/978-3-032-02496-1_2

reported that more than 90% of oil and gas resources were located in the Phanerozoic, while the oil and gas resources in the Proterozoic, which spans 2 billion years, accounted for less than 0.2% (Klemme and Ulmishek 1991). Recently, with improvements in petroleum exploration, commercial accumulations of oil and gas have been discovered on almost all continents. Examples include the West Siberian and Volga–Ural regions of Russia, the Salt Basin in Oman, and the Middle–Upper Proterozoic to Lower Cambrian oil and gas in India and Pakistan, which have been industrially developed (Craig et al. 2013). Proterozoic strata in regions such as the McArthur Basin and Adelaide region in northern Australia, the Taoudeni Basin in West Africa, the Great Lakes region in North America, and the São Francisco Basin in Brazil have become exploration targets in recent years (Craig et al. 2013; Hlebszevitsch et al. 2009). In the McArthur Basin, geological reserves of more than 400 billion m^3 of shale gas have been discovered in Paleoproterozoic Barney Creek (1.64 billion years ago) (Ahmad et al. 2013), and the Mesoproterozoic Middle Velkerri (1.4 billion years ago) shale gas has geological reserves exceeding tens of billions of tons of oil and gas equivalents (Cox et al. 2022). In China, the Dengying Formation in the Weiyuan gas field in the Sichuan Basin has produced more than 15 billion m^3 of natural gas, and the discovery of the Anyue gas field has led to continuous increases in proven reserves in the Ediacaran–Cambrian formations in the Moxi–Gaoshiti areas of the central Sichuan Basin, reaching 81.02 trillion m^3, with the Dengying Formation accounting for 36.98 trillion m^3 (Du et al. 2014; Zou et al. 2014).

The Mesoproterozoic–Cambrian period was an important time in Earth's history for biological evolution and environmental changes, including Neoproterozoic Snowball Earth events and Cambrian global greenhouse events (Fairchild and Kennedy 2007; Hoffman et al. 1998; Kuang et al. 2019), which involved dramatic changes in global climate, oceanic environments, and biological evolution (Hoffman et al. 2017). During this critical period in Earth's history, multiple sets of significant high-quality hydrocarbon source

rock systems, including the Mesoproterozoic Chuanlinggou Formation, Ediacaran Doushantuo Formation, and the Lower Cambrian Yuertusi Formation, were deposited in the North China Craton, Yangtze Craton, and Tarim Craton (Zhao et al. 2018a, b; Ma et al. 2020; Li et al. 2021). Hydrocarbon source rocks serve as the foundation for oil and gas exploration, making the prediction of their distribution crucial for successful exploration efforts (Zhao et al. 2018a, b). In this study, we systematically examine the sedimentary fill processes of these hydrocarbon source rocks, describe the paleoclimatic and paleoenvironmental backgrounds in which they formed, and ultimately investigate the models responsible for organic matter accumulation in each source rock.

2.2 Geological Background

China contains three major cratonic blocks: the North China Block, the South China Block, and the Tarim Block (Fig. 2.1) (Zhao et al. 2019). The formation and evolution of these blocks are closely related to the convergence and breakup of multiple supercontinents during the Paleoproterozoic, particularly the Columbia Supercontinent (ca. 2.1–1.4 Ga) and the Rodinia Supercontinent (ca. 1.2–0.6 Ga) (Li et al. 2008; Zhao et al. 2004). The Yanliao Basin is a continental rift in the Yanshan–Liaoxi region of the north–central North China Craton and is notable for its substantial accumulation of thick Mesoproterozoic sediments, which are considered possible records of the break-up of the Columbia Supercontinent (Li et al. 2019a, b). The Mesoproterozoic strata within the Yanliao Basin are composed of three distinct systems: the Changcheng System, the Jixian System and an unnamed system (Zhai et al. 2016). The Changcheng System and Jixian System are dominated by shallow marine clastic sedimentary and carbonate rocks. The Changcheng System can be further divided into four formations: the Changzhougou, Chuanlinggou, Tuanshanzi, and Dahongyu Formations. Samples collected from the Kuancheng area (Member 2 and Member 3) indicate that the Chuanlinggou Formation is the oldest source rock in China and was deposited in

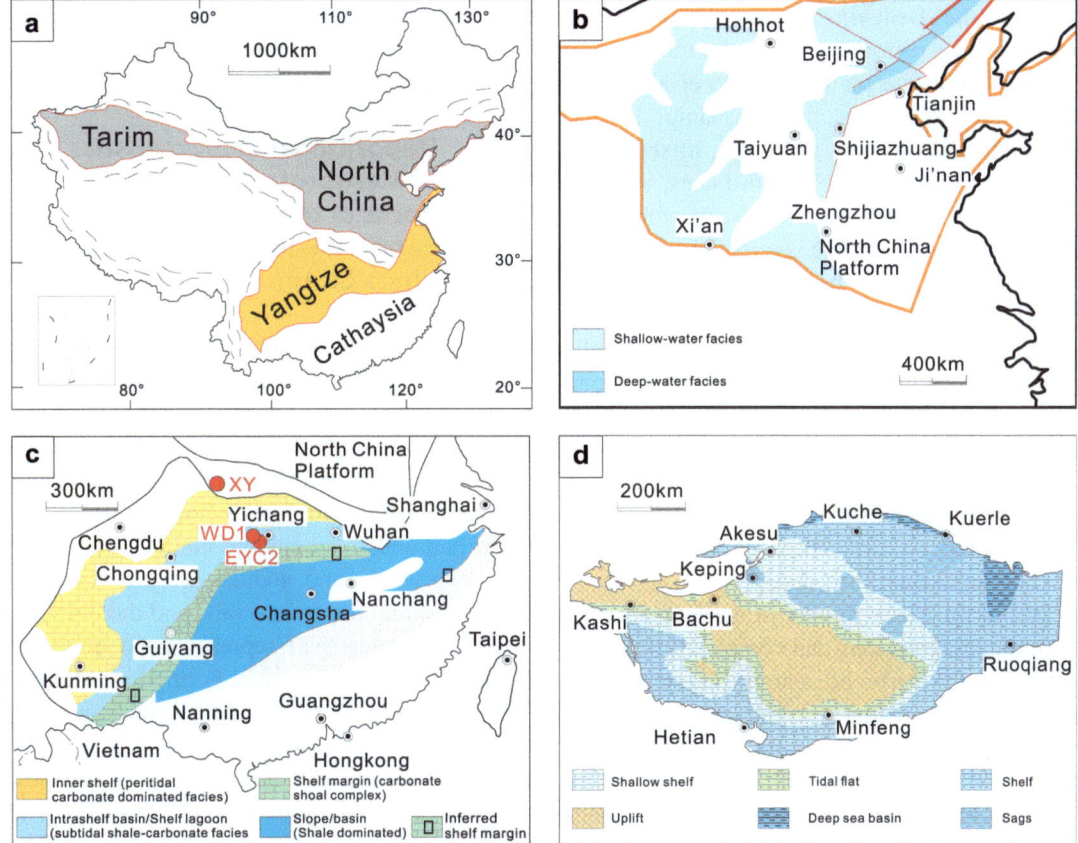

Fig. 2.1 **a** Distribution and location of North China, Yangtze and Tarim blocks; **b** paleogeography of Mesoproterozoic Chuanlingou Formation in the North China Block; **c** paleogeography of Ediacaran Doushantuo Formation in the Yangtze Block (modified after Jiang et al. 2011), XY = Xiaoyang section; **d** paleogeography of Cambrian Yuertusi Formation

a shallow-sea tidal flat–shelf environment (Luo et al. 2016; Liu et al. 2019b; Cai et al. 2021).

During the early Neoproterozoic, the amalgamation and collision of the Yangtze Block and the Cathaysia Block formed the South China Block (Fig. 2.1) (Wang and Li 2003). The Ediacaran Yangtze Platform in South China is considered a passive continental margin that formed along the southeastern margin of Rodinia during its breakup (~ 820 Ma) (Wang and Li 2003; Li et al. 1999). The Ediacaran Yangtze Platform gradually developed on the basis of Nantuo glacial paleotopography and features a rimmed carbonate platform with various vertical and lateral lithofacies combinations (Zhu et al. 2003, 2007, 2013; Jiang et al. 2011; Vernhet 2007; Vernhet

and Reijmer 2010). Following the late Neoproterozoic Marinoan "Snowball Earth" glaciation event, the Doushantuo Formation was extensively deposited on the Yangtze Platform (Fig. 2.1) (Jiang et al. 2011; Zhu et al. 2003, 2007). Wang et al. (2019) conducted lithofacies paleogeographic reconstructions of the Middle-to-Upper Yangtze region (including the Sichuan Basin and surrounding areas), revealing that paleouplifts and marginal depressions significantly controlled the sedimentary filling sequences and stratigraphic distribution of the Doushantuo Formation. Therefore, in the Sichuan Basin and its vicinity, the lithological sequence of the Doushantuo Formation is typically incomplete and is primarily composed of siliciclastic sediments until the deposition of the Dengying Formation, when the entire Yangtze region experienced a transition to a typical carbonate platform (Xiao et al. 2021). In the paleo-uplift areas, the Doushantuo Formation is characterized by shoreface, mixed shelf, and atypical carbonate platform facies, whereas in the marginal depression areas, the formation shows relatively complete layers with significant thickness, dominated by deep-water shelf and restricted basin facies (Wang et al. 2019). Jiang et al. proposed four major sedimentary facies belts from northwest to southeast across the Ediacaran Yangtze Platform: the inner shelf, intrashelf basin, shelf margin, and slope/basin (Fig. 2.1) (Jiang et al. 2011; Zhu et al. 2007). The studied Xiaoyang section is located in the Chuanbei Depression, approximately 10 km southeast of Zhenba County in southern Shaanxi Province.

The Tarim Block formed in the late Archean and has undergone long and complex drift evolution since the early Ediacaran, and it was an independently drifting ancient land mass near the equator during the early Paleozoic Cambrian (Chen et al. 2019; Yang et al. 2021). The Tarim Basin is surrounded by the following boundaries: the northern boundary is the South Tianshan Northern Fault Zone, the southwestern boundary is the West Kunlun Fault Zone, and the southeastern boundary is the Arjin Fault Zone (Huang et al. 2008). The Tarim Basin can be divided into nine primary tectonic units: the "four uplifts" are the Tabei uplift, the Bachu uplift, the Taichung uplift and the Southeast uplift; the "five depressions" are the Kuche depression, the Northern depression, the Tanggu depression, the Southeast depression and the Southwest depression (Zhang et al. 2023). The early Cambrian Tarim Basin experienced two large-scale sea floods, which led to the extensive development of a set of shale and siliciclastic combinations of black shale formations in the Tarim Basin, namely, the Yurtus Formation hydrocarbon source rocks, which are the most important hydrocarbon source rocks in the Cambrian and the Lower Paleozoic strata of the Tarim Basin (Zhu et al. 2016).

2.3 Sedimentary Background and Sedimentary Filling Processes in Marine Basins

2.3.1 Depositional Characteristics of Mesoproterozoic Source Rocks of the Chuanlinggou Formation in the North China Craton

The Chuanlinggou Formation is composed of fine-grained clastic rocks that are more than 800 m thick, including iron-rich sandstones at the lower interval, illite mudstones in the middle interval, and dolomitic shale in the upper interval. Many thin interlayers and lenses are visible in the lower part and are composed mainly of siltstone. Dolomite lenses and nodules were observed in the middle and upper intervals, and the dolomite in the uppermost part represents the transition to the overlying Tuanshanzi Formation. Member 1 is composed of iron-bearing sandstone, with ooidal and stromatolitic ironstones in the northwestern Yanliao Basin. It is regarded as a shallow marine depositional setting on the shoreline with water depths below the mean low tide surface and above the wave base surface (< 100 m) (Tang et al. 2015; Lin et al. 2019). Member 2 is dominated mainly by illite shale and siltstone with small wave cross-laminae in some lenses (Liu et al. 2022). Member 3 consists of dark silty shale with fine interlayers of siltstone and occasional carbonate nodules and dolomitic shale in the uppermost layer (Li et al. 2015). The lower interval of the Chuanlinggou Formation is dominated by hematite-rich medium- to fine-grained sandstones with a gradual increase of clay content upward. The middle interval is dominated by gray–dark illite shale. In the upper interval, manganese dolomite precipitated with a gradual decrease of clastic content upward. The overall formation indicates a large transgression–regression cycle in the Chuanlinggou Formation, with the largest transgression corresponding to the deposits of dark organic-rich shale in Member 3.

Using collected samples from the Chuan-linggou Formation, in one study, the bulk mineral contents, carbonate minerals, silicate minerals and clay minerals were calculated with the Brumsack model (Brumsack 1989). Based on the mineralogy and composition, the lithofacies were classified into four lithofacies groups and twelve lithofacies. The samples were subdivided into six lithofacies: siliceous mudstone, clay-rich siliceous mudstone, argillaceous/siliceous mixed mudstone, siliceous/calcareous mixed mudstone, calcareous mudstone, and silicate-rich argillaceous mudstone. Calcareous mudstone and siliceous/calcareous mixed mudstone were observed mainly in upper Member 3 (Fig. 2.2). Clay-rich siliceous mudstone was deposited mainly in the lower Member 3, whereas silicate-rich argillaceous mudstone was deposited mainly in the upper Member 2. Argillaceous/siliceous

mixed mudstone was mostly found in the lower Member 3.

Microscopic observations revealed that six primary types of structures were present in the samples: massive structures, horizontal bedding, graded bedding, sand vein structures, lenticular bedding and wave ripples. The origin of sand vein structures may be earthquake-driven soft sediment deformation during continental breakup (Qiao and Gao 2007), or the structures may be related to methane emission events in shallow-sea environments (Shi et al. 2008, 2016; Tang et al. 2009). The sedimentary structure in the lower member was dominated by horizontal bedding and occasional graded bedding. Massive mudstone was predominant in upper Member 2 with some lenticular bedding. Compared with the upper Member 2, the lower Member 2 was charac-terized by more horizontal bedding and more

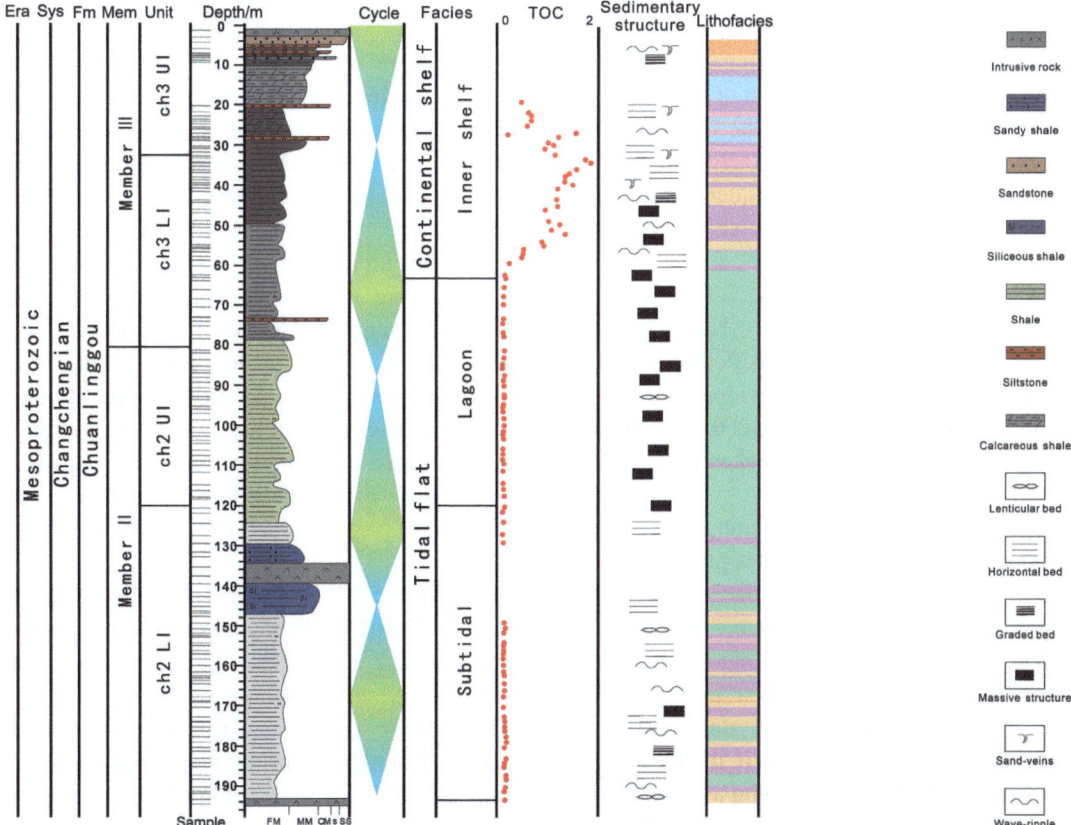

Fig. 2.2 Synthetic stratigraphic column of the Chuanlinggou Formation. Sys: system; Fm: formation; Mem: member; LA: lithofacies assemblage; FM: fine mud; MM: medium mud; CM: coarse mud; S: silt; SS: sandstone

event gravity flow deposits, indicating that more storm events occurred in the lower Member and brought more sand into the massive, horizontal mudstone. Member 2 was deposited in an intertidal–subtidal muddy flat environment with flaser/lenticular bedding. Member 3 was deposited in a shallow shelf environment with more horizontal silt laminae, graded bedding and wave ripple, which suggests that the water depth was near the storm wave base.

2.3.2 Depositional Characteristics of Neoproterozoic Doushantuo Source Rocks and Sedimentary Filling of the Sichuan Basin

During the Doushantuo stage, the Chuanbei Depression, as a platform margin depression in the Sichuan Basin, has a sediment thickness of 120–300 m and is considered a potential hydrocarbon generation center of the Doushantuo Formation. The "skip-like" tectonic pattern of the depression and hydrothermal activity in the Chengkou area jointly determine the lateral heterogeneity of the stratigraphic structure of the sequence. Combined with previous research, a correlation section was established from northwest to southeast across the Chuanbei Depression (Fig. 2.3) (Wang et al. 2019, 2020; Yang et al. 2016; He 2022; Duan et al. 2019). The Doushantuo Formation in the Chuanbei Depression can be divided into three and a half third-order sequences, and the sedimentary filling evolution of each sequence unit is detailed as follows:

Sequence SQ1 is a lower stratigraphic unit of the Doushantuo Formation and is situated between SB1 and SB2 (Fig. 2.3). In the Nanjiang area, which serves as an erosion source, this sequence unit is absent; however, sedimentation is observed from Zhenba and Lianghekou to the Gaozhu area in Wuxi. Sequence SQ1 starts during the initial stages of sedimentation in the Doushantuo Formation, which coincides with a global climate warming period. The melting of the Marinoan ice sheet led to a rise in absolute sea level on the Yangtze Plate, resulting in the widespread

deposition of cap dolostones in the mid-lower Yangtze region, which is far from the provenance area (Wang et al. 2015; Jiang et al. 2011; Zhao et al. 2012). During the early development of Sequence SQ1 in the Chuanbei Depression, the Xiaoyang section was located close to the Hannan oldland. At this time, the sea level did not surpass the northwest slope fold belt of the Chuanbei depression. The supply rate of sediment from the provenance exceeded the rate of sea level rise, leading to a decrease in relative sea level. Consequently, approximately 40 m of progradational turbidite deposits were deposited at the base of the Xiaoyang section (Fig. 2.3), forming the Lowstand System Tract (LST1). As the sea level continued to rise above the northwest slope fold belt of the depression, the sediment input capacity in the Xiaoyang region weakened. However, the relative sea level remained in a declining state, corresponding to the sedimentary expression of the diminishing scale of the progradational parasequence set within the LST1 system tract. With a further increase in absolute sea level, when the sediment supply rate became lower than the rate of absolute sea level rise, the relative sea level rose, resulting in the development of the Transgressive System Tract (TST1). The Xiaoyang section exhibited thin-bedded turbidite facies as evidence of retrogradation during this phase (Fig. 2.3). When the transgression reached its maximum, the sediment input rate in the Xiaoyang region was at its minimum, resulting in underdeveloped turbidite facies. In the context of a gentle slope, slumping and deformed siltstone-rich mudstone replaced the turbidite facies, indicating the onset of the Highstand System Tract (HST1) development stage.

Sequence SQ2 represents the lower to middle part of the Doushantuo Formation, situated between SB2 and SB3 (Fig. 2.3). It exhibits lateral continuity similar to that of Sequence SQ1 and has a consistent thickness of approximately 20–30 m. During the initial stages of Sequence SQ2, the sea level gradually decreased, leading to the development of a Lowstand System Tract (LST2) characterized by turbidite deposition at Lianghekou and in the Xiaoyang region

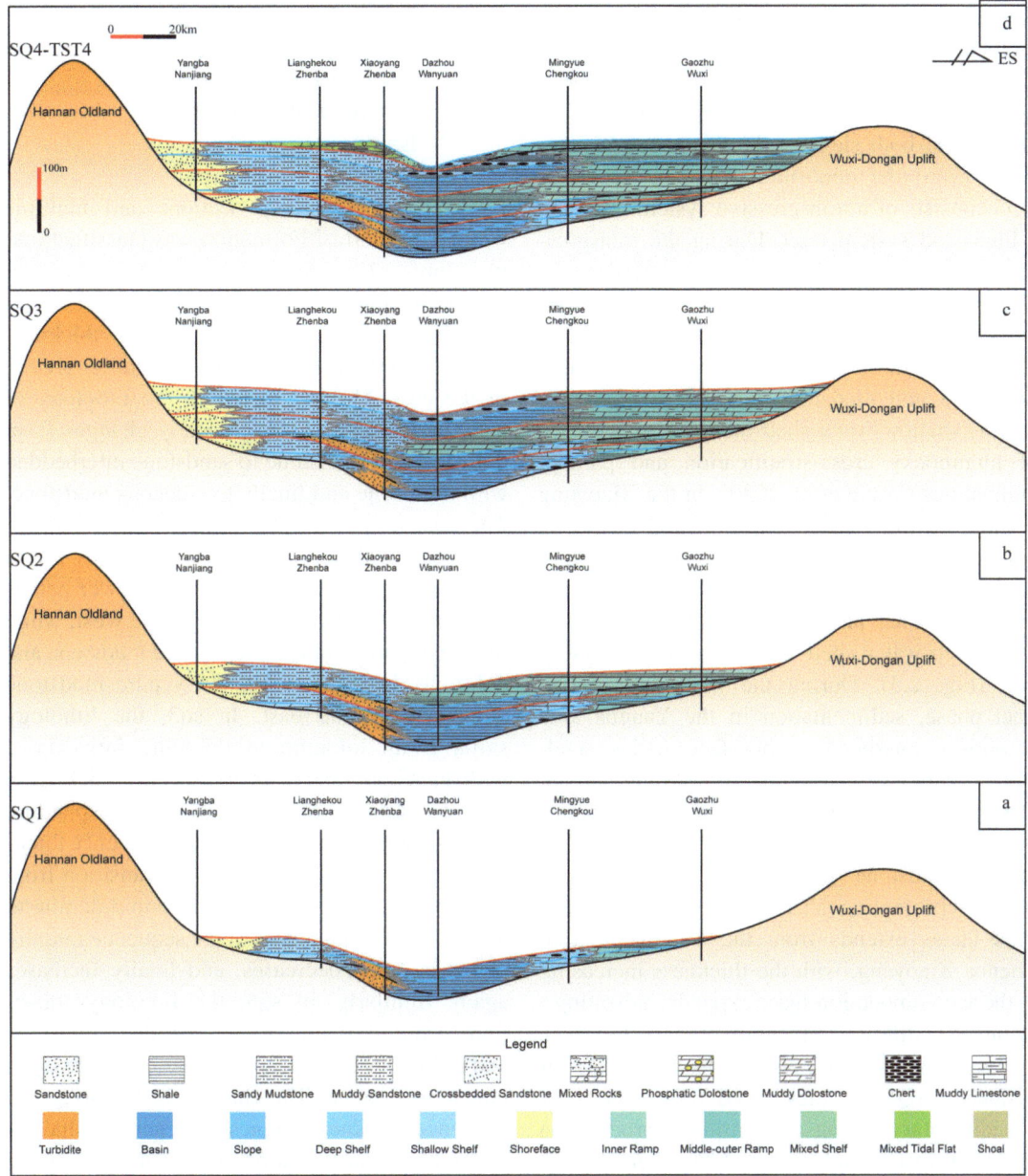

Fig. 2.3 Sedimentary evolution model of Doushantuo Formation in the depression on the northern margin of the Yangtze Block

(Fig. 2.3). During the Transgressive System Tract (TST2), the detrital inputs weakened, resulting in the predominance of slumping and deformed siltstone-rich mudstone on the northwestern slope of the depression (Fig. 2.3). In the Highstand System Tract (HST), the Xiaoyang section transitions from deep-water shelf sandstone with laminated mudstone to rhythmic, fine-grained sandy mudstone and then to sand-rich, shallow-water shelf sandstone, reflecting the progradational characteristics of the late-stage Highstand System Tract (Fig. 2.3).

Sequence SQ3 is situated between SB3 and SB4 and occurs in the middle to upper part of the Doushantuo Formation. It has the widest regional distribution, extending to the Yangba area in Nanjiang. The lateral thickness variation is significant (Fig. 2.3), gradually thickening from both sides of the Chuanbei Depression toward the depositional center. Sequence SQ3 consists of a transgressive system tract and a highstand system tract. During the transgressive system tract, sedimentation occurred in the Yangba area, resulting in the deposition of approximately 25 m of shoreface facies mudstone (Wang et al. 2019). In the Lianghekou area, interbedded sandstone and mudstone are deposited, representing shallow-water shelf facies, with evidence of hummocky cross-stratification and parallel laminations (Wang et al. 2020). In the Xiaoyang section, 40 m of sand-rich laminated mudstone interbedded with thin-bedded, rhythmic, fine-grained sandy developed, forming multiple retrogradational parasequence sets. This represents deep-water shelf transgressive sedimentation (Fig. 2.3). During the highstand system tract phase, sedimentation in the Yangba area comprises interbedded shoreface facies sandstone with cross-stratification (Wang et al. 2019). In the Lianghekou and Xiaoyang areas, the dominant sedimentation is characterized by muddy sandstone and interbedded sandy siltstone, representing shallow-water shelf facies. This facies extends from the Yangba area to Zhenba Xiaoyang, with the thickness increasing as the accommodation space expands, indicating a filling and infilling sedimentary process (Fig. 2.3).

The transgressive system tract (TST4) of Sequence SQ4 was developed on the basis of a sedimentary filling and infilling process on the northwestern side of the Chuanbei Depression (Fig. 2.3). Consequently, TST4 has a relatively thin lateral thickness of approximately 15 m (Fig. 2.3). In the Lianghekou area, phosphatic dolomite is deposited in mixed tidal flat facies. In the Xiaoyang section, the sedimentation is dominated by mixed tidal flat and mixed shelf facies, characterized by the interbedding of shallow-water mixed lithofacies and deep-water mixed lithofacies (Fig. 2.3).

2.3.3 Depositional Characteristics of Early Paleozoic Source Rocks of the Yuertusi Formation in the Tarim Basin

By comparing outcrop sections and multiple wells, the Yuertusi Formation was classified into two third-order sequences and five fourth-order sequences (sq1–sq5) (Fig. 2.4). Well correlation studies have revealed that the thickness of the Yuertusi Formation is notably greater in the east. Moving eastward, the lithology of the Yuertusi Formation gradually changes from dolomite and sandstone to sandstone interbedded with mudstone and finally to siliceous mudstone. The eastern part of the formation is relatively complete, with an increase in thickness and a darker mudstone color. In sq2, the lower part is predominantly dark mudstone in the west, while the upper part contains interbedded mudstone and dolomite. In contrast, relatively pure mudstone is present in the east. In sq3, the lithology shifts from dolomite to dolomite intercalated with mudstone and then to siliceous rock from west to east, accompanied by a corresponding increase in mudstone content and sequence thickness. In sq4, there is a lithologic transition from interbedded dolomitic mudstone and dolomite to mudstone. The thickness of the sequence initially increases, then decreases, and finally increases again. Similarly, in sq5, the lithology transitions from dolomite to interbedded mudstone and dolomite, indicating a continuous increase in mudstone content. The changes in lithology and sequence thickness suggest a generally shallow water depth on the northwestern margin of the Tarim Basin. Specifically, sq2 has a relatively small influence range, representing gradual sea level changes, and is characterized by thick and continuous sedimentary mudstone. Sq4 was deposited over a large area, with a rapid rise in sea level. This resulted in an increase in the deep-water sedimentary area in the northeast of the basin, and compared with the previous period, the depositional center shifted southwestward.

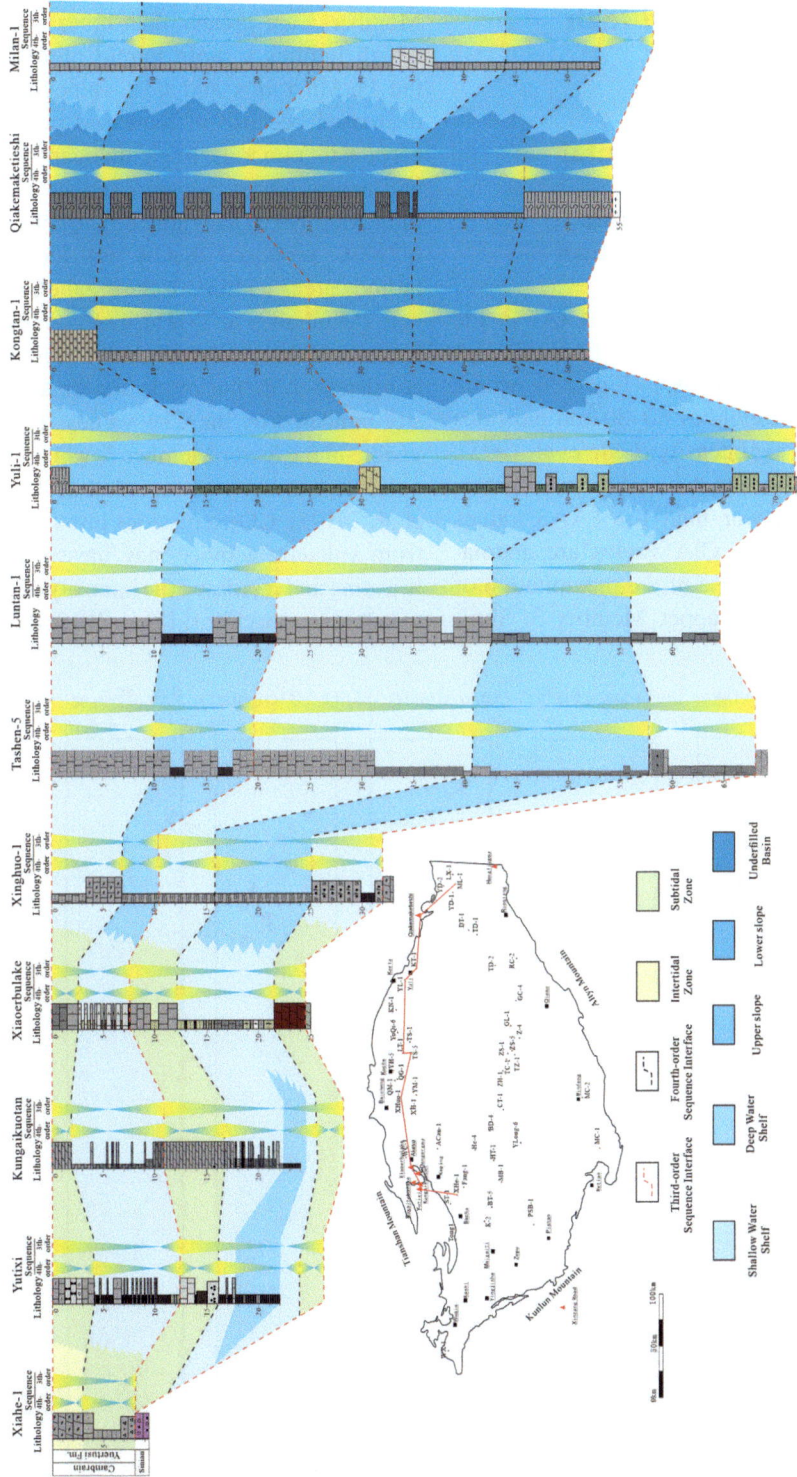

Fig. 2.4 The depositional characteristics of the fourth-order sequences of the Yuertusi Formation in the northwestern Tarim Basin

2.4 Paleoclimate and Paleoenvironmental Reconstruction and Their Characteristics

2.4.1 Paleoclimatic and Paleoenvironmental Reconstruction of the Mesoproterozoic Chuanlinggou Formation

The chemical index of alteration (CIA) in the Chuanlinggou Formation ranges from 60.6 to 71.8, with an average of 68.1 (Fig. 2.5), which falls within the threshold of 60–80 and indicates moderate weathering intensity (Fedo et al. 1995). The CIA values in the Chuanlinggou Formation are slightly lower than the PAAS values (~ 70.4). Notably, the CIA values are anomalously low (< 70) because of the precipitation of calcareous minerals in upper Member 3. Simi-

larly, Fe/Mn ratios are regarded as a viable proxy to track paleoclimate, and the ratios in Member 3 are greater than those in Member 2, which implies a more humid climate. Notably, the Fe/Mn ratios and CIA values anomalously decreased because of increased precipitation of carbonate minerals (e.g., manganese dolomite) and deposition of calcareous shale in the uppermost Member 3. Research on the paleogeography of NCC proposed that it was located in low-latitude areas within 30° during the Mesoproterozoic (Mitchell et al. 2021). The commonly prevalent warm and moist climate at low latitudes is consistent with the moderate weathering intensity indicated by the CIA values. Tuff rocks are visible in the lower Member 3, which implies that volcanic activity was recorded at that time (Liu et al. 2019a). CO_2 emissions into the atmosphere from volcanic activity may have contributed to warming and increased the intensity of continental weathering (Joachimski et al. 2012; Romano et al.

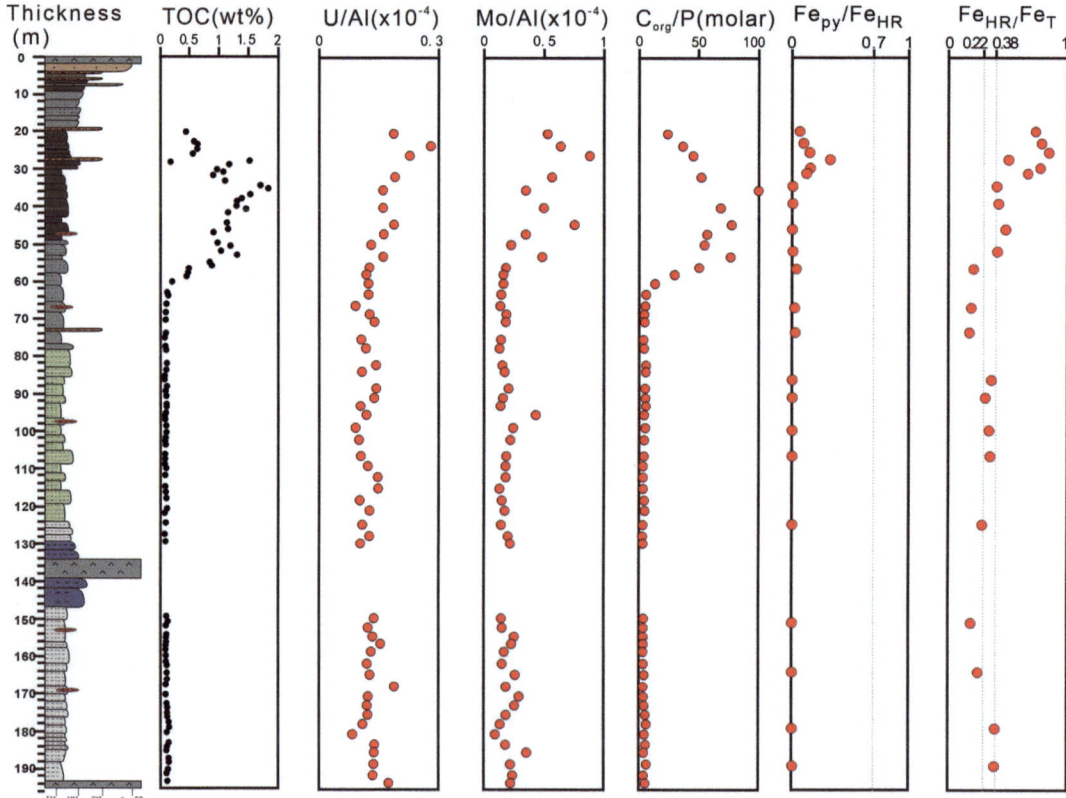

Fig. 2.5 Stratigraphic variation in paleoclimate and redox proxies of the Chuanlingou Formation

2013). This may have produced a warmer and wetter climate and enriched the accumulation of organic matter in Member 3.

Redox conditions are among the most significant components in paleomine settings, and abundant sedimentary geochemical proxies, including trace metal enrichment factors, Fe speciation, and Corg/P ratios, are available (Algeo and Li 2020).

Most redox-sensitive elements in the Chuanlinggou Formation are less abundant than UCC (Mo ~ 1.1 ppm, U ~ 2.7 ppm, V ~ 140 ppm) (Rudnick and Gao 2014), with depleted Mo (0.32 ppm on average), U (1.17 ppm on average) and V (82 ppm on average). Similarly, the enrichment factors, including V_{EF}, Mo_{EF}, and U_{EF}, are less than 1, with lean V_{EF} (0.53–1.21, 0.77 on average), Mo_{EF} (0.11–0.97, 0.28 on average), and U_{EF} (0.25–0.84, 0.41 on average). The Corg/P ratios are distributed between 3.36 and 100.21, with an average of 16.56. The ratios reveal eight points with ratios > 50, corresponding to high TOC contents of 0.89–1.83. The enrichment factors and Corg/P ratios suggest that the Chuanlinggou Formation was deposited in oxic–suboxic seawater. However, the FeHR/FeT ratios range from 0.18 to 0.84, with an average of 0.43, and the FePy/FeHR ratios range from 0.05 to 0.3, indicating oxic–ferruginous anoxic enrichment. Overall, the Chuanlinggou Formation was deposited mainly in a suboxic–ferruginous anoxic environment. A comprehensive analysis of the Chuanlinggou Formation in the context of paleogeography suggested that the Mo and U concentrations decreased because of mild continental weathering with semirestricted–restricted conditions in the Yanliao Basin. The northern part of the Yanliao Basin may have been adjoined to open seawater during the deposition of the Chuanlinggou Formation. It is assumed that the Kuancheng area was farther from the proto-Asian Sea and that the surrounding paleo-continent was at a lower elevation. The oxic–suboxic environment in Member 2 and the ferruginous anoxic environment in Member 3 suggest a transgression–regression cycle in the Chuanlinggou Formation, and the dark shale corresponds to the peak period of transgression.

2.4.2 Paleoclimatic and Paleoenvironmental Reconstruction of the Neoproterozoic Doushantuo Formation

The CIA_{corr} values of the Doushantuo Formation in the Xiaoyang section of Zhenba range from 75.62 to 86.78, with an average of 83.06 (Fig. 2.6), reflecting warm, humid to hot, moist climatic conditions in the source area and moderate to strong weathering intensity. The average CIA_{corr} of the Xiaoyang section is significantly greater than the PAAS value of 70, indicating relatively intense weathering (Taylor and McLennan 1985). Although uncertainties still exist in the global paleogeographic reconstructions of the Ediacaran period, many researchers currently place the South China Block in the mid-low latitude region (Zhao et al. 2018a, b; Kidder and Worsley 2004; Zhang et al. 2015). If the presumed position of the South China Block is correct, then the sedimentation period of the Doushantuo Formation likely experienced a temperate–subtropical–tropical paleoclimate characterized by higher precipitation and annual mean temperatures. Therefore, the paleoweathering intensity and paleoclimate background of the source area indicated by CIA_{corr} in this study are consistent with the inferred low-latitude position of the South China Block and the expected paleoclimate conditions. The CIA_{corr} values for Doushantuo Member 2 range from 75.62 to 83.83, with an average of 80.27, primarily reflecting moderate weathering intensity under a warm and humid climate. The CIA_{corr} values for Doushantuo Member 3 range from 75.82 to 86.78, with an average of 83.68, indicating warm, humid and hot, moist climatic conditions in the source area, as well as moderate to strong weathering intensity. The CIA_{corr} values for Doushantuo Member 4 range from 81.36 to 85.31, with an average of 83.83, indicating moderate weathering intensity and a climate dominated by warm and humid conditions (Fig. 2.6). The CIA_{corr} values in the Xiaoyang section show an overall increasing trend from Doushantuo Member 2 to Member 4, with relatively lower values and larger fluctuations in Member 2 and overall higher and less

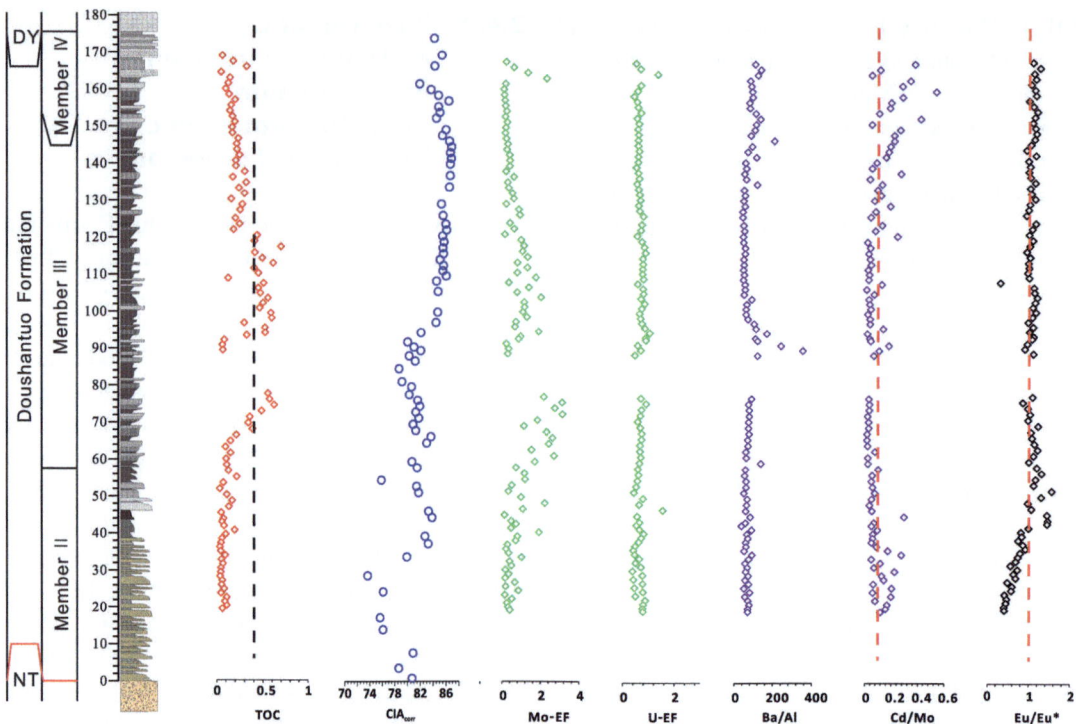

Fig. 2.6 Stratigraphic variation in paleoclimate and redox proxies of the Doushantuo Formation

fluctuation in Member 3, particularly with CIA_{corr} values above 85 in the upper part of Member 3, indicating extreme weathering intensity under hot and humid climatic conditions. The CIA_{corr} values for Member 4 are relatively lower than those for Member 3 (Fig. 2.6). Therefore, in general, during Doushantuo Member 2 in the Upper Yangtze region, the prevailing climate was relatively warm and humid, with significant climatic fluctuations. The climate gradually transitioned to a stable hot and humid climate during Member 3, and during Member 4, it gradually returned to warm and humid conditions but was more stable compared to Member 2 (Fig. 2.6).

The TOC contents of the Doushantuo Formation in the Xiaoyang section range from 0.03 to 0.70, with the highest TOC contents occurring in the lower part of Doushantuo Member 3 (mostly higher than 0.5). The Mo_{EF} of Doushantuo Member 2 is relatively low, ranging primarily from 0.12 to 2.23 (average = 0.58) (Fig. 2.6).

These features suggest relatively oxic conditions. In contrast, the Mo_{EF} in the lower part of Doushantuo Member 3 range from 0.13 to 3.13 (average = 1.38), indicating relatively anoxic conditions (Fig. 2.6). The Mo_{EF} in upper Member 3 and Member 4 return to low values (average = 0.42), suggesting that redox conditions return to oxic conditions (Fig. 2.6). Ba/Al ratios, as indicators of paleoproductivity, are relatively stable across the section, with an average value of 78.8 ppm, suggesting that paleoproductivity was stable during the whole Doushantuo period. The highest values (up to 353.2 ppm) are observed in the lower part of Member 3, indicating that paleoproductivity was relatively high in this interval. Cd/Mo ratios are generally elevated (> 0.1) in sediment deposited in upwelling settings (Sweere et al. 2016). In the Xiaoyang section, Cd/Mo values are relatively high in Member 2 and Member 4, implying that upwelling events were relatively significant during these two periods.

2.4.3 Paleoclimatic and Paleoenvironmental Reconstruction of the Early Paleozoic Yuertusi Formation

The Sugetbulak section in the Tarim Basin is characterized by different sequences in the Yuertusi Formation, and the elemental contents and their ratios are used to explore the characteristics of the depositional environments associated with the four

sequences. The sq2 of the Yuertusi Formation is characterized by a V/(V + Ni) mass fraction ratio of 0.96 and a U_{EF} value of 64.75, indicating an oxygen-poor environment. The CIA curve shows a high average value of 67.77, indicating a warm and humid climate (Fig. 2.7). Compared with sq3–sq5, sq2 has much higher Ba/Al ratios and ex-Ba contents, indicating high productivity during sq2.

Compared with sq2, sq4 has a lower TOC content, with a V/(V + Ni) ratio between 0.14 and 0.84, indicating oxygen-deficient conditions

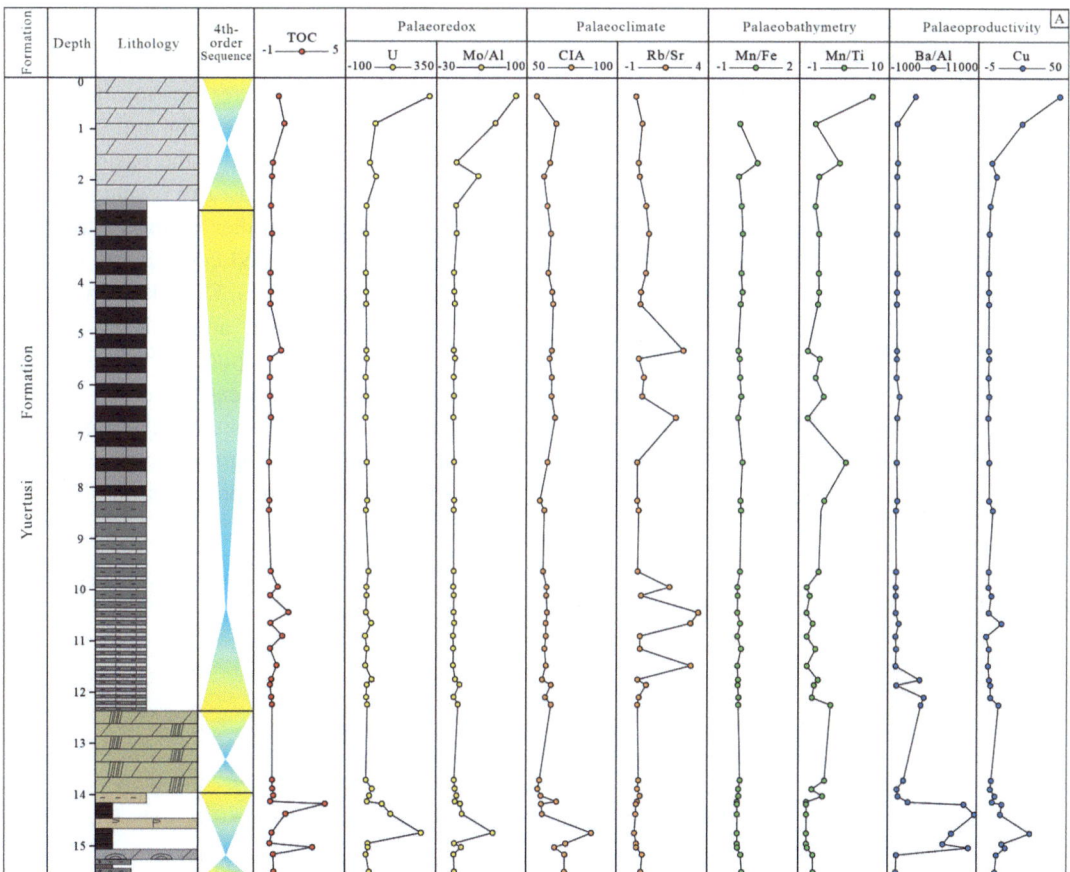

Fig. 2.7 The Sugetbulak section of the Yuertusi Formation exhibits depositional environments variations in the following proxies: TOC content, paleo-redox (EF(U), V/ (V + Ni)), paleoclimatic (CIA, Bb/Sr), sediment input rate (Si/Al, Ti/Al), and paleo-productivity (Ba/Al, Cu)

alternating with suboxic conditions. The CIA curve indicates moderate weathering, and the Rb/Sr ratio indicates deposition in a warm and humid environment. The Si/Al and Ti/Al curves exhibit no obvious changes, suggesting a lower terrigenous supply (Fig. 2.7).

2.5 Global Geological Events and Their Effects on Organic Matter Enrichment

2.5.1 Organic Matter Accumulation in the Mesoproterozoic Chuanlinggou Formation

The combination of Cd, Mn, Co and Mo can be used to estimate the relative importance of productivity versus preservation and has the potential to reconstruct environmental conditions during the deposition of organic-rich sediments (Sweere et al. 2016). An analysis of samples from the lower Member 3 and upper Member 2 indicated upwelling settings, whereas an analysis of samples from the other parts indicated restricted settings (Fig. 2.8). These results demonstrate that the accumulation of organic matter in the Chuanlinggou Formation was dominated by primary

productivity according to the Cd/Mo and Co*Mn proxies, and the higher the primary productivity was, the higher the TOC.

As a consequence, a gradual opening process with a broad transgression during the Chuanlinggou Formation can be inferred (Fig. 2.9). In Member 2, the Yanliao Basin was in the early period of rifting with low sea level. Member 2 was deposited in an intertidal–subtidal muddy flat with oxic–suboxic seawater. A cooler and drier paleoclimate may have led to weaker weathering, which resulted in insufficient nutrients and subsequently reduced production of organic matter. Moreover, oxic–suboxic environments were not conducive to the preservation of organic matter. By the period of Member 3, open seawater was injected southward into the Yanliao Basin (Yan and Liu 1998), which led to the transition in the basin environment from a restricted basin to a semirestricted basin. Moreover, the further rifting of the basin may have led to improved connections between the basin and the open sea. Upwelling, a warmer and wetter paleoclimate, and volcanic activity increased the nutrient input to the basin. This increase in nutrient content increased primary productivity and then promoted the enrichment of organic matter. A ferruginous anoxic environment is beneficial for the preservation of organic matter.

Fig. 2.8 Cd/Mo versus Co*Mn (inferred conditions based on modern settings, after Sweere et al. 2016)

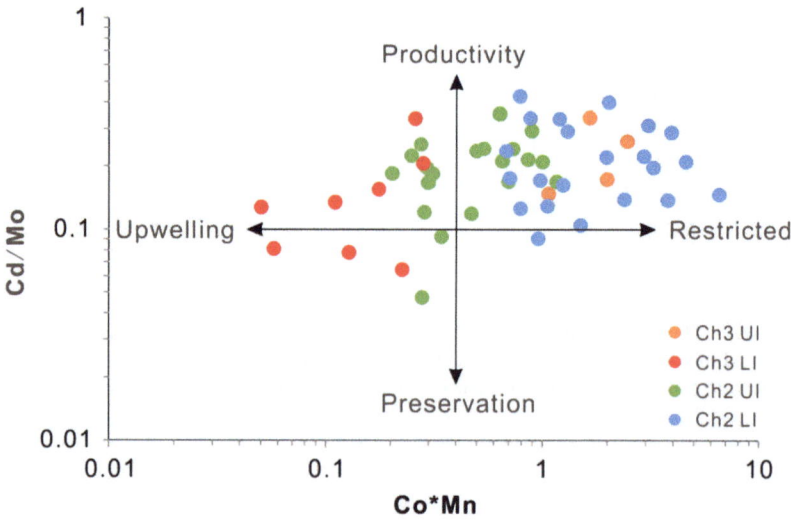

Fig. 2.9 The depositional model of the Chuanlinggou Formation in North China Craton

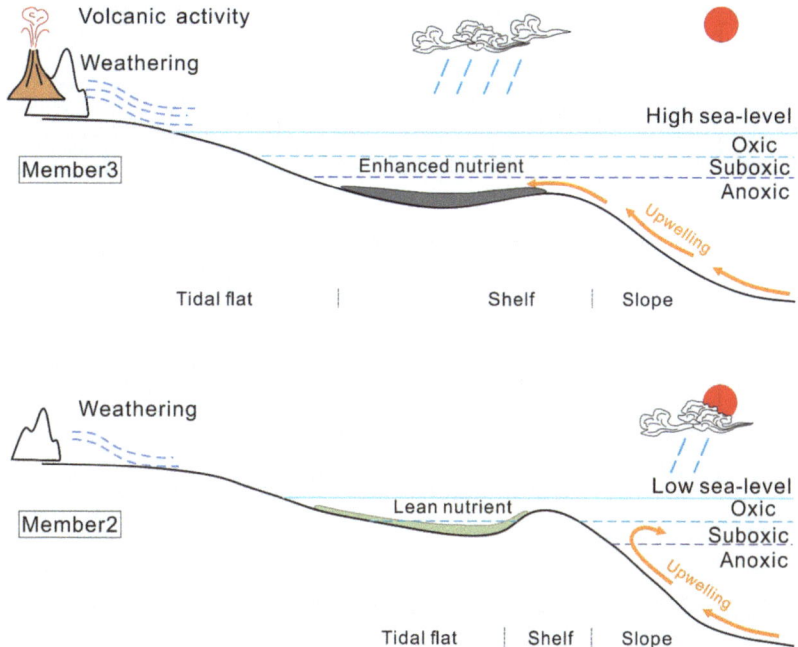

2.5.2 Organic Matter Accumulation in the Neoproterozoic Doushantuo Formation

Following the termination of the Marinoan glaciation event, the climate gradually transitioned from an icehouse to a greenhouse state during the Doushantuo Member 1 period. The sea level was relatively low, and the overall marine environment was oxygenated, leading to limited organic matter development (Fig. 2.10). During this period, sedimentation was absent in Zhenba County, which served as an erosion area. During the Doushantuo Member 2, the climate experienced two prominent warm–cold cycles. The warm periods resulted in increased input of terrigenous clastics, stimulating primary productivity in the euphotic zone. This, in turn, triggered the release of dissolved organic carbon from the deep sea, along with associated sulfide events, facilitating organic matter enrichment in the intraplatform basin (western Hubei Province) and shelf-margin slope (western Hunan Province and northern Guizhou Province). However, the strata in the Chuanbei Depression are dominated by turbidite deposits, and the redox conditions are oxic. Therefore, organic matter

accumulation is relatively low in the Chuanbei Depression (Fig. 2.10). Three additional warm–cold cycles occurred during the Doushantuo Member 3 phase, including the Gaskiers glaciation event (Chen et al. 2020). The sea level was relatively high, and the interval was dominated by laminated mudstone. The amalgamation of the Gondwana supercontinent led to intensified continental weathering (Halverson et al. 2010), resulting in a continued increase in clastic input. Furthermore, this increased ocean oxidation and triggered biodiversity events, leading to organic matter enrichment during warm periods. Suboxic–anoxic bottom water conditions also increased the preservation of organic matter. At the end of Doushantuo Member 4, strong upwelling currents and anoxic conditions facilitated the accumulation of organic matter.

2.5.3 Organic Matter Accumulation in the Early Paleozoic Yuertusi Formation

The Al-(Co × Mn) cross-plotting technique has been found to be effective in distinguishing the

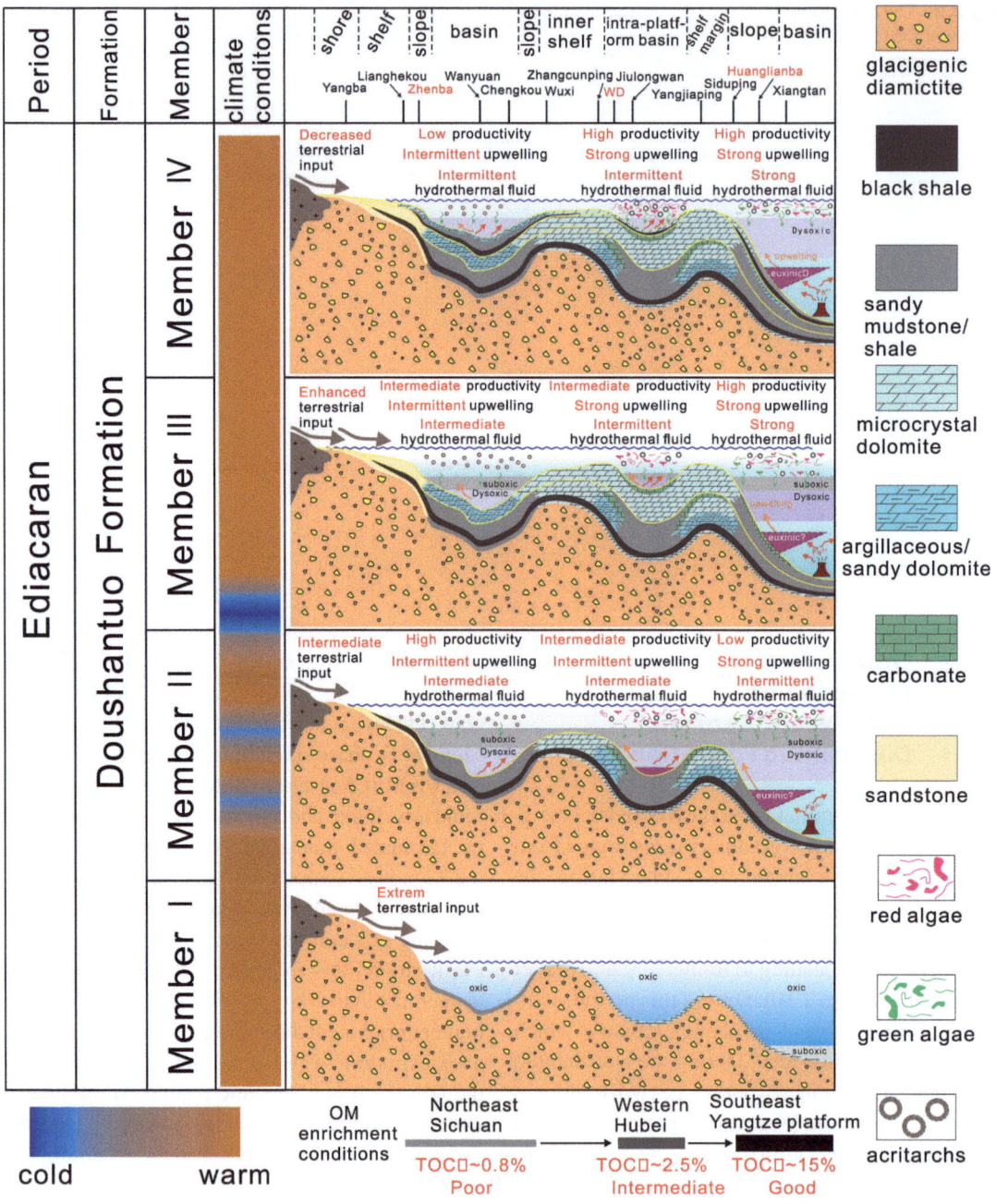

Fig. 2.10 Organic matter accumulation model for the Doushantuo Formation

properties of water columns (Sweere et al. 2016). The data show that the (Co × Mn) values of sq2 and sq4 are low, and only a few samples were plotted in the higher values casting-restricted region, indicating that the organic matter enrichment was controlled mainly by upwelling during these intervals (Fig. 2.11). A rendezvous plot correlation analysis was conducted to determine the relationships between organic matter enrichment and various indicators. The results revealed that the sq2 layer had the highest correlation (R^2 = 0.4426) between TOC and paleoproductivity (ex-Ba) (Fig. 2.11). On the other hand, weaker correlations were observed between the TOC content and various indicators for sq4 and sq5 and, in some cases, were even negative (Fig. 2.11), suggesting that paleoproductivity was the main factor controlling organic matter enrichment in the study area.

Based on the analysis of the above various indicators, it can be concluded that the early Cambrian period was characterized by high sedimentation rates and strong upwelling, which brought abundant nutrients to the Yuertusi Formation. The warm and humid environment further increased the paleoproductivity of the sq2 of the Yuertusi Formation, resulting in the formation of hydrocarbon-generating material. The presence of an oxygen-deficient reducing environment provided ideal preservation conditions for the deposited organic matter, leading to TOC enrichment. A high-quality hydrocarbon source rock depositional model was established under the influence of hydrothermal activity, upwelling currents, and productivity, as shown in Fig. 2.12. The sq2 period of the Yuertusi Formation, which featured high paleoproductivity due to upwelling current, and the oxygen-deficient reducing environment provided excellent preservation conditions for deposited organic matter, resulting in significant TOC enrichment. The sq4 period of the Yuertusi Formation was affected by weakly restricted water conditions with low sediment input rates and nutrient supply. Although the paleoproductivity was lower than that of the sq2 layer, the suboxic sedimentary environment was conducive to preserving the generated sedimentary organic matter. In contrast, sq3 was also influenced by upwelling but was subjected to a relatively hot and dry climate. This had a negative impact on the growth and reproduction of hydrocarbon-generating parent materials, and oxygen-rich conditions were not conducive to preserving organic matter. Sq5 is influenced mostly by weak water conditions, low productivity in ancient times, and frequent changes in redox conditions, which are not conducive to the generation and preservation of organic matter.

2.6 Potential Source Rock–Reservoir Patterns

The Mesoproterozoic Chuanlinggou Formation in the North China Craton contains widely distributed organic-rich shales that represent potential source rocks. Together with overlying Paleozoic and Cenozoic strata that provide reservoir and seal intervals, they form a speculative petroleum system, although no commercial discoveries have yet been reported (Zhao et al. 2018a, b). In the Yangtze Craton, the Neoproterozoic Doushantuo Formation is one of the most important source rocks for gas accumulation in the Ediacaran Dengying Formation (Zou et al. 2014). In the northwestern part of the craton, reservoirs in the Dengying Formation are developed mainly in platform–margin shoal complexes, consisting of algal–clot dolostone, stromatolitic dolostone, and algal clastic dolostone. Reservoir spaces are dominated by intergranular pores, intercrystalline dissolution pores, and cavities. The second and fourth members of the Dengying Formation were modified by the first and second phases of the Tongwan orogeny, respectively, generating two large-scale regional unconformities. Subsequent epigenetic karstification led to the development of widely distributed karst-cavity reservoirs (Yang et al. 2023). The platform-margin shoal reservoirs of the Dengying Formation are underlain by high-quality source rocks of the Doushantuo Formation and overlain by source rocks of the Cambrian Maidiping and Qiongzhusi formations, forming a "sandwich-type" source–reservoir–seal assemblage favorable for hydrocarbon accumulation (Yang et al. 2023). Gas reservoirs in the second and fourth members of the Dengying Formation are dual-source systems, as they are charged by hydrocarbons derived from both the Doushantuo Formation and the Maidiping–Qiongzhusi Formation (Xie et al. 2021). In particular, the Doushantuo source rocks and platform-margin shoal reservoirs of the Dengying Formation establish a favorable "lower-source, upper-reservoir" configuration (Yang et al. 2023).

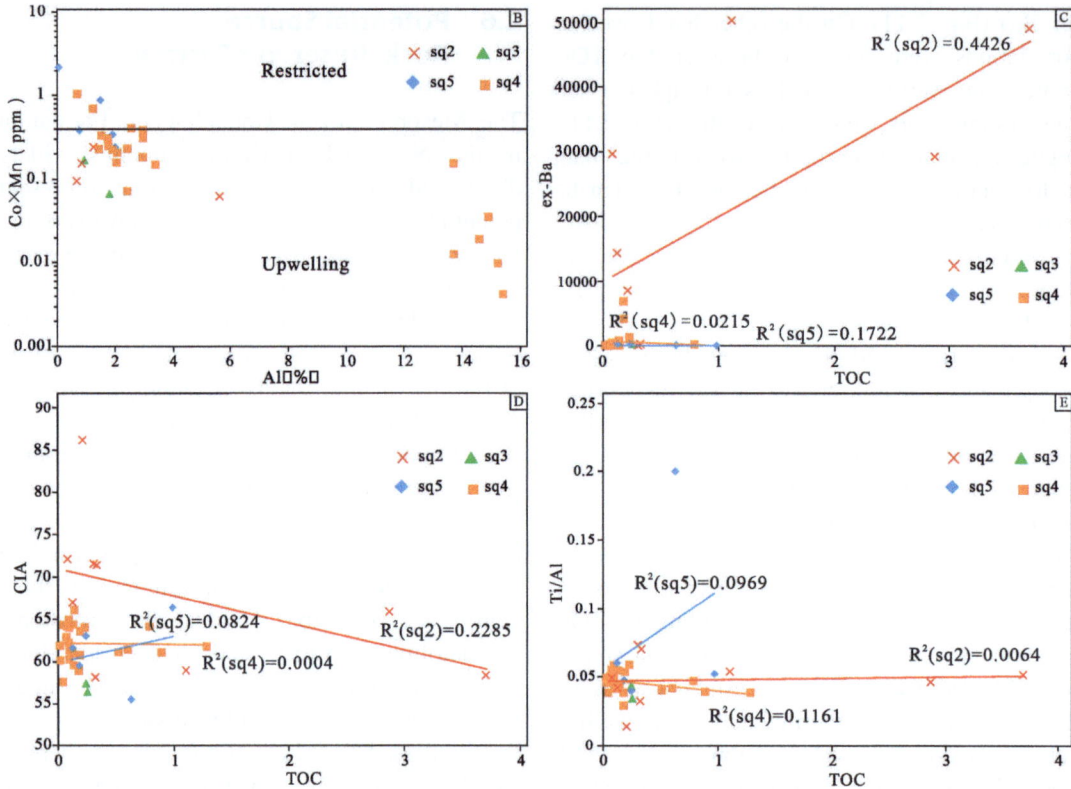

Fig. 2.11 Plots of the main controlling factors of the organic matter enrichment of the Yuertusi Formation, including ex-Ba, CIA, and Ti/Al

In the Tarim Basin, organic-rich shales of the Cambrian Yuertusi Formation are present mainly in the northern depression, its margins, and the Maigaiti Slope. These shales are the principal source rocks for petroleum systems of the Ediacaran, Cambrian, and Ordovician strata in platform–basin regions (Wang et al. 2024). The reservoirs in these intervals are predominantly carbonates, including vuggy dolostone and karst-fracture–cavity limestones. In the upper Ediacaran, the dolostone reservoirs of the Qigebulake Formation are controlled by high-energy grain shoal facies, enhanced by karstification, and dominated by vuggy porosity. In the Cambrian, dolostone reservoirs occur in the Xiaoerbulake, Wusongger, and Shayilik formations. The Xiaoerbulake reservoirs are associated mainly with gentle-slope shoal and high-energy grain shoal facies, whereas the Wusongger and Shayilik reservoirs are related to weakly rimmed to rimmed

carbonate platforms, developing both platform-margin reef–shoal and platform-interior shoal facies, dominated by vuggy and fracture–vuggy types. In the Ordovician reservoirs, the karst-fracture–cavity limestone reservoirs of the Yingshan, Yijianfang, and Lianglitage formations are widely distributed in northern Tarim, central Tarim, the Gucheng area, and the Maigaiti Slope, controlled jointly by karstification and strike-slip faulting (Li et al. 2021). The Yuertusi Formation, together with the underlying Ediacaran dolostone and the overlying Cambrian Series 2 dolostone and evaporites, forms a "stacked" source–reservoir–seal assemblage (Jia and Zhang 2023). Furthermore, it has established a "long-distance conduit-type" source–reservoir–seal system with Middle–Lower Ordovician reservoirs and Upper Ordovician tight limestone and argillaceous shales (Jia and Zhang 2023).

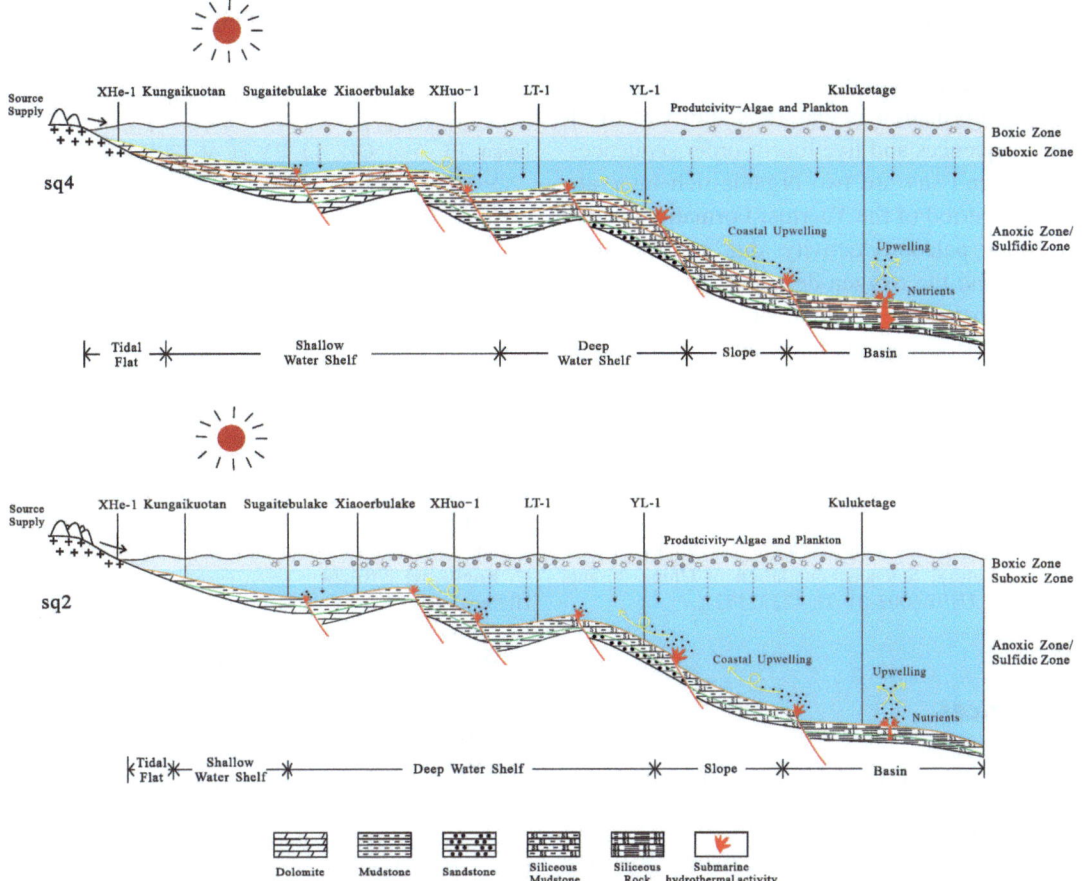

Fig. 2.12 The depositional model of the high-quality hydrocarbon source rocks in the Yuertusi Formation, which was formed under the influence of hydrothermal activity upwelling currents, and productivity. **a** Sq4 of the Yuertusi Formation, **b** sq2 of the Yuertusi Formation

2.7 Summary

Three selected hydrocarbon source rocks in cratonic basins are analyzed in this study. Organic-rich Member 3 of the Mesoproterozoic Chuanlingou Formation in the Kuancheng section is dominated by transgressive dark shale, which corresponds to the inner shelf. The accumulation of organic matter in the Chuanlinggou Formation is predominantly related to primary productivity. Upwelling, a warmer and wetter paleoclimate, and volcanic activity increased the nutrient input to the basin. This increase in nutrient content increased primary productivity and then promoted the enrichment of organic matter. The Ediacaran Doushantuo Formation underwent a sedimentary evolution process of "steep slope–slope–shelf–mixed platform". The lower part of the Doushantuo Formation is characterized by steep slope facies composed of turbidite and dark laminated mudstone facies. The turbidite sequences are overlain by a gentle slope facies dominated by slumped and folded siltstone. In the middle part, deep-water shelf facies characterized by interbedded sandy laminated mudstone and rhythmic sandy mudstone are present. The upper-middle part evolves into a shallow-water shelf facies dominated by sand-rich interbedded siltstone. At the top, there is a mixed-facies zone between shallow-water tidal flat and deep-water mixed-shelf facies. Doushantuo Member 3 is an organic-rich interval that is triggered by a warm climate and increased clastic inputs. Redox conditions are the main factor controlling the accumulation of organic matter. The Yuertusi Formation is classified into two third-order sequences and five fourth-order sequences, and sq2 and sq4 are two organic-rich intervals. The sq2 period of the Yuertusi Formation, which has high paleoproductivity due to upwelling current, and the oxygen-deficient reducing environment provided excellent preservation conditions for deposited organic matter, resulting in significant TOC enrichment. The suboxic sedimentary environment of sq4 is also conducive to preserving the generated sedimentary organic matter.

Acknowledgements This research was funded by the National Natural Science Foundation of China (Grants U24B6001, U19B6003-01-02, 42272173).

References

Ahmad M, Dunster JN, Munson TJ (2013) McArthur Basin. In: Ahmad M, Munson TJ (eds) Geology and mineral resources of the Northern Territory: Darwin, Australia, Northern Territory Geological Survey Special Publication 5, 72 pp

Algeo TJ, Li C (2020) Redox classification and calibration of redox thresholds in sedimentary systems. Geochim Cosmochim Acta 287:8–26

Brumsack H-J (1989) Geochemistry of recent TOC-rich sediments from the Gulf of California and the Black Sea. Geol Rundsch 78(3):851–882

Cai J, Luo S, Guan Y, Dong P (2021) Main controlling factors of organic matter enrichment of the Mesoproterozoic source rocks from the North China Craton. Geol J 56(1):422–433

Chen YQ, Yan W, Han CW, Yan L, Ran QG, Kang Q, He H, Ma Y (2019) Structural and dendimentary basin transformation at the Cambrian/Neoproterozoic interval in Tarim Basin: implication to subsalt dolostone exploration. Nat Gas Geosci 30:40–48

Chen C, Wang J, Wang Z et al (2020) Variation of chemical index of alteration (CIA) in the Ediacaran Doushantuo Formation and its environmental implications. Precambr Res 347:105829

Cox GM, Collins AS, Jarrett AJM et al (2022) A very unconventional hydrocarbon play: the Mesoproterozoic Velkerri Formation of northern Australia. AAPG Bull 106(6):1213–1237

Craig J, Biffi U, Galimberti RF et al (2013) The palaeobiology and geochemistry of Precambrian hydrocarbon source rocks. Mar Pet Geol 40:1–47

Du J, Zou C, Xu C et al (2014) Theoretical and technical innovations in strategic discovery of a giant gas field in Cambrian Longwangmiao Formation of central Sichuan paleo-uplift, Sichuan Basin. Pet Explor Dev 41(3):268–277

Duan JB, Mei QH, Li BS et al (2019) Sinian-early Cambrian tectonic-sedimentary evolution in Sichuan Basin. Earth Sci (Chin Engl Abs) 44(3):738–755

Fairchild IJ, Kennedy MJ (2007) Neoproterozoic glaciation in the earth system. J Geol Soc 164(5):895–921

Fedo CM, Wayne Nesbitt H, Young GM (1995) Unraveling the effects of potassium metasomatism in sedimentary rocks and paleosols, with implications for paleoweathering conditions and provenance. Geology 23(10):921–924

Halverson GP, Wade BP, Hurtgen MT et al (2010) Neoproterozoic chemostratigraphy. Precambr Res 182(4):337–350

He D (2022) Multi-cycle superimposed sedimentary basins in China: formation, evolution, geologic framework and hydrocarbon occurrence. Earth Sci Front (Chin Engl Abs) 29(6):24

Hlebszevitsch JC, Gebhard I, Cruz CE et al (2009) The 'Infracambrian System' in the southwestern margin of Gondwana, southern South America. Geol Soc Lond Spec Publ 326(1):289–302

Hoffman PF, Kaufman AJ, Halverson GP et al (1998) A Neoproterozoic snowball earth. Science 281(5381):1342–1346

Hoffman PF, Abbot DS, Ashkenazy Y et al (2017) Snowball earth climate dynamics and Cryogenian geology-geobiology. Sci Adv 3(11):e1600983

Huang BC, Zhou YX, Zhu RX (2008) Discussions on Phanerozoic evolution and formation of continental China, based on paleomagnetic studies. Earth Sci Front 15:348–359

Jia C, Zhang S (2023) The formation of marine ultra-deep petroleum in China. Acta Geol Sin 97:2775–2801

Jiang G, Shi X, Zhang S et al (2011) Stratigraphy and paleogeography of the Ediacaran Doushantuo Formation (ca. 635–551 Ma) in South China. Gondwana Res 19(4):831–849

Joachimski MM, Lai X, Shen S, Jiang H, Luo G, Chen B, Chen J, Sun Y (2012) Climate warming in the latest Permian and the Permian-Triassic mass extinction. Geology 40(3):195–198

Kidder DL, Worsley TR (2004) Causes and consequences of extreme Permo-Triassic warming to globally equable climate and relation to the Permo-Triassic extinction and recovery. Palaeogeogr Palaeoclimatol Palaeoecol 203(3–4):207–237

Klemme HD, Ulmishek GF (1991) Effective petroleum source rocks of the world: stratigraphic distribution and controlling depositional factors. AAPG Bull 75(12):1809–1851

Kuang HW, Liu YQ, Geng YS et al (2019) Important sedimentary geological events of the Meso-Neoproterozoic and their significance. J Palaeogeogr (Chin Engl Abs) 21(1):1–30

Li ZX, Li XH, Kinny PD et al (1999) The breakup of Rodinia: did it start with a mantle plume beneath South China? Earth Planet Sci Lett 173(3):171–181

Li ZX, Bogdanova SV, Collins AS et al (2008) Assembly, configuration, and break-up history of Rodinia: a synthesis. Precambr Res 160(1):179–210

Li C, Planavsky NJ, Love GD, Reinhard CT, Hardisty D, Feng LJ, Bates SM, Huang J, Zhang QR, Chu XL, Lyons TW (2015) Marine redox conditions in the middle Proterozoic ocean and isotopic constraints on authigenic carbonate formation: insights from the Chuanlinggou Formation, Yanshan Basin, North China. Geochim Cosmochim Acta 150:90–105

Li M, Hinnov L, Kump L (2019a) Acycle: time-series analysis software for paleoclimate research and education. Comput Geosci 127:12–22

Li SZ, Li XY, Wang GZ, Liu YM, Wang ZC, Wang TS, Cao XZ, Guo XY, Somerville I, Li Y, Zhou J, Dai LM, Jiang SH, Zhao H, Wang Y, Wang G, Yu S (2019b) Global Meso-Neoproterozoic plate reconstruction and formation mechanism for Precambrian basins: constraints from three cratons in China. Earth Sci Rev 198:28

Li J, Tao X, Bai B et al (2021) Geological conditions, reservoir evolution and favorable exploration directions of marine ultra-deep oil and gas in China. Pet Explor Dev 48(1):60–79

Lin YT, Tang DJ, Shi XY, Zhou XQ, Huang KJ (2019) Shallow-marine ironstones formed by microaerophilic iron-oxidizing bacteria in terminal Paleoproterozoic. Gondwana Res 76:1–18

Liu D, Wang X, Zhang H, Shi C (2019a) Zircon SHRIMP U-Pb age of the Chuanlinggou Formation of the Changcheng Group, North China and the stratigraphic implications. Earth Sci Front 26(3):183–189

Liu XG, Li SZ, Zhang J, Li XY, Zhao SJ, Dai LM, Wang GZ (2019b) Meso-Neoproterozoic strata and target source rocks in the North China Craton: a review. Precambr Res 334:22

Liu SQ, Zhu XY, Su WB, Zhao TP, Qiu YF, He YT, Zhang HF (2022) Black shales of the Chuanlinggou Formation in the northern North China Craton and its constraints on redox state of seawater during the Early Mesoproterozoic. Acta Petrol Sin 38(6):1667–1684

Luo QY, George SC, Xu YH, Zhong NN (2016) Organic geochemical characteristics of the Mesoproterozoic Hongshuizhuang Formation from northern China: implications for thermal maturity and biological sources. Org Geochem 99:23–37

Ma Y, Li M, Cai X et al (2020) Mechanisms and exploitation of deep marine petroleum accumulations in China: advances, technological bottlenecks and basic scientific problems. Oil Gas Geol 41(4):655–672

Mitchell RN, Kirscher U, Kunzmann M, Liu Y, Cox GM (2021) Gulf of Nuna: astrochronologic correlation of a Mesoproterozoic oceanic euxinic event. Geology 49(1):25–29

Qiao X, Gao L (2007) Mesoproterozoic palaeoearthquake and palaeogeography in Yan-Liao Aulacogen. J Palaeogeogr 9(4):337–352

Romano C, Goudemand N, Vennemann TW, Ware D, Schneebeli-Hermann E, Hochuli PA, Brühwiler T, Brinkmann W, Bucher H (2013) Climatic and biotic upheavals following the end-Permian mass extinction. Nat Geosci 6(1):57–60

Rudnick RL, Gao S (2014) Composition of the continental crust, treatise on geochemistry. Elsevier, pp 1–51

Shi X, Jiang G, Zhang C, Gao L, Liu J (2008) Sand veins and MISS from the Mesoproterozoic black shale (ca. 1.7 Ga) in North China: implication for methane degassing from microbial mats. Sci China Ser D Earth Sci 51(11):1525–1536

Shi C, Liu D, Cui Y, Zhang C, Qiu Y (2016) Soft-sediment structures developed in northern part of North China paleocontinent and their constraint on geodynamic environment of sedimentary basin. Geol Rev 62(1):37–53

Sweere T, van den Boorn S, Dickson AJ, Reichart GJ (2016) Definition of new trace-metal proxies for the controls on organic matter enrichment in marine sediments based on Mn Co, Mo and Cd concentrations. Chem Geol 441:235–245

Tang D, Shi X, Liu J, Wang X, Pei Y (2009) Authigenic carbonate precipitation and pyrites from sand-veins in the Chuanlinggou Formation of the North China platform—evidence for anaerobic oxidation of methane in the Mesoproterozoic. J Palaeogeogr 11(4):361–374

Tang D, Shi X, Liu D, Lin Y, Zhang C, Song G, Wu J (2015) Terminal Paleoproterozoic ooidal ironstone from North China: a sedimentary response to the initial breakup of Columbia supercontinent. Earth Sci 40(2):290–304

Taylor SR, McLennan SM (1985) The continental crust: its composition and evolution, 312 pp

Vernhet E (2007) Paleobathymetric influence on the development of the late Ediacaran Yangtze platform (Hubei, Hunan, and Guizhou provinces, China). Sed Geol 197(1):29–46

Vernhet E, Reijmer JJ (2010) Sedimentary evolution of the Ediacaran Yangtze platform shelf (Hubei and Hunan provinces, Central China). Sed Geol 225(3–4):99–115

Wang J, Li Z (2003) History of Neoproterozoic rift basins in South China: implications for Rodinia break-up. Precambr Res 122(1):141–158

Wang ZJ, Wang J, Jiang XS et al (2015) New progress for the stratigraphic division and correlation of Neoproterozoic in Yangtze Block, South China. Geol Rev 61(1):1–22

Wang Z, Liu J, Jiang H et al (2019) Lithofacies paleogeography and exploration significance of Sinian Doushantuo depositional stage in the middle-upper Yangtze region, Sichuan Basin, SW China. Pet Explor Dev 46(1):41–53

Wang H, Li Z, Liu S et al (2020) Ediacaran extension along the northern margin of the Yangtze platform, South China: constraints from the lithofacies and geochemistry of the Doushantuo Formation. Mar Pet Geol 112:104056

Wang Q, Xu Z, Zhang R et al (2024) New fields, new types of hydrocarbon explorations and their resource potentials in Tarim Basin. Acta Petrol Sin 45(1):15

Xiao D, Cao J, Luo B, Tan X, Xiao W, He Y, Li K (2021) Neoproterozoic postglacial paleoenvironment and hydrocarbon potential: a review and new insights from the Doushantuo Formation Sichuan Basin, China. Earth Sci Rev 212:103453

Xie Z, Li J, Yang C (2021) Geochemical characteristics of Sinian-Cambrian natural gas in central Sichuan paleo-uplift and exploration potential of Taihe gas area. Nat Gas Ind 41(7):1–14

Yan YZ, Liu ZL (1998) The relationship between biocommunities and paleoenvironments in Changcheng Period of the Yanshan Basin, North China. Acta Micropalaeontol Sin 03:29–46

Yang Y, Wen L, Luo B et al (2016) Sedimentary tectonic evolution and reservoir-forming conditions of the Dazhou-Kaijiang paleo-uplift, Sichuan Basin. Nat Gas Ind B 3(6):515–525

Yang XQ, Li Z, Gao B, Zhou YQ (2021) The Cambrian Drumian carbon isotope excursion (DICE) in the Keping area of the northwestern Tarim Basin, NW China. Palaeogeogr Palaeoclimatol Palaeoecol 571

Yang Y, Wen L, Zhou G et al (2023) New fields, new types and resource potentials of hydrocarbon exploration in Sichuan Basin. Acta Petrol Sin 44(12):2045

Zhai M, Zhao Y, Zhao T (2016) Main tectonic events and metallogeny of the North China Craton. Springer

Zhang S, Li H, Jiang G et al (2015) New paleomagnetic results from the Ediacaran Doushantuo Formation in South China and their paleogeographic implications. Precambr Res 259:130–142

Zhang X, Gao Z, Maselli V, Fan T (2023) Tectono-sedimentary characteristics and controlling factors of the lower-middle Cambrian gypsum-salt rocks in the Tarim Basin, Northwest China. Mar Petrol Geol 151:106189

Zhao G, Cawood PA (2012) Precambrian geology of China. Precambr Res 222:13–54

Zhao G, Sun M, Wilde SA et al (2004) A Paleo-Mesoproterozoic supercontinent: assembly, growth and breakup. Earth Sci Rev 67(1):91–123

Zhao G, Wang Y, Huang B et al (2018a) Geological reconstructions of the East Asian blocks: from the breakup of Rodinia to the assembly of Pangea. Earth Sci Rev 186:262–286

Zhao W, Hu S, Wang Z et al (2018b) Petroleum geological conditions and exploration importance of Proterozoic to Cambrian in China. Pet Explor Dev 45(1):1–14

Zhao W, Wang X, Hu S, Zhang S, Wang H, Guan S, Ye Y, Ren R, Wang T (2019) Hydrocarbon generation characteristics and exploration prospects of Proterozoic source rocks in China. Sci China Earth Sci 62:909–934. https://doi.org/10.1007/s11430-018-9312-4

Zhu M, Zhang J, Yang A et al (2003) Sinian-Cambrian stratigraphic framework for shallow-to deep-water environments of the Yangtze Platform: an integrated approach. Prog Nat Sci 13(12):951–960

Zhu M, Zhang J, Yang A (2007) Integrated Ediacaran (Sinian) chronostratigraphy of South China. Palaeogeogr Palaeoclimatol Palaeoecol 254(1–2):7–61

Zhu M, Lu M, Zhang J et al (2013) Carbon isotope chemostratigraphy and sedimentary facies evolution of the Ediacaran Doushantuo Formation in western Hubei, South China. Precambr Res 225:7–28

Zhu GY, Chen FR, Chen ZY, Zhang Y, Xing X, Tao XW, Ma DB (2016) Discovery and basic characteristics of the high quality source rocks of the Cambrian Yuertusi Formation in Tarim Basin. Nat Gas Geosci 27:9–19

Zou C, Du J, Xu C et al (2014) Formation, distribution, resource potential, and discovery of Sinian-Cambrian giant gas field, Sichuan Basin, SW China. Pet Explor Dev 41(3):306–325

Tectonic–Sedimentary Differentiation of Cratonic Basins in China and Its Controlling Effects on the Development of Large-Scale Carbonate Reservoirs

Shuangjian Li, Yingqiang Li, Shoutao Peng, Yunqing Hao, Juntao Zhang, Jingbin Wang, and Xuhui Xu

Abstract

The Tarim, Yangtze, and North China cratons in China are characterized by small scales, poor tectonic stability, and strong tectonic–sedimentary differentiation. Based on the investigations of the prototypes and paleogeographical evolution of the Tarim, Sichuan, and Ordos Basins, this study classified tectonic–sedimentary differentiation and analyzed its controlling effects on large-scale reservoirs. In this study, the main factors controlling the development of three types of large reservoirs, i.e., the reef-shoal reservoirs of the Sinian Dengying Formation in the Sichuan Basin, the intraplatform grain shoal reservoirs of the Cambrian Longwangmiao Formation and the Xixiangchi Group, and the Ordovician paleokarst reservoirs in the Tabei paleo-uplift in the Tarim Basin, were analyzed. Based on the analytical results, this study summarized the controlling mechanisms of tectonic–sedimentary differentiation on the development of large-scale

reservoirs and obtained the following conclusions: (1) The orderly superimposition of rift basins, basins with passive continental margins, cratonic depression basins, and foreland basins in the extensional–convergent tectonic cycles of marine basins controls the regular development of reef-shoal reservoirs, intraplatform grain shoal reservoirs, and paleokarst reservoirs. (2) The development of reef-shoal reservoirs is controlled by the formation of intracratonic and marginal rifts in a strong extensional setting, whereas the development of intraplatform grain shoal reservoirs is controlled by the formation of intracratonic depressions in a weak extensional–weak compression setting, and the development of karst reservoirs is controlled by paleo-uplifts that formed during the strong compression period. (3) The development of high-quality reservoirs in the Sinian Dengying Formation and the Cambrian Longwangmiao Formation and Xixiangchi Group in the Sichuan Basin was controlled by high-energy facies belts, early dolomitization, and early karstification. High-quality reservoirs show a zonal distribution under strong tectonic differentiation and a planar discontinuous distribution under weak tectonic differentiation. (4) The development of Ordovician karst in the Tabei paleo-uplift of the Tarim Basin was controlled by the lithology of the parent rocks, paleoclimate, tectonic landforms, and exposure time. Karst reservoirs are distributed in

S. Li (✉) · Y. Li · S. Peng · Y. Hao · J. Zhang · J. Wang
Key Laboratory of Geology and Resources in Deep Strata, SINOPEC, Beijing, China
e-mail: lishuangjian.syky@sinopec.com

S. Li · Y. Li · Y. Hao · J. Zhang · J. Wang
Petroleum Exploration and Production Research Institute, SINOPEC, Beijing, China

X. Xu
Geophysical Exploration Technology Research Institute, SINOPEC, Nanjing, China

Y. Ma et al. (eds.), *Petroleum Geology and Exploration of Deep Marine Strata in China*, Advances in Oil and Gas Exploration & Production, https://doi.org/10.1007/978-3-032-02496-1_3

the form of rings under tectonic differentiation.

Keywords

Chinese cratonic basin · Tectonic differentiation · Carbonate reservoir · Reef-shoal facies · Grain shoal · Karst

3.1 Introduction

Tectonic differentiation activities such as extensional rifting, uplifting and denudation, and basement fault activation occurred in marine cratonic basins under the influence of dynamic mechanisms such as the stress field at plate tectonic boundaries, preexisting fabrics, and mantle plume activity. Consequently, tectonic units such as intracratonic rifts, paleo-uplifts, and deep-seated faults formed, causing sedimentary differentiation and further controlling the development of basic accumulation conditions such as source rocks, reservoirs, and cap rock (Liu et al. 2013; Ma et al. 2017; Wang et al. 2017; He et al. 2019a). With respect to the controlling effects of marine basin structures on the formation and distribution of oil and gas fields, previous evaluations focused on the effects of structural stability, such as long-term inherited paleo-uplifts and stable passive continental margins (Budd et al. 1995, He et al. 2008, 2010b; Jin 2014). However, an increasing number of exploration practices and studies have shown that moderate tectonic differentiation has a constructive effect on the elements critical for hydrocarbon accumulation in cratonic basins, especially the formation and distribution of high-quality reservoirs (Ma et al. 2010; Zhao et al. 2012; Zou et al. 2014). With the discovery of the Kaijiang–Liangping continental shelf and the Mianyang–Changning extensional trough in the Sichuan Basin, the formation and development mechanisms of high-quality reservoirs along both sides of intracratonic rifts have become hot research topics since 2000 (Ma 2006; Liu et al. 2016; He et al. 2017). Recently, evidence for the development of paleorifts in paleocratons has been found in the deep layers of the Ordos and Tarim Basins (Wang et al. 2005; Zhou et al. 2015; Guan et al. 2017), providing an opportunity to search for deep exploration targets. Compared with the cratons in regions such as North America, Africa, and Siberia, the Tarim, Yangtze, and North China cratons in China have small scales, poor tectonic stability, and strong tectonic–sedimentary differentiation. However, some basic geological problems associated with these small cratons are not yet clear and require further study, including the types of tectonic–sedimentary differentiation, its controlling effects on the development of large-scale reservoirs, and the regularity of the controlling effects.

As indicated by the statistics of large-scale carbonate oil and gas fields worldwide, carbonate reservoirs are dominated by organic reefs, grain shoals, dolomites, and karst weathering crusts, whose reserves account for more than 90% of the total carbonate reserves (Luo et al. 2008). The carbonate reservoirs discovered in China in the early stage include mainly early Paleozoic karst weathering crust reservoirs, such as those in the Renqiu oil field in the Bohai Bay Basin, the Jingbian gas field in the Ordos Basin, and the Tabei paleo-uplift in the Tarim Basin (He 2002; Du et al. 2002; Hou et al. 2011; Zhang et al. 2012b; Guo 2013; Zhao et al. 2013). Since 2000, significant progress has been made in the exploration of reef-shoal and grain-shoal dolomite reservoirs on platform margins. Gas fields such as Puguang, Yuanba, Gaoshiti, and Moxi in the Sichuan Basin, as well as the Tazhong No. 1 oil and gas field in the Tarim Basin, have been discovered (Ma et al. 2005; Hu 2011; Jiao et al. 2011; Du et al. 2014; Zou et al. 2014). The formation and development of large-scale reservoirs in these oil and gas fields are closely related to the tectonic–sedimentary differentiation of basins. To elucidate the basic development and distribution patterns of large-scale carbonate reservoirs, this study systematically organized the results of the prototype evolution and lithofacies paleogeography of the Tarim, Sichuan, and Ordos Basins. On this basis, the types of tectonic–sedimentary differentiation of the three major cratonic basins are summarized, and the types of tectonic–sedimentary differentiation corresponding to different tectonic evolution stages of different basins are determined.

Moreover, in this study, the development characteristics and main controlling factors of high-quality reservoirs in key strata are analyzed using stratigraphic correlation, core observations, and reservoir geochemistry, and development models of large-scale reservoirs under different types of tectonic–sedimentary differentiation are established to provide a basis for the prediction of reservoir distribution.

3.2 Division and Correlation of the Tectonic Evolution Stages of Three Major Cratons in China

The Chinese continent is composed of multiple blocks, including the Tarim Craton, the North China Craton, the South China Craton, and many other microcontinents (Ren et al. 1999; Zhang et al. 1995). Since the Neoproterozoic, the Chinese continent has been jointly controlled by three global dynamic systems, i.e., the Paleo-Asian Ocean, the Tethyan Ocean, and the Paleo-Pacific Ocean, during the break-up and amalgamation of global supercontinents such as Rodinia and Pangaea. After the three tectonic evolution stages of the Proto-, Paleo-, and Neo-Tethys Oceans, five tectonic layers generally developed in sedimentary basins (Fig. 3.1), forming multicycle superimposed basins (He et al. 2010a).

(1) Sinian–Silurian tectonic layers: Approximately 850 Ma, the Rodinia paleocontinent broke up and formed the Proto-Tethys Ocean, controlling the Early Paleozoic ocean–continent pattern in China. The Sinian–Silurian experienced a complete extensional–compressional tectonic cycle, forming various types of basins, including basins of passive continental margins, intracratonic fault depression basins, depression basins, and peripheral foreland basins, in sequence. The tectonic evolution of the Tarim and Sichuan Basins was similar to that of the Nanhuaan–Sinian rift basins on the margin of cratonic basins to the Cambrian–Ordovician intracratonic depression basins and then the Silurian foreland basins. Affected by uplift and denudation, only Cambrian–Ordovician intracratonic depression basins were preserved in the Ordos Basin.

(2) Devonian–Permian tectonic layers: During the Late Paleozoic, the Paleo-Tethys Ocean expanded, and the Paleo-Asian Ocean closed, resulting in extensive tectonic compression activity in the S–N direction. As the major plates converged and merged with each other, various types of basins, such as intracontinental rift basins, depression basins, basins of passive continental margins, and peripheral foreland basins, developed in three major cratonic basins in China. The tectonic evolution of the Tarim and Ordos Basins was highly similar. Specifically, both basins featured marine–continental transitional facies and continental facies and were dominated by intracratonic depression basins under a weak extension–compression background. Marine sediments developed in the Sichuan Basin, which was dominated by craton margin rift basins under a strong extensional background.

(3) Middle–Upper Triassic tectonic layers: During the Middle–Late Triassic (Indosinian period, approximately 230–210 Ma), the Paleo-Tethys Ocean closed, and the Pangaea supercontinent formed (Zhang et al. 1995). Peripheral foreland basins developed in China's three major cratonic basins. After the Indosinian movement, the three major cratonic basins entered the continental sedimentary stage.

(4) Jurassic–Cretaceous tectonic layers: After the Indosinian, as the Pangaea supercontinent broke up, the world entered the evolution stage of the modern plate tectonic system. The Chinese continent was affected by the extension and closure of the Neo-Tethys Ocean in the southwest, the extension and extinction of the (paleo and present) Pacific Ocean in the east, and the closure of the Okhotsk Ocean in the north (Ren et al. 1999). Under the convergence of the three tectonic dynamic systems, large intracontinental depression basins and reactivated foreland basins developed on the Chinese continent.

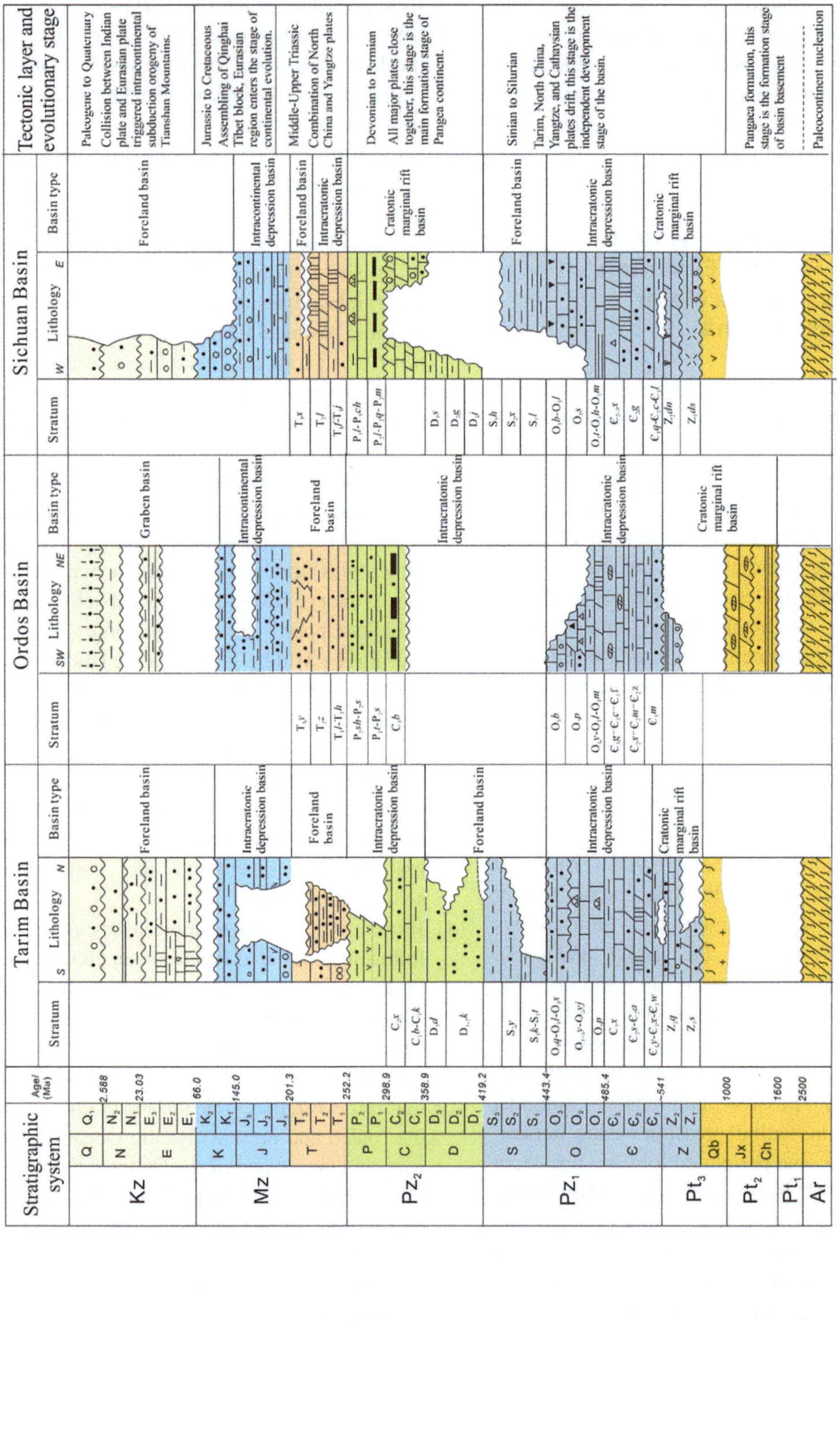
Stratigraphic system
Age/(Ma)
Tarim Basin
Ordos Basin
Sichuan Basin
Stratum
Lithology
Basin type
Tectonic layer and evolutionary stage
Foreland basin
Intracratonic depression basin
Cratonic marginal rift basin
Graben basin
Intracontinental depression basin
Paleogene to Quaternary
Collision between Indian plate and Eurasian plate triggered intracontinental subduction orogeny of Tianshan Mountains.
Jurassic to Cretaceous
Assembling of Qinghai Tibet block, Eurasian region enters the stage of continental evolution.
Middle-Upper Triassic
Combination of North China and Yangtze plates
Devonian to Permian
All major plates close together, this stage is the main formation stage of Pangea continent.
Sinian to Silurian
Tarim, North China, Yangtze, and Cathaysian plates drift, this stage is the independent development stage of the basin.
Pangaea formation, this stage is the formation stage of basin basement
Paleocontinent nucleation
Shale
Mudstone
Coal bed
Silty mudstone
Sandy mudstone
Siltstone
Sandstone
Conglomerate
Glutenite
Limestone
Argillaceous limestone
Edgewise limestone
Breccia limestone
Grainstone
Dolomite
Argillaceous dolomite
Doloarenite
Granular dolomite
Breccia dolomite
Anhydrite dolomite
Anhydrite
Granite
Andesite
Chlorite schist
Unconformity surface

◂**Fig. 3.1** Tectonic evolution stages of marine cratonic basins in China. *Note* Chronostrata: Ar—Archaeozoic; Pt$_1$—Paleoproterozoic; Pt$_2$—Mesoproterozoic; Pt$_3$—Neoproterozoic; Pz$_1$—Lower Paleozoic; Pz$_2$—Upper Paleozoic; Mz—Mesozoic; Kz—Cenozoic; Ch—Changchengian; Jx—Jixianian; Qb—Qingbaikou Formation; Z—Sinian; Z$_1$—Lower Sinian; Z$_2$—Upper Sinian; Є—Cambrian; Є$_1$—Lower Cambrian; Є$_2$—Middle Cambrian; Є$_3$—Upper Cambrian; O—Ordovician; O$_1$—Lower Ordovician; O$_2$—Middle Ordovician; O$_3$—Upper Ordovician; S—Silurian; S$_1$—Lower Silurian; S$_2$—Middle Silurian; S$_3$—Upper Silurian; D—Devonian; D$_1$—Lower Devonian; D$_2$—Middle Devonian; D$_3$—Upper Devonian; C—Carboniferous; C$_1$—Lower Carboniferous; C$_2$—Upper Carboniferous; P—Permian; P$_1$—Lower Permian; P$_2$—Upper Permian; T—Triassic; T$_1$—Lower Triassic; T$_2$—Middle Triassic; T$_3$—Upper Triassic; J—Jurassic; J$_1$—Lower Jurassic; J$_2$—Middle Jurassic; J$_3$—Upper Jurassic; K—Cretaceous; K$_1$—Lower Cretaceous; K$_2$—Upper Cretaceous; E—Paleogene; E$_1$—Paleocene; E$_2$—Eocene; E$_3$—Oligocene; N—Neogene; N$_1$—Miocene; N$_2$—Pliocene; Q – Quaternary. Strata in the Tarim Basin: Z$_1s$—Sugaibulake Formation; Z$_2q$—Qigebulake Formation; Є$_1y$—Yuertusi Formation; Є$_1x$—Xiaoerbulake Formation; Є$_1w$—Wusonger Formation; Є$_2s$—Shayilike Formation; Є$_2a$—Awatage Formation; Є$_3x$—Lower Qiulitage Formation; O$_1p$—Penglaiba Formation; O$_{1\text{-}2}y$—Yingshan Formation; O$_2yj$—Yijianfang Formation; O$_3q$—Qiaerbake Formation; O$_3l$—Lianglitage Formation; O$_3s$—Sangtamu Formation; S$_1k$—Kepingtage Formation; S$_1t$—Tataaiertage Formation; S$_2y$—Yimuwutagan Formation; D$_{1\text{-}2}k$—Keziertage Formation; D$_3d$—Donghetang Formation; C$_1b$—Bachu Formation; C$_1k$—Kalashayi Formation; C$_2x$—Xiaohaizi Formation. Strata in the Ordos Basin: Є$_1m$—Mantou Formation; Є$_2x$—Xuzhuang Formation; Є$_2m$—Maozhuang Formation; Є$_2z$—Zhangxia Formation; Є$_3g$—Gushan Formation; Є$_3c$—Changshan Formation; Є$_3f$—Fengshan Formation; O$_1y$—Yeli Formation O$_1l$—Liangjiashan Formation; O$_1m$—Majiagou Formation; O$_2p$—Pingliang Formation; O$_3b$—Beiguoshan Formation; C$_3b$—Benxi Formation; P$_1t$—Taiyuan Formation; P$_1s$—Shanxi Formation; P$_2sh$—Shihezi Formation; P$_2s$—Shiqianfeng Formation; T$_1l$—Liujiagou Formation; T$_1h$—Heshanggou Formation; T$_2z$—Zhifang Formation; T$_3y$—Yanchang Formation. Strata in the Sichuan Basin: Z$_1ds$—Doushantuo Formation; Z$_2dn$—Dengying Formation; Є$_1q$—Qiongzhusi Formation; Є$_1c$—Canglangpu Formation; Є$_1l$—Longwangmiao Formation; Є$_2g$—Gaotai Formation; Є$_{2\text{-}3}x$—Xixiangchi Group; O$_1t$—Tongzi Formation; O$_1h$—Honghuayuan Formation; O$_1m$—Meitan Formation; O$_2s$—Shizipu Formation; O$_3b$—Baota Formation; O$_3l$—Linxiang Formation; S$_1l$—Longmaxi Formation; S$_1x$—Xiaoheba Formation; S$_1h$—Honghuayuan Formation; S$_2x$—Shizipu Formation; S$_3h$—Hanjiadian Formation; D$_1j$—Jinbaoshi Formation; D$_2g$—Guanwushan Formation; D$_3s$—Shawozi Formation; P$_2l$—Liangshan Formation; P$_2q$—Qixia Formation; P$_2m$—Maokou Formation; P$_3l$—Longtan Formation; P$_3ch$—Changxing Formation; T$_1f$—Feixianguan Formation; T$_1j$—Jialingjiang Formation; T$_2l$—Leikoupo Formation; T$_3x$—Xujiahe Formation

(5) Paleogene–Quaternary tectonic layers: In the Cenozoic, the collision between the Indian Plate and the Eurasian Plate triggered the intracontinental subduction orogeny of Tianshan, Longmenshan, and Qinling. Consequently, strongly folded orogenic belts formed on the margins of large cratonic basins, transforming the marginal sedimentation of most marine cratons.

The multicycle superposition of different directions, properties, and intensities led to the multicycle superimposed development of sedimentary basins on the Chinese continent. The remaining marine sedimentary strata are distributed mainly in the Tarim, Ordos, and Sichuan basins, which retain Sinian–Ordovician, Cambrian–Ordovician, and Sinian–Middle Triassic strata, respectively, in their main bodies (Fig. 3.1).

3.3 Tectonic–Sedimentary Differentiation Types and Large-Scale Reservoir Ages of Three Major Cratons in China

The tectonic differentiation of cratonic basins mainly refers to the differential tectonic deformations in cratonic basins and their regular changes under the influence of factors such as tectonic stress and preexisting fabrics. Tectonic differentiation is generally manifested through block faulting, uplift and denudation, and multiphase activation of basement faults; the formation of tectonic units such as intracratonic rifts, paleo-uplifts, paleodepressions, and deep-seated faults; and the significant control of stratigraphic sequences, sedimentation, lithofacies paleogeography, and elements critical for hydrocarbon accumulation (Wang et al. 2017). Tectonic–sedimentary differentiation in cratonic basins refers to sedimentary differentiation caused by the tectonic processes of basement faults, synsedimentary faults, and paleo-uplifts (Zhang et al. 2020).

Marine cratonic basins in China experienced multiphase tectonic cycles of strong extension–weak extension–weak compression–strong compression during the Sinian–Triassic. Under different tectonic stresses, the tectonic–sedimentary differentiation of the basins showed similar characteristics and controlled similar large-scale reservoirs. Tectonic–sedimentary differentiation can be divided into three basic types according to the tectonic stress state of cratonic basins (Fig. 3.2):

(1) Tectonic–sedimentary differentiation controlled by synsedimentary faults in an extensional or local extensional setting. This type of tectonic–sedimentary differentiation developed during the tectonic evolution stage of strong extension in the early formation stage of cratonic basins and controlled the development of reef–shoal reservoirs, such as those in the Upper Sinian–Lower Cambrian and Upper Permian–Lower Triassic strata in the Sichuan Basin and the Sinian strata in the Tarim Basin.

(2) Tectonic–sedimentary differentiation is controlled by synsedimentary paleo-uplifts and intracratonic depressions in a weak extension–compression setting. This type of tectonic–sedimentary differentiation developed during the stable development stage of cratonic basins and controlled the development of the grain-shoal reservoirs and the tidal flat–lagoon facies dolomite reservoirs on the margin of paleo-uplifts, such as the Middle–Upper Cambrian and Middle–Lower Triassic reservoirs in the Sichuan Basin, the Middle–Upper Cambrian reservoirs in the Tarim Basin, and the Middle Cambrian–Lower Ordovician reservoirs in the Ordos Basin.

(3) Tectonic–sedimentary differentiation is controlled by denudation-type paleo-uplifts and foreland depressions in a strongly compressional setting. This type of tectonic–sedimentary differentiation developed during the shrinking stage of cratonic basins and controlled the development of carbonate karst and clastic reservoirs, such as the Upper Ordovician–Lower Silurian clastic reservoirs in the Sichuan Basin, the Lower

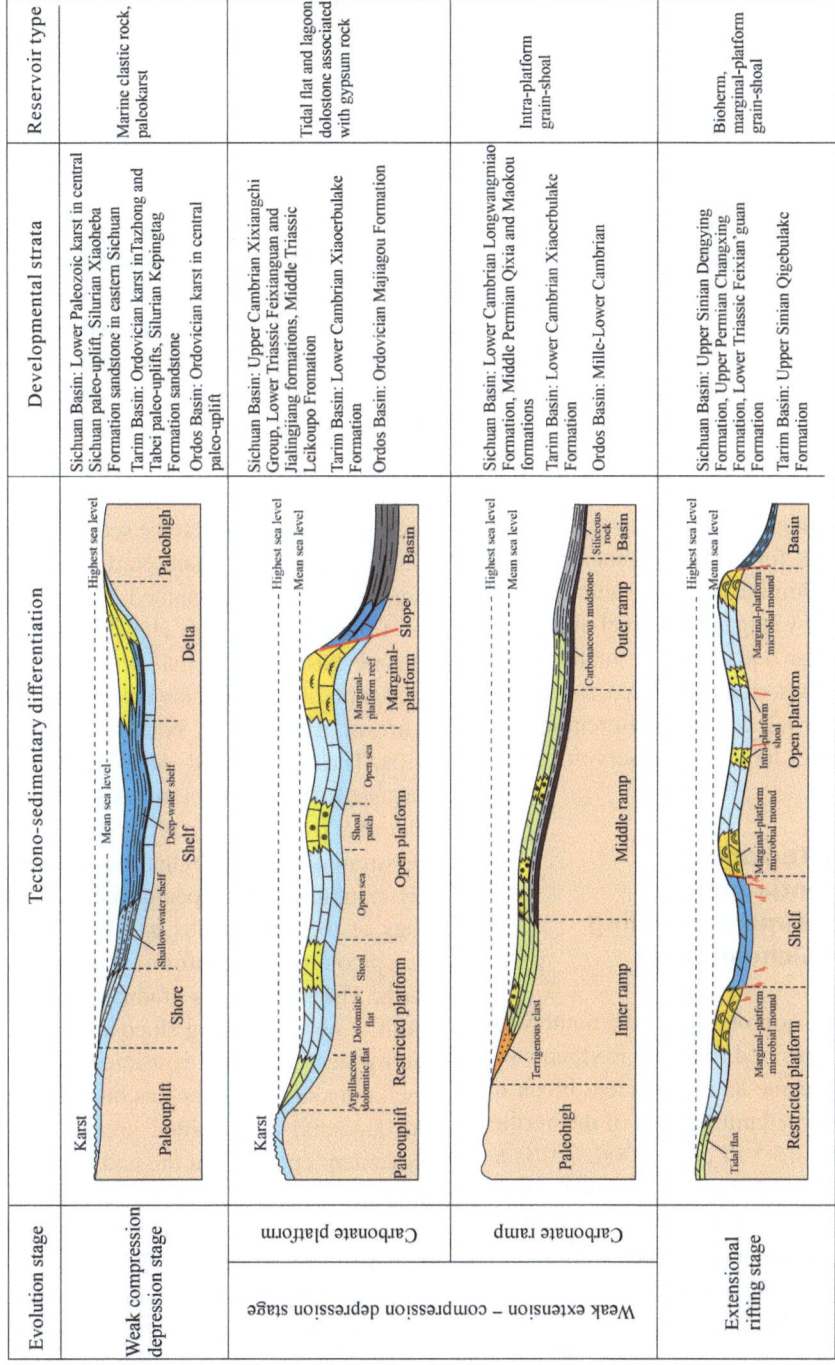

Evolution stage	Tectono-sedimentary differentiation	Developmental strata	Reservoir type
Weak compression depression stage		Sichuan Basin: Lower Paleozoic karst in central Sichuan paleo-uplift, Silurian Xiaoheba Formation sandstone in eastern Sichuan Tarim Basin: Ordovician karst in Tazhong and Tabei paleo-uplifts, Silurian Kepingtag Formation sandstone Ordos Basin: Ordovician karst in central paleo-uplift	Marine clastic rock, paleokarst
Weak extension – compression depression stage (Carbonate platform)		Sichuan Basin: Upper Cambrian Xixiangchi Group, Lower Triassic Feixianguan and Jialingjiang formations, Middle Triassic Lcikoupo Fromation Tarim Basin: Lower Cambrian Xiaoerbulake Formation Ordos Basin: Ordovician Majiagou Formation	Tidal flat and lagoon dolostone associated with gypsum rock
Weak extension – compression depression stage (Carbonate ramp)		Sichuan Basin: Lower Cambrian Longwangmiao Formation, Middle Permian Qixia and Maokou formations Tarim Basin: Lower Cambrian Xiaoerbulake Formation Ordos Basin: Mille-Lower Cambrian	Intra-platform grain-shoal
Extensional rifting stage		Sichuan Basin: Upper Sinian Dengying Formation, Upper Permian Changxing Formation, Lower Triassic Feixian guan Formation Tarim Basin: Upper Sinian Qigebulakc Formation	Bioherm, marginal-platform grain-shoal

Fig. 3.2 Tectonic–sedimentary differentiation types of marine cratonic basins in China

Silurian clastic reservoirs in the Tarim Basin, and the karst reservoirs in the central Sichuan, Tazhong, Tabei, and Ordos central paleo-uplifts (Fig. 3.2).

3.4 Controlling Effects of Tectonic–Sedimentary Differentiation on the Development of Large-Scale Reservoirs in an Extensional or Local Extensional Setting

Rimmed platforms tend to develop in cratonic basins. They have external high-energy margins, and their slopes significantly increase from several degrees to 60 degrees or more after they stretch into deep-water basins. Platform margins are subjected to strong seawater wave action and are favorable sites for the development of large-scale reservoirs dominated by organic reefs and grain shoals. By investigating the reef-shoal reservoirs of the Sinian Dengying Formation in the Sichuan Basin, this study analyzed the controlling effects of strong tectonic differentiation on the development of large-scale reservoirs.

3.4.1 Sinian Tectonic Differentiation and Sedimentary Filling of the Sichuan Basin

During the Late Neoproterozoic, the South China continent was separated from the Rodinia supercontinent and became an independent block that is located in the low-latitude zone of the northern hemisphere ($10°$–$15°$; Li et al. 2008, 2013) and to the north of the Laurentia continent and is permanently submerged by seawater (Powell et al. 1994; Li 1999; Wang and Li 2003). Along with the extensional rifting within the South China continental plate (Zhang et al. 2013), rifting and tectonic thermal events occurred on the margin of the Yangtze Craton, and many pre-Sinian rifts formed in the craton (Dong et al. 2013; Zhong

et al. 2013; Yang et al. 2014; Gu and Wang 2014; Guan et al. 2017). The pre-Sinian extension in the Yangtze Craton laid the foundation for the formation of the Late Sinian–Early Cambrian rift zone. The reactivation of early rifts affected the overlying strata during the late extension (Zhong et al. 2013; Yang et al. 2014; Zhao et al. 2017). More importantly, the weak tectonic zones that formed on the margins of these early rifting zones largely determined the distribution of late rifts, such as the Mianyang–Changning rift trough (Liu et al. 2013; Wei et al. 2015; Du et al. 2016).

The sedimentary filling of the Sinian Dengying Formation was significantly affected by paleorifts (Fig. 3.3). During the deposition of the second member of the Dengying Formation (the Deng 2 Member for short; Fig. 3.3a), large open oceanic basins formed on the west, north, and east sides of the Yangtze Block, and deep faults on the basin margins controlled the sedimentary patterns of the rimmed carbonate platforms. During this stage, two paleorifts, named Guangyuan–Anyue and Chengkou–Western Hubei, developed in the oceanic basins, and shallow-water high-energy algal mound–shoal facies belts developed on the platform margins on both sides of the paleorifts. The algal mound–shoal facies belts mainly consisted of granular dolomites and dolomites with microbial frameworks consisting of foam sponges and stromatolite. These algal mound–shoals generally exhibited a U-shaped distribution along both sides of the rift troughs and were connected to the epicontinental platform–margin shoals in the north. The rift troughs gradually deepened from south to north, forming slope and basin facies in deep-water areas and restricted platform facies (including subfacies such as intraplatform shoals and lagoons) in nonrifted areas. The Tongwan Movement episode I at the end of the deposition stage of the Deng 2 Member caused the extensive withdrawal of seawater. Consequently, the high-lying platform margin shoals underwent meteoric water leaching and even formed dissolution vugs, thus becoming excellent oil and gas reservoirs. During the deposition of the fourth member of the Dengying Formation (the Deng 4 Member for short; Fig. 3.3b), the regional extension further

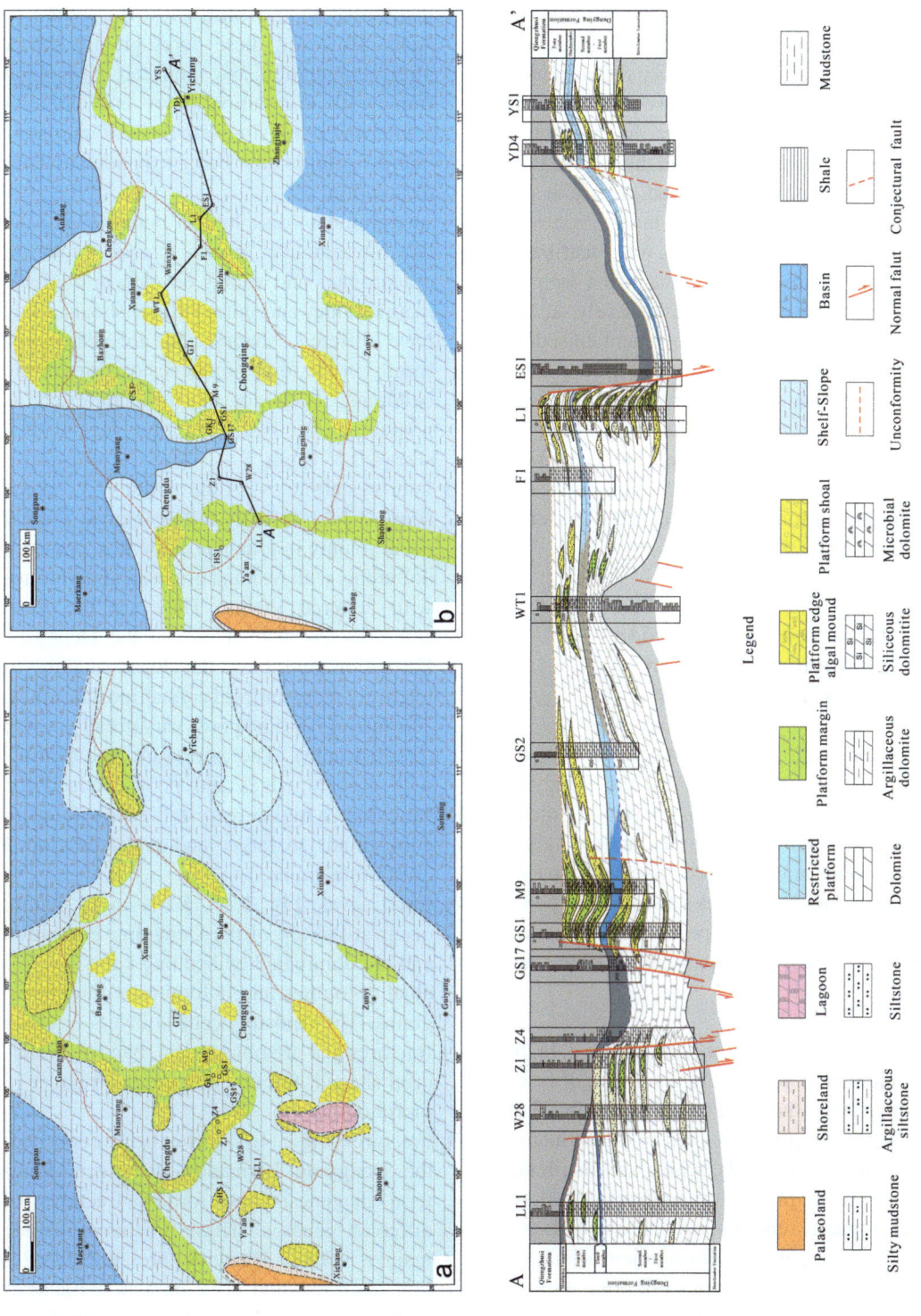

Fig. 3.3 Lithofacies Paleogeographic Map of the Dengying Formation in the Sinian System, Sichuan Basin

intensified, and the rift boundaries expanded continuously. Moreover, the Guangyuan–Anyue rift extended to southern Sichuan; thus, rifts run through the basin in the north–south direction. Although the types of sedimentary facies belts at this stage were the same as those in the deposition stage of the Deng 2 Member, the distribution of the platform margin zones and the sedimentary characteristics with rifts changed greatly at this stage. Specifically, the platform margin zones on the east and west sides of the rifts were distributed in the north–south direction and in the form of an arc protruding eastward, respectively, at this stage. During the Tongwan Movement episode II at the end of the deposition stage of the Deng 4 Member, the entire Upper Yangtze Craton was uplifted again, and the top of the Deng 4 Member generally underwent denudation and karstification, resulting in the formation of a regional weathering crust unconformity (Zou et al. 2014; Li et al. 2018; Yang et al. 2019).

3.4.2 Reservoirs of the Sinian Dengying Formation in the Sichuan Basin

As indicated by the observation results of typical outcrop sections and drill cores, the Dengying Formation reservoirs mainly consist of microbial dolomites, granular dolomites, and crystalline dolomites and have diverse reservoir spaces, mainly including framework (dissolution) pores, intergranular (dissolution) pores, intercrystalline (dissolution) pores, and dissolution fractures and vugs.

(1) Framework (dissolution) pores (Fig. 3.4a, b). Framework pores are the main type of reservoir space in the Dengying Formation. Their formation is closely related to the formation of mounds and shoals, which are distributed mainly in the platform margin zones. Microbial mounds continuously bond and trap argillaceous particles during their formation, forming numerous framework pore systems. The dissolution pores in the soluble particles and cements in the intergranular pores are generally 0.1–0.2 mm in size, which is affected by the particle size. They are distributed in the form of strips, grids, or spots. However, these pores greatly shrink after cementation and then can form algal framework dissolution pores and interclot framework pores similar to those of intergranular dissolution after being expanded by later dissolution.

(2) Intergranular (dissolution) pores (Fig. 3.4c, d). Intergranular pores mostly occur in doloarenites and widely develop in platform margin zones and intraplatform shoals. They have different shapes and are distributed locally in the form of patches or strips along beddings. They are mostly manifested as nonfabric-selective dissolution vugs with a size of 0.3–1 mm.

(3) Intercrystalline (dissolution) pores (Fig. 3.4e). Intercrystalline pores mainly occur mainly in granular dolomites, powdery dolomites with residual grains, and dolomites with adhered grains. They are the most common reservoir spaces in the Dengying Formation and have developed mainly in intensely recrystallized crystalline dolomites. The intercrystalline pores vary greatly in size (0.1–0.4 mm). The pores are polyhedrons with straight edges and poor connectivity. Moreover, a higher recrystallization intensity corresponds to larger and better-developed crystalline grains.

(4) Karst caves (Fig. 3.4f). Karst caves mainly occur at the top of the Dengying Formation and are dominated by bedding karst caves of algal dolomites and dolomites with adhered grains. They mostly occur in layers or along fractures and are oblate, oval, banded, water-drop shaped, fracture shaped, or irregular in morphology.

(5) Fractures (Fig. 3.4g–i). After being deposited, the Dengying Formation underwent long-term diagenesis and formed fractures of different geneses. These fractures include mainly tectonic fractures, pressure dissolution fractures, and dry-contraction fractures and may occur in different sedimentary facies belts.

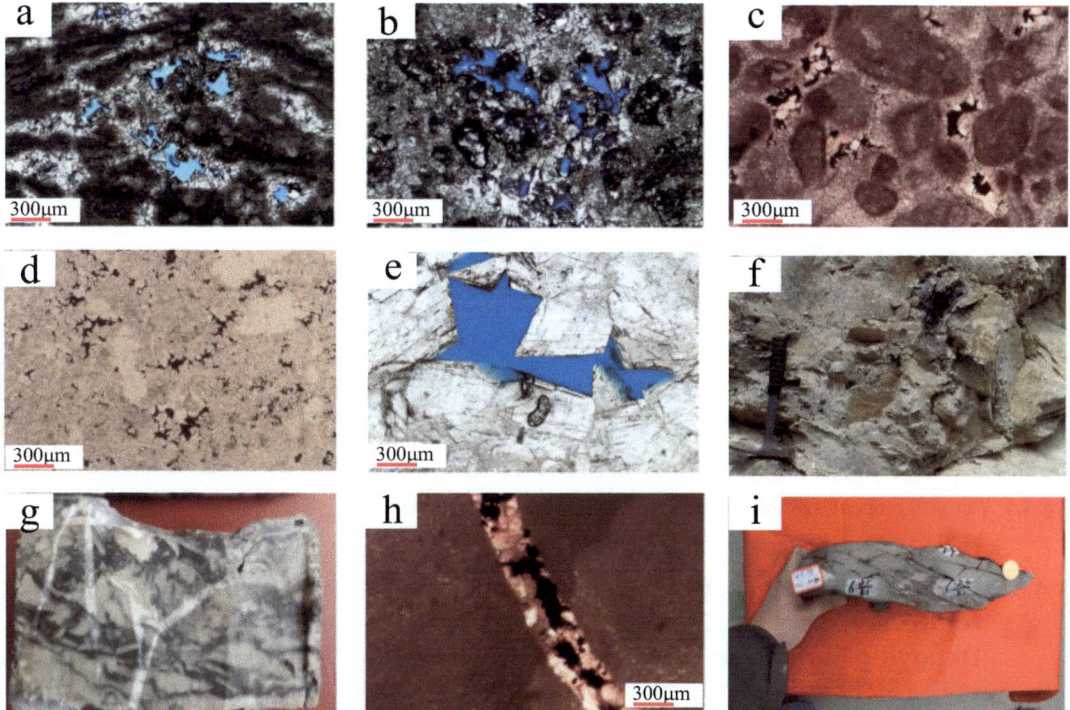

Fig. 3.4 Petrological characteristics of reservoirs in the Sinian Dengying Formation in the Sichuan Basin. **a** Well HS1, depth: 5233.59 m, Z_2dn^4, microbial pane pores, 5× ; **b** Well GK1, depth: 5443.86 m, Z_2dn^2, interclot dissolution pores, 5×; **c** Well Chuanshen 1, depth: 8192 m, Z_2dn^4, algal dolomite intergranular pores filled with asphalt; **d** Yangba section, Z_2dn^4, intercrystalline pores and intercrystalline dissolution pores filled with asphalt; **e** Hujiaba section, Z_2dn^2, dolomite intercrystalline pores in clotted limestones; **f** Yanhe section, Z_2dn^2, karstization-related karst caves filled with asphalt; **g** Well Chuanshen 1, depth: 8161.90 m, Z_2dn^4, breccia dolomites rich in asphalt; **h** Well Lin 1, depth: 3513.07 m, Z_2dn^4, dolomicrite fractures filled with dolomites and asphalt; **i** Well TX1, depth: 2190.26 m, Z_2dn^4, well-developed tectonic fractures

The reservoirs of the Dengying Formation in the Sichuan Basin are strongly heterogeneous, and their reservoir performance is significantly affected by sedimentary facies belts and karstification. Compared with the small cylindrical samples, the full-diameter core samples from this formation generally had more favorable physical properties. Specifically, their porosities were predominantly 2–6% (average: 3.41%), and those with porosities greater than 6% accounted for 13% of all the samples (Fig. 3.5). Their horizontal permeability was predominantly $(0.1–1) \times 10^{-3}$ μm^2 (average: 2.55×10^{-3} μm^2), and those with a horizontal permeability greater than 1×10^{-3} μm^2 accounted for 26.5% of all the samples (Fig. 3.5). The porosity of small-diameter samples was mostly less than 2% (average: 1.99%), and those with porosities greater than 6% accounted for less than 4% of all the samples (Fig. 3.5). Small-diameter samples mostly had horizontal permeabilities of less than $(0.1–1) \times 10^{-3}$ μm^2 (average: 0.9×10^{-3} μm^2), and those with horizontal permeabilities greater than 1×10^{-3} μm^2 accounted for 15% of all the samples (Fig. 3.5). In general, the physical properties of the Deng 2 Member are better than those of the Deng 4 Member, but the porosity and permeability are poorly correlated (Fig. 3.5). These results indicate that the reservoirs of the two members are highly heterogeneous, are affected by fractures to varying degrees and are fracture–vug reservoirs.

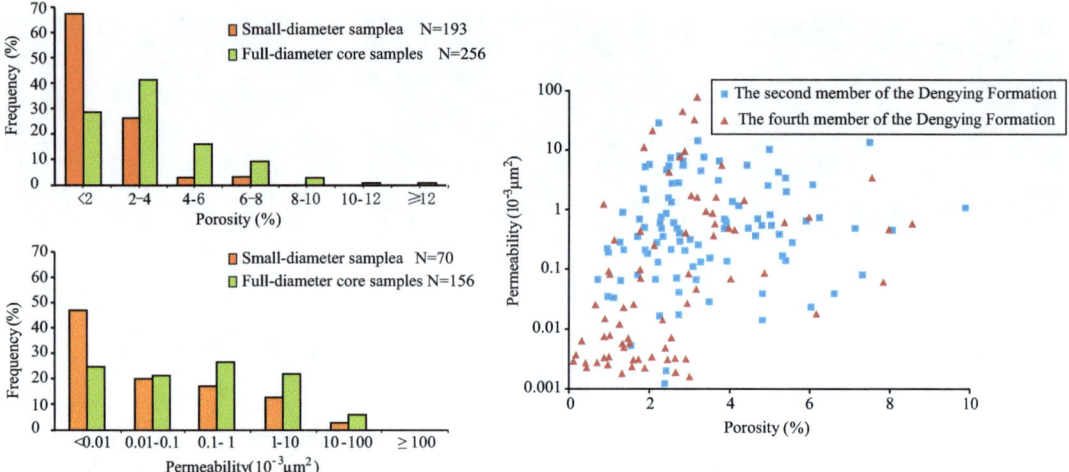

Fig. 3.5 Physical properties of the Sinian Dengying Formation reservoirs

3.4.3 Main Controlling Factors for Reservoirs of the Sinian Dengying Formation in the Sichuan Basin

The development of high-quality reservoirs in the Dengying Formation is affected by deposition and diagenesis, such as lithofacies, dolomitization, penecontemporaneous dissolution, supergene karstification, and hydrocarbon filling. High-energy facies belts, early dolomitization, and penecontemporaneous karstification are the main controlling factors.

(1) High-energy facies belts lay the foundation for the development of high-quality reservoirs.

During the deposition of the Dengying Formation, the study area generally featured an uplift-depression sedimentary pattern and was in a restricted carbonate platform environment composed of sedimentary microfacies, including algal mounds, grain shoals, tidal flats, and lagoons. The reservoirs in the platform-margin algal mound–shoal facies belt mainly consist of algal frameworks/algal stromatolites, algal clots, algal doloarenites, spongy algae, and algae-bonded dolomites. These reservoirs feature high sedimentary water-body energy and high initial

rock porosity, and their reservoir spaces are related mainly to algal structures, followed by intergranular pores. The intraplatform shoal facies reservoirs mainly consist of algal-laminar and algal-clastic dolomites. These reservoirs have lower sedimentary water-body energy and initial rock porosity than platform margin zones do. The dolomitic flats and lagoons mainly consist of dolomicrites. They feature low sedimentary water-body energy and low initial rock porosity and can form effective reservoirs only after dissolution or fracturing. As suggested by the statistics of more than 10 key exploratory wells in central and northern Sichuan, the high-quality reservoirs of the Dengying Formation are composed of algal stromatolites, algal clot dolomites, and algal doloarenites and have abundant layered pores and three-dimensional pane pores. The physical property statistics reveal that the algal stromatolites, algal clot dolomites, and algal doloarenites have average porosities of up to 5.41%, 3.81%, and 3.87%, respectively, indicating that they have the most favorable reservoir physical properties and that the reservoirs are significantly controlled by the physical properties of the rocks (Fig. 3.6).

(2) Early dolomitization and multiphase karstification are key to the development of high-quality reservoirs

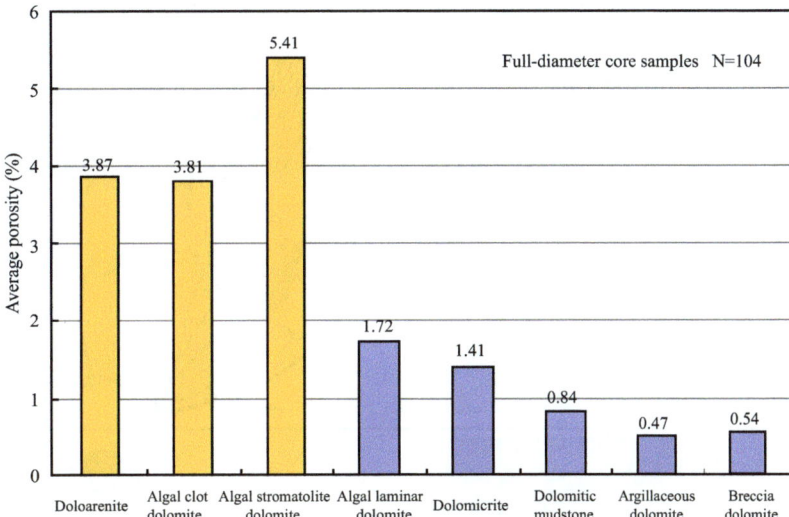

Fig. 3.6 Statistical histogram showing the physical properties of rocks in the Sinian Dengying Formation

Under the action of microorganisms, the seepage reflux of seawater on the platform margin, and evaporative pumping on restricted platforms, the Sinian Dengying Formation generally experienced penecontemporaneous dolomitization, in which microbial dolomitization occurred globally. During the Late Sinian, the global paleoclimate shifted from an icehouse climate to a greenhouse climate (Coppold and Powell 2000) with low oxygen partial pressure and high CO_2 partial pressure (Partin et al. 2000; Planavsky et al. 2014). The microbial mat covered the seafloor from the tidal flat area to the euphotic zone (Grotzinger and Knoll 1999), forming a widespread microorganism-dominated tidal flat environment. Owing to the persistently low partial pressure of oxygen, microorganisms might have contributed much more to organic mineralization than they do today (Knoll 2015). Therefore, microbial carbonates dominated during the Sinian under the organic mineralization of microorganisms. Many experiments have indicated that the regulatory effect of microorganisms can overcome the dynamic obstacles associated with dolomitization. With the participation of some anaerobic bacteria, the formation of dolomites does not require a supersaturated solution with a high magnesium ion concentration. Therefore, the development of microorganisms is favorable for early dolomitization. In addition, the

fibrous cements in the grape-like dolomites of the Deng 2 Member have bundle-shaped and radial dolomite aggregates with optical characteristics of positive elongation (Fig. 3.7; Wang et al. 2020) and well-preserved primary structures—banded growth zones and cathodoluminescent (CL) zones—which are distinct from the crystal characteristics of aragonites and calcites (Hood et al. 2011; Hood and Wallace 2012). These results reveal that these fibrous cements are primary dolomites directly deposited in the contemporaneous seawater environment, further revealing that the Sinian Dengying Formation dolomites formed during deposition. Early dolomitization of carbonate rocks is conducive to the formation of intercrystalline pores, fluid infiltration in the supergene stage, and pore retention in the deep burial stage, strongly guaranteeing the development of high-quality reservoirs.

The penecontemporaneous supergene karstification and the supergene karstification caused by the Tongwan Movement are key to the formation of reservoirs in the Dengying Formation, especially the penecontemporaneous karstification. According to the statistics of the pore space types of reservoirs in the Deng 4 Member, the reservoir spaces are significantly fabric selective and mainly include interalga fenestral vugs and framework vugs, which are flat, occur along the beddings, and are mainly millimeter- and centimeter-scale

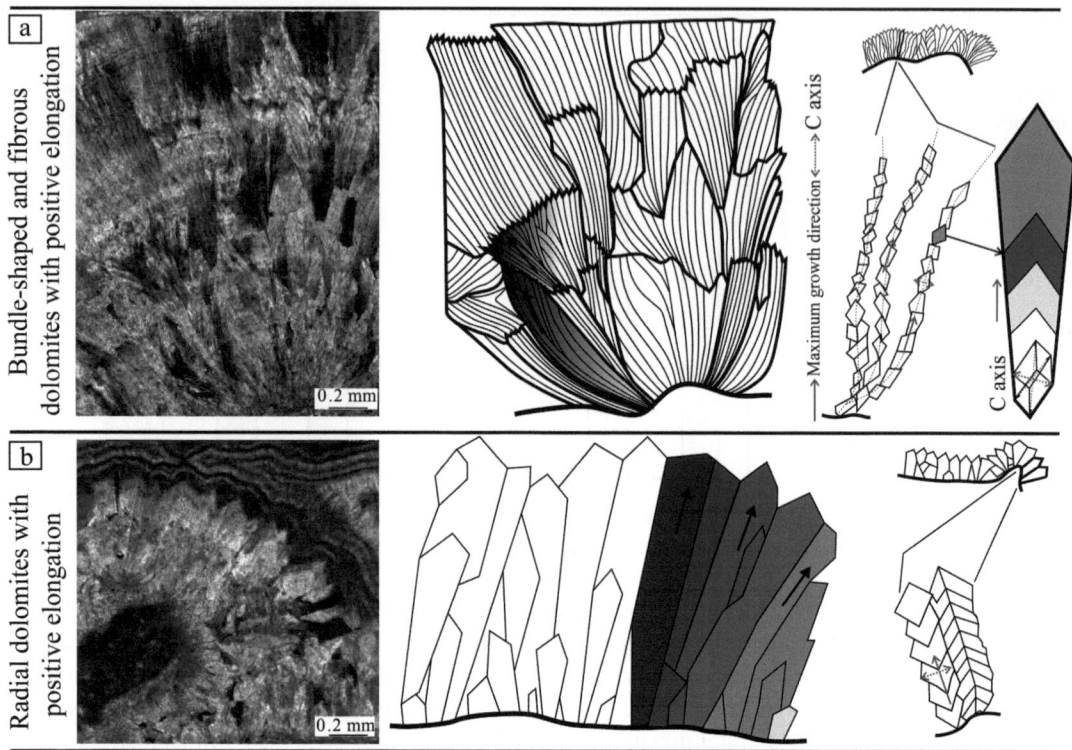

Fig. 3.7 Petrological characteristics of fibrous dolomite cements in the Deng 2 Member and schematic diagrams of the bundle (**a**) and radial (**b**) dolomite cements (modified after Wang et al. 2020)

pores and fractures. These characteristics significantly differ from the nonselective characteristics of the karst pores formed during the supergene stage and the dissolution pores formed during the burial stage. Moreover, the first-generation cements in the vugs are fibrous, foliated aragonites and high-magnesium calcites cemented on the seabed, and their in situ U–Pb dating results approximate the sedimentary epoch of the strata (Shen et al. 2021). These findings indicate that the vugs were formed in the penecontemporaneous stage before seafloor cementation. In addition, the vertical distribution of the dissolution pores is significantly different in a single section of the Dengying Formation reservoirs. In a single algal mound sedimentary cycle, the vugs are the most developed in the middle and upper parts, and their density and physical properties decrease downward. In contrast, the vugs in the supergene karsts are concentrated mainly within a certain depth range near an unconformity, and their density gradually decreases with increasing depth (Zhu

et al. 2014). All the evidence indicates that the dissolution of reservoirs was strongly affected by penecontemporaneous sea level fluctuations.

3.4.4 Reservoir Development Pattern of the Sinian Dengying Formation in the Sichuan Basin

During the deposition of the Dengying Formation, the rise in the relative sea level against the background of intraplatform rifting differentiation created favorable conditions for the formation of microbial mound–shoal suites in platform margin zones. Overall, the Dengying Formation featured shallow-water algal dolomite deposition and platform margin mound–shoal suites composed of stromatolites, clotted (lumpy) limestones, oolites, foam sponge dolomites, and doloarenites. During this period, numerous pores were formed in the primary microbial skeleton.

In the penecontemporaneous stage, the sea level decreased frequently. Accordingly, the algal mounds in the high parts of micropaleo-landforms were frequently and shortly exposed and underwent the dissolution of meteoric water, forming multiple sets of thick, large, and superimposed segments of bedding fabric-selective dissolution vugs and fractures longitudinally. The algal mounds located from the lower intertidal zone to the upper subtidal zone showed the development of interalga seafloor cements that were composed mainly of unstable high-magnesium calcites and aragonites because of the relatively high water-body energy. When the sea level decreased, the algal mounds were exposed to a meteoric water environment and were prone to dissolve. As a result, many bedding inter-algae fenestral and framework vugs with fabric-selective dissolution characteristics formed, with the density increasing in parts closer to the shortly exposed surfaces. When the sea level rose again, the high parts of the micropaleo-landforms received algal mound sediments again, thus forming thick and multilayered superimposed algal mound systems vertically. When the sea level decreased again, the algal mounds in the high parts of the micropaleo-landforms were shortly exposed and underwent the dissolution of meteoric water again. Finally, multiple sets of thick superimposed bedding fabric-selective dissolution pore–fracture segments formed vertically (Fig. 3.8).

3.5 Tectonic–Sedimentary Differentiation in a Weak Extension–Weak Compression Setting and Its Controlling Effects on the Development of Large-Scale Reservoirs

Cratonic basins in a stable tectonic setting feature weak tectonic differentiation and weakly active faults, rifts, and paleo-uplifts. As a result, two sedimentary systems, i.e., carbonate ramps and carbonate platforms, were formed mainly in these cratonic basins.

A carbonate ramp is an assemblage of a set of genetically related carbonate sedimentary facies formed in the shallow-water environment of a shelf ramp with an extremely low slope (less than 1°) from the coast to the sedimentary surface of a basin. Sedimentation on carbonate ramps varies with changes in water depth, water temperature, and wave and tide intensity. The main sedimentary facies include inner, middle, and outer slope facies and basin facies. Among them, the inner and middle ramp subfacies are favorable for reservoir development. As the product of basin shrinkage in a weak compressional setting, a carbonate platform sedimentary system formed during the middle–late stages of a large-scale tectonic movement cycle. Tectonic uplift and sea level drop formed large-scale evaporative platforms, where extremely thick evapora-

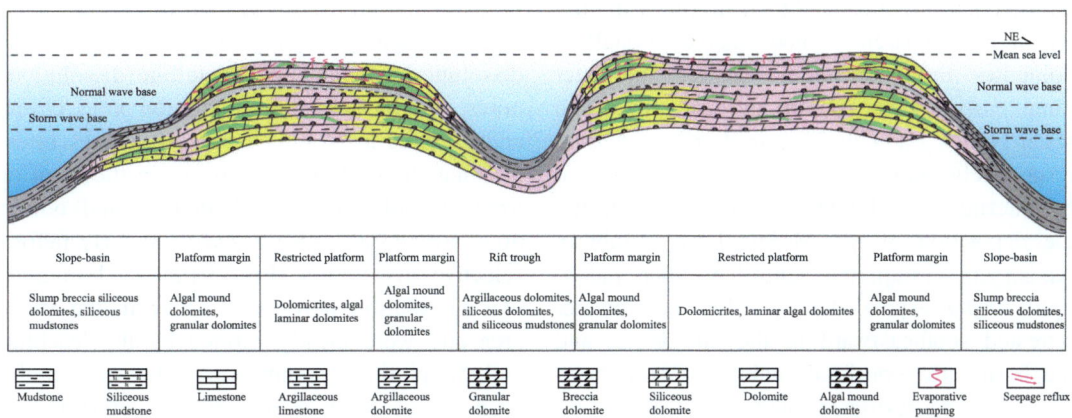

Fig. 3.8 Reservoir development pattern of the Sinian Dengying Formation in the Sichuan Basin under an intensely extensional tectonic–sedimentary differentiation setting

tive dolomites and anhydrite-bearing rocks were deposited. The reservoirs on carbonate ramps mainly consist of dolomites of the tidal flat facies and shoal facies, and they transitioned to high-quality reservoirs after multiphase exposure and dissolution. Based on an investigation of the Cambrian Longwangmiao Formation and Xixiangchi Group in the Sichuan Basin, this study analyzed the main factors controlling the development of high-quality reservoirs on ramps and evaporative platforms.

3.5.1 Cambrian Tectonic Differentiation and Sedimentary Activity of the Sichuan Basin

The sedimentary pattern of the Middle–Upper Yangtze region changed greatly during the Early Cambrian. The Sinian intracratonic rifts in the Middle–Upper Yangtze region were filled with clastic sediments during the Canglangpu stage, and the region developed carbonate ramps during the Longwangmiao stage. The region was high in the northwest and low in the southeast and dipped gently southward overall during the Early Cambrian. The uplifted platform area in the west had relatively shallow and high-energy water bodies and sufficient clastic sources. It transitioned from a nearshore mixed tidal flat to a carbonate inner ramp eastward (Fig. 3.9a). Numerous grain shoals were deposited in the uplifted platform area, and their lithology mainly comprises doloarenites and coarse crystalline dolomites. From the middle ramp to the outer ramp eastward, the depth and energy of the water bodies gradually increased and decreased, respectively, and the lithology was dominated by micrites and dolomitic limestones. During the deposition of the Middle–Late Cambrian Xixiangchi Group, the central Sichuan paleo-uplift further expanded eastward as the seawater retreated southeastward in the late stage, and the basin area consisted mainly of carbonate-restricted–semirestricted platforms (Fig. 3.9b). The restricted platforms were distributed at the periphery of the central Sichuan–northern

Sichuan paleo-uplifts. Their rock assemblages consisted of dolomites, argillaceous dolomites, and calcareous dolomites, and their sedimentary subfacies included mainly tidal flats and intraplatform grain shoals (oolitic shoals and arenaceous shoals), with saline lagoons developing locally (Fig. 3.9b). The grain shoals in the restricted platforms developed from the high areas in the paleolandforms and showed a banded distribution in the SW–NE direction at the periphery of the paleo-uplift. The semirestricted platforms were distributed mainly in eastern and southeastern Sichuan. Their rock assemblages consisted of medium–thick laminated–massive dolomites, dolomitic limestones, and limestones (Fig. 3.9c). Shoals were also visible on the southeastern margin of the Sichuan Basin, and their rock types included mainly bioclastic limestones, bioclastic dolomites, oolitic limestones, and reef limestones.

3.5.2 Reservoirs of the Cambrian Longwangmiao Formation and Xixiangchi Group in the Sichuan Basin

As revealed by drill cores and outcrop sections, the reservoirs of the Longwangmiao Formation and the Xixiangchi Group are dominated by residual doloarenites, oolitic dolomites, and powdery–finely crystalline dolomites, followed by powdery dolomites, (sandy) (residual) doloarenites, and partially dolomitized granular limestones. The reservoir spaces mainly include intergranular dissolution pores and intercrystalline (dissolution) pores, with intragranular dissolution pores, caves, and fractures developing to varying degrees. Intergranular dissolution pores occur in the form of irregular polygons in doloarenites and oolitic dolomite and are mostly related to early fabric-selective dissolution (Fig. 3.10a–d). Intragranular dissolution pores also occur in the study area and were mostly formed by the leaching and dissolution of meteoric water due to the drop in the relative sea level. Intercrystalline pores occur in granular dolomites and should have resulted from the transformation of the

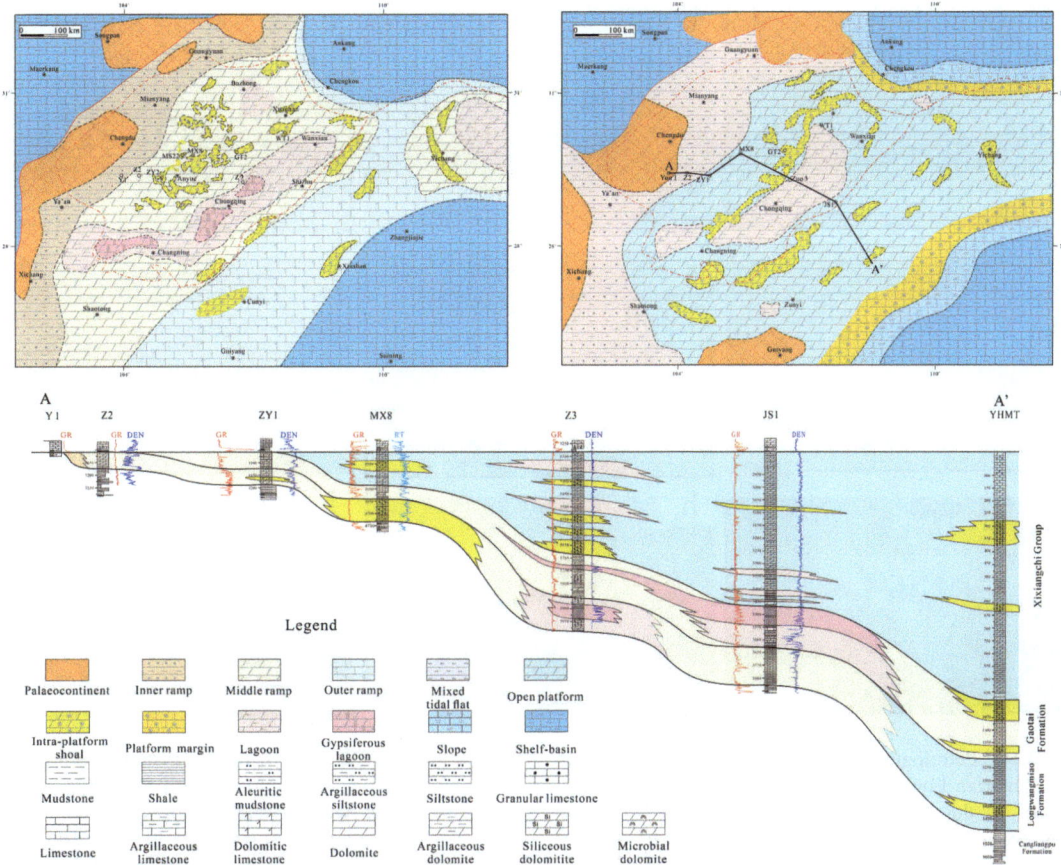

Fig. 3.9 Lithofacies paleogeographic map of the Cambrian Longwangmiao Formation and Xixiangchi Group in the Sichuan Basin and its surrounding areas

early pore system by dolomitization (Fig. 3.10c). Early pore systems are important for the formation of dolomites with coarse crystalline grains (such as powdery–finely crystalline dolomites and mesocrystalline dolomites), as chemical water–rock interactions and dolomitization cannot occur without early pore systems. Intercrystalline pores can form intergranular dissolution pores after being dissolved in the late stage (Fig. 3.10d–f). The common microfractures in the reservoirs in the Longwangmiao Formation and the Xixiangchi Group include tectonic fractures related to tectonic movements and stylolites related to diagenesis. Multiphase tectonic fractures have developed in the reservoirs of the Longwangmiao Formation (Fig. 3.10h, i), and they cut each other. These tectonic fractures are extremely significant for the increase in permeability despite their

limited reservoir capacity. The physical properties of some tectonic fractures decreased after they were filled with late calcite, dolomite, or asphalt.

The 201 small-diameter samples from the Longwangmiao Formation had an average porosity of 2.84% (mostly 0.5–5%) and an average permeability of $0.052 \times 10^{-3} \ \mu m^2$, and 81.6% of them had a permeability of $< 0.1 \times 10^{-3} \ \mu m^2$. Overall, the Longwangmiao Formation is characterized by medium–low porosity and low permeability, with a weak positive correlation between porosity and permeability (Fig. 3.11), indicating strong reservoir heterogeneity. As indicated by the statistics of the relationship between rock types and porosity, the porosity of rocks is in the order of granular dolomites > powdery–finely crystalline dolomites > granular limestones > dolomicrites > micrites. Among them, granular dolomites have

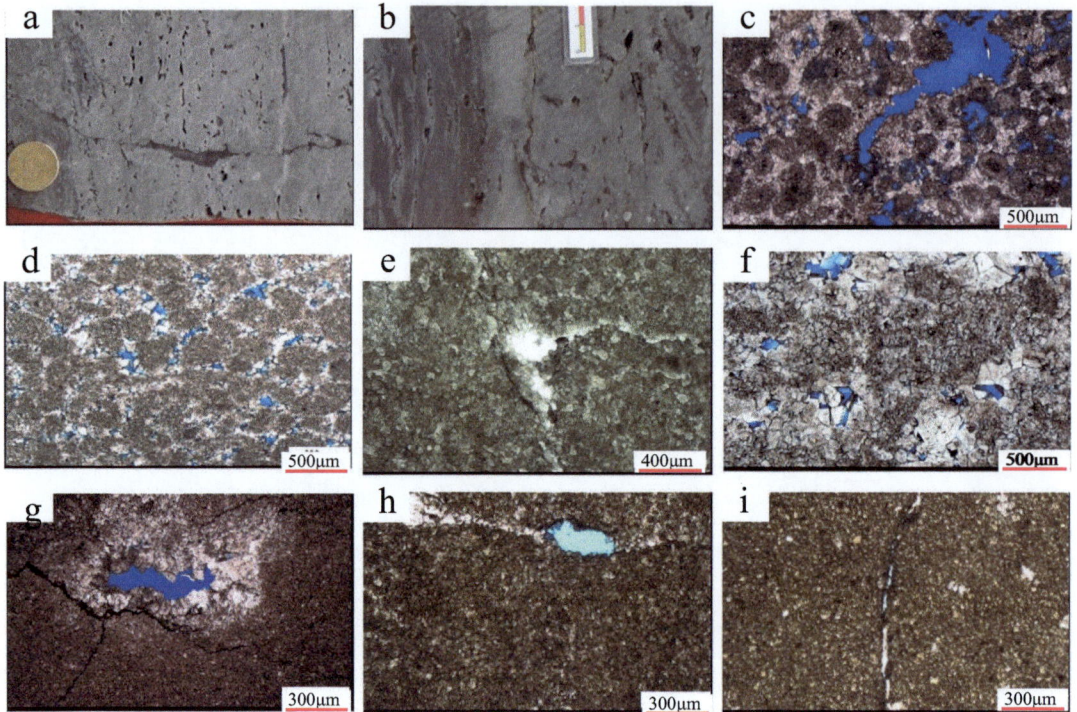

Fig. 3.10 Petrological characteristics of reservoirs in the Cambrian Longwangmiao Formation and Xixiangchi Group in the Sichuan Basin. **a** Well MS22, depth: 4944.58 m, $\in_1 l$, grayish-white doloarenites with dissolution pores along beddings; **b** Well GT2, depth: 5343.20 m, $\in_{2-3} x$, gray doloarenites with dissolution pores along beddings; **c** LQH section, $\in_1 l$, doloarenites with intergranular dissolution pores; **d** Well H12, depth: 4710 m, $\in_{2-3} x$, doloarenites with intergranular dissolution pores; **e** Well WT1, depth: 6893.8 m, $\in_1 l$, finely crystalline dolomites with intercrystalline dissolution pores; **f** Well PQ2, depth: 3338.6 m, $\in_{2-3} x$, finely crystalline dolomites with intercrystalline dissolution pores; **g** LQH section, $\in_1 l$, dolomicrites with intercrystalline dissolution pores; **h** XS section, $\in_{2-3} x$, doloarenites with dissolution and expansion along stylolites; **i** XS section, $\in_1 l$, dissolution fractures with powdery dolomites

porosities of 2–4% and permeabilities of 0.01–1×10^{-3} μm^2, indicating the optimal reservoir physical properties.

The 506 small-diameter samples from the Xixiangchi Group had an average porosity of 2.57%, and 79.2% of them had porosities ranging from 1.5–4% (Fig. 3.11). These samples had an average permeability of 0.064×10^{-3} μm^2, and most of them had a permeability of $< 0.1 \times 10^{-3}$ μm^2. The porosity and permeability of the Xixiangchi Group reservoirs greatly vary, reflecting that the reservoirs are strongly heterogeneous and are affected by fractures. As indicated by the data analysis results, coarsely textured dolomites and pinhole dolomites formed in the karst area, and the high-energy grain shoal environment has relatively high porosity and permeability.

3.5.3 Main Controlling Factors for Reservoirs of the Cambrian Longwangmiao Formation and Xixiangchi Group in the Sichuan Basin

A locally restricted sedimentary environment under the tectonic–sedimentary differentiation of uplifts and depressions is conducive to the development of large-area dolomite reservoirs. The main factors controlling the development of dolomite reservoirs include high-energy facies belts, early dolomitization, and early dissolution.

(1) High-energy grain shoals in the background of intraplatform uplift–depression differenti-

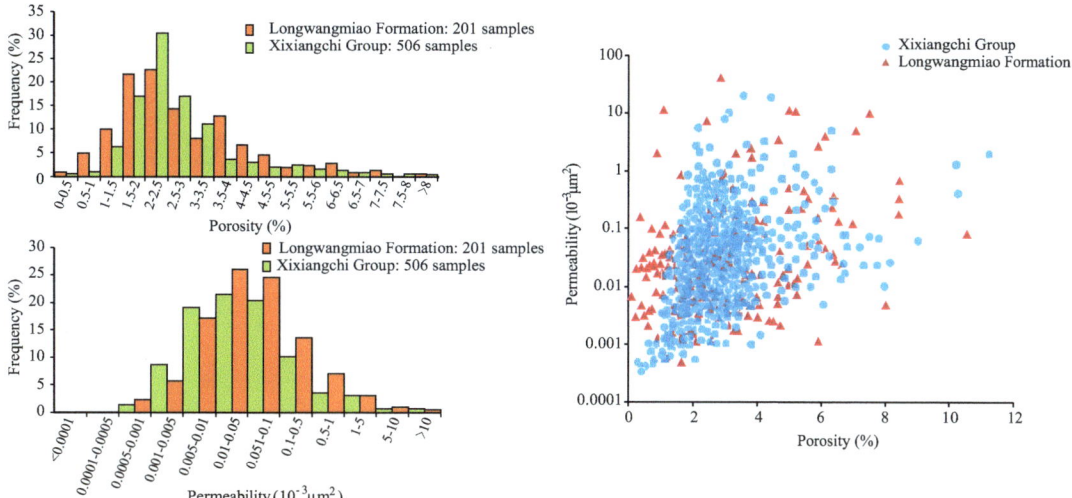

Fig. 3.11 Physical properties of the reservoirs in the Cambrian Longwangmiao Formation and Xixiangchi Group in the Sichuan Basin

ation laid the foundation for reservoir development.

Carbonate reservoirs of high-energy sedimentary facies laid the foundation for the development of primary matrix pore–cavity reservoirs. During the deposition of the Longwangmiao Formation–Xixiangchi Group, the whole Sichuan Basin was present as a deposited carbonate ramp that was high in the northwest and low in the southeast. Owing to the poor water circulation and the arid climate, restricted anhydrite-bearing lagoons were formed in secondary sags that developed in the middle ramp areas in the east-central part of the Sichuan Basin (Zhou et al. 2014; Wang et al. 2016). Numerous grain shoals formed in the paleolandforms' high parts on the margin of the lagoons because of the high energy of shallow seawater (Zhang et al. 2018; Yang et al. 2015; Zhou et al. 2014). Reservoirs were formed mainly from granular dolomites, granular limestones subjected to selective dolomitization, and granular (mainly powdery–finely crystalline) dolomites. As revealed by relevant studies, the powdery–finely crystalline dolomites contained residual grain apparitions, their original sedimentary facies were determined to be grain shoals and tidal flats (lagoons), and their lithology mainly comprised penecontemporaneous dolomicrites and granular

limestones without undergoing dolomitization, which had poor porosity or contained no pores.

(2) Early dolomitization was key to the formation of high-quality reservoirs.

As shown by the analytical results of the different lithologies and physical properties of the Longwangmiao Formation, compared with limestone, dolomite has significantly better reservoir performance, and reservoirs mostly develop at horizons with high dolomite contents. Locally restricted high-salinity seawater laid the foundation for early dolomitization in a setting of weak tectonic differentiation. Specifically, the seawater continuously evaporated and formed high-salinity concentrated seawater in a lagoon. Owing to high salinity and a small number of organisms, sediments were formed mainly by the chemical precipitation of aragonites and/or high-magnesium calcites (Zhao and Zhu 2001; Chen et al. 2010; Zhang et al. 2012a, b). Owing to the limited production capacity of carbonate rocks, the lime mudstone layers were very thin. With the precipitation of aragonites and high-magnesium calcite, massive amounts of calcium ions but a limited number of magnesium ions were consumed, leading to an increase in the Mg^{2+}/Ca^{2+} ratio. When the Mg^{2+}/Ca^{2+} ratio increased to a certain level, the

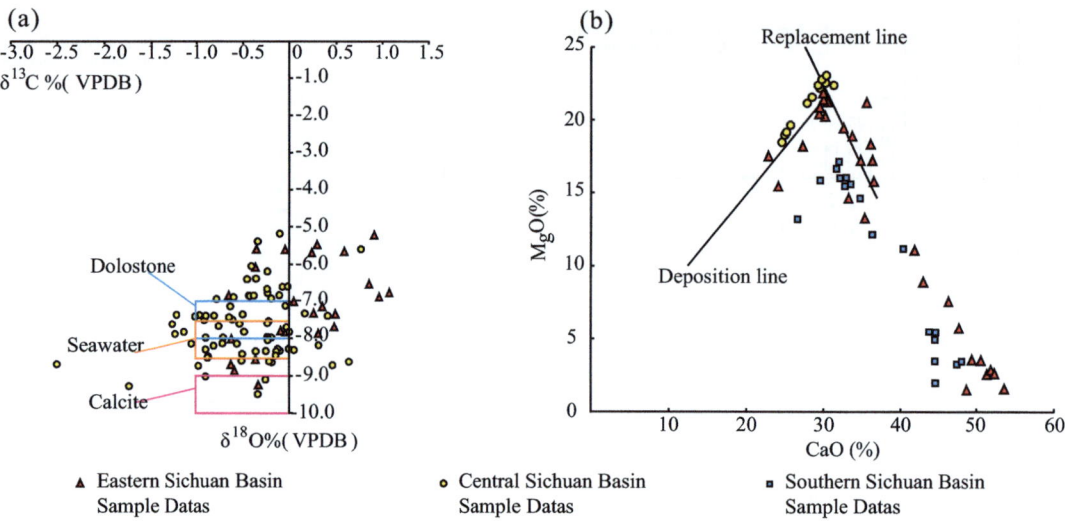

Fig. 3.12 Cross plots of C–O isotopes (**a**) and Mg–CaO (**b**) of the Longwangmiao Formation in the Sichuan Basin

particles, mud particles, and lime mud began to be dolomitized. Moreover, the high-magnesium brine had a high relative density and inevitably flowed back downward. In particular, extensive and intense seepage reflux dolomitization occurred in shoals on the lagoon margin because of the high permeability of the high-magnesium brine.

It is generally believed that the penecontemporaneous metasomatic dolomites related to the evaporitic environment retain the micritic–powdery textures of the protoliths. The buried dolomites usually have crystalline granular textures, and their crystalline grain size is positively correlated with the protolith grain size and recrystallization time. As indicated by the analytical results of the petrological and geochemical characteristics of the Longwangmiao Formation, in addition to micritic-powdery dolomites, doloarenites and oolitic dolomites on the inner ramp were also formed by penecontemporaneous metasomatism. The residual protolith textures indicate that the dolomites hardly underwent superimposed transformation by recrystallization when they were buried. The dolomites of the inner ramp facies in central Sichuan mostly have the same range of $\delta^{18}O$ values as the contemporaneous marine dolomites do, reflecting that the diagenetic fluids are the contemporaneous evapo-

rating seawater. Moreover, there is a linear positive correlation between the MgO and CaO contents, indicating an evaporite mode. In contrast, there is a negative correlation between the MgO and CaO contents of granular dolomites in eastern Sichuan, and the granular dolomites have higher $\delta^{18}O$ values than the contemporaneous seawater, indicating the significant influence of fluids subjected to seepage reflux dolomitization (Fig. 3.12). Therefore, the anhydrite-related Cambrian dolomite reservoirs in the Sichuan Basin are mainly the products of penecontemporaneous seepage reflux dolomitization and evaporative pumping dolomitization. Moreover, the high-salinity seawater environment of the inner ramps and restricted platforms laid the foundation for early metasomatic dolomitization.

(3) Karstification further improved the reservoir and permeability performance of dolomites

Dolomites of the grain shoal facies constitute the main body of reservoirs in the Longwangmiao Formation. However, not all of them act as reservoirs, as some granular (psammitic and oolitic) dolomites are very tight. In the vertical direction, porous granular dolomites, tight granular dolomites, and tight dolomicrites formed multiphase cycles and were superimposed on each other. According to the statistics, effective reservoirs are generally located at the top of sedimentary cycles. For example, the Longwangmiao Formation at Well Moxi 21 has a thickness of 120 m, including 70-m-thick doloarenites. Three-phase cycles composed of the lower dolomicrites, the middle tight doloarenites, and the upper porous doloarenites with thicknesses decreasing upward have developed in the vertical direction. Reservoir development is significantly related to the exposure and dissolution of shoals during their deposition. As verified by simulation experiments on the influence of mineral composition on dissolution intensity, the penecontemporaneous dissolution of meteoric water occurring after dolomitization contributed greatly to pore development (Chen et al. 2020). The residual soluble components, such as undolomitized calcareous materials and gypsum, were dissolved by meteoric water and formed fabric-selective dissolution pores. The residual soluble components with certain contents formed honeycomb dissolution pores. Moreover, insoluble dolomite formed solid frameworks, which were conducive to the preservation of dissolution pores. The dissolution of supergene meteoric water occurred at the end of the Silurian. Dissolution pores and caves along or vertical to beddings have developed under the influence of paleo-landforms and changes in sea level, significantly affecting the strata around the central Sichuan paleo-uplift. Dissolution vug layers with thicknesses of tens of centimeters or several meters are visible under the sedimentary discontinuities. The dissolution vugs are present mainly as elliptic or nearly elliptic polygons with smooth edges. These vugs generally have diameters of 0.3–5 cm, and most of them are isolated, leading to poor connectivity. They are locally connected by dissolution fractures, forming a small fracture–cavity system. Dissolution in rocks is not controlled by petrofabrics. The most significant karst characteristic is the presence of large elongated dissolution vugs along beddings. These vugs are distributed stably in the form of layers, with a low degree of filling and strong connectivity.

3.5.4 Reservoir Development Pattern of the Cambrian Longwangmiao Formation and Xixiangchi Group in the Sichuan Basin

High-quality reservoirs on carbonate ramps or restricted platforms have developed mainly in highly dolomitized grain shoal facies. The sediments of the grain shoal facies are conducive to pore formation and have high porosity and permeability. In contrast, the physical properties of the micritic–powdery dolomites and micrites deposited in a tidal flat or lagoon environment are poor. A shallow-water evaporitic environment, such as a tidal flat or a lagoon, is favorable for dolomitization. Continuous changes in sea level can cause shoals to be exposed and suffer from meteoric water leaching, creating an environment conducive to the dolomitization of grain shoals. Thinly laminated evaporates can ensure the full but not excessive dolomitization of the underlying shoal–facies sediments. Moreover, shoal–facies sediments have high initial porosity and permeability and can ensure the downward infiltration of high-magnesium fluids, increasing the thickness of dolomitization. Otherwise, it is difficult for high-magnesium fluids to infiltrate downward if the underlying strata are composed of micrites or argillaceous micrites deposited in intershoal zones and limestone flats, thus greatly restricting the thickness of dolomitization.

High-quality reservoirs on carbonate ramps or restricted platforms were generally formed in three stages, namely, grain shoal deposition, grain shoal dolomitization, and the dissolution of dolomite grain shoals. Shoals were deposited in the high parts of paleo-landforms, followed by a series of seabed cementation. When the water level dropped, the shoals were exposed quickly, and penecontemporaneous dissolution or dissolution related to low-order sequence boundaries occurred, forming the initial reservoir space. Dissolution was the most common in this stage. In the burial stage, the exposure of the overlying strata and the concentration of seawater caused

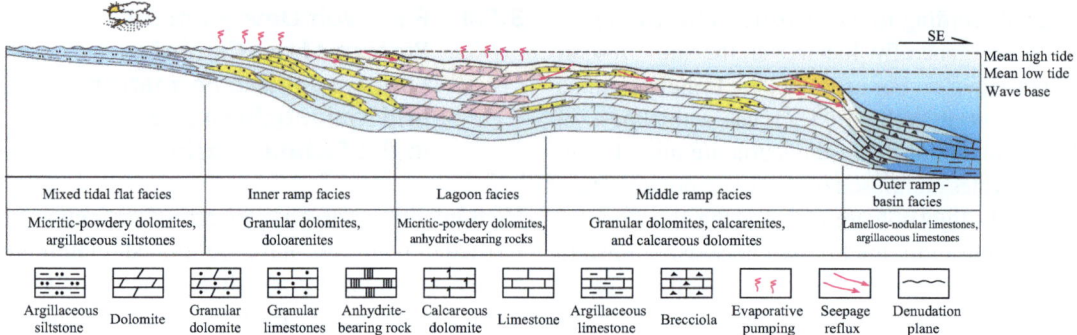

Fig. 3.13 Reservoir development pattern of the Cambrian Longwangmiao Formation in the Sichuan Basin in the setting of tectonic–sedimentary differentiation in a weak extension background

dolomitization during the penecontemporaneous and shallow burial periods, which maintained and improved the initial pore system. Dolomite strata in grain shoals were affected by supergene karstification, as some areas were affected by tectonic uplift, and the continuous leaching of atmospheric precipitation resulted in non-fabric-selective dissolution along paleo-uplifts and their surrounding areas. As a result, karst reservoirs formed along beddings (Fig. 3.13).

3.6 Controlling Effects of Tectonic Differentiation on the Development of Large-Scale Reservoirs in a Strong Compression Setting

Marine basins are deposited and shrink against the background of strong compression. Orogenesis on the margin of the basins becomes active and brings in terrigenous clastics, and denuded paleo-uplifts tend to develop within the basins. Therefore, a strong compression background is favorable for the development of karst reservoirs. Large-scale paleo-uplifts have developed in the three major marine basins in China, and karst reservoirs have formed on the periphery of these paleo-uplifts, such as the Tabei–Tazhong paleo-uplift in the Tarim Basin; the central paleo-uplift in the Ordos Basin; and the

Leshan–Longnüsi, Kaijiang, and Luzhou paleo-uplifts in the Sichuan Basin. The karst reservoirs in marine basins in China vary greatly due to the influence of various factors, such as tectonic paleo-landforms, hydrological characteristics, bedrock properties, fault distribution, exposure time, and paleoclimate. Overall, the Sichuan Basin contains mainly juvenile karst, small-scale paleo-uplifts, and slightly different paleolandforms. This basin features short-term exposure, with karrens, clints, and dolines developing near unconformities. The northern Tarim Basin contains mainly young and adult karsts, large-scale paleo-uplifts, and greatly different paleolandforms (Fig. 3.14). This basin was exposed for a long time and was subjected mainly to the dissolution of surface water and underground rivers, with the eventual development of clints, karrens, large caves, cave roof collapses, cones, and ponors. Karst reservoirs have developed at a certain depth below unconformities in this basin. The Ordos Basin contains mainly senile karst, large-scale paleo-uplifts, and slightly different paleolandforms. This basin features long-term exposure and was subjected to peneplanation. The main landforms in this basin include isolated peaks, residual mounds, and eluviums. Taking the Tabei paleo-uplift in the Tarim Basin as an example, this paper introduces the controlling effects of tectonic differentiation on the development of karst reservoirs against the background of compression.

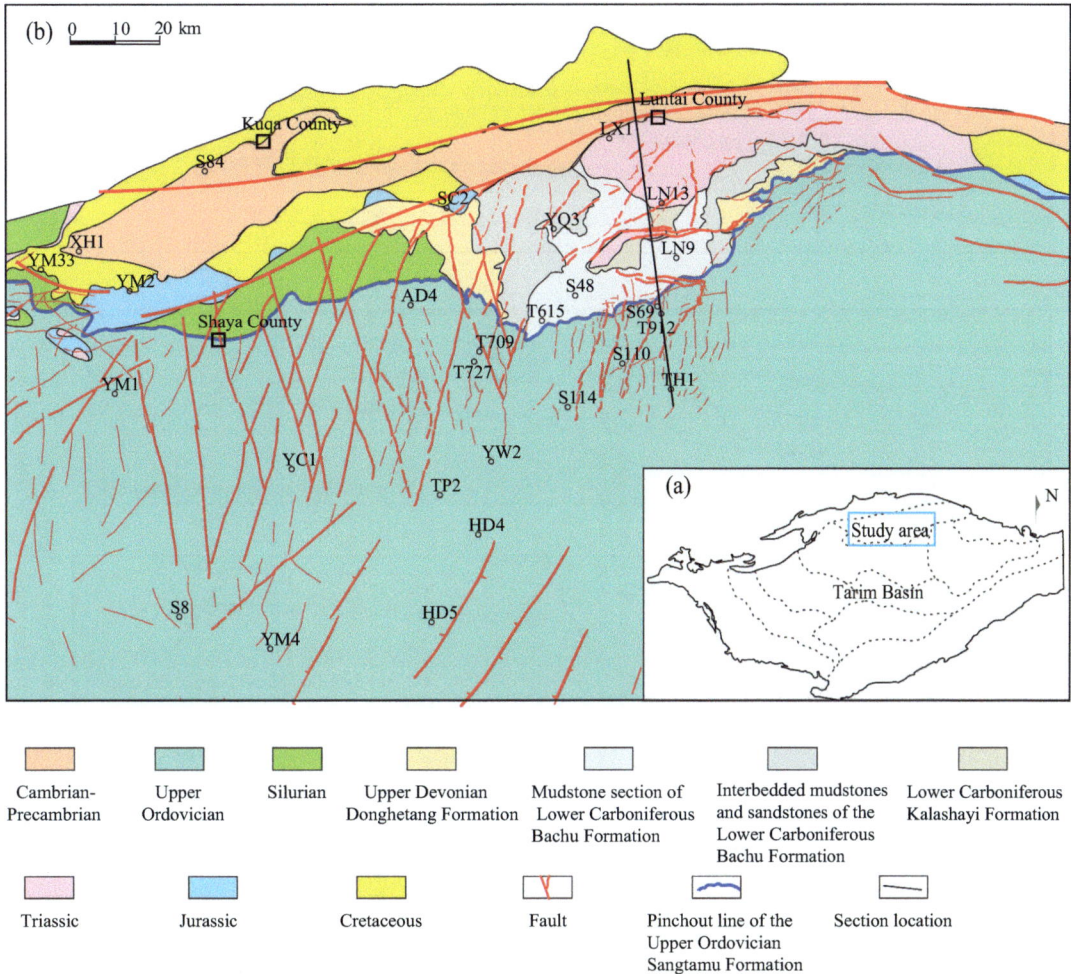

Fig. 3.14 Map showing the distribution of the overlying strata on the top of the Middle-Lower Ordovician carbonate rocks in the Tabei paleo-uplift

3.6.1 Formation and Evolution of the Tabei Paleo-Uplift

The Early Paleozoic tectonic pattern of the Tarim Basin was controlled by the surrounding tectonic environment. During the Cambrian–Middle Ordovician, an extensional tectonic setting developed in the Tarim Basin, and the Tabei area in the basin was located on a neritic carbonate platform on the passive continental margin. As the water level rose, the structural and paleogeographic setting in the Tabei area transitioned from a restricted platform environment during the Cambrian to a normal open platform environment during the Middle Ordovician. As

a result, the Tabei area began to widely accept sediments, including dolomitic limestones and siliceous carbonate rocks. Among them, the main carbonate sedimentary sequences included evaporitic platforms, restricted carbonate platforms, open carbonate platforms, and slopes.

At the end of the Middle Ordovician, the tectonic setting of the Tarim Basin transitioned from an extensional system to a compressional environment, and the compressive and thrusting orogeny around the basin led to strong differentiation and changed the tectonic pattern of the basin. As a result, the northern Tarim (also referred to as Tabei) and middle-southwest Tarim uplift zones, which were distributed nearly west–east

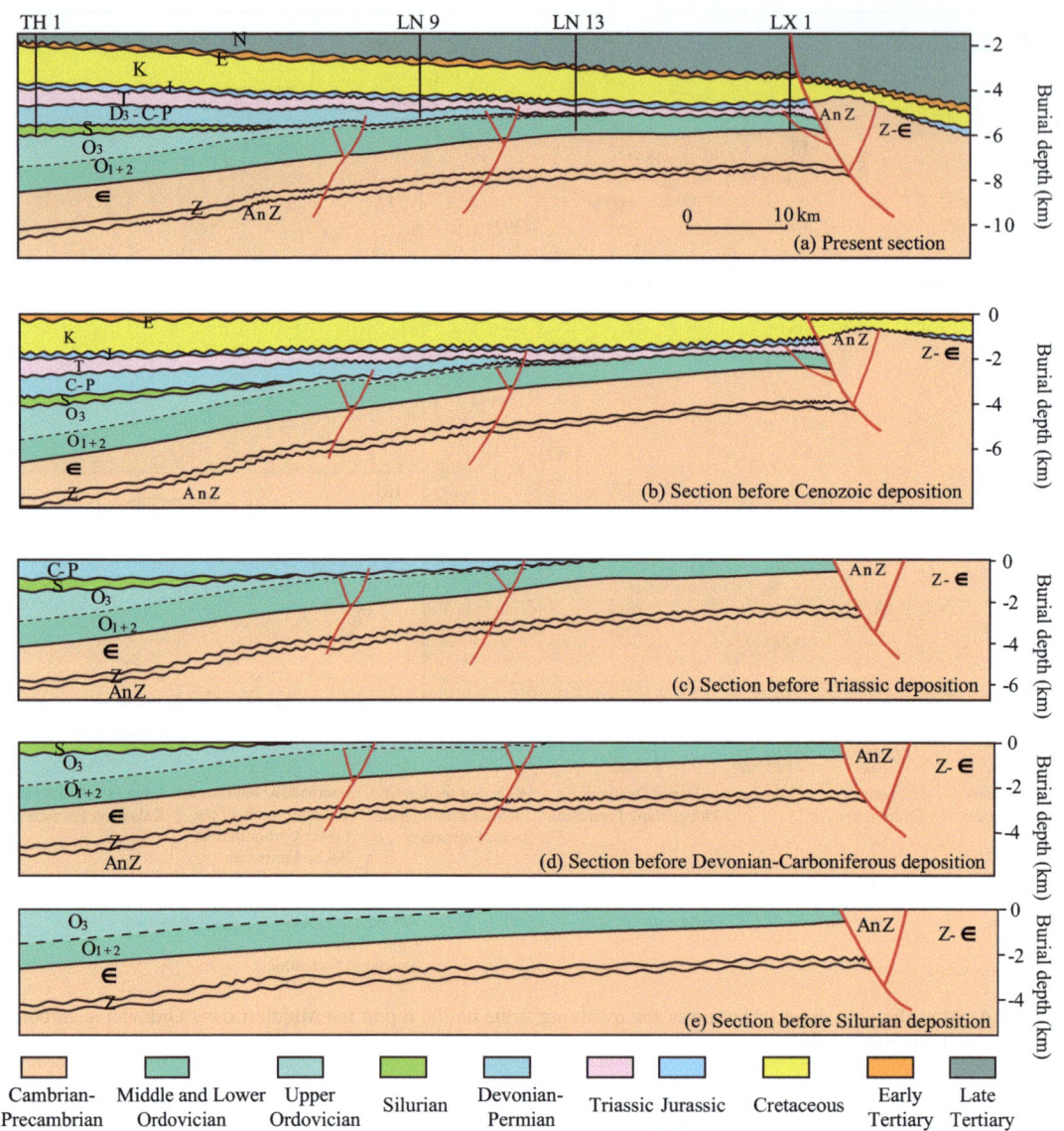

Fig. 3.15 Structural evolution section of the Tabei paleo-uplift in the Tarim Basin

under a compression background, formed, and the strata in the uplift areas were uplifted and denudated (Lin et al. 2008). The Tabei paleo-uplift extended in the nearly west–east direction along the inner axis of the craton and had a steep northern wing, well-developed faults, a wide and gentle southern wing, and alternately distributed concavities and convexities in the west–east direction, appearing as an asymmetrical uplift overall (Fig. 3.15). Orogenic compressive tectonism at the end of the Late Ordovician

further intensified the uplift and denudation of the two major uplifts in the Tarim Basin. From the end of the Late Ordovician to the Carboniferous, marine clastics, carbonate rocks, and evaporites were deposited on the southern slope of the Tabei paleo-uplift. During the Permian–Jurassic, clastic rock formations were deposited in the Tabei paleo-uplift. However, the high parts of the uplift lay in the uplifted and denudated zone for a long time and were not completely buried until the Cretaceous, creating favorable paleoland-

Fig. 3.16 Typical diagrams of pores, vugs, and fractures in Middle-Lower Ordovician carbonate rocks in the Tabei paleo-uplift. **a** granular limestones of the Yijianfang Formation at Well T727, with intragranular dissolution, intragranular dissolution pores, and local micro-fractures; **b** Yingshan Formation at Well S69, with dissolution pores developing along high-angle fractures; **c** Yijianfang Formation at Well S114, with dissolution pores distributed along micro-fractures in silty micrites; **d** Yijianfang Formation at Well S114, with porphyritic-lobate asphalt distribution, reflecting the development of porous reservoirs in the early stage; **e** Yingshan Formation at Well YQ3, with limestones and argillaceous breccias from the collapse of caves; **f** Yingshan Formation at Well T615, with caves filled with oil-immersed fine-grained sandstones with clear beddings

forms for the development of carbonate karst reservoirs.

3.6.2 Karst Reservoirs in the Tabei Paleo-Uplift, Tarim Basin

The karst reservoirs in the Tabei paleo-uplift have developed mainly in the Lower Ordovician strata, which have a sedimentary environment consisting of platform margin shoals, open and wide platforms, and semirestricted or restricted platforms and lithologies comprising micrites, granular limestones, biological limestones, dolomitic limestones, dolomites, and karst rocks. Among them, the main rock types include micrites and granular limestones, which are widely distributed vertically and horizontally. The lithological characteristics indicate that the Lower Ordovician strata are prone to undergo karstification and structural rupture and form high-quality reservoirs (Zhu et al. 2009; Peng et al. 2010).

Owing to the small effects of meteoric water during the early diagenetic stage, intragranular dissolution pores and moldic pores have developed mainly in the Lower Ordovician carbonate rocks in the Tabei paleo-uplift (Fig. 3.16a). Fabric selective dissolution indicates the presence of meteoric water effects due to short-term exposure in the process of penecontemporaneous diagenesis. The matrix pores formed during the early diagenetic stage generally have low porosity and permeability and contribute little to the formation of reservoirs. As shown by the statistics of the test results of the Middle–Lower Ordovician full-diameter core samples, the porosities of the 700 samples from the Yingshan Formation ranged from 0.1–13.8% (average: 1.93%, dominant range: 1–3%), and the permeabilities ranged from 0.0008 to 144×10^{-3} μm^2 (average: 2.22 $\times 10^{-3}$ μm^2). The statistics also reveal that the porosities of 1420 samples from the Yijianfang Formation ranged from 0.1 to 7.0% (average: 1.93%, dominant range: 1–3%), and the permeabilities ranged from 0.0015 to 7484×10^{-3} μm^2 (average: 9.34×10^{-3} μm^2) (Fig. 3.17).

Effective reservoir spaces depend mainly on secondary dissolution vugs and dissolution-enlarged fractures related to karstification, tectonic fractures, and complex pore–fracture–

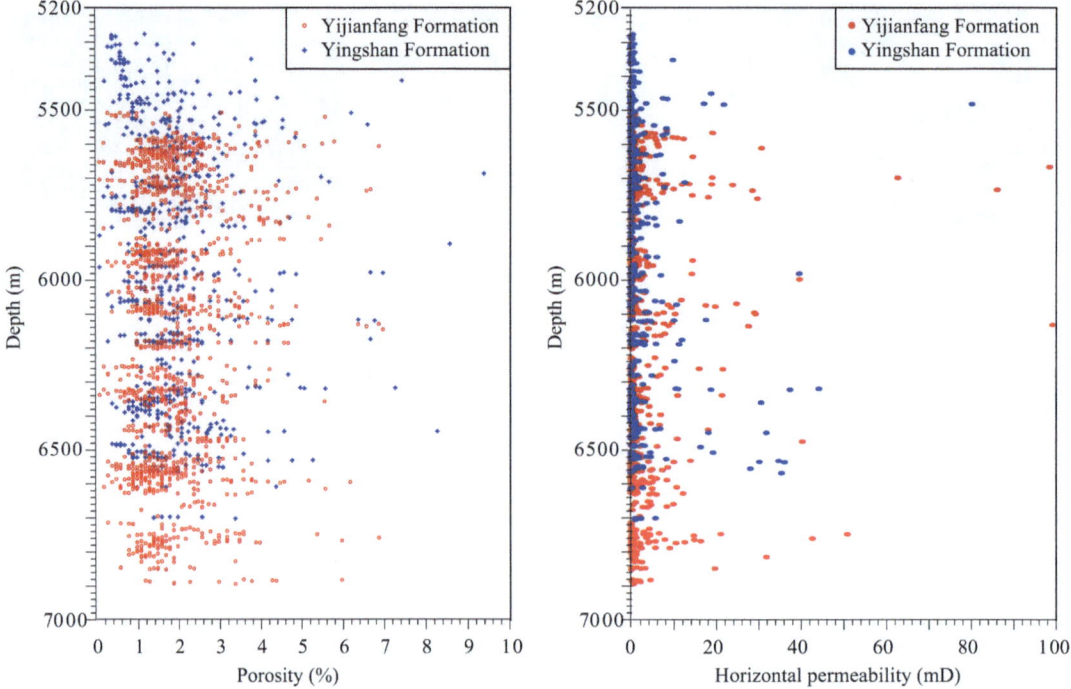

Fig. 3.17 Depth-varying distribution of the porosity and permeability of Middle-Lower Ordovician full-diameter core samples from the Tabei paleo-uplift

vug networks. According to core observations, collapse breccias and sandstones and mudstones deposited in underground rivers are common in the north of the Tahe area (Fig. 3.16b–e). The mechanical interstitial materials and deposits in some caves usually have high porosity and oil-bearing capacity. For example, the largest fully filled karst cave discovered in the core of Well T615 is up to 20 m high (height: 5535–5555 m). This cave is filled with light brown–gray oil-immersed fine-grained sandstones and has yielded industrial oil flow (Fig. 3.16f). In the southern Tahe area, fault-controlled fractures and vugs predominate, and they are commonly filled with chemical fillings such as calcites. With large-scale unfilled karst caves as the main reservoir space, the karst reservoirs in the area are character-ized by enlarged borehole diameters and increased sonic interval transit times on log curves and by high-amplitude reflection structures in the shape of collapse columns in the seismic time section (Xiao et al. 2003). Moreover, circulation loss and venting frequently occurred during drilling. As

shown by the statistics, among more than 2500 wells drilled in the Middle and Lower Ordovician strata, the drilling of up to 1200 wells revealed lost circulation, venting, or filling and a direct probability of penetration of large-scale fracture–vug bodies of up to 48%. Moreover, most of the 1200 wells yielded high–relatively high oil and gas productivity after tests.

3.6.3 Main Controlling Factors for Paleokarst Reservoirs in the Tabei Paleo-Uplift, Tarim Basin

The formation, evolution, and macroscopic distri-bution characteristics of karst reservoirs, as well as the development and preservation of reser-voir spaces, are controlled mainly by four factors, i.e., parent rock lithology, paleoclimate, structural landforms, and exposure time.

During the Middle Caledonian, the Tarim plate was in the tropical region between the equator

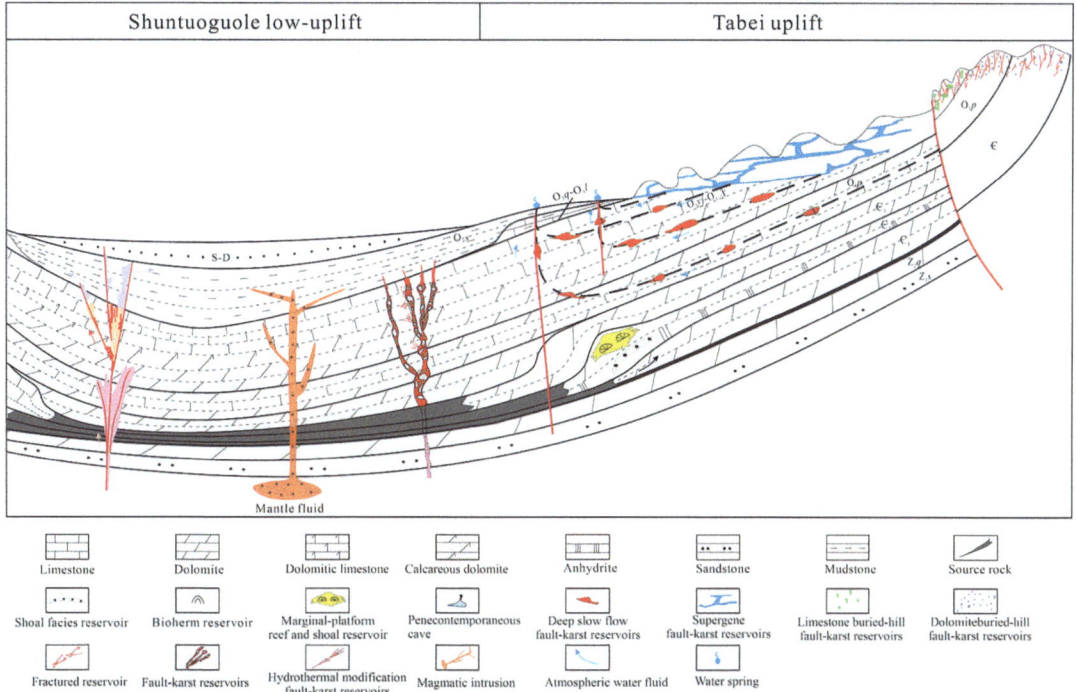

Fig. 3.18 Development pattern of karst reservoirs in the Tabei paleo-uplift in the Tarim Basin in the setting of tectonic–sedimentary differentiation in a strong extension background (stratum code description shown in Fig. 3.14; from Lv et al. 2021)

and the Tropic of Capricorn (Jia 1997). Owing to the high temperature and rainfall, the Tarim plate was hot and humid, leading to rapid dissolution, which was conducive to karstification. The dissolution had a wide range of influence. However, it was small in scale and was concentrated along paleofaults due to the limit of the short sedimentary hiatus. During the Late Caledonian–Early Hercynian, the Tarim plate lay in the mid-latitude arid climate zone of the northern hemisphere (Jia 1997). The paleostructural and geomorphic differentiation and the sedimentary hiatus time caused by intensive Early Hercynian movement were far greater during this period than during the Middle Caledonian. In the rainy season during this period, the flood flow on the surface or downward infiltration into the ground carried large amounts of terrigenous clastics, which experienced extremely strong chemical dissolution and mechanical impact and erosion, resulting in the formation of surface peak clusters, valleys, and large and complex underground karst fracture–

vug systems in the Middle–Lower Ordovician denudated areas.

Controlled by the differences in the Middle Caledonian–Early Hercynian tectonic activities of paleo-uplifts, the karst reservoirs in different areas show significantly different distribution patterns (Fig. 3.18).

In the high part of the Tabei paleo-uplift, the Middle–Lower Ordovician strata in the main body of the Tahe area underwent strong deformation and large-scale uplift and denudation, as well as the superimposition of multiphase unconformities and high-intensity karstification. As a result, typical karsts have developed in the Yingshan Formation (Xie et al. 2013). They are strata-bound karst fracture–vug bodies and are distributed mainly along paleochannels and monadnocks. On the structural slope of the Tabei paleo-uplift, fault-controlled reservoirs with deep fluid circulation have developed in the Yingshan and Yijianfang formations from the south of the Tahe area to the Tuoputai–Yuejin area. These reservoirs differ from those in the main body area and were formed as follows. The meteoric freshwater from the high part of the Tabei paleo-uplift continuously converged southward and southwestward and infiltrated downward along the faults under the pressure of the groundwater head. Then, in the form of subsurface and slow-flow dissolution, the meteoric freshwater dissolved the Ordovician strata and even the Cambrian strata along faults and sequence boundaries in the deep parts, influencing the depth range of up to hundreds to thousands of meters. Locally, it broke through the pressure of the Upper Ordovician strata and discharged to the surface in the form of ascending springs. Such karstification with deep fluid circulation reached the lower strata of the Yingshan Formation vertically and the Shunbei No. 1 zone horizontally, forming dissolution fracture–vug-type reservoirs, which are linearly distributed on a certain scale (He et al. 2019b). In the Shunbei area of the Shuntuoguole low uplift and the north slope of the central Tarim uplift, the Ordovician strata are not directly exposed, and their overlying Paleozoic clastics have a thickness greater than 3000 m. Moreover, these strata are relatively gentle and have small structural slopes, leading to weak karstification with deep fluid circulation. Reservoirs of multiple genetic types have developed in these areas, mainly including approximately tubular caves, fracture bodies, or vugs in breccia zones, which were controlled by the evolution and structures of strike-slip faults. In addition, local formation brine and deep hydrothermal fluids migrated upward along fault zones and interacted with the dolomites of the Penglaiba Formation and limestones of the Yingshan Formation, forming structurally controlled hydrothermal reservoirs (Lv et al. 2021).

3.7 Conclusions

(1) The strong tectonic–sedimentary differentiation in the marine cratonic basins of China controls the development and distribution of large-scale carbonate reservoirs. The orderly superimposition of rift basins, basins of passive continental margins, cratonic depression basins, and foreland basins in the extensional–convergent tectonic cycles of the marine basins controls the regular occurrence of reef–shoal reservoirs, intraplatform grain shoal reservoirs, and paleokarst reservoirs.

(2) The development of reef–shoal reservoirs is controlled by intracratonic and marginal rifts formed in a strong extensional setting. The development of intraplatform grain shoal reservoirs is controlled by the formation of intracratonic depressions formed in a weak extensional–weak compression setting. The development of karst reservoirs is controlled by paleo-uplifts formed during the strong compression period.

(3) The development of high-quality reservoirs of platform–margin reef–shoal facies and intraplatform grain shoal facies is generally controlled by high-energy facies belts, early dolomitization, and early karstification. High-quality reservoirs exhibit a zonal distribution under strong tectonic differentiation and a planar discontinuous distribution under weak tectonic differentiation. The development of karst reservoirs is controlled by the lithology of the parent rocks, paleoclimate, tectonic landforms, and exposure time. Karst reservoirs are distributed in the form of rings under tectonic differentiation.

(4) In addition to controlling the distribution of high-quality reservoirs, tectonosedimentary differentiation also governs the configuration of source–reservoir assemblages. Strong tectonic differentiation leads to sedimentary differentiation between platform margins and deep-water shelves, enabling the formation of source–reservoir assemblages with lateral contact. In contrast, reservoirs formed under weak tectonic differentiation can only connect with the underlying source rocks via faults, resulting in lower-generation and upper-reservoir assemblage configurations.

Acknowledgements The authors acknowledge the financial support of Natural Science Foundation of China (Grants U24B6001, U19B6003 and U20B6001).

References

Budd DA, Saller AH, Harris PM et al (1995) Unconformities and porosity in carbonate strata. AAPG 63

Chen LQ, Shen ZG, Hou FH et al (2010) Formation environment of Triassic evaporate rock basin and dolostone reservoirs in the Sichuan Basin. Pet Geol Exp 32(4):334–346

Chen YN, Zhang JY, Li WZ et al (2020) Lithofacies paleogeography reservoir origin and distribution of Cambrian Longwangmiao Formation in Sichuan Basin. Mar Petrol Geol 25(2):171–180

Coppold M, Powell W (2000) A geoscience guide to the burgess shale: geology and paleontology in Yoho National Park Burgess Shale Geoscience Foundation. B.C 2000

Dong SW, Gao R, Yin A et al (2013) What drove continued continent-continent convergence after ocean closure? Insights from high-resolution seismic-reflection profiling across the Daba Shan in central China. Geology 41:671–674

Du JH, Zou WH, Fei BS et al (2002) Buried hill composite hydrocarbon accumulation zone of Jizhong depression. Science Press

Du JH, Zou CN, Xu CC et al (2014) Theoretical and technical innovations in strategic discovery of a giant gas field in Cambrian Longwangmiao Formation of central Sichuan paleo-uplift Sichuan Basin. Pet Explor Dev 41(3):268–277

Du JH, Wang ZC, Zou CN et al (2016) Discovery of intra-cratonic rift in the Upper Yangtze and its coutrol effect on the formation of Anyue giant gas field. Acta Petrol 37(1):1–16

Grotzinger JP, Knoll AH (1999) Stromatolites in Precambrian carbonates: evolutionary mileposts or environmental dipsticks? Annu Rev Earth Planet Sci 27(1):313–358

Gu ZD, Wang ZC (2014) The discovery of Neoproterozoic extensional structures and its significance for gas exploration in the Central Sichuan Block Sichuan Basin South. Sci China Earth Sci 44(10):2210–2220

Guan SW, Wu L, Ren R et al (2017) Distribution and petroleum prospect of Precambrian rifts in the main cratons China. Acta Petrol. Sin 38(1):9–22

Guo Q (2013) Research on natural gas enrichment rule of the weathering crust karst reservoir of the Lower Paleozoic era theme in western area of Jingbian Gasfield. Northwest University

He FQ (2002) Karst weathering crust oil-gas field on carbonate unconformity: An example from the Tahe Oilfield in the Ordovician reservoir in the Tarim Basin. Int Geol Rev 4:391–397

He DF, Li DS, Tong XG et al (2008) Accumulation and distribution of oil and gas controlled by paleo-uplift in poly-history superimposed basin. Acta Petrol Sin 4:475–488

He DF, Li DS, Tong XG (2010a) Stereoscopic exploration model for multi-cycle superimposed basins in China. Acta Petrol Sin 31(5):695–709

He ZL, Peng ST, Zhang T (2010b) Controlling factors and genetic pattern of the Ordovician reservoirs in the Tahe area Tarim Basin. Oil Gas Geol 31(6):743–752

He DF, Ma YS, Cai XY et al (2017) Comparison study on controls of geologic structural framework upon hydrocarbon distribution of marine basins in western China. Acta Petrol Sin 33(4):1037–1057

He DF, Ma YS, Liu B et al (2019a) Main advances and key issues for deep-seated exploration in petroliferous basins in China. Earth Sci Front 26(1):1–12

He ZL, Yun L, You DH (2019b) Genesis and distribution prediction of the ultra-deep carbonate reservoirs in the transitional zone between the Awati and Manjiaer depressions Tarim Basin. Earth Sci Front 26(1):13–21

Hood AVS, Wallace MW (2012) Synsedimentary diagenesis in a Cryogenian reef complex: ubiquitous marine dolomite precipitation. Sediment Geol 255:56–71

Hood AVS, Wallace MW, Drysdale RN (2011) Neoproterozoic aragonite-dolomite seas? Widespreadmarine dolomite precipitation in Cryogenian reef complexes. Geology 39:871–874

Hou FH, Fang SX, He J et al (2011) Distribution characters and comprehensive evaluation of Middle Ordovician Majiagou 51–54 submembers reservoirs in Jingbian gas field area Ordos Basin. Mar Pet Geol 16(1):1–13

Hu DF (2011) Differences in reef-bank reservoir features between Puguang and Yuanba gas fields and their reasons. NGI 31(10):17–21+114–115

Jia CZ (1997) Tentonic characteristics and Petroleum Tarim Basin. Petroleum Industry Press, Beijing, pp 1–425

Jiao WW, Lu XX, Zhou YY et al (2011) Main controlling factors of the Ordovician carbonate reservoirs in Tazhong area the Tarim Basin. Oil Gas Geol 32(2):199–206

Jin ZY (2014) TA study on the distribution of oil and gas reservoirs controlled by source-cap rock assemblage in unmodified foreland region of Tarim Basin. Oil Gas Geol 35(6):763–770

Knoll AH (2015) Life on a young planet. Princeton University Press, Princeton

Li XH (1999) U–Pb zircon age of granites from the southern margin of Yangtze Block and the timing of Neoproterozoic Jinning Orogeny in SE China: termination of Rodinia assembly? Precambrian Res 97:43–57

Li ZX, Bogdanova SV, Collins AS (2008) Assembly configuration and break-up history of Rodinia. Precambrian Res 160(1–2):179–210

Li ZX, Evans DAD, Halverson GP (2013) Neoproterozoic glaciations in a revised global palaeogeography from the breakup of Rodinia to the assembly of Gondwanaland. Sediment Geol 294:219–232

Li SJ, Gao P, Huang BY et al (2018) Sedimentary constraints on the tectonic evolution of Mianyang-Changning trough in the Sichuan Basin. Oil Gas Geol 39(05):889–898

Lin CS, Yang HJ, Liu JY et al (2008) Paleohigh geom orphology and paleogeographic frame work and their controls on the formation and distribution of stratigraphic traps in the Tarim Basin. Oil Gas Geol 02:189–197

Liu SG, Sun W, Luo ZL et al (2013) Xingkai tafrogeny and petroleum from Upper Sinian to Cambrian strata in Sichuan Basin, China. J Chengdu Univ Technol Sci Technol Ed 40(5):511–520

Liu SG, Wang YG, Sun W et al (2016) Controlling effects of extensional troughs on marine oil and gas distribution in Sichuan Basin. J Chengdu Univ Technol Sci Technol Ed 43(1):1–23

Luo P, Zhang J, Liu W et al (2008) Basic characteristics of marine carbonate reservoirs in China. Earth Sci Front 15(1):36–50

Lv HT, Han J, Zhang JB et al (2021) Development characteristics and formation mechanism of ultra-deep carbonate fault-dissolution body in Shunbei area Tarim Basin. Pet Geol Exp 43(1):14–22

Ma YS (2006) Cases of discovery and exploration of marine fields in China (part6) Puguang gas field in Sichuan basin. Mar Pet Geol 11(2):35–40

Ma YS, Mou CL, Guo TL et al (2005) Sequence stratigraphy and reservoir distribution of the Changxing Formation in northeastern Sichuan Basin. Earth Sci Front 12(3):179–185

Ma YS, Cai XY, Zhao PR et al (2010) Distribution and further exploration of the large-medium sized gas fields in Sichuan Basin. Acta Petrol Sin 31(3):347–354

Ma YS, He DF, Cai XY et al (2017) Distribution and fundamental science questions for petroleum geology of marine carbonate in China. Acta Petrol Sin 33(4):1007–1020

Partin CA, Bekker A, Planavsky NJ et al (2000) Large-scale fluctuations in Precambrian atmospheric and oceanic oxygen levels from the record of U in shales. Earth Planet Sci Lett 369:284–293

Peng ST, He ZL, Ding Y et al (2010) Characteristics and major controlling factors of carbonates reservoir in the Middle Ordovician Yijianfang formation, Tuofutai Area, Tahe Oilfield. Pet Geol Exp 32(2):108–114

Planavsky NJ, Reinhard CT, Wang X et al (2014) Low Mid-Proterozoic atmospheric oxygen levels and the delayed rise of animals. Sci 346(6209):635–638

Powell CM, Preiss WV, Gatehouse CG et al (1994) South Australian record of a Rodinian epicontinental basin and its mid-Neoproterozoic break-up (approximately 700 Ma) to form the Palaeo-Pacific Ocean. Tectonophysics 237:113–140

Ren JS, Niu BG, Liu ZG (1999) Soft collision superposition orogeny and polycyclic suturing. Earth Sci Front 6(3):85–93

Shen AJ, Zhao WZ, Hu AP et al (2021) The dating and temperature measurement technologies for carbonate minerals and their application in hydrocarbon accumulation research in the paleo-uplift in central Sichuan Basin SW China. Pet Explor Dev 48(3):476–487

Wang J, Li ZX (2003) History of Neoproterozoic rift basins in South China: implications for Rodinia break-up. Precambrian Res 122:141–158

Wang SC, Fu ST, Li XZ et al (2005) The influence to the accumulation of oil and gas of the development of the arcuate structure zone in the transition zone from the plat to the geosyncline in the west of Ordos Basin. Nat Gas Geosci 16(4):421–427

Wang SL, Zheng MP, Zhang Z et al (2016) Carbon and oxygen isotopic compositions of Cambrian marine carbonates in Sichuan Basin China: implications for sedimentary evolution and potash finding. Earth Sci Front 23(5):202–220

Wang ZC, Zhao WZ, Hu SY et al (2017) Control of tectonic differentiation on the formation of large oil and gas fields in craton basins: a case study of Sinian-Triassic of the Sichuan Basin. NGI 37(1):9–23

Wang JB, He ZL, Zhu DY et al (2020) Petrological and geochemical characteristics of the botryoidal dolomite of Dengying Formation in the Yangtze Craton South China: Constraints on terminal Ediacaran "dolomite seas." Sediment Geol 406:1–17

Wei GQ, Yang W, Du JH et al (2015) Geological characteristics of the Sinian-Early Cambrian intracratonic rift Sichuan Basin. NGI 35(1):24–35

Xiao YR, He FY, Sun YM (2003) Reservoir characteristics of paleocave carbonates—a case study of Ordovician paleocave in Tahe Oilfield Tarim Basin. Oil Gas Geol 1:75–80+86

Xie DQ, Zheng ML, Jiang HS et al (2013) Formation and evolution of the Shaya uplift and constraints on oil and gas distribution in the Tarim Basin. Geotecton et Metallog 37(3):398–409

Yang ZR, Wang XJ, Feng XK et al (2014) Geological research and significance of a rift valley in the Presinian period in central Sichaun Basin. NGI 34(3):80–85

Yang XF, Wang XZ, Tang H et al (2015) Research sedimentary microfacies of the Longwangmiao formation in Moxi area central Sichuan basin. Acta Sedimentol Sin 33(5):972–982

Yang YM, Yang Y, Yang G et al (2019) Gas accumulation conditions and key exploration and development technologies of Sinian and Cambrian gas reservoirs in Anyue gas field. Acta Petrol 40(4):493–508

Zhang J, He Z, Xu HB et al (2012a) Petrological characteristics and origin of Permian Fengcheng formation dolomitic rocks in Wuerhe-Fengcheng area Junggar basin. Acta Sedimentol Sin 30(5):859–867

Zhang XF, Li M, Chen ZY et al (2012b) Characteristics and karstification of the Ordovician carbonate reservoir Halahatang area northern Tarim Basin. Acta Petrol 28(3):815–826

Zhang X, Luo W, Wen L et al (2018) Sedimentary facies evolution characteristics and petroleum geological significance of Cambrian Group in Sichuan Basin. Fault-Block Oil Gas Field 148(4):15–21

Zhang XX, Chen AQ, Dang N et al (2020) Tectono-sedimentary differentiation of lower Palaeozoic carbonate rock in Ordos basin NW China and its implications for hydrocarbon-play generation. Carsol Sin 39(2):215–224

Zhang GW, Meng QR, Lai SC (1995) Tectonics and structure of Qinling orogenic belt. Sci China Ser B 38(11):1379–1394

Zhang GW, Guo AL, Wang YJ et al (2013) Tectonics of South China continent and its implications. Sci China Earth Sci 56:1804–1828

Zhao DL, Zhu XM (2001) Sedimentary Petrology. Petroleum Industry Publishing House, Beijing

Zhao WZ, Wang ZC, Hu SY et al (2012) Large-scale hydrocarbon accumulation factors and characteristics of marine carbonate reservoirs in three large onshore cratonic basins in China. Acta Petrol 33(S2):1–10

Zhao WZ, Shen AJ, Pan WQ et al (2013) A research on carbonate karst reservoirs classification and its implication on hydrocarbon exploration: cases studies from Tarim Basin. Acta Petrol 29(9):3213–3222

Zhao WZ, Wei GQ, Yang W et al (2017) Discovery of Wanyuan-Dazhou Intracratonic Rift and its exploration significance in the Sichuan Basin, SW China. Pet Explor Dev 44(5):659–669

Zhong Y, Li YL, Zhang XB et al (2013) Features of extensional structures in pre-Sinian to Cambrian strata Sichuan Basin, China. J Chengdu Univ Technol Sci Technol Ed 40(5):498–510

Zhou JG, Fang C, Ji HC et al (2014) A development rule of lower cambrian longwangmiao grain beaches in the Sichuan basin. Nat Gas Ind 34(8):27–36

Zhou XB, Li JH, Wang HH et al (2015) The type of prototypic basin and tectonic setting of Tarim Basin formation from Nanhua to Sinian. Earth Sci Front 22(3):290–298

Zhu GY, Zhang SC, Wang HH et al (2009) The formation and distribution of deep weathering crust in north Tarim Basin. Acta Petrol 25(10):2384–2398

Zhu DY, Jin ZJ, Zhang RQ et al (2014) Characteristics and developing mechanism of Sinian Dengying Formation dolomite reservoir with multi-stage karst. Earth Sci Front 21(6):335–345

Zou CN, Du JH, Xu CC et al (2014) Formation distribution resource potential and discovery of the Sinian-Cambrian giant gas field Sichuan Basin, SW China. Pet Explor Dev 41(3):278–293

Structural Transformation in Cratonic Basins and Influence on Oil and Gas Accumulation in Deep Marine Strata

4

Xuhui Xu, Jianlin Lu, Chenyu Zhang, Mingzhe Deng, Shang Deng, Peng Deng, Shuaiqiang Shan, and Zongxin Zuo

Abstract

Guided by the idea of a "superposition-controlled reservoir", the technical route of geologic analysis, seismic data interpretation, analogue modelling and basin digital modelling, geometry, kinematics, and dynamics are adopted in this paper to carry out systematic research on the deformation characteristics and evolution processes of different reformation types and their control on oil and gas accumulation and the depletion of deep marine strata in the Palaeozoic. In this paper, four stages of tectonic evolution and five types of intracontinental transformation since the Late Triassic are established, the time frame of transformation deformation is constructed, dynamic hydrocarbon accumulation characteristics and models under the control of different types of superimposed transformation are established, and favourable hydrocarbon accumulation zones and targets are proposed.

Keywords

Superposition-controlled reservoir · Tectonic evolution · Tectonic deformation · Dynamic hydrocarbon accumulation

X. Xu
Sinopec Geophysical Research Institution, Nanjing, China

J. Lu (✉) · C. Zhang · M. Deng · S. Deng · P. Deng · S. Shan · Z. Zuo
Sinopec Petroleum Exploration and Production Research Institute, Beijing, China
e-mail: lujl.syky@sinopec.com

4.1 Introduction

Based on the study results of the deformation characteristics of the unified landmass, which formed during the Late Triassic and was subjected to intracontinental differential compression, the Meso-Cenozoic superimposed reformation process of the three cratonic basins can be divided into four evolutionary stages—T3–J2, J3–K1, K2–E, and N–Q—and five major reformation types, namely, deformation in the piedmont, intraplatform decollement fold, differential subsidence–uplift, intrabasinal strike-slip fault, and tectonic–thermal event. The time–space frames of intracontinental reformation and deformation under the far-field effect of supercontinental convergence are constructed, and the space–time distributions of the five main reformation types and their control on oil and gas enrichment and depletion are determined.

4.2 Temporal and Spatial Framework of Structural Deformation in the Three Cratons

4.2.1 Tectonic Evolution of Cartons in the Meso-Cenozoic

The basement landmass of the large basin in central and western China is the result of mutual splicing and accretion of many microcontinents scattered between the north continent and Gondwanaland in the Palaeozoic. These microconti-

© The Author(s) 2026
Y. Ma et al. (eds.), *Petroleum Geology and Exploration of Deep Marine Strata in China*, Advances in Oil and Gas Exploration & Production, https://doi.org/10.1007/978-3-032-02496-1_4

nents gradually pieced to the southern margin of the Eurasian continent during multiple tectonic events and became part of the Eurasian continent at the end of the Late Palaeozoic. Since the Late Triassic, the lithospheres of the Chinese mainland have been affected by the continuous splicing of north continents and polycyclic convergence and accretion of the south plate margin in the Tethys–Pacific tectonic domain, and each landmass has a different response under polycyclic compression and convergence in three directions: north, east and west (Liu et al. 2015). Under the interaction of the three tectonic domains of the Palaeo-Asian Ocean, Tethys Ocean, and Pacific Ocean, the Chinese mainland has undergone four tectonic deformation stages under a constantly uplifting tectonic setting (Xu et al. 2005).

The first tectonic deformation stage occurred from the Late Triassic to the Middle Jurassic. The Palaeo-Tethys Ocean in the north of the Qiangtang terrane began to close in the Late Triassic (Chen et al. 2015) and finally closed in the Early Jurassic, whereas the Meso-Tethys Ocean continued to expand (Fig. 4.1a). In the Early Jurassic, the Lhasa terrane was separated from Gondwanaland and drifted northwards. The rift zone expanded rapidly, and the Neo-Tethys Ocean began to open. The Palaeo-Pacific plate began to subduct to the Eurasian continent in the Late Triassic and gradually subducted from the Early–Middle Jurassic. The Mongolia–Okhotsk oceanic plate in the north of the Chinese mainland continued to subduct to the south and closed in a scissor-like shape from southwest to northeast in the Middle Jurassic.

The second tectonic deformation stage occurred from the Late Jurassic to the Early Cretaceous. The Meso-Tethys Ocean began to subduct in both directions in the Late Jurassic and closed in the Early Cretaceous (Zhang et al. 2016). Accompanied by the rapid expansion of the Neo-Tethys Ocean, the Lhasa terrane drifted northwards and merged into the Asiatic landmass in the Early Cretaceous (Zhu et al. 2021; Fig. 4.1b). At this stage, the Palaeo-Pacific plate subducted and rolled back, and the east margin of the Chinese continent completely transformed from the Palaeo-Asian ocean tectonic domain to the marginal Pacific tectonic domain and entered the active Andean continental margin stage, in which the left lateral strike-slip pull-apart basin developed.

The third tectonic deformation stage occurred from the Late Cretaceous to the Palaeogene. In the Late Cretaceous, the Pacific plate, in place of the Palaeo-Pacific plate, began to subduct down to the Eurasian continent in the SWW direction and began to subduct and retreat (Fig. 4.1c). From the Late Cretaceous to the Middle Palaeocene, the Neo-Tethys Ocean Basin continued to subduct to the southern margin of the Eurasian continent to accumulate volcanic arc magma at the Gangdese–Nyenchen Tanglha continental margin and the flysch sedimentation in front of the Xigaze forearc in the middle and south Lhasa landmass. In the Early Eocene, the Indian landmass collided with the Eurasian continent (Yin and Harrison 2000), and the Neo-Tethys Ocean was finally closed to form a south-trending Himalayan superimposed orogenic belt.

The fourth tectonic deformation stage occurred from the Neogene to Quaternary. In the Oligocene–Miocene stage, the Himalayas did not experience strong uplift, and the molasse assemblage of orogenic facies is missing. Since the end of the Miocene, the Indian landmass has continued to thrust northwards (Zhu et al. 2021), and intracontinental subduction has advanced northwards along the Siwalik belt at the south foot of the Himalayas, resulting in rapid uplift of the Himalayas (Figs. 4.1d and 4.2).

4.2.2 Temporal and Spatial Framework of Cratonic Basins During Intracontinental Deformation

The corresponding continental basins have been developed in the Chinese mainland under the four tectonic deformation frameworks mentioned above. This monograph uses the Tarim Basin, Sichuan Basin, and Ordos Basin as typical basins for analysis (Fig. 4.2).

In the first to second tectonic deformation stage (T_3–K_1), the closure of the Nujiang Ocean

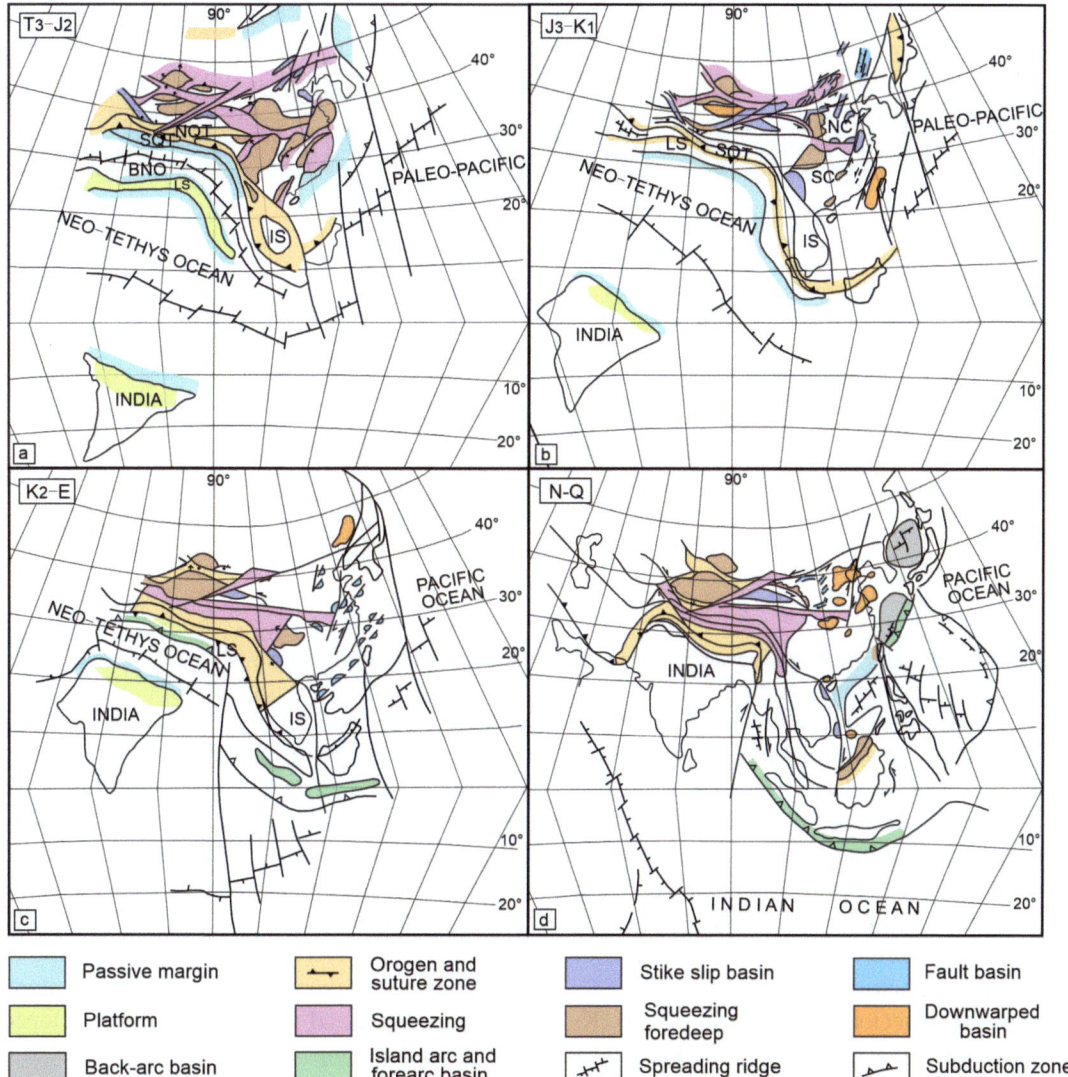

Fig. 4.1 Schematic paleotectonic maps of China and its neighbours. *SQT* South Qiangtang Terrane, *NQT* North Qiangtang Terrane, *LS* Lhasa Terrane, *BNO* Bangong- Nujiang Ocean, *IS* Indochina Block, *NC* North China, *SC* South China

during the Tethys evolution caused the land-masses to collide, as did the oblique closure of the Mongolia–Okhotsk Ocean south of the Siberian continent to the Palaeo-Asian continent. More-over, influenced by the transform fault because of oceanic crustal activity of the eastern Pacific plate, the eastern margin of the continent triggered the reactivation of early landmass aggregation in the central and western areas and strong intra-continental subduction in the Palaeozoic orogenic belt (Huang et al. 2017; Table 4.1). The Tianshan

Mountains in the central and western areas first reversed as a whole, and intracontinental fore-land basins began to develop in the mountain fronts in the north and south from the Triassic to the Jurassic. The previously formed back-arc fold belt on the south side of the South Tianshan Mountains thrust the Tarim landmass. The east-wards movement of the Tarim landmass induced the creation of the Altyn strike-slip fault, and the eastwards movement of small landmasses such as Songpan and Alashan between the east and west

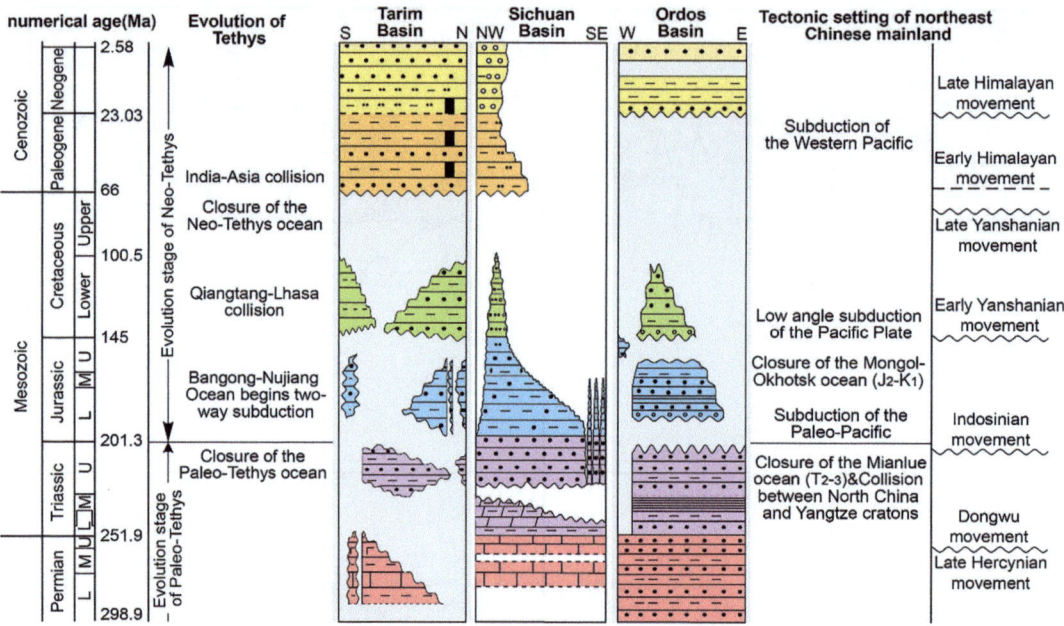

Fig. 4.2 Unconformity distribution and corresponding tectonic events of the three Cratons

suture zones caused the previously formed N–S-trending rift valley between them to reverse, and the compression produced from the west to the east led to the creation of the compresso-shear strike-slip basin at the front of Altyn Mountain and the flexural basin at the front of West Sichuan and Helan Mountain, respectively, with a structural pattern presented as a gentle slope inclined to the orogenic belt (Fig. 4.3).

In the third tectonic deformation stage (K_2–E), from the Late Cretaceous to the Palaeogene, the main body of the Yarlung Zangbo Neo-Tethys ocean basin gradually closed, the Indian plate converged reversely with the Chinese mainland, the Lhasa–Qiangtang landmass continued to be compressed under aggregation, and the Kunlun and Tianshan Mountains again underwent strong compression in the nearly N–S direction. Because the eastwards expansion of the Philippine plate turned clockwise to the east, the Pacific Ocean was forced to retreat to the east and change directions. The stresses acting upon the North and South China plates altered from the previous compression to extension, thus causing the landmass of the Chinese mainland to begin to separate and repel eastwards and form strong differential ascending activity in the west and descending

activity in the east. On both sides of the Tianshan Mountains, basinwards thrust faults and foredeeps developed again. Oblique overthrusts were formed by Kunlun Mountain on the southwest margin of the Tarim Basin. The sedimentary centre migrated to the foredeep basin, and the south of the central uplift zone was strongly tilted to the south. Moreover, the eastwards migration of the landmass triggered strike-slip activity along the Altyn fault, and strike-slip basins developed and were superimposed in the Southeast Tarim Basin. The eastwards movement of the Songpan–Ganzi landmass and the continuous compression and inversion of the South Qinling rift zone led to the successive development of foreland basins in the western and northeastern Sichuan regions. Under the unbalanced compression and repulsion of surrounding landmasses, the Ordos landmass underwent left-lateral movement around the rigid basement and was uplifted as a whole. In addition, extensional fault depression basins developed around the landmass.

In the fourth tectonic deformation stage (N–Q), owing to the collision between the Indian and Asiatic landmasses and the continuous intracontinental subduction, as well as the WNW-trending subduction of the Philippine oceanic

Table 4.1 The prototype basin differences among the three basins in Meso-Cenozoic

Chronostratigraphy			Tarim Basin		Sichuan Basin		Ordos Basin	
			Basin Prototype	Distribution	Basin prototype	Distribution	Basin prototype	Distribution
Cenozoic	Quaternary Neogene		Squeezing foredeep and stike-slip	Kuche-Northern Tarim squeezing foredeep / Southwest Tarim squeezing foredeep / Southeast Tarim stike-slip	–	Integral uplift	–	Integral uplift
	Paleogene		Squeezing foredeep and stike-slip	Kuche-Northern Tarim squeezing foredeep / Southwest Tarim squeezing foredeep / Southeast Tarim stike-slip	Squeezing foredeep	Northern Sichuan / Western Sichuan / Southern Sichuan	–	Integral uplift
Mesozoic	Creatceous	Upper			Squeezing foredeep and Intracontinental depression	Northern Sichuan squeezing foredeep / Western Sichuan squeezing foredeep / Southern Sichuan intracontinental depression	Squeezing foredeep	Whole area
		Lower	Squeezing foredeep	Kuche-Northern/ Southeast/ Southwest Tarim				
	Jurassic	Upper	Squeezing foredeep and Intracontinental depression and Stike-slip	Kuche squeezing foredeep / Southwest Tarim squeezing foredeep / Southeast	Squeezing foredeep and Intracontinental depression	Northern Yangtze squeezing foredeep / Intracontinental depression	Squeezing foredeep and Intracontinental depression	Southwest Ordos squeezing foredeep / Eastern Ordos depression
		Middle		Tarim squeezing foredeep / Northern Tarim intracontinental depression / Wuqia stike-slip				
		Lower						
	Triassic	Upper	Squeezing foredeep	Northern Tarim / Southwest Tarim				

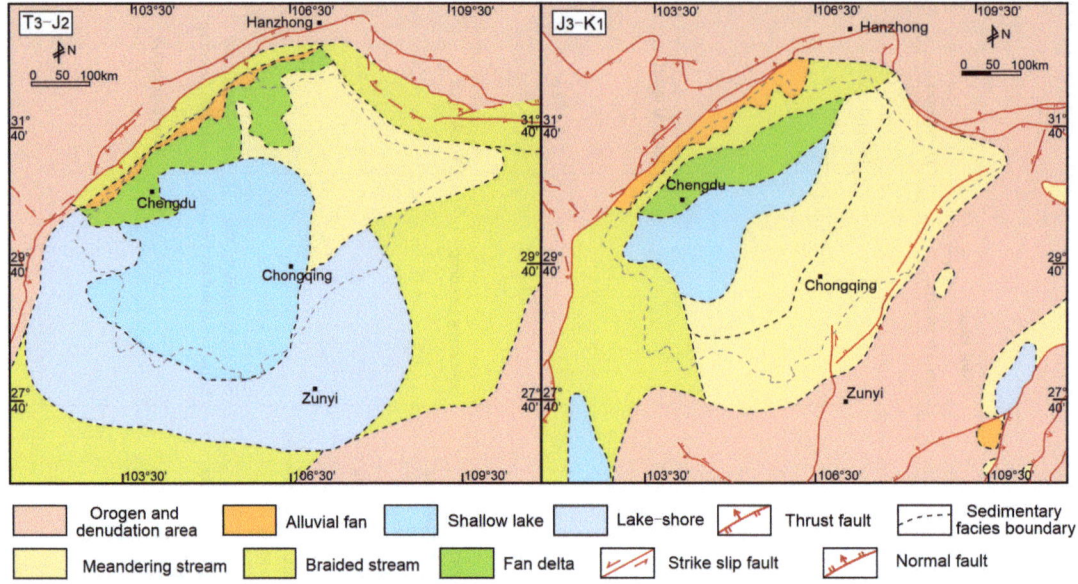

Fig. 4.3 Basin types and depositional distribution of Sichuan Basin in two tectonic stages

crust in the West Pacific, the Chinese landmass was compressed in dual directions. The crust of the Qinghai–Tibet region was strongly uplifted and thickened. These basins are characterized mainly by N–S-trending extensional fault depressions and east–west compression–shear faults. The Tianshan and Kunlun Mountains overthrust continuously into the basin, and the Tarim Basin developed two foredeeps in the north and south, expanding into the basin. The Sichuan and Ordos Basins inherited the previous movement pattern. The Sichuan Basin uplifted largely as a whole and contracted to the southwest of Sichuan; the Ordos Basin developed strike-slip extensional fault depressions around its periphery due to the rotation of the landmass.

4.3 Structural Deformation Types and Mechanisms in Cratons During the Meso-Cenozoic Period

The oceans between continental plates disappeared from the Mesozoic to the Cenozoic. After continental collision, the remote effects of plate activity (Liu et al. 2015; Shi et al. 2015), continental subduction influenced by plate drag (Dong et al. 2016; Shu et al. 2017), and plate interactions caused by material heterogeneity between plates were the main dynamic mechanisms that caused further tectonic deformation at plate edges and inside. Multiple dynamic forces of different natures acted superimposed on the plate boundary and inside from different directions, which not only caused the plate to flex, uplift, and tilt (He et al. 2008) but also induced the formation of complex geological structures under the action of a regional detachment system (Li et al. 2020) and triggered tectonic–thermal events in the basin under the condition that plate subduction disturbed the deep heat flow field. Subsequently, the basic structural features of marine sedimentary basins in the Chinese continental plate fully reformed, and multiple reformation types, namely, deformation in the piedmont, multiple detachment folding, differential subsidence-uplift, intrabasin high-angle strike-slip fault, and tectonic-thermal events, occurred.

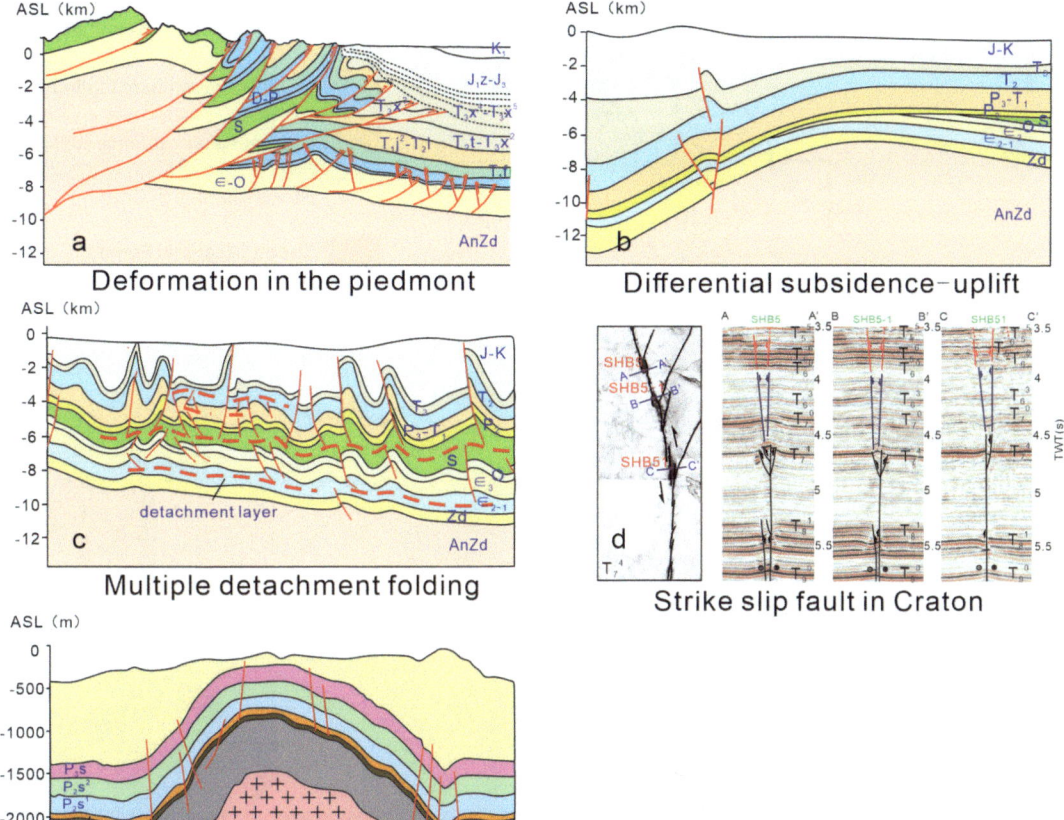

Fig. 4.4 Structure diagram of five reformation types, "J1z" stands for "Ziliujing Formation", "T3x" stands for "Xujiahe Formation", "T2t" stands for "Tianjingshan Formation", "T2l" stands for "Leikoupo Formation", "T1j" stands for "Jialingjiang Formation", "T1f" stands for "Feixianguan Formation", "P3s" stands for "Shiqianfeng Formation", "P2s" stands for "Shihezi Formation", "P1s-P1t" stands for "Shanxi-Taiyuan Formation", "P1b" stands for "Benxi Formation"

4.3.1 Deformation Type and Distribution

4.3.1.1 Basic Structures of Different Deformation Types

The deformation in the piedmont is characterized mainly by strong thrust deformation, generally involving the basement, and the crust is significantly thinner. The different strata and basement thicknesses involved in structural deformation can generally be divided into two basic styles, i.e., "thick-skinned structures" and "thin-skinned structures". The transition area from a "thick-skinned structure" to a "thin-skinned structure" can be further divided into transition zone I and transition zone II (Xu et al. 2016). Intraplatform multiple detachment folding is characterized mainly by the development of multiple sets of detachment layers. The deformation between the upper and lower walls is disharmonious, and the caprock has slipped off. According to the different widths of synclines and anticlines in terms of fold deformation, the fold can be divided into two basic styles, i.e., a trough-like style (wide anticlines–narrow synclines) and a partition style (narrow anticlines–wide synclines). Different subsidence–uplift reformation refers to the simultaneous occurrence of subsidence and

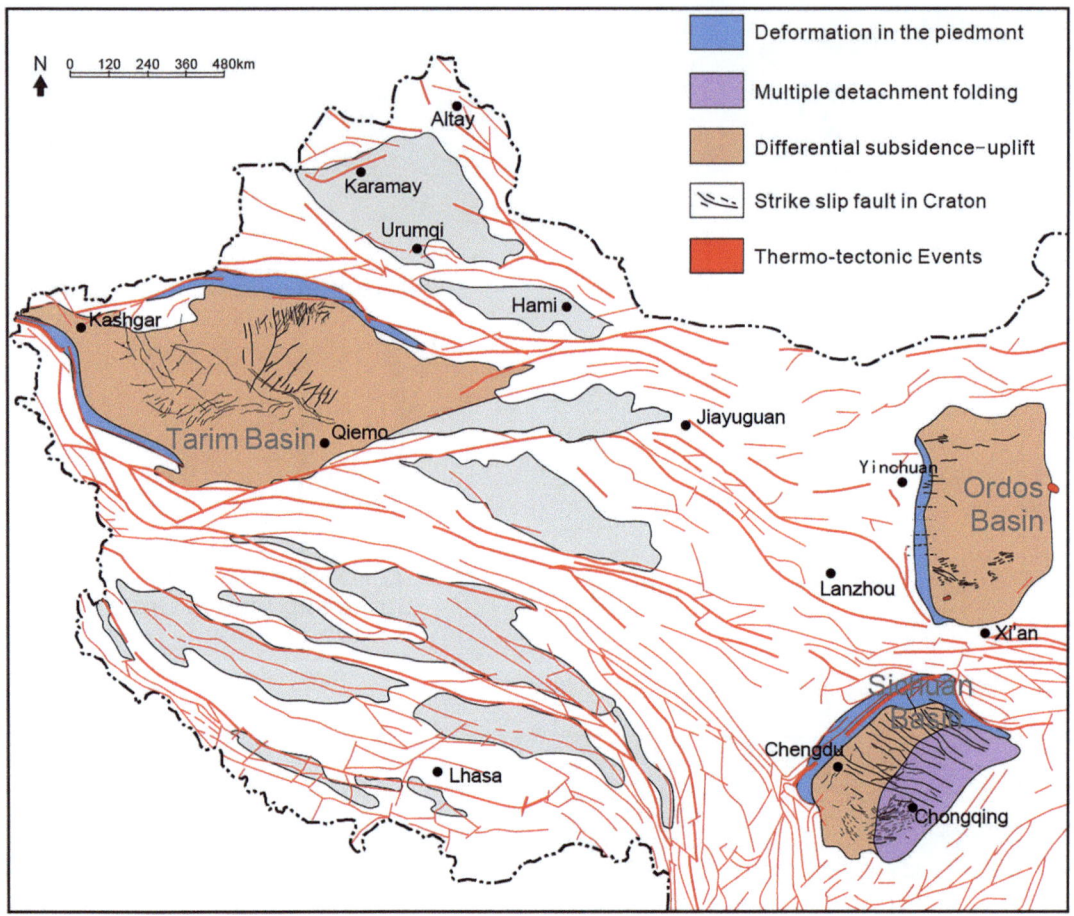

Fig. 4.5 Space distribution of five reformation types

uplift zones in the same basin during the same period. There may be not only continuous subsidence but also continuous uplift in some areas in the same basin, while multistage subsidence and uplift activities occur in other areas. Intrabasinal strike-slip fault reformation refers to the occurrence of multiple sets of strike-slip faults in the basin; the slip distance of the fault is generally small, the strike is stable, the plane shape is straight and segmented, and the deformation is stratified longitudinally. Tectonic–thermal events refer to reformation mainly by volcanic rocks and hydrothermal fluids intruding into the stratum, yet volcanic eruption is short and widespread (Fig. 4.4).

4.3.1.2 Time–Space Distribution of Different Deformation Types

The main transformation types in the Sichuan Basin include piedmont deformation, differential subsidence uplift transformation, multiple detachment folds, and strike-slip transformation. Examples of piedmont deformation include the Longmen Mountain piedmont deformation zone, the Daba Mountain piedmont deformation zone on the basin's north edge, and the Micang Mountain piedmont deformation zone on the basin's northeast edge (Fig. 4.5). Examples of differential subsidence uplift transformation include the northern and western edge of the basin, which is a subsidence area, whereas the interior of the basin is a relatively uplifted area. However, in the subsidence area formed by the tectonic load of the Longmen piedmont on the western edge of the basin, multiple anticlinal structural belts have developed, forming local uplifted structures. Multiple detachment folds are distributed mainly on the eastern margin of the Sichuan Basin (Fig. 4.5), and strike-slip transformation occurs within the craton, such as that in the Gaoshiti Moxi area in the central Sichuan Basin (Fig. 4.5).

The main transformation types in the Tarim Basin include piedmont deformation, differential subsidence uplift transformation, and strike-slip transformation. Examples of piedmont deformation include the north margin of the basin of the South Tianshan Mountains and the south margin of the Kunlun Mountains (Fig. 4.5). Areas of differential subsidence uplift transformation include the Tazhong uplift of the Bachu uplift, where the main subsidence area is located on the north and southwest edges of the basin (Fig. 4.5). Strike-slip transformation occurred within the craton, such as in the Shunbei area of the Tarim Basin.

The main transformation types in the Ordos Basin include piedmont deformation, differential subsidence uplift transformation, strike-slip transformation, and tectonic thermal transformation. Piedmont deformation occurred on Liupan Mountain and Helan Mountain in the western margin of the Ordos Basin (Fig. 4.5). Differential subsidence uplift transformation occurred in the depression on the western edge of the basin, which is superimposed on the ancient uplift in the centre of the basin (Fig. 4.5); the basin is inclined to the west as a whole, the Yimeng uplift is located in the north of the basin, and the Weibei uplift and Weibei depression are developed in the south of the basin. Strike-slip transformation occurred within the craton, such as dense parallel strike-slip faults that developed in the Zhenjing area. Tectonic thermal transformation is distributed mainly in the east of the basin (Fig. 4.5).

Different reformation types clearly differ in structure, but in terms of evolution, these are all the responses of the basin to the activities at the peripheral structure zone; as a result, they are closely related to the evolution of the time–space relationship.

In the Sichuan Basin, the relationships among the three reformation types of thrust reformation in the piedmont, differential subsidence–uplift, and multiple detachment folding are as follows: in the Late Triassic, piedmont deformation occurred in the piedmont of the northwestern margin of the Upper Yangtze platform, subsidence occurred in the central area of the piedmont of the Longmen Mountain, and the central area of the Sichuan Basin was relatively uplifted. In the Middle Jurassic, piedmont deformation occurred in the piedmont of Daba Mountain, which corresponds to the occurrence of subsidence in the north area of the Sichuan Basin, and the uplift amplitude in the centre of the basin was small. From the Late Jurassic to the Early Cretaceous, a trough-like fold zone gradually formed in the southeastern margin of the Sichuan Basin, and multiple detachments induced fold deformation to be layered up and down, exhibiting obvious incongruous characteristics. Piedmont thrust deformation developed in western and northern Sichuan, where the middle and north of the piedmont of Longmen Mountain and Micang and Daba Mountain subsided, and the central and southern Sichuan regions continued to rise. From the Late Cretaceous to the Palaeogene, the piedmonts of Longmen, Micang and Daba Mountain underwent deformation and reformation, the northern Sichuan Basin was uplifted as a whole and experienced denudation, the southern Sichuan area partially subsided, the multiple detachment folding in eastern Sichuan was further strengthened, and a partition-style fold zone developed in eastern Sichuan.

The spatiotemporal relationship of piedmont deformation at the basin margin is closely related to the intrabasin differential subsidence in the Tarim Basin. In the Early–Middle Jurassic, fault depressions developed at the margin of the Tarim Basin, such as the Kuqa fault depression, Southwest Tarim fault depression, East Tarim depression and Southeast Tarim fault depression. In the

Late Jurassic, compression activity intensified at the basin margin, as did deformation in the Kuqa piedmont. In the Early Cretaceous rifting stage, the basin was high in the west and low in the east. The deposition distribution was wider than that in the Jurassic, including the Kuqa depression, East Tarim depression and Southwest Tarim rift. In the Late Cretaceous, compression occurred at the basin margin again, and the basin was uplifted on a large scale. In the Palaeocene–Early Eocene rifting/depression stage, the Kuqa–Awati depression, East Tarim depression and Southwest Tarim rift developed; in the Miocene–Quaternary, compression shear foreland basins developed, the basin margin was thrusted, the flexural structure subsided and filled with very thick deposits, the alluvial fan river sedimentary system was well developed, the transpression and fault activities were strong, and the basin uplift–depression pattern was finalized.

Reformation in the Ordos Basin is relatively simple and mainly manifested as continuous activity in the piedmont zone at the western margin of the basin during the Mesozoic and the continuous accumulation of sediments in the Tianhuan depression, which formed in the corresponding piedmont zone. A typical tectonic–thermal event occurred at the east margin of the basin in the Cretaceous, which was the origin of the Zijinshan rock mass.

4.3.2 Deformation Mechanism

The reformation in basins in central and western China had typical "intracontinental" characteristics in the Meso-Cenozoic, and the parts where reformation took place were far from the plate collision zone. Controlled by the difference in the nature and distribution of cratonic basement materials and the difference in the lithology assemblage of basin sediments and under the interaction of plates, a regional structure with different continental rheological properties formed in the basin, which dominated the internal deformation process of different types of reformation (Fig. 4.6).

Deformation in the piedmont: The geotectonic settings of piedmont deformation in central and western China are the same and fall within the tectonic system around the Qinghai–Tibet Plateau in the Tethys tectonic domain (Jia 2005). Intracontinental piedmont thrust deformation develops far from the plate collision zone and is controlled by the detachment layers at different levels and the geometric form and location of the composite landmass. Composite intracontinental piedmont thrust is a tectonic deformation caused by stepped detachment thrust along the detachment layer at different depths from the orogenic wedge to the basin under compressional stress at the plate margin (Xu et al. 2016). The piedmont deformation is concentrated at the margin of the basin. Owing to the difference in material composition in different basins, the deformation intensity and depth also differ greatly, thus resulting in multiple structural styles of piedmont deformation.

Differential subsidence–uplift: As with piedmont deformation, differential subsidence–uplift in the basin is controlled by crustal flexure, which is caused by the tectonic sedimentary load due to foreland thrust activity at the basin margin. The regional subsidence centre formed by flexure and the corresponding relative structural height (i.e., caused by uplift) causes the local stratum thickening, thinning and denudation. The differential subsidence–uplift in the Mesozoic–Cenozoic in the basins of central and western China is controlled mainly by the splicing of the Chinese continental plate since the Late Triassic and the collision of the Indian Eurasian plate in the Cenozoic. The basement of the Sichuan Basin consists of multiple rock masses with different lithologies, which corresponds to the structural features of multiple uplifts and depressions in the basin. The basement of the Tarim Basin is divided into two rock masses, one in the north and one in the south, and the centre of the basin is a palaeo-uplift. The basement of the Ordos Basin is relatively rigid as a whole, and the deformation of the uplift depression in the basin is characterized mainly by westwards dipping. Therefore, the differential subsidence–uplift pattern in the basin is controlled not only by the activity of peripheral structural belts but also by the lithology of the basin basement. The difference in the plane distribution of different basement properties in the basins of central and

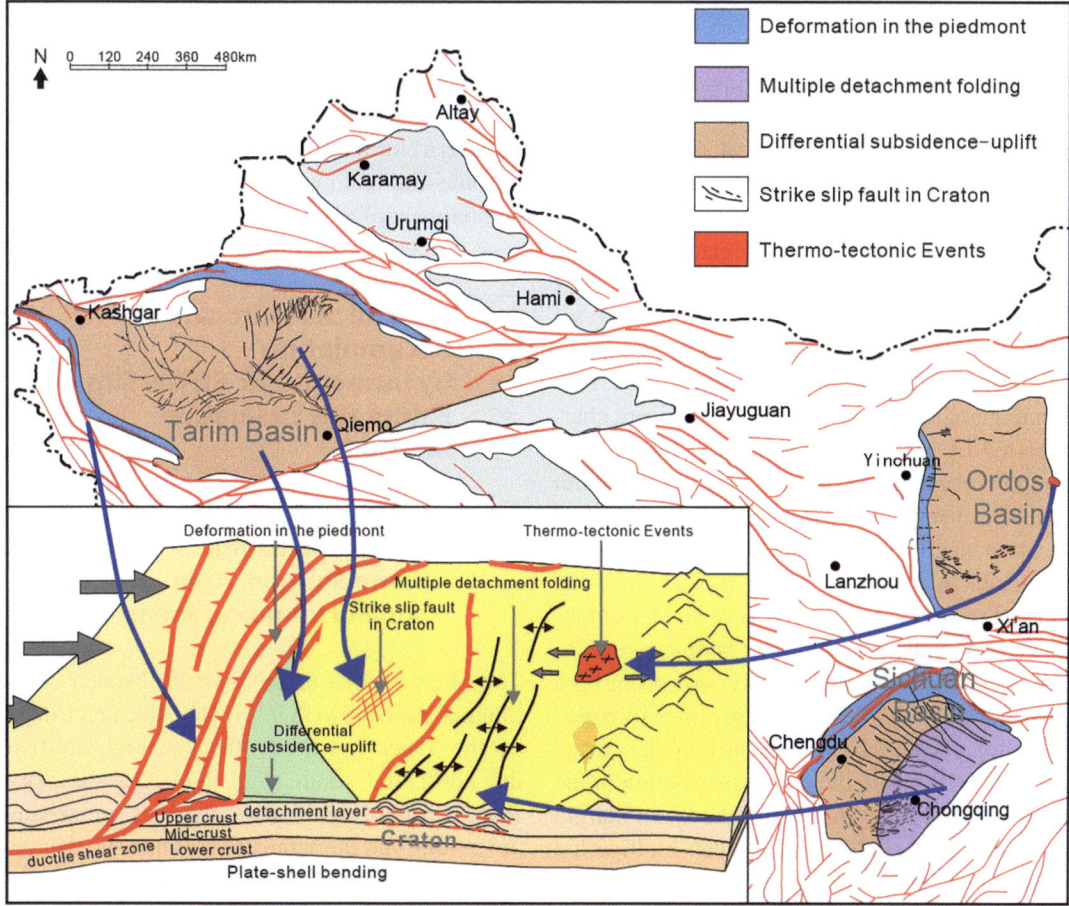

Fig. 4.6 Schematic diagram of the genetic mechanism of five deformations

western China determines the difference in differential subsidence–uplift in the basin.

Multiple detachment folding: The detachment layer plays an important role in influencing and controlling structural deformation styles, especially in areas where thin-skinned thrust structures develop (Scharer et al. 2004; Sherkati et al. 2005; Brogi 2008; Yan et al. 2009). Two or more tectonic deformation systems can be divided in general by taking the detachment layer as the boundary. The nondeformability of the upper detachment layer manifests as a parallel fold geometry, whereas the weak lower detachment layer often manifests as inconsistent folding and piercing deformation. Continuous tectonic pressure may lead to an increase in fold amplitude or a more tightly closed geometry. The geolog-

ical parameters of the detachment layer determine the structural style of the thrust-fold zone. These parameters include the rock mechanical properties of the detachment layer (e.g., compressive strength, tensile strength, shear strength, Young's modulus, Poisson's ratio, internal friction angle), the distribution of the detachment layer and its 3D structural form (e.g., thickness change, relative depth, continuity of plane distribution, space combination), and the rheological behaviour of the detachment layer and the vertical coupling relationship between the detachment layer and the competent nondeformability layer (Li et al. 2020).

Strike-slip fault in the craton: Taking the Tarim Basin as an example, the development of a strike-slip fault is controlled by multistage stresses at the periphery of the Tarim Basin. The Middle

Caledonian is the prototype stage of the development of strike-slip faults in the Tarim Basin. At the end of the Middle Ordovician, affected by the West Kunlun orogeny and under remote compression from the southwest margin of the basin, a NE-trending strike-slip fault prototype formed in the Central Tarim and Shuntuoguole areas. From the Middle Ordovician to the Late Ordovician, under the joint action of the West Kunlun Mountains and Altyn orogeny, the previously formed prototype NE-trending strike-slip fault continued to be active to create the NE-trending main strike-slip faults, which were characterized by simple shear activities. Under joint compression from the two directions of the South Talim Basin and limited by the north boundary of the basin, nearly N–S-trending compressive stress developed in the North Tarim Basin. Since the North Tarim uplift was relatively stable, conjugate shear joints or small-scale conjugate strike-slip faults formed under compression from both the south and north during this period. Moreover, as the stress extends from the basin boundary to the interior of the basin, the stress distribution in the basin is strong on both sides of the basin margin and weak in the abdomen. Under multiperiod orogenesis at the basin margin in the Hercynian, Indosinian–Yanshanian, and Himalayan periods, the development and evolution of strike-slip faults in the Central Tarim and Shuntuoguole areas also show splicing from the north and south sides to the middle abdomen and become the current fault distribution pattern. In short, the creation of strike-slip faults in the Tarim Basin is the result of the multiperiod activities of the NS structural system and the final splicing under noncoaxial compression in the basin.

Tectonic–thermal events: The underlying cause of the occurrence of tectonic–thermal events in the Ordos Basin in the Early Cretaceous lies in intensified deep thermal activity in the basin lithosphere, upwelling of deep asthenosphere materials, and thinning of the lithosphere (Ren et al. 2020). Fundamentally, the deep subduction of the Pacific plate from east to west induced extensive and strong extensional activity in the eastern North China craton in the Early Cretaceous, and a series of metamorphic core complexes, exten-

sional domes, extensional basins, and magmation of mainly intermediate–acidic rocks in planar distribution occurred successively, which corresponds to the peak period when the tectonic system in eastern North China was at its turning point. The tectonic–thermal event in the Ordos Basin is the embodiment of the turn of this tectonic system.

4.4 Control of Hydrocarbon Accumulation by Intracratonic Strike-Slip Faults and Evaluation Methods

4.4.1 Distribution of Intracratonic Strike-Slip Faults in the Shunbei Area and Surrounding Areas

The Shunbei area is geologically located in the Shuntuoguole Lower Uplift, a tectonic unit fringed by the Tabei Uplift and the Tazhong Uplift to the north and south (Fig. 4.7). Over the past few decades, significant advances in seismic processing and the acquisition of new 3D datasets have revealed the existence of a spectacular strike-slip fault system that developed across Shunbei and its surrounding areas (Fig. 4.7). Based on the orientation and distribution of the fault sets, there are at least three representative strike-slip fault systems (Fig. 4.7): (1) The NNE-trending strike-slip fault system consisting of SB1, SB8, SB12 fault sets, etc., (2) the NNW-trending strike-slip fault system consisting of SB7, SB11 and SB13 fault sets, etc., and (3) the SB5 fault and the associated NNE-oriented fault sets (e.g., the SB1 fault).

4.4.2 Controlling Factors and Evaluation Methods for Hydrocarbon Accumulation

Previous studies have demonstrated that the segmentation structural style, the structure of the salt layer located between the Lower Cambrian source rocks and Middle–Lower Ordovician

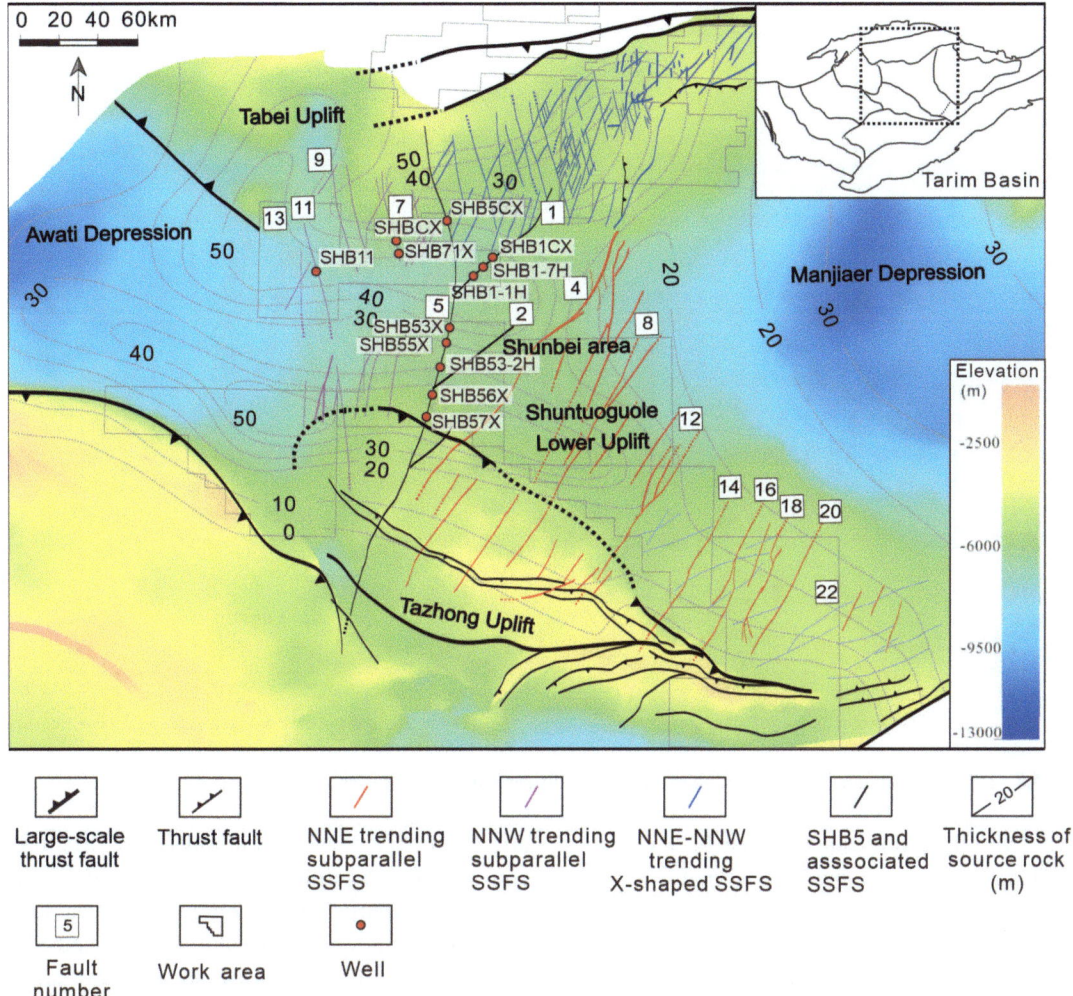

Fig. 4.7 Distribution of intracratonic strike-slip faults in Shunbei area and the surroundings. In the legend, "SSFS" stands for "strike-slip fault system"

carbonates, and the movement activities of strike-slip faults during the hydrocarbon accumulation period are key factors controlling hydrocarbon abundance in strike-slip faults (Deng et al. 2022; Bian et al. 2022). In this section, we present a summary of these controlling factors and introduce the discoveries of SB4 and SB8 faults, which are rich in hydrocarbons.

4.4.2.1 Segmentation of Structural Styles

The strike-slip faults in Shunbei and adjacent areas display subvertical segments at depth, and en-echelon normal faults are relatively shallow (Deng et al. 2019). The underlying subvertical segments, rooted through surface $T_9{}^0$ and extending upwards across surface $T_7{}^4$, commonly display segmented geometry on surface $T_7{}^4$. Three types of typical structural styles are mainly identified on the top

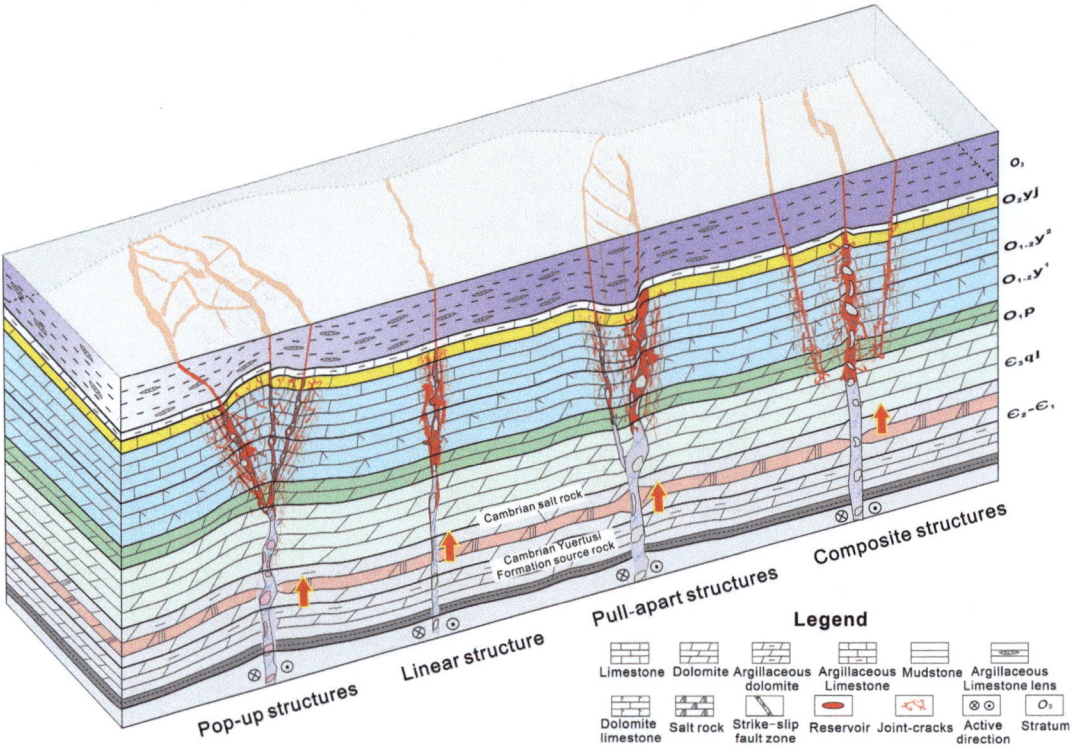

Fig. 4.8 Different structural styles and the associated hydrocarbon accumulation

of the Middle Ordovician (surface T_7^4), i.e., pull-apart structures and push-up structures in the stepovers and pure strike-slip segments with a relatively straight, throughgoing trace (Fig. 4.8). In addition to these three structural styles, a composite structural style occurs in the southern segment of the SB5 fault (Deng et al. 2021a, b). The composite structural style is characterized by the aforementioned three segmentation styles overprinted by grabens on both sides of the underlying subvertical segments (Fig. 4.8). The underlying subvertical segments in all four structural styles link source rocks at depth with the reservoirs in the carbonates beneath surface T_7^4. Strike-slip faults play important roles in hydrocarbon accumulation in the Ordovician carbonates. However, drilling reveals that different structural styles have significantly different influences on hydrocarbon accumulation.

Compared with pure strike-slip segments, larger-scale reservoirs are developed in pull-apart structures and push-up structures (Deng et al. 2022). Flower structures with multiple fault planes

are commonly developed in pull-apart structures and push-up structures, whereas pure strike-slip segments display only one single subvertical fault plane. Exploration practices in the SB1 and SB5 faults suggest that hydrocarbons are more abundant in the pull-apart structures and push-up structures than in the strike-slip segments (Deng et al. 2022).

The reservoir connectivity differs between pull-apart structures and push-up structures. Previous studies suggest that dilational breccia with high porosity and permeability can form in pull-apart structures. Drilling practices in pull-apart structures also reveal good reservoir connectivity and hydrocarbon abundance (Deng et al. 2022). Compared with pull-apart structures, which formed under internal tensile stress, push-up structures formed under internal compressive stress. In this case, the internal structures are characterized by a lesser degree of opening. Drilling practices suggest that push-up structures have poor reservoir connectivity (Deng et al. 2022). In the southern segment of the SB5 fault

(Fig. 4.7), the underlying subvertical segments are overprinted by grabens and display great structural complexity (Deng et al. 2021a, b; Yun and Deng 2022). Because of the layered structure, the grabens are not directly connected to the source rocks (Fig. 4.8). Accordingly, the reservoir size associated with the graben fault planes is smaller than that associated with the subvertical segments.

4.4.2.2 Strike-Slip Salt Tectonics and Their Influence on Hydrocarbon Migration

In the Shunbei area and its surroundings, the strike-slip faults are rooted through surface T90 and extend upwards across surface T_7^4. In map view, the strike-slip faults commonly display segmented geometry on surface T_7^4 (Deng et al. 2019). Push-up structures are developed in the right-stepping stepovers, and pull-apart structures are developed in the left-stepping stepovers. In the stepover regions, transpressional and transtensional stress states occurred. The gypsum salt rocks in the Middle Cambrian strata (above surface T_8^3) were widely distributed in the evaporative platform in the Shunbei area (Fig. 4.9). The stress response patterns of the gypsum salt rocks to transpression and transtension in the underlying subvertical segments of the strike-slip faults contrast those in the upper carbonate layers (Bian et al. 2022). In rigid carbonate strata, the transpression exhibited a positive flower structure, whereas the transtension was characterized by a negative flower structure. However, in the gypsum salt rocks, the transpressional surface was concave downwards, whereas the transtensional surface was convex upwards. During the transpression, the top of the upper carbonate (surface T_7^4) was uplifted, whereas the top of the gypsum salt rocks (surface T_8^3) was depressed due to salt escape. The salt migrated from the high-stress area in the fault zone to the low-stress area, resulting in the thinning of the Middle Cambrian strata. Conversely, during transtension, the top of the upper carbonate (surface T_7^4) was depressed, whereas the top of the gypsum salt rocks (surface T_8^3) was uplifted due to salt diapirs. The salt also tended to migrate from the high-stress area to the low-stress area in the fault zone, resulting in the

thickening of the Middle Cambrian strata. During a pure strike-slip fault, the salt thickness remained stable, and the vertical movements in both the upper carbonate and the gypsum salt rocks were minimal (Bian et al. 2022).

The gypsum salt rocks in the Middle Cambrian strata strongly influence petroleum migration. In the SB5 fault, the wells drilled in the transpressional segments were characterized by high oil production. The oil production in pure strike-slip segments and transtensional segments is relatively low (Bian et al. 2022). In the seismic sections, the gypsum salt rocks are characterized by significant thinning in the transpressional segments, whereas the gypsum salt rocks are significantly thickened in the transtensional segments. The fault connectivity in the transpressional segments is greater than that in the transtensional segments for hydrocarbon migration.

4.4.2.3 Influence of Strike-Slip Fault Activity on Hydrocarbon Accumulation

The evolution of the strike-slip fault system in the Shunbei area and surrounding areas can be divided into at least three stages due to variations in palaeostress environments. The first stage is the III episode of the Middle Caledonian stage (Qiu et al. 2019; Yun and Deng 2022). At this stage, the NNE-trending and NNW-trending strike-slip faults exhibit sinistral and dextral strike-slip activity, respectively. In the cross section, pressure ridges occur along the strike-slip segment, indicating a transpressional stress environment. The second stage is the Late Caledonian period. At this stage, the strike-slip segments at depth were reactivated, forming right-stepping echelon normal faults over the NNE-trending strike-slip faults and left-stepping echelon normal faults over the NNW-trending strike-slip faults, indicating sinistral and dextral strike-slip activities, respectively (Figs. 4.9 and 4.10). Because most of the echelon fault angles are less than 45°, the strike-slip faults were active in a transtensional stress environment (Deng et al. 2019). The third stage is the middle–late Hercynian stage. During this stage, the NNE-trending strike-slip faults were reactivated again, whereas the NNW-trending faults remained

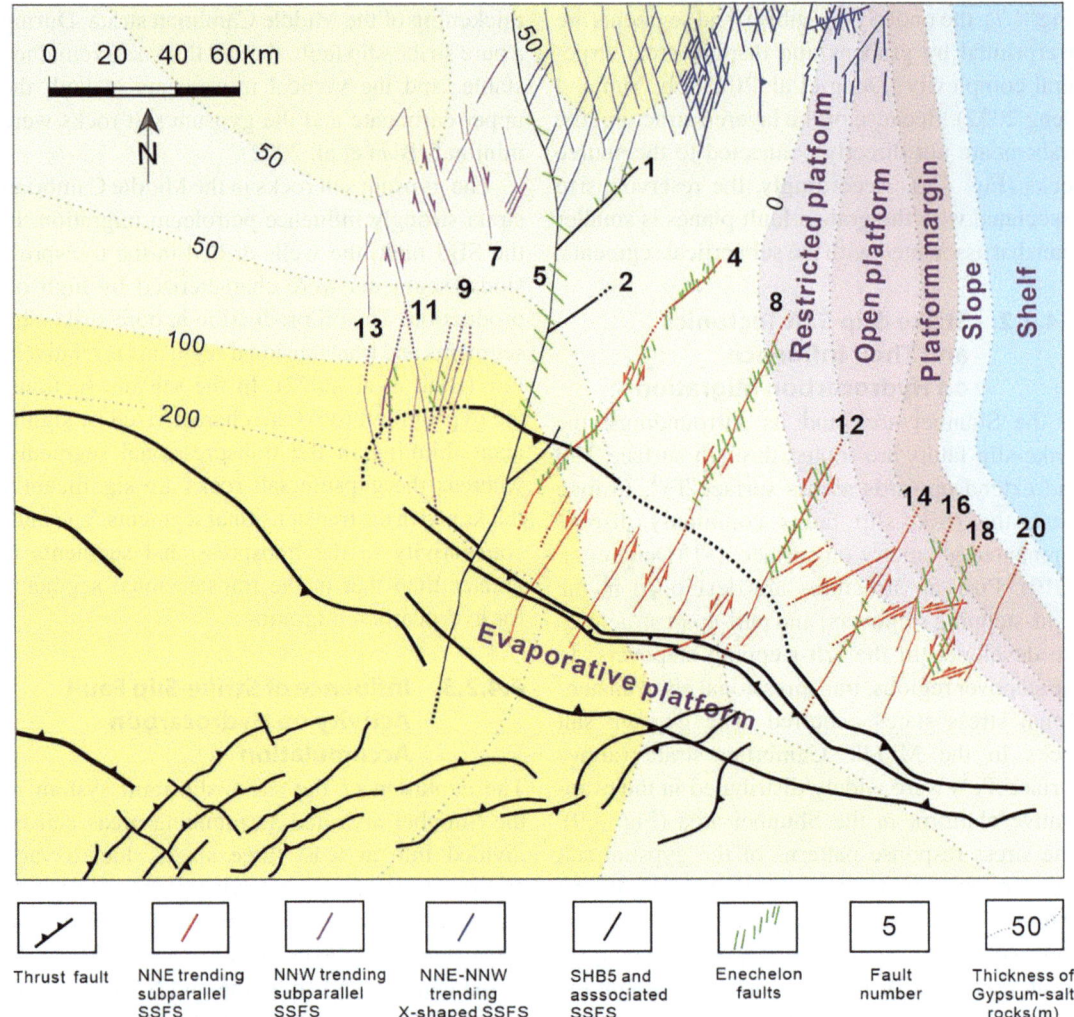

Fig. 4.9 Distribution characteristics of the Middle Cambrian gypsum-salt rocks across Shunbei area and the surroundings. The strike-slip faults overprinted with en- echelon normal faults in Permian strata are presented, indicating fault activities in late Hercynian period

inactive. The echelon normal faults formed in this stage developed in the strata above the $T_6{}^3$ interface and broke through the $T_5{}^6$ interface, some of which may have broken into the Permian igneous rock strata (Qiu et al. 2022). The major oil accumulation period in the Shunbei area is likely the late Hercynian (Qi 2020; Yang et al. 2022). Because the NNE-trending strike-slip faults were reactivated in the late Hercynian, whereas the NNW-trending strike-slip faults remained inactive, it is estimated that the hydrocarbon abundance is higher in the former than in the latter.

4.4.3 The Discovery of SB4 and SB8 Faults with High Degrees of Hydrocarbon Abundance

Through the structural characterization of strike-slip faults and the analyses of the factors controlling hydrocarbon accumulation as described above, it is inferred that the exploration potential of the SB4 and SB8 faults was high because of the thinning of salt layers and strong activity during the Hercynian period. Exploration practices have later proven that faults SB4 and SB8 are rich in hydrocarbons.

On March 11, 2021, the first high-production well, SHB42X, deployed in the Shunbei No. 4 fault, obtained 1000 tons (oil and gas equivalent) of oil and gas flows per day (Fig. 4.10). Later, six high-production wells (SHB42X, SHB41X, SHB44X, SHB43X, SHB45X and SHB46X) were drilled (Fig. 4.10). Among these wells, the daily oil and gas equivalent of SHB43X is the highest. Under the condition of a 12 mm nozzle, the daily crude oil output is 888 m^3, and the daily natural gas output is 61.51×10^3 m^3; the daily crude oil equivalent can reach 1188.1 t. After the discovery of the SB4 fault, well SHB8X deployed on the SB8 fault zone was discovered, with the daily crude oil equivalent reaching 1126 t.

At present, the whole SB4 fault has been put into production. Among the high-production wells discovered thus far, three wells are located in the pull-apart structure, two are distributed in the push-up structure, and one is distributed in the pure strike-slip segment (Fig. 4.10). The 1000 t exploratory well SHB8X in the No. 8 fault zone is distributed in the pull-apart segment. The SB4 and SB8 fault zones are characterized by strong fault activity in the late stage, and en-echelon normal faults with considerable lengths can be found on surface T_7^0 (Fig. 4.10). Current exploration practices suggest that fault SB4 is rich in hydrocarbons as a whole fault zone. Integrated with the comprehensive evaluation of oil and gas accumulation conditions based on the structural characterization of strike-slip faults presented above, many more

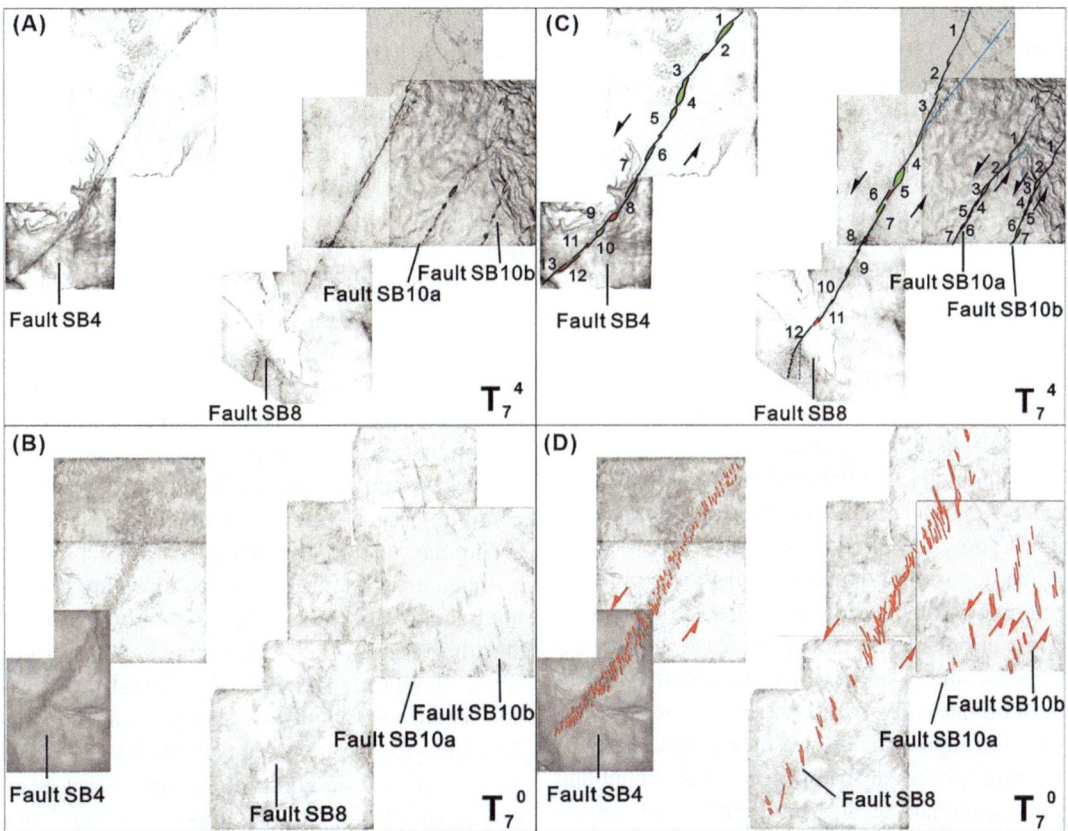

Fig. 4.10 Uninterpreted coherence slice mosaics (**a, b**) and interpreted coherence slice mosaics (**c, d**) of surfaces T74 and T70, respectively, of the same area in which faults SB4, SB8, SB6 and SB10 occur (Fig. 4.7). The segments of faults SB4, SB8, SB6 and SB10 are labeled. In (**d**), the en echelon normal faults are colored in red. High-production wells deployed on faults SB4 and SB8 are labeled

high-production wells are expected to be discovered in the SB8 fault zone and faults with similar characteristics located nearby.

4.5 Influence of Deformation Type on Petroleum Occurrence in Deep Marine Strata

4.5.1 Characteristics and Main Controlling Factors of Deep Marine Oil and Gas Accumulation

4.5.1.1 Deep Oil and Gas Accumulation Conditions in Central and Western China

In addition to the splitting–accretion–convergence cycle of the global supercontinent, the three Chinese cratons of Tarim, Yangtze, and North China also underwent a cyclic evolution of separation and convergence of palaeoplates and the growth and disappearance of palaeo-oceanic crust. The Tarim, Sichuan, and Ordos marine basins are all multicycle superimposed basins developed based on the aforementioned three Precambrian small cratons (Jia et al. 2007). Corresponding to the global supercontinent cycle, it is believed that since the Proterozoic, the three cratonic marine basins have generally undergone four geological history stages, namely, the proto-China landmass, the Palaeo-China landmass, the Palaeo-Asian landmass and the Neo-Asian landmass. The first three stages all comprise the marine deposit buildup stage. The corresponding prototype basins and their parallel superposition combination are developed in each geological history evolution stage, which controls the orderly development of hydrocarbon source rocks, reservoirs, and caprocks, presents the characteristics of their multiperiod superimposed development and reformation, and creates multiple sets of marine accumulation assemblages (He et al. 2022).

In the creation stage of the "Proto-China landmass" (i.e., pre-Sinian), basins developed mainly at the margins of the three cratons and were characterized mainly by continental margin rift, discrete continental margin depression, and a back-arc spreading basin that developed in an extensional environment. In the creation stage of the "Palaeo-Chinese landmass" (i.e., Sinian–Silurian), the three palaeocratons of Tarim, North China, and Yangtze were developed mainly with basin prototypes such as rifts and depressions (intraplatform or platform margin) and were deposited mainly with gypsum salt and carbonate rocks. In the creation stage of the "Palaeo-Asian landmass" (i.e., the Devonian and Middle Triassic), the whole area was in a discrete tectonic environment, and discrete continental margin depressions developed juxtaposed with rifts, which controlled the development and distribution of multiple types of source and reservoir conditions. During this period, the sedimentary filling in the basin changed from mainly marine facies to mainly marine–continental transitional facies clastic rocks.

The basin prototype controls the development of basic hydrocarbon accumulation conditions such as source rocks and reservoirs (Xu et al. 2022). In the rift sedimentary system, the source rocks are mainly deep-water shelf facies sediments and thick, dark shale. The extension range of the rift controls the distribution of source rocks. If carbonate platforms are developed on both sides of the rift, excellent-quality reef beach facies reservoirs may be created. If it is a palaeolandmass, sandbodies of rivers, deltas, and shores are often deposited at its margin. Both of the above cases may form excellent source–reservoir configurations of "in situ source rock and adjacent accumulation". In the intraplatform depression sedimentary system, the source rocks are mainly mud shale, siliceous shale, and carbonate rock, and the reservoirs can be composed of various types of intraplatform shoal, dolomitized granular shoal, microbial rock, gypsum salt rock and dissolution dolomite, which are often interbedded alternately to form good source–reservoir assemblages of "lower source upper reservoir" and "in situ source and accumulation". The passive continental margin sedimentary system can be divided into steep slope and gentle slope types. The lithology of the source rocks in the steep slope system is mainly carbonaceous shale and marlstone, which are distributed in belts along the periphery of the platform margin. The inner belt of the platform margin often develops with platform margin and intraplatform shoal facies deposits. These reservoirs can develop into various types of reservoirs of different sizes to constitute a source–reservoir configuration of "in situ source rock and adjacent accumulation" or "lower source upper reservoir" after later exposure, karstification, and fault reformation by faults. In the gentle slope system, the source rocks are distributed in the basin facies and the outer slope facies, mainly the marine carbonaceous shale and siliceous shale. The reservoirs include dolomite, grain beach, and microbial mound beach on the inner gentle slope and the middle gentle slope, which form a good source–reservoir configuration of "lower source upper reservoir" or "source–reservoir in one" with the source rocks.

Therefore, the basin prototype formed in the process of basin creation and evolution is the most direct controlling factor for the creation and distribution of source rocks and reservoirs and the creation of accumulation assemblages.

4.5.1.2 Main Characteristics and Controlling Factors of Hydrocarbon Accumulation and Enrichment

The hydrocarbons in the marine carbonate rocks of the three craton basins underwent complex multiperiod accumulation processes, and the hydrocarbon accumulation period was generally early. In the Mesozoic and Cenozoic, the basins underwent strong reformations of multiple tectonic movements. Most of them have become basin residues. The previously formed multilevel hydrocarbon accumulation system experienced strong denudation after uplift and has undergone reorganization and reaccumulation; thus, a coexistence pattern of primary and secondary oil and gas reservoirs has formed (Ma et al. 2017; Jin 2012).

The hydrocarbon accumulation modes and characteristics of the three craton basins are similar. The following accumulation modes generally occur: "source from shelf and accumulation at the margin, reservoir near source laterally", "lower source upper reservoir and fault channelling", "source reservoir in one, in situ accumulation in source", "source in new formation and accumulation in old formation, source–reservoir connection, structural adjustment and enrichment control". The above four types of hydrocarbon enrichment modes are found in marine carbonate rocks in southern China (Sichuan). An enrichment control mode is also found in the Shunbei ultradeep fault–karst strike-slip fault zone, which is characterized by a "multistage hydrocarbon supply from Cambrian, deep fault–karst reservoirs, in situ vertical channelling, mainly late accumulation and strike-slip fault-controlled enrichment". In general, the distribution of oil and gas reservoirs is controlled by high-quality source rock zones, and the enrichment of oil and gas is controlled by high-quality carbonate rock reservoirs, whereas the migration, accumulation, and preservation of oil and gas are controlled by the fault system and differential subsidence–uplift patterns formed by tectonic movement. In short, an enrichment law with the characteristics of "near-source accumulation, high reservoir enrichment, dynamic adjustment, and continuous preservation" exists.

4.5.2 Control of Mesozoic–Cenozoic Tectonic Movement on Deep Hydrocarbon Accumulation

4.5.2.1 Overall Tectonic Characteristics

Owing to the special location of the Chinese mainland, the basins in China were squeezed mainly by the expanding edge of the Mesozoic Ocean in the west, subducted by the Pacific Plate in the east, and blocked by the Palaeozoic orogenic belt in the north during the Mesozoic–Cenozoic. Under these geodynamic conditions, basins are formed when continental crust is compressed, but their manifestations are diverse (Zhu et al. 1983). During this process, the West China continent experienced (1) subduction of Palaeo-Tethyan and Early Tethyan oceanic crust from the end of the Triassic to the end of the Jurassic; (2) the final closure of the Tethys Ocean from the end of the Cretaceous to the Palaeogene; and (3) the collision of the Indian plate in the Neogene.

Tectonic deformation during the three stages strongly affects the early-stage basin morphology, especially the occurrence of various materials in the basin generated by sedimentary filling and evolution. The main structural reformation types include deformation in the piedmont, intraplatform detachment folding, differential subsidence–uplift, intrabasinal strike-slip fault, and tectonic–thermal events. These reformation types affect the oil and gas reservoirs formed in the early stage of the basin and their accumulation conditions to some extent, thus changing the state of oil and gas occurrence, resulting in the remigration, accumulation, enrichment, and depletion of oil and gas. Different types of reformation have different effects on oil and gas. Research on the control of these structural deformations on oil and gas accumulation is important for evaluating deep formations, especially favourable exploration zones in the marine facies field of the superimposed basin.

4.5.2.2 Control of Differential Uplift–subsidence on Hydrocarbon Accumulation

(1) Typical examples of differential subsidence–uplift controlling fluid migration and accumulation conditions

Taking the Kaijiang uplift zone in the Sichuan Basin as an example, the evolution of the Kaijiang uplift zone has experienced four stages: the Jialingjiang Formation sedimentary period, which is the prototype stage; the Leikoupo Formation sedimentary period, which is the development stage; the Xujiahe Formation sedimentary period, which is the decline stage; and the Yanshanian–Himalayan period, which is the reformation stage. Two sets of source rocks, two sets of reservoirs and two sets of caprocks have mainly developed in this area, from which two sets of accumulation assemblages have formed. The accumulation assemblages are the Huanglong Formation accumulation assemblage and the Changxing–Feixian-

guan Formation accumulation assemblage. High-quality source rocks underlie the Huanglong Formation. Reservoirs are well developed in eastern Sichuan, and the source reservoirs are well configured. Bitumen fillings are found in the Huanglong Formation, indicating that palaeoreservoirs once existed. The Changxing–Feixianguan Formation is developed with shoal facies reservoirs, the underlying Longtan Formation is rich in hydrocarbon sources, and the overlying Jialingjiang Formation has a gypsum layer, where the capping conditions are good, and the source–reservoir–cap assemblage is well configured.

The uplifting of the Kaijiang area provides structural conditions for the creation of large-scale stratigraphic–palaeo structural composite traps at the upwards dip boundary of the underlying Carboniferous Huanglong Formation, which control the secondary migration and early accumulation of oil and gas in the reservoir. There is a good configuration relationship between the creation of palaeotraps and the massive generation of oil and gas, and large palaeogas reservoirs have developed around the palaeo-uplift.

The uplift of the Kaijiang area controlled the secondary migration and the early oil and gas accumulation. The creation of the palaeo-uplift led to the uplift of the strata, the upwards dip of the reservoir of the Carboniferous Huanglong Formation, and the migration and accumulation of oil and gas towards the palaeo-uplift. The late tectonic movement destroyed the palaeoreservoir, and the oil and gas were redistributed to form large-stratum palaeostructure composite oil and gas reservoirs.

(2) Accumulation characteristics and reservoir control mode of differential subsidence–uplift

Differential subsidence-uplift in a basin is often characterized by multiperiod and multistage development. The spatiotemporal relationship between the reformation process of this structure and the accumulation conditions of deep marine strata (Fig. 4.11a) determines the final enrichment zone of deep marine oil and gas. The characteristics of oil and gas accumulation in differential subsidence–uplift areas should consider not only the

c. After subsidence and uplift deformation

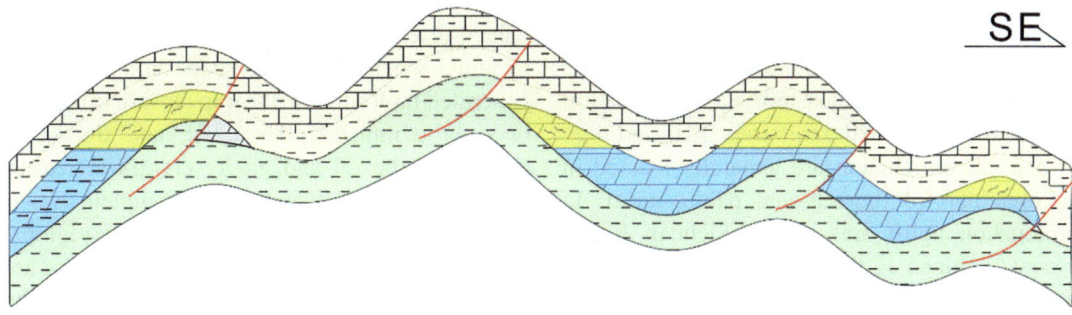

b. During subsidence and uplift deformation

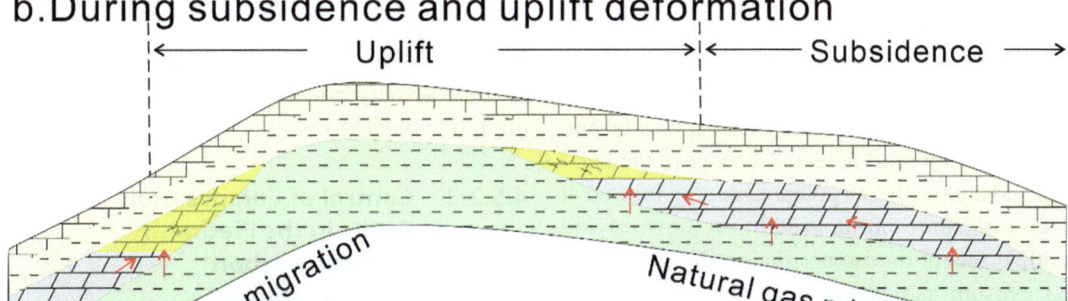

a. Before subsidence and uplift deformation

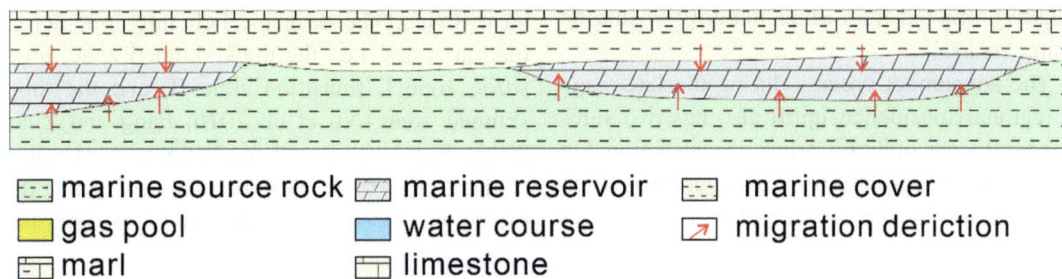

▦ marine source rock	▨ marine reservoir	▦ marine cover
▨ gas pool	▨ water course	⬈ migration deriction
▦ marl	▦ limestone	

Fig. 4.11 Reservoir control **mode** of differential subsidence-uplift

spatiotemporal relationship and the nature of the subsidence–uplift reformation but also the accumulation process in which the source rock matures and oil and gas migrate to deep marine strata.

Previous studies have shown that against the background of Mesozoic basin reformation, the control of regional subsidence–uplift on oil and gas accumulation conditions can be attributed to the fact that when regional subsidence–uplift occurs, it mainly controls the direction of oil and gas accumulation; the postadjustment process of regional subsidence–uplift controls oil and gas redistribution.

Regional subsidence–uplift has dual effects on hydrocarbon accumulation. The denudation effect caused by crustal uplift results in the loss or destruction of some local deep marine sources, reservoirs, and cap rocks; however, the local structural highlands formed during uplift (Fig. 4.11b) can also be used as regional fluid migration and accumulation zones to form early oil and gas traps. As noted in the discovered gas reservoirs in southeastern Sichuan, the oil and gas reservoirs with large trap areas and high oil and gas enrichment are generally located in the high part of the regional palaeo-uplift. In the Luzhou uplift area of the Sichuan Basin, a karst highland is formed in the high part of the uplift, which becomes an area where vertical dissolution pores and fractures develop. Through vertical microfracture migration, lateral palaeo-fault communication and microdialysis in pore throats, large amounts of hydrocarbons gradually accumulate from the marginal depression and slope area with high potential energy to the karst highland area with low potential energy located at the core of palaeo-uplift. In addition, the overlying multiple sets of mudstone, shale, silty mudstone, and gypsum salt layers function as seals, thus forming a distribution zone of oil and gas reservoirs. Notably, the rapid deposition and continuous burial of the overlying clastic sediments in the subsidence area during the accumulation period deepened the burial depth of the underlying strata and warmed the strata rapidly, which greatly promoted the further maturity of marine source rocks.

The adjustment process during the late stage of regional subsidence–uplift also has two sides to the influence of hydrocarbon accumulation. On the one hand, regional subsidence–uplift is often accompanied by the activity of early basement faults, resulting in a series of associated faults, which have destructive effects on preexisting palaeo-reservoirs. Some reservoirs are reformed by faults; when oil and gas escape along the newly generated fault surface, the trap closure is reduced or even completely destroyed, and the water content of the oil and gas reservoirs increases, or the spaces may even become complete "water reservoirs". On the other hand, owing to regional compression activities, the newly formed traps are mainly structural (Fig. 4.11c). Furthermore, the microfractures generated in areas with concentrated local tectonic stress create new reservoir spaces, and a series of local faults formed by tectonic activity can also become new oil and gas migration channels.

4.5.2.3 Control of Compression Faults in the Piedmont on Hydrocarbon Accumulation

(1) Control of compression faults in the piedmont on hydrocarbon migration and key accumulation conditions

In recent years, we have elucidated the piedmont thrust structure in central and western China and have divided the orogenic belt into four deformation zones, i.e., the thick-skinned zone, transition zone I, transition zone II, and the thin-skinned zone (Xu et al. 2016; Fang and Zhao 2019) (Fig. 4.12). Based on a better understanding of the zoning of the deformation structure in the piedmont, we focus on the study of deforma-

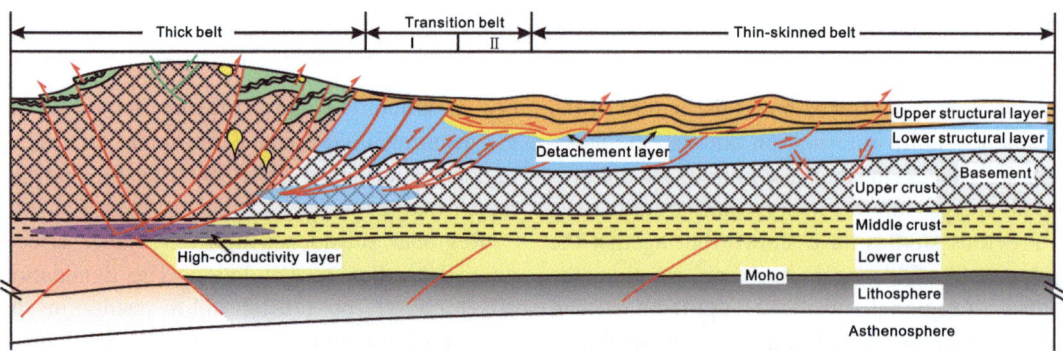

Fig. 4.12 The structural belts within the Piedmont in central and western China. Modified from Xu (2019)

tion characteristics and hydrocarbon control in the Longmen and Micang piedmonts at the periphery of the Sichuan Basin and obtain the following conclusions.

① Controlling effect of thrust deformation of the "separated layer and imbricated structure" on oil and gas in the Longmen Piedmont

The Longmen piedmont features all four deformation zones, namely, thick-skinned, transition I, transition II, and thin-skinned zones (Fig. 4.13). The thick-skinned zone is heavily deformed, and the Beichuan–Yingxiu boundary fault detaches away from the deep middle–upper crust (the detachment depth is 20–40 km). In transition zone I, the main faults of Huanglianqiao and Xiaoba detach away near the basement surface and merge into the Beichuan–Yingxiu deep fault in the direction of the orogenic belt, resulting in an inverted "γ" deformation pattern (Fig. 4.13). Moreover, the detachment layers of the middle and shallow strata in the caprock control the development of the Leiguzhen fault, Nanmuyuan fault, and secondary faults in transition zone II, and they also exhibit an inverted "γ" deformation pattern with their respective main faults (Fig. 4.13). Therefore, the Longmen piedmont is affected by different levels and types of detachment layers from its back edge to its front edge and from deep to shallow,

reflecting the characteristics of thrust deformation of "separated layer and imbricated structure" at different levels (Deng et al. 2021a, b).

Under the control of the "separated layer and imbricated structure" thrust deformation mode, the upper and lower walls of the thrust nappe fault in the Longmen Piedmont show significant structural shortening and deformation differences, which control the different tectonic-fluid activities on both sides and form two relatively independent tectonic–fluid systems on the upper and lower sides. In general, the strong shortening deformation and uplift of the upper wall led to large-scale fluid discharge from the orogenic belt to the basin and from deep to shallow, after which the oil and gas migrated, accumulated, and dissipated, whereas the weak shortening and uplift of the lower wall caused the fluid to flow mainly between structural layers, and the oil and gas accumulated and formed reservoirs near the hydrocarbon source (Fig. 4.13).

② Control effect of the "detachment–reformation" deformation of the Micang piedmont on hydrocarbon migration and accumulation

Thick-skinned, transition zone II, and thin-skinned zones are present in the Micang piedmont; transition zone I is missing (Fig. 4.14). Based on field geological outcrop profiles, thermal

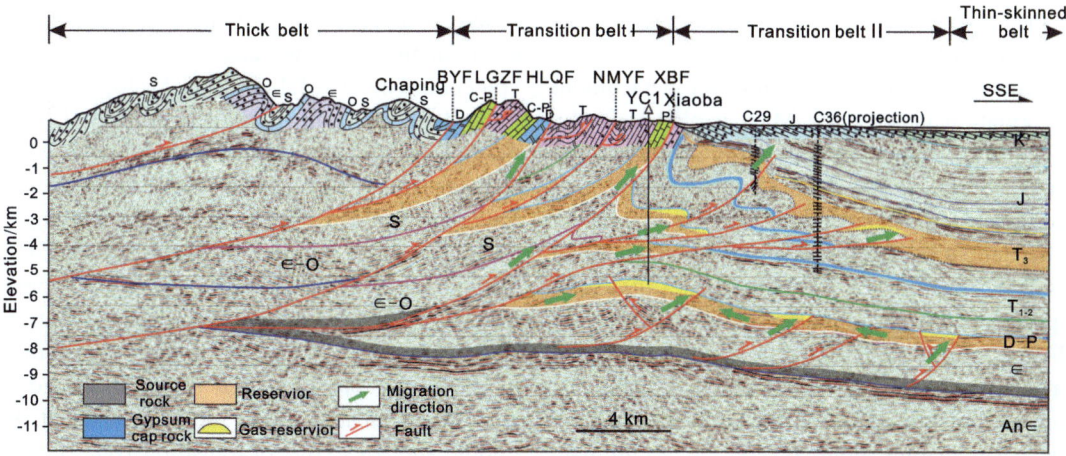

Fig. 4.13 "Separated layer and imbricated structure" and hydrocarbon migration and accumulation in Longmen piedmont

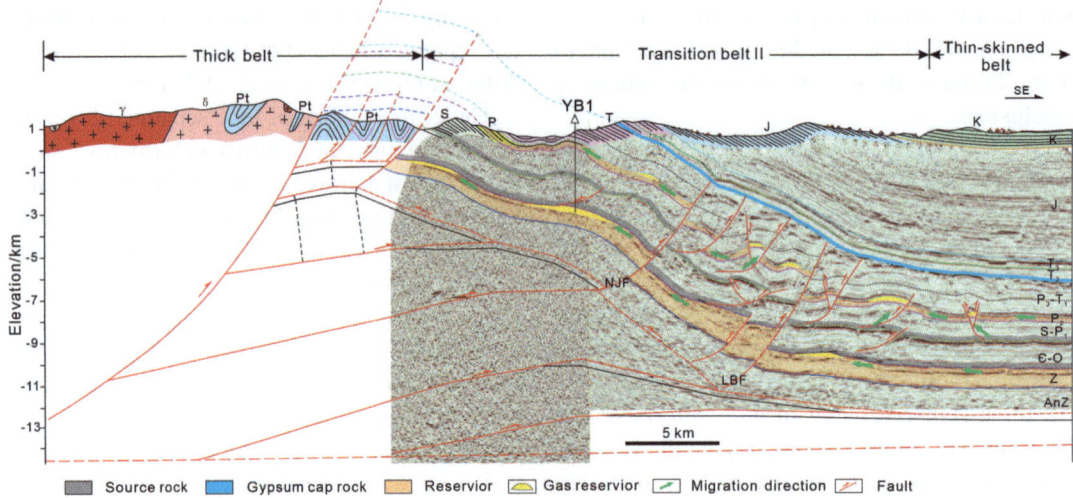

Fig. 4.14 "Detachment-reformation" thrust structure and hydrocarbon migration-accumulation in Micang piedmont

chronology analysis, fine structure analysis, and seismic profile interpretation, the upper and lower layers of the Micang piedmont are not characterized by "detachment–reformation" deformation (Fig. 4.14), i.e., the Micang piedmont is characterized by multilayer detachment deformation, interlayer detachment folds and well-developed faults in its early stage (corresponding to the middle and late Indochina period), and the deformation is not coordinated, whereas in its late stage (corresponding to the middle Yanshanian period), the basement structural wedge is strongly compressed and thrust and is stacked vertically, thus resulting in the passive uplift and tilting of the sedimentary cover as a whole.

Under the control of multilayer detachment deformation in the early Micang piedmont, the regional detachment layers such as Cambrian mud shale, Silurian mudstone, and Triassic gypsum salt rock control and separate several sets of relatively independent structure–fluid systems vertically. In its late stage, under the control of tilting and uplifting, the large pressure gradient structure in the formation creates dynamic conditions for the large-scale bedding flow of fluid, which causes the large-scale stratification fluid to discharge from the basin margin along layers and accumulate inside the basin (Fig. 4.14). There are slight differences in the different assemblages. The plasticity

of the Triassic gypsum salt in the upper assemblage causes low pressure locally, thus increasing the difference in vertical pressure and causing some of the fluids to be discharged vertically. The oil and gas are adjusted and escape accordingly; the stable physical properties of mudstone in the lower assemblage prevent the fluid from flowing vertically.

(2) Reservoir control mode in the piedmont deformation zone

Based on the above analysis of piedmont structural deformation and its control on hydrocarbon accumulation, we constructed four types of reservoir control modes (Table 4.2): ① Prototype mode: located in the front of transition zone II, with multiple detachment layers in the early stage, stable in the late stage, and in situ oil and gas accumulation; ② Adjustment and reformation mode: located at the front edge of transition zone I or the middle of Zone II, with high stress anomalies, tilting and thrust lately, strong reformation, and continuous oil and gas migration laterally to the lower pressure area of the back edge; ③ Adjustment and filling mode: located at the middle and back edge of transition zone II, with multiple detachment layers in the early stage, fold amplification or uplifting as a whole in the late stage, weak reformation, and continuous oil and gas filling in the high-pressure area of the receiving basin; and ④ Destruction mode: located at the back edge of transition zone II or the middle and back of transition zone I, undergoing uplifting and exposure, with oil and gas escaping vertically.

Through the dissection of typical oil and gas reservoirs, combined with the study of piedmont tectonic deformation processes and fluid response patterns and a comparative analysis both domestically and internationally, it is believed that source-trap matching, late-stage preservation, source-fault-cap matching, and late-stage charging conditions are the key factors for hydrocarbon accumulation and enrichment in piedmont deformation zones (transition zones I and II, respectively). On this basis, five comprehensive evaluation indicators are proposed, including deformation zone type, deformation zone scale, fault-cap combination, reservoir-forming conditions, and detachment layer properties.

Based on the comprehensive division of deformation patterns, 22 deformation zones were classified for the piedmont belts around the three cratonic basins. By analysing the hydrocarbon distribution and the main controlling factors of zonal enrichment in the aforementioned composite intracontinental piedmont deformation zones, the evaluation index system was clarified, and a quantitative comprehensive evaluation model was constructed using fuzzy mathematics to achieve rapid and effective grading and classification evaluation. By applying the newly established grading and classification quantitative evaluation technology, five favourable zones were comprehensively selected. Within the optimized favourable zones in Longmen Mountain and Micang Mountain, two risk wells, YC1 (Fig. 4.13) and YB1 (Fig. 4.14), were demonstrated and submitted. The target areas have resource scales of 251.6 billion cubic metres and 189 billion cubic metres, respectively, and are expected to achieve new breakthroughs in the deep marine realm of piedmont deformation zones.

4.5.2.4 Control of Multidetachment Layers in Hydrocarbon Accumulation

(1) Control of multidetachment layers in hydrocarbon accumulation conditions

Table 4.2 Summary of hydrocarbon accumulation modes in piedmont zone at the periphery of Sichuan Basin

Reservoir type	Structure location	Deformation characteristics and evolution	Accumulation mode	Controlling factors	Well
① Prototype	Front of zone II	Multiple detachment layers in the early and stable in the late	Multi-source hydrocarbon supply and in-situ accumulation	Source cap configuration	Shuangyushi wells
② Adjustment and filling	Middle and back of zone II	Multiple detachment layers in the early and fold amplification or uplift as a whole in the late	Bedding channeling, in-situ adjustment and continuous filling	Filling intensity and preservation condition in late stage	/
③ Adjustment and reformation	Front of Zone I or middle of Zone II	Concealed tilting and thrust, strong reformation	Fault channeling, reformed residual	Lateral sealing and preservation condition	Well PZ1, JX2, LB1
④ Destruction	Middle and back of zone I	Continuous thrust, uplift and strong denudation	Escape vertically alone fault or in uplift and denudation	Preservation condition	Well LS1, TX1

The large-scale multilayer detachment structures developed in the eastern Sichuan area result in layered differential deformation in the vertical direction to form the vertical superposition of multiple petroleum systems, which constitutes the basis of three-dimensional exploration. Extensive studies have shown that most gas fields in the eastern Sichuan area have experienced conversion, adjustment, and redistribution from large-scale oil accumulation to large-scale gas accumulation. In this process, the deformation of the multilayer detachment structure strongly affects the conditions of oil and gas accumulation in the area (Ma et al. 2018; Guo et al. 2014).

Multilayer detachment deformation in eastern Sichuan has experienced two main evolutionary processes: multistage uplift in the middle and late Yanshan periods and overall uplift in the Himalayan period (Qiu et al. 2020; Li et al. 2021), which play important roles in controlling the adjustment of oil and gas reservoirs formed in the early stage and the creation and preservation of traps (especially shale gas) in the later stage. Analysis of hydrocarbon generation and evolution in the Wufeng and Longmaxi Formation shale in the Fuling area revealed that, in the Early Cretaceous, the Ro values mostly exceeded 2.0%, the thermal evolution was in the overmature stage, and the pyrolysis gas from liquid hydrocarbons was dry gas. The structure uplifting after the Late Cretaceous resulted in adjustment and reformation of the gas reservoir. Analysis of palaeo-pressure also revealed that staged uplift and pressure relief since the Late Cretaceous were the main driving forces for the late expulsion of hydrocarbons from shale gas in the area (according to 2020 data from the Jianghan Oil Field Branch), and late structural uplift resulted in pressure relief and discharge of early-stage shale oil and gas, which migrated through faults created by tectonic activity.

The detachment layer between different structural layers is also a regional caprock, which vertically separates different source–reservoir–cap assemblages. In the process of upwards thrusting of the fault, the detachment layer shields the fault from continuing to pass upwards and prevents damage to the overlying source–reservoir–cap assemblage so that oil and gas can be preserved in the same structural layer. Between the two sets of detachment layers, fault activity causes the formation to deform and creates a trap with a certain structural amplitude. Many structural traps, such as dual structures, fault-related folds, and anticlines, have developed in the structural layers of the Middle–Upper Cambrian and Lower Silurian Longmaxi Formations in eastern Sichuan.

Late tectonism is characterized by enrichment and a high yield of shale gas. On the one hand, a large number of fractures can be generated to improve shale reservoir performance. Triaxial stress experiments and simulations of Longmaxi Formation shale reveal that when the burial depth of shale increases from 3000 m to 1000–1500 m, the confining pressure decreases from 50 MPa to approximately 15 MPa; when the mud shale begins to change from plastic to brittle, microfractures begin to open on a large scale, the porosity and permeability of the shale greatly improve (Ma et al. 2018), and the shale reservoir properties improve, which is conducive to enrichment and high yield.

On the other hand, faults and fractures caused by tectonism may also damage shale sealing, resulting in the loss of shale gas. The effects of fault opening and closing on shale gas single-well production are much more obvious. For example, the Wujiang fault zone that developed in the southwestern Jiaoshiba area is a thrust fault with Silurian strata as multiple thrust-nappe detachment layers. The fracture zone formed by a multistage fault connects to the surface. The wells drilled near the fault zone are characterized by low-pressure coefficients and low single-well production; however, the single-well production near the fault with good sealing in the same block is not affected by the fault.

Moreover, the difference in structural deformation also results in the planar zoning characteristics of shale gas enrichment in eastern Sichuan. Under multistage tectonism, organic-rich shale can be divided into several tectonic blocks. Owing to the differences in fracture development, fracture nature, and structural burial history, the gas-bearing properties, pressure coefficients, and single-well production after the fracturing of shale differ among these different blocks (Ma et al. 2018). In areas with relatively stable structures and good preservation conditions, the gas well production is usually high, such as in wells JY-1HF, DY-2HF, and DY-1HF, whereas in areas with similar shale development but strong structural reformation and relatively poor preservation conditions, the gas well production is generally low (such as in wells HY-1, YQ-1, and YY-1) (Guo 2014).

Recent exploration practices have shown that oil and gas accumulation can also occur between different structural layers under good source-reservoir and channelling conditions. Exploratory breakthroughs of Well PQ-1 reveal that against the background of a multilayer detachment structure in eastern Sichuan, the late structural uplift led to communication between the high-quality Wufeng–Longmaxi Formation shale and the reservoirs of the Cambrian Xixiangchi group, thus forming good hydrocarbon supply conditions and channelling systems. Conventional gas reservoirs are present in Cambrian structural traps. An industrial gas flow of 25.13×10^4 m^3/d was obtained in the well test.

(2) Multilayer detachment reservoir control mode

Based on the exploration and development practices and previous studies, the hydrocarbon accumulation modes controlled by multilayer detachment structural deformation can be classified into two categories and five types (Fig. 4.15).

① Conventional oil and gas reservoirs

"New source–old reservoir, lateral hydrocarbon supply" mode

This mode is represented by Well PQ-1 located at the Pingqiao fault anticline in the high and steep fold belt in eastern Sichuan. The gas reservoir is a deep conventional gas reservoir with the characteristics of "a source in a new formation and accumulation in an old formation". The source rock is the shale of the Upper Ordovician Wufeng Formation and Lower Silurian Longmaxi Formation, and the reservoir is the porous dolomite of the Upper Cambrian Xixiangchi group. The main channelling system is the late-formed source rock connected to faults and fractures, and the source and reservoir are connected laterally through the fault. The uplift caused by the deformation of the multilayer detachment structure during the later period leads to the release of overpressured shale gas, which constitutes the driving force for hydrocarbon expulsion. The released gas migrates to the structural traps of the Xixiangchi group through faults and fractures to accumulate and form reservoirs (Fig. 4.15).

② Unconventional oil and gas

Unconventional gas reservoirs are characterized by self-generation and self-accumulation, and late preservation conditions are important for enrichment and high yields. Because of the difference in multilayer detachment reformation, the main controlling factors of preservation conditions at different locations are significantly different, thus resulting in different hydrocarbon accumulation modes.

Syncline enrichment mode

This type of accumulation mode is common in eastern Sichuan; it is generally developed in residual synclines, with the characteristics of annular retention and central enrichment, and most of the gas reservoirs are the normal pressure type. Under the influence of compression, "A"-type microfractures are well developed in shale formations with good physical properties, and gas production is affected by the burial depth of the formation and the distance from the denudation area. Taking the Pengshui area as an example, the closer to the syncline core, the deeper the burial depth, the higher the formation pressure coefficient, and the higher the single-well production (Fig. 4.15).

Anticline enrichment mode

This kind of gas reservoir is located in a positive structure, where natural fractures are developed in the shale formation, as well as "V" type microfractures, with good physical properties. The proportion of free gas in gas reservoirs is high and char-

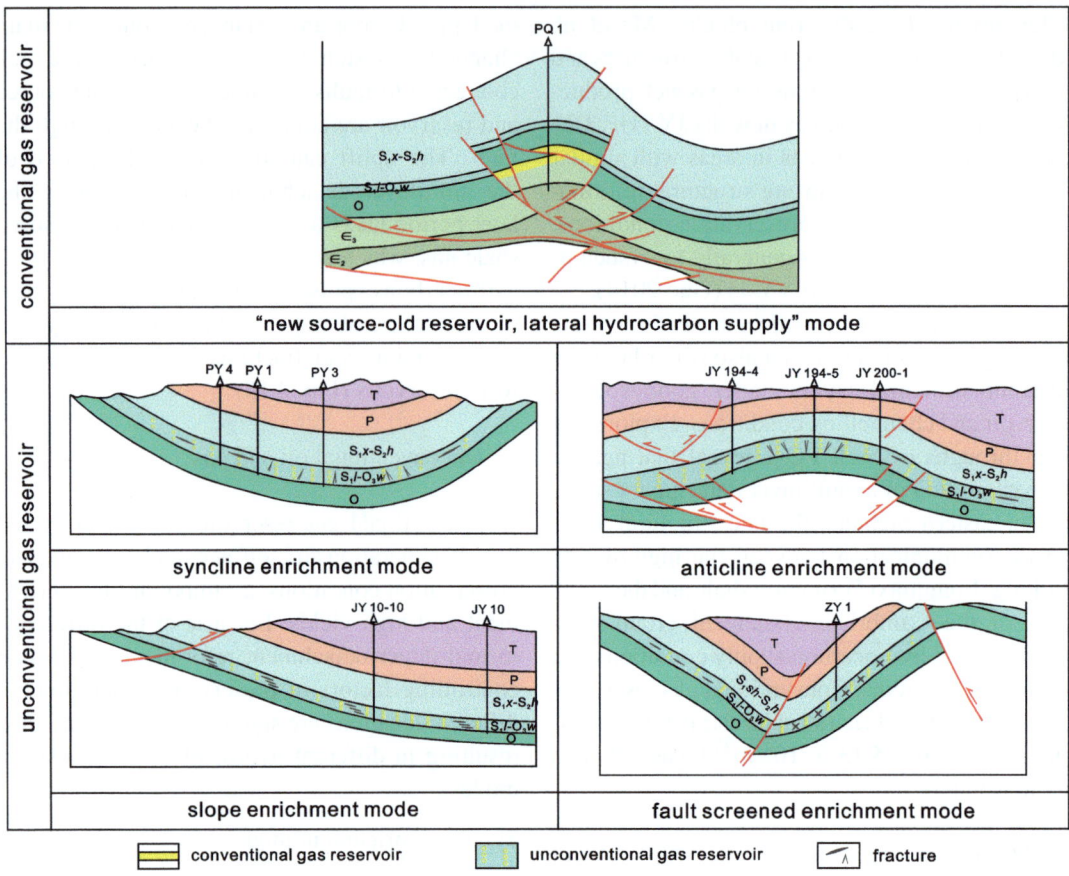

Fig. 4.15 Reservoir control mode of multi-layer detachment in East Sichuan

acterized by short-distance migration and accumulation. This type of gas reservoir is an overpressured gas reservoir, with a large formation pressure coefficient and high single-well production, which is present mostly in the Jiaoshiba and Yongchuan areas.

Slope enrichment mode

Affected by formation uplift, the stress in the slope area is released, where "E" type bedding fractures develop in the shale formation. The shale gas can easily escape laterally, and its enrichment is controlled mainly by lateral sealing conditions. This type of gas reservoir is relatively developed in the southern Sichuan area (Fig. 4.15). This type usually transitions from overpressure to normal pressure.

Fault-screened enrichment mode

Under this accumulation mode, the shale gas is trapped in the lower wall of the fault due to lateral sealing of the reverse fault. "X" type oblique crossing fractures are developed in the shale formation, which makes it easy to form complex fracture networks. Most of these gas reservoirs are normal pressure reservoirs, represented by Well ZY-1 at the Daozhen syncline (Fig. 4.15).

4.5.2.5 Influence of Tectonic–Thermal Events on Hydrocarbon Accumulation

(1) General characteristics of the thermal event in Ordos

The Mesozoic–Cenozoic tectonic–thermal events in the Ordos Basin involved three main types of thermal reform. At the end of the Late Triassic and Early–Middle Jurassic, the tectonic–thermal event affected mainly the southwestern basin. At this time, the magmatic activity intensity was weak, and tuff could be seen in the Upper Triassic and Lower Jurassic strata. A tectonic–thermal event in the Early Cretaceous affected the whole basin, and magmatic and hydrothermal activity was strong during this time. Evidence of basalt ejection during this period was found in the inner and southern parts of the basin, and the alkaline complex represented by the Zijinshan rock mass formed in the east of the basin.

The Zijinshan rock mass is exposed east of the Ordos Basin and has a length of 7.5 km, a width of 4 km, an area of approximately 23.3 km^2, and a subsurface distribution range of approximately 400 km^2. In recent years, hundreds of billions of cubic metres of tight sandstone gas have been produced around the rock mass. The gas reservoir distribution and production are different from those of other gas fields in the basin, and the abundance of tight gas in blocks close to the rock mass is much greater than that in the blocks on the north side. According to Li et al. (2019), two magmatism stages, namely, intrusion and eruption, since the Early Cretaceous affected and controlled the accumulation of regional tight gas. Magmatism helps generate hydrocarbons, and faults and fractures redistribute gas reservoirs vertically, improve local reservoir conditions, and dominate the creation of distant gas reservoirs. Moreover, the rock mass also has a certain destructive effect, which leads to the escape of gas from the source or near the source to a distant location and the surface.

(2) Summary of hydrocarbon accumulation

Multiperiod tectonic–thermal events in the Mesozoic and Cenozoic played important roles in hydrocarbon accumulation in the Ordos Basin. The tectonic–thermal event at the end of the Late Triassic promoted the extensive development of tuff at the bottom of the Yanchang Formation. The shale and volcanic ash in the Chang-7 members are interbedded in the Tongchuan area in the southern basin. The tuff interlayer content is 5–7%, and the total organic carbon

(TOC) content of the tuff shale is high, reaching 20%. In addition, many hydrothermal minerals are present in the Chang 7 members, and their contents are positively correlated with the abundance of organic matter in the source rocks. The analysis revealed that this tectonic–thermal event may have promoted the flourishing growth of organisms, which is conducive to the enrichment of organic matter in the shale.

The Ordos Basin experienced a tectonic–thermal event in the Early Cretaceous that had the most profound influence on hydrocarbon accumulation. During this period, the occurrence of magmatic underplating caused the substances in the asthenosphere to upwell into the mantle such that the lithosphere clearly thinned, which led to a great increase in the geothermal field of the basin. The geothermal gradient increased rapidly from approximately 2.4 °C/100 m at the end of the Late Triassic to 4 °C/100 m in the Early Cretaceous. Studies have shown that the starting times of hydrocarbon generation differ among the three sets of Lower Palaeozoic, Upper Palaeozoic, and Mesozoic source rocks in the basin, but the peak time of hydrocarbon generation is in the Early Cretaceous. Studies of the Zijinshan rock mass at the east margin of the basin reveal that the intrusion of the rock mass can promote hydrocarbon generation in the surrounding source rocks, and its thermal influence range reaches 13.5 km, which is approximately 1.2 times the radius of the rock mass. The arc-shaped and radial faults and fracture zones distributed around the rock mass formed by upswelling of the rock mass are beneficial to oil and gas channelling and the improvement of reservoir physical properties (Li et al. 2019). A study on the origin of organic nanopores in the Shanxi Formation in the east of the basin revealed that abnormal thermal events concentrated from the Late Jurassic to the end of the Early Cretaceous controlled the intensity and range of the pyrolysis gas generated. During this process, many organic nanopores were produced by asphaltization, which increased the porosity of the shale (Liu et al. 2022). In addition, the dating of illite in the reservoir reveals that the accumulation period of Upper Palaeozoic gas reservoirs and Mesozoic oil reservoirs was mainly Early Cretaceous (Ren et al. 2020). In summary, tectonic–thermal events strongly influence hydrocarbon generation and expulsion from source rocks, reservoir creation, and oil and gas migration and accumulation in the Ordos Basin.

4.6 Summary

As the unified landmass formed in central and western China during the Late Triassic is affected by intracontinental differential compression deformation, the superimposed reformation process of the three craton basins can be divided into four evolution stages, namely, T_3–J_2, J_3–K_1, K_2–E, and N–Q, and five major reformation types, namely, deformation in the piedmont, intraplatform decollement fold, differential subsidence–uplift, intrabasinal strike-slip fault, and tectonic–thermal event. Five types of spatiotemporal frames of intracontinental reformation and deformation under the far-field effect of supercontinental convergence are constructed; and the superimposed reformation process and deformation style play important roles in controlling and reforming the deep marine oil and gas accumulation conditions, which results in the creation of different types of oil and gas enrichment and depletion modes.

Acknowledgements The authors acknowledge the financial support of Natural Science Foundation of China (Grants U24B6001 and U19B6003).

References

Bian Q, Deng S, Lin HX et al (2022) Strike-slip salt tectonics in the Shuntuoguole Low Uplift, Tarim Basin, and the significance to petroleum exploration. Mar Petrol Geol 139:105600

Brogi A (2008) Kinematics and geometry of Miocene low-angle detachments and exhumation of the metamorphic units in the hinterland of the northern Apennines (Italy). J Struct Geol 30(1):2–20

Chen XH, Gehrels G, Yin A et al (2015) Geochemical and Nd–Sr–Pb–O isotopic constrains on Permo-Triassic magmatism in eastern Qaidam Basin, northern Qinghai-Tibetan plateau: implications for the evolution of the Paleo-Tethys. J Asian Earth Sci 114:674–692

Deng S, Li HL, Zhang ZP et al (2019) Structural characterization of intracratonic strike-slip faults in the central Tarim Basin. AAPG Bull 103(1):109–137

Deng P, Fang CM, Deng MZ et al (2021a) New insights into the structural character in the central-northern segment of the Longmen Mountains: implications for the multiple-decollement deformation. Petrol Geol Exp 43(1):45–55

Deng S, Liu YQ, Liu J et al (2021b) Structural styles and evolution models of intracratonic strike-slip faults and the implications for reservoir exploration and appraisal: a case stury of the Shunbei area, Tarim Basin. Geotecton Metall 45(6):1111–1126

Deng S, Zhao R, Kong QF et al (2022) Two distinct strike-slip fault networks in the Shunbei area and its surroundings, Tarim Basin; hydrocarbon accumulation, distribution, and controlling factors. AAPG Bull 106(1):77–102

Dong YP, Yang Z, Liu XM et al (2016) Mesozoic intracontinental orogeny in the Qinling Mountains, central China. Gondwana Res 30:144–158

Fang CM, Zhao L (2019) Compound intracontinental piedmont thrust structure transformation and its formation mechanisms. Petrol Geol Exp 41(6):791–799

Guo XS (2014) Rules of two-factor enrichment for marine shale gas in southern China: Understanding from the Longmaxi Formation shale gas in Sichuan Basin and its surrounding area. Acta Geol Sin 88(7):1209–1218

Guo XS, Hu DF, Wen ZD et al (2014) Major factors controlling the accumulation and high productivity in marine shale gas in the Lower Paleozoic of Sichuan Basin and its periphery: a case study of the Wufeng-Longmaxi Formation of Jiaoshiba area. Geol China 41(3):893–901

He DF, Li DS, Tong XG et al (2008) Accumulation and distribution of oil and gas controlled by paleo-uplift in poly-history superimposed basin. Acta Petrol Sin 29(4):325–329

He ZL, Lu JL, Lin JH et al (2022) Marine basins in China—a prototype-reconstruction analyses and ordered hydrocarbon accumulation patterns. Earth Sci Front 29(6):60–72

Huang ZG, Fang CM, Gao CL et al (2017) The basinal prototype sequence and clastic hydrocarbon reservoirs in four main basins in Central and Western China. Petroleum Industry Press, Beijing

Jia CZ (2005) Foreland thrust-fold belt features and gas accumulation in Midwest China. Petrol Explor Dev 32(4):9–15

Jia CZ, Li BL, Zhang XY et al (2007) Formation and evolution of the Chinese marine basins. Chin Sci Bull 52(S1):1–8

Jin ZJ (2011) Formation and accumulation of oil and gas in marine carbonate strata in Chinese sedimentary basins. Sci China Earth Sci. https://doi.org/10.1007/s11430-011-4264-4

Li LT, Wu KQ, Liang JS et al (2019) The domination of tight sand gas accumulation by the Zijinshan thermal event in the block B, eastern Ordos Basin. Nat Gas Geosci 30(10):1430–1438

Li YQ, He DF, Li SJ (2020) Deformation characteristics and structural model of multi-layer detachment structures in the western Hunan and Hubei to eastern Chongqing fold belt. Chin J Geol 55(3):894–908

Li CX, He DF, Lu G et al (2021) Multiple thrust detachments Multiple thrust detachments hydrocarbon accumulation in the northeastern Sichuan Basin, southwestern China. AAPG Bull 105(2):357–390

Liu SF, Li WP, Wang K et al (2015) Late Mesozoic development of the southern Qinling-Dabieshan foreland fold-thrust belt, Central China, and its role in continent–continent collision. Tectonophysics 644–645(3):220–234

Liu HL, Zou C, Mei J et al (2022) Genesis and geological significance of organic matter nanopores in transitional facie strata: a case study of the Shanxi formation in eastern Ordos basin. Acta Geol Sin 96(7):2562–2572

Ma YS, He DF, Cai XY et al (2017) Distribution and fundamental science questions for petroleum geology of marine carbonate in China. Acta Petrol Sin 33(4):1007–1020

Ma YS, Cai XY, Zhao PR (2018) China's shale gas exploration and development: understanding and practice. Petrol Explor Dev 45(4):561–574

Qi LX (2020) Characteristics and inspiration of ultra-deep fault-karatreservoir in the Shunbei area of Tarim Basin. China Petrol Explor 25(1):102–111

Qiu HB, Deng S, Cao ZC et al (ù) The evolution of the complex anticlinal belt with crosscutting strike-slip faults in the central Tarim Basin, NW China. Tectonics 38(6):2087–2113

Qiu NS, Feng QQ, Borjigin T et al (2020) Yanshanian-Himalayan differential tectono-thermal evolution and shale gas preservation in Dingshan area, southeastern Sichuan Basin. Acta Petrol Sin 41(12):1610–1622

Qiu HB, Deng S, Zhang JB et al (2022) The evolution of a strike-slip fault network in the Guchengxu High, Tarim Basin (NW China). Mar Petrol Geol 140:105655

Ren ZL, Qi K, Liu RC et al (2020) Dynamic background of Early Cretaceous tectonic thermal events and its control on various mineral accumulation such as oil and gas in the Ordos Basin. Acta Petrol Sin 36(4):1212–1234

Scharer KM, Burbank DW, Chen J et al (2004) Detachment folding in the Southwestern Tian Shan-Tarim foreland, China: shortening estimates and rates. J Struct Geol 26(11):2119–2137

Sherkati S, Molinaro M, Lamotte DFD et al (2005) Detachment folding in the central and eastern Zagros fold-belt (Iran): salt mobility, multiple detachments and late basement control. J Struct Geol 27(9):1680–1696

Shi W, Dong SW, Zhang YQ et al (2015) The typical large-scale superposed folds in the central South China: implications for Mesozoic intracontinental deformation of the South China Block. Tectonophysics 664(28):50–66

Shu LS, Yin HW, Faure M et al (2017) Mesozoic intracontinental underthrust in the SE margin of the North China Block: Insights from the Xu-Huai thrust-and-fold belt. J Asian Earth Sci 141(A):161–173

Xu XH, Lu JL, Wang BH et al (2022) Marine basins in China: petroleum resource dynamic evolution and exploration directions. Earth Sci Frontiers 29(6):73–83

Xu XH, Fang CM, Liu JL et al (2016) Deformation zoning model of piedmont thrust, western China, and its petroleum response. Petrol Geol Exp 41(6):779–790+820

Xu XH, Gao CL, Huang ZG et al (2005) Three stages of tectonic movements in formation of petroliferous basins in China. Oil Gas Geol 26(2):155–162

Yan DP, Zhang B, Zhou MF et al (2009) Constraints on the depth, geometry and kinematics of blind detachment faults provided by fault-propagation folds: an example from the mesozoic fold belt of South China. J Struct Geol 31(2):150–162

Yang S, Wu GH, Zhu YF et al (2022) Key oil accumulation periods of ultra-deep fault-controlled oil reservoir in northern Tarim Basin, NW China. Petrol Explor Dev 49(2):285–299

Yin A, Harrison MT (2000) Geologic evolution of the Himalayan-Tibetan orogen. Annu Rev Earth Pl Sc 28:211–280

Yun L, Deng S (2022) Structural styles of deep strike-slip faults in Tarim Basin and the characteristics of their control on reservoir formation and hydrocarbon accumulation: a case study of Shunbei oil and gas field. Acta Petrol Sin 42(6):770–787

Zhang YX, Li ZW, Zhu LD et al (2016) Newly discovered eclogites from the Bangong Meso-Tethyan suture zone (Gaize, central Tibet, western China): mineralogy, geochemistry, geochronology, and tectonic implications. Int Geol Rev 58(5):574–587

Zhu X, Chen HJ, Sun ZC et al (1983) The Mesozoic-Cenozoic tectonics and petroliferous basins of China. Acta Geol Sin 57(2):235–242

Zhu RX, Zhao P, Zhao L (2021) Tectonic evolution and geodynamics of the Neo-Tethys Ocean. Sci China Ser D Earth Sci 65(1):1–24

Geothermal and Geopressure Regimes in Ultra-Deep Sedimentary Strata, Tarim and Sichuan Basins

Nansheng Qiu, Jian Chang, Anlai Ma, Yifeng Liu, Qianqian Feng, and Dan Li

Abstract

The temperature field fundamentally governs the process and total volume of hydrocarbon generation from source rocks, as well as the resultant phase state of fluids within basin reservoirs. A lack of clarity regarding the thermal regime of ultradeep strata in China's Tarim and Sichuan Basins currently impedes hydrocarbon exploration efforts. This research addresses this deficiency by summarizing and contrasting the current thermal and pressure characteristics in the two basins, with a specific analysis of the hydrocarbon phases in the Tarim Basin's Ordovician strata. The modern thermal settings show a distinct contrast: the average heat flux for the Tarim Basin is 42.5 ± 7.6 mW/m^2, compared to 53.8 ± 7.6 mW/m^2 for the Sichuan Basin, thereby affirming their characterization as "cold" and "warm" basins, respectively. The thermal contrast is sustained across depth: the geothermal gradients in the Tarim Basin decrease slightly with depth (e.g., 21.6 ± 2.9 °C/km for 0–5000 m to 19.6 ± 2.8 °C/km for 0–7000 m),

while those in the Sichuan Basin show a slight increase (e.g., 21.9 ± 2.3 °C/km for 0–5000 m to 23.3 ± 2.4 °C/km for 0–7000 m). Ultradeep formation temperatures are spatially controlled primarily by basement architecture, fault systems, and hydrothermal convection. Furthermore, thermal modeling established distinct paleoheat flow histories for the basins: the Tarim Basin's internal heat flux generally declined after the Early Cambrian, punctuated solely by an abrupt Early Permian peak. Conversely, the Sichuan Basin's thermal history unfolded in three stages—a steady state from the Cambrian to Early Permian, followed by a rapid rise at the close of the Early Permian, and a subsequent cooling phase after the Late Permian. The consistently higher thermal regime in the Sichuan Basin resulted in differential hydrocarbon generation and conservation in their ultradeep strata. The pressure evolution path of the main reservoirs was reconstructed using numerical simulation constrained by fluid inclusions. The Upper Ordovician reservoir in the Tarim Basin largely maintained normal pressure or mild overpressure throughout its geological history. Conversely, the Cambrian Formation in the Sichuan Basin experienced an abnormal high-pressure stage, peaking during the Late Cretaceous when the strata reached its maximum burial depth. Furthermore, the maturity evolution of the main Paleozoic source rocks were reconstructed utilizing the Easy%R$_o$ model,

N. Qiu (✉) · J. Chang · Q. Feng · D. Li
China University of Petroleum, Beijing, China
e-mail: qiunsh@cup.edu.cn

A. Ma
Zhejiang University, Zhoushan, China

Y. Liu
Petroleum Exploration and Production Research Institute, SINOPEC, Beijing, China

Y. Ma et al. (eds.), *Petroleum Geology and Exploration of Deep Marine Strata in China*, Advances in Oil and Gas Exploration & Production, https://doi.org/10.1007/978-3-032-02496-1_5

constrained by the thermal history data. The final analysis included a discussion of the hydrocarbon preservation phase state in the deep Tarim Basin reservoirs, informed by the established temperature and pressure data. This comprehensive study significantly advances the theoretical understanding of ultradeep geothermal and pressure fields, providing a vital framework for accurately assessing ultradeep source rock maturation and the resultant hydrocarbon preservation state in both the Tarim and Sichuan Basins.

Keywords

Thermal history · Pressure evolution · Maturation · Oil and gas phase · Ultradeep strata

5.1 Introduction

Driven by escalating energy demands and the diminishing returns from shallower fields, the ultradeep strata (> 6000 m) of sedimentary basins have emerged as the paramount focus for hydrocarbon exploration and production in China (Jin, 2011; Zou et al. 2018). ecent successes include the discovery of major accumulations, such as the Anyue gas field in the Sichuan Basin and the Shunbei oil field in the Tarim Basin (Jiao 2018; Wang et al. 2014; Zou et al. 2014). A critical observation is the contrast in fluid phases: ultradeep Lower Paleozoic reservoirs in the Tarim Basin still yield liquid hydrocarbons, whereas equivalent ultradeep Sinian and Cambrian reservoirs in the Sichuan Basin exclusively contain gaseous hydrocarbons. This sharp difference in the hydrocarbon phase state across the two basin's deep systems raises a fundamental question regarding the geological controls on fluid preservation.

The thermal pressure fields of a basin are among the key factors governing hydrocarbon accumulation and the maturation histories of source rocks. Accurately reconstructing thermal pressure evolution is crucial for studying hydrocarbon accumulation processes and evaluating the potential of oil, gas, and geothermal resources.

Therefore, clarifying the thermal and pressure regimes of the Tarim and Sichuan Basins is essential to resolving the question of differential hydrocarbon phase states. Previous studies have indeed investigated the current geothermal field and thermal history of these basins (Chang et al. 2017, 2022a, 2022b; Li et al. 2010a, b, 2022; Liu et al. 2016, 2018, 2020; Qiu et al. 2012, 2022; Xu et al. 2011, 2018; Zhu et al. 2022). However, research specifically detailing the characteristics of the ultradeep strata of the Tarim Basin—at depths ranging from 6000 to 10,000 m—remains absent. This study addresses that gap by first introducing and analyzing the differences in the present-day heat flow, geothermal gradients, and ultradeep formation temperatures of both the Tarim and Sichuan Basins. Subsequently, the paleo-heat flow histories of the two basins are reconstructed based on paleogeothermometer data and basin modeling. In addition, by integrating numerical simulations and evidence from fluid inclusions, the pressure evolution history of the main reservoirs was reconstructed. Finally, we discuss the influencing factors on the geothermal fields, source rock maturation histories, and differential hydrocarbon phases within the ultradeep strata of the Tarim and Sichuan Basins. This paper offers crucial new insights into the thermal and pressure regimes of the Tarim and Sichuan Basins, supporting subsequent ultradeep exploration and resource evaluation.

5.2 Thermal History of Ultradeep Strata

5.2.1 Present-Day Heat Flow of the Tarim and Sichuan Basins

Based on extensive data, the present-day horizontal heat flow distribution across the Tarim and Sichuan Basins was systematically analyzed, revealing a stark thermal contrast between the two (Qiu et al. 2022; Fig. 5.1). Pecifically, the heat flow for the Tarim Basin spans a range of 27.4–66.5 mW/m^2, averaging a low value of 42.5 $\pm$ 7.6 mW/m^2 (Fig. 5.1a). This low average clearly

Fig. 5.1 Current horizontal heat flow distribution in the Tarim and Sichuan basins Qiu et al. (2022)

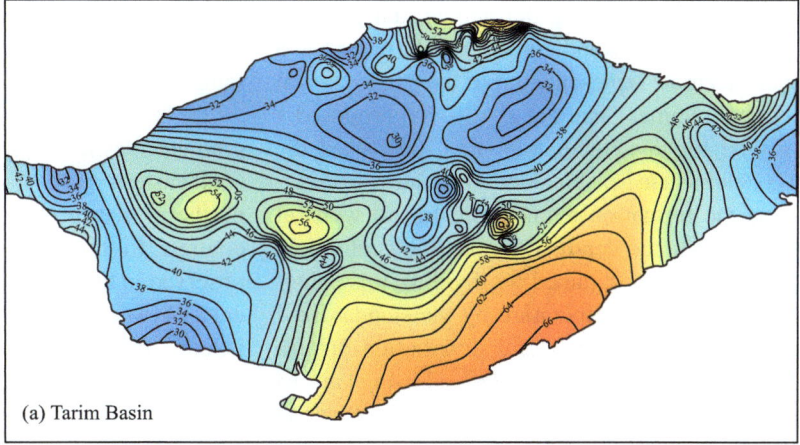

(a) Tarim Basin

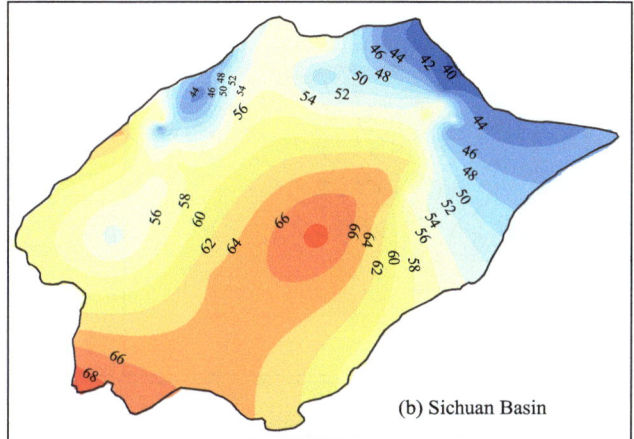

(b) Sichuan Basin

designates the Tarim Basin as a typical "cold" basin. In general, elevated heat flow is observed in uplifted regions characterized by a shallow basement, such as the southeastern Tarim Basin, where values exceed 60 mW/m². Conversely, depressed areas containing thick Cenozoic sediment successions–specifically the Southwest and Northern Depressions–exhibit substantially lower heat flow, typically ranging from 30 to 40 mW/m². The anomalously high thermal flux in the western margin of the Bachu Uplift and the Kuqa Depression is likely attributable to thermal disturbances associated with Late Cenozoic thrust faulting and strike-slip movements, respectively.

The Sichuan Basin exhibits an average heat flow of 53.8 ± 7.6 mW/m² (Fig. 5.1b). Elevated thermal values, ranging between 60 and 70 mW/m², characterize the Central Sichuan Uplift and

the Southeastern Gentle Structural Zone. In sharp contrast, a considerably lower thermal regime, typically 40–54 mW/m², prevails across the northern sector of the Eastern Steep Structural Zone, the Northern Gentle Structural Zone, and the north-central Western Depression.

5.2.2 Present-Day Geothermal Gradient of the Tarim and Sichuan Basins

The geothermal gradient is defined as the rate at which subsurface temperature increases per unit of burial depth within a basin. Prior investigations have established that this gradient varies with depth across different sedimentary basins (Chang et al. 2016a; Feng et al. 2009; Liu et al. 2020;).

Consequently, utilizing the geothermal gradient calculated at a standardized depth provides a more accurate representation of the basin's actual thermal regime. Chang et al. (2016a) proposed a method for this calculation: first, they used the 1D steady-state heat conduction equation to determine the formation temperature for all wells at a chosen unified depth. The resulting geothermal gradient is then obtained by dividing this calculated formation temperature by the standardized depth. The resulting lateral distribution pattern of the geothermal gradient, derived using various standardized depths in the Tarim and Sichuan Basins, is illustrated in Fig. 5.2.

Calculated for a unified depth interval of 0–5000 m, the Tarim Basin's geothermal gradient ranges from 14.9 to 31.9 °C/km (average 21.6 ± 2.9 °C/km) (Fig. 5.2a). Spatially, the thermal gradient reaches its peak (28–30 °C/km) in the southeastern part of the basin and along the northern margin of the Kuqa Depression, in contrast to minimum values of 12–16 °C/km observed in the Kalpin Uplift. Intermediate gradients, between 16 and 20 °C/km, are characteristic of the Manjiaer Depression, the northern Shuntuoguole Low Uplift, the Kongquehe slope, and the western margin of the Southwest Depression. Notably, the comparatively high gradients

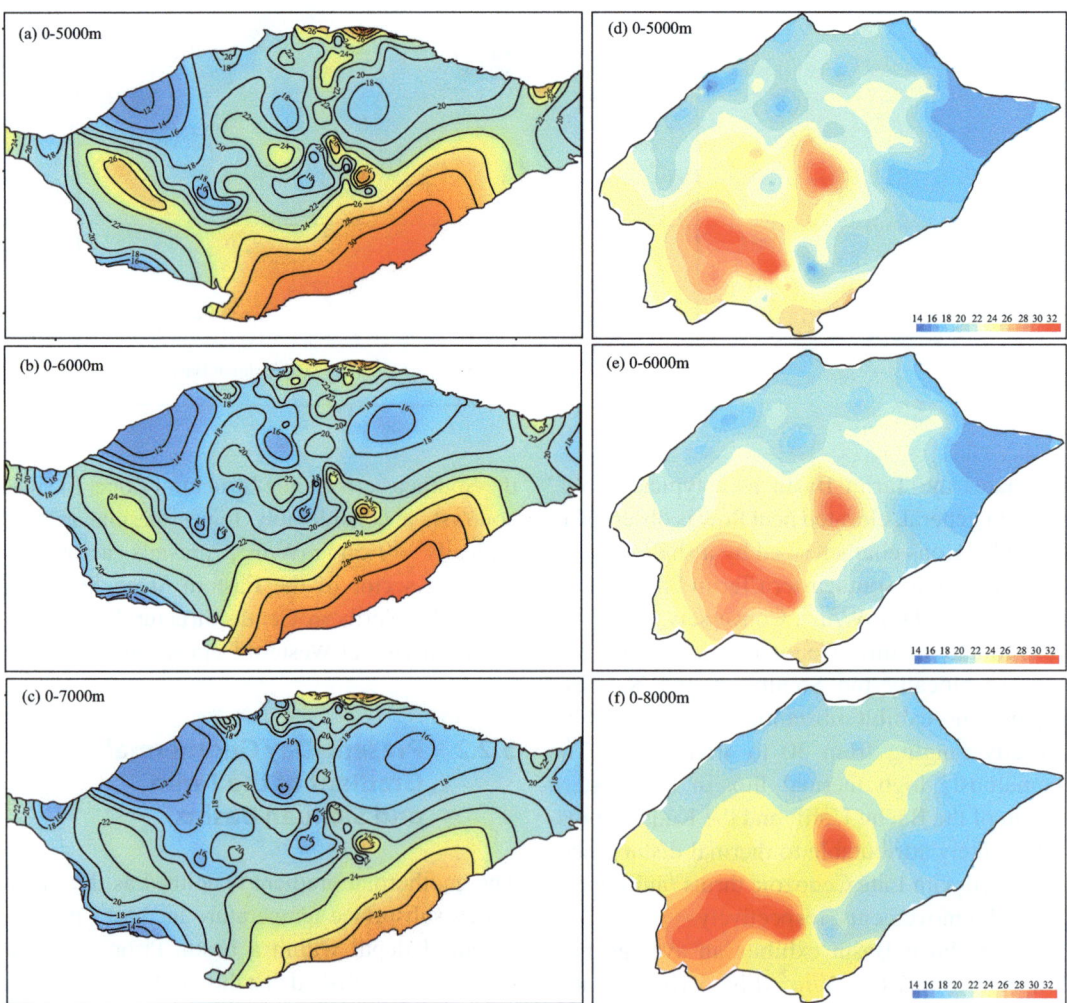

Fig. 5.2 Spatial variation of geothermal gradients at unified depth intervals in the Tarim and Sichuan basins

(22–26 °C/km) observed in the eastern Southwest Depression and the southern Shuntuoguole Low Uplift are attributed to deep hydrothermal upwelling (Liu et al. 2020). Across the Tarim Basin, the average geothermal gradients at unified depths are low: 20.5 ± 2.8 °C/km (ranging from 14.0 to 31.7 °C/km) for the 0–6000 m interval, and 19.6 ± 2.8 °C/km (ranging from 13.3 to 31.4 °C/km) for the 0–7000 m interval (Fig. 5.2b and c). While the spatial pattern remains consistent with that of the 0–5000 m depth interval, the geothermal gradient values exhibit a clear inverse relationship with depth.

For the 0–5000 m depth interval, the Sichuan Basin's geothermal gradient ranges from 14.8 to 30.3 °C/km (average 21.9 ± 2.3 °C/km) and typically increases from north to south (Fig. 5.2d). Higher thermal gradients (22–32 °C/km) characterize the Central Sichuan Uplift and Southeastern Gentle Structural Zone, in contrast to the cooler regimes (14–20 °C/km) of the basin's northern, western, and eastern sectors. Significantly, examination of deeper intervals (0–6000 m and 0–8000 m) reveals that the average gradient strengthens with depth, reaching 22.1 ± 2.5 °C/km (16.1–31.0 °C/km) and 23.3 ± 2.4 °C/km (18.0–33.6 °C/km), respectively (Fig. 5.2e and f). This behavior is diametrically opposed to the trend identified in the Tarim Basin.

5.2.3 Paleoheat Flow Evolution of the Tarim and Sichuan Basins

(1) Method and Theory of Thermal History Reconstruction

The thermal history of a sedimentary basin is typically defined using two primary approaches: tectono-thermal modeling (e.g., the stretching model of Mckenzie (1978)) and analysis of thermal maturity indicators (e.g., apatite fission tracks (AFT) and vitrinite reflectance). The latter method is generally favored as it is directly constrained by empirical data from samples. While vitrinite reflectance (R_o), along with apatite/zircon fission

track and (U-Th)/He data, are routinely employed to constrain thermal evolution, their application is limited. Specifically, vitrinite macerals are absent in Early Paleozoic strata, and suitable apatite or zircon minerals are often lacking in marine carbonate successions. To address this challenge, a suite of alternative paleotemperature proxies has been developed, including vitrinite-like reflectance (R_v), bitumen reflectance (R_b), graptolite reflectance (R_g), and the conodont alteration index (CAI), among others. A particularly robust solution involves converting the reflectance of vitrinite-like macerals, bitumen, and graptolites into an Equivalent Vitrinite Reflectance (R_{equ}), which has become a widely accepted and effective thermal indicator.

The thermal history reconstruction in this research was robustly constrained using a variety of geological indicators, including Apatite and Zircon (U-Th)/He ages, AFT data (both track lengths and ages), and equivalent vitrinite reflectance (R_o) data. Modeling of these thermal indicators was performed using specific kinetic models: the Easy%R_o model (Sweeney and Burnham 1990) was employed for reflectance data, the fan model (Laslett et al. 1987) for AFT annealing, and specific models for both apatite (Farley 2000) and zircon (Reiners et al. 2004) were used for He ages. To execute the 1D forward and inverse thermal models, the study relied on the HeFTy (version 2009) and Thermodel for Windows (version 2011) software packages. A key methodological distinction exists among the software: Thermodel is suitable for analyzing the evolution of AFT and R_o profiles downhole (i.e., multiple depths in one well). In contrast, both HeFTy and QTQt integrate all three primary datasets (AFT, (U-Th)/He, and R_o); however, HeFTy is typically applied to single samples, while QTQt facilitates the simultaneous modeling of multiple samples within the same wellbore. The timing of tectonic thermal events and temperature paths, alongside the burial history, can be constrained by AFT data. This allows for determining the average paleothermal gradient (G_t) from the surface to the sample depth using the modeled paleotemperature (T_t) and its corresponding paleoburial depth (Z_t), following the relationship: $G_t = (T_t - T_{surface})/Z_t$. The required Z_t is derived directly from the sample's burial history. This calculated G_t represents an average across the entire depth interval. Finally, paleoheat flow values are modeled by integrating this calculated G_t with the thermal conductivities specific to each stratum.

(2) Tarim Basin

Previous research into the basin's thermal evolution presents a duality of interpretations. One school of thought posits a significant cooling trend, evidenced by a decrease in the thermal gradient from high values, possibly around 40 °C/km or higher heat flow, during the Early Paleozoic to a current 20–25 °C/km (Chang et al., 2017). Conversely, other studies suggest that the basin has maintained a consistently cold thermal regime, reporting Early Paleozoic thermal gradients around 20 °C/km –a value nearly identical to current conditions. The subsequent section introduces the thermal history derived from a single well.

We selected Well TZ1 (located in the Central Uplift) for thermal history modeling constrained by R_{equ} data. The resulting profile clearly shows an unconformity between the Upper Ordovician and Carboniferous units, corresponding to a distinct offset in the R_{equ}-depth relationship between the Lower Ordovician and Carboniferous strata. Crucially, the R_{equ}-depth curve slope is steeper in the Cambrian–Lower Ordovician section than in the overlying strata. This difference signifies a higher thermal gradient during the Early Paleozoic compared to the subsequent Late Paleozoic and Mesozoic eras. The proposed thermal history path was validated by simulating the required thermal indicators (AFT and R_{equ}) until they aligned well with measured values. The modeled R_{equ} demonstrated an excellent fit with the measured data in Well TZ1 (Fig. 5.3). These results indicate that the Cambrian and Ordovician units reached their peak temperature during the Late Ordovician, followed by gradual cooling after the Caledonian event, causing thermal gradients to decrease progressively to their present-day values.

This study illustrates mainly the paleogeothermal histories of the northern Shuntuoguole low uplift, Tabei uplift, and Tazhong uplift (Fig. 5.4). Despite variations among individual wells, the thermal evolution in these three units generally shows a decreasing heat flow trend since the Early Cambrian. This overall cooling trend is punctuated by a sudden thermal maximum observed in the northern Shuntuoguole low uplift, northwest Tazhong uplift, and western Tabei uplift, likely resulting from the thermal influence of a Large Igneous Province (LIP). The northern Shuntuoguole low uplift also distinguishes itself by possessing a much cooler thermal regime

Fig. 5.3 Burial and thermal history of well TZ1 in the Tarim Basin. The measured (+) and modeled R_{equ} values (black line) are shown in the right figure Qiu et al. (2022)

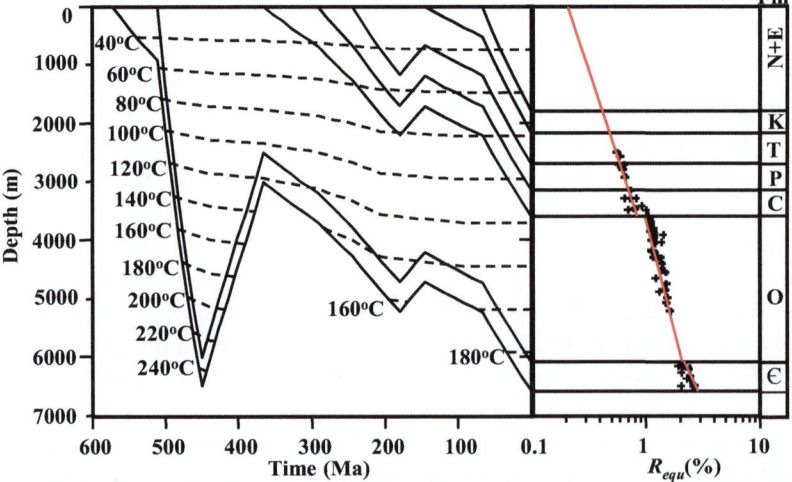

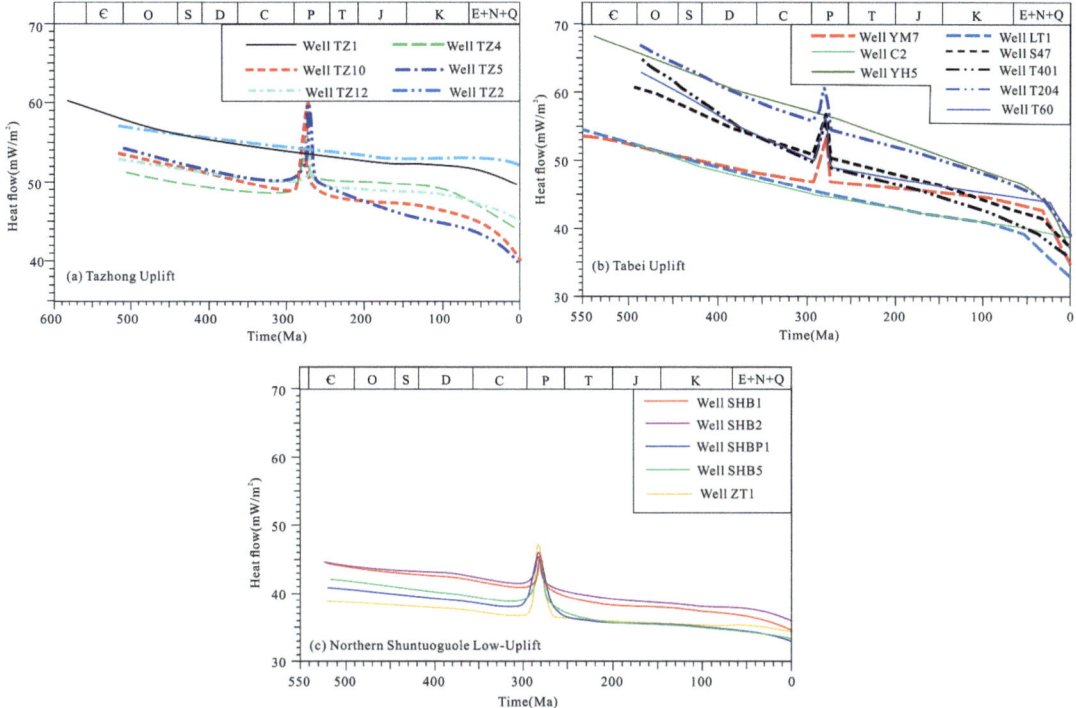

Fig. 5.4 Heat flow histories of the Tazhong uplift, Tabei uplift and northern Shuntuoguole low uplift in the interior of the Tarim Basin Li et al. (2022)

compared to the Tazhong and Tabei uplifts. This comparatively lower heat flow is a consequence of the unit's long-term tectonic stability throughout its geological history.

(3) Sichuan Basin

Since the 1980s, investigations into the thermal regime of the Sichuan Basin have been ongoing (Xu et al. 2011). Recently, the proliferation of deep boreholes has yielded a wealth of deep formation temperature measurements, significantly advancing the basis for geothermal research in the region. Consequently, numerous recent publications have focused on thermal history reconstruction (He et al. 2011, 2014; Huang et al. 2012; Jiang et al. 2018; Liu et al. 2018; Xu et al. 2018; Zhu et al. 2010, 2016), allowing researchers to successfully document the basin's thermal evolution since the Sinian.

The thermal history modeling results for a single well are presented in Fig. 5.5. Well

MX13's R_o simulation results showed an excellent match with measured data, confirming the modeling's high credibility. This well's history indicates it experienced a 56 mW/m^2 heat flow, which remained lower and more stable from the Sinian through the Early Permian. Subsequently, the heat flow peaked at the end of the Early Permian, exceeding 80 mW/m^2, before finally declining to its present-day value of 63 mW/ m^2.

Though exhibiting regional differences, the thermal histories of the Sichuan Basin's tectonic units share a common trajectory (Fig. 5.6): a period of thermal stability prior to the Early Permian, followed by a pronounced spike at the close of the Early Permian, and a subsequent, prolonged cooling phase from the Late Permian onward. This dramatic thermal spike is best illustrated in the Western Depression, where heat flow rose sharply from its Early Paleozoic baseline of 50–55 mW/m^2 to a peak of 80–115 mW/m^2 at the Permian boundary. This maximum thermal

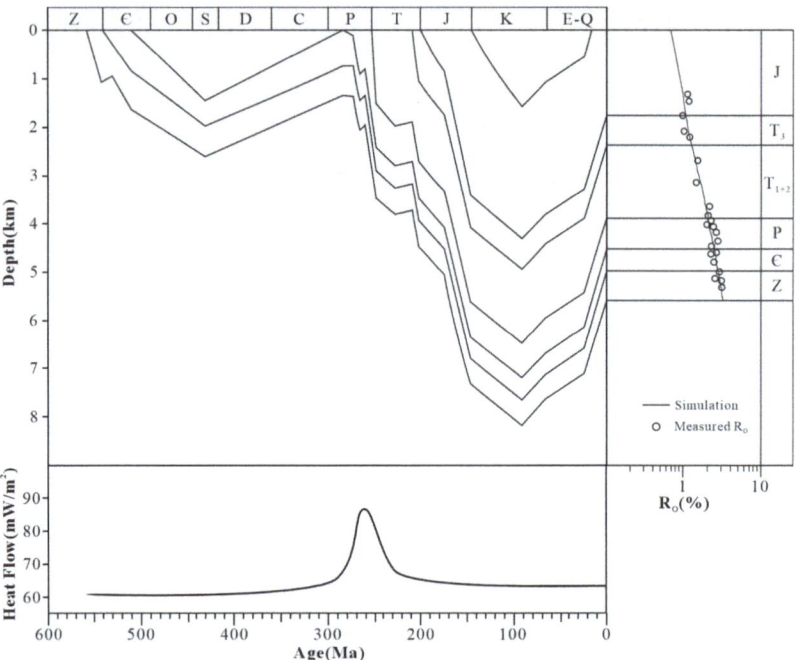

Fig. 5.5 Burial and thermal history for well MX13 in the Sichuan Basin Liu et al. (2018)

flux is a known consequence of the Emeishan plume activity (He et al. 2014; Zhu et al., 2016), followed by a swift decline. The central basin showed a parallel, albeit slightly muted, history, increasing from an initial 55–60 mW/m² to a peak of 82–95 mW/m². Less intense thermal pulses were recorded in the northern and eastern basins, with their peaks constrained to

62–75 mW/m². Distinctly, the southern Sichuan region deviates from this pattern, maintaining a relatively stable thermal profile. Its heat flow only modestly increased in the Middle and Late Permian before settling at current values of 50–60 mW/m².

The Emeishan mantle plume (LIP) induced a major Permian thermal pulse across the

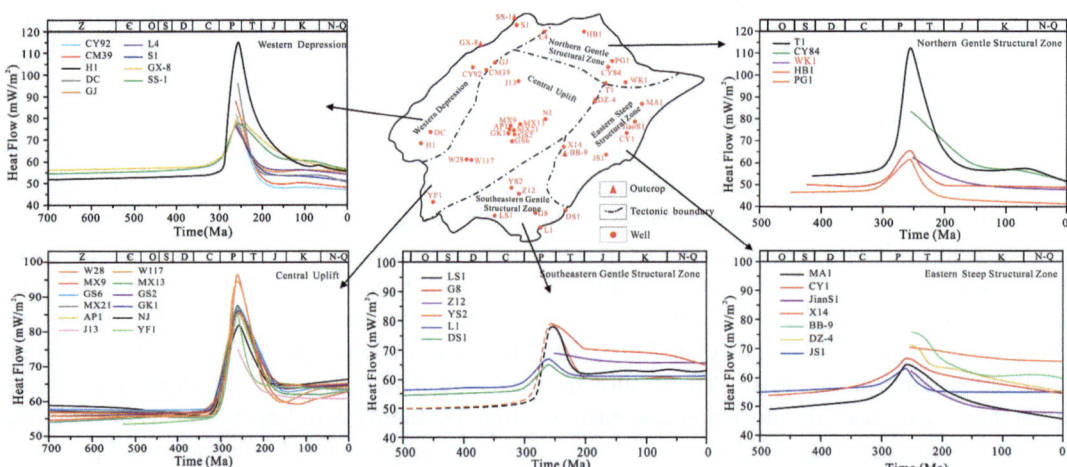

Fig. 5.6 Thermal histories of different tectonic units of the Sichuan Basin Qiu et al. (2022)

Sichuan Basin. However, the intensity of this thermal event varies spatially as a function of its distance from the plume center. Southwest Sichuan, being proximal to the center, is characterized by pronouncedly elevated heat flow peaks. This contrasts sharply with Northeast Sichuan, which is distal to the plume, where the peak heat flow values are considerably suppressed or entirely absent. (He et al. 2011, 2014; Huang et al. 2012; Jiang et al. 2018; Liu et al. 2018; Xu et al. 2018; Zhu et al. 2010, 2016).

5.3 Pressure Evolution of Typical Ultradeep Strata

Overpressures are pervasive features across sedimentary basins. Understanding their distribution and genesis is fundamental for pre-drilling pore pressure prediction, which is critical for ensuring drilling safety. Beyond operational concerns, pore pressure—particularly overpressure—significantly impacts crucial subsurface processes, including porosity evolution and the generation, migration, accumulation, and preservation of petroleum (Bloch et al. 2002; Hao et al. 2007; Mello and Karner 1996; Nguyen et al. 2013). Most encountered overpressures are primarily attributed to disequilibrium compaction. This occurs because the rate of pore fluid expulsion cannot keep pace with the increasing lithostatic load during burial, causing the fluid to bear a portion of the extra stress. (Neuzil 1995; Osborne and Swarbrick 1997). The second major category is fluid expansion overpressure, which encompasses several mechanisms that cause the relative volume of pore fluid to increase within a constrained rock volume (Bowers 1995; Mouchet and Mitchell 1989; Osborne and Swarbrick 1997). These mechanisms include aquathermal expansion, diagenesis (e.g., of gypsum, kaolinite or smectite), gas generation, and fluid transfer (Mouchet and Mitchell 1989; Neuzil 1995; Osborne and Swarbrick 1997; Swarbrick and Osborne 1998; Yardley and Swarbrick 2000).

5.3.1 Methods for Pore Pressure Analysis and Paleopressure Construction

Evaluating pore pressure requires either direct field measurement or indirect calculation. Direct measurements, acquired using devices like wireline repeat formation testers (RFTs) and drill-stem tests (DSTs), enable the direct determination of pressure in permeable rock units. However, for fine-grained lithologies (e.g., thick shales), pressure values must be derived solely from indirect indicators. Recognition of overpressure in the subsurface typically involves monitoring a comprehensive array of logging and drilling indicators, including data derived from geophysical logs (such as sonic, resistivity, and density), operational parameters like the drilling mud's weight and viscosity and the penetration rate, physical clues from the shale cutting characteristics, and subtle geochemical shifts (Ortoleva 1994).

Fluid inclusions are crucial micro-analytical tools, offering reliable PVTx data at the time of their entrapment (Aplin et al. 1999; Bourdet et al. 2010; Munz et al. 2004; Pironon and Bourdet 2008;). Trapping conditions are established by finding the intersection point of isochores derived from co-existing hydrocarbon and brine (aqueous) inclusions. Aplin et al. (1999) outlined an iterative calibration technique for hydrocarbon isochores. We initiated PVTx modeling by using the current reservoir oil composition and applying the SRK equation with Péneloux volume correction (Péneloux et al. 1982). Using PVTsim, the total molar volume and saturation pressure are calculated at the homogenization temperature. The critical step is adjusting the initial composition—by adding or removing gas—until the calculated vapor volume fraction (at room temperature and constant molar volume) matches the fraction measured by Confocal Laser Scanning Microscopy (CLSM) and 3D reconstruction. This calibrated mixture is the "effective composition," from which the fluid's isochore and phase envelope are generated. The brine inclusions present a distinct challenge due to their complex geochemistry. Munz et al. (2004) simplified the problem by suggesting that the homogenization temperature of brine inclusions directly equates to the trapping temperature, citing the low solubility of hydrocarbons in water. For modeling purposes, the NaCl–H$_2$O system, as modeled by Mao and Duan (2008), is commonly used to construct the aqueous isochore, a method previously shown to yield results comparable to more complex NaCl–H$_2$O–CH$_4$ systems (Bourdet et al. 2008; Pironon and Bourdet 2008).

Raman spectroscopy is a versatile tool for fluid inclusion (FI) studies because the spectral characteristics of Raman-active species are sensitive to their local molecular environment. This method is advantageous, offering fast, nondestructive analysis with confocal spatial resolution reaching 2–3 μm. CH$_4$ and CO$_2$ are common vapor components in hydrocarbon FIs, particularly those trapped at high (T–P) conditions (Dutkiewicz et al. 2003; Lüders et al. 2012; Pironon et al. 2001). A key methane feature is its C–H symmetric stretching band (υ_1), where the peak position systematically shifts to lower wavenumbers with increasing pressure (Lu et al. 2007; Seitz et al. 1993, 1996). Similarly, the bands of the Fermi diad in the spectrum of CO$_2$ shift to lower frequencies (or wavenumbers) with increasing density (Rosso and Bodnar 1995). Furthermore, the Raman spectral behavior of CO$_2$–H$_2$O, CH$_4$–H$_2$O, CH$_4$–CO$_2$, and N$_2$–CO$_2$–CH$_4$ mixtures was analyzed to determine the internal pressure of fluid inclusions (Lu et al. 2007; Qiu et al. 2020).

Basin modeling can be used to forward model pore pressure and provides information on the timing of overpressure development (Yardley and Swarbrick 2000). The PRA BasinMod and Schlumberger PetroMod are two of the most widely used software programs for basin modeling.

5.3.2 Ordovician Pore Pressure in the Tazhong Area

A broom-like structure, defined by numerous northwest and east–west trending basement faults, characterizes the Tazhong area in the center of the Tarim Basin's Central Uplift. Drilling operations generally employed low-density mud, with weights ranging from 1.1 to 1.27 g/cm^3. In Well TZ1, Repeat Formation Tester (RFT) measurements fall below the freshwater hydrostatic gradient (9.8 MPa/km below 2500 m), and these measured pressures are distinctly lower than those derived from the mud weight. Across the region, Drill-Stem Test (DST) data generally align well with the mud weight, with exceptions occurring where DST pressures were measured underbalanced (e.g., the Upper Ordovician Formation in Well TZ12). Based on the drilling mud weights, formations in the Tazhong area are typically interpreted as being normally pressured or mildly underpressured. he pore pressure at the top of the Ordovician strata ranges from 35 to 62 MPa. This pressure is primarily a function of burial depth and systematically increases from the southeast to the northwest. Conversely, the overpressures observed at the top of the Ordovician (spanning 0–10 MPa) show no correlation with depth. Spatially, two high-overpressure zones exist: the southwestern sector and the tract from the eastern Central Fault Horst to the North Slope. Normal pressure is confined to the southeast and the No. 1 Fault belt.

Petroleum and aqueous inclusions were extensively found within calcite-filled hydraulic fractures. Under UV-epifluorescence, two distinct types of petroleum inclusions were differentiated by color: one exhibiting yellow or green–yellow fluorescence, and the other displaying blue or white–blue fluorescence. The composition and isochores of both the blue and yellow fluorescent oil inclusions were subsequently reconstructed using PVTsim software. These specific inclusions were deliberately selected for modeling because their fluorescence colors, low homogenization temperatures (T$_h$), and minimal gas volume percentages designated them as the most representative samples preserving the initial fluid composition. In wells TZ62 and TZ166, the Ordovician reservoirs are dominated by dry gases with only minor liquid hydrocarbons, a situation likely attributable to their proximity to the

Fig. 5.7 Pore pressure evolution of the O_3 reservoir in well TZ62

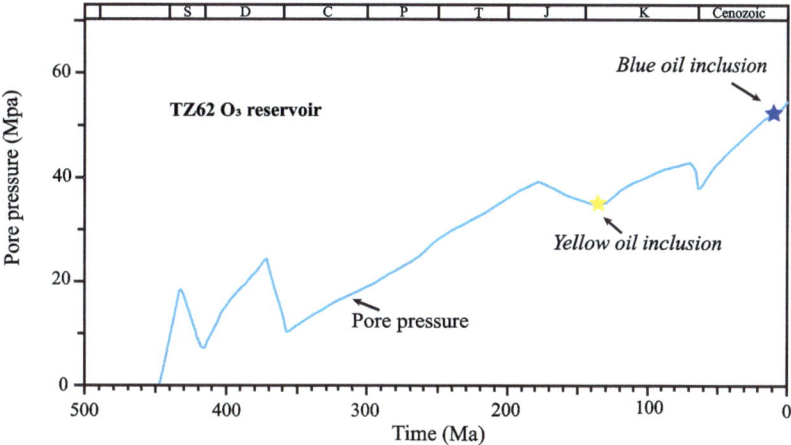

fracture system and subsequent intensive late gas charging. The derived trapping conditions for the yellow fluorescence inclusion assemblage were 34 MPa and 100 °C, while those for the blue fluorescence inclusion assemblage were 43 MPa and 130 °C. When evaluated against the thermal-burial history, these conditions indicate that the trapping pressures and temperatures were nearly normal at the time of entrapment. These trapping events are temporally constrained, with the blue and yellow oil inclusions being trapped around 10 Ma and 135 Ma, respectively. By integrating these fluid inclusion constraints with numerical simulation, the complete pressure evolution path of the Upper Ordovician (O_3) reservoir can be reconstructed (Fig. 5.7). The analysis of the TZ12 sample, however, introduced a methodological challenge: unlike the TZ62 and TZ166 samples, it contains many single liquid-phase hydrocarbon inclusions at room temperature—a characteristic also observed by Liu et al. (2013) in other Ordovician samples with homogenization temperatures as low as 10 °C. Crucially, because these single liquid-phase inclusions lack a measurable vapor fraction at room temperature, the stated PVTx method cannot accurately reconstruct the trapping pressure. Nevertheless, hydrocarbon inclusions with a low T_h are still considered valuable indicators of a overpressure system (Liu et al. 2013; Pironon and Bourdet 2008).

5.3.3 Pore Pressure in the Lower Paleozoic Strata in the Sichuan Basin

Overpressured formations spanning the Cambrian to Jurassic have been documented across numerous Sichuan Basin subtectonic units. Notably, some maximum pore pressures approach the lithostatic limit. (Gao et al. 2004; Guo et al. 2012; Jiang et al. 2003; Hao et al. 2010; Shi et al. 2009). The Central Paleouplift of the Sichuan Basin is favorable for petroleum accumulation due to its sustained geological history. Pressure data analysis reveals multiple distinct vertical pressure systems in the Moxi–Gaoshiti area (Fig. 5.8). System I (Lower Jurassic and overlying strata): Normally pressured, with a slight deviation noted between 900 and 1300 m. System II (Base of Lower Jurassic to top of Middle Triassic Leikoupo): The first true overpressure compartment. A sharp increase in DST and mud weight near 3000 m indicates intense overpressure localized directly beneath the gypsum layer. S System III (Middle Triassic to Permian): Characterized by a high pressure ratio (r) greater than 2.0. System IV (Cambrian): Clearly differentiated from upper zones by sealing layers and gas distribution, with a high pressure ratio ($r \approx 1.6$) (Precise partitioning of Lower Paleozoic overpressure is limited by data). System V (Below 5000 m): Characterized by almost normal pressure conditions.

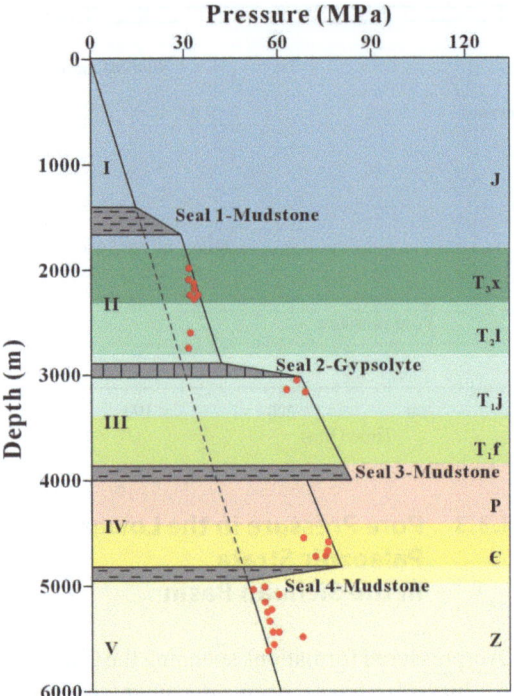

Fig. 5.8 Pressure systems and seals profile of the Gaoshiti–Moxi area. Specifically, Systems I and V are categorized as normally pressured, while Systems II, III, and IV exhibit overpressured conditions. The analysis is constrained by direct field measurements (DSTs), represented by the red solid dots collected from wells. The reference hydrostatic pressure gradient is depicted by the dashed line, clearly contrasting it against the measured pore pressures to identify and quantify the zones of overpressure

The Upper Triassic Xujiahe Formation data plots near the loading curve (Fig. 5.9), strongly suggesting disequilibrium compaction is a key generating factor. This mechanism is supported by the region's rapid subsidence rate (> 200 m/ Ma peak) from the Early Permian to the Late Cretaceous, which is highly conducive to creating compaction overpressure in low-permeability sequences. Slight deviations from the loading curve (Fig. 5.9) imply a secondary contribution from fluid expansion. Concurrent gas generation from Upper Triassic coal seams would have intensified these overpressures by increasing pore fluid volume. However, later unloading due to uplift complicated the thermal history and subsequently reduced the overpressure magnitude. Consequently, while gas generation's precise contribution to present-day overpressure is unquantifiable,

petrophysical properties clearly indicate disequilibrium compaction is the primary source for the Upper Triassic overpressures.

Analysis using the ratio of the effective stress offset (LC) to the measured overpressure (Fig. 5.9) indicates that fluid expansion contributes substantially to the Cambrian overpressure, estimated at 70–100%. Conversely, the role of aquathermal expansion in the Central Paleouplift is likely insignificant due to the low prevailing geothermal gradient (~ 20 °C/km). Theoretically, hydrocarbon generation can significantly increase fluid volume within sealed rock, potentially creating overpressure up to the lithostatic pressure (Barker 1990; Osborne and Swarbrick 1997). Given the geological context, the thermal cracking of oil is the most likely primary origin for the observed Cambrian overpressure.

We constructed a comprehensive pressure evolution model for the Cambrian Formation in the Central Paleouplift by integrating its overpressure mechanisms with burial and hydrocarbon generation histories. From the Cambrian to the Devonian, slow sedimentation and shallow burial kept the pore fluid generally normally pressured. Even when Cambrian source rocks entered the oil window at the end of the Ordovician, only mild overpressure likely developed in the reservoirs, as significant leakage—exceeding 25%— prevents overpressure generation from oil (Guo et al. 2011). Moreover, the preservation of any initial overpressure was unlikely, given the uplift of Cambrian strata toward the surface by the close of the Carboniferous period. Subsequently, the Central Paleouplift underwent a phase of swift sedimentation during the Early Permian, a condition highly conducive to the development of disequilibrium compaction overpressure within the shallower Permian and Upper Triassic Formations. Crucially, obvious overpressure started to develop in the Cambrian Formation itself and intensified significantly as the amount of oil cracking to gas increased during the Jurassic period; this late-stage thermal cracking of oil to gas is considered the most critical factor for overpressure development in these ancient strata. Furthermore, disequilibrium compaction overpressure might also have originated within

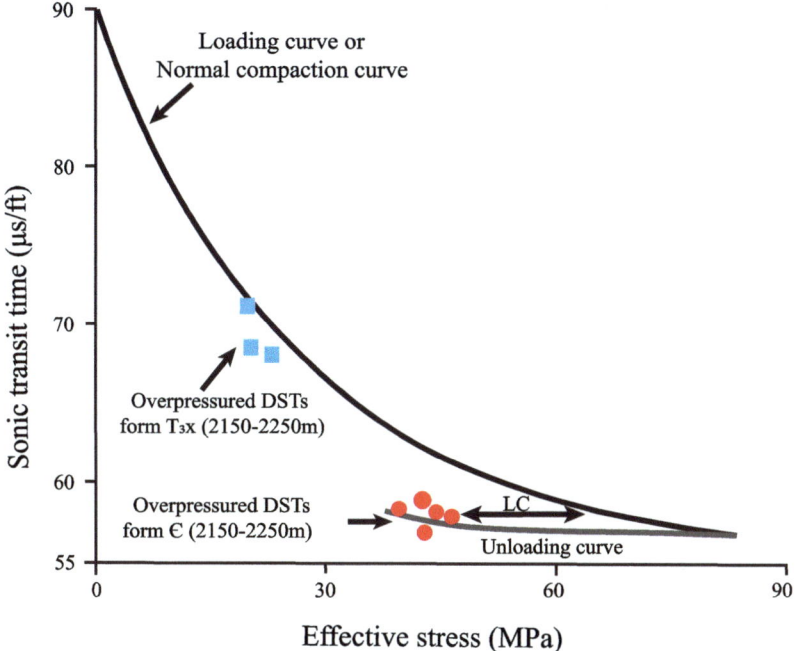

Fig. 5.9 Sonic transit time versus effective stress in Moxi–Gaoshiti. The loading curve is derived from sequences interpreted as having experienced only normal hydrostatic pressure. The plot highlights pressure anomalies: the solid dots (ϵ, Cambrian) and squares (T_3x, Late Triassic Xujiahe Formation) represent overpressured units. Notably, the Cambrian overpressured formation plots with a significant deviation from the normal loading curve. The horizontal distance labeled LC quantifies this deviation, reflecting the contribution of fluid expansion to the generation of the overpressure

the sealing strata due to heightened vertical stress during deep burial. Maximum pore pressure was attained during the Late Cretaceous deep burial peak (90 Ma). Prior work corroborates this maximum, showing high paleopressure ($1.8 < r < 2.4$) n the Sinian Formation via fluid inclusion state equations (Liu et al. 2008). Post-90 Ma, uplift and corresponding unloading caused both pore pressure and overpressure to gradually decline to their current levels. The present-day pressure regime exhibits spatial heterogeneity (Fig. 5.10a): the Weiyuan structure, having undergone the most extensive denudation, reverted to normal fluid pressure, whereas the Moxi–Gaoshiti area retains moderate overpressure.

Excess pressure in the Dengying Formation's second member (Z_2dn^2) commenced during the Triassic (Fig. 5.10). After this inception, the pressure rose steadily and slowly from the late Triassic through the Middle Jurassic. A subsequent phase, marked by rapid strata subsidence (Middle Jurassic to end of Early Cretaceous), led to a significant amplification of this excess pressure. Post-Late Cretaceous, the Z_2dn^2 unit experienced swift uplift; this tectonic change prompted the excess pressure to progressively wane until the Late Neogene, when it dissipated completely (diminished to zero).

5.4 Maturation of the Main Deep Petroleum Source Rocks

Organic matter maturity is primarily controlled by its temperature and heating duration. The Easy%R_o chemical kinetic model (Sweeney and Burnham 1990) commonly depicts this relationship. Utilizing measured maturity data, we employed the Easy%R_o model within BasinMod 1D software to reconstruct the temperature and maturity histories of the source rocks.

Fig. 5.10 Pressure evolution and dominant overpressure mechanisms for the Central Paleouplift Cambrian Formation in the Sichuan Basin. **a** Moxi–Gaoshiti pressure evolution and **b** Weiyuan pressure evolution. Maximum Late Cretaceous (90 Ma) pressure cited from Liu et al. (2008)

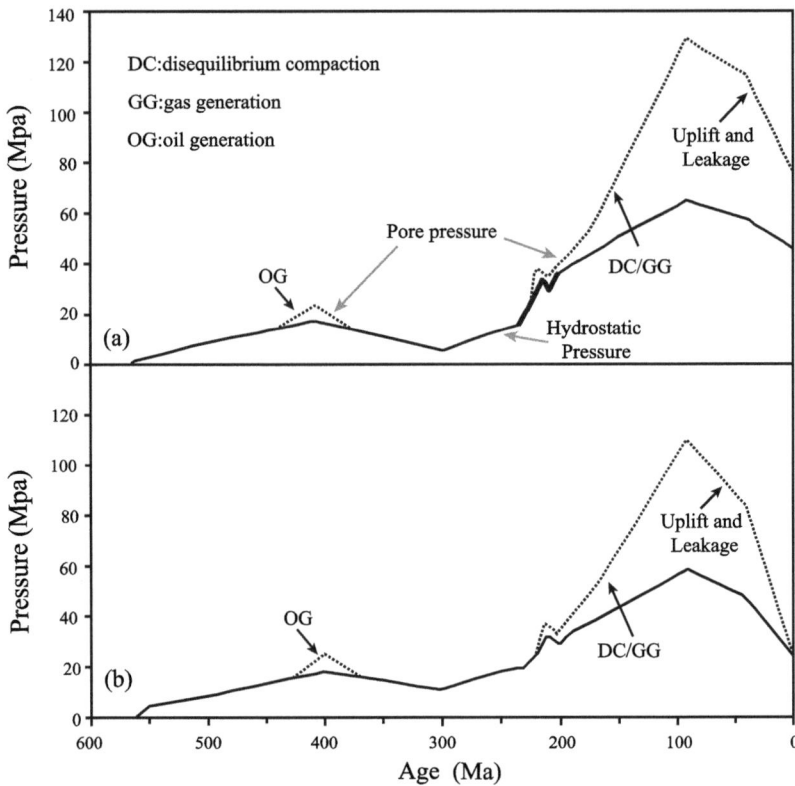

5.4.1 Maturation Evolution of the Lower Paleozoic Source Rocks in the Tarim Basin

We analyzed the maturation histories of the Lower Ordovician Yingshan and Lower Cambrian Yuertusi Formation source rocks. Figure 5.11 illustrates the maturation evolution for Paleozoic source rocks in Well LT1 (Tabei Uplift): Cambrian source rocks were in the high-mature oil-generating stage from the Late Ordovician to the Early Cretaceous. They subsequently entered the gas-generating stage from the Late Cretaceous to the present, attaining a current maturity over 1.3%. Lower Ordovician source rocks were in the low-mature oil-generating stage during the Late Ordovician–Early Cretaceous, transitioning to the middle-mature stage from the Late Cretaceous to the Paleogene. Due to rapid Neogene subsidence, the current maturity exceeds 1.0%, placing them in the high-mature oil generation stage.

The maturation histories of the Lower Ordovician Yingshan Formation and Lower Cambrian Yuertusi Formation source rocks of the typical wells have been reconstructed in northern Shuntuoguole Low Uplift (Fig. 5.12), a key area for oil and gas exploration. Notably, since the drilling failed to reach the Lower Cambrian and Lower Ordovician source rocks, we used only seismic data to infer the thickness of the lower stratum. Combined with the heat flow evolution trend, the early heat flow evolution history can be inferred. The Lower Ordovician source rock has maintained a low- to high-mature stage ($0.5\% < R_o < 1.3\%$) since the Silurian, aligning with the oil and gas phase of the Middle Ordovician reservoirs (Fig. 5.12a). Conversely, Lower Cambrian source rocks generally began generating oil in the Early Ordovician ($R_o > 0.5\%$) (Fig. 5.12b). Due to differences in thermal evolution, the Shunbei area entered the gas generation stage ($R_o > 1.3\%$) from east to west during the Middle Silurian–Late Carboniferous (Fig. 5.12b). Although the Lower Cambrian

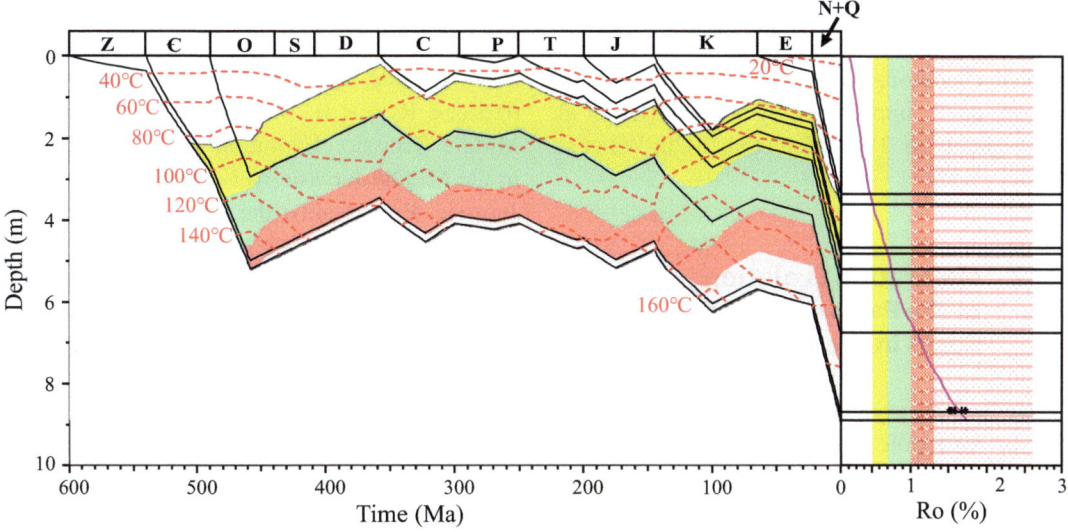

Fig. 5.11 Burial, thermal history and maturation of source rock in well LT1

Fig. 5.12 Maturation history of Lower Ordovician Yingshan and Lower Cambrian Yuertusi source rocks in the Shunbei area Li et al. (2022)

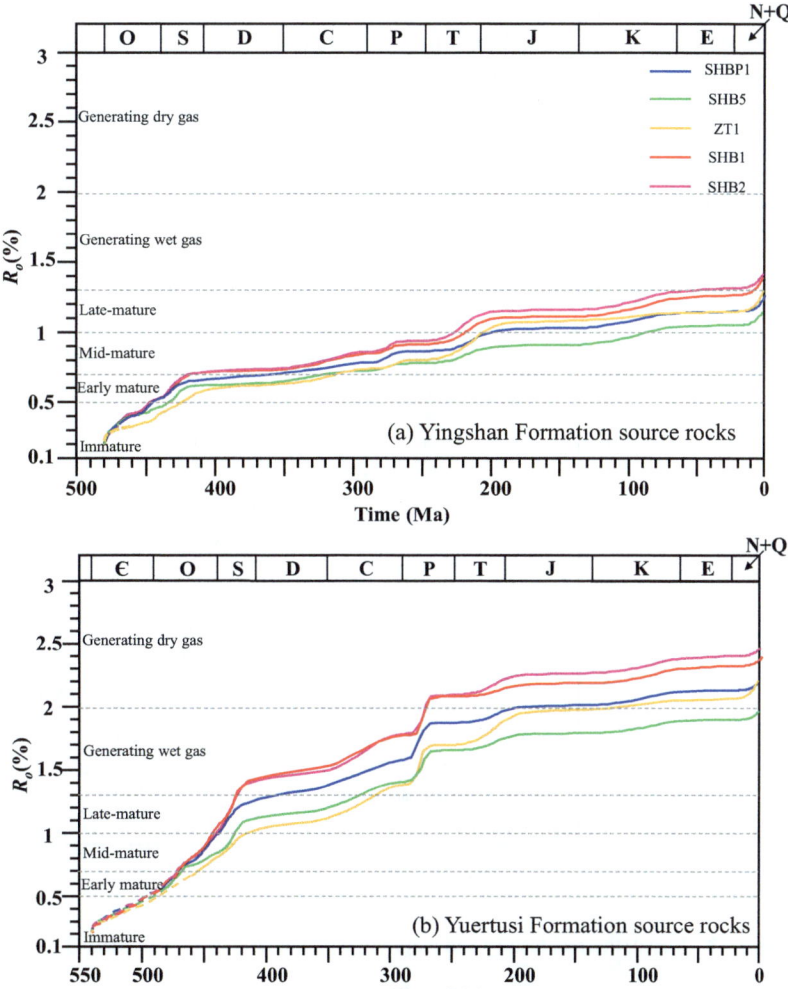

source rocks reached the gas stage, the deep burial and high-pressure environment inhibited their thermal evolution, raising the liquid oil generation window's upper limit by ~ 0.5% (Gu et al. 2019; McTavish. 1998; Ren et al. 2020; Wu et al. 2016).

5.4.2 Maturation Evolution of the Paleozoic Source Rocks in the Sichuan Basin

Figure 5.13 clearly depicts the maturation evolution of Paleozoic source rocks in Well MS1 (northern Sichuan Basin). Lower Cambrian Qiongzhusi Formation source rocks initiated maturity in the Ordovician, but the Caledonian tectonic uplift subsequently slowed their thermal evolution. They then underwent a phase of rapid evolution beginning in the Middle and Late Triassic, eventually reaching the late-mature stage ($1.3\% < R_o < 2.6\%$). The strata reached the main gas generation stage ($R_o > 2.6\%$) during the Jurassic, driven by continuous burial and rapidly increasing temperature. Maturation culminated during the Middle and Late Cretaceous, halting completely thereafter due to subsequent strata uplift and cooling. Conversely, Permian source

rock maturation tracked a distinct, multi-stage trajectory. They developed swiftly during the Early Triassic, attaining the early maturity stage ($0.5\% < R_{equ} < 0.7\%$), but then paused in the Middle Triassic. Evolution rapidly reactivated in the Middle Jurassic, yet decelerated post-Late Jurassic. Remarkably, the Permian source rocks endured two temporary cessation phases during the Triassic and attained an exceptionally high peak maturity of 3.6%.

The Permian mantle plume extensively perturbed the thermal regime by the close of the Early Permian, potentially influencing Paleozoic source rock maturity in the Sichuan Basin (Fig. 5.14). Most source rocks (e.g., in Wells MS1, DS1, JS1, and the Lower Permian/Longmaxi Formation) achieved normal maturation by the Early Permian's end, their evolution primarily driven by Mesozoic sedimentation, not plume heat. This thermal resilience occurred either in the northern/eastern Sichuan Basin due to lower heat flow anomalies, or in the central uplift/southwest, where shallow source rock burial precluded a substantial temperature rise, despite high heat flow. Conversely, the Lower Cambrian Qiongzhusi Formation in Wells W28, GS6, and YF1 matured swiftly, reaching the late stage ($R_o = 1.0\%–1.3\%$), clearly indicating the plume's

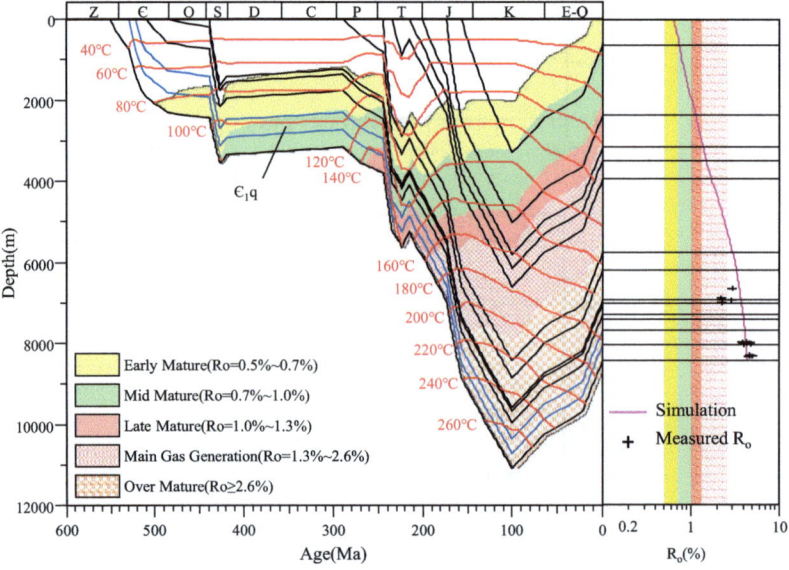

Fig. 5.13 Burial, thermal history and maturation of source rock in well MS1 Qiu et al. (2021)

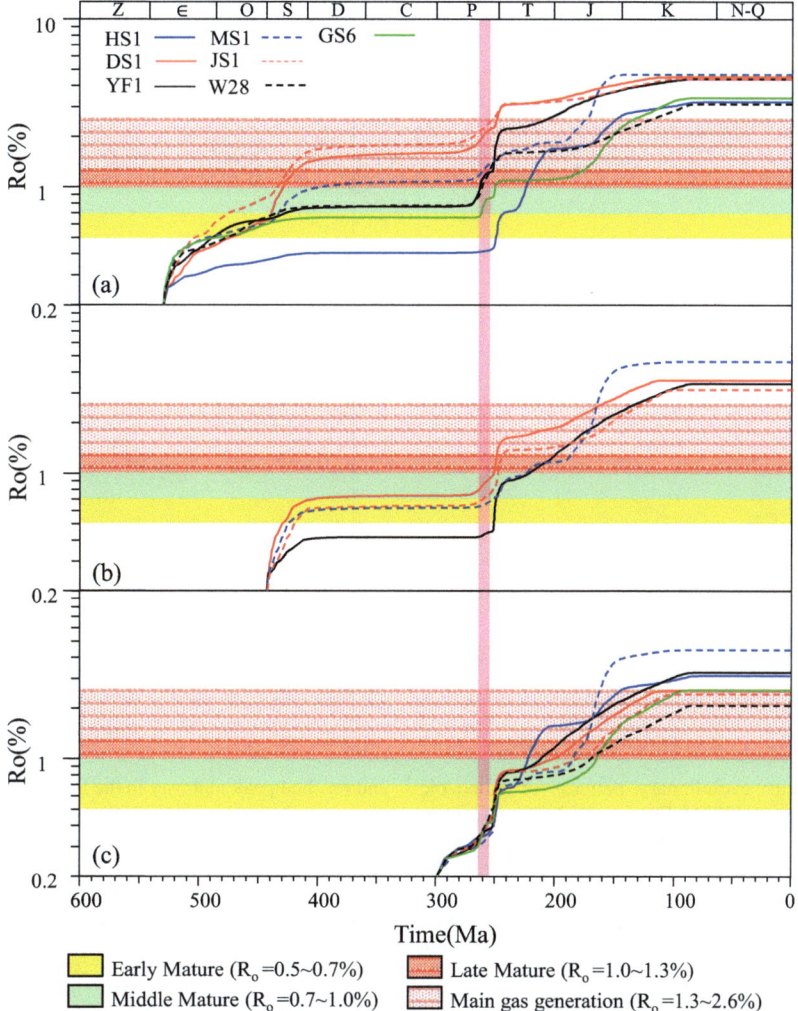

Fig. 5.14 Maturation histories of **a** the Lower Cambrian Qiongzhusi Formation, **b** the Lower Silurian Longmaxi Formation, and **c** the Lower Permian source rocks in wells HS1, JS1, YF1, MS1, DS1, W28 and GS6 Feng et al. (2021)

intense thermal impact on this unit. Well HS1 was an anomaly: its Qiongzhusi Formation remained unaffected despite high heat flow. This is ascribed to significant Caledonian erosion during Central Paleouplift formation, which removed overlying strata (Liu et al., 2021) and kept the Qiongzhusi Formation at a critically shallow burial depth during the Early Permian plume event. These varied patterns collectively demonstrate that the Emeishan mantle plume's thermal effect on maturity was fundamentally contingent on the co-occurrence of a high heat flow anomaly and adequate burial depth toward the end of the Early Permian.

By comparing differences in maturity evolution, we identify distinct maturation patterns for both the Lower Permian and the Lower Cambrian Qiongzhusi Formation source rocks (Fig. 5.15). Three specific maturity evolution patterns are classified for the Qiongzhusi Formation: (1) continuous Heating: Observed in southern Sichuan (Fig. 5.15a); (2) Silurian–Permian Heating Cessation: Found in western Sichuan and northwestern central Sichuan (Fig. 5.15b); and (3) complex Silurian–Permian and Triassic Heating Cessation: Prevalent across the central uplift, northern structural zone, and the western part of the eastern structural zone (Fig. 5.15c). Similarly, the Lower

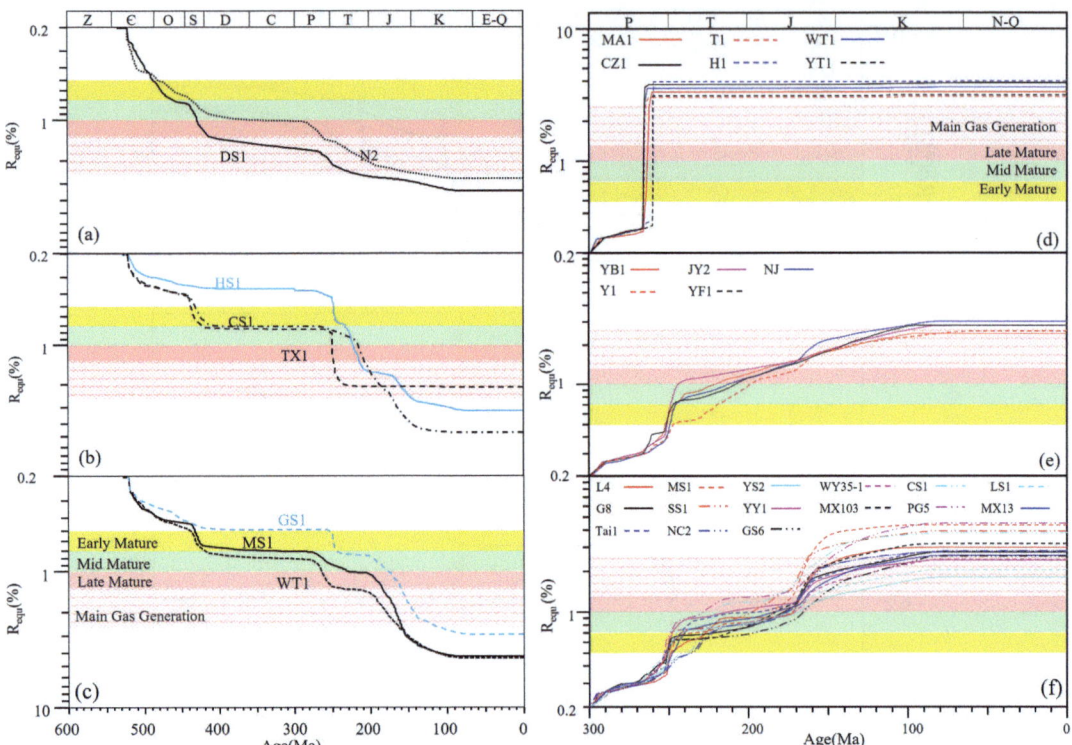

Fig. 5.15 Maturity evolutionary patterns of Paleozoic source rocks Qiu et al. (2021), Feng et al. (2022). **a** Continuous heating pattern of the Qiongzhusi Formation source rock. **b** Silurian–Permian heating cessation pattern of the Qiongzhusi Formation source rock. **c** Silurian–Triassic heating cessation patterns of the Qiongzhusi Formation source rock. **d** Permian source rocks show rapid maturation and reached peak maturity in the Early Triassic. **e** Permian source rocks underwent rapid changes during the Early Triassic, followed by continuous evolution. **f** Permian source rocks experienced two periods of rapid evolution (Early Triassic and Middle Jurassic), with significantly slower maturation during the intervening Middle Triassic

Permian source rocks exhibit three maturity evolutionary patterns: Pattern I, where rocks that experienced magmatic intrusions matured rapidly and reached peak maturity during the Middle Permian (Fig. 5.15d); Pattern II, Found in southwestern Sichuan Basin, this pattern shows rapid Early Triassic evolution followed by a slower rate (Fig. 5.15e); and Pattern III (Most Common): Characterized by two rapid evolutionary periods (Late Permian and Middle Jurassic), separated by a temporary Middle Triassic maturation cessation (Fig. 5.15f). Overall, Mesozoic sedimentation, uplift, and the high heat flow from the Permian Emeishan basalt eruption were the main controls on the thermal maturity of these Paleozoic source rocks.

5.5 Oil and Gas Preservation Phase State in Ultradeep Reservoirs of the Tarim Basin

The reservoir fluid phase is linked to the source rock hydrocarbon generating phase and the subsequent secondary alteration of the oil post-filling. In the Shuntuoguole area of the Tarim Basin, the ultradeep Ordovician exhibits a complex fluid mix where light oil, volatile oil, condensate, and dry gas coexist (Fig. 5.16). Figure 5.17 illustrates the typical PVT phase for the Shuntuguole reservoir. Reservoirs within the No. 5 Fault (F5) demonstrate the most complex fluid phase, varying latitudinally: the fluid transitions from light oil in the northern section to volatile oil in the middle

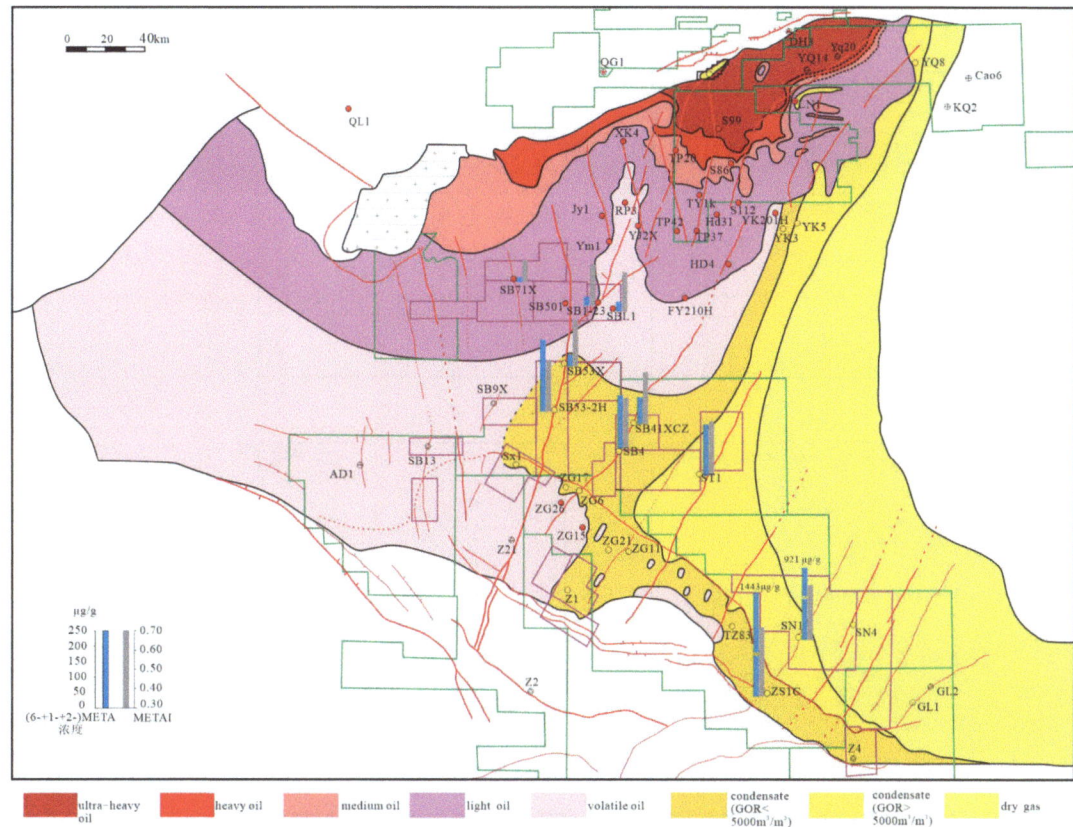

Fig. 5.16 Reservoir fluid phase of ultradeep Ordovician strata in Shuntuoguole and adjacent areas

section, concluding with the condensate phase in the southern section.

5.5.1 Geochemical Characteristics and Secondary Alteration of Oil from the Ultradeep Ordovician

(1) Geochemical Characteristics of Ultradeep Ordovician Oil

Whole-oil chromatograms from the Shuntuoguole area show major n-alkane peaks at low carbon numbers (Fig. 5.18), strongly indicating high oil maturity. Additionally, all oils exhibit high methylcyclohexane (MCH) concentrations and low aromatic hydrocarbon content. The sole exception is oil from Well SB4, notable for its relatively elevated toluene and benzene levels.

The pristane/phytane (Pr/Ph) ratio of the oil ranged from 0.93 to 2.25, which clearly suggests the source rocks were deposited in a slightly reducing sedimentary environment. Furthermore, the oil generally displayed relatively low Pr/nC_{17} ratios (0.10–0.38) and Ph/nC_{18} ratios (0.12–0.47). As shown in Fig. 5.19, these ratios are diagnostic, indicating that the source rocks primarily comprised Type II organic matter derived from marine algae. A contrasting signature was observed in oil samples from Well SB71X, which also had relatively low Pr/nC_{17} and Ph/nC_{18} ratios, but these values reflect a source rock containing mixed Type II/III organic matter. Finally, some oil from wells ST1 and SN1, as well as oil recovered from the No. 4 and No. 5 south faults, exhibited particularly low Pr/nC_{17} and Ph/nC_{18} ratios. In these specific cases, the low values are interpreted not as a source rock indicator, but as a signature of high thermal maturity.

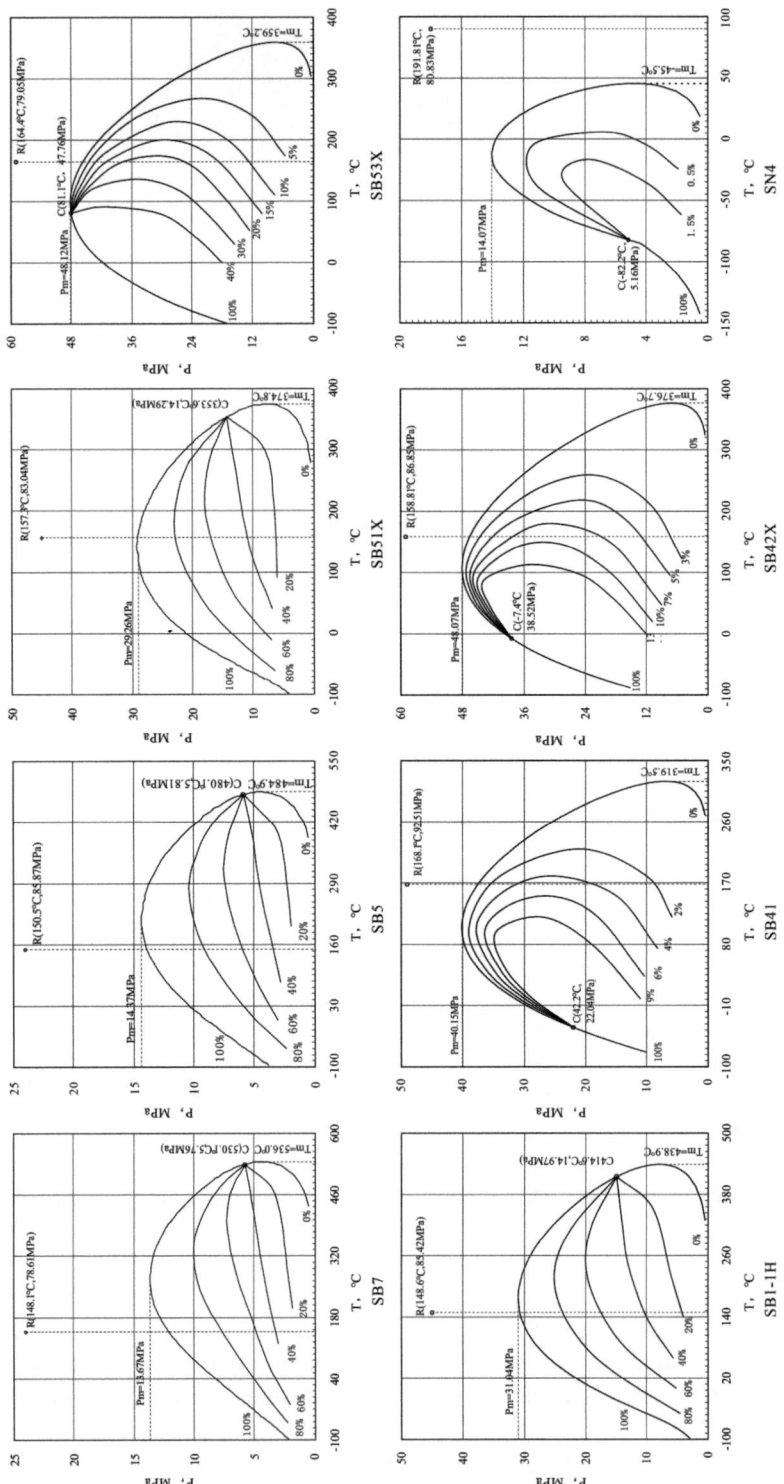

Fig. 5.17 Typical PVT fluid phase diagram of the Ordovician reservoir in the Shuntuoguole area

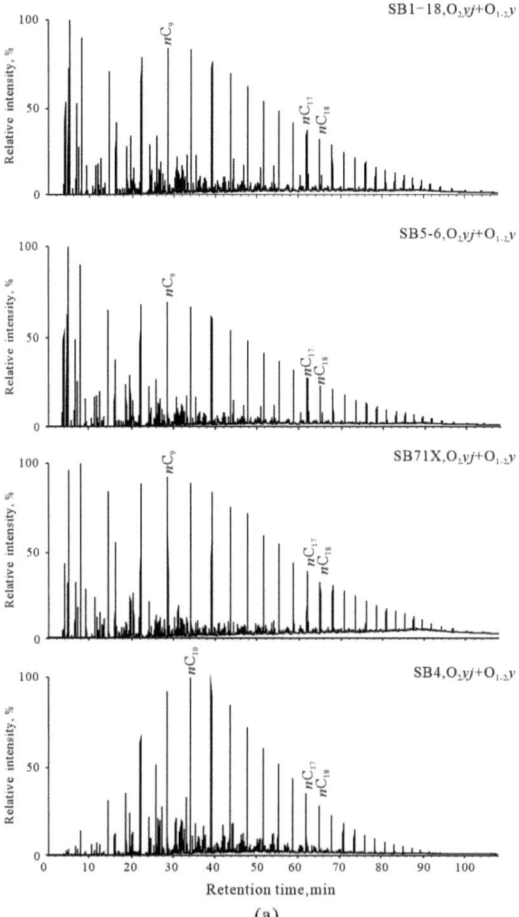

Fig. 5.18 **a** Whole-oil gas chromatograms and **b** $C_2–C_8$ light hydrocarbon gas chromatograms of oil samples from ultradeep Ordovician strata in the North Shuntuoguole area. nC_3 = n-propane; nC_4 = n-butane; nC_5 = n-pentane; $2-MC_5$ = 2-methylpentane; $3-MC_5$ = 3-methylpentane; nC_6 = n-hexane; MCYC5 = methylcyclopentane; Bz = benzene; CYC_6 = cyclohexane; $2-MC_6$ = 2-methylhexane; $3-MC_6$ = 3-methylhexane; 1c3-DMCYCY5 = 1,*cis*,3-dimethylcyclopentane; 1t3-DMCYC5 = 1,*trans*,3-dimethylcyclopentane; 1t2-DMCYC5 = 1,*trans*,2-dimethylcyclopentane; nC_7 = nheptane; $MCYC_6$ = methylcyclohexane; Tol = toluene; $2-MC_7$ = 2-methylheptane; nC_8 = n-octane

Owing to its high maturity, ultradeep Ordovician oil from the Shuntuoguole area has a relatively low biomarker. Oil from the No. 7, No. 1, and No. 5 north faults has a complete series of tricyclic terpane and sterane (Fig. 5.20). The oil from the No. 4 and No. 8 faults has no detectable biomarkers. Oil with detectable biomarkers featured $C_{23}TT$ greater than $C_{21}TT$ and a lower content of C_{28} sterane, similar to marine oil from the Tazhong and Tabei Uplift.

(2) **Secondary Alteration of the Deep Oil**

① **Biodegradation**

Biodegradation is the most frequent secondary process impacting oil reservoirs. For instance, Ordovician heavy oil in the Tahe and West Lunnan Oilfields (Tarim Basin) is linked to Late Hercynian biodegradation. Nevertheless, the Ordovician oil in the Shuntuoguole area contrasts sharply

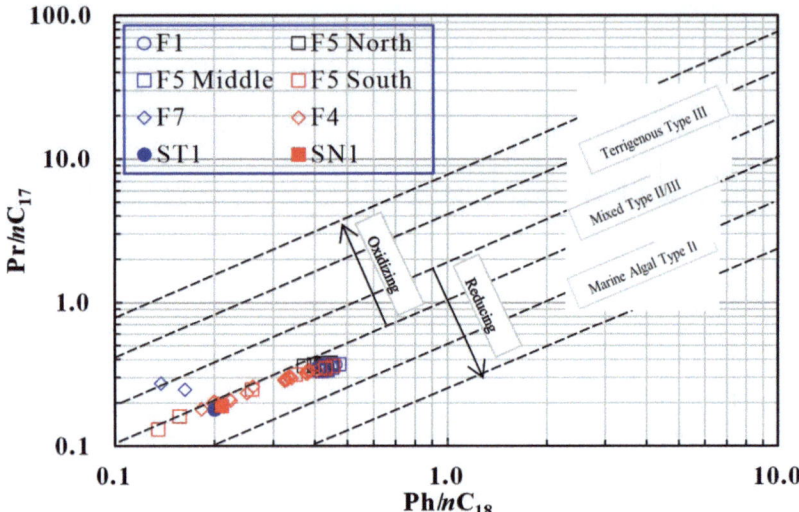

Fig. 5.19 Plot of Pr/nC_{17} versus Ph/nC_{18} of oil from the Shuntuotuole area, Tarim Basin

with this heavy oil. The Shuntuoguole oil mostly comprises light oil, volatile oil, and condensate oil. Critically, it contains few Unresolved Complex Mixtures (UCMs) and no detectable 25-norhopane, both of which are common indicators of bacterial alteration. This compositional evidence strongly suggests that the oil in the Shuntuoguole area has undergone low, or negligible, biodegradation.

The resistance to biodegradation of 2-methylalkanes (2 M) with five and six carbons (C_5 and C_6) is inherently lower than that of the corresponding 3-methylalkanes (3 M). Consequently, the 2 M-C_5/3 M-C_5 and 2 M-C_6/3 M-C_6 ratios decrease with increasing biodegradation. Furthermore, as n-alkanes are more susceptible to biodegradation than iso-alkanes, relevant ratios like i-C_5/n-C_5 and 3 M-C_5/n-C_6 increase proportionally with biodegradation. The Ordovician oil from the Shuntuoguole area consistently displayed relatively high 2 M-C_5/3 M-C_5 ratios (1.41–1.81) and 2 M-C_6/3 M-C_6 ratios (0.79–1.09), alongside relatively low i-C_5/n-C_5 ratios (0.31–0.90) and 3 M-C_5/n-C_6 ratios (0.16–0.37) (Fig. 5.21). The Shuntuoguole oil's chemical signature contrasts sharply with two biodegraded Ordovician oil samples from the Tahe Oilfield's Aiding (AD) block. The AD samples exhibited relatively low 2 M-C_5/3 M-C_5 (1.55–1.61) and 2 M-C_6/3 M-C_6 ratios (0.49–0.51), yet high i-

C_5/n-C_5 (0.90–0.98) and M-C_5/n-C_6 ratios (0.35–0.46). These diagnostic ratio differences conclusively indicate that the Shuntuoguole oil underwent no biodegradation.

The Ordovician Yijianfang and Yingshan reservoirs (O2yj + O1-2y) in North Shuntuoguole lack the conditions for biodegradation. The overlying strata—including the Upper Ordovician (O3q, O3l, O3s) and Lower Silurian (S1k, S1t) ormations—created a substantial seal, exceeding 1800 m thick before S2y deposition. Although the Middle Silurian Yimugantawu Formation (S2y) was completely eroded at the end of the Late Caledonian movement, the Ordovician reservoirs remained protected and unsusceptible to biodegradation due to the overlying Sangtamu Formation (O3s) mudstone, which maintained a thickness of 700–1000 m and served as a persistent cap rock. Furthermore, despite the Upper Carboniferous Xiaohaizi Formation (C2x) being eroded during the Late Hercynian, the Lower Carboniferous Bachu Formation mudstone (C1b) served as an additional, effective cap rock for the reservoirs during that subsequent period.

② **Evaporative Fractionation**

Evaporative fractionation, alternatively termed gas washing or phase-controlled fractionation, is a widely adopted mechanism for describing

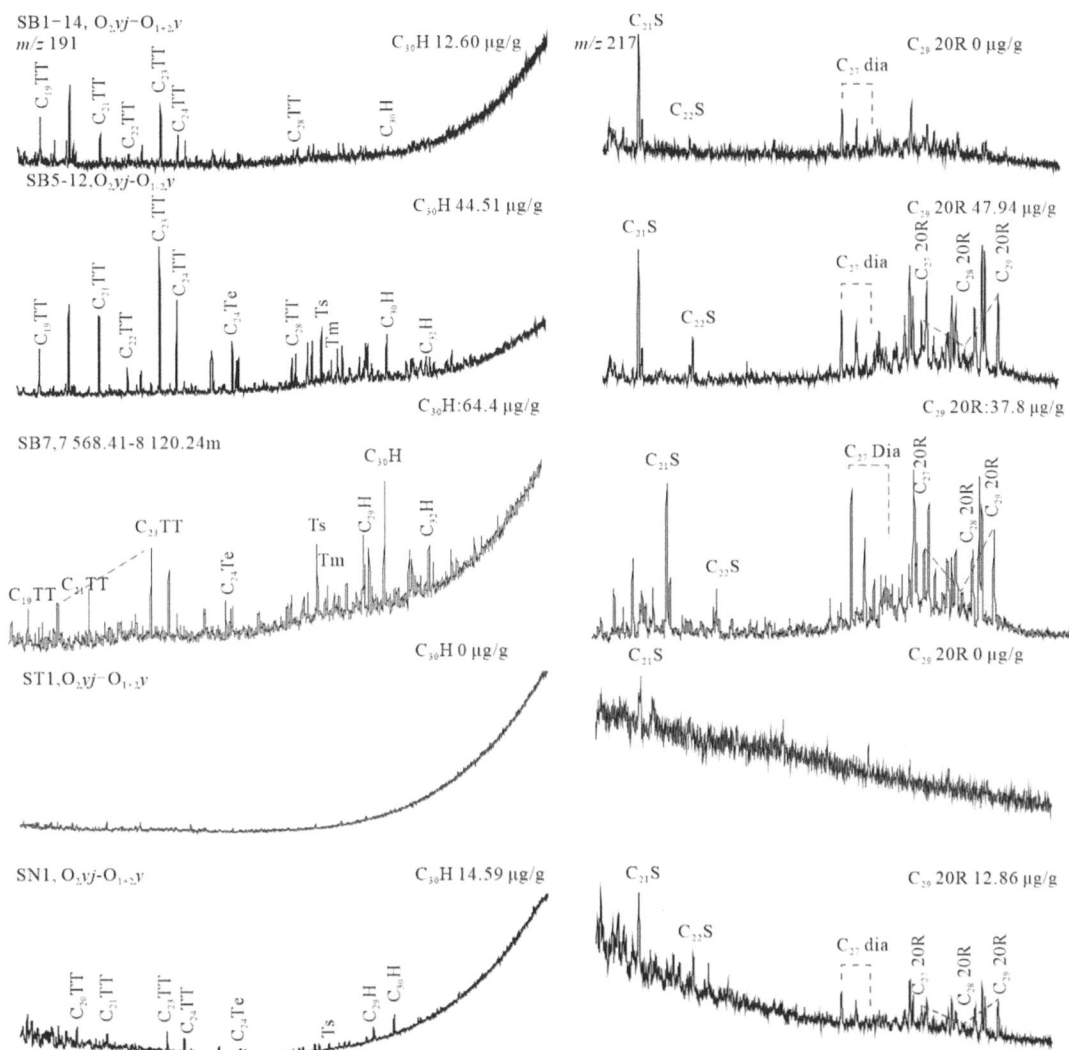

Fig. 5.20 Mass chromatograms of terpanes and steranes of ultradeep Ordovician oil from the Shuntuoguole area

oil diversity within complex sedimentary basins. Specifically, Zhang et al. (2011) concluded that the occurrence of waxy oil and condensate oil in East Lungu, as well as the condensate found in the east of the Tazhong Uplift, are direct results of this evaporative fractionation process.

The degree of evaporative fractionation is conventionally determined by plotting the logarithm of the molar concentration of n-alkanes Ln[MC(i)] against their corresponding carbon number. An analysis of molar fraction plots for typical oils across various Shuntuoguole fault systems (Fig. 5.22) revealed that almost all oils

exhibited linear relationships between Ln[MC(i)] and carbon number. This linearity is strongly supported by high regression coefficients (R^2) consistently exceeding 0.98. This compelling statistical evidence indicates that no evaporative fractionation occurred in the Shuntuoguole oil, consistent with prior work.

Oils that have undergone evaporative fractionation typically exhibit high toluene/n-heptane ratios and low n-heptane/MCH ratios. However, the analysis of oils from the Shuntuoguole area generally shows the opposite pattern: almost all samples have low toluene/n-heptane ratios (0.10–

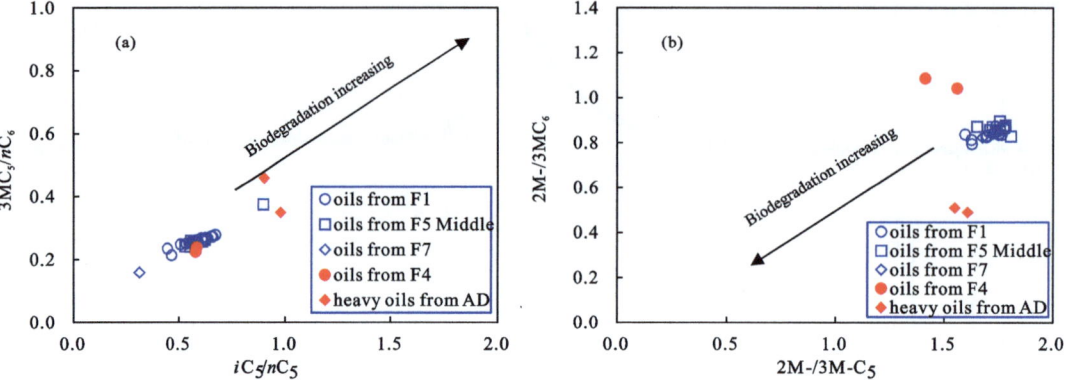

Fig. 5.21 **a** 3 M-C$_5$/nC$_6$ versus iC$_5$/nC$_5$ and **b** 2 M-/3 M-C$_6$ versus 2 M-/3 M-C$_5$ of oil samples from ultradeep Ordovician strata in the Shuntuoguole area, Tarim Basin

0.38, averaging 0.29) and low n-heptane/MCH ratios (1.50–1.80, averaging 1.71) (Fig. 5.23). The only exception is the highly mature oil in the southern Shuntuoguole area, which displays high values for both the toluene/n-heptane and n-heptane/MCH ratios. The consistently low values observed in the majority of the samples further confirm the conclusion that most of the oil experienced no evaporative fractionation.

③ **Oil Cracking**

Methyldiamantane concentration serves as a chemical indicator for assessing the degree of oil cracking (Dahl et al. 1999). Crucially, determining the appropriate methyldiamantane baseline for the oil from a specific basin is essential for ensuring the accuracy of this method. For instance, Zhang et al. (2011) suggested the methyldiamantane baseline for Tarim Basin marine oil is approximately 20 ug/g. They further suggested that any higher observed baseline values are likely attributable to differences in the analytical procedures and quantitative methods employed by various laboratories.

Using D$_{16}$-adamantane as an internal standard, the diamondoid concentrations of 27 diamondoid isomers of oil from the Shuntuoguole area were quantified (Fig. 5.24). The (4- + 3-) methyldiamantane [(4- + 3-)MD] concentrations of oil from wells SB7 and SB71X on the No. 7 fault were the

lowest, with values of 0–9.41 µg/g. From north to south, the (4- + 3-)MD concentrations of oil from the No. 5 fault gradually increase. The average (4- + 3-)MD concentrations of oil from the north section, middle section and south section were 38.43 (8 samples), 45 (9 samples) and 144.16 µg/g (7 samples), respectively. The (4- + 3-)MD concentration of oil from the No. 1 fault varied from 14.05 to 37.74 µg/g, with an average value of 26.72 µg/g (17 samples). The condensate oil in the No. 4 fault had an average (4- + 3-)MD concentration of 96 µg/g (8 samples). The (4- + 3-)MD concentrations of oil from wells ST1 and SN1 were higher, with values of 641 and 1598 µg/g (Fig. 5.25), respectively.

The (4- + 3-)MD concentrations of oil from different faults reveal that from the northwest to the southeast, the (4- + 3-)MD concentration gradually increases, suggesting that oil cracking increases.

④ **Thermochemical Sulfate Reduction (TSR)**

TSR is the process where petroleum hydrocarbons react with inorganic sulfate in high-temperature (80–200 °C) gypsum-bearing reservoirs, yielding CO$_2$, H$_2$S, and solid asphalt.4 Under extreme conditions, oil can be fully oxidized to CO$_2$ and H$_2$S. Mango (1997) noted that TSR significantly increases the oil's K$_1$. In the Shuntuoguole area, the Ordovician oil K$_1$ ratio 0.91–1.14 suggests only weak TSR activity. An H$_2$S content greater

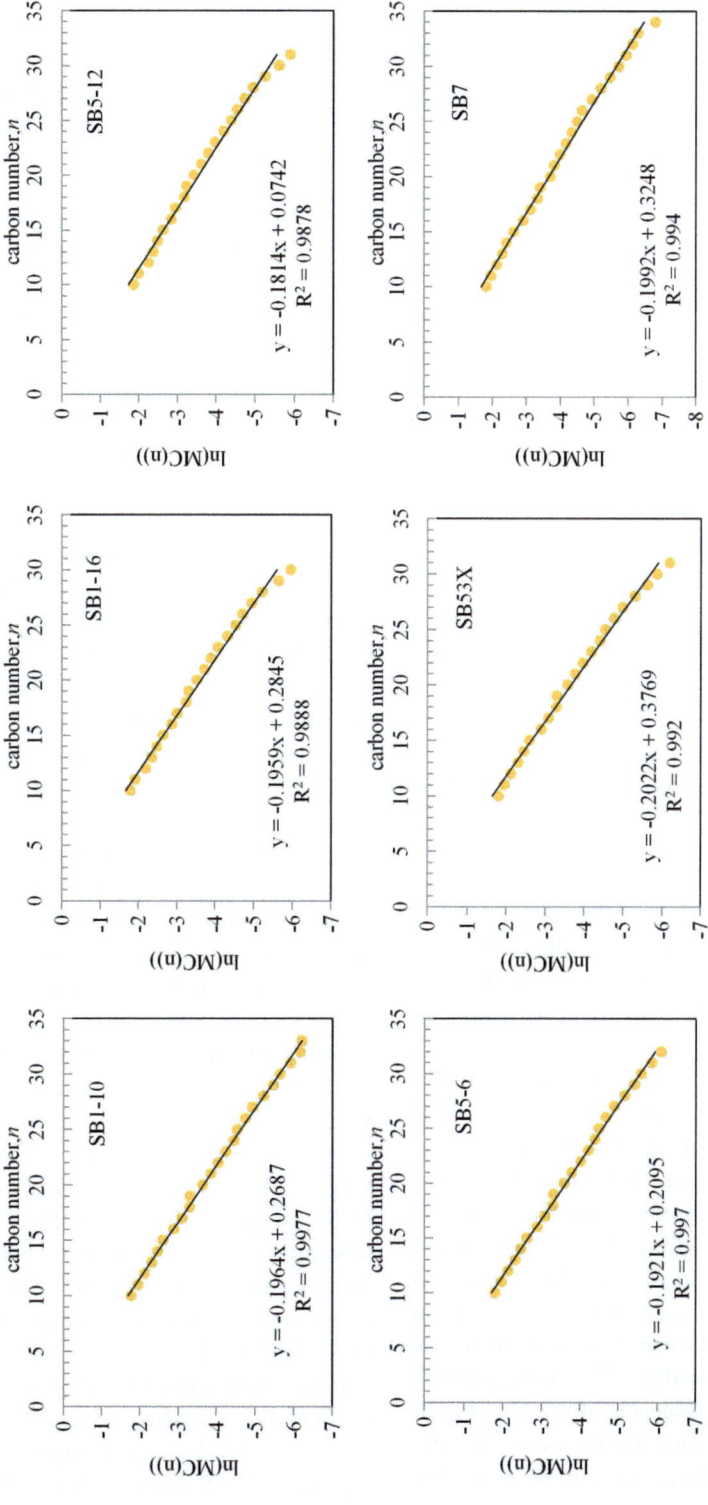

Fig. 5.22 Relationships between the molar fractions of *n*-alkanes and the carbon numbers of oil in the Shuntuoguole area

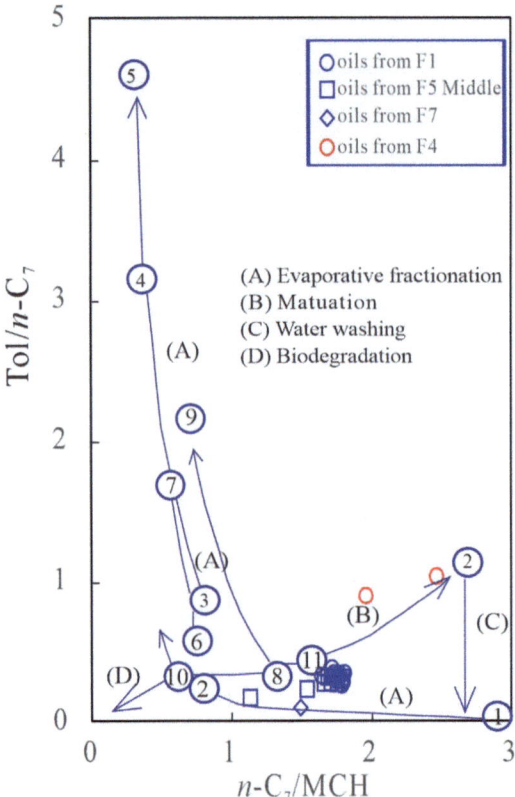

Fig. 5.23 Toluene/nC_7 versus nC_7/MCC$_6$ ratios of ultra-deep Ordovician oil samples from the Shuntuoguole area, Tarim Basin

than 3% generally indicates the occurrence of TSR. In the Shuntuoguole, only the natural gas from well SB4 has a H$_2$S content greater than 3%, suggesting that the TSR in most Ordovician reservoirs is weak.

Thiadiamondoids (thiaadamantane, thiadia-mantane, thiatriadamantane) are detectable in most Shuntuoguole oils (Fig. 5.26). Using D$_{16}$-adamantane w as the internal standard, the oil's thiadiamondoid concentration ranged from 0 to 80 μg/g, with most concentrations below 20 μg/g, indicating a low degree of TSR. Generally, the thiadiamondoid concentration of oil from the No. 4 fault is higher than that of oil from the other faults, suggesting that a slightly higher TSR degree occurs in the reservoir of the No. 4 fault, which is consistent with the H$_2$S content from natural gas from Ordovician reservoirs in the No. 4 fault.

For the oil from well SN1, although the H$_2$S content is low, not only can lower thidiamondoids be detected with a concentration of 80 μg/g but also thiatetraamantane and tetramantanthoil can be detected (Fig. 5.27), suggesting that the oil from well SN1 experienced moderate TSR.

5.5.2 Geochemical Characteristics and Origin of Natural Gas

(1) Chemical Composition and Isotopic Characteristics of Natural Gas

The ultradeep Ordovician natural gas in the North Shuntuoguole (SB) area is n-alkane dominated. Methane (C1) content ranges from 46.89 to 84.18%, and heavy hydrocarbon gas (C2–C5) spans 2.93–43.59%. Although the initial test gas from vertical well SB1 had a high dryness coefficient (C1/C1-5) of 0.97, all other SB wells range from 0.52 to 0.88, classifying them as wet gas (below 0.95). In contrast, the Ordovician natural gas found in the GuLong (GL), Gucheng (GC), Shuntuoguole (ST), and South Shuntuoguole (SN) areas exhibits significantly different characteristics, with relatively low heavy hydrocarbon contents (less than 2%) and dryness coefficients consistently greater than 0.95. Consequently, the natural gas from these four regions is classified as dry gas. Notably, the Gucheng area holds the highest dryness coefficient, with a value of approximately 0.998.

Ordovician natural gas in the SB area exhibits relatively light carbon isotopic composition. The δ^{13}C1(methane) values range from − 49.6‰ to − 44.7‰, with δ^{13}C2 and δ^{13}C3 values spanning − 39.3‰ to−− 32.5‰ and _35.6‰ to_ 30.6‰, respectively. This methane carbon isotope signature is notably lighter than that of the natural gas found in the ST, SN, GL, and GC areas, where δ^{13}C1 values range from − 40.2‰ to − 30.2‰. The carbon isotopic distribution for the Ordovician natural gas in the SB area exhibits a clear positive series distribution (δ^{13}C1 < δ^{13}C2 < δ^{13}C3 < δ^{13}C4), strongly indicating an organic origin. Similarly, most Ordovician natural gas from the ST, SN, GL, and GC areas also

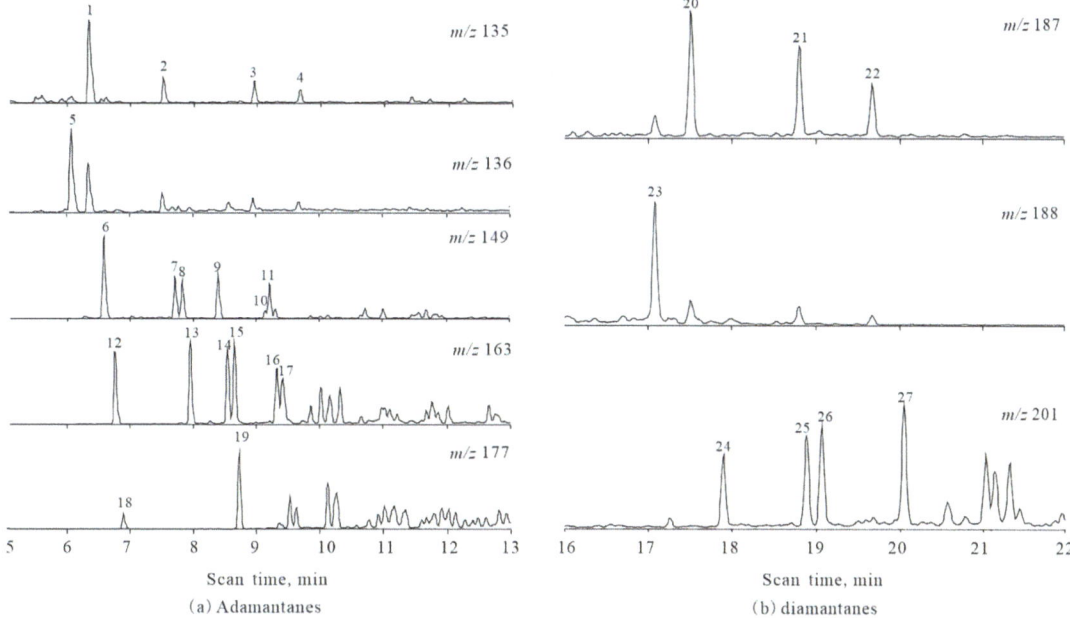

Fig. 5.24 Mass chromatograms of diamondoids of oil from well SB41X-C

Fig. 5.25 Plot of the concentrations of (4- + 3-) MD versus C_{29} $\alpha\alpha\alpha20R$ of oils from the North Shuntuoguole area

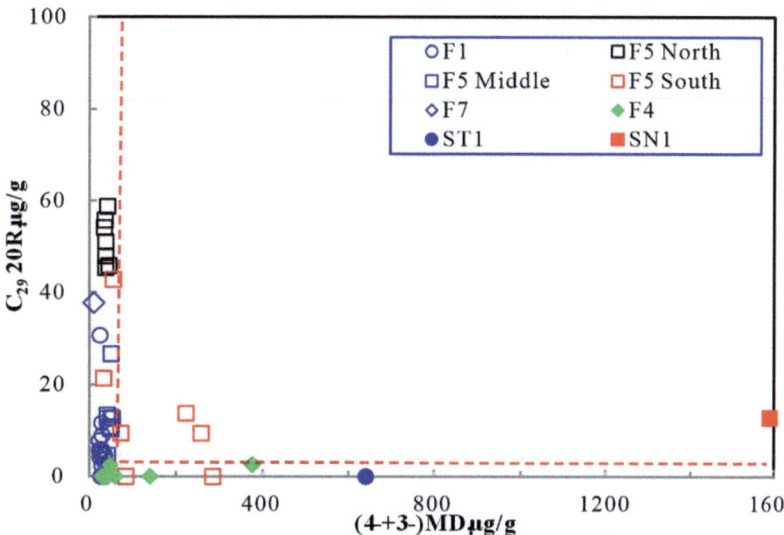

shows a positive series distribution. However, a few samples from these regions display partial inverse carbon isotope distributions. For instance, the $\delta^{13}C1$, $\delta^{13}C2$ and $\delta^{13}C3$ values of the gas from well GC 6 are 33.3‰, 37.6‰ and 34.9‰, respectively, while the corresponding values for gas from well GC 11 are 30.8‰, 30.7‰ and 31‰.

(2) **Origin of Natural Gas**

The majority of Ordovician natural gas in the Shuntuoguole area exhibits a positive series distribution of carbon isotopes, which suggests the gas is entirely of organic origin. All natural gas in the area is definitively thermogenic, as demonstrated by its placement in both Bernard's

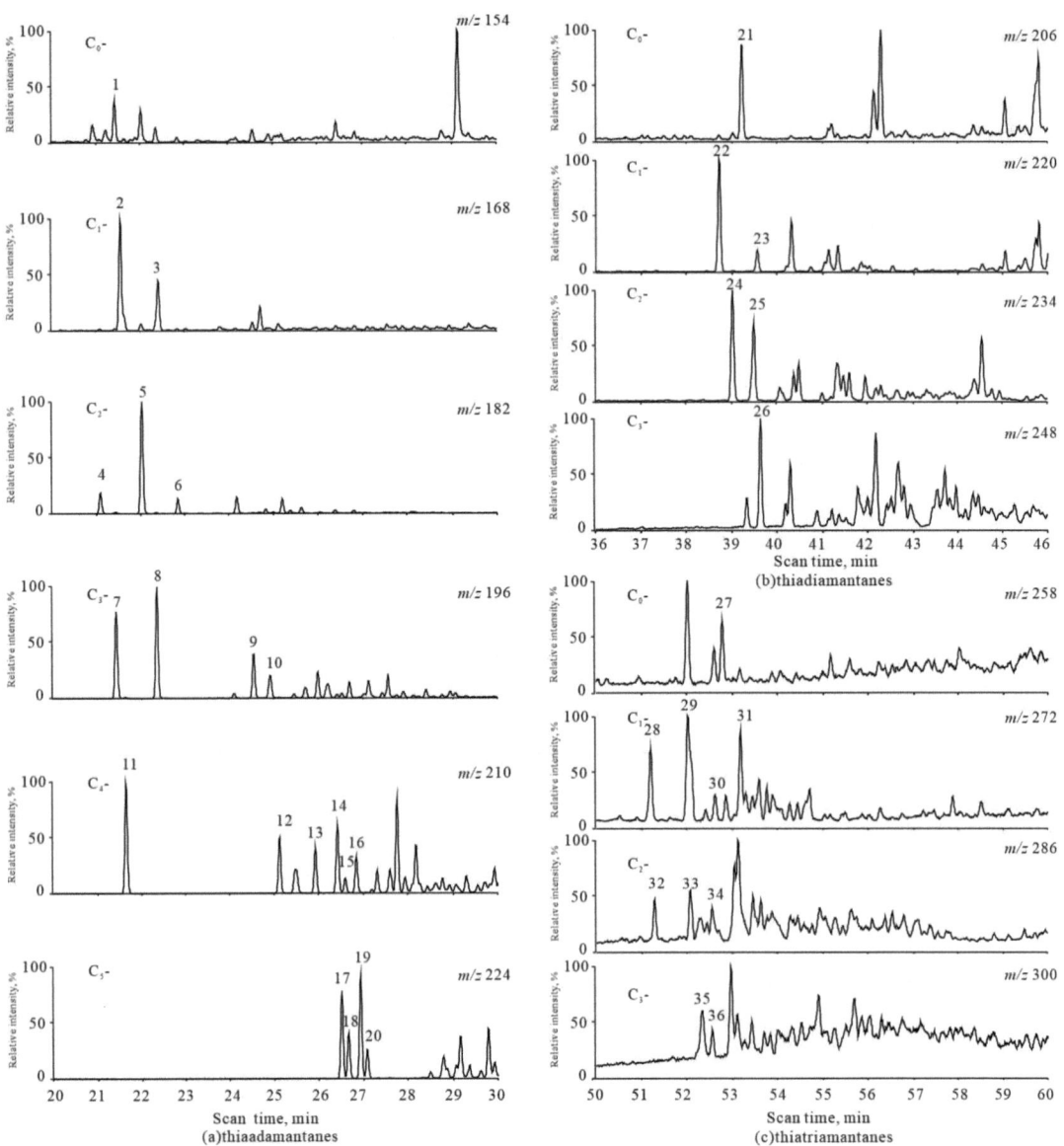

Fig. 5.26 Mass chromatograms of thiadiamondoids of oil from well SB41X-C

(Fig. 5.28a) and Molikv's (Fig. 5.28b) diagrams relating wetness and $\delta^{13}C1$. The $\delta^{13}C2$ (carbon isotopic value of ethane) can classify gas types, with values typically around 29–28‰ ifferentiating oil-type from coal-type gas in China. Since the Shuntuoguole Ordovician reservoir's $\delta^{13}C2$ values are less than 32‰, all this gas is classified as oil-type gas. Furthermore, the $\delta^{13}C1$–$\delta^{13}C2$ diagram (Fig. 5.29) shows that the Ordovician natural gas in Shuntuoguole is compositionally similar to natural gas generated from Type II kerogen in the Delaware/Val Verde Basin.

(3) **Two Types of Natural Gas in the Shuntuoguole Area**

Li et al. (2012) utilized the Ln(C1/C2) vs. Ln(C2/C3) diagram to differentiate between kerogen-cracking and oil-cracking gases in sapropelic organic matter. In the SB area Fig. 5.30, the

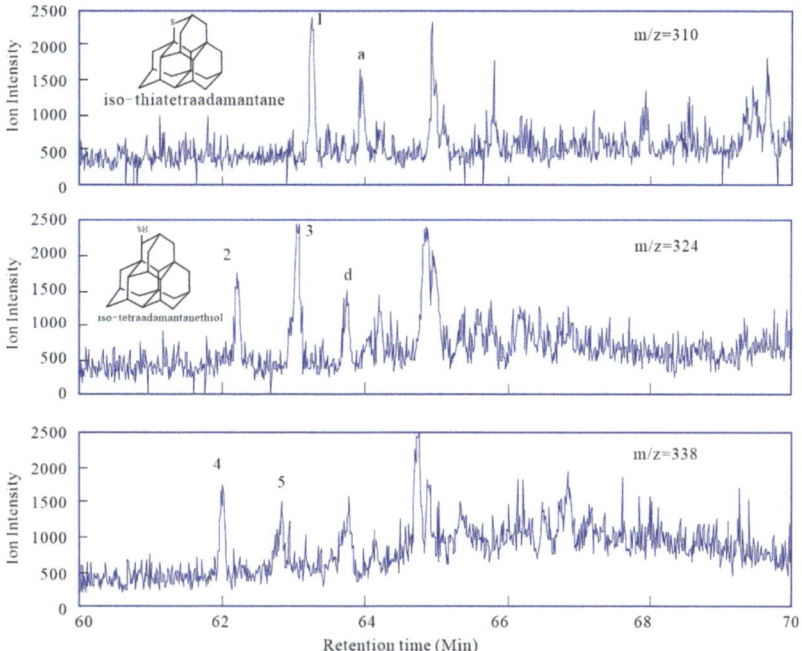

Fig. 5.27 Mass chromatograms of thiatetramantane and tetramantanthoil of oil from well SN1

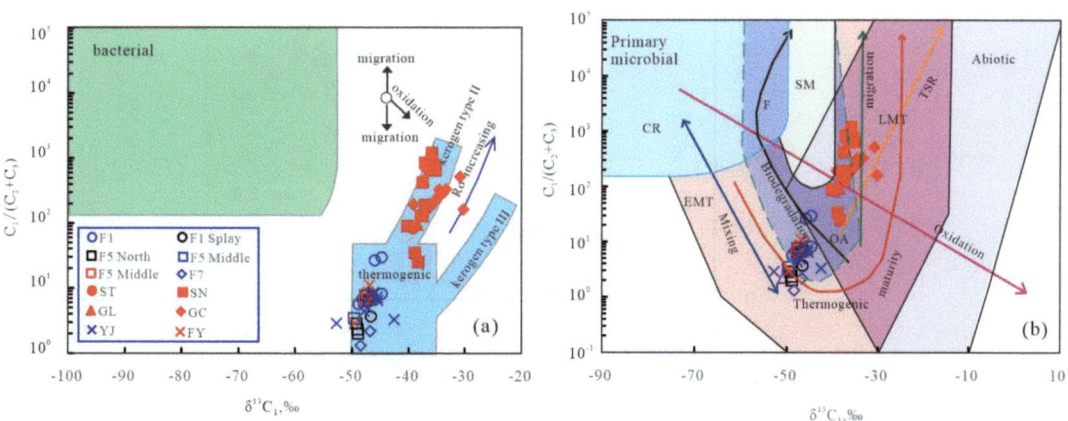

Fig. 5.28 Genetic classification of gas in the Shuntuoguole area using the $C_1/(C_2 + C_3)$ and $\delta^{13}C$ values of methane

Ordovician gas Ln(C1/C2) values range from 0.81 to 3.55, and Ln(C2/C3) from 0.35 to 1.65. Most gas maturity is below 1.0%. Analyzed samples plot between the kerogen-cracking and oil-cracking fields, suggesting a mixed gas combined with oil-cracking gas. Conversely, most natural gas samples from the ST, SN, GL, and GC areas exhibit Ln(C1/C2) values greater than 4 and widely varying Ln(C2/C3) values (0.55–3.4). Critically, these samples plot directly on the oil-cracking gas curve, indicating the gas is predominantly sourced from oil cracking. Thermal maturity in the ST, SN, and GL areas ranges from 1.3 to 2.0%, while the GC area maturity is significantly higher, exceeding 2.0%.

Fig. 5.29 Carbon isotope ratios of methane and ethane in ultradeep Ordovician natural gas from the Shuntuoguole area

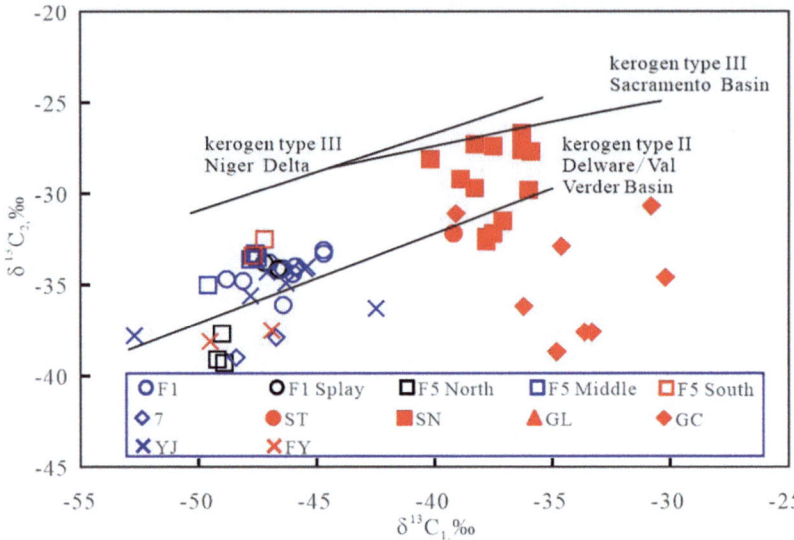

Fig. 5.30 Ln(C1/C2) versus Ln(C2/C3) plot for Ordovician gas in Shuntuoguole and adjacent areas

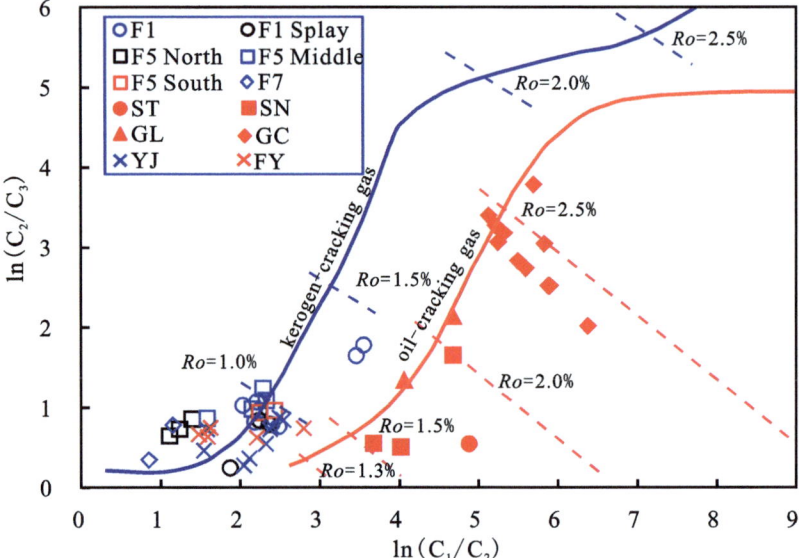

5.5.3 Preservation of the Oil Phase in Ultradeep Ordovician Strata

Oil-cracking kinetic parameters effectively predict the crucial geological temperature for oil phase preservation. Various researchers have calculated the stability temperatures of separate oil phases using these parameters (Waples 2000; Tian et al. 2006; Ma 2015; Li et al. 2021). For example, Waples (2000) proposed that the separate oil phase can be maintained at 179–182 °C

under a slow heating rate 2 °C/Ma, whereas the oil remains stable at a higher temperature of 194 °C when the heating rate increases significantly to 20 °C/Ma. This is supported by Ma (2015), who specifically calculated that Tarim Basin marine oils (Tahe Oilfield) remain preserved as a separate oil phase up to 178 °C.

In the SB area, the paleogeothermal history and current Ordovician ultradeep temperatures remained below 180 °C. This lower thermal regime permitted the preservation of light and volatile oil phases in the reservoirs. Moreover, the natural gas here is primarily kerogen-cracking gas, with only a minor contribution from early-stage oil-cracking gas. Conversely, the ST, SN, GL, and GC areas experienced paleogeothermal temperatures exceeding 180 °C, and their present-day temperatures are significantly higher. This intense heat caused large-scale oil cracking, restricting the preservation window to condensate and dry gas phases only. The natural gas composition in these areas is consistent with later-stage oil cracking.

5.6 Summary

This chapter summarizes the thermal and pressure fields in ultradeep strata across the Tarim and Sichuan Basins by highlighting five key contrasts: (1) Thermal Regime: The Tarim Basin is characterized by "cold" heat flow and thermal gradients, while the Sichuan Basin is typically "warm". (2) Paleoheat Flow Evolution: Thermal modeling indicates that the Tarim Basin interior's paleoheat flow diminished after the Early Cambrian, with a single abrupt peak during the Early Permian. Conversely, the Sichuan Basin's paleoheat flow sustained three steady states from the Cambrian to Early Permian, followed by a rapid rise at the end of the Early Permian and a subsequent decline post-Late Permian. (3) Pressure History: Throughout geological time, the Upper Ordovician reservoir pressure in the Tarim Basin was primarily normal or mildly overpressured. In contrast, the Cambrian Formation pore pressure in the Sichuan Basin underwent an abnormal high-pressure stage, culminating in its maximum value during the deepest burial phase in the Late Cretaceous. (4) Source Rock Maturation: Maturity histories were reconstructed for the main Paleozoic source rocks in both basins. Furthermore, the maturity evolutionary patterns for three distinct sets of Paleozoic source rocks in the Sichuan Basin were successfully summarized. (5) Tarim Basin Fluid Phase and Alteration: The ultradeep Ordovician rocks in the Tarim Basin possess complex fluid phases, including light oil, volatile oil, condensate, and dry gas. The oil within this ultradeep Ordovician unit

underwent various secondary alterations, specifically biodegradation, evaporative fractionation, oil cracking, and TSR. It was determined that the geological temperature for oil phase preservation can reach 180 °C.

Acknowledgements This work was supported by the National Natural Science Foundation of China (Grants U24B6001 and U19B6003). We gratefully acknowledge the SINOPEC Southwest Oil & Gas Company, the SINOPEC Northwest Oilfield Company and the PetroChina Southwest Oil & Gas Field Company for providing geological data and core samples for this study.

References

Aplin AC, MacLeod G, Larter SR et al (1999) Combined use of confocal laser scanning microscopy and PVT simulation for estimating the composition and physical properties of petroleum in fluid inclusions. Mar Petrol Geol 16:97–110

Aziz K, Settari A (1979) Petroleum reservoir simulation. Elsevier Applied Science Publication, London

Barker C (1990) Calculated volume and pressure changes during the thermal cracking of oil to gas in reservoirs. AAPG Bull 74(8):1254–1261

Bethke CM (1985) A numerical model of compaction-driven groundwater flow and heat transfer and its application to the paleohydrology of intracratonic sedimentary basins. J Geophys Res-Sol Ea 90:6817–6828

Bloch S, Lander RH, Bonnell L (2002) Anomalously high porosity and permeability in deeply buried sandstone reservoirs: origin and predictability. AAPG Bull 86(2):301–328

Bourdet J, Pironon J, Levresse G et al (2008) Petroleum type determination through homogenization temperature and vapour volume fraction measurements in fluid inclusions. Geofluids 8:46–59

Bourdet J, Pironon J, Levresse G et al (2010) Petroleum accumulation and leakage in a deeply buried carbonate reservoir, Níspero field (Mexico). Mar Petrol Geol 27:126–142

Bowers GL (1995) Pore pressure estimation from velocity data: accounting from overpressure mechanisms besides undercompaction. SPE Drill Completion 10(2):89–95

Cao HY, Zhu CQ, Qiu NS (2016) Maximum paleotemperature of main Paleozoic argillutite in the Eastern Sichuan Basin. Chinese Chinese J Geophys-Ch 59(03):1017–1029 (NB: in Chinese with English abstract)

Chang J, Qiu NS, Xu W (2017) Thermal regime of the Tarim Basin, Northwest China: a review. Int Geol Rev 59(1):45–61

Chang J, Qiu NS, Li CX et al (2022a) Zircon He diffusion kinetics models and its implications on the thermal history reconstruction of the ancient strata in the craton basins. Chinese J Geophys-Ch 65(2):711–725 (NB: in Chinese with English abstract)

Chang J, Qiu NS, Song XY et al (2016a) Multiple cooling episodes in the Central Tarim (Northwest China) revealed by apatite fission track analysis and vitrinite reflectance data. Int J Earth Sci 105(4):1257–1272

Chang J, Qiu NS, Zhao XZ et al (2018) Mesozoic and Cenozoic tectono-thermal reconstruction of the western Bohai Bay Basin (East China) with implications for hydrocarbon generation and migration. J Asian Earth Sci 160:380–395

Chang J, Qiu NS, Zhao XZ et al (2016a) Present-day geothermal regime of the Jizhong depression in Bohai Bay basin, East China. Chinese J Geophys-Ch 59(3):1003–1016 (NB: in Chinese with English abstract)

Chang J, Yang X, Qiu NS et al (2022b) Zircon (U-Th)/He thermochronology and thermal evolution of the Tarim Basin, Western China. J Asian Earth Sci 230:105210

Dahl JE, Moldowan JM, Peters KE et al (1999) Diamondoid hydrocarbons as indicators of natural oil cracking. Nature 399(6731):54–57

Deng B, Liu SG, Li Z et al (2013) Differential exhumation at eastern margin of the Tibetan Plateau, from apatite fission-track thermochronology. Tectonophysics 591:98–115

Dutkiewicz A, Ridley J, Buick R (2003) Oil-bearing CO_2–CH_4–H_2O fluid inclusions: oil survival since the Palaeoproterozoic after high temperature entrapment. Chem Geol 194:51–79

Farley KA (2000). Helium diffusion from apatite: general behavior as illustrated by Durango fiuorapatite. J Geophys Res 105(B2):2903–2914

Feng CG, Liu SW, Wang LS (2009) Present-day geothermal regime in Tarim Basin, northwest China. Chinese J Geophys-Ch 52(11):2752–2762 (NB: in Chinese with English abstract)

Feng QQ, Qiu NS, Fu XD et al (2022) Maturity evolution of Permian source rocks in the Sichuan Basin, southwestern China: the role of the Emeishan mantle plume. J Asian Earth Sci 132:105180

Feng QQ, Qiu NS, Fu XD et al (2021) Permian geothermal units in the Sichuan Basin: implications for the thermal effect of the Emeishan mantle plume. Mar Petrol Geol 132:105226

Gao SL, Yao WH, Zhu GS (2004) Pressure features of Upper Triassic series of middle-west Sichuan bas in and the assemble of oil and nature gas. Northwest Geol 37(1):75–79

Goldstein RH (2001) Fluid inclusions in sedimentary and diagenetic systems. Lithos 55(1–4):159–193

Gu Y, Wan YL, Huang JW et al (2019) Prospects for ultra-deep oil and gas in the "deep burial and high pressure" Tarim Basin. Petrol Geol Exp 41:157–164 (NB: in Chinese with English abstract)

Gu Y, Zhao YQ, Jia CS et al (2012) Analysis of hydrocarbon resource potential in Awati Depression of Tarim Basin. Petrol Geol Exp 34:257–266

Guo XW, He S, Liu KY et al (2011) Quantitative estimation of overpressure caused by oil generation in petroliferous basins. Org Geochem 42(11):1343–1350

Guo YC, Pang XQ, Chen DX et al (2012) Evolution of continental formation pressure in the middle part of the Western Sichuan Depression and its significance on hydrocarbon accumulation. Petrol Explor Dev+ 39(4):457–465

Hao F, Zou HY, Gong ZS et al (2007) Hierarchies of overpressure retardation of organic matter maturation: case studies from petroleum basins in China. AAPG Bull 91(10):1467–1498

Hao GL, Liu GD, Xie ZY et al (2010) Distribution and origin of abnormal pressure in Upper Triassic Xujiahe Formation reservoir in central and southern Sichuan. Glob Geol 29(2):298–304 (NB: in Chinese with English abstract)

He LJ, Xu HH, Wang JY (2011) Thermal evolution and dynamic mechanism of the Sichuan Basin during the Early Permian-Middle Triassic. Sci China Earth Sci 54:1948–1954

He LJ, Huang F, Liu QY et al (2014) Tectono-thermal evolution of Sichuan basin in early Paleozoic. J Earth Sci Environ 36(2):10–17

Huang F, Liu QY, He LJ (2012) Tectono-thermal modeling of the Sichuan Basin since the Late Himalayan period. Chinese J Geophys 55:3742–3753 (NB: in Chinese with English abstract)

Jiang QC, Hu SY, Jiang H et al (2015) Calculation and inducement of lacuna in the Mid-Permian Maokou Formation of the Sichuan Basin. Nat Gas Ind 38:21–29 (NB: in Chinese with English abstract)

Jiang Q, Qiu NS, Zhu CQ (2018) Heat flow study of the Emeishan large igneous province region: implications for the geodynamics of the Emeishan mantle plume. Tectonophysics 724–725:11–27

Jiang X, Zuo YA, Xu RF et al (2003) The distribution characteristics of the formation pressure within the Feixianguan Formation in the northeastern Sichuan Basin. Nat Gas Explor Dev 26(2):1–5 (NB: in Chinese with English abstract)

Jiao FZ (2018) Significance and prospect of ultra-deep carbonate fault-karst reservoirs in Shunbei Area, Tarim Basin. Oil Gas Geol 39(2):207–216 (NB: in Chinese with English abstract)

Jin ZJ (2011) Formation and accumulation of oil and gas in marine carbonate strata in Chinese sedimentary basins. Sci China Earth Sci 41(7):910–926

Rosso KM, Bodnar RJ (1995) Microthermometric and Raman spectroscopic detection limits of CO_2 in fluid inclusions and the Raman spectroscopic characterization of CO_2. Geochim Cosmochim Ac 59(19):3961–3975

Laslett GM, Green PF, Duddy IR (1987) Thermal annealing of fission tracks in apatite 2. A quantitative analysis. Chem Geol 65:1–13

Li SM, Amrani A, Pang XQ et al (2015a) Origin and quantitative source assessment of deep oils in the Tazhong Uplift, Tarim Basin. Org Geochem 78:1–22

Li D, Chang J, Qiu NS et al (2022) The thermal histories in Sedimentary basins: a case study of the central Tarim Basin, Western China. J Asian Earth Sci 229:105149

Li JH, Zhou XB, Li WB et al (2015) Preliminary reconstruction of tectonic paleogeography of Tarim Basin and its adjacent areas from Cambrian to Triassic, NW China. Geol Rev 61:1225–1234 (NB: in Chinese with English abstract)

Li MJ, Wang TG, Lillis PG et al (2012) The significance of 24-norcholestanes, triaromatic steroids and dinosteroids in oils and Cambrian-Ordovician source rocks from the cratonic region of the Tarim Basin, NW China. Appl Geochem 27:1643–1654

Li MJ, Wang TG, Chen JF et al (2010a) Paleo-heat flow evolution of the Tabei Uplift in Tarim Basin, northwest China. J Asian Earth Sci 37(1):52–66

Li SM, Pang XQ, Jin ZJ et al (2010b) Petroleum source in the Tazhong Uplift, Tarim Basin: new insights from geochemical and fluid inclusion data. Org Geochem 41:531–553

Lin F, Sum AK, Bodnar RJ (2007) Correlation of methane Raman ν1 band position with fluid density and interactions at the molecular level. J Raman Spectrosc 38(11):1510–1515

Liu KY, Bourdet J, Zhang BS et al (2013) Hydrocarbon charge history of the Tazhong Ordovician reservoirs, Tarim Basin as revealed from an integrated fluid inclusion study. Pet Explor Dev 40(2):171–180 (NB: in Chinese with English abstract)

Liu SG, Wang H, Sun W et al (2008) Energy field adjustment and hydrocarbon phase evolution in Sinian-Lower Paleozoic Sichuan Basin. J China Univ of Geosci 19(6):700–706

Liu SG, Yang Y, Deng B et al (2021) Tectonic evolution of the Sichuan Basin. Southwest China. Earth-Sci Rev 213:103470

Liu SW, Lei X, Feng CG et al (2016) Estimation of subsurface formation temperature in the Tarim Basin, northwest China: implications for hydrocarbon generation and preservation. Int J Earth Sci 105:1329–1351

Liu W, Qiu NS, Xu QC et al (2018) Precambrian temperature and pressure system of Gaoshiti-Moxi block in the central paleo-uplift of Sichuan Basin, southwest China. Precambrian Res 313:91–108

Liu YC, Qiu NS, Li HL et al (2020) Terrestrial heat flow and crustal thermal structure in the northern slope of Tazhong uplift in Tarim Basin. Geothermics 83:101709

Lu WJ, Chou IM, Burruss RC et al (2007) A unified equation for calculating methane vapor pressures in the CH_4–H_2O system with measured Raman shifts. Geochim Cosmochim Ac 71(16):3969–3978

Lüders V, Plessen B, di Primio R (2012) Stable carbon isotopic ratios of CH_4–CO_2-bearing fluid inclusions in fracture-fill mineralization from the Lower Saxony Basin (Germany)—a tool for tracing gas sources and maturity. Mar Petrol Geol 30:174–183

Ma AL (2015) Kinetics of oil-cracking of different types of marine oils from Tahe oilfield, Tarim Basin, NW China. Nat Gas Geosci 26(6):1120–1128

Mao SD, Duan ZH (2008) The P,V, T,x properties of binary aqueous chloride solutions up to T = 573 K and 100 MPa. J Chem Thermodyn 40(7):1046–1063

Mango FD (1997) The light hydrocarbons in petroleum: a critical review. Org Geochem 26(7–8):417–440

Matthew D, Sublett J, Eszter S et al (2021) Raman spectral behavior of N_2, CO_2, and CH_4 in N_2–CO_2–CH_4 gas mixtures from 22 degrees C to 200 degrees C and 10 to 500 bars, with application to other gas mixtures. J Raman Spectrosc 52(3):750–769

Mckenzie D (1978) Some remarks on the development of sedimentary basins. Earth Planet Sci Lett 40:25–32

McTavish RA (1998) The role of overpressure in the retardation of organic matter maturation. J Petrol Geol 21:153–186

Mello UT, Karner GD (1996) Development of sediment overpressure and its effect on thermal maturation; application to the Gulf of Mexico Basin. AAPG Bull 80(9):1367–1396

Mouchet J, Mitchell AF (1989) Abnormal pressures while drilling: origins, prediction, detection, evaluation, vol 2. Elf Aquitaine, Boussens, France

Munz IA, Wangena M, Girardb J et al (2004) Pressure-temperature- time-composition (P-T-t-X) constraints of multiple petroleum charges in the Hild field, Norwegian North Sea. Mar Petrol Geol 21(8):1043–1060

Neuzil CE (1995) Abnormal pressures as hydrodynamic phenomena. Am J Sci 295(6):742–786

Nguyen BTT, Jones SJ, Goulty NR et al (2013) The role of fluid pressure and diagenetic cements for porosity preservation in Triassic fluvial reservoirs of the Central Graben. North Sea. AAPG Bull 97(8):1273–1302

Ortoleva PJ (1994) Basin compartments and seals. AAPG Mem 61:477

Osborne MJ, Swarbrick RE (1997) Mechanisms for generating overpressure in sedimentary basins: a reevaluation. AAPG Bull 81(6):1023–1041

Péneloux A, Rauzy E, Fréze R (1982) A consistent correction for Redlich-Kwong-Soave volumes. Fluid Phase Equilibr 8(1):7–23

Pironon J, Thiéry R, Ougougdal MA et al (2001) FT-IR measurements of petroleum fluid inclusions: methane, n-alkanes and carbon dioxide quantitative analysis. Geofluids 1:2–10

Pironon J, Bourdet J (2008) Petroleum and aqueous inclusions from deeply buried reservoirs: experimental simulations and consequences for overpressure estimates. Geochim Cosmochim Ac 72(20):4916–4928

Qiu Y, Wang XL, Liu X et al (2020) In situ Raman spectroscopic quantification of CH_4–CO_2 mixture: application to fluid inclusions hosted in quartz veins from the Longmaxi Formation shales in Sichuan Basin, southwestern China. Petrol Sci 17:23–35

Qiu NS, Chang J, Zhu CQ et al (2022) Thermal regime of sedimentary basins in the Tarim, Upper Yangtze and North China Cratons, China. Earth-Sci Rev 224:103884

Qiu NS, Chang J, Zuo YH et al (2012) Thermal evolution and maturation of lower Paleozoic source rocks in the Tarim Basin, northwest China. AAPG Bull 96(5):789–821

Qiu NS, Liu W, Fu XD et al (2021) Maturity evolution of Lower Cambrian Qiongzhusi Formation shale of the Sichuan Basin. Mar Petrol Geol 128:105061

Ren ZL, Cui JP, Qi K et al (2020) Control effects of temperature and thermal evolution history of deep and ultra-deep layers on hydrocarbon phase state and hydrocarbon generation history. Nat Gas Ind 40:22–30 (NB: in Chinese with English abstract)

Reiners PW, Spell TL, Nicolescu S et al (2004) Zircon (U-Th)/He thermochronometry: He diffusion and comparisons with 40Ar/39Ar dating. Geochim Cosmochim Acta 68(8):1857–1887

Richardson NJ, Densmore AL, Seward D et al (2008) Extraordinary denudation in the Sichuan Basin: insights from low-temperature thermochronology adjacent to the eastern margin of the Tibetan Plateau. J Geophys Res-Sol Ea 113:B04409

Sclater JG, Christie PAF (1980) Continental stretching: an explanation of the post-mid-cretaceous subsidence of the central North Sea basin. J Geophys Res-Sol Ea 85:3711–3739

Seitz JC, Pasteris JD, Chou IM (1993) Raman spectroscopic characterization of gas mixtures; I, Quantitative composition and pressure determination of CH_4, N_2 and their mixtures. Am J Sci 293:297–321

Seitz JC, Pasteris JD, Chou IM (1996) Raman spectroscopic characterization of gas mixtures. II. Quantitative composition and pressure determination of the $CO2$–$CH4$ system. Am J Sci 296(6):577–600

Shi JN, Zheng RC, Xu FB et al (2009) Characteristics and accumulation effects of temperature-pressure fields in Changxing Formation, eastern Sichuan Basin, China. J Chengdu Univ Technol (Sci & Technol Ed) 36(5):531–537 (NB: in Chinese with English abstract)

Swarbrick RE, Osborne MJ (1998) Mechanisms that generate abnormal pressures: an overview. In: Law B E, Ulmishek G F, Slavin I (eds) Abnormal pressures in hydrocarbon environments, vol 70. AAPG Memoir, pp 13–34

Swarbrick RE, Osborne MJ, Yardley GS (2002) Comparison of overpressure magnitude resulting from the main generating mechanisms. In: Huffman AR, Bowers GL (eds) Pressure regimes in sedimentary basins and their prediction, vol 76. AAPG Memoir, pp 1–12

Sweeney JJ, Burnham AK (1990) Evaluation of a simple model of vitrinite reflectance based on chemical kinetics. AAPG Bull 74:1559–1570

Tian H, Wang ZM, Xiao ZY et al (2006) Oil cracking to gases: kinetics modeling and geological significance. Chinese Sci Bull 51(22):2763–2770

Wang FY, Zhang SC, Zhang BM et al (2003) Maturity and its history of Cambrian marine source rocks in the Tarim Basin. Geochim Cosmochim Ac 32:461–468

Wang ZM, He AD (2009) Hydrocarbon enrichment and exploration domains in Mid-Western Tabei Uplift, Tarim Basin. XinJiang Pet Geol 30:153–156 (NB: in Chinese with English abstract)

Wang ZM, Xie HW, Chen YQ et al (2014) Discovery and exploration of Cambrian subsalt dolomite original hydrocarbon reservoir at Zhongshen-1 Well in Tarim Basin. China Pet Explor 19(2):1–13 (NB: in Chinese with English abstract)

Waples DW (2000) The kinetics of in-reservoir oil destruction and gas formation: constraints from experimental and empirical data, and from thermodynamics. Org Geochem 31(6):553–575

Wu H, Qiu NS, Feng QQ et al (2020) Reconstruction of tectonic uplift process with thermo-kinematic method. Chinese J Geophys-Ch 63:2329–2344 (NB: in Chinese with English abstract)

Wu YD, Zhang ZN, Ji LM et al (2016) The changes of hydrocarbon yield and Ro for source rock in the semi-open simulation with increasing of fluids pressure. Nat Gas Geosci 27:883–891(NB: in Chinese with English abstract)

Xu HL, Wei GQ, Jia CZ et al (2012) Tectonic evolution of the Leshan-Longnüsi paleo-uplift and its control on gas accumulation in the Sinian strata, Sichuan Basin. Petrol Explor Dev+ 39:406–416

Xu M, Zhu CQ, Tian YT et al (2011) Borehole temperature logging and characteristics of subsurface temperature in the Sichuan Basin. Chin J Geophys-Chin Ed 54(4):1052–1060 (NB: in Chinese with English abstract)

Xu QC, Qiu NS, Liu W et al (2018) Thermal evolution and maturation of Sinian and Cambrian source rocks in the central Sichuan Basin, Southwest China. J Asian Earth Sci 164:143–158

Yang FL, Wang TG, Li MJ (2016) Geochemical study of Cambrian source rocks in the cratonic area of Tarim Basin, NW China. Nat Gas Geosci 27:861–872

Yardley GS, Swarbrick RE (2000) Lateral transfer: a source of additional pressure? Mar Petro Geol 17:523–537

Yu S, Pan CC, Wang JJ et al (2011) Molecular correlation of crude oils and oil components from reservoir rocks in the Tazhong and Tabei Uplifts of the Tarim Basin, China. Org Geochem 42:1241–1262

Yuuki H, Junji T, Junji Y (2020) Pressure measurement and detection of small H_2O amounts in high-pressure H_2O–CO_2 fluid up to 141 MPa using Fermi diad splits and bandwidths of CO_2. J Raman Spectrosc 51(6):1003–1018

Zhang SC, Su J, Wang XM, et al (2011) Geochemistry of Paleozoic marine petroleum from the Tarim Basin, NW China: part 3. Thermal cracking of liquid hydrocarbons and gas washing as the major mechanisms for deep gas condensate accumulations. Org Geochem 1394–1410

Zhang SC, Huang HP (2005) Geochemistry of Palaeozoic marine petroleum from the Tarim Basin, NW China: part l Oil family classification. Org Geochem 36:1204–1214

Zhang SC, Liang DG, Li MW (2002) Correlation between molecular fossils and oil sources in Tarim Basin. Chin Sci Bull 47:16–23 (NB: in Chinese with English abstract)

Zhang SC, Moldowan JM, Graham SA et al (2000) Paleozoic oil-source rock correlations in the Tarim Basin, NW China. Org Geochem 31:273–286

Zhou CX, Yu S, Huang WY et al (2021) Oil maturities, mixing and charging episodes in the cratonic regions of the Tarim Basin, NW China: insight from biomarker and diamondoid concentrations and oil bulk properties. Mar Pet Geol 126:104903

Zhu CQ, Hu SB, Qiu NS et al (2016) The thermal history of the Sichuan Basin, SW China: evidence from the deep boreholes. Sci China Earth Sci 59:70–82

Zhu CQ, Xu M, Yuan YS et al (2010) Palaeo-geothermal response and record of the effusing of Emeishan basalts in Sichuan basin. China Sci Bull 55:474–482 (NB: in Chinese with English abstract)

Zhu CQ, Xu T QNS et al (2022) Distribution Characteristics of the deep geothermal field in the Sichuan Basin and its main controlling factors. Front Earth Sci 10:824056

Zou CN, Du JH, Xu CC et al (2014) Formation, distribution, resource potential and discovery of the Sinian-Cambrian giant gas field, Sichuan Basin, SW China. Pet Explor Dev 41(3):278–293 (NB: in Chinese with English abstract)

Zou CN, Yang Z, He DB et al (2018) Theory, technology, and prospects of conventional and unconventional natural gas. Pet Explor Dev 45(4):575–587 (NB: in Chinese with English abstract)

Diversity and Effectiveness of Deep Marine Source Rocks in the Sichuan and Tarim Basins

6

Maowen Li, Enze Wang, Xiaoxiao Ma, Qianglu Chen, Liyu Zhang, and Kun Yuan

Abstract

Deep marine petroleum systems are critical to the energy supply in China, making them a key focus of petroleum exploration and research. A comprehensive understanding of the diversity and effectiveness of deep marine source rocks is essential for objectively evaluating their resource potential. However, significant challenges remain, including the unclear heterogeneity of source rocks, ambiguous hydrocarbon origins, and uncertainties regarding source rock effectiveness, all of which hinder accurate resource assessments. To address these issues, this study examines the geological characteristics of deep marine source rocks in two major petroliferous basins in China—the Sichuan Basin and the Tarim Basin. Drawing on field outcrop profiles and newly acquired drilling data, five geological models are established to characterize the formation and distribution of deep marine source rocks. Furthermore, stable carbon isotope and biomarker analyses of different source rock types underscore their distinct roles in hydrocarbon

accumulation within deep marine petroleum systems. This study also provides a preliminary discussion of the unconventional petroleum resource potential of carbonate source rocks. The results highlight the pronounced heterogeneity of deep marine source rocks, which is largely governed by sedimentary facies distributions. Multiple lines of evidence suggest that the stable carbon isotopes of kerogen are minimally affected by thermal maturity and can still retain valuable paleoenvironmental information even during high-overmature stages. Although hydrocarbon cracking at high maturity reduces the absolute concentration of biomarkers, certain proxies remain effective for characterizing source rock properties and performing oil–source correlation. During the black oil stage, specific biomarkers can still provide insights into source rock characteristics. In the Tarim Basin, deep marine petroleum systems are not solely sourced from organic-rich shales but are also significantly influenced by carbonate source rocks. Two primary hydrocarbon origins are identified: (1) source rocks of varying lithofacies within the different sedimentary facies in the Yuertusi Formation and (2) contributions from source rocks in other formations, including the Ordovician and Sinian. The resource potential of carbonate source rocks is substantial, with hydrocarbons not only migrating into conventional reservoirs but also potentially remaining trapped within the source rocks. Under favor-

M. Li · E. Wang (✉) · X. Ma · Q. Chen · L. Zhang · K. Yuan
State Key Laboratory of Shale Oil and Gas Enrichment Mechanisms and Efficient Development, SINOPEC, Beijing, China
e-mail: wangenze9939@163.com

Petroleum Exploration and Production Research Institute, SINOPEC, Beijing, China

© The Author(s) 2026
Y. Ma et al. (eds.), *Petroleum Geology and Exploration of Deep Marine Strata in China*, Advances in Oil and Gas Exploration & Production, https://doi.org/10.1007/978-3-032-02496-1_6

able economic conditions, these carbonate reservoirs could become promising targets for unconventional petroleum exploration and development. Therefore, potential exploration targets for unconventional petroleum resources include the Middle Permian Maokou Formation in the Sichuan Basin and its adjacent areas, as well as the marlstone reservoirs of the Lower Cambrian Yuertusi Formation and the Lower Xiaoerbulak Formation in the Tarim Basin.

Keywords

Marine source rock · Deep strata · Source rock development models · Source rock effectiveness

6.1 Introduction

Deep marine petroleum systems are crucial components of the energy supply in China (Ma et al. 2020, 2024). Petroleum resources at depths greater than 4500 m account for 40% of the total petroleum resources in China, with major accumulations in the Tarim, Sichuan, and Ordos basins (Ma et al. 2020). Deep marine source rocks in China were deposited primarily during the Middle to Late Proterozoic and Paleozoic eras and are notably distributed in major petroliferous basins such as the Tarim and Sichuan Basins (Liu et al. 2020; Wang et al. 2022b; Qiu et al. 2023; Zhang et al. 2024a, b). Traditionally, source rocks have been predominantly considered to be mudstones and shales. However, owing to variations in tectonic and sedimentary settings, the geological characteristics of marine source rocks can differ significantly, even within the same strata. Current research often overlooks the heterogeneity within marine source rocks, which limits the understanding of critical aspects such as hydrocarbon origins, resource evaluation, and accumulation mechanisms in major petroliferous basins (Wei et al. 2014; Huang et al. 2016; Zhao et al. 2021).

The origins of deep petroleum systems are diverse. Hydrocarbons can be derived from early-stage hydrocarbon generation by organic matter (Tissot et al. 1974, 1987; Pepper and Corvi 1995) or from the thermal cracking of highly mature kerogen and organic matter under high-temperature conditions (Tissot and Welte 1984; Zhao et al. 2005; Spigolon et al. 2015; Wang et al. 2022a). This diversity complicates the accurate assessment of the resource potential of deep marine petroleum systems. Additionally, source rocks from different sedimentary-tectonic settings exhibit distinct hydrocarbon-generating biological assemblages (Pepper and Corvi 1995; Liu et al. 2020). Theoretical geochemical approaches—such as elemental, molecular, and isotopic analyses—can be employed to characterize source composition (Li et al. 1992; Larter et al. 1996; Peters et al. 2005), providing potential insights into hydrocarbon origins and the identification of effective source rocks. However, because deep petroleum systems are typically highly mature, the reliability of geochemical indicators under high-overmature conditions remains a subject of debate (Huang et al. 2016), making it challenging to trace hydrocarbon sources and objectively assess source rock effectiveness. A systematic characterization of the geological attributes of deep marine source rocks is urgently needed to enhance our understanding of their diversity and to accurately identify effective source rocks for objective resource assessment.

Given the unclear heterogeneity of deep marine source rocks, the uncertainty surrounding deep petroleum sources, and the ongoing debate over source rock effectiveness, this study reviews the geological characteristics of deep marine source rocks in two major petroliferous basins in China—the Sichuan Basin and the Tarim Basin. By utilizing field outcrop data and newly acquired drilling results from both basins, five geological models for deep marine source rock development are established. Furthermore, stable carbon isotope and biomarker analyses of various source rock types highlight their crucial roles in hydrocarbon accumulation within deep marine petroleum systems. This study also provides a preliminary assessment of the unconventional resource potential of carbonate source rocks.

These findings offer critical insights into the effectiveness of deep marine source rocks in China, expanding both the scale and exploration scope of deep marine petroleum resources.

6.2 Current Understanding of the Deep Marine Source Rocks in the Sichuan and Tarim Basins

As noted earlier, the deep petroleum resource potential of the Sichuan and Tarim Basins is substantial, primarily because of the presence of high-quality deep marine source rocks. In this section, previous research is synthesized to review the geological characteristics of deep marine source rocks in these two basins, as well as a prior understanding of hydrocarbon origins.

6.2.1 Characteristics of Deep Marine Source Rocks in the Sichuan and Tarim Basins

Multiple sets of high-quality marine source rocks have developed in the Sichuan Basin (Zhu et al. 2006; Liang et al. 2008; Zou et al. 2011; Liu et al. 2020); from bottom to top: the Sinian Doushantuo Formation, Cambrian Qiongzhusi Formation, Silurian–Ordovician Wufeng–Longmaxi Formation, Permian Qixia–Maokou Formation, Permian Dalong Formation, and Triassic Leikoupo Formation (Zhu et al. 2006, 2021; Liang et al. 2008; Xia et al. 2010; Zou et al. 2011; Yang 2016; Liu et al. 2020; Wang et al. 2022b; Guo et al. 2023). The characteristics of these source rocks are summarized in Table 6.1.

The Sinian Doushantuo Formation source rocks are distributed primarily in the northwestern and southeastern depressions of the Sichuan Basin and were deposited in a transgressive–regressive depositional setting. These source rocks consist predominantly of mudstone with high total organic carbon (TOC) content and are concentrated mainly in the second and fourth members of the formation (Zhu et al. 2022). In the Sichuan

Basin, the effective thickness of the Doushantuo Formation can reach 233 m, with an average TOC content of 1.85%. The kerogen is primarily Type I, and the vitrinite reflectance (Ro) values indicate that these source rocks have reached an overmature stage (Zou et al. 2011). The Cambrian Qiongzhusi Formation is widespread across the basin and displays lithological variations because of differences in sedimentary environments. It is predominantly composed of black shale and siltstone (Liu et al. 2020; Guo et al. 2023). The Qiongzhusi Formation source rocks have an average thickness of approximately 200 m and an average TOC content of 3.5%, with kerogen also dominated by Type I kerogen. Like the Doushantuo Formation, the Qiongzhusi Formation has reached an overmature stage (Zou et al. 2011; Liu et al. 2020). The Wufeng–Longmaxi Formation is present primarily in the eastern and southern parts of the Sichuan Basin. It was deposited in an anoxic deep-water shelf environment influenced by upwelling, forming a sequence of black and dark gray shales. The Wufeng–Longmaxi Formation shales in the Sichuan Basin have an average TOC content of 2.59%, with Type I kerogen predominant. The equivalent Ro values range from 1.6 to 3.6%, indicating that the source rock is generally in the high-mature to overmature stage (Zou et al. 2011; Wang et al. 2022b; Feng et al. 2023). The Permian strata in the Sichuan Basin contain two key sets of source rocks: the Qixia–Maokou Formation and the Dalong Formation. The Qixia–Maokou Formation consists primarily of carbonate rocks with relatively low TOC contents (approximately 0.4%) but substantial thickness (200–400 m). In contrast, the Dalong Formation has more complex lithologies, including black mudstone, marlstone, and silicalite. Although its average TOC content is relatively high (5.86%), the formation is relatively thin, with a maximum thickness of 55 m. Both formations have reached the high-mature to overmature stage (Zhu et al. 2006; Liang et al. 2008; Xia et al. 2010). The Triassic Leikoupo Formation in the Sichuan Basin is composed mainly of marlstone and muddy dolomite, with TOC contents ranging from 0.4 to 0.6%. The kerogen types are primarily Type II_1 and II_2, and the formation has

Table 6.1 The geological features of deep marine source rocks in the Tarim and Sichuan basins

Basin	System	Formation	Lithology	TOC (%)	Kerogen type	Thickness (m)	Thermal maturity	References
Sichuan Basin	Sinian	Doushantuo	Balck shale and carbonates	0.67–3.02% (1.85%)	I	32–233 (100)	Ro = 2.67–4.50%	Zhu et al. (2006), Zou et al. (2011), Zhu et al. (2021)
	Cambrian	Qiongzhusi	Black shale and silty shale	0.43–22.15% (3.50%)	I	46–445 (200)	Ro = 2.30–5.20%	Zhu et al. (2006), Zou et al. (2011), Guo et al. (2023)
	Silurian–Ordovician	Wufeng-Longmaxi	Black and dark-gray shale	0.51–25.73% (2.59%)	I	23–847 (203)	Ro = 1.60–3.60%	Zhu et al. (2006), Zou et al. (2011), Wang et al. (2022a; b)
	Permian	Qixia-Maokou	Carbonate	Average value is approximately 0.4%	I, II	200–400	Ro more than 2.0%	Zhu et al. (2006)
	Permian	Dalong	Black shale, marlstone, and silicalite	Maximum 24.31% (average 5.86%)	III	5–55	Ro = 1.51–2.18% (1.75%)	Liang et al. (2008), Xia et al. (2010)
	Triassic	Leikoupo	Marlstone and muddy dolomite	0.4–0.6%	III1-II2	250–350	Ro more than 2.0%	Yang (2016)
Tarim	Cambrian	Yuertusi	Shale, mudstone and marlstones	1.2–5.0%	I-II	120–415	Ro more than 2.0%	Zhang et al. (2000), Cai et al. (2009), Li et al. (2010), Huang et al. (2016), Zhu et al. (2022)
	Ordovician	Heituao	Carbonaceous and siliceous mudstone	Mainly greater than 1.0%, with a maximum value of 7.6%	II-I	50–200	Ro = 1.7–2.2%	Zhang et al. (2005), Zhao et al. (2008), Cai et al. (2009), Huang et al. (2016)
	Ordovician	Yijianfang and Lianglitage	Shale, argillaceous limestone and marlstone	0.3–2.9%	I and III	20–100	Ro = 0.8–1.3%	Zhang et al. (2000), Zhang et al. (2005), Zhao et al. (2008), Li et al. (2010)

a thickness of 250–350 m. These source rocks have also reached the overmature stage (Yang 2016).

In the Tarim Basin, the primary deep marine source rocks include the Cambrian Yuertusi Formation, Lower Ordovician Heituao Formation, and Middle to Upper Ordovician Yijianfang and Lianglitage formations (Zhang et al. 2000, 2005; Zhao et al. 2008; Cai et al. 2009; Li et al. 2010; Huang et al. 2016; Zhu et al. 2022). The Cambrian Yuertusi Formation was deposited in basin–tidal flat environments and is primarily composed of black shale, mudstone, and marlstone. It is widely distributed across the Tarim Basin, with thicknesses reaching 415 m in the western region. The TOC content ranges from 1.2 to 5.0%, and the kerogen types are predominantly Type I–II. Ro values generally exceed 2.0%, indicating an overmature stage (Zhang et al. 2000; Cai et al. 2009; Li et al. 2010; Huang et al. 2016; Zhu et al. 2022). This formation is widely recognized as a high-quality source rock. The Lower Ordovician Heituao Formation is distributed mainly in the eastern Tarim Basin, including the Manjiaer Depression and Tadong area. Deposited in deepwater to shallow-water shelf environments, its lithology consists predominantly of carbonaceous and siliceous mudstones. Compared with the Yuertusi Formation, the Heituao Formation is significantly thinner, with thicknesses ranging from 50 to 200 m. The TOC content generally exceeds 1.0%, reaching a maximum of 7.6%, while the kerogen types are primarily Type II–I. Ro values range between 1.7 and 2.2%, indicating a high to overmature stage (Zhang et al. 2005; Zhao et al. 2008; Cai et al. 2009; Huang et al. 2016). The Middle to Upper Ordovician Yijianfang and Lianglitage formations were deposited primarily in shallow-water shelf and ramp environments. These source rocks are present mainly in the Tadong area and in the Tazhong and Tabei uplifts. Their thicknesses range from 20 to 100 m, with TOC contents varying between 0.3 and 2.9%. The kerogen types are predominantly Type I and Type III, while the Ro values range from 0.8 to 1.3%, indicating that these source rocks have reached the mature to high-mature stage.

6.3 Hydrocarbon Origins of Deep Marine Petroleum Systems in the Sichuan and Tarim Basins

Owing to the great burial depths and high geothermal conditions of deep marine petroleum systems, hydrocarbons often undergo various secondary alterations, such as thermochemical sulfate reduction (TSR) and thermal maturity effects. These processes complicate the determination of hydrocarbon origins. Extensive research has been conducted on the Permian and Silurian source rocks in the Sichuan Basin, particularly their contributions to conventional and unconventional petroleum accumulations. The role of these source rocks in major discoveries, such as the Puguang, Yuanba, and Fuling gas fields, is well documented and is not discussed further here (Ma 2006; Guo 2011; Guo et al. 2014; Guo et al. 2016).

The origin of hydrocarbons in the Sinian to Lower Paleozoic gas reservoirs of the Sichuan Basin remains a subject of debate, with multiple competing hypotheses proposed (Zhu et al. 2006; Wei et al. 2014). Specifically, four primary perspectives exist regarding the source of Sinian natural gas: (1) self-generation and self-storage (Xu et al. 1989), (2) source from the Cambrian Qiongzhusi Formation (Dai 2003; Zhu et al. 2006), (3) mixed contributions from Sinian and Cambrian source rocks (Chen 1992; Zhao et al. 2021), and (4) inorganic origin (Wang 1982). For instance, on the basis of gas composition and methane isotopic data, Zhu et al. (2006) suggested that the Sinian gas reservoirs in the Weiyuan gas field, southern Sichuan, are primarily sourced from Cambrian source rocks. Zhao et al. (2021) quantitatively estimated the relative contribution of Sinian source rocks to Dengying Formation gas reservoirs using δ^2H_{CH4} values. Additionally, strong secondary alterations pose significant challenges to hydrocarbon origin characterization. For example, Zhu et al. (2006) proposed that the carbon isotope inversion of methane and ethane in the Sinian reservoirs of the Weiyuan gas field is influenced by TSR, whereas Wei et al. (2014) attributed the isotopic inversion observed in Sinian and Cambrian gas reservoirs primarily to high paleogeothermal conditions. These uncertainties highlight the need for further research on the hydrocarbon origins of deep marine source rocks in the Sichuan Basin.

The origin of hydrocarbons in the Tarim Basin has long been a subject of interest. Since the breakthrough at the SC2 well, the hydrocarbon origins of the Lower Paleozoic petroleum systems in the basin have attracted significant scholarly attention (Gu et al. 1994; Jia 1997; Zhang et al. 2000, 2002a, 2023; Zhang and Huang 2005; Zhao et al. 2008; Cai et al. 2009; Li et al. 2010; Tian et al. 2012; Huang et al. 2016; Cai 2018). Early biomarker studies suggested that petroleum in the Lower Paleozoic of the Tarim Basin primarily originated from Middle to Upper Ordovician source rocks (Zhang et al. 2000, 2002a; Zhang and Huang 2005). However, given the relatively low thickness and TOC content of these source rocks compared with their Cambrian counterparts, many researchers argue that Cambrian and Lower Ordovician source rocks are the predominant contributors to Lower Paleozoic reservoirs (Kang 2003). In support of this view, alkane carbon isotope analyses have also indicated greater contributions from Cambrian and Lower Ordovician source rocks (Li et al. 2010). Beyond conventional approaches such as biomarker and organic carbon isotope analyses, advanced techniques—such as organic sulfur isotope studies— have been employed to investigate hydrocarbon sources in the Lower Paleozoic strata of the Tarim Basin. For instance, Cai (2018) reported that the organic sulfur isotopic composition of Paleozoic crude oil in the Tarim Basin closely matches that of Lower Cambrian source rocks. However, no study has yet systematically explained the discrepancies between results derived from organic sulfur isotopes, organic carbon isotopes, and biomarkers. Therefore, further in-depth investigations are needed to clarify the hydrocarbon origins of the Lower Paleozoic petroleum systems in the Tarim Basin.

6.4 Diversity and Sedimentary Models of Deep Marine Source Rocks in the Sichuan and Tarim Basins

Traditional perspectives suggest that marine source rocks in marine depositional systems exhibit relatively low heterogeneity (Katz and Lin 2014). However, the results of this study demonstrate that deep-marine source rocks, which formed in different depositional facies, display significant heterogeneity. Using the Cambrian Yuertusi Formation in the Tarim Basin as a case study, the source rock properties across various depositional facies were characterized on the basis of the latest drilling data (Fig. 6.1). The findings indicate that the Yuertusi Formation was deposited in five distinct facies: paleouplift, tidal flat, shallow-water shelf, deep-water shelf, and basin facies. In terms of the source rock thickness, the deep-water shelf facies is the thickest, typically exceeding 10 m, with a maximum recorded thickness of 32 m in the TS5 well. In contrast, the source rocks in the shallow-water shelf and tidal flat facies are generally less than 5 m thick, whereas those in the basin facies are intermediate in thickness, averaging approximately 10 m in the KT1 well. From a lithological perspective, source rocks in the deep-water shelf and basin facies are relatively homogeneous, primarily consisting of black shale, siliceous mudstone, and silicalite, except in the Xiaoerbulake well, where gray mudstone is also present. In comparison, the shallow-water shelf and tidal flat facies exhibit similar lithological characteristics, both being dominated by gray mudstone. In terms of the TOC content, the source rocks in the basin facies present the highest values, ranging from 1.6 to 33.1%, with an average of 11.8%. In the deep-water shelf facies, the TOC content varies between 1.9 and 22.4%, with an average exceeding 6% across all the boreholes. Conversely, the TOC content in source rocks from the shallow-water

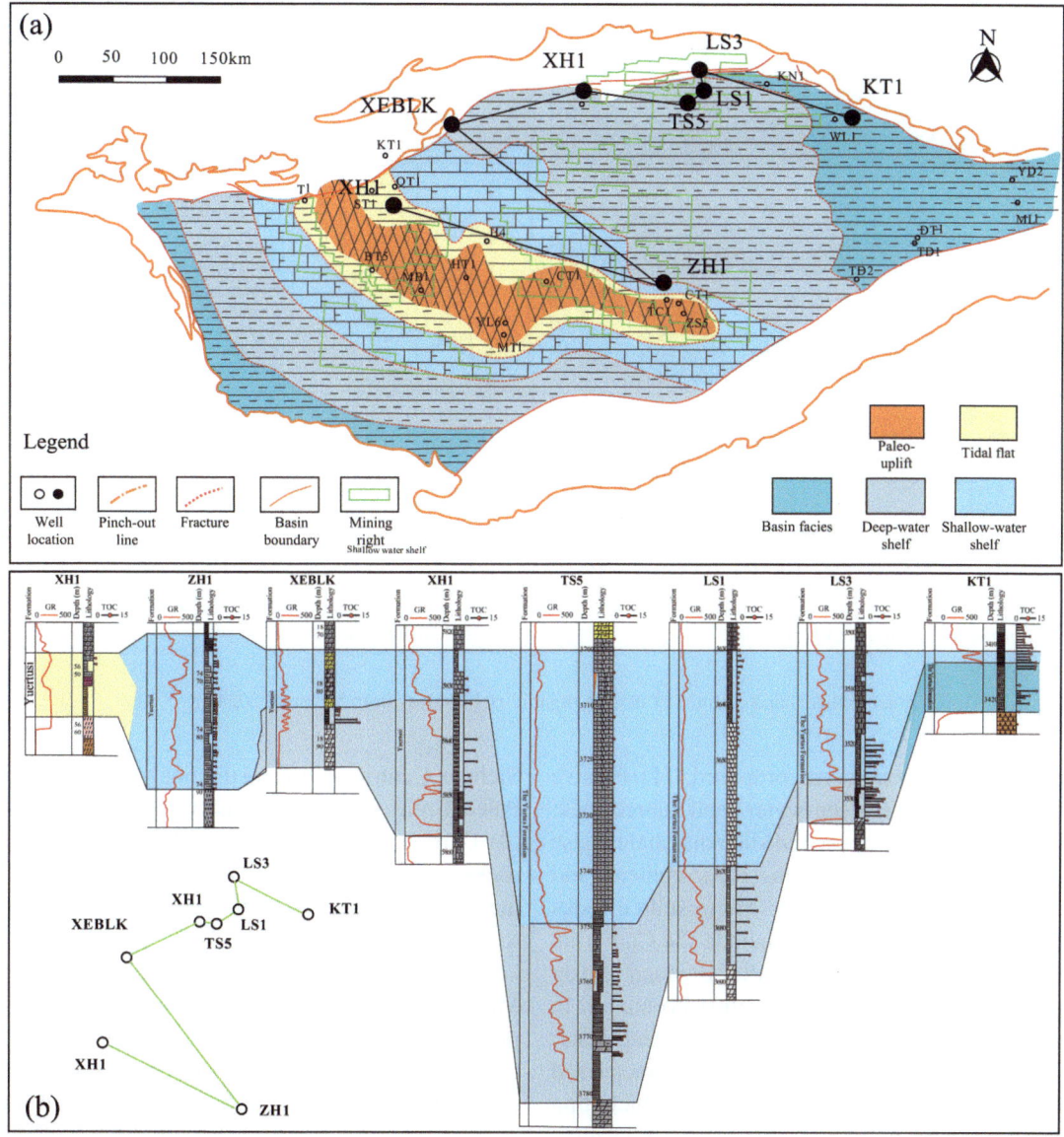

Fig. 6.1 The sedimentary facies distribution and profiles of the Yuertusi Formation in the Tarim Basin. **a** The sedimentary facies distribution of the Yuertusi Formation in the Tarim Basin. **b** Profiles showing lithology and sedimentary facies of the Yuertusi Formation in the Tarim Basin

shelf and tidal flat facies is generally less than 1%, with average values less than 0.5%. These results highlight the critical role of regional sedimentary facies in controlling both lithological characteristics and organic matter accumulation in source rocks.

Building on the fundamental influence of depositional facies on source rock properties, this study further delineates the development models of deep-marine source rocks in the Sichuan and Tarim Basins (Fig. 6.2). These findings indicate that passive continental margins, intraplate rift zones, and intracratonic basins provide favorable structural conditions for the accumulation of high-quality source rocks. On the basis of these geological settings, five distinct source rock

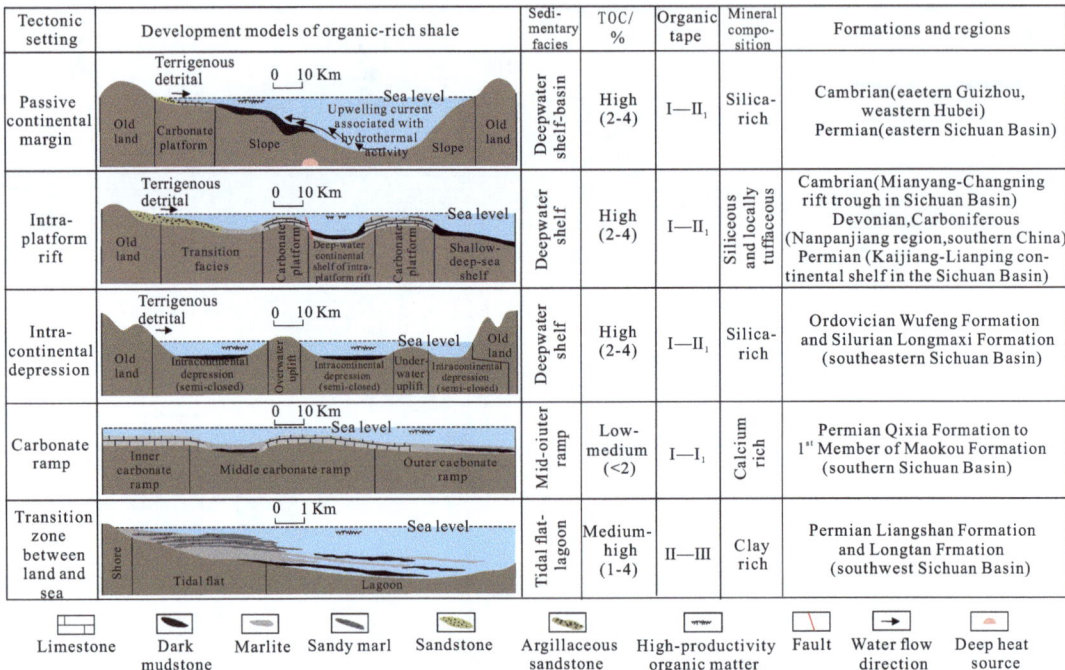

Tectonic setting	Development models of organic-rich shale		Sedimentary facies	TOC/%	Organic tape	Mineral composition	Formations and regions
Passive continental margin	Terrigenous detrital 0 10 Km Sea level Upwelling current associated with hydrothermal activity Old land, Carbonate platform, Slope, Old land		Deepwater shelf-basin	High (2-4)	I—II₁	Silica-rich	Cambrian(eaetern Guizhou, weastern Hubei) Permian(eastern Sichuan Basin)
Intra-platform rift	Terrigenous detrital 0 10 Km Sea level Old land, Transition facies, Carbonate platform, Deep-water continental shelf of intra-platform rift, Carbonate platform, Shallow-deep-sea shelf		Deepwater shelf	High (2-4)	I—II₁	Siliceous and locally tuffaceous	Cambrian(Mianyang-Changning rift trough in Sichuan Basin) Devonian,Carboniferous (Nanpanjiang region,southern China) Permian (Kaijiang-Lianping continental shelf in the Sichuan Basin)
Intra-continental depression	Terrigenous detrital 0 10 Km Sea level Old land, Old land, Intracontinental depression (semi-closed), Overwater uplift, Intracontinental depression (semi-closed), Underwater uplift, Intracontinental depression (semi-closed)		Deepwater shelf	High (2-4)	I—II₁	Silica-rich	Ordovician Wufeng Formation and Silurian Longmaxi Formation (southeastern Sichuan Basin)
Carbonate ramp	0 10 Km Sea level Inner carbonate ramp, Middle carbonate ramp, Outer caebonate ramp		Mid-oiuter ramp	Low-medium (<2)	I—I₁	Calcium rich	Permian Qixia Formation to 1ˢᵗ Member of Maokou Formation (southern Sichuan Basin)
Transition zone between land and sea	0 1 Km Sea level Shore, Tidal flat, Lagoon		Tidal flat-lagoon	Medium-high (1-4)	II—III	Clay rich	Permian Liangshan Formation and Longtan Frmation (southwest Sichuan Basin)

Limestone | Dark mudstone | Marlite | Sandy marl | Sandstone | Argillaceous sandstone | High-productivity organic matter | Fault | Water flow direction | Deep heat source

Fig. 6.2 Development models and features of marine organic-rich shales (from Ma et al. 2024)

development models are proposed: (1) Passive continental margin deep-water shelf source rock: Predominantly composed of siliceous shale, these source rocks exhibit high TOC content and contain Type I–II₁ kerogen. These source rocks are widely distributed and include the Cambrian Qiongzhusi Formation in the Sichuan Basin and the Lower Cambrian Yuertusi Formation in the Tarim Basin. (2) Intraplate rift deep-water shelf source rock: This rock is characterized by significant lithofacies variations, including siliceous mudstone and calcareous mudstone, with elevated TOC contents and types I to II₁ kerogen. Representative examples include the Cambrian Maidiping–Qiongzhusi Formation and the Upper Permian Wujiaping–Dalong Formation, which are influenced by the Mianyang–Changning extensional fault and the Kaijiang–Liangping rift. (3) Foreland Basin deep-water shelf source rocks: Composed predominantly of siliceous shale, these source rocks exhibit high TOC contents and contain Type I to II₁ kerogen. A representative example is the Upper Ordovician Wufeng Formation–Lower Silurian Longmaxi Formation, which developed between the Sichuan Central Uplift and the Jiangnan–Xuefeng Paleo-Uplift in the southeastern Sichuan Basin. (4) Carbonate ramp source rock: These source rocks primarily consist of calcareous mudstone, muddy limestone, and marlstone and contain low to moderate TOC contents and Type I to II₁ kerogen. A notable example is the Middle Permian Maokou Formation in the Sichuan Basin, which predominantly developed along the inner and outer carbonate ramps. (5) Marine–continental transitional lagoonal–tidal flat source rock: With characteristics of argillaceous mudstone, these source rocks exhibit moderate to high TOC contents and contain type II–III kerogen. A representative example is the Upper Permian Longtan Formation.

6.5 Hydrocarbon Origins and Effectiveness of Deep Marine Source Rocks

Deep-marine source rocks display considerable heterogeneity, making determination of their hydrocarbon origins and assessment of their effectiveness in hydrocarbon generation essential. A comprehensive understanding of these factors is critical for evaluating the resource potential of deep marine petroleum systems.

6.5.1 Hydrocarbon Origins and Effectiveness of Deep Marine Source Rocks

The study of hydrocarbon generation and oil–source correlation in overmature source rocks presents significant challenges. Owing to their low absolute concentrations, molecular biomarkers are highly susceptible to thermal evolution and secondary alterations (Huang et al. 2016). In contrast, stable carbon isotopes offer valuable insights into paleogeographical, paleoclimatic, and paleoenvironmental conditions during source rock deposition. The isotopic compositions of hydrocarbon components vary significantly depending on their source, making them

powerful tools for determining hydrocarbon origin. However, before organic carbon isotopes are applied to hydrocarbon origin analysis, assessment of the impact of thermal maturity on isotopic composition is crucial.

To evaluate the influence of thermal maturation on organic carbon isotopes, this study incorporated data from both hydrocarbon generation thermal simulation experiments and geological profile samples. Three source rock samples were selected for thermal simulation: the Cambrian Alum Shale from Sweden and the Cretaceous shale from the Western Canada Basin, both of which are classified as black shales, along with a Devonian marlstone sample from the Western Canada Basin. The results of these thermal simulation experiments (Fig. 6.3) reveal consistent carbon isotope trends across different lithologies. As thermal maturity increases (with equivalent Ro values ranging from 0.7 to 3.0%), the $\delta^{13}C$ values of kerogen remain relatively stable, whereas the $\delta^{13}C$ values of methane become progressively

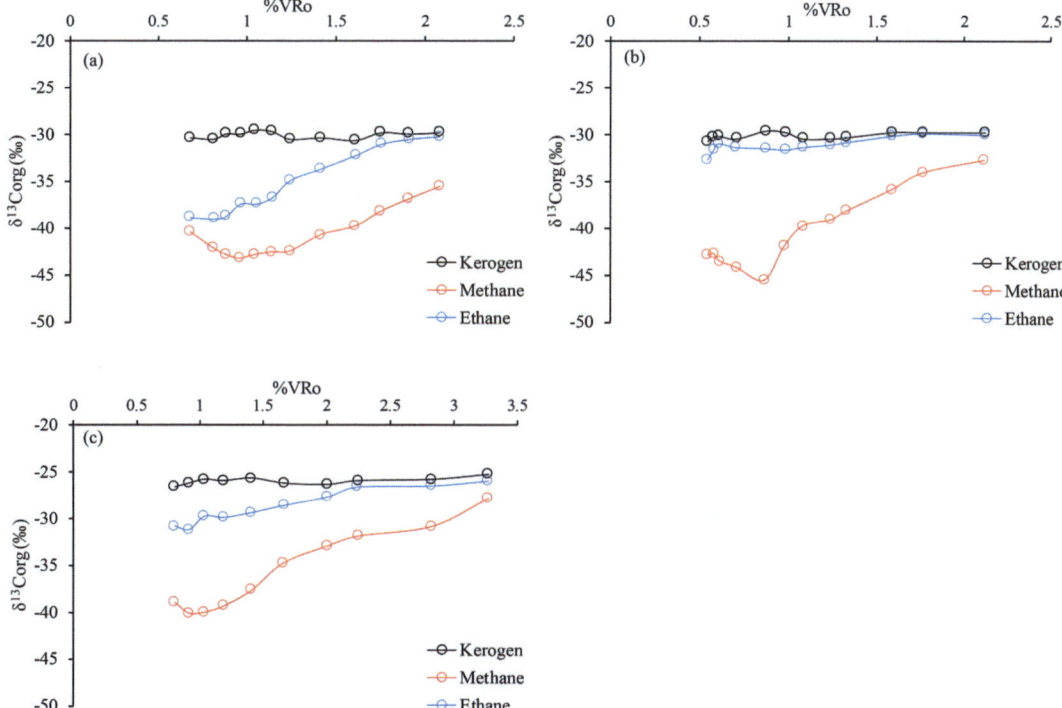

Fig. 6.3 Organic carbon isotope evolution features of kerogen, methane, and ethane of different thermal maturity samples obtaining by simulation experiments. **a** Cambrian Alum Shale from Sweden; **b** Cretaceous shale from the Western Canada Basin; **c** Devonian marlstone from the Western Canada Basin

enriched in ^{13}C. In contrast, the δ^{13}C values of ethane initially become lighter before shifting to heavier values, with an inflection point occurring at approximately Ro = 1.0%. Additionally, the δ^{13}C values of kerogen in Cambrian and Cretaceous black shales fluctuate around − 30‰, whereas the δ^{13}C value of the Devonian lagoonal marlstone is approximately − 25.5‰. These findings suggest that stable carbon isotopes, in addition to the compositional characteristics of overmature source rocks, provide crucial insights into the paleosedimentary environments of source rocks.

Data from geological profile samples (Fig. 6.4) further support these conclusions. The Eagle Ford Shale in the United States, which is primarily composed of laminated organic-rich carbonate rock with carbonate mineral contents exceeding 60%, has kerogen δ^{13}C values ranging from − 23‰ to − 28‰. The δ^{13}C enrichment from the immature to overmature stages does not exceed 2‰. Additionally, the δ^{13}C values of the corresponding saturated hydrocarbon and aromatic fractions are 1.5‰ and 2.5‰ lower than those of kerogen, respectively. These results align with expected trends, where compared with their corresponding kerogen, hydrocarbon components at different thermal evolution stages exhibit lower δ^{13}C values. These observations indicate that thermal maturity has a minimal influence on the δ^{13}C values of kerogen and heavier hydrocarbon fractions.

On the basis of the combined evidence from hydrocarbon generation thermal simulation experiments and geological profiles, it can be concluded that the stable carbon isotopes of kerogen are minimally affected by thermal maturity and can serve as reliable proxies of paleoenvironmental conditions in source rocks. Consequently, organic carbon isotopes represent a robust and effective tool for determining oil–source correlations in overmature systems.

Previous studies have demonstrated that many biomarkers are highly susceptible to thermal maturity (Seifert and Moldowan 1978; Peters et al. 1990, 2005). Huang et al. (2016) emphasized that the variability observed in biomarkers in studies of hydrocarbon origins in Paleozoic reservoirs of the Tarim Basin is largely attributed to ultrahigh thermal maturity. Therefore, caution is necessary when biomarkers are employed for oil–source correlation analysis in overmature systems. While thermal maturity alters both the absolute concentration and distribution trends of certain biomarkers, some biomarkers still preserve critical information regarding source rock characteristics. Their preservation is not solely governed by thermal maturity but is also influenced by geological factors such as mineral composition and the petroleum generation environment.

Fig. 6.4 Organic carbon isotope evolution features of kerogen, saturated hydrocabron, and aromatic hydrocarbon of different thermal maturity samples obtaining by geological samples

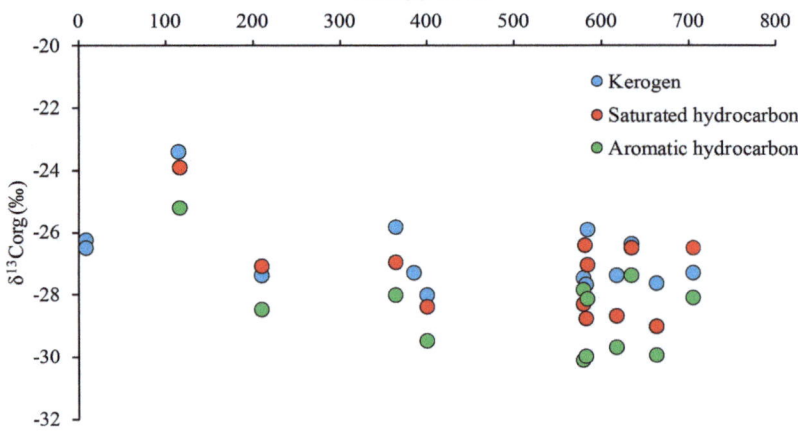

For instance, Brønsted and Lewis acid sites within clay minerals play crucial roles in hydrocarbon cracking and decarboxylation by donating hydrogen cations (H^+) and accepting electron pairs, thereby facilitating the thermal evolution of organic matter (Johns 1979; Geatches et al. 2010; Tong et al. 2010; Wu et al. 2012; He et al. 2022). Wang et al. (2024) further reported that the transformation of steranes and hopanes into dihaponoids and diasteranes under high thermal maturity occurs only when sufficient clay minerals are present. Although biomarker concentrations decrease because of hydrocarbon cracking at highly overmature stages, certain biomarkers continue to retain source rock characteristics, with their preservation influenced by variations in the hydrocarbon generation environment. Thus, during the black oil phase, selecting biomarkers remains valuable for characterizing source rock properties and conducting oil–source correlations.

In the Western Canada Basin, biomarker data from crude oils derived from different Devonian source rocks are presented in Fig. 6.5.

Oils from the Middle Devonian Rainbow-Zama-Shekilie subbasins are believed to originate from organic-rich carbonates of the Lower Keg River Formation, as well as the carbonates and laminated deposits of the Muskeg Formation. In contrast, oils from the Upper Devonian Rimbey–Meadowbrook reef tend to have originated from organic-rich shale within the Duvernay Formation. The results reveal distinct biomarker correlations among Ts/(Ts+Tm), diasterane/sterane, C_{29}-Ts/C_{29}-hopane, common terpenoid distributions, and regular sterane ratios between oils derived from evaporite/carbonate and shale source rocks. These findings indicate that source rocks with differing lithologies exhibit unique geochemical characteristics and follow distinct diagenetic and maturation pathways. Furthermore, an analysis of terpenoid ratios suggests that although thermal maturation and TSR can influence the distributions of common terpenoids, the primary geochemical signatures inherited from the original source rocks remain discernible.

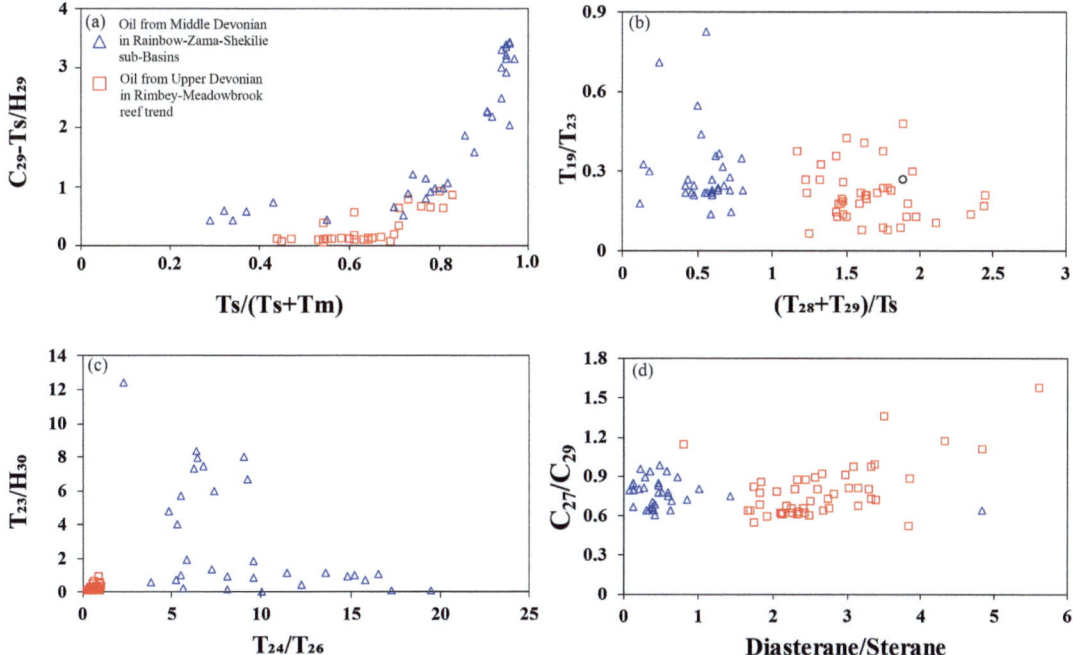

Fig. 6.5 Molecular geochemical parameters of crude oils derived from marlstone/evaporite and black shale source rocks in the Western Canada Basin

6.5.2 Variations in Organic Carbon Isotopes in Deep Marine Source Rocks Deposited in Different Facies

Organic carbon isotopes serve as reliable indicators for investigating the origins of deep marine petroleum systems. The Yangtze Platform preserves one of the most well-preserved successions of Ediacaran to Early Cambrian strata globally, containing a series of continuous profiles that transition from shallow-water to deep-water environments. Examining the organic carbon isotopes of source rocks across multiple depositional settings—ranging from platform to ramp (transitional) to basin facies—not only enhances our understanding of paleoenvironmental conditions but also provides a critical basis for interpreting the formation mechanisms of source rocks. Accordingly, in this study, a statistical analysis of organic carbon isotopic compositions

in source rocks of varying lithologies and sedimentary facies across different strata within the Yangtze Platform was conducted (Fig. 6.6). The results reveal significant variations in the distribution of organic carbon isotopes among source rocks of different lithologies. Specifically, the $\delta^{13}C_{org}$ values for black shale, silicalite, and carbonate source rocks range from $-37.67‰$ to $-27.99‰$ (average of $-34.21‰$), $-36.63‰$ to $-30.78‰$ (average of $-33.56‰$), and $-35.36‰$ to $-26.27‰$ (average of $-30.24‰$), respectively. These findings indicate that black shale has the lightest organic carbon isotopic signatures, followed by silicalite, whereas carbonate source rocks have the heaviest $\delta^{13}C_{org}$ values.

In the Tarim Basin, the $\delta^{13}C_{org}$ values of Cambrian–Ordovician source rocks vary according to depositional facies. In the deep-water shelf zone of the Keping outcrop, Lower Cambrian source rocks have $\delta^{13}C_{org}$ values ranging from $-34.0‰$ to $-35.9‰$. In contrast, source rocks from the undercompensated basin facies in the

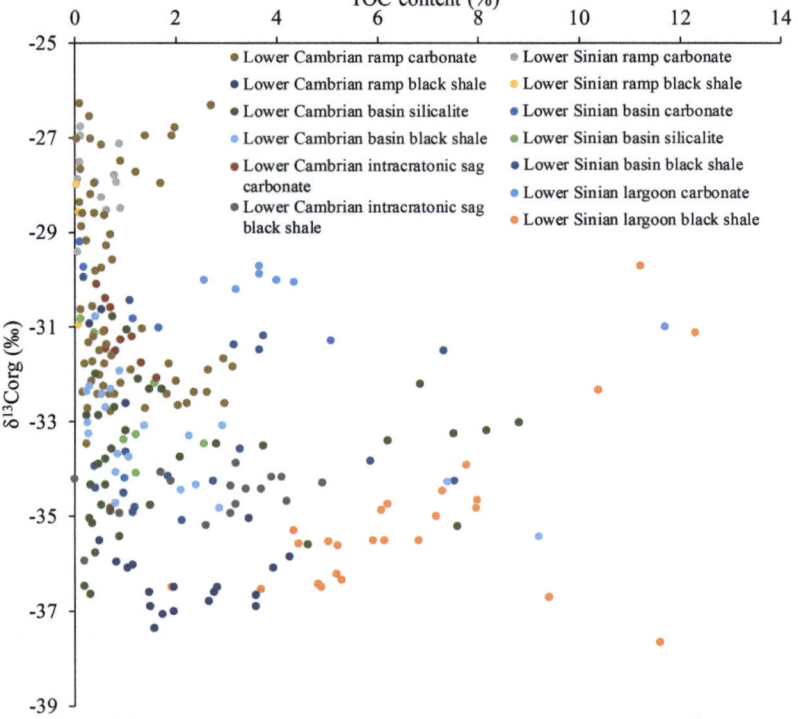

Fig. 6.6 Organic carbon isotope features in source rocks of different lithologies and depositional facies from the Ediacaran–Cambrian strata of the Yangtze Platform

Tadong area range from − 30.5‰ to − 32.0‰, whereas those from the Lower Cambrian source rocks deposited in the restricted platform facies of the Bachu area range from − 28.6‰ to − 30.0‰. These data suggest that lithofacies and depositional facies exert primary control over the isotopic composition of kerogen in source rocks. Notably, within the deep-water shelf facies, the δ^{13}Corg values of kerogen in the Yuertusi Formation systematically decrease from the LT1 well to the Keping outcrop (i.e., from east to west), indicating regional environmental variations within the Yuertusi Formation. Moreover, compared with their Cambrian counterparts, Ordovician source rocks generally exhibit heavier δ^{13}Corg values, with most values exceeding − 30‰. On the basis of organic carbon isotope data from source rocks of different sedimentary facies in the Tarim Basin, an organic carbon isotope distribution model for marine depositional systems has been developed (Fig. 6.7). The model indicates that the δ^{13}Corg values of kerogen in siliceous shale are generally less than − 34‰, whereas those in calcareous shale range from − 34‰ to − 30‰. In contrast, carbonate source rocks associated with platform facies, which typically exhibit low TOC contents, display δ^{13}Corg values exceeding − 30‰. These patterns reflect the influence of

carbon cycling processes in open-water environments. The observed variations in organic carbon isotopic compositions across various lithofacies may be attributed to differences in microbial contributions, influenced by strong oceanic stratification and fluctuating redox conditions. Higher δ^{13}Corg values (>− 30‰) in samples deposited in shallow-water environments resemble isotopic shifts observed in the Phanerozoic and may reflect isotopic fractionation primarily driven by photosynthesis. Conversely, lower δ^{13}Corg values in samples from deep-water facies likely indicate a significant contribution of chemotrophic or methane-oxidizing microorganisms to organic matter accumulation under anoxic conditions (Wang et al. 2014).

6.5.3 Origins of Deep Marine Hydrocarbons in the Tarim Basin

Organic geochemical indicators (biomarkers) continue to play a crucial role in oil–source correlation during the black oil stage. Therefore, this section addresses the controversial issue of deep petroleum system origins in the Tarim Basin, employing organic geochemical methods

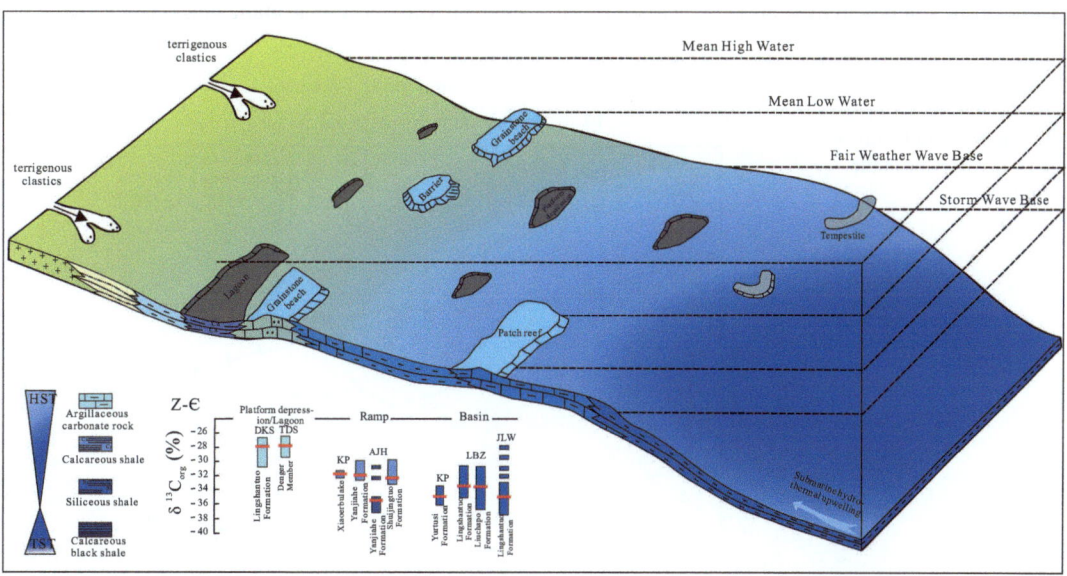

Fig. 6.7 Stable carbon isotope features of various type deep marine source rocks in the Tarim and Sichuan basins

to systematically analyze hydrocarbon origins and evaluate the effectiveness of deep-marine source rocks.

The δ^{13}Corg values of n-alkanes from different wells in the Tarim Basin are presented in Fig. 6.8. The distribution patterns of n-alkane carbon isotopes in Tarim crude oils reveal the complex interplay between hydrocarbon origins and postaccumulation alterations. On the basis of data from multiple wells in the Tarim Platform area, these n-alkane δ^{13}C distributions can be categorized into five distinct types. For instance, in the YG2 and YG2-1 wells, the δ^{13}C values of n-alkanes in Ordovician reservoir crude oil are approximately $-$ 37‰, with minimal isotopic variation among different molecular-weight n-alkanes. This pattern reflects the primary characteristics of black siliceous shale, with biological contributions from the ramp-basin facies (Type C). Crude oil samples from the YM2 well (Ordovician) and the ZS1 well (6841–6897 m, Cambrian) exhibit n-alkane δ^{13}C values ranging from $-$ 34‰ to $-$ 32‰ (Type B), with negligible differences between high- and low-molecular-weight n-alkanes, which are indicative of black shale from basin facies. In the source rock of the YD2 well (Ordovician Heituao Formation), the δ^{13}C values of n-alkanes range from $-$ 32‰ to $-$ 30‰ (Type A), with an increasing trend in δ^{13}C values for low-molecular-weight n-alkanes. This pattern

reflects carbon isotope fractionation controlled by the thermal degradation of kerogen in black shale from basin facies. In contrast, crude oils from the ZS1C well and YN2 well exhibit δ^{13}C differences of up to $-$ 10‰ or more between various n-alkanes (Type D), suggesting the influence of TSR on the carbon isotopic composition of crude oils derived from black shale in basin facies. Moreover, crude oils from wells Q6 and Q7 (Devonian) show distinct δ^{13}C variations between different n-alkanes (Type E). The lower δ^{13}C values of low-molecular-weight n-alkanes suggest contributions from aquatic organisms, whereas the higher δ^{13}C values of high-molecular-weight n-alkanes likely indicate inputs from terrigenous higher plants or carbonate source rocks associated with platform facies. The δ^{13}C distribution pattern of Type E suggests contributions from multiple source rocks or multiple phases of hydrocarbon generation. Overall, variations in kerogen isotopic compositions and the diverse patterns of δ^{13}C values in n-alkanes highlight the heterogeneity of sedimentary environments in Lower Paleozoic source rocks within the Tarim Basin. These findings underscore the complexity of crude oil origins and the multiple alteration processes influencing petroleum accumulation.

Peters et al. (2005) reported significant differences in the C_{29}/C_{30} hopane and C_{34}-22S hopane/

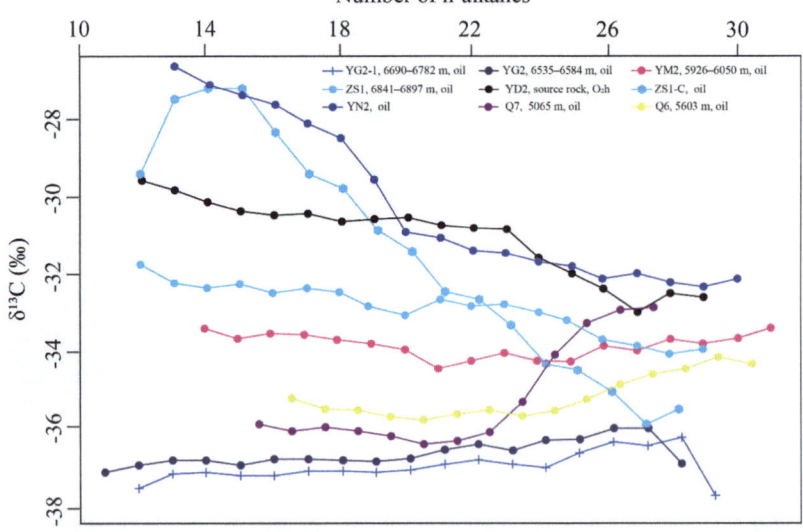

Fig. 6.8 Distribution characteristics of n-alkane δ^{13}C values in crude oils of Paleozoic reservoir in the Tarim Basin

C_{35}-22S hopane ratios between marine carbonate source rocks and crude oils generated in shale. The distributions of these ratios in crude oils from the Tazhong oil field are presented in Fig. 6.9. The results indicate that only a limited number of samples fall within the marine shale zone, whereas the majority are associated primarily with carbonate and marlstone zones.

By integrating the $\delta^{13}C$ values of n-alkanes with biomarker data, the hydrocarbon origins in the Tarim Basin can be further elucidated. Beyond black shales, the contribution of carbonate source rocks to deep marine petroleum systems cannot be overlooked. This suggests that the extensively developed black shale of the Yuertusi Formation alone is unlikely to have been the sole contributor to the Lower Paleozoic deep petroleum system in the Tarim Basin. Consequently, two possible hydrocarbon source scenarios emerge for the deep marine petroleum system in this region. The first scenario posits that both source rocks deposited in shallow-water shelf or tidal flat facies and the black shales of the Yuertusi Formation jointly function as effective source rocks. This interpretation highlights the pronounced heterogeneity of deep marine source rocks. Alternatively, the second scenario suggests that, in addition to the Yuertusi Formation, source rocks from other strata—such as the Middle to Upper Ordovician—also contributed to deep petroleum accumulations.

6.6 Carbonate-Rich Shales as Effective Contributors to Conventional and Unconventional Petroleum Resource Potential in the Tarim Basin and Sichuan Basin

Numerous Sinian–Cambrian outcrop profiles in southern China reveal significant vertical inheritance between Cambrian basin-facies fine-grained sediments and their Sinian counterparts. Laterally, the organic-rich black shales of basin facies transition sequentially into organic-rich marlstones of ramp facies and organic-lean carbonate rocks of platform-margin facies (e.g., Wang et al. 2016; Jiang et al. 2007). Consequently, compared with basin-facies black shales, ramp-facies organic-rich marlstones adjacent to high-energy carbonate reservoirs of platform margin facies exhibit supe-

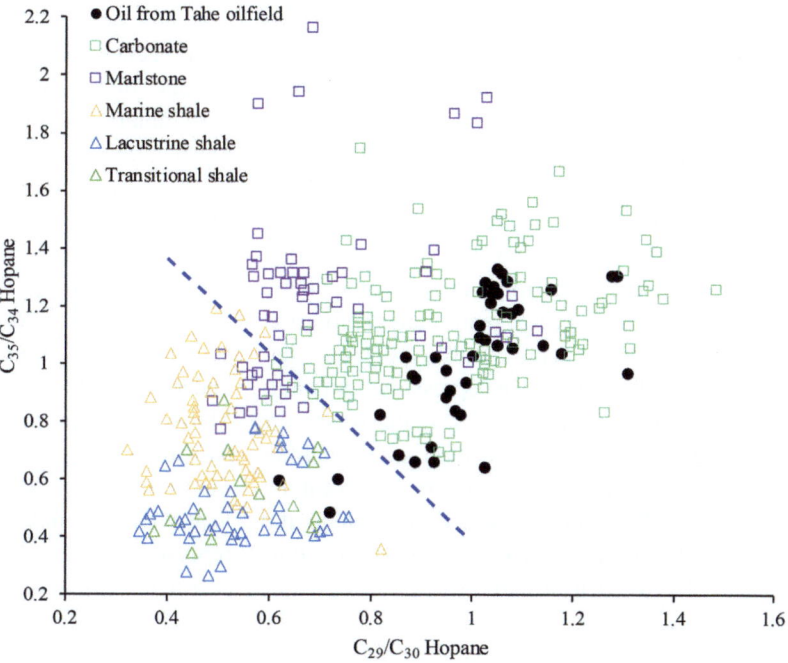

Fig. 6.9 C_{29}/C_{30} hopan and C_{34}-22S hopane/ C_{35}-22S hopane ratios of the crude oils of Paleozoic reservoir in the Tarim Basin

rior lateral hydrocarbon expulsion conditions, potentially demonstrating higher expulsion efficiency.

Research on the origins of deep marine petroleum in the Tarim Basin highlights the significant contribution of deep marine carbonate source rocks to hydrocarbon accumulation, emphasizing their resource potential. However, the TOC content of carbonate source rocks is considerably lower than that of black shales (Zhang et al. 2002b; Liang et al. 2008). Theoretically, this would result in a much lower hydrocarbon generation potential than that of black shales. Nevertheless, the strong mechanical rigidity of carbonate minerals enables carbonate source rocks to develop a compaction-resistant framework, allowing for them to better retain primary pores at similar burial depths relative to black shales. Furthermore, the chemically reactive nature of carbonate minerals makes them susceptible to interactions with organic acids released during kerogen maturation, leading to the formation of secondary pores (Ma et al. 2010, 2019). Therefore, carbonate source rocks generally exhibit well-developed pore systems. However, owing to their relatively low hydrocarbon generation capacity, most hydrocarbons remain confined within the inorganic pore spaces of the carbonate matrix. Only carbonate formations with sufficiently high TOC contents can generate hydrocarbons in quantities adequate to meet absorption thresholds and drive hydrocarbon expulsion (Wang et al. 2022a, 2025). This suggests that a substantial proportion of hydrocarbons within carbonate source rocks remains within their pore systems. Under favorable economic conditions, these formations could become viable targets for unconventional petroleum exploration and development. This study proposes that carbonate-rich reservoirs in depressions and ramp facies function primarily as "source–reservoir integrated" unconventional reservoirs, making them promising exploration targets. For example, the Middle Permian Maokou Formation in the Sichuan Basin and its adjacent areas was predominantly deposited in a carbonate ramp environ-

ment characterized by extensive marlstone and calcareous mudstone development. The marlstone has moderate to good TOC levels and high to overmature thermal maturity (Guo 2021). Detailed sedimentary microfacies and geochemical analyses indicate that the dominant clay minerals in the Maokou Formation marlstone are sepiolite and talc (rich in Mg and Si), which are derived from contemporaneous seawater, with their distribution controlled by depositional conditions (Song et al. 2024). The clay mineral content follows the sequence: outer ramp (5–70%) > deep-water lagoon, middle ramp (3–47%) > shallow-water lagoon-flat (10–13%). Organic matter enrichment is closely associated with the clay mineral content, with the highest TOC values observed in deep-water lagoons and middle ramp facies. Moreover, effective porosity is influenced by both the clay mineral content and TOC content. The deep-water lagoon and middle ramp facies not only exhibit elevated TOC values but also feature well-developed pore systems. Consequently, these facies, whose TOC and porosities are superior to those of the outer ramp, represent the most favorable targets for unconventional petroleum exploration. The specific surface area is significantly correlated with the micropore and mesopore volumes. At the high-overmature stage (Ro ranging from 1.54 to 2.17%), the specific surface area reaches its peak, predominantly comprising organic matter pores, intercrystalline pores in clay minerals and pyrite, and microfractures. At lower thermal maturity stages, intercrystalline pores in clay minerals and pyrite are less developed. This suggests that at the high-overmature stage, intercrystalline pores in clay minerals and pyrite constitute the primary mesopore types and play a crucial role in enhancing the specific surface area and promoting natural gas adsorption and accumulation. Therefore, the Middle Permian Maokou Formation in the Sichuan Basin and its surrounding regions, along with the marlstones of the Lower Cambrian Yuertusi Formation and the Lower Xiaoerbulak Formation in the Tarim Basin, represent key targets for future unconventional petroleum exploration.

6.7 Conclusion

This study provides a review of the geological characteristics of deep marine source rocks in two major petroliferous basins in China—the Sichuan Basin and the Tarim Basin. By integrating field outcrop observations and newly acquired drilling data, five geological models for the development of deep marine source rocks were established. Through the analysis of stable carbon isotopes and biomarkers from various source rock types, this study highlights their respective roles in deep marine petroleum systems and preliminarily evaluates the unconventional

1. The heterogeneity of marine source rocks is a crucial factor that cannot be overlooked. Sedimentary facies significantly influence the lithofacies and TOC content of source rocks, leading to substantial variations in their properties. Recognizing this heterogeneity objectively is essential for accurate evaluation of hydrocarbon generation potential.
2. Multiple lines of evidence suggest that the stable carbon isotopes of kerogen remain largely unaffected by thermal maturation. Even at high-overmature stages, they continue to provide valuable paleoenvironmental information about the source rocks. Although hydrocarbon cracking at high-overmature stages reduces the absolute concentration of biomarkers, certain geochemical proxies still retain key characteristics of the source rock. Therefore, during the black oil phase, some biomarkers remain effective for characterizing source rock properties and conducting oil–source correlations.
3. The deep marine petroleum system in the Tarim Basin does not originate solely from organic-rich shale; carbonate source rocks also significantly contribute to hydrocarbon accumulation, and their resource potential needs to be reassessed objectively. Two possible hydrocarbon origins for the deep marine petroleum system in the Tarim Basin are proposed: first, petroleum may originate from source rocks with different lithologies deposited in various sedimentary facies of the Yuertusi Forma-

tion; second, source rocks from other stratigraphic intervals, such as the Ordovician and Sinian, also play a substantial role in petroleum generation and accumulation.
4. The resource potential of carbonate source rocks is considerable, and most hydrocarbons generated within these formations may still be retained in their pore systems. Under favorable economic conditions, these source rocks could become promising targets for unconventional petroleum exploration. On the basis of this understanding, the Middle Permian Maokou Formation in the Sichuan Basin and its surrounding areas, the marlstone of the Lower Cambrian Yuertusi Formation, and the Lower Xiaoerbulak Formation marlstone in the Tarim Basin are identified as key targets for future unconventional petroleum exploration.

Acknowledgements The authors acknowledges the financial support from the National Natural Science Foundation of China (Grants U24B6001 and U19B6003).

References

Cai CF (2018) Application of organic sulfur isotopic composition of petroleum origin and evolution: a review. Nat Gas Geosci 29(2):159–167

Cai CF, Li KL, Ma AL, Zhang CM, Xu ZM, Worder RH, Wu GH, Zhang BS, Chen LX (2009) Distinguishing Cambrian from Upper Ordovician source rocks: evidence from sulfur isotopes and biomarkers in the Tarim Basin. Org Geochem 40(7):755–768

Chen WZ (1992) Discussing again on the source of Sinian gas pool in Weiyuan gasfield in Sichuan Basin. Nat Gas Ind 12(6):28–32

Dai JX (2003) Pool-forming periods and gas source of Weiyuan gasfield. Pet Geol Exp 25(5):473–480

Feng Y, Xiao XM, Gao P, Wang EZ, Hu DF, Liu RB, Li G, Lu CG (2023) Restoration of sedimentary environment and geochemical features of deep marine Longmaxi shale and its significance for shale gas: a case study of the Dingshan area in the Sichuan Basin, South China. Mar Pet Geol 151:106186

Geatches DL, Clark SJ, Greenwell HC (2010) Role of clay minerals in oil-forming reactions. J Phys Chem A 114(10):3569–3575

Gu J, Chou J, Yan H (1994) Sedimentary facies and petroleum accumulations: the petroleum exploration in the Tarim Basin, vol 3. Petroleum Industry Press, Beijing

Guo TL (2011) Basic characteristics of deep reef-bank reservoirs and major controlling factors of gas pools in the Yuanba gas field. Nat Gas Indust 31(10):12–16 (in Chinese with English abstract)

Guo TL (2021) Geological characteristics and exploration prospect of carbonate source rock gas in Sichuan Basin. J Southwest Pet Univ (Science and Technology Edition). 43(1):1–16

Guo XS, Guo TL, Huang RC, Duan JB (2014) Cases of discovery and exploration of marine fields in China (Part 16): Yuanba gas field in Sichuan Basin. Marine Origin Petrol Geol 19(4):57–64 (in Chinese with English abstract)

Guo XS, Hu DF, Wei ZH, Li YO, Wei XF (2016) Discovery and exploration of Fuling shale gas field. China Petrol Explor 21(3):24–37 (in Chinese with English abstract)

Guo TL, Xiong L, Ye SJ, Dong XX, Wei LM, Yang YT (2023) Theory and practice of unconventional gas exploration in carrier beds: insight from the breakthrough of new type of shale gas and tight gas in Sichuan Basin, SW China. Pet Explor Dev 50(1):24–37

He M, Wang ZY, Moldwan MJ, Peters K (2022) Insights into catalytic effects of clay minerals on hydrocarbon composition of generated liquid products during oil cracking from laboratory pyrolysis experiments. Org Geochem 163:104331

Huang HP, Zhang SC, Su J (2016) Palaeozoic oil–source correlation in the Tarim Basin NW China: a review. Org Geochem 94:32–46

Jia CZ (1997) Tectonic features and oil and gas in Tarim Basin, China. Petroleum Industry Press, Beijing

Jiang GQ, Kaufman AH, Christie-Blick N, Zhang SH, Wu HC (2007) Carbon isotope variability across the Ediacaran Yangtze platform in South China: implications for a large surface-to-deep ocean $\delta^{13}C$ gradient. Earth Planet Sci Lett 261(1–2):303–320

Johns WD (1979) Clay mineral catalysis and petroleum generation. Annu Rev Earth Planet Sci 7:183–198

Kang YZ (2003) Geological characteristics of the formation of the large Tahe oilfield in the Tarim Basin and its prospects. China Geology 30(3):315–319

Katz BJ, Lin F (2014) Lacustrine basin unconventional resource plays: Key differences. Mar Pet Geol 56:255–265

Later SR, Bowler BFJ, Li M, Chen M, Brincat D, Bennett B, Noke K, Donohoe P, Simmons D, Kohnen M, Allan J, Telnaes N, Horstad I (1996) Molecular indicators of secondary oil migration distances. Nature 383:593–597

Li MW, Larter SR, Stoddart D, Bjoroey M (1992) Liquid chromatographic separation schemes for pyrrole and pyridine nitrogen aromatic heterocycle fractions from crude oils suitable for rapid characterization of geochemical samples. Anal Chem 64(13):1337–1344

Li MW, Ma XX, Jin ZJ, Li ZM, Jiang QG, Wu SQ, Li Z, Xu ZX (2022) Diversity in the lithofacies assemblages of marine and lacustrine shale strata and significance for unconventional petroleum exploration in China. Oil Gas Geol 43(1):1–25 (in Chinese with English abstract)

Li SM, Pang XQ, Jin ZJ, Yang HJ, Xiao ZY, Gu QY, Zhang BS (2010) Petroleum source in the Tazhong Uplift, Tarim Basin: new insights from geochemical and fluid inclusion data. Org Geochem 41(6):531–553

Li MW, Ma XX, Jin ZJ, Li ZM, Jiang QG, Wu SQ, Li Z, Xu ZX (2022) Diversity in the lithofacies assemblages of marine and lacustrine shale strata and significance for unconventional petroleum exploration in China. Oil Gas Geol 43(1):1–25

Liang DG, Guo TL, Chen JP, Bian LZ, Zhao Z (2008) Some progresses on studies of hydrocarbon generation and accumulation in marine sedimentary regions, Southern China (Part 1): distribution of four suits of regional Marine Source Rocks. Mar Orig Pet Geol 1–16

Liu SG, Deng B, Sun W, Song JM, Jiao K, Ye YH, Xie GL (2020) May Sichuan Basin be a Super Petroliferous Basin? J Xihua University (Natural Science Edition). 39(5):20–35

Ma YS (2006) Cases of discovery and exploration of marine fields in China (Part 6): Puguang gas field in Sichuan Basin. Marine Origin Petrol Geol 11(2):35–40 (in Chinese with English abstract)

Ma YS, Cai XY, Zhao PR, Zhang XF (2010) Formation mechanism of deep-buried carbonate reservoir and its model of three-element controlling reservoir: a case study from the Puguang Oilfield in Sichuan. Acta Geol Sin 84(8):1087–1094

Ma YS, He ZL, Zhao PR, Zhu HQ, Han J, You DH, Zhang JT (2019) A new progress in formation mechanism of deep and ultra-deep carbonate reservoir. Acta Petrolei Sinica. 40(12):1415–1425

Ma YS, Li MW, Cai XY, Xu XH, Hu DF, Qu SL, Li GS, He DF, Xiao XM, Zeng YJ, Rao Y (2020) Mechanisms and exploitation of deep marine petroleum accumulations in China: advances, technological bottlenecks and basic scientific problems. Oil Gas Geol 41(4):655–672

Ma YS, Cai XY, Li MW, Li HL, Zhu DY, Qiu NS, Pang XQ, Zeng DQ, Kang ZJ, Ma AL, Shi KB, Zhang JT (2024) Research advances on the mechanisms of reservoir formation and hydrocarbon accumulation and the oil and gas development methods of deep and ultra-deep marine carbonates. Pet Explor Dev 51(4):692–707

Pepper AS, Corvi PJ (1995) Simple kinetic models of petroleum formation. Part I: oil and gas generation from kerogen. Mar Pet Geol 12(3):291–319

Peters KE, Moldowan JM, Sundararaman P (1990) Effects of hydrous pyrolysis on biomarker thermal maturity parameters: monterey Phosphatic and Siliceous Members. Org Geochem 15:249–265

Peters KE, Walters CC, Moldowan JM (2005) The biomarker guide, biomarkers and isotopes in petroleum exploration and earth hisory. Cambridge University Press, Cambridge

Qiu NS, Chang J, Feng QQ, Zeng S, Liu XY, Li HL, Ma AL (2023) Maturation history of deep and ultra-deep source rocks, central and western basins, China. Earth Sci Front 30(6):199–212

Seifert WK, Moldowan JM (1978) Applications of steranes, terpanes and monoaromatics to the maturation, migration and source of crude oils. Geochim Cosmochim Acta 42:77–95

Song JM, Wang JR, Liu SG, Li ZW, Luo P, Jiang QC, Jin X, Yang D, Huang SP, Fan JP, Ye YH, Wang JK, Deng HS, Wang B, Guo JX (2024) Types, composition and diagenetic evolution of authigenic clay minerals in argillaceous limestone of sepiolite-bearing strata: a case study of Mao-1 Member of Middle Permian Maokou Formation, eastern Sichuan Basin, SW China. Pet Explor Dev 51(2):351–363

Spigolon ALD, Lewan MD, de Barros Penteado HL, Coutinho LFC, Mendonça Filho JG (2015) Evaluation of the petroleum composition and quality with increasing thermal maturity as simulated by hydrous pyrolysis: a case study using a Brazilian Source Rock with Type I Kerogen. Org Geochem 83–84:27–53

Tian YK, Yang CP, Liao ZW, Zhang HZ (2012) Geochemical quantification of mixed marine oils from Tazhong area of Tarim Basin, NW China. J Petrol Sci Eng 90–91:96–106

Tissot B, Welte D (1984) Petroleum formation and occurrence 219. Springer, New York, p 699

Tissot B, Durand B, Espitalie J, Combaz A (1974) Influence of the nature and diagenesis of organic matter in the formation of petroleum. AAPG Bull 58:499–506

Tissot BP, Pelet R, Ungerer PH (1987) Thermal history of sedimentary basins, maturation indices, and kinetics of oil and gas generation. AAPG Bull 71:1445–1466

Tong DS, Zhou CH, Li MY, Yu WH, Beltramini J, Lin CX, Xu ZP (2010) Structure and catalytic properties of Sn-containing layered double hydroxides synthesized in the presence of dodecylsulfate and dodecylamine. Appl Clay Sci 48(4):569–574

Wang XB (1982) Natural gas from deep earth sources. Chin Sci Bull 17:1069–1071

Wang XQ, Shi XY, Jiang GQ, Tang DJ (2014) Organic carbon isotope gradient and ocean stratification across the late Ediacaran-Early Cambrian Yangtze Platform. Sci China Earth Sci 57:919–929

Wang XQ, Jiang GQ, Shi XY, Xiao SH (2016) Paired carbonate and organic carbon isotope variations of the Ediacaran Doushantuo Formation from an upper slope section at Siduping, South China. Precambrian Res 273:53–66

Wang EZ, Li MW, Ma XX, Qian MH, Cao TT, Li ZM, Yang WW, Jin ZJ (2024) Diahopane and diasterane as the proxies for paleoenvironment, hydrocarbon generation condition, and shale oil accumulation. Chem Geol 670:122447

Wang EZ, Fu YX, Guo TL, Li MW (2025) A new approach for predicting oil mobilities and unveiling their controlling factors in a lacustrine shale system: Insights from interpretable machine learning model. Fuel 379:132958

Wang EZ, Feng Y, Guo TL, Li MW (2022a) Oil content and resource quality evaluation methods for lacustrine shale: a review and a novel three-dimensional quality evaluation model. Earth-Sci Rev 232:104134

Wang EZ, Guo TL, Li MW, Li CR, Dong XX, Zhang NX, Feng Y (2022b) Exploration potential of different lithofacies of deep marine shale gas systems: Insight

into organic matter accumulation and pore formation mechanisms. J Nat Gas Sci Eng 102:104563

Wei GQ, Xie ZY, Bai GL, Li J, Wang ZH, Li AG, Li ZS (2014) Organic geochemical characteristics and origin of natural gas in the Sinian-Lower Paleozoic reservoirs, Sichuan Basin. Natural Gas Ind 34(3):44–49

Wu LM, Zhou CH, Keeling J, Tong DS, Yu WH (2012) Towards an understanding of the role of clay minerals in crude oil formation, migration and accumulation. Earth Sci Rev 115:373–386

Xia ML, Wen L, Wang YG, Hong HT, Fan Y, Wen YC (2010) High-quality source rocks in trough facies of Upper Permian Dalong Formation of Sichuan Basin. Pet Explor Dev 37(6):654–662

Xu YC, Shen P, Li YC (1989) The oldest gas pool of China—Weiyuan Sinian gas pool, Sichuan Province. Acta Sedimentol Sin 7(4):3–12

Yang KM (2016) Hydrocarbon potential of source rocks in the Middle Triassic Leikoupo Formation in the Western Sichuan Depression. Pet Geol Exp 38(3):366–374

Zhang SC, Huang HP (2005) Geochemistry of Palaeozoic marine petroleum from the Tarim Basin, NW China. Part 1 Oil Family Classification. Org Geochem 36:1204–1214

Zhang SC, Hanson AD, Moldowan JM, Graham SA, Liang DG, Chang E, Fago F (2000) Paleozoic oil–source rock correlations in the Tarim Basin, NW China. Org Geochem 31:273–286

Zhang K, Su J, Chen YQ, Ma SH, Zhang HZ, Yang CL, Fang Y (2023) The biogeochemical features of the Cambrian-Ordovician source rock and origin of ultra-deep hydrocarbon in the Tarim Basin. Acta Geol Sin 97(6):2026–2041

Zhang SC, Liang DG, Li MW, Xiao ZY, He ZH (2002a) Molecular fossils and oil–source rock correlations in Tarim Basin, NW China. Chin Sci Bull 47:20–27

Zhang SC, Liang DG, Zhang DJ (2002b) Evaluation criteria for Paleozoic effective hydrocarbon source rocks. Pet Explor Dev 29(2):8–12

Zhang SC, Huang HP, Xiao ZY, Liang DG (2005) Geochemistry of Palaeozoic marine petroleum from the Tarim Basin, NW China. Part 2: Maturity assessment. Org Geochem 36(8):1215–1225

Zhang SC, Wang HJ, Su J, Wang XM, He K, Liu YK (2024a) Control of Earth system evolution on the formation and enrichment of marine ultra-deep petroleum in China. Pet Explor Dev 51(4):870–885

Zhang ZY, Zhu GY, Chen WY, Wu L, Ren R, Zhang CL (2024b) Cryogenian–Cambrian tectono-sedimentary evolution, paleoclimate and environment effects, and formation of Petroleum resources in the Tarim Block. Earth Sci Rev 248:104632

Zhao WZ, Wang ZY, He HQ, Zhang MJ, Wang HJ, Wang YP, Qin Y (2005) Gas formation mechanism of marine carbonate source rocks in China. Science in China Series D: Earth Science 48:441–453

Zhao MJ, Wang ZM, Pan WQ, Liu SB, Qin SF, Han JF (2008) Lower Palaeozoic source rocks in Manjiaer Sag, Tarim Basin. Pet Explor Dev 35(4):417–423

Zhao WZ, Xie ZY, Wang XM, Shen AJ, Wei GQ, Wang ZC, Wang K (2021) Sinian gas sources and effectiveness of primary gas-bearing system in Sichuan Basin, SW China. Pet Explor Dev 48(6):1089–1099

Zhu GY, Zhang SC, Liang YB, Ma YS, Dai JX, Li J, Zhou GY (2006) The characteristics of natural gas in Sichuan basin and its sources. Earth Sci Front 13(2):234–248

Zhu GY, Zhao K, Li TT, Zhang ZY (2021) Formation mechanism and distribution prediction of source rocksin the Ediacaran Doushantuo Formation. South China. Acta Geologica Sinica. 95(8):2553–2574

Zhu GY, Hu DF, Chen YQ, Xue N, Zhao K, Zhang ZY, Li TT, Chen ZY (2022) Geochemical characteristics and formation environment of source rock of the Lower Cambrian Yuertusi Formation in well Luntan 1 in Tarim basin. Acta Geol Sin 96(6):2116–2130

Zou CN, Dong DZ, Yang H, Wang YM, Huang JL, Wang SF, Fu CX (2011) Conditions of shale gas accumulation and exploration practices in China. Natural Gas Ind 26–39

Multiple Types of Deep Carbonate Reservoirs in China: Consensuses and New Developments

7

Dongya Zhu, Jingbin Wang, Donghua You, Kaibo Shi, Juntao Zhang, Chongyang Wu, Qian Ding, Huili Li, Xunyu Cai, Quanyou Liu, Zhiliang He, and Bo Liu

Abstract

Oil and gas exploration in the Tarim Basin and Sichuan Basin has been gradually expanding to deep–ultradeep carbonate strata (> 8000 m). Three types of large-scale high-quality carbonate reservoirs have been found, including fault-dissolution reservoirs associated with fault–fluid coupling alteration, Cambrian subsalt dolomite and microbialite reservoirs, which are the most important hydrocarbon reservoirs to be discovered after karst fracture–cave and reef–shoal facies reservoirs. The deep–ultradeep Ordovician carbonates in the Tarim Basin were first broken and fractured by strike–slip faults and then partly dissolved by diagenetic fluids, such as meteoric water, hydrothermal fluids, and fluids derived from thermochemical sulfate reduction (TSR), leading to the formation of fault-dissolution reservoirs with coupled fault and fluid alterations. The granular dolomites of the Cambrian subsalt shoal facies are prone to leaching by quasi contemporaneous meteoric water to form high-quality dolomite reservoirs, which are rich in intergranular, intragranular, and intercrystalline pores and dissolution vugs. Microbialite reservoirs occur widely in the Upper Ediacaran Dengying Formation in the Sichuan Basin and the Qigebulak Formation and Lower Cambrian Xiaoerbulak Formation in the Tarim Basin. The microbialite reservoirs are mainly microbial thrombolites, stromatolites, and microbial reef mounds, which generally contain abundant framework pores, with porosities as high as 11.29%. A series of geological factors, such as oil and gas charging, high concentrations of CO_2 and H_2S derived from TSR, and fluid overpressure in subsalt, ensure that the porosity in ultradeep carbonates can be well maintained. One or more of these types of ultradeep carbonate reservoirs are predicted to exist at depths > 10,000 m.

D. Zhu (✉) · J. Wang · D. You · J. Zhang · C. Wu ·
Q. Ding · H. Li · Z. He
Petroleum Exploration and Production Research Institute
of SINOPEC, Beijing, China
e-mail: zhudy.syky@sinopec.com

D. Zhu · J. Wang · J. Zhang · C. Wu · Q. Ding · H. Li
Key Laboratory of Geology and Resources in Deep
Stratum of SINOPEC, Beijing, China

K. Shi · Q. Liu · B. Liu
Peking University, Beijing, China

X. Cai
China Petroleum and Chemical Corporation (Sinopec),
Beijing, China

Keywords

Deep–ultradeep carbonate · Development and preservation · Fault–fluid coupling alteration · Microbialite reservoir · Subsalt dolomite

© The Author(s) 2026
Y. Ma et al. (eds.), *Petroleum Geology and Exploration of Deep Marine Strata in China*, Advances in Oil and Gas Exploration & Production, https://doi.org/10.1007/978-3-032-02496-1_7

7.1   Introduction

Recently, increasing attention has been given to the exploration and development of oil and gas in deep–ultradeep carbonates. Consequently, the reserves and production of ultradeep oil and gas in North America, the Middle East, and the Asia-Pacific region have shown an increasing trend year by year. Deep–ultradeep carbonates are also important targets for oil and gas exploration and development in China (Ma et al. 2019). Wells Tashen 1 and Luntan 1 in the Tarim Basin (Yang et al. 2020; Zhu et al. 2015) and well Mashen 1 in the Sichuan Basin all have high-quality carbonate reservoirs with good oil and gas reserves in ultra-deep strata (> 8000 m). Industrial oil and gas reserves and production have also been found in ultradeep Ordovician carbonate reservoirs at depths of 7000–8000 m in the Shunbei area (Ma et al. 2022). The deepest well, Yuejin 3-3XC, in the Shunbei area (up to 9432 m) has a daily production of 200 tons of oil and 50,000 m^3 of natural gas from the Ordovician carbonate reservoir.

However, a key bottleneck factor restricting the success rate of hydrocarbon exploration is whether large-scale high-quality reservoirs can be found in ultradeep (7000–8000 m and even deeper) environments. Ehrenberg et al. (2009) concluded that the porosity of carbonate rocks gradually decreases with increasing burial depth and geological age. On the basis of the successful explorations of deep–ultradeep carbonate reservoirs in the Tarim, Sichuan and Ordos Basins in China (Fig. 7.1), many scholars agree that high-quality carbonate reservoirs can exist in deep–

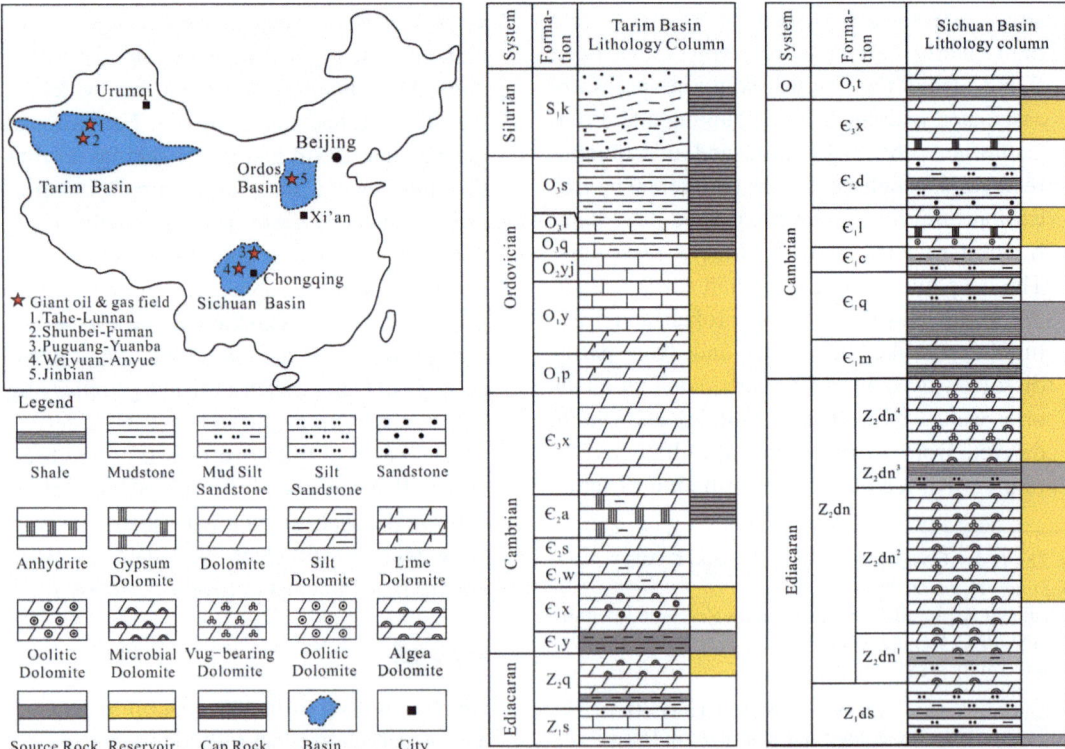

Fig. 7.1 Major Paleozoic marine carbonate basins and stratigraphic columns in western China. Formation names in Sichuan Basin: Tongzi (O$_1$t), Xixiangchi (Є$_3$x), Douposi (Є$_2$d), Longwangmiao (Є$_1$l), Canglangpu (Є$_1$c), Qiongzhusi (Є$_1$q), Maidiping (Є$_1$m), Dengying (Z$_2$dn) including four members (Z$_2$dn^1, Z$_2$dn^2, Z$_2$dn^3, Z$_2$dn^4), Doushantuo (Z$_1$ds). Formation names in Tarim Basin: Kepingtag (S$_1$k), Sangtamu (O$_3$s), Lianglitag (O$_3$l), Qalbak (O$_3$q), Yijianfang (O$_2$yj), Yingshan (O$_1$y), Penglaiba (O$_1$p), Xiaqiulitag (Є$_3$x), Awatage (Є$_2$a), Shay-ilik (Є$_2$s), Wusonger (Є$_1$w), Xiaoerbulak (Є$_1$x), Yuertusi (Є$_1$y), Qigebulak (Z$_2$q), Sugaitebulak (Z$_1$s)

ultradeep strata under the influences of lithofacies, tectonics, faults and diagenetic fluids (He et al. 2017; Ma et al. 2010; Zhu et al. 2019).

The types of deep–ultradeep carbonate reservoirs are diverse and can generally be classified into five major types: reef–shoal, karst fracture–cave below unconformity, fault-dissolution, subsalt dolomite and ancient microbialite reservoirs. Among them, reef–shoal and karst fracture–cave reservoirs have a long history of petroleum exploration. In addition, their features, controlling factors and geological models have been well studied. Typical reef–shoal gas fields include the Puguang, Yuanba and Anyue gas fields in the Sichuan Basin (Ma et al. 2007; Shen et al. 2015), and typical karst fracture–cavity oil and gas fields include the Tahe–Lunnan oil and gas field in the northern Tarim Basin (Ma et al. 2011; Zhao et al. 2014). In the process of advancing to deep–ultradeep carbonate strata in recent years, some new types of carbonate reservoirs, such as fault-dissolution, subsalt dolomite and ancient microbialite reservoirs, have proven to be important reservoir types. The characteristics and controlling factors of the development of these new carbonate reservoirs have not been studied in detail. The main purpose of this paper is therefore to summarize the main controlling factors and mechanisms of deep–ultradeep carbonate reservoirs based on typical field outcrop profiles, ultradeep drilling cores, geochemical data and the latest hydrocarbon exploration results. The results of this summary will provide basic knowledge for expanding hydrocarbon exploration into even deeper realms (> 10,000 m).

7.2 Fault–Fluid Coupling Alteration of Deep Carbonate Reservoirs

7.2.1 Fault–Fluid Coupling Types

The development and preservation of porosity in deep–ultradeep carbonate reservoirs are generally influenced by a series of factors, such as sedimentary facies, tectonic activities, faults, and diagenetic fluids (He et al. 2017; Ma et al. 2010). Fault–fluid coupling alteration is a critical factor in the formation of ultradeep carbonate reservoirs (Zhu et al. 2019). Strike–slip faults and related fractures can not only directly contribute to the development of ultradeep fault-controlled breccia spaces but also serve as key conduits for various types of diagenetic fluids migrating upward or downward. In particular, hydrothermal fluid migrating upward along fault–fracture conduits can strongly dissolve and alter carbonate reservoirs (Davies and Smith 2006; Stacey et al. 2021).

Large-scale high-quality carbonate reservoirs associated with fault–fluid coupling alteration are currently targets of hydrocarbon exploration in deep–ultradeep carbonates around the world. Specifically, hydrocarbon exploration is now focusing on fault–fluid coupling ultradeep reservoirs in the Sichuan and Tarim Basins in China, the western Canadian Basin in Canada, the Mississippi River Basin in the United States, and Oman in the Middle East (Davies and Smith 2006; Stacey et al. 2021). Accurately distinguishing and identifying fluid types migrating along faults and learning dissolution and alteration rules are key issues in studying development and preservation mechanisms as well as predicting deep–ultradeep carbonate reservoirs. By assessing the petrography of diagenetic minerals, carbon, oxygen and strontium isotopic compositions (δ^{13}C, δ^{18}O and ^{87}Sr/^{86}Sr), elemental compositions, and fluid inclusions from more than 300 deep–ultradeep carbonate samples in basins in China, three endmember geochemical indicators have been determined (Table 7.1), and geochemical identification diagrams have been constructed (Fig. 7.2). Using these geochemical indicators and diagrams, the main diagenetic fluids can be accurately distinguished.

The types of diagenetic fluids are often closely related to tectonic or fault activity, and the coupling relationships include tectonic uplift–meteoric water, fault–hydrothermal fluids, and deep burial–TSR-derived fluids (Zhu et al. 2019). Carbonates, which are affected by tectonic uplift or associated sea level fluctuations, tend to be exposed above seawater for a period of time and consequently suffer meteoric water karst and alteration. As a result, meteoric water macrocrys-

Table 7.1 The end-member geochemical indicators for different types of diagenetic fluids in deep–ultradeep carbonate reservoir

Geochemical indicator	Meteoric water	Hydrothermal	TSR-derived fluid	Formation water
$\delta^{18}O_{V\text{-PDB}}$	− 17.2‰	− 9.3‰	− 10.2‰	− 8.0‰
$\delta^{13}C_{V\text{-PDB}}$	− 1.4‰	− 1.2‰	− 15.7‰	0.5‰
Homogenization temperature (Th)	53.5–64.7 °C	155.9–199.1 °C	150.3–218.2 °C	50.2–90.2 °C
$^{87}Sr/^{86}Sr$	0.710340	0.709513	0.707638	0.708347
δEu	0.92	76.03	10.9	0.96
Typical reservoir, well	The Ordovician carbonate reservoir in the northern Tarim Basin Well AD3, S85, TS3, T904	The Ordovician carbonate reservoir in the center of Tarim Basin Well Zhong 4, Zhong 16, Tazhong 12	The Permian-Triassic carbonate reservoir in the Puguang, Sichuan Basin Well PG1, PG10, YB2	The Ordovician carbonate reservoir in the Tahe oilfield, Tarim Basin Well S91, S94, S119, TP2

talline calcite often fills in karst caves. Meteoric calcite has a strongly negative oxygen isotope value ($\delta^{18}O_{V\text{-PDB}}$: − 17.2‰) and a strongly high strontium isotope ratio ($^{87}Sr/^{86}Sr$: 0.710340). The calcite precipitated from hydrothermal fluids generally has a high strontium isotope ratio ($^{87}Sr/^{86}Sr$: 0.709513) (Fig. 7.2) and a significantly high positive Eu anomaly (δEu: 76.03). The homogenization temperature (Th) of fluid inclu-

sions in calcite can reach 155.9–202 °C, which is much higher than the formation temperature. Hydrothermal fluid may have originated from meteoric water migrating downward into the basement strata. After heating, the fluid migrates upward as hydrothermal fluid (Katz et al. 2006). The calcite precipitated from such hydrothermal fluid should have very negative oxygen isotope values, e.g., − 25 to − 15‰ (Westphal et al. 2004)

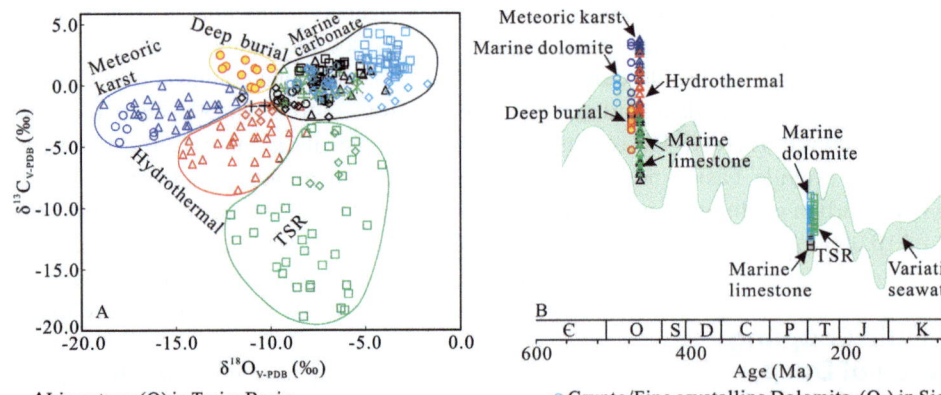

△ Limestone (O) in Tarim Basin
✕ Seawater calcite (O) in Tarim Basin
△ Shallow burial formation water calcite (O) in Tarim Basin
△ Meteoric mega-crystalline calcite (O) in Tarim Basin
△ Hydrothermal calcite (O) in Tarim Basin
○ Limestone (O) in Sichuan Basin
＋ Seawater calcite (O) in Sichuan Basin
○ Deep burial formation water calcite (O) in Sichuan Basin
○ Meteoric mega-crystalline calcite (O) in Sichuan Basin

○ Crypto/Fine crystalline Dolomite (O_1) in Sichuan Basin
□ Limestone (T_1f) in Sichuan Basin
□ Crypto/Fine crystalline Dolomite (T_1f) in Sichuan Basin
□ TSR calcite (T_1f) in Sichuan Basin
◇ Limestone (O) in Ordos Basin
◇ Crypto/Fine crystalline Dolomite (O) in Ordos Basin
◇ Hydrothermal dolomite (O) in Ordos Basin
◇ TSR calcite (O) in Ordos Basin

Fig. 7.2 Carbon, oxygen and strontium isotope identification diagrams for different types of fluids in the deep–ultra-deep marine carbonate reservoir (Modified from Zhu et al. 2019); Variation of seawater $^{87}Sr/^{86}Sr$ is from Veizer et al. (1999)

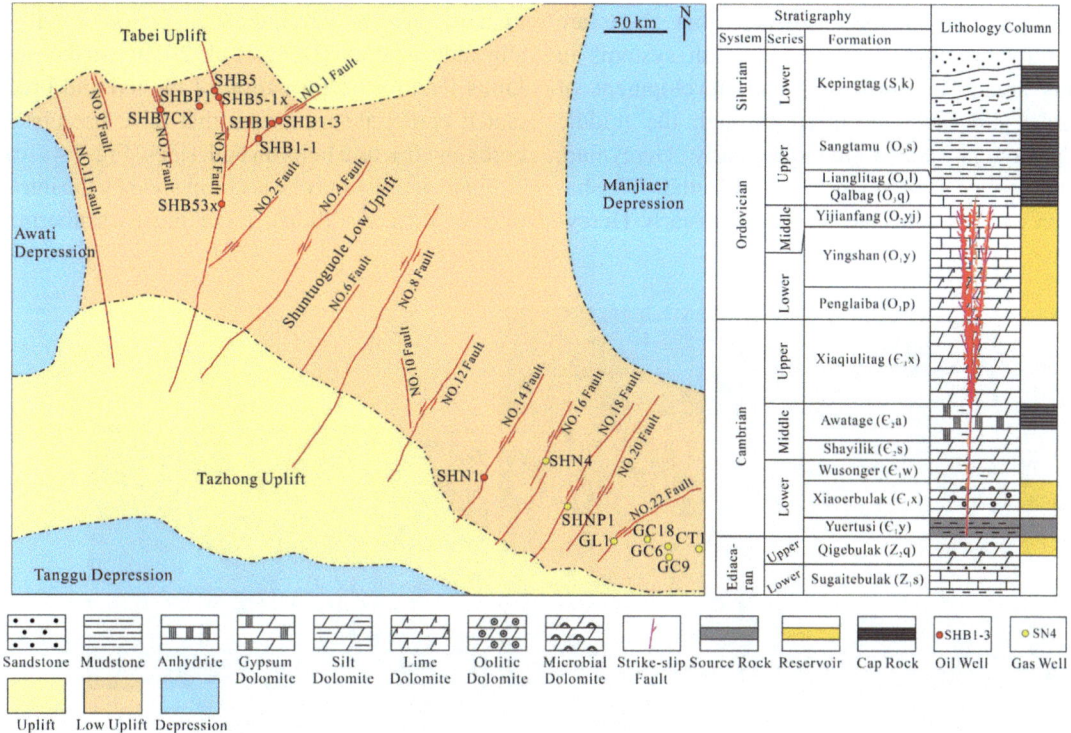

Fig. 7.3 Tectonic units and strike-slip faults in Shunbei area, Tarim Basin

or − 26.5 to − 15.1‰ (Katz et al. 2006), as well as low salinity (Katz et al. 2006).

As the burial depth and temperature gradually increase (close to or above 175 °C), thermochemical sulfate reduction (TSR) often occurs in deep–ultradeep marine gypsum-bearing carbonate reservoirs, which generally leads to the formation of minerals such as pyrite, calcite, and pyrobitumen. The calcite precipitated in TSR-derived fluids typically has a significantly light carbon isotopic composition ($\delta^{13}C_{V\text{-}PDB}$: − 15.7‰) with strontium isotope ratios consistent with those of contemporaneous formation water ($^{87}Sr/^{86}Sr$: 0.707638). The Th values of fluid inclusions range between 150.3 and 218.2 °C. Because of high fluid temperature, TSR-related calcite has a moderately negative oxygen isotopic composition ($\delta^{18}O_{V\text{-}PDB}$: − 10.2‰) and a high positive Eu anomaly (δEu: 10.9).

7.2.2 Fault-Dissolution Reservoirs of the Shunbei Oil and Gas Field in the Tarim Basin

7.2.2.1 An Overview of the Shunbei Oil and Gas Field

The Shunbei oil and gas field, which is located in the Shuntuoguole low uplift area (the transitional zone between the Awati and Manjiaer depressions) (Fig. 7.3), has experienced the most important oil and gas discovery in the Tarim Basin over the past 10 years. This event is a landmark achievement in the oil and gas exploration and development of ultradeep marine carbonate rocks (Ma et al. 2022). The vertical burial depth of oil and gas reservoirs typically exceeds 7200 m. Well SHB84X revealed oil and gas reservoir with a vertical depth of more than 8937.77 m, and the daily oil production from the well is over 1000 tons.

Structural analysis indicates that there are multiple groups of strike–slip fault systems in this area. The formation and development of strike–slip faults occurred through the middle Caledonian, late Caledonian–early Hercynian, middle–late Hercynian, and Indosinian–Yanshanian periods. The late Caledonian–early Hercy-

nian period is, however, the main period of strike–slip fault activity (Deng et al. 2019). Strike–slip faults lead to the formation of faulted and brecciated carbonate reservoirs that have brecciated cores and fracture belts on both sides (Fig. 7.4a–c).

Fault activity may be accompanied by the intrusion of different fluids, including hydrothermal

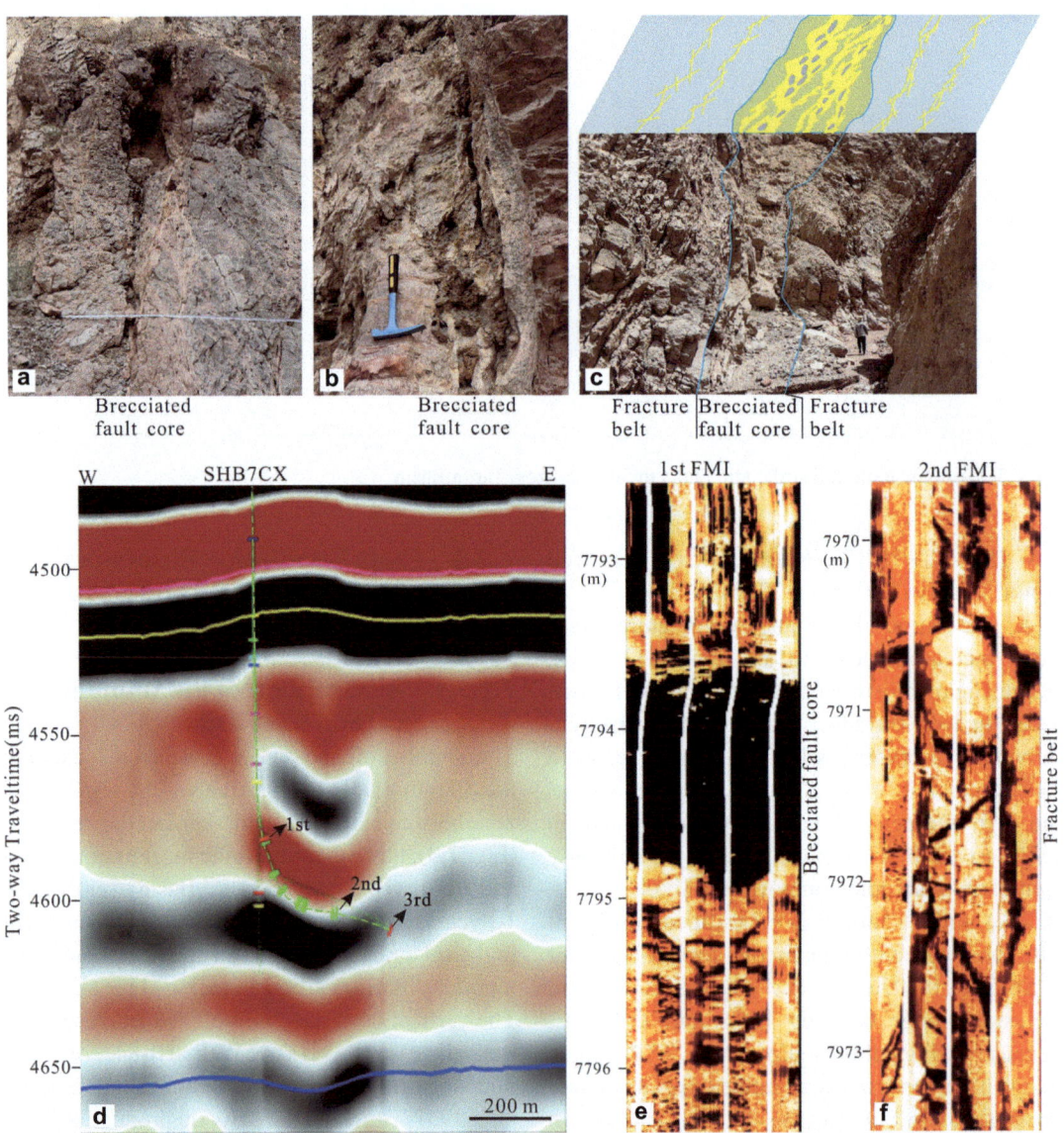

Fig. 7.4 Features of fault-dissolution reservoir in Shunbei area, Tarim Basin. **a, b** Core of the strike-slip fault are filled with carbonate breccia and cavities, southern party of the Piqiang fault, O_{1-2}; **c** fault-dissolution reservoir has a structure of "brecciated fault core- fracture belt on both sides", southern party of the Piqiang fault, O_{1-2}; **d** drilling trajectory of well SHB7CX in seismic profile; 1st, 2nd and 3rd black arrows indicating the locations of mud loss during drilling; **e, f** imaging logging characteristics of Ordovician Yingshan Formation in Well SHB7CX

fluids from Cambrian or Precambrian basements and hydrocarbon-bearing fluids from lower Cambrian sources. Fault–fluid coupling alterations promote the formation of fault-dissolution cavity reservoirs along the strike–slip fault zone, which are spatially "high-steep and plate-like" and layered and stacked vertically (Fig. 7.3).

7.2.2.2 Cases of Fault-Dissolution Reservoirs in the Shunbei Oil and Gas Field

(1) Well SHB7CX

Well SHB7CX is located in the NW-trending strike–slip fault zone. It is the result of horizontal sidetrack drilling in the direction of 85.75° from the northeast of straight well SHB7. During the sidetrack drilling process through the Yingshan Formation, three drilling mud leaking locations were revealed at 7787.00 m, 7979.70 m and the bottom of the well. The first mud loss is located near the main fault zone (brecciated fault core), and the imaging log shows the presence of caves. The second and third leakages are located in the fracture belt based on the imaging log (Fig. 7.4d–f).

Four samples of coarse macrocrystalline calcite cement were obtained from the cuttings near the first leakage point. These calcite cements have very negative $\delta^{18}O$ values and are rich in radiogenic ^{87}Sr. These geochemical features are similar to those of the calcite cements associated with karst in the Tabei uplift. This similarity may suggest the presence of relatively low-temperature meteoric fluids in the carbonate strike–slip fault zone in the southern coverage of the Tabei uplift.

(2) Well SHBP1

Well SHBP1 is located to the west of the No. 5 fault zone, approximately 11 km southwest of well SHB5. High-quality dolomite reservoirs are present in the lower member of the Yingshan Formation and are located below 8400 m. This section of dolomite has incredible porosity at depths greater than 8400 m (Fig. 7.5a, b). For example, the porosity at 8404–8407.50 m is 8.8% (the porosity is calculated according to the empir-

ical formula of the relationship between resistivity and porosity), and the porosity at 8429–8437 m is 10.9–17.7%. Cores at 8450–8451.55 m at the bottom of the well moreover exhibit abundant pores in dolomites. Overall, the reservoir quality varies vertically, while the size and distribution of the pores are heterogeneous. The development of reservoirs is controlled by faults and their interactions with fluids.

(3) Well SHB1-3

Well SHB1-3 is located in the No. 1 fault zone and has the greatest cumulative oil and gas production in the Shunbei oilfield. A large amount of bitumen occurs in the cored sections, and the bitumen layer contains gravels of mud, limestone and other mixtures (Fig. 7.5c, d). The XRD results of the mud–gravel samples reveal that they are dominated by clay minerals (67–68%), followed by quartz (20–23%), pyrite (4–7%), and other minerals. There are obvious differences in the bitumen content, which is unevenly distributed in the mud gravels. A small amount of cubic pyrite is dispersed in the mixture of clay and bitumen. Vein-like irregular calcite cements are locally developed and have groups of liquid phase inclusions. The mixed volcanic rock debris shows secondary fractures and pores, which were filled mainly by calcite and limonite. The secondary structural characteristics of the volcanic rock fragments indicate that they originate from the weathering and denudation zone and are filled with micritic calcite and limonite in the fractures and cavities.

(4) Well SHB53X

Well Shunbei 53X is located in the southern segment of the No. 5 fault zone. Only one core sample, with a depth of 7750.00–7750.12 m (length 12 cm), was obtained at the top of the Yijianfang Formation. Three fractures with different inclinations and dips are present in this core, which is heavily silicified (Fig. 7.5e). The matrix part is typically characterized by euhedral quartz. Under cathodoluminescence, the quartz does not emit light and appears black; the calcite

Fig. 7.5 Dissolution-filling characteristics related fault/ fracture in cores of some wells in the Shunbei oil and gas field. **a** Obvious orientation of pores and vugs in dolomite together with giant crystal calcite cement, O_1y, 8450.30–8450.43 m, Well SHBP1; **b** dolomite breccia and pores extremely developed in the dolomite fault zone, O_1y, 8451.00–8451.15 m, Well SHBP1; **c** reservoir bitumen, mud strips, and pyrite, O_2yj, 7265.21–7265.28 m, Well SHB1-3; **d** interlayer and deformation characteristics of mud and bitumen, O_2yj, 7265.83–7265.87 m, Well SHB1-3; **e** multiple sets of fractures (F1, F2 and F3) and secondary calcite cements in the silicified limestone, O_2yj, 7750.00–7750.12 m, Well SHB53x; **f** the pores, vugs and secondary calcite cements controlled by the fractures in dolomite, $\mathcal{C}_3q$, 7627.05–7627.23 m, Well SHNP1

between quartz crystals exhibits uneven intensity with an orange–yellow luminescence. The granular quartz in the fractures also does not emit light and is black. The calcite cements in the fractures are different. Some calcites exhibit high-intensity luminescence and are bright yellow, whereas other calcites exhibit nonluminous-dark luminescence with a ring zone.

(5) Well SHNP1

Well SHNP1 is located in the No. 18 fault zone. It is a fracture–cavity dolomite reservoir that was altered by hydrothermal fluids in the Cambrian period (Fig. 7.5f). CT scanning analysis of full-diameter cores reveals that fracture–cavity dolomite reservoirs have higher porosity and permeability. The matrix dolomite is fine to medium crystalline and subhedral dolomite. The development of vugs/pores is clearly controlled by fractures. Fracture–cavity edges are lined with euhedral dolomite cements, followed by quartz and calcite cements. The calcite cement is patchy and is controlled mainly by the fracture system in terms of three-dimensional pore space distribution.

In addition, well SHN4, which is located in the No. 16 fault zone, has a silicified carbonate reservoir that formed under the activity of silica-rich hydrothermal fluid; it is typically characterized by high quartz content and heterogeneous pore development, reaching 20.5% (You et al. 2018). Well GL1, located in the No. 22 fault zone, reveals the transformation of limy dolomite by silicon-rich hydrothermal fluids in the transition section between dolomite and limestone in the Yingshan Formation. Wells GC6, GC9, GC18 and CT1 also reveal similar silicified carbonate reservoirs in the Yingshan Formation.

7.2.2.3 Geological Model of Fault-Dissolution Reservoirs

On the basis of the sedimentary facies, petrology, alterations in different diagenetic fluids (meteoric water karst, hydrothermal fluid dissolution, TSR-derived fluid alteration), regional structural evolution and activities of the strike–slip faults in the Shunbei area, a geological model of the development of fault-dissolution reservoirs under fault–fluid coupling alteration is proposed (Fig. 7.6).

During the middle Caledonian orogenic period, regional tectonic movement led to the formation of uplifts in the Tazhong and Tabei areas. At the same time, strike–slip faults in different zones began to develop. In the late Caledonian–early Hercynian period, regional tectonic movement reactivated and further strengthened the strike–slip faults and led to the widespread development of echelon normal faults. The strike–slip faults in turn caused the development of fault–cavity reservoirs (Fig. 7.6), which commonly have a "brec-

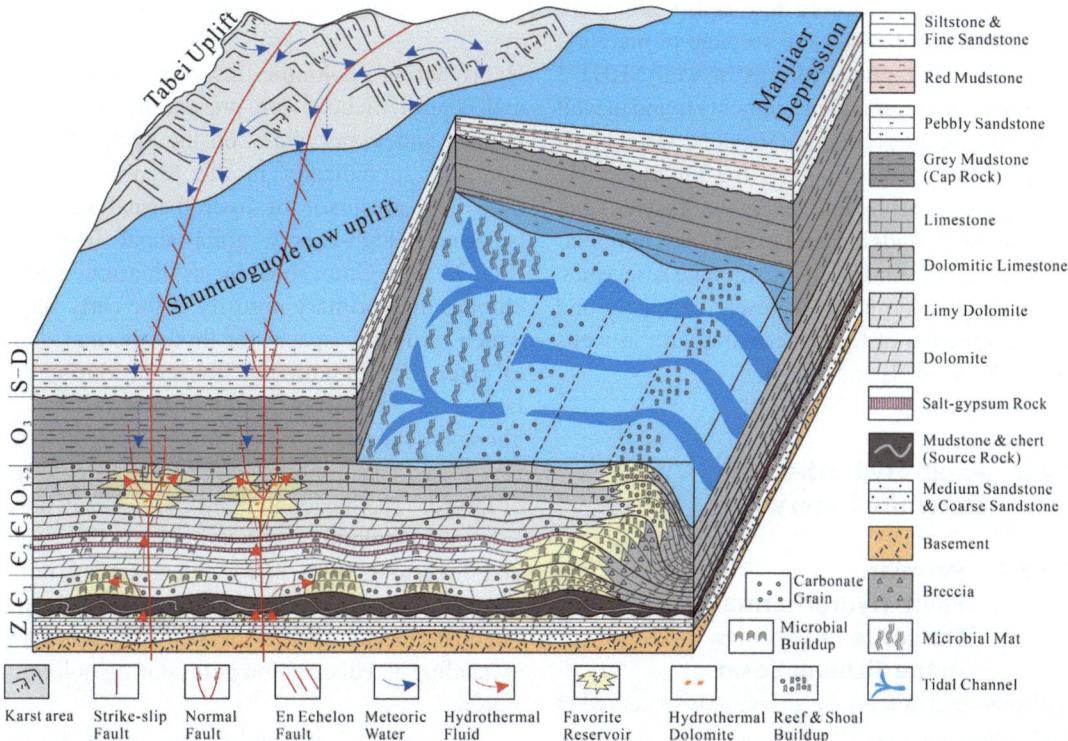

Fig. 7.6 Geological model of ultradeep reservoir under the alterations of meteoric water and hydrothermal fluid controlled by strike-slip faults

ciated fault core–fracture belt" structure along the strike–slip faults (Fig. 7.4).

The tension-twisted strike–slip faults cutting downward through Cambrian strata into the Precambrian basement provided conduits for the migration of both diagenetic fluids and oil and gas. The diagenetic fluids included mainly meteoric water, hydrothermal fluids, and TSR-derived fluids (Table 7.1, Fig. 7.2). Owing to the seal of the tight carbonate and mudstone caprocks in the Upper Ordovician, the alteration of fluids is limited to specific areas, such as near the strike–slip fault zones and in reef–shoal dolomites and microbialites that have high original porosity (Fig. 7.6).

During the Silurian–Carboniferous periods, the Shuntuoguole low uplift was lower than that of the Shaya uplift. Karstification in the late Caledonian–early Hercynian occurred mainly in the northern Shaya uplift, where the Ordovician carbonates were exposed above the surface. The overall pooling of meteoric water is from north to south. The strike–slip faults in the northern part of the Shuntuoguole low uplift are more likely to be channels for the seepage of meteoric water. The mud–gravel fillings in well SHB1-3 and low-temperature calcite cements found in well SHB7 may be the products of meteoric water alteration. In the southern part of the Shuntuoguole low uplift, considerable evidence reveals that hydrothermal fluids with relatively high temperatures altered the Ordovician carbonate reservoirs, leading to the formation of reservoir spaces and the precipitation of hydrothermal quartz in wells SHB53X, SHNP1, and SHN4.

7.2.3 Fault–Hydrothermal Dolomite Reservoir

7.2.3.1 Permian Fault–Hydrothermal Dolomite Reservoirs in the Sichuan Basin

Hydrothermal dolomite is commonly observed in the Permian Qixia Formation and Maokou Formation in the Sichuan Basin. Although the formation of dolomites is related to dolomitization by hydrothermal fluids along faults and fractures, there are significant differences in the characteristics of hydrothermal dolomitization in different areas of the basin. According to the contact relationship between the hydrothermal dolomite and the surrounding rock and the degree of dolomitization, the development patterns of the Permian hydrothermal dolomite can be categorized into three types (Fig. 7.7): (1) granular hydrothermal dolomite, which is derived from high-energy granular shoal facies limestone and is distributed mainly in the middle and upper parts of the Qixia Formation in the western and central Sichuan Basin; (2) karst breccia-derived hydrothermal dolomite, which is the result of hydrothermal dolomitization of the karst limestone breccia and is distributed mainly in the Maokou Formation in the central Sichuan Basin; and (3) tight limestone-derived hydrothermal dolomite, which is the result of the transformation of tight limestone deposited in low-energy facies and is distributed mainly in the southeastern Sichuan Basin.

(1) Granular hydrothermal dolomite

Hydrothermal dolomite reservoirs in western and central Sichuan are present mainly in the middle and upper parts of the Qixia Formation and sporadically in the Maokou Formation. The precursor of hydrothermal dolomite is usually high-energy grain shoal limestone (Fig. 7.7a–c). Grain shoal facies limestone generally has high primary porosity in the early stage, which provides space for hydrothermal fluid to enter the dolomitizing limestone during the later stage. In hydrothermal dolomite, residual or ghost structures of grains, oolites, foraminifera, bryozoans, gastropods, and algae are often observed. The residual matrix limestone adjacent to the dolomite is also dominated by granular limestone (Fig. 7.7b). Intergranular, intragranular, dissolution and molding pores in the granular limestone precursor are inherited as intercrystalline and dissolution pores during dolomitization.

The lateral distribution of hydrothermal dolomite reservoirs is unstable, and both the thickness and degree of dolomitization vary greatly.

Fig. 7.7 Hydrothermal dolomite of Permian Qixia and Maokou formations in the Sichuan Basin. **a** Granular limestone is partially dolomitized along fractures, Qixia Formation, Guangyuan Xibeixiang outcrop; **b** granular limestone is partially dolomitized, Qixia Formation, Guangyuan Xibeixiang outcrop; **c** zebra-like hydrothermal dolomite along fractures, Qixia Formation, Baoxing Wulong outcrop; **d** hydrothermal dolomite in karst breccia, Maokou Formation, Baoxing Lianghekou outcrop; **e**, **f** hydrothermal dolomite along fracture in the Maokou Formation tight limestone, well Tailai 6

The medium–thick layered and massive dolomite is gray to light gray and almost completely hydrothermally dolomitized. The crystals are relatively coarse and essentially do not contain limestone residues. Pinhole-shaped dissolved pores and vugs are often present and are often filled with late-stage hydrothermal minerals, such as calcite, dolomite, quartz, pyrite, illite and barite. Typical hydrothermal-related hydraulic fractures and zebra structures are often present in hydrothermal dolomite, and large amounts of coarse-crystalline saddle dolomite crystals fill the pores and fractures (Fig. 7.7c). The porphyritic dolomite is generally gray to dark gray, and the degree of dolomitization is incomplete. Along the fractures, the porphyritic dolomite is composed of white dolomitized areas and a gray limestone matrix (Fig. 7.7a).

(2) Karst breccia-derived hydrothermal dolomite

The middle Permian hydrothermal dolomite in the central Sichuan Basin is present mainly in the middle and upper parts of the Maokou Formation and is mostly breccia dolomite (Fig. 7.7d). The thickness of the dolomite section is generally several meters to more than 10 m, and the maximum thickness can reach 31 m (well Guangtan 2). Breccia dolomite usually occurs at locations of faults, fractures, and karst collapse breccias. Dolomite breccias are mostly angular–subangular and vary in size. Locally, residual matrix limestone breccias often "float" in hydrothermal dolomite. The interbreccia vugs and fractures are filled with coarse-megacrystalline saddle dolomite and small amounts of calcite and bitumen. Compared with matrix limestone, saddle dolomite usually emits stronger luminescence under cathode rays and has an obvious ringzone structure. In addition, hydrothermal minerals such as fluorite and pyrite can be observed in the intergranular vugs.

At the end of the middle Permian period, affected by the tectonic uplift of the Dongwu movement, the entire Sichuan Basin was exposed above the surface and eroded by meteoric water, forming a leopard-like and layered karst system. Afterward, at the beginning of the late Permian

period, the strong basement fault and related expansion activities caused the deep hydrothermal fluid to migrate upward into the karst breccia and fracture–cavity system of the Maokou Formation. As a result, the karst carbonate breccias underwent hydrothermal dolomitization. A large number of the saddle dolomite cements precipitated in the karst fracture–cave system along the basement faults.

(3) Tight   limestone-derived   hydrothermal
      dolomite

In the Dianjiang–Fuling–Zhongxian area of southeastern Sichuan, there are layer-like, lenticular and irregular dolomite patches in the upper part of the middle Permian Maokou Formation, and the surrounding carbonate consists mainly of tight limestone, including nodular limestone and micrite limestone. Owing to the poor porosity and absence of early karst, hydrothermal dolomite was replaced mainly along faults and fractures (Fig. 7.7e, f), and the placement was closely related to hydrothermal fluids along the faults.

According to field outcrops and well core observations, the thickness of dolomite sections generally ranges from several meters to more than twenty meters. On the basis of its occurrence, structure and microscopic characteristics, hydrothermal dolomite can be divided into three categories: crystal–fine crystalline matrix dolomite, fine–medium crystalline matrix dolomite, and coarse-crystalline saddle dolomite cement. Zebra structures, hydraulic fractures and burst breccias, and dissolution vugs are often seen in these types of dolomite. The formation sequence is matrix dolomite, saddle dolomite and calcite cement. In addition, columnar quartz with a perfect crystal morphology occurs in many dissolution vugs.

7.2.3.2 Comparison of Fault-Hydrothermal Dolomite in Different Areas

Statistics show that the thickness of hydrothermal dolomite in the Qixia Formation and Maokou Formation in the Sichuan Basin is positively correlated with the development of faults and volcanic rocks. Hydrothermal dolomite is relatively thick, reaching 50 m thick near the Permian volcanic rocks in the western Sichuan Basin. It is distributed mainly along the NW-trending basement faults in the southwestern and northwestern Sichuan Basin. The intrusion of the Emeishan mantle plume in the early late Permian significantly promoted the convective circulation of deep magnesium-rich hydrothermal fluids along basement faults. Hydrothermal fluids not only led to hydrothermal dolomitization in the Qixia and Maokou Formations but also strongly dissolved and altered the carbonates along faults and fractures. The peak homogenization temperatures of the fluid inclusions in saddle hydrothermal dolomite range mainly from 140 to 150 °C. Saddle dolomite has a very negative oxygen isotopic composition ($\delta^{18}O_{V-PDB}$: − 15.3‰) and a high $^{87}Sr/^{86}Sr$ ratio (up to 0.709905). The U–Pb dating results of saddle dolomite reveal that its formation age is approximately 235–255 Ma, which is the same time as or immediately after the Emeishan mantle plume.

Examples of world-famous oil and gas fields with hydrothermal dolomite as the main reservoir include the Clarke Lake middle Devonian gas field in Canada (Lonnee and Machel 2006), the middle Cambrian–middle Devonian gas field in the Western Canada Basin (Stacey et al. 2021), and the Upper Ordovician hydrothermal dolomite gas field in New York, USA (Smith 2006). The features of the hydrothermal fluids in these oil and gas fields are similar. Hydrothermal fluids originate from hot brines in the deep basin basement. Thermal convection drives large-scale hydrothermal fluids to convect and circulate upward along the conduits of faults and fractures to shallow limestones, and meteoric water precipitation at nearby topographic high point seepage along the faults to the basement, supplying basement brines. Large-scale hydrothermal convective circulation generally leads to the formation of large-scale hydrothermal dolomite reservoirs. The Permian rocks in the Sichuan Basin share similarities with the rocks in these regions, resulting in the formation of large-scale hydrothermal dolomitization reservoirs. In the Ordovician carbonate rocks in Tabei, Shunbei and other areas of the

Tarim Basin, some hydrothermal dolomite can also be seen along strike–slip faults and fractures, but the overall scale is small, which is related to the limited circulation scale of hydrothermal fluid.

7.3 Origin and Distribution of the Deep Subsalt Dolostone Reservoirs

7.3.1 Global Deep Subsalt Carbonate Reservoirs

Subsalt oil and gas reservoirs are covered and sealed under gypsum or salt layers (Fu et al. 2020). There have been a series of breakthroughs in the exploration of deep subsalt oil and gas reservoirs worldwide. Industrial oil and gas flows have been discovered in deep subsalt traps in the Permian Basin offshore of Brazil, the Gulf Basin in Mexico, the Great Campos Basin in South America, the Zagros Basin in the Middle East, the Caspian Basin in Russia, and the Po Valley Basin and Aquitaine Basin in Europe. Proven oil and gas reserves and production have been increasing rapidly (Zhang et al. 2015). In China, the discovery of the Anyue Ediacaran–Cambrian giant gas field in the Sichuan Basin (Zou et al. 2014) and the Cambrian subsalt reservoir of well Zhongshen 1 in the Tarim Basin (Wang et al. 2014) indicate great prospective reserves in the deeply buried subsalt strata of marine cratonic basins in China.

Most reservoirs in the subsalt traps are marine or lacustrine reef and shoal carbonates (Table 7.2). For example, the reservoir of the Lula Field in the Santos Basin is a lower Cretaceous lacustrine buildup (Wang et al. 2017). The reservoir of the Libra giant field in the Brazilian presalt province contains both lacustrine reef facies and coquina facies (Carlotto et al. 2017). The subsalt reservoirs in the Zagros Basin are dominated by thick-layered Cretaceous shells and clam limestones, which are strongly affected by meteoric karst (Du et al. 2016). In the Tarim Basin and Sichuan Basin, the deep subsalt reservoirs are dominated by dolomitized reefs and shoals, which constitute an effective reservoir–caprock combi-

Table 7.2 Typical deep subsalt carbonate reservoirs around the world

Tectonic background	Basin	Source rock		Reservoir			Caprock		Typical oil/gas field	References
		Sys.	Petrology	Sys.	Petrology	Pressure	Sys.	Petrology		
Passive continental edge	Mexico Bay Basin	/	/	J-K	Carbonates	/	/	/	Bermudez	Zhang (2000)
	Santos Basin	K	Shale	K	Reef, shoal carbonates	Over-pressure	K	Gypsum, salt	Lula	Wang et al. (2012), Wang et al. (2017)
Foreland	Zagros basin	/	Shale	K	Reef, chalk limestones	/	N_1	Gypsum, salt	Papileh	Zhang (2000), Du et al. (2016)
	Po Valley Basin	T_2	Shale, limestone	T	Carbonates	Over-pressure	/	/	Malossa	Wang et al. (2012)
Craton	Tarim Basin	ϵ_1	Shale	ϵ_1	Reef, shoal carbonates	Normal pressure	ϵ_2	Gypsum, salt	Central Tarim	Wang et al. (2012)
	Sichuan Basin	O-P	Mudstone	ϵ-P	Reef, shoal carbonates	Normal pressure	T	Gypsum, salt	Puguang	Wang et al. (2012)
	Caspian Sea Basin	D_2-P_1	Shale, silt limestone	D_3-C_2	Reef, shoal carbonates	Normal pressure	P	Shale, gypsum, salt	Kashagan, Tengiz	Wang et al. (2012), Zhang (2000)

nation with the overlying gypsum and mudstone (Jiang et al. 2021; Zhang 2000). Reef and shoal dolomites are often located at depositional highs and are consequently susceptible to exposure when sea levels fall. Meteoric leaching and karst due to high-frequency exposure generally lead to further improvement in the porosity of subsalt dolomite reservoirs. Rapid burial of the salt forms a regional caprock that prevents large-scale inflow of diagenetic fluids into subsalt carbonates and consequently prevents heavy cementation of the porosity. In addition, overpressure allows the pores of the subsalt reservoir to be maintained continuously (Table 7.2).

7.3.2 Deep Cambrian Subsalt Dolomite Reservoirs

The Tarim Basin was located near the equator during the Cambrian period, and the middle–lower Cambrian strata consist mainly of evaporite and dolomite successions that were deposited in a restricted evaporative environment (Fig. 7.8). In recent years, a number of risk drillings have been performed in deep Cambrian subsalt dolomite reservoirs, revealing good oil and gas accumulation conditions and various oil and gas shows. Well Xinghuo 1 revealed and confirmed high-quality hydrocarbon source rocks in the lower Cambrian Yuertusi Formation for the first time in the basin area. The well core of 8408 m from Well Tashen 1 revealed high-quality Cambrian dolomite reservoirs with porosities of up to 9.1% and the presence of light oil. Well Zhongshen 1 revealed industrial oil and gas flow in Cambrian subsalt dolomite below 6400 m (Wang et al. 2014). In 2020, an oil test in well Luntan 1 revealed high-yield industrial oil and gas flow in Cambrian subsalt dolomites below 8200 m (Fig. 7.8), representing a new breakthrough in oil and gas exploration in ultra-deep carbonate reservoirs in the platform area of the Tarim Basin and setting a new world record for oil and gas discovery depth in the Craton Basin (Yang et al. 2020). These exploratory wells all contain evaporative gypsum–salt rocks from the middle and lower Cambrian periods, providing high-quality regional sealing conditions.

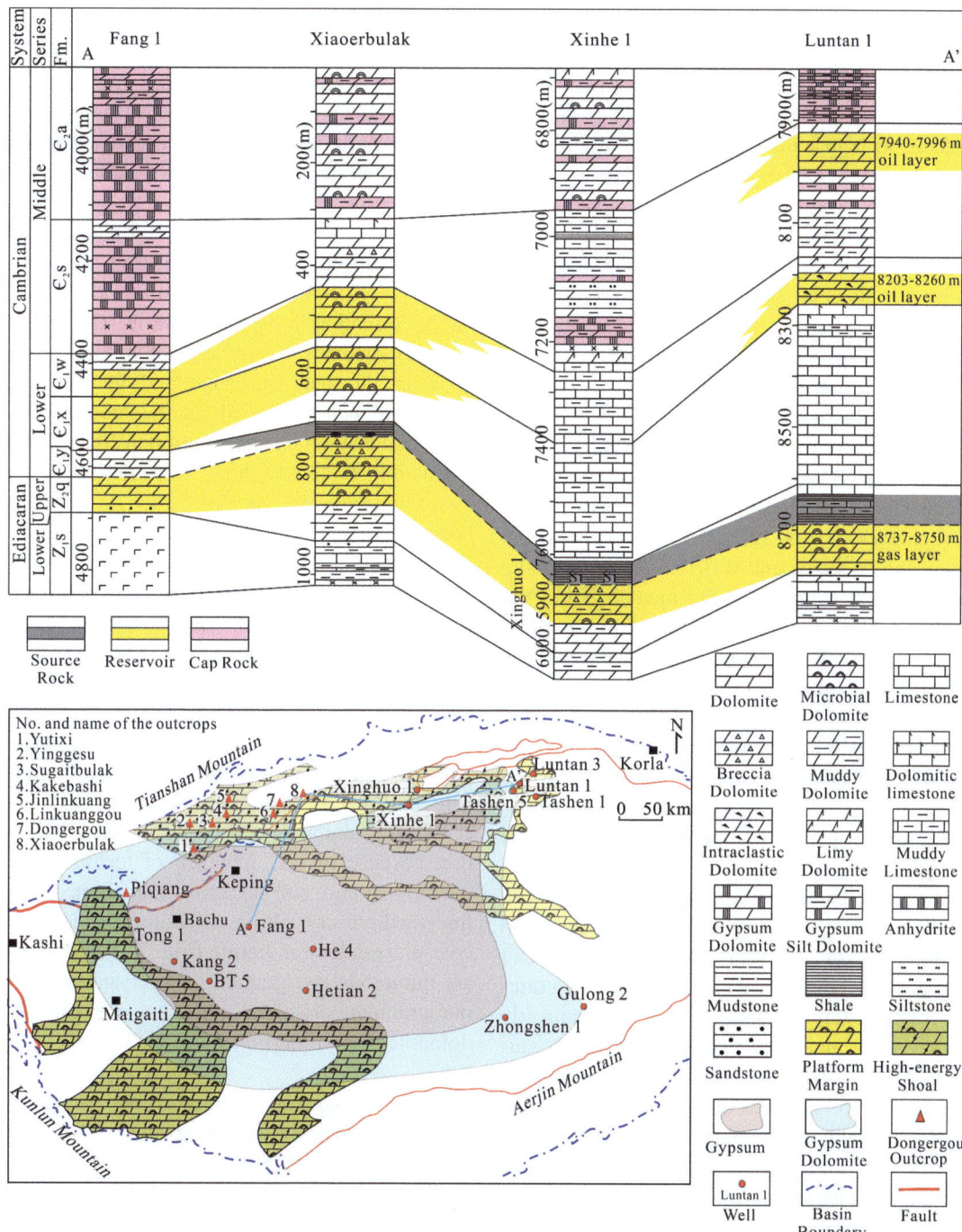

Fig. 7.8 Stratigraphy and distribution of the Ediacaran and Lower-Middle Cambrian strata. The distribution of the Cambrian gypsum and Qigebulak Formation high-energy shoal and platform margin facies are modified from Yang et al. (2020)

The above exploration practices indicate that the deep–ultradeep Cambrian subsalt strata in the platform area have good combinations of source rocks, reservoirs and caprocks (Fig. 7.8), leading to high hydrocarbon exploration potential under the subsalt in the Tarim Basin. The deep Cambrian subsalt strata in the Sichuan Basin have a similar combination of source rocks, reservoirs and caprocks as well as a high exploration potential (Hu et al. 2016).

7.3.2.1 Types of Deep Cambrian Subsalt Dolomite Reservoirs

The Cambrian subsalt dolomite reservoirs in both the Sichuan Basin and Tarim Basin have been studied for a long time, including the classification and characterization (Zhao et al. 2012) and the sedimentary facies and distribution of dolomite reservoirs (Jiang et al. 2022; Liu et al. 2011). On the basis of both previous studies and this research, this paper briefly summarizes the characteristics of subsalt dolomite reservoirs, focusing mainly on the development and preservation mechanism of porosity in dolomite reservoirs.

Middle Cambrian gypsum salt rocks in the Sichuan Basin are located mainly in the southeastern area, and the thickness can reach several kilometers. Middle–lower Cambrian gypsum salt rocks occur extensively in the Tarim Basin, with a thickness of 100–450 m and an area of up to 143,000 km^2. Thin interbeds of gypsum salt and dolomite and brownish-red medium- to thin-layered gypsum-bearing dolomite are frequently observed in field outcrops around the basins (Fig. 7.9a, b). Thick-layered microbial dolomite, algal dolomite, granular dolomite, and medium–coarse crystalline dolomite are present among the middle Cambrian gypsum salt layers as well as in the lower Cambrian layers (Fig. 7.9c–h). During a short-term sea level decrease, the gypsum-bearing dolomite successions would suffer from short-term exposure of the contemporaneous period and would be eroded by meteoric water, forming erosion surfaces after the exposure (Fig. 7.9d, e). Consequently, abundant dissolution vugs have developed in the dolomites below the exposed erosion surface (Fig. 7.9e–h).

According to the sedimentary facies, structure and pore characteristics, Cambrian subsalt dolomite reservoirs can be divided into two categories. The first category includes the granular shoal dolomite reservoir in the high-energy tidal flat and platform margin gentle slope facies, which includes algae grains, sand grains and medium- to coarse-crystalline dolomites in the middle Cambrian Shaylik Formation, the lower Cambrian Wusongger Formation and the Xiaoerbulak Formation in the Tarim Basin. The second category includes the microbial dolomites developed in the platform edge and gentle slope facies and is represented by the mound–shoal microbialites in the lower Cambrian Xiaoerbulak Formation. Granular shoal dolomites are often associated with microbialites, such as thrombolites and stromatolites. This section focuses on the characteristics of grain-shoal dolomite reservoirs. The characteristics of microbialite reservoirs are described in detail in subsequent sections.

The middle–lower Cambrian dolomite reservoirs are mostly medium–thick layered silty–fine crystalline dolomite (Fig. 7.9f) with residual ghosting of the grain structure. The grain types are algal-coated grains, intraclasts and oolites (Fig. 7.10a, b). The grain-bearing silty–fine crystalline dolomites underwent recrystallization because of the influence of tectonic–fluid alterations and subsequently transformed into fine–medium crystalline dolomite and medium–coarse crystalline dolomite (Fig. 7.10c–l). There are numerous intergranular, intercrystalline, and intragranular dissolved pores in the granular dolomite reservoirs (Fig. 7.10a–c).

7.3.2.2 Diagenesis Events of the Deep Cambrian Subsalt Dolomite Reservoirs

The deep Cambrian subsalt dolomite reservoirs in the Tarim Basin have undergone complex and multistage diagenetic evolution. According to the time sequence relative to the pressure solution and different cement types and fabrics (Fig. 7.10) (Flügel 2013; Scholle and Ulmer-Scholle 2003), diagenesis can be further divided into early diagenesis and late diagenesis, which occurred in the

◀**Fig. 7.9** Characteristics of the Cambrian subsalt dolomite reservoirs. **a** Thin interbeds of dolomite and anhydrite, $\mathcal{C}_2$, Shapingzi outcrop in Leibo, Sichuan Basin; **b** brown-red medium-thin layered gypsum-bearing dolomite, $\mathcal{C}_2$a, Xiaoerbulak outcrop, Tarim Basin; **c** thick layered microbial dolomites in gypsum-bearing dolomite strata, $\mathcal{C}_2$a, Xiaoerbulak outcrop, Tarim Basin; **d** synsedimentary eroding surface in gypsum-bearing dolomite and dissolution pores and vugs under the eroding surface, $\mathcal{C}_2$s, Huanghe outcrop in Shizhu County, Sichuan Basin; **e** synsedimentary eroding surface in gypsum-bearing dolomite and dissolution pores and vugs under the eroding surface, $\mathcal{C}_2$, Banqiao outcrop in Tongzi County, Sichuan Basin; **f** thick-layered granular dolomite contains numerous dissolution vugs, $\mathcal{C}_1$x, Xiaoerbulak outcrop, Tarim Basin; **g** medium-coarse crystalline dolomite contains numerous dissolution vugs, $\mathcal{C}_1$x, well Hetian 2, Tarim Basin; **h** medium-coarse crystalline dolomite contains numerous inter-crystalline and dissolution pores, $\mathcal{C}_1$x, well Hetian 2, Tarim Basin

contemporaneous near-surface to shallow burial stage and middle to deep burial stage, respectively (Fig. 7.11).

Early diagenesis refers to a series of diagenetic events that occur prior to pressure dissolution, including mainly seafloor diagenetic environments, meteoric water diagenetic environments and shallow burial diagenetic environments. The depth of early diagenesis is approximately tens of meters to hundreds of meters. Consequently, early diagenesis does not occur completely separated from the depositional environment. The chemical properties of diagenetic fluids are still controlled by the surface hydrodynamic system and are still in an oxidative–suboxidative environment (Machel 2004; Machel and Buschkuehle 2008). Early diagenesis of the deep Cambrian subsalt dolomite reservoirs in the Tarim Basin includes micritization, microbial calcification (MC), isopachous fibrous cement (IFC), meteoric water dissolution, crescent cement and seepage silty sand, isopachous dental cement (IDC), early dolomitization, and block anhydrite cement (BAC) (Fig. 7.10).

Late diagenesis refers to a series of diagenetic events that occur after large-scale pressure dissolution, mainly in the middle–deep burial diagenetic environment. In this environment, the diagenetic fluid completely separated from the original sedimentary background, and as a result, its properties completely changed. It is characterized by formation brine with complex chemical compositions because of long-term water–rock interactions under high-temperature and high-pressure conditions and is generally observed in reducing conditions (Machel 2004; Machel and Buschkuehle 2008). The migration and circulation of diagenetic fluid is no longer controlled by surface hydrodynamics, and its scale is limited. Late diagenesis includes euhedral dolomite cement (EDC), silicification and siliceous cementation (Q1, Q2 and Q3), hydrocarbon charging and associated dissolution, block and saddle dolomite cement (BDC and SD), block calcite cement (BCC), and fault-related fragmentation (Fig. 7.10g–l).

7.3.2.3 Development Model of the Deep Cambrian Subsalt Dolomite Reservoirs

The development of high-quality deep Cambrian subsalt dolomite reservoirs in the Tarim Basin is controlled mainly by sedimentary facies and early diagenesis (Fig. 7.11). High-energy shoal granular limestones and microbial mounds developed under shallow seawater conditions. The granular limestones and microbial mounds were dolomitized because of the evaporation of seawater and subsequent reflux of high-salinity brine (Fig. 7.11a). The granular shoal dolomites of high-energy facies and microbial mound dolomites commonly contain numerous primary pores (intercrystalline, intergranular, and microbial framework pores). Consequently, they constitute the matrix for high-quality dolomite reservoirs. Contemporaneous meteoric water dissolution owing to high-frequency sea level fall and exposure further improve the porosity and permeability, which is key to the formation of high-quality dolomite reservoirs (Fig. 7.11b). A similar enhancement in reservoir quality has also been observed in the dolomites of the deepest gas field in the U.S. (Smith et al. 2004).

The effective preservation of the early-developed porosity in the subsalt dolomite reser-

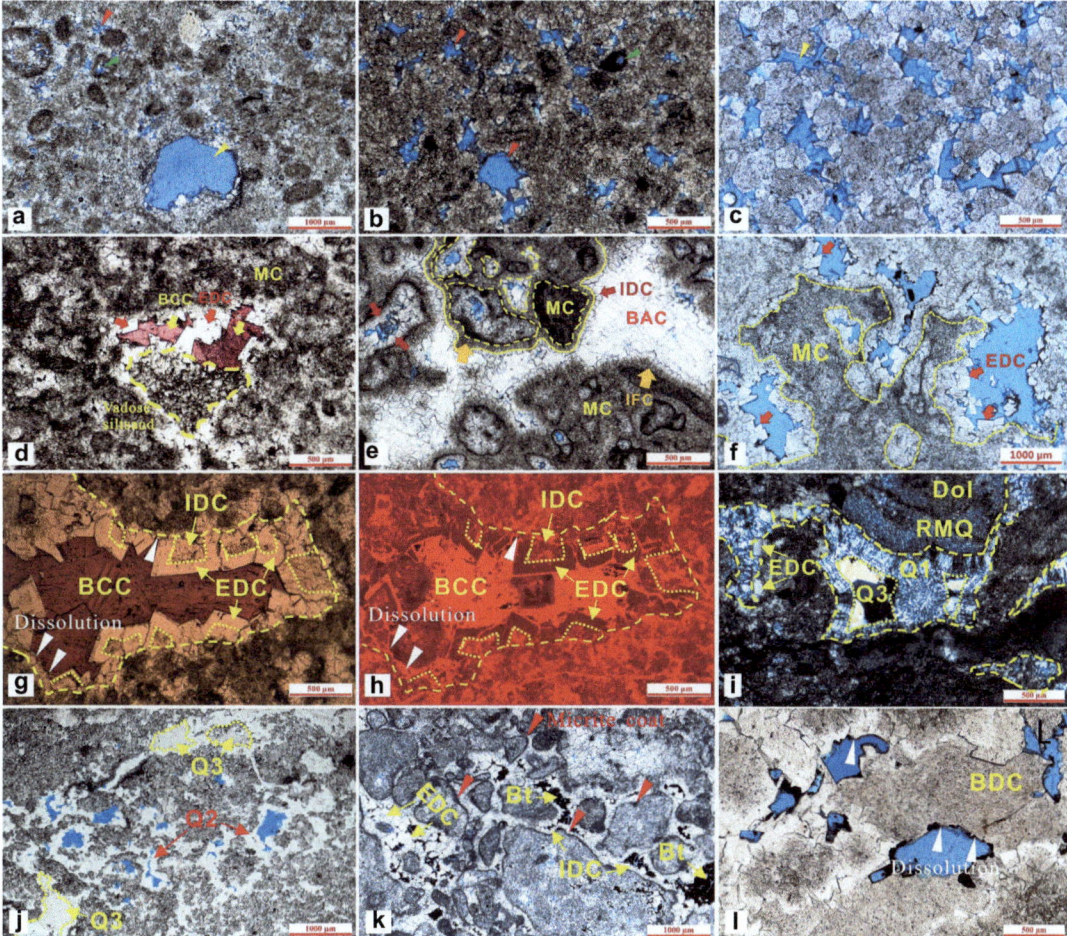

Fig. 7.10 Characteristics of the pores and diagenetic minerals in the Cambrian subsalt dolomite reservoirs in the Tarim Basin (Modified from Jiang et al. 2021). **a** Fine crystalline dolomite with residual grain structure, intergranular pores (red arrow), intragranular pores (green arrow) and dissolution vugs (yellow arrow), Kakebashi section; **b** grain dolomite with intergranular pores (red arrow), Jinlinkuang section; **c** residual fine-grained medium crystalline dolomite with intergranular pores and intercrystalline pores (yellow arrow), Well Batan 5, 5784.30 m; **d** dissolution vugs related to meteoric leaching with seepage silt sands, euhedral dolomite cement (EDC), and massive calcite cement (BCC) in thrombolite dolomite, Yingersu section; **e** Isopachous fibrous cement (IFC) is partially eroded by meteoric water, leading to formation of framework cavities which are gradually filled with isopachous dental cement (IDC) and block anhydrite cement (BAC), sponge dolomite, Well Fang 1, 4598.5 m; **f** Euhedral dolomite cement (EDC) partially fills in dissolution pores among clots, thrombolite dolomite, Yingersu section; **g**, **h** residual of thick isopachous dental cement (IDC) after meteoric water dissolution, and euhedral dolomite cement (EDC) and block calcite cement (BCC) formed in the middle-deep burial environment, thrombolite dolomite, Kungaikuotan section; **i** rimed microcrystalline quartz (RMQ) replaces early micritic dolomite, and framework pores are successively filled by euhedral dolomite (EDC), fibrous quartz (Q1) and cluster quartz (Q3), stromatolite dolomite, Well Kang 2, 5631.79 m; **j** pore is partially or completely filled with granular quartz (Q2) and clustered quartz (Q3) in turn, very fine crystalline dolomite, Shiairike section; **k** micrite coat surrounding intraclast and intergranular pores are filled with isopachous dental cement (IDC), euhedral dolomite cement (EDC) and bitumen (Bt) in turn, Yingersu section; **l** block dolomite cement (BDC) filled in large dissolved vugs undergoes dissolution due to organic acids, micrite dolomite, Yutixi section

Fig. 7.11 Formation and preservation model of the deep Cambrian subsalt dolomite reservoirs in the Tarim Basin

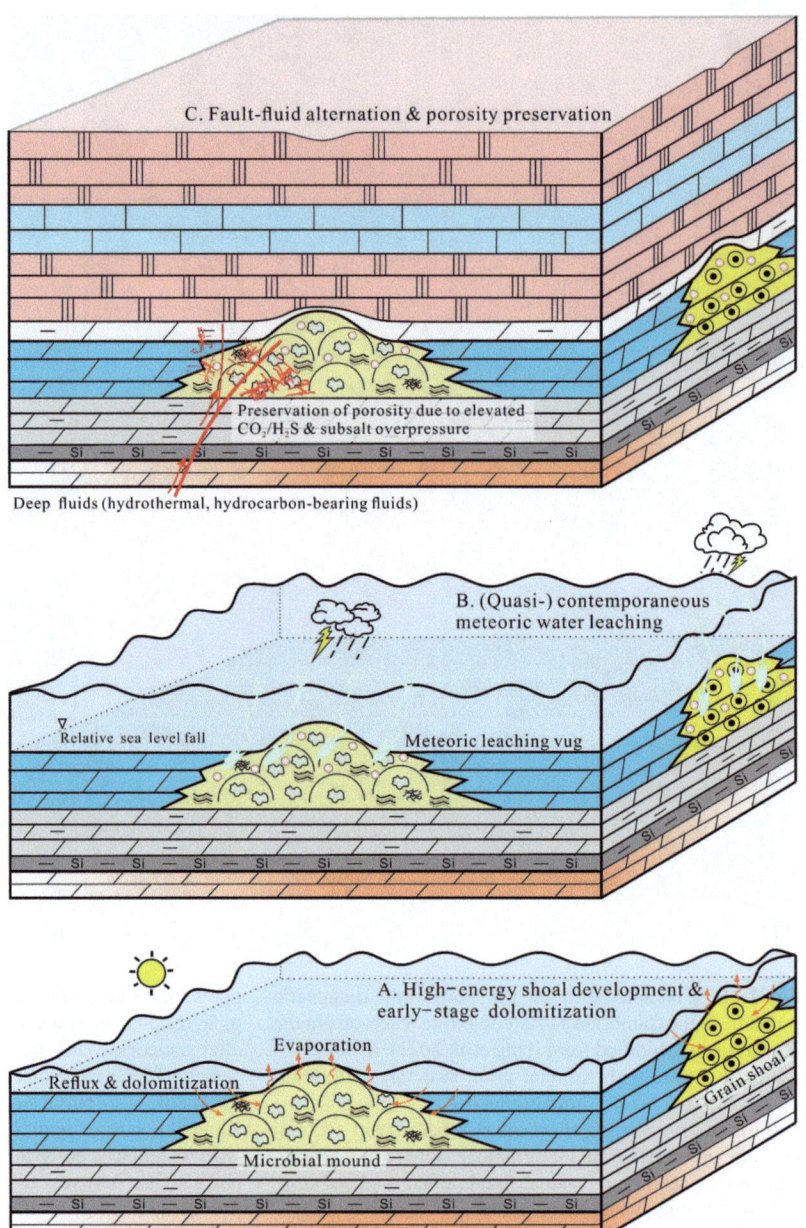

Fig. 7.11 Formation and preservation model of the deep Cambrian subsalt dolomite reservoirs in the Tarim Basin

voirs is due to early dolomitization, rapid sealing of thick gypsum–salt rocks, oil and gas charging and elevated levels of sour gases (CO_2 and H_2S) (Fig. 7.11c). Early dolomitization enhanced the ability of carbonate rocks to resist compaction–compression dissolution in the deeply buried diagenetic environment. Thick gypsum–salt rock has a high sealing capacity and high thermal conductivity. The gypsum–salt rock with a high sealing capacity can effectively block the descending diagenetic fluid, reduce the diagenetic fluid flux into the underlying dolomite strata, and form fluid overpressure in the subsalt dolomite reservoirs, which not only weakens compaction and pressure solutions but also diminishes cementation in the dolomite reservoir. High thermal conductivity can effectively reduce the diagenetic temperature of the underlying dolomite formation and reduce the diagenetic intensity. In addition, the fluid activity connected by faults causes local dolomite strata to undergo recrystallization, dissolution, cementation and other burial diagenetic transformations, which lead to the reconstruction of the pore structure of the reservoir and aggravate the heterogeneity of the reservoir.

7.4 Microbialite Reservoirs in Ultradeep Ancient Strata

7.4.1 Evolution of Precambrian Conditions for Microbialites

7.4.1.1 Precambrian 'Dolomite Sea' Environments

Microbialites are defined as carbonate sediments that are captured or bonded to detrital sediments by benthic microbial activities or formed in situ by the inorganic–organic-induced mineralization associated with microbial activities (Riding 2000). At present, the earliest fossil record of life dates back to 4.1 billion years (Riding 2000). The large-scale prosperity of eukaryotic communities (e.g., cyanobacteria) subsequently occurred during the Proterozoic period, which contributed to the formation of widespread Precambrian stromatolitic strata that can be traced back 3.7 billion years ago (Nutman et al. 2016).

Stromatolite is the most common type of microbial carbonate. Its abundance peaked at 1250 Ma in the Proterozoic period, then declined in the Neoproterozoic period, and exhibited a large-scale decline in the Phanerozoic period, with the exception of a brief recovery in the Cambrian period (Fig. 7.12). Recent investigations have demonstrated that stromatolites can be considered both fossils and sedimentary structures depending on the energy in the depositional environment and that well-preserved stromatolites can provide insight into both biological and environmental factors in ancient ecosystems (Suosaari et al. 2019a). In general, the declining trend of the abundance of stromatolites corresponded to the flourishing and diversity of heterotrophic metazoans (Fig. 7.12) (Riding 2006). The grazing and chewing interference of these heterotrophic animals signifi-

cantly affected the living space of microorganisms (e.g., cyanobacteria), leading to the restriction of stromatolite development and prosperity (Garrett 1970). In other words, most stromatolites can form through the mediation of these benthic microbial communities with photosynthesis in environments where metazoan reproduction is inhibited (Pomar and Hallock 2008). For example, heterotrophic metazoans had not yet emerged during the Precambrian period, and the microorganisms represented by cyanobacteria widely flourished, leading to the formation of widely distributed stromatolites (Monty 1973). However, since the late Neoproterozoic (i.e., the Ediacaran period), the well-known Ediacaran-type complex and giant metazoan communities have gradually flourished (Xiao and Laflamme 2009). The interference of grazing, drilling and eating resulted in the destruction of the living space of microorganisms and bacteria (Pomar and Hallock 2008), thus inhibiting the development and prosperity of stromatolites. Since then, stromatolites have been mostly confined to some special environments, such as the extreme and high-salinity environments in Shark Bay, Australia (Papineau et al. 2005; Riding 2006), and their abundance in terms of both scale and diversity is much lower than that in the Precambrian period (Fig. 7.12).

7.4.1.2 Ancient Microbialite Dolomitization by 'Dolomite Sea' Environments

Under favorable seawater conditions, microbial-mediated precipitation of dolomite plays a major role in the total number of Precambrian dolomite sequences. At present, microbial-mediated dolomitization experiments using sulfate-reducing bacteria in which microorganisms can precipitate ordered dolomite with a superreflective structure under anaerobic conditions (Vasconcelos et al. 1995) are publicly accepted by geologists. Subsequently, Vasconcelos and McKenzie (1997) found dolomite in the black organic-rich sediments of the Lagoa Vermelha lagoon, Brazil, similar to the dolomite precipitated by experiments. Furthermore, they established the formation mechanism and typical

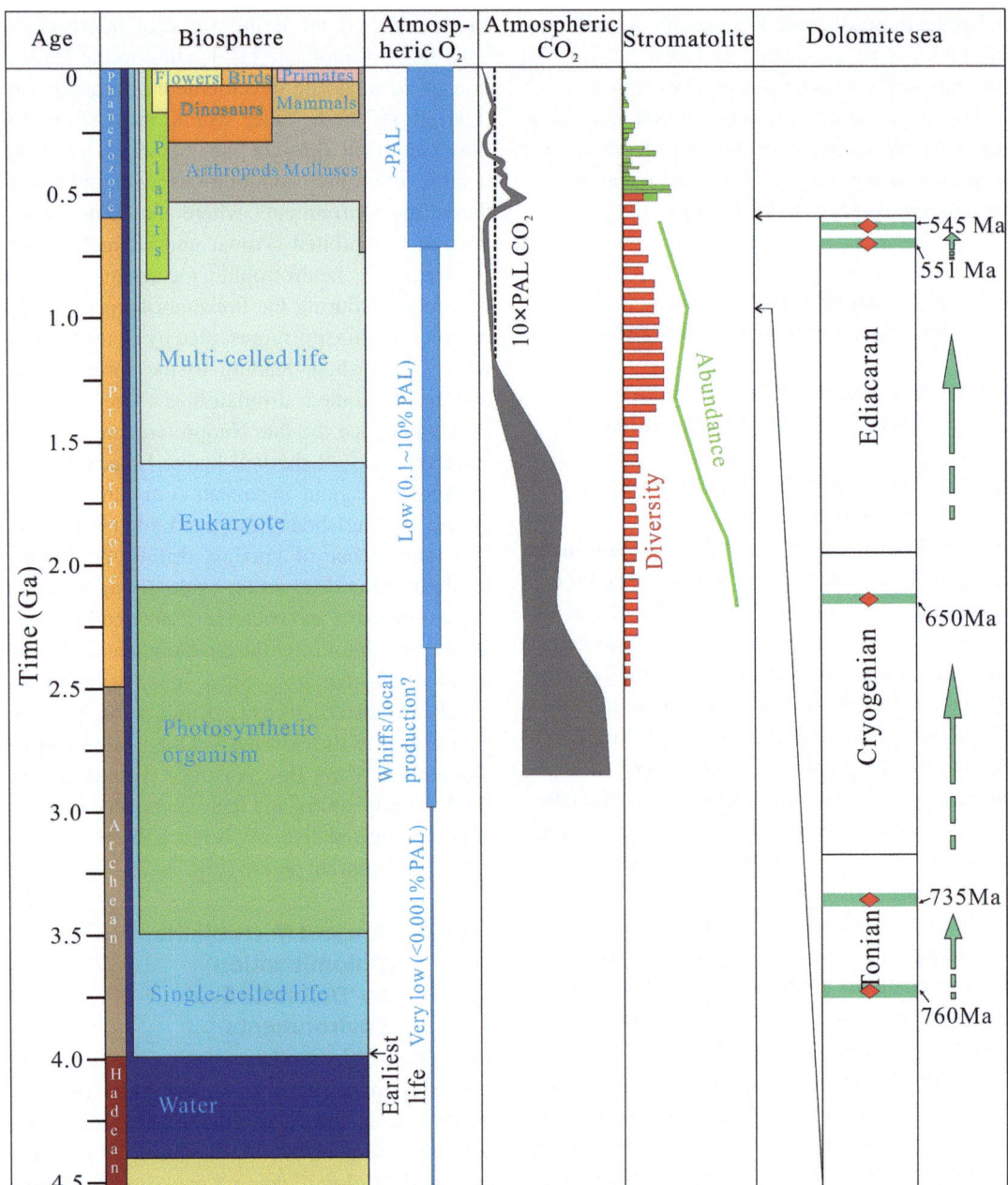

Fig. 7.12 Evolutionary diagrams of microbial activity, stromatolites, and 'dolomite sea' environments in geological history, modified from Hood and Wallace (2012) and Wang et al. (2020)

microbial dolomite model at natural sea surface temperatures based on the isotopic data and hydrological model of the lagoon (Vasconcelos and McKenzie 1997).

Moreover, the consensus is that the physicochemical conditions of Precambrian seawater are significantly different from those of Phanerozoic seawater. Hood et al. (2011) proposed a special Neoproterozoic 'dolomite sea' environment with high Mg/Ca ratios (> 3.0–10.0), low sulfate concentrations, and anoxic (or sulfurized) microconditions, which could directly precipitate primary dolomite or promote dolomitization with the role of microorganisms. This environment can account for the occurrence of abundant Neoproterozoic microbial dolomite sequences worldwide despite the influence of metazoans on the subsistence of microorganisms. Combined with previous studies (Hood et al. 2011; Wang et al. 2020), the unique Neoproterozoic 'dolomite sea' environment in the deep-time system was ubiquitous but did not persist, extending from the Tonian (ca. 760 Ma) to the late Ediacaran (ca. 551–545 Ma) periods (Fig. 7.12). Microbial mediation does not necessarily precipitate dolomites but leads to the formation of calcites. Microbial mediation in the 'dolomite sea' was likely the key process controlling the precipitation of microbial dolomites in the Precambrian period. This view provides a new window for understanding the 'dolomite problem' and widespread microbial dolomite strata in the Precambrian period.

For example, the thrombolites from the upper Ediacaran Dengying Formation in the Sichuan Basin consist of very tiny rhombic dolomite crystals several millimeters in diameter. The formation of these thrombolites was related to microbial activity under 'dolomite sea' conditions that occasionally occurred during the late Ediacaran period (Wang et al. 2020). Abundant fossils or relics related to microbial mediation, such as snowflake-shaped, long strip, spherical and dumbbell-shaped crystals, are observed among dolomite crystals several millimeters or nanometers in size (Zhu et al. 2022). They are generally considered to be traces of extracellular polymers (EPS) secreted by microorganisms (e.g., cyanobacteria). EPS can not only provide nucleation sites for dolomite formation but also overcome the kinetic obstacles associated with dolomite precipitation (Sánchez-Román et al. 2008; Vasconcelos and McKenzie 1997; You et al. 2014).

7.4.2 Development Characteristics of Ancient Microbialite Reservoirs

Abundant oil and gas resources have been discovered within ultradeep ancient microbialite reser-

voirs, such as the upper Ediacaran Dengying Formation in the Sichuan Basin, the Qigebulak Formation, and the lower Cambrian Xiaoerbulak Formation in the Tarim Basin. These microbialite reservoirs have high hydrocarbon exploration potential. The Dengying Formation microbialite reservoirs are distributed mainly along the platform margins of the Mianyang–Changning and Exi rifting troughs in the Sichuan Basin and the surrounding areas. The scale and thickness of these reservoirs are relatively larger than those in the interior platform (Figs. 7.13 and 7.14). In addition, abundant natural gas reserves and production have been obtained from the Dengying Formation microbialite reservoirs, which have been widely studied by different researchers (Wei et al. 2015; Zhao et al. 2017). Recently, the newly drilled well Luntan 1 in the Tarim Basin has also revealed some gas-bearing layers in the Qigebulak Formation microbialites at depths > 8700 m (Fig. 7.8). In addition, a high-porosity microbialite reservoir with bitumen can be observed from the Qigebulak Formation and Xiaoerbulak Formation in some outcrops (Figs. 7.15 and 7.16), indicating favorable exploration potential in these ancient microbialite reservoirs.

7.4.2.1 Types of Microbialites

The main types of microbialites in these ancient strata include stromatolites, thrombolite, dendrolite, leiolite, oncolite, and laminite. At present, live stromatolites are still growing in some extreme environments, such as the high-salinity lagoon (salinity, up to 70‰) of Shark Bay, Australia (Fig. 7.15a) (Bauld 1984; Papineau et al. 2005), and the backreef lagoon (salinity, up to 94–120‰) of Exuma, Bahamas (Dupraz and Visscher 2005). Observations linking stromatolite morphology to physiography in modern microbial systems in Hamelin Pool, Western Australia, suggest that both biology and the environment control stromatolite morphology (Suosaari et al. 2019b). When physiography leads to a high-energy regime, environmental control is the main factor determining stromatolite morphology. In contrast, when physiography promotes a low-energy environment, the response of biological communities becomes the main driver of macroscale stro-

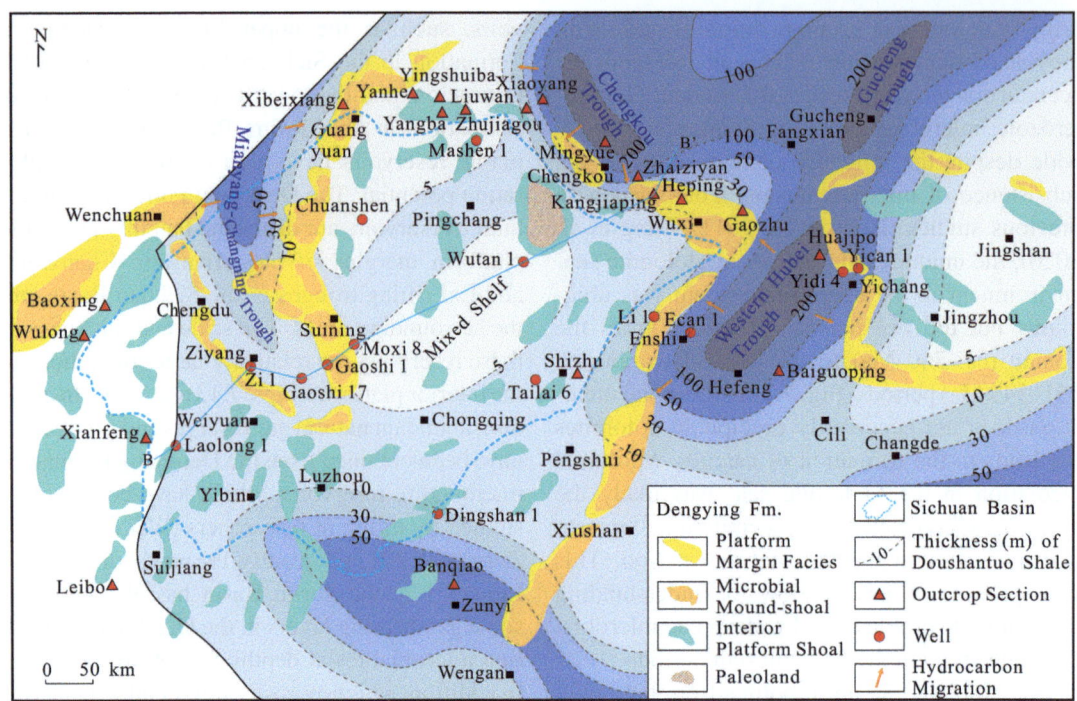

Fig. 7.13 Map of Doushantuo formation hydrocarbon source rocks and Dengying formation microbialite reservoirs in Sichuan Basin and its periphery, modified from Wang et al. (2019), Wang et al. (2020), and Zhu et al. (2022)

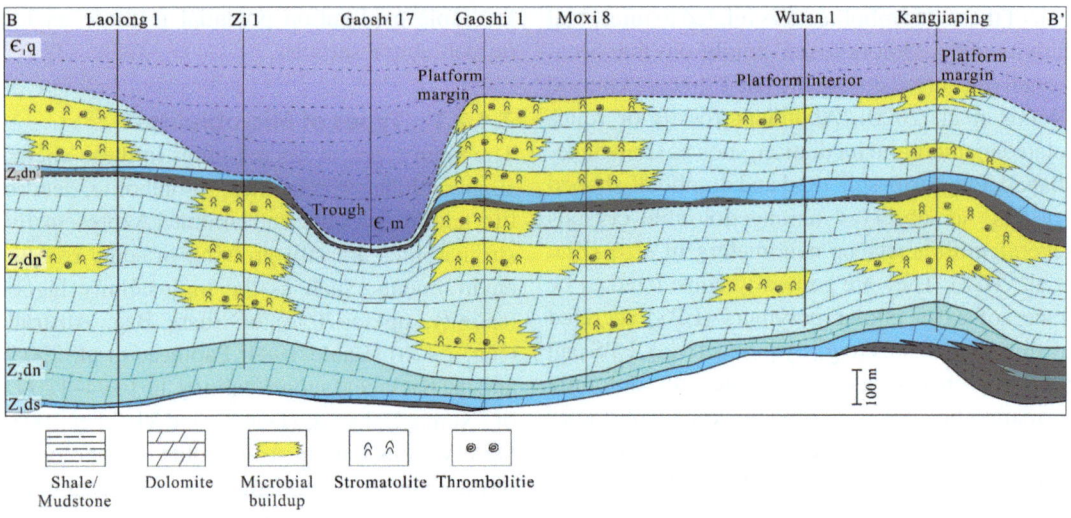

Fig. 7.14 Profile of the Dengying formation microbialite reservoirs in the Sichuan Basin

matolite morphology. The Hamelin Pool can be seen as a microbial carbonate factory, which was constructed through lithifying microbial mats to form microbialites and the erosion and degradation of weakly lithified microbial mats, resulting in the extensive production of sand-sized micritic sediments (Suosaari et al. 2022).

Stromatolites can be considered potential economic oil and gas reservoirs in ancient ultra-deep strata because of the high preservation of

Fig. 7.15 Characteristics of microbialite reservoirs in the Upper Ediacaran. **a** Modern stromatolites in Shark Bay, Australia; **b, c** sinuous short columnar stromatolite, Z_2q, Linkuanggou outcrop, Tarim Basin; **d** horizontal layered stromatolites, Z_2dn^2, Wuxi Heping outcrop, Sichuan Basin; **e** dome-shaped stromatolites, Z_2dn^2, Wangcang Gucheng outcrop, Sichuan Basin; **f** microbial thrombolite containing framework pores, partially filled by dolomite cement, Z_2dn^2, Yangba outcrop, Sichuan Basin; **g** botryoidal dolomite, Z_2dn^2, Yangba outcrop, Sichuan Basin; **h** botryoidal dolomite, Z_2q, Dongergou outcrop, Tarim Basin

Fig. 7.16 Characteristics of microbialite reservoirs in the Lower Cambrian Xiaoerbulak formation. **a** Microbial mounds (yellow arrows) with progradational structures, $\mathcal{C}_1$x, Sugaitebulak outcrop, Tarim Basin; **b** features of pores within these microbial reefs, $\mathcal{C}_1$x, Sugaitebulak outcrop, Tarim Basin; **c** features of pores within the microbial thrombolite, $\mathcal{C}_1$x, Sugaitebulak outcrop, Tarim Basin; **d** features of layered framework pores in the microbial reefs, filled by bitumen, $\mathcal{C}_1$x, Sugaitebulak outcrop, Tarim Basin

primary porosity and thus have become the main targets of oil and gas exploration in sedimentary basins in recent years (Bhat et al. 2012; Carlotto et al. 2017; Hu et al. 2020). In the Keping–Aksu area on the northwestern margin of the Tarim Basin, different types of stromatolites with numerous pores and vugs occur widely within the dolomite strata of the upper Ediacaran Qige-bulak Formation (Yan et al. 2019). Medium–thick layered stromatolites are distributed mainly in the middle and upper parts of the Qigebulak Forma-tion (Fig. 7.15b). The stromatolite morphologies are mainly conical, curved columnar (Fig. 7.15b, c), dome, stratified, and wavy. In the Sichuan Basin, stratified, wavy and dome stromatolites are commonly seen in the Dengying Formation (Fig. 7.15d, e).

Thrombolites are a type of microbial carbonate rock formed from microbial frameworks mixed with the microbial binding of carbonate particles (Fig. 7.15f) and often have a clump microbial structure. Microbial clots are usually surrounded by a dark gray thin shell rich in clay or organic matter. There are many pores among microbial clots, and some of the pores are filled with calcite or dolomite cements.

Botryoidal dolomite commonly occurs in the upper Edicaran strata; for example, Dengying Formation microbialites (especially the Deng 2 Member) are present in nearly the entire Yangtze Plate and the Qigebulak Formation in the Tarim Basin (Fig. 7.15g, h). It is generally a few mm to 1–2 cm in size, with a multilayer concentric ring structure, and the thickness of a single-layer ring is approximately 1–2 mm. Most botryoid dolomite covers the surface of the thick microbialite layer, growing in a grape shape upward from the surface and filling in the pores between the layers. The origin of these botryoidal dolomites in the Dengying Formation is controversial. Shi et al. (2011) argued that their formation is related to short-term exposure to meteoric water during the synsedimentary period. However, our recent study suggested that these botryoidal dolomites preserve original growth zones and original fascicular crys-tals very well; hence, they were likely directly precipitated from special 'dolomite sea' envi-ronments during the deposition of the Dengying Formation microbialites (Wang et al. 2020).

Moreover, different types of microbialites can be mixed to form thick microbial reefs. For example, three periods of microbial reefs were found in the lower Cambrian Xiaoerbulak Formation on the northwestern margin of the Tarim Basin (Fig. 7.16), reflecting typical retrograding sequences. Multiple layers rich in microbial rocks grow superimposed and deaccumulate in the forward (sea) upward direction. These complicated microbial reefs have disorderly internal structures, showing features of upward growth (Fig. 7.16a, b). Microbial reefs are composed of microbial dolomite, mucoclastic dolomite, clastic dolomite and mud–silty dolomite, and microbial dolomite can be further subdivided into laminite, thrombolite, stromatolite, and algal dolomite. These reefs contain abundant layered pores, and some bitumen can be seen within some pores (Fig. 7.16c). These microbes, together with the high-quality source rocks of the lower Cambrian Yuertusi Formation, can form favorable petroleum systems. There is a good relationship between the generation and the reservoir, so oil and gas (bitumen) can be seen in the reservoir pores of the microbialite (Fig. 7.16d).

7.4.2.2 Characterization of Microbial Reservoir Spaces

The types of storage spaces in microbial reservoirs include mainly primary and secondary pores (Fig. 7.17). The primary pores are composed of pores within and between microbial frameworks, such as the fenestral and framework pores in the thrombolites (Fig. 7.17d, h), and the interlayer pores developed along the lamina in stromatolites (Fig. 7.17g). The secondary dissolution pores are related mainly to meteoric karstification. The upper Ediacaran strata in both the Sichuan Basin and Tarim Basin were subjected to tectonic uplift and subsequent meteoric karstification at the end of the Ediacaran period, leading to the formation of karst pores in the microbialites. The presence of an unconformity between the Deng 2 and Deng 3 members indicates the influence of karstification (Fig. 7.17a). There are abundant karst pores and vugs in the Deng 2 member microbialites below the unconformity (Fig. 7.17a–c), and the secondary karst pores and vugs are filled with bitumen (Fig. 7.17b, c). Similarly, bitumen filled in these breccia pores is also commonly seen in

the upper part of the Qigebulak Formation, Tarim Basin (Fig. 7.17e, f).

The reservoir spaces of the Dengying Formation stromatolites are characterized through CT analysis (Fig. 7.17j, l). The results show that there are abundant pores within stromatolites distributed along the laminations of the stromatolites. The minimum pore diameter of the stromatolites is 9.3 μm, the maximum is 498.5 μm, and the average is 30.1 μm. The total number of pores obtained by a CT scan is 26,672, of which 8921 are isolated. The measured porosity is 11.29%, and the permeability is 1.420×10^{-3} μm^2, indicating good reservoir properties.

7.4.3 Preservation Mechanism of Microbialite Reservoirs

The primary and secondary pores in microbialite reservoirs need to be effectively preserved during a long process of burial diagenesis so that they can contribute to the main part of the current oil and gas storage spaces. Several factors ensure that these early pores can avoid being cemented and destroyed (Zhu et al. 2020). First, the microbial dolomite framework formed by microbial bonding and microbial dolomitization gives the reservoir a strong ability to resist mechanical compaction and chemical pressure solutions (Croize et al. 2013). Second, no large-scale fluid cementation occurs during deep burial (Leach et al. 1991). Third, hydrocarbon charging events in microbialite reservoirs, such as oil, bitumen, and natural gas, not only cover the surface of the particles around the pores but also change the properties of the formation fluids, effectively inhibiting the transmission of Ca^{2+}, Mg^{2+}, and CO_3^{2-} ions and hindering the growth of cementation minerals (Ehrenberg et al. 2012). Fourth, hydrocarbon generation from kerogen (Behar et al. 1995) and thermochemical sulfate reduction (TSR) reactions (Ma et al. 2007) gradually increase the CO_2 and H_2S concentrations in microbialite reservoirs, which reduces the pH value of the formation water to create acidic conditions (Lerman and Mackenzie 2018). In such acidic formation fluids, carbonate minerals are more prone to dissolve instead of precipitate.

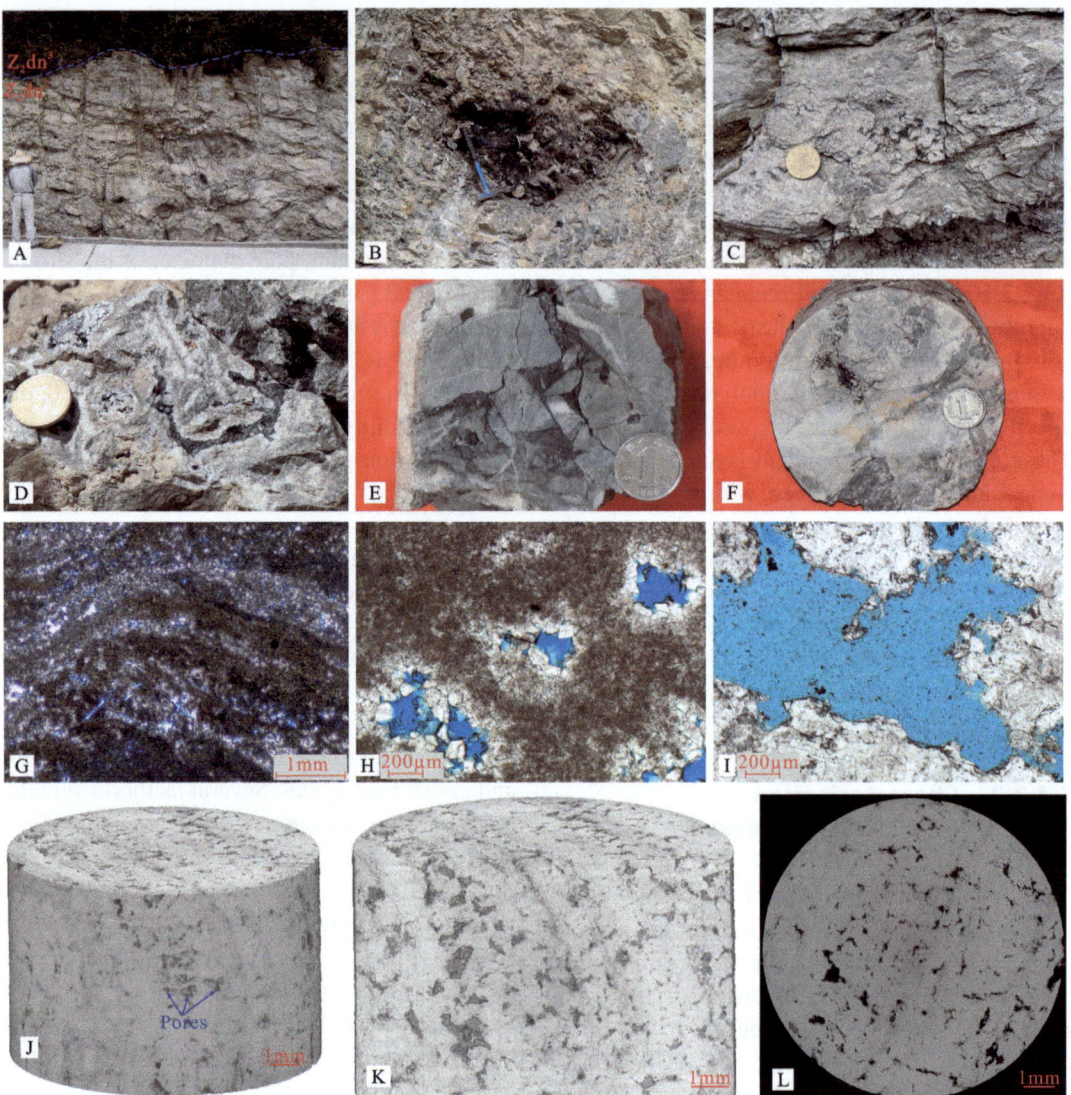

Fig. 7.17 Characteristics of pores and bitumen in the Ediacaran microbialite reservoirs. **a** Karst unconformity between Z_2dn^2 microbialites and Z_2dn^3 siliceous mudstones, Guangwushan town Yingshuiba section; **b** vugs of microbialite are filled with bitumen, Z_2dn^2, Wangcang Yanhe section; **c** pores in microbial thrombolite are filled with bitumen, Z_2dn^2, Nanjiang Liuwan section; **d** framework pores in the microbial thrombolites are filled with bitumen, Z_2q, Dongergou outcrop, Tarim Basin; **e** pores between karst breccias of the microbial thrombolites are filled with bitumen, Z_2q, Luntan 3 well, Tarim Basin; **f** pores between karst breccias of the microbial thrombolites are filled with bitumen, Z_2q, Qitan 1 well, Tarim Basin; **g** matrix pores along laminae in stromatolites, polarized light, $\times 25$, Z_2dn, Wuxi Heping section; **h** fenestral pores in thrombolite, cast thin section, polarized light, $\times 50$, Z_2dn, Yudu section; **i** karst cavities in thrombolite, cast thin section, polarized light, $\times 50$, Z_2dn^2, Zhenba Zhujiagou section; **j–l** micron CT characteristics of the stromatolites, Z_2dn^2, Zhujiagou outcrop, Sichuan Basin

Whether the increasing concentrations of H_2S and pCO_2 in carbonate reservoirs can continuously dissolve carbonate rocks and cause the porosity to continuously increase has long been debated. Although high CO_2 and H_2S contents result in acidic fluid environments that can dissolve carbonate rocks, many researchers have argued that high CO_2 and H_2S contents cannot significantly increase the porosity of carbonate reservoirs (Biehl et al. 2016). According to a case study of Zechstein carbonates in the Lower Saxony Basin, NW Germany, carbonate dissolution features and increased porosity along fractures were expected from acidic CO_2-rich fluids but were not observed petrographically. Therefore, the impact of elevated levels of CO_2 on reservoir quality is low, as fluids quickly become saturated with Ca^{2+}, and porosity creation during burial is completely negligible (Biehl et al. 2016). Hao et al. (2015) believed that the high concentrations of H_2S and CO_2 related to TSR would not cause dissolution of the carbonate reservoir; instead, they contended that TSR might lead to the precipitation of calcite. Net porosity creation by acidic formation brines is very rarely if ever possible according to petrography (Ehrenberg et al. 2012) because the brines are very limited owing to the lack of inflow and outflow circulation.

Although TSR-derived fluid with high concentrations of CO_2 and H_2S might not increase new porosity in microbialite reservoirs, acidic fluid should be better for the long-term preservation of porosity because it does not precipitate calcite or dolomite to fill the existing pores (Liu et al. 2016). The solubility of $CaCO_3$ in solution increases with increasing pCO_2. Hence, precipitation of $CaCO_3$ does not take place in an environment with a high level of CO_2.

The Dengying Formation microbialite reservoirs commonly have high concentrations of H_2S and CO_2, which are associated with organic diagenesis and TSR in the deep burial stage (Liu et al. 2016). For example, the CO_2 and H_2S concentrations are 14.19% and 0.85%, respectively, in well GS1; 4.35% and 2.75%, respectively, in well GS2; and 6.23% and 1.03%, respectively, in well MX8. In acidic formation water, due to the presence of H_2S and CO_2, cement minerals (calcite/dolomite) are less likely to precipitate. As a result, the porosity could be well preserved.

7.5 Reservoir Potential at a Depth of 10,000 m

Based on the previous exploration results, these deep and ultradeep carbonate reservoirs can be summarized into five main types and

models, including reef shoal, karst fracture–cave below unconformity surface, fault–fluid coupling alteration (fault-dissolution and hydrothermal dolomite), subsalt dolomite and ancient microbialite (Fig. 7.18). The characteristics of these reservoir space types, development patterns, and representative oil and gas fields/wells are summarized in detail in Fig. 7.18.

Owing to the influence of burial compaction and diagenesis, the total porosity of carbonate rocks usually decreases with increasing burial depth and age. Both primary and secondary porosities in carbonate reservoirs decrease gradually with increasing depth during normal burial compaction and cementation (Schmoker and Hally 1982). In the early shallow burial stage (< 300 m), more than 50% of the porosity in carbonate should be reduced. The porosity loss in carbonate mud is more strongly affected by compaction, whereas in carbonate sands, it is more strongly affected by cementation (Goldhammer 1997). In terms of carbonate reservoirs from the Precambrian to Silurian periods, the porosity generally decreased to < 10% when they were buried at a depth up to 6000 m, and the median porosity (P50) was close to 5% (Ehrenberg et al. 2009). Therefore, hydrocarbon exploration has traditionally been limited to depths of approximately 5000–6000 m.

However, owing to the limitation of drilling depth, the above trend of porosity variation is only captured to a depth of 6000 m. In recent years, China has drilled a series of ultradeep wells greater than 8000 m in depth in both the Tarim Basin and Sichuan Basin and has found abundant high-quality carbonate reservoirs and oil and gas reserves in the ultradeep realm, thus surpassing the exploration depth limit of 6000 m in the early stage. Well Zhongshen 1 in the Tarim Basin reveals that the deep Cambrian subsalt Xiaoerbulak Formation from 6760 to 6810 m consists of high-porosity granular dolomite and microbial dolomite reservoirs, which are gas layers with a gas production of 158,545 m^3/day (Wang et al. 2014). In well Tashen 1, a high-quality Cambrian dolomite reservoir with a porosity of up to 9.1% was discovered at a depth of 8407.56 m, where a small amount of condensed oil was obtained.

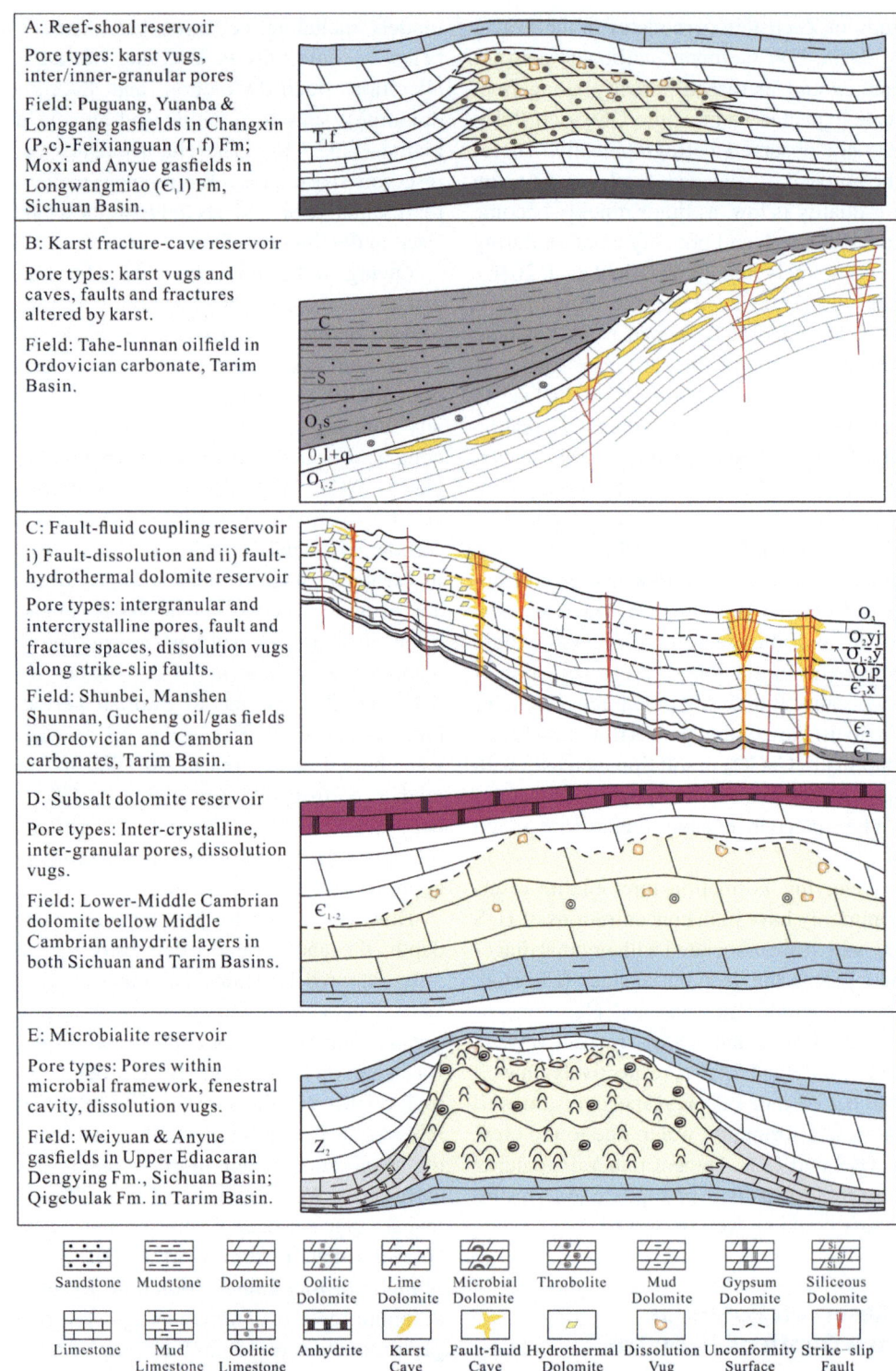

Fig. 7.18 Types, features, and geological models of deep–ultra-deep carbonate reservoirs

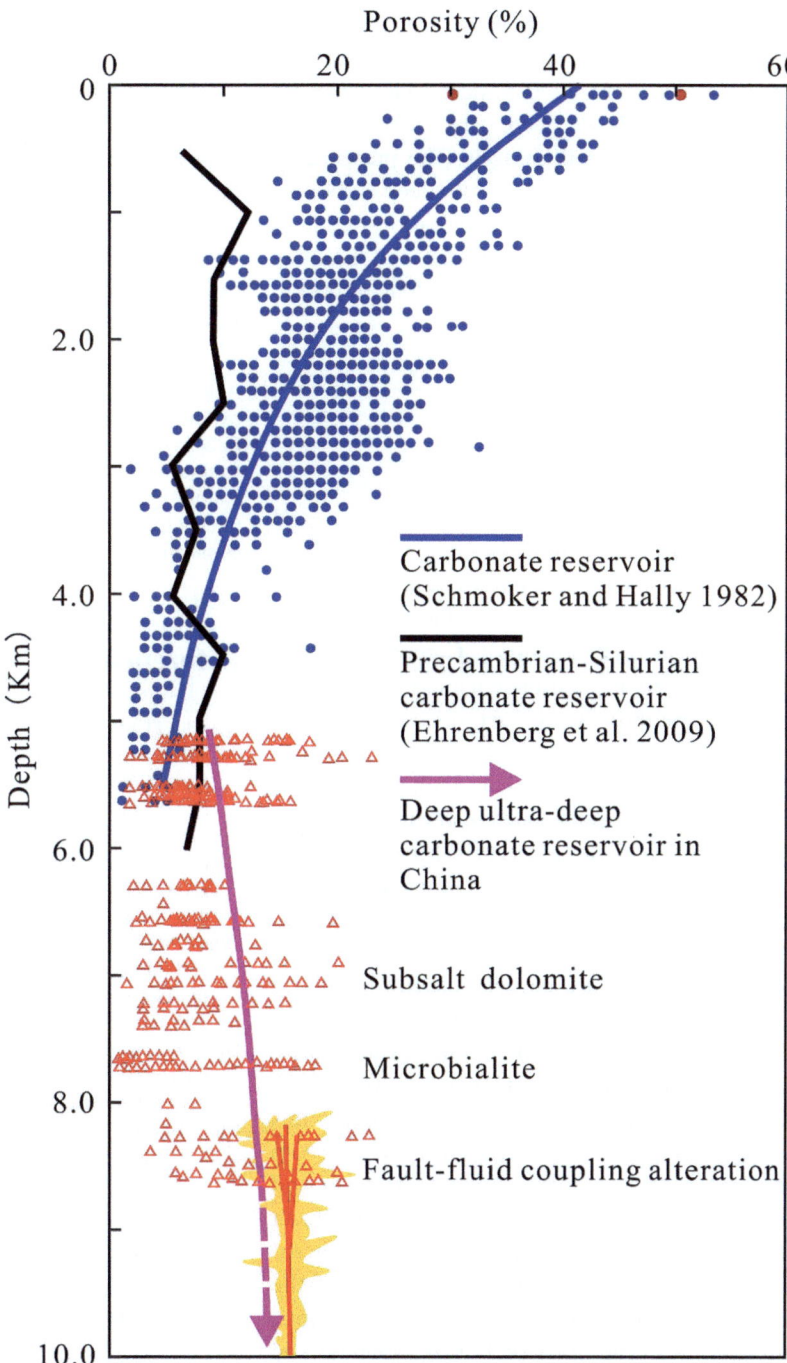

Fig. 7.19 Prediction diagram of deep–ultra-deep carbonate reservoirs extending to a depth of 10,000 m

The lower Cambrian Wusongger Formation granular dolomites from 8203 to 8260 m in well Luntan 1 are high oil- and gas-producing reservoirs with a gas production of 45,917 m³/day and an oil production of 134 m³/day (Yang et al. 2020). The upper Ediacaran Qigeburak Formation microbialites from 8737 to 8750 m have a porosity of 4.0% and are tested as gas layers. The Ordovician fault-dissolution reservoirs are the major reservoirs in the Shunbei and Fuman areas of the Tarim Basin. The oil and gas production fault-dissolution reservoirs are generally 7000–8000 m deep, and some are more than 9000 m deep. The annual output has exceeded one million tons. The presence of these multiple types of deep–ultradeep carbonate reservoirs may enable the exploration depth of oil and gas in the Tarim Basin and Sichuan Basin to further increase to more than 10,000 m (Fig. 7.19).

7.6 Summary

In the deep–ultradeep Ediacaran, Cambrian and Ordovician carbonate strata in the Tarim and Sichuan Basins, in addition to the karst fracture–cave and reef–shoal reservoirs, fault–fluid coupling alteration reservoirs (fault-dissolution, fault-hydrothermal dolomite), Cambrian subsalt dolomite reservoirs and ancient microbialite reservoirs are the three main reservoir types revealed by recent exploration.

The activities of strike–slip faults and related fractures in deep carbonates result in the formation of faulted-cavity reservoirs. Meteoric water, hydrothermal fluids and other diagenetic fluids further dissolve and alter carbonates along strike–slip faults to form fault–fluid coupling alteration reservoirs, which are key exploration targets in Ordovician carbonates at depths of more than 7000 m or even 8000 m in the Tarim Basin. The dolomites formed in shallow evaporative lagoon environments, especially the dolomites of high-energy grain shoal facies, are prone to leaching by synsedimentary meteoric water and consequently developing into high-quality subsalt dolomite reservoirs. Ancient Precambrian microbialite reservoirs are widely observed in the Edicaran strata, mainly including microbial clots, stromatolites, microbial reefs and mounds, and contain abundant primary framework pores and karst vugs. Dolomitization and microbial frameworks with strong resistance to compaction and pressure dissolution, the presence of gypsum seals, oil and gas charging, and high concentrations of CO_2 and H_2S are key factors in the preservation of porosity in deep–ultradeep dolomite and microbialite reservoirs.

High-quality hydrocarbon reservoirs of fault dissolution, subsalt dolomite and ancient microbialite have been proven to exist at depths of more than 8000 m and are expected to potentially exist at depths of more than 10,000 m. These reservoirs are important potential oil and gas exploration targets in ultradeep carbonates in the Tarim Basin and Sichuan Basin.

Acknowledgments This study was financially supported by the National Natural Science Foundation of China (U24B6001, U19B6003 and U2244209) and projects from SINOPEC (YK-2023-10 and KLP24015). Professor Hairuo Qing at the University of Regina and Professor Gregor Eberli at the University of Miami gave constructive comments to improve this article.

References

Bauld J (1984) Microbial mats in marginal marine environments: Shark Bay, Western Australia, and Spencer Gulf, South Australia. In: Cohen Y, Castenholz RW (Eds) HalvorsonMicrobial mats: stromatolites. Alan Liss, New York

Behar F, Vandenbroucke M, Teermann SC et al (1995) Experimental simulation of gas generation from coals and a marine kerogen. Chem Geol 126:247–260

Bhat GM, Craig J, Hafiz M et al (2012) Geology and hydrocarbon potential of Neoproterozoic–Cambrian Basins in Asia: an introduction. Geol Soc Lond Spec Publ 366:1–17

Biehl BC, Reuning L, Schönherr J et al (2016) Do CO_2-charged fluids contribute to secondary porosity creation in deeply buried carbonates? Mar Pet Geol 76:176–186

Carlotto MA, da Silva RCB, Yamato AA et al (2017) Libra: a newborn giant in the Brazilian Presalt Province. In: Merrill RK, Sternbach CA (eds) Giant fields of the decade 2000–2010. The American Association of Petroleum Geologists, Tulsa

Croize D, Renard F, Gratier JP (2013) Compaction and porosity reduction in carbonates: a review of observations, theory, and experiments. Adv Geophys 54:181–238

Davies GR, Smith LB (2006) Structurally controlled hydrothermal dolomite reservoir facies: an overview. AAPG Bull 90:1641–1690

Deng S, Li H, Zhang Z et al (2019) Structural characterization of intracratonic strike-slip faults in the central Tarim Basin. AAPG Bull 103:109–137

Du YM, Zheng SF, Gong X et al (2016) Characteristics and main controlling factor of the Upper Cretaceous Sarvak Reservoir, SouthWest Iran. Acta Sediment Sinica 34:137–148

Dupraz C, Visscher PT (2005) Microbial lithification in marine stromatolites and hypersaline mats. Trends Microbiol 13:429–438

Ehrenberg SN, Nadeau PH, Steen Ø (2009) Petroleum reservoir porosity versus depth: influence of geological age. AAPG Bull 93:1281–1296

Ehrenberg SN, Walderhaug O, Bjørlykke K (2012) Carbonate porosity creation by mesogenetic dissolution: reality or illusion? AAPG Bull 96:217–233

Flügel E (2013) Microfacies of carbonate rocks: analysis, interpretation and application. Springer Science and Business Media, Berlin

Fu L, Li JZ, Xu WL et al (2020) Characteristics and main controlling factors of Ordovician deep subsalt reservoir in central and eastern Ordos Basin. J Nat Gas Geosci 31:1548–1561

Garrett P (1970) Phanerozoic stromatolites: noncompetitive ecologic restriction by grazing and burrowing animals. Science 169:171–173

Goldhammer RK (1997) Compaction and decompaction algorithms for sedimentary carbonates. J Sediment Res 67:26–35

Hao F, Zhang XF, Wang CW et al (2015) The fata of CO_2 derived from thermochemical sulfate reduction (TSR) and effect of TSR on carbonate porosity and permeability, Sichuan Basin, China. Earth Sci Rev 141:154–177

He Z, Zhang J, Ding Q et al (2017) Factors controlling the formation of high-quality deep to ultra-deep carbonate reservoirs. Oil Gas Geol 34:633–644 (in Chinese with English abstract)

Hood AvS, Wallace MW (2012) Synsedimentary diagenesis in a Cryogenian reef complex: ubiquitous marine dolomite precipitation. Sediment Geol 15:56–71

Hood AvS, Wallace MW, Drysdale RN (2011) Neoproterozoic aragonite-dolomite seas? Widespread marine dolomite precipitation in Cryogenian reef complexes. Geology 39:871–874

Hu SY, Shi SY, Wang TS et al (2016) Effect of gypsum-salt environment on hydrocarbon generation, reservoir-forming and hydrocarbon accumulation in carbonate strata. China Pet Explor 21:20–27

Hu Y, Cai C, Pederson CL et al (2020) Dolomitization history and porosity evolution of a giant, deeply buried Ediacaran gas field (Sichuan Basin, China). Precambrian Res 338:1–21

Jiang QC, Li JZ, Wang ZC et al (2022) Quantitative thickness prediction of Cambrian gypsum-salt rocks in Eastern Sichuan Basin and its petroleum significance. Nat Gas Ind 42:34–46

Jiang W, Luo Q, Shi K et al (2021) Origin of a microbial-dominated carbonate reservoir in the Lower Cambrian Xiaoerbulake Formation, Bachu-Tazhong area, Tarim Basin, NW China. Mar Pet Geol 133:105254

Katz DA, Eberli GP, Swart PK et al (2006) Tectonic hydrothermal brecciation associated with calcite precipitation and permeability destruction in Missis-

sippian Carbonate Reservoirs, Montana and Wyoming. AAPG Bull 90:1803–1841

Leach DL, Plumlee GS, Hofstra AH et al (1991) Origin of late dolomite cement by CO_2-saturated deep basin brines: evidence from the Ozark region, central United States. Geology 19:348–351

Lerman A, Mackenzie FT (2018) Carbonate minerals and the CO_2-carbonic acid system. Encyclopedia of geochemistry; Encyclopedia of earth sciences series. Springer, Cham, Switzerland

Liu W, Zhang GY, Pan WQ et al (2011) Lithofacies palaeogeography and sedimentary evolution of the Cambrian in Tarim area. J Palaeogeogr 13:529–538

Liu Q, Zhu D, Jin Z et al (2016) Coupled alteration of hydrothermal fluids and thermal sulfate reduction (TSR) in ancient dolomite reservoirs—an example from Sinian Dengying Formation in Sichuan Basin, Southern China. Precambrian Res 285:39–57

Lonnee J, Machel HG (2006) Pervasive dolomitization with subsequent hydrothermal alteration in the Clarke Lake gas field, Middle Devonian Slave Point Formation, British Columbia, Canada. AAPG Bull 90:1739–1761

Ma Y, Guo X, Guo T et al (2007) The Puguang gas field: New giant discovery in the mature Sichuan Basin, southwest China. AAPG Bull 91:627–643

Ma YS, Cai XY, Zhao PR et al (2010) Formation mechanism of deep buried carbonate reservoir and its model of three element controlling reservoir: a case study from the Puguang oil field in Sichuan. Acta Geol Sinica 84:1087–1094

Ma Y, Cai X, Zhao P (2011) The research status and advances in porosity evolution and diagenesis of deep carbonate reservoir. Earth Sci Front 18:181–192 (in Chinese with English abstract)

Ma YS, He ZL, Zhao PR et al (2019) A new progress in formation mechanism of deep and ultra-deep carbonate reservoir. Acta Petrolei Sinica 40:1415–1425 (in Chinese with English abstract)

Ma YS, Cai XY, Yun L et al (2022) Practice and theoretical and technical progress in exploration and development of Shunbei ultra-deep carbonate oil and gas field, Tarim Basin, NW China. Pet Explor Dev 49:1–20

Machel HG (2004) Concepts and models of dolomitization: a critical reappraisal. Geol Soc Lond Spec Publ 235:7–63

Machel HG, Buschkuehle BE (2008) Diagenesis of the Devonian Southesk-cairn carbonate complex, Alberta, Canada: marine cementation, burial dolomitization, thermochemical sulfate reduction, anhydritization, and squeegee fluid flow. J Sediment Res 78:366–389

Monty C (1973) Precambrian background and Phanerozoic history of stromatolitic communities, an overview. Ann Société Géologique Belg 96:585–624

Nutman AP, Bennett VC, Friend CR et al (2016) Rapid emergence of life shown by discovery of 3,700-million-year-old microbial structures. Nature 537:535–538

Papineau D, Walker JJ, Mojzsis SJ et al (2005) Composition and structure of microbial communities from stromatolites of Hamelin Pool in Shark Bay, Western Australia. Appl Environ Microbiol 71:4822–4832

Pomar L, Hallock P (2008) Carbonate factories: a conundrum in sedimentary geology. Earth Sci Rev 87:134–169

Riding R (2000) Microbial carbonates: the geological record of calcified bacterial–algal mats and biofilms. Sedimentology 47:179–214

Riding R (2006) Microbial carbonate abundance compared with fluctuations in metazoan diversity over geological time. Sediment Geol 185:229–238

Sánchez-Román M, Vasconcelos C, Schmid T et al (2008) Aerobic microbial dolomite at the nanometer scale: implications for the geologic record. Geology 36:546–549

Schmoker JW, Hally RB (1982) Carbonate porosity versus depth: a predictable relation for South Florida. AAPG Bull 66:2561–2570

Scholle PA, Ulmer-Scholle DS (2003) A color guide to the petrography of carbonate rocks: grains, textures, porosity, diagenesis. AAPG Memoir 77, American Association of Petroleum Geologists, Tulsa

Shen A, Zhao W, Hu A et al (2015) Major factors controlling the development of marine carbonate reservoirs. Pet Explor Dev 42:597–608

Shi Z, Liang P, Wang Y et al (2011) Geochemical characteristics and genesis of grapestone in Sinian Dengying Formation in south-eastern Sichuan Basin. Acta Petrol Sinica 27:2263–2271

Smith LBJ (2006) Origin and reservoir characteristics of Upper Ordovician Trenton–Black River hydrothermal dolomite reservoirs in New York. AAPG Bull 90:1691–1718

Smith Jr LB, Eberli GP, Sonnenfeld M (2004) Sequence-stratigraphic and paleogeographic distribution of reservoir-quality dolomite, Madison Formation, Wyoming and Montana. In: Grammer GM, Harris PM, Eberli GP (eds) Integration of outcrop and modern analogs in reservoir modeling: AAPG Memoir 80. American Association of Petroleum Geologist, Tulsa

Stacey J, Corlett H, Holland G et al (2021) Regional fault-controlled shallow dolomitization of the Middle Cambrian Cathedral Formation by hydrothermal fluids fluxed through a basal clastic aquifer. GSA Bulletin 133:2355–2377

Suosaari EP, Reid RP, Andres MS (2019a) Stromatolites, so what?! A tribute to Robert N. Ginsburg. Depos Rec 5:486–497

Suosaari EP, Reid RP, Oehlert AM et al (2019b) Stromatolite Provinces of Hamelin Pool: physiographic controls on stromatolites and associated lithofacies. J Sediment Res 89:207–226

Suosaari EP, Reid RP, Mercadier C et al (2022) The microbial carbonate factory of Hamelin Pool, Shark Bay, Western Australia. Sci Rep 12:12902

Vasconcelos C, McKenzie JA (1997) Microbial mediation of modern dolomite precipitation and diagenesis under anoxic conditions (Lagoa Vermelha, Rio de Janeiro, Brazil). J Sediment Res 67:378–390

Vasconcelos C, McKenzie JA, Bernasconi S et al (1995) Microbial mediation as a possible mechanism for natural dolomite formation at low temperatures. Nature 377:220–222

Veizer J, Ala D, Azmy K et al (1999) $^{87}Sr/^{86}Sr$, $\delta^{13}C$ and $\delta^{18}O$ evolution of Phanerozoic seawater. Chem geol 161:59–88

Wang Y, Su J, Wang K et al (2012) Distribution and accumulation of global deep oil and gas. Nat Gas Geosci 23:526–534 (in Chinese with English abstract)

Wang ZM, Xie HW, Chen YQ et al (2014) Discovery and exploration of Cambrian subsalt dolomite original hydrocarbon reservoir at Zhongshen-1 Well in Tarim Basin. China Petrol Explor 19:1–13

Wang Y, Wang X, Kang H et al (2017) Genetic analysis of microbial reef-bank of Cretaceous lacustrine carbonate in Santos Basin, Brazil. J Chengdu Uni Tech (Scie Tech Ed) 44:67–75

Wang Z, Liu J, Jiang H et al (2019) Lithofacies paleogeography and exploration significance of Sinian Doushantuo depositional stage in the middle-upper Yangtze region, Sichuan Basin, SW China. Pet Explor Dev 46:41–53

Wang JB, He ZL, Zhu DY et al (2020) Petrological and geochemical characteristics of the botryoidal dolomite of Dengying Formation in the Yangtze Craton, South China: constraints on terminal Ediacaran "dolomite seas". Sediment Geol 406:105722

Wei G, Du J, Xu C et al (2015) Characteristics and accumulation modes of large gas reservoirs in Sinian-Cambrian of Gaoshiti-Moxi region, Sichuan Basin. Acta Petrolei Sinica 36:1–12

Westphal H, Eberli GP, Smith LB et al (2004) Reservoir characterization of the Mississippian Madison Formation, Wind River Basin, Wyoming. AAPG Bull 88:405–432

Xiao S, Laflamme M (2009) On the eve of animal radiation: phylogeny, ecology and evolution of the Ediacara biota. Trends Ecol Evol 24:31–40

Yan W, Yang G, Yi Y et al (2019) Characteristics and genesis of Upper Sinian dolomite reservoirs in Keping area, Tarim Basin. Acta Petrolei Sinica 40:295–307, 321 (in Chinese with English abstract)

Yang H, Chen Y, Tian J et al (2020) Great discovery and its significance of ultra-deep oil and gas exploration in well Luntan-1 of the Tarim Basin. China Petrol Explor 25:62–72 (in Chinese with English abstract)

You XL, Sun S, Zhu JQ (2014) Significance of fossilized microbes from the Cambrian stromatolites in the Tarim Basin, Northwest China. Sci China Earth Sci 57:2901–2913

You D, Han J, Hu W et al (2018) Characteristics and formation mechanisms of silicified carbonate reservoirs in well SN4 of the Tarim Basin. Energy Explor Exploit 36:820–849

Zhang GY (2000) Formation and evolution of the Paleozoic cratonic basin in Tarim and associated oil and gas. Geological Publishing House, Beijing

Zhang GY, Ma F, Liang YB et al (2015) Domain and theory-technology progress of global deep oil & gas exploration. Acta Petrolei Sinica 36:1156–1166 (in Chinese with English abstract)

Zhao W, Wang Z, Hu S et al (2012) Large-scale hydrocarbon accumulation factors and characteristics of marine carbonate reservoirs in three large onshore cratonic basins in China. Acta Petrolei Sinica 33:1–10 (in Chinese with English abstract)

Zhao W, Shen A, Qiao Z et al (2014) Carbonate karst reservoirs of the Tarim Basin, northwest China: types, features, origins, and implications for hydrocarbon exploration. Interpretation 2:65–90

Zhao WZ, Wei GQ, Yang W et al (2017) Discovery of Wanyuan-Dazhou Intracratonic Rift and its significance for gas exploration in Sichuan Basin, SW China. Pet Explor Dev 44:697–707

Zhu DY, Meng QQ, Jin ZJ et al (2015) Formation mechanism of deep Cambrian dolomite reservoirs in the Tarim basin, northwestern China. Mar Pet Geol 59:232–244

Zhu DY, Liu QY, Zhang JT et al (2019) Types of fluid alteration and developing mechanism of deep marine carbonate reservoirs. Geofluids 2019:1–18

Zhu DY, Liu QY, He ZL et al (2020) Early development and late preservation of porosity linked to presence of hydrocarbons in Precambrian microbialite gas reservoirs within the Sichuan Basin, southern China. Precambrian Res 342:105694

Zhu DY, Liu QY, Wang JB et al (2022) Transition of seawater conditions favorable for development of microbial hydrocarbon source—reservoir assemblage system in the Precambrian. Precambrian Res 347:106649

Zou C, Du J, Xu C et al (2014) Formation, distribution, resource potential and discovery of the Sinian-Cambrian giant gas field, Sichuan Basin, SW China. Pet Explor Dev 41:306–325

Dynamic Mechanism and Prediction Methods for Deep Marine Hydrocarbon Accumulation

8

Xiongqi Pang, Hong Pang, Tao Hu, and Junqing Chen

Abstract

Abundant deep hydrocarbon resources have been discovered in deep marine strata worldwide, whereas the dynamic mechanism and prediction methods for deep marine reservoirs are still unclear, resulting in great challenges in selecting favourable exploration targets. Compared with middle-shallow hydrocarbon reservoirs, deep reservoirs exhibit significant differences in terms of media conditions and temperature and pressure environments and feature multiphase accumulation characteristics in multirock masses, multisource and stage accumulation characteristics, multidynamic and category accumulation characteristics and wide distribution and variable production capacity. The dynamic process and the variation characteristics of oil and gas expulsion from source rocks were divided into four stages. The CPD did not contribute at the first stage before entering the HET, contribution varied from 25 to 50% at the second stage, contribution was more than 50% at the third stage in oil/gas expulsion, and contributed little at the fourth stage for the reformation of the reservoirs formed in the earlier stages due to tectonic movements. Generally, an effective CPD is the most important driving force for deep tight oil/gas reservoir formation. The HET of the source rock, buoyancy-driven hydrocarbon accumulation depth, hydrocarbon accumulation depth limit, and ASDL are the accumulation dynamic boundaries for deep hydrocarbon reservoir formation. Finally, dynamic models and predictive evaluation methods for deep hydrocarbon reservoirs were proposed.

Keywords

Deep marine strata · Hydrocarbon accumulation · Dynamic mechanism · Quantitative prediction methods

Abbreviations

CPD	Capillary Pressure Displacement
HET	The hydrocarbon expulsion threshold
ASDL	The maximum burial depth of active source rocks
mD	MilliDarcy
hm	Hundred meters
MPa	Megapascal
Q_p	The total hydrocarbon amount generated in kg/m^3 for oil or m^3/m^3 for gas
TOC	The organic matter content

X. Pang (✉) · H. Pang · T. Hu · J. Chen
State Key Laboratory of Petroleum Resources and Engineering, China University of Petroleum (Beijing), Beijing, China
e-mail: pangxq@cup.edu.cn

College of Geosciences, China University of Petroleum (Beijing), Beijing, China

© The Author(s) 2026
Y. Ma et al. (eds.), *Petroleum Geology and Exploration of Deep Marine Strata in China*, Advances in Oil and Gas Exploration & Production, https://doi.org/10.1007/978-3-032-02496-1_8

KTI	Kerogen-type index	Q_{ed}	The hydrocarbon amount expelled as a diffusion phase, kg/km^2 or m^3/km^2
Ro	Thermal evolution degree		
Z	Buried depth		
D	Density	Q_{eog}	The gas amount expelled from the source rocks as an oil-solution phase, kg/km^2 or m^3/km^2
R_p	The hydrocarbon amount generated in kg/t_{toc} or m^3/t_c		
Q_{ro}	The total hydrocarbon amount remaining, kg/m^3 or m^3/km^3	Q_{es}	The hydrocarbon expulsion amount as a free phase, kg/km^2 or m^3/km^2
Φn	The source rock porosity with normal compaction in %	V_e	Velocity of hydrocarbon expulsion
$\triangle \Phi$	The extra porosity of the source rock with abnormal compaction, %	S_e	Rate of hydrocarbon expulsion
		R_e	Efficiency of hydrocarbon expulsion
ρ_o	The crude oil density in t/m^3	SRI	Source rock index
S_1	The amount of free hydrocarbons already present in the rock sample	V_e	The hydrocarbon expulsion velocity or the hydrocarbon amount expelled per square meter of the source rock
"A"	Soluble organic matter extracted from rock samples by chloroform solvents		
		Ksr	The expansion coefficient of the skeleton content
B_k	The factor of light hydrocarbon compensation, which is the ratio of lost-to-residual light hydrocarbons quantitatively related to the major factors established before	$\triangle V_{f3}$	The fluid volume related to F_3
		Φ	Porosity
		H	The source rock thickness
		μi	Fluid viscosity
Q_{rg}	The total hydrocarbon amount remaining, kg/m^3 or m^3/km^3	Pc	The capillary pressure in the source rock
		r	The throat radius of the surrounding rocks
Q_{rgb}	The amount of adsorbed gas		
Q_{rgo}	The amount of oil-dissolved gas	R	The throat radius of reservoirs
Q_{rgw}	The amount of water-dissolved gas	θ	The wetting angle of hydrocarbon/water
Q_e	The total hydrocarbon amount expelled, kg/m^3 for oil or m^3/m^3 for gas	F_{9o}	The capillary pressures for oil in the rocks
R_e	The hydrocarbon amount expelled, $kg/T\ toc$ or m^3/Tc	F_{9g}	The capillary pressures for gas in the rocks
Q_{rm}	The residual hydrocarbon amount	Q_{e1}	The hydrocarbon amount of Q_{ed} was contributed by F_1
		D	The diffusion coefficient
Q_{ew}	The hydrocarbon amount expelled as water solution, kg/km^2 or m^3/km^2	dc/dz	Hydrocarbon concentration gradient
		S	Diffusion area

φ	Source rock porosity
t	Diffusion period
Q_{ei}	The oil/gas amount expelled by each of the other eight driving forces
HC	Hydrocarbon
LHED	Liquid hydrocarbon expulsion depth
OM	The hydrocarbons generated from organic matter
Q_r	The amounts of residual hydrocarbon
S_2	The hydrocarbon content of the source rock from the pyrolysis of kerogen
HCI_o	The original hydrocarbon generation potential index
HCI_p	The residual hydrocarbon generation potential index
q_e	The hydrocarbon expulsion rate: the amount of hydrocarbon expelled per unit organic carbon after the source rock reaches the HET
E_{hc}	The hydrocarbon expulsion intensity, t/km^2
$q_e(Z)$	Hydrocarbon expulsion rate per unit TOC, mg/g
$\rho(Z)$	The source rock density
S (n)	The source rock area, m^2
GT	The variation rates of the geothermal gradient
HFU	Heat flow unit
BHAD	The buoyancy-driven hydrocarbon accumulation depth
P_e	The hydrocarbon expansion pressure
P_w	The overlying hydrostatic pressure
P_c	The capillary pressure
HADL	The hydrocarbon accumulation depth limit
H/C	The atomic ratios of hydrogen to carbon
O/C	The atomic ratios of oxygen to carbon
("$S_1 + S_2$"/TOC)	Hydrocarbon generation potential index

Ve	The hydrocarbon expulsion rate
Ke	The cumulative amount ratio of the hydrocarbons expelled from the source rocks to the cumulative amounts of generated hydrocarbons
HI	The hydrogen index
ASDLg	The ASDLs for natural gas
ASDLo	The ASDLs for natural oil
Qr	The amount of residual hydrocarbon in the source rock
Qrm	The minimal amount of hydrocarbon needed to saturate the source rock porosity
Qew	The amount of hydrocarbon dissolved in water
Qed	The amount of hydrocarbon diffused out of the source rock
Qes	The amount of hydrocarbon expelled in a separate phase
Qe'	The amount of hydrocarbon expelled in early expulsion
Qew'	The amount of hydrocarbon dissolved in water in early expulsion
Qed'	The amount of hydrocarbon diffused out of the source rock in early expulsion

8.1 Medium Conditions and the Temperature and Pressure Environment of Deep Hydrocarbon Accumulation

Deep hydrocarbons refer to the oil and gas resources in the stratum with a burial depth of more than 4500 m (Dyman et al. 2001a, b; Pang et al. 2015a, b, c). Abundant deep and ultra-deep hydrocarbon resources have been discovered worldwide, and have become important for increasing reserves and production. However, deep reservoirs feature complex geological conditions and face great challenges in exploration and development. Compared with middle-shallow hydrocarbon reservoirs, deep reservoirs exhibit significant differences in terms of the medium conditions and temperature and pressure environments reflected in the aspects presented below.

8.1.1 Medium Conditions of Deep Reservoirs

8.1.1.1 Stratigraphic Age

Deep hydrocarbon reservoirs are distributed in strata of different ages. According to the stratigraphic ages of the discovered deep hydrocarbon reservoirs, they are distributed mainly in older Neogene, Cretaceous, and Upper Palaeozoic strata. The stratigraphic ages of 1462 deep hydrocarbon reservoirs, which are distributed mainly in five strata, namely, the Neogene, Palaeogene, Cretaceous, Jurassic, and Upper Palaeozoic, have been statistically analysed (Fig. 8.1a). From the perspective of reserves, the 2P (probable) recoverable reserves of deep hydrocarbons in the Neogene, Palaeogene, Cretaceous, Jurassic, and Upper Palaeozoic account for 12.8%, 22.3%, 18.3%, 12.8%, and 22.2% of the global total, respectively (Bai et al. 2014) (Fig. 8.1b).

The proportion of deep gas reserves gradually increases with increasing reservoir age and burial depth.

The stratigraphic ages of deep hydrocarbon reservoirs in China are quite different. Statistical analysis revealed that deep hydrocarbon reservoirs in western China are distributed mainly in Palaeozoic strata, including the main production areas of Proterozoic Sinian, Early Palaeozoic Cambrian, and Ordovician strata in the Tarim and Sichuan Basins (Xu et al. 2014; Luo et al. 2015; Deng et al. 2018, 2019; Zhao et al. 2019; Zhu et al. 2019; Li et al. 2019a, b; Li et al. 2019a, b; Zhao et al. 2020). However, deep hydrocarbon reservoirs in eastern China are distributed mainly in the Palaeogene and Cretaceous strata of the Bohai Bay and Songliao Basins. Deep hydrocarbons are produced mainly in the Cenozoic Palaeogene and Cretaceous (Ran et al. 2021; Bai et al. 2018; Ma et al. 2012; He et al. 1999), but bedrock reservoirs are distributed mainly in Precambrian metamorphic and volcanic rocks.

8.1.1.2 Density

Density is among the important physical properties of rocks and is closely related to the physical and electrical properties of a reservoir (Wang

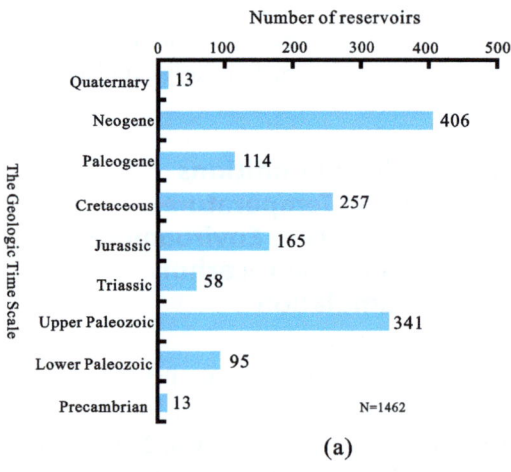

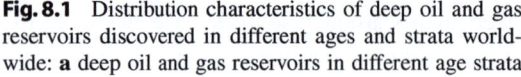

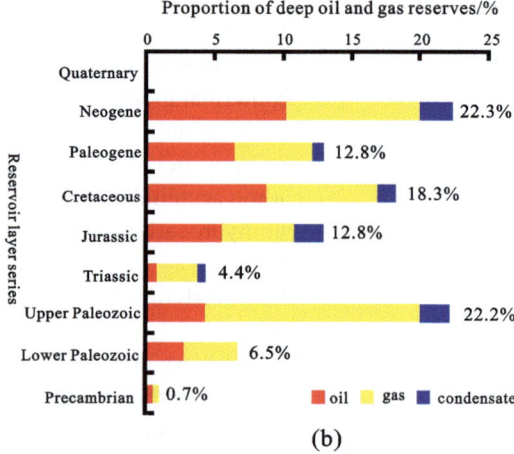

(a)

(b)

Fig. 8.1 Distribution characteristics of deep oil and gas reservoirs discovered in different ages and strata worldwide: **a** deep oil and gas reservoirs in different age strata worldwide and **b** global deep reservoir reserves and phase states in different age strata

et al. 2009). Understanding density and its variation is highly important for gravity data interpretation and deep hydrocarbon exploration (Xu et al. 2016). The rock types found in deep hydrocarbon reservoirs are mainly clastic and carbonate rocks, with a small number of volcanic and metamorphic rocks (Pang et al. 2015a, b, c). The reservoir density is generally high and gradually increases with increasing burial depth. It is generally greater than 2.5 g/cm³ at burial depths exceeding 4500 m. Compared with the shallow strata, the average rock density increases by 0.3 g/cm³, a 12% increase. In accordance with the core and outcrop data of the Subei Basin, Xu et al. (2016) measured the density of the strata at different depths and divided the density interface. The results revealed that rock density tends to gradually increase with increasing burial depth (or ageing) (Table 8.1).

8.1.2 Physical Reservoir Characteristics

8.1.2.1 Porosity

Porosity is an important physical parameter for characterizing reservoir rock. It is the main spatial type of reservoir hydrocarbon enrichment. The porosity of deep hydrocarbon reservoirs is generally low due to deep depth and strong compaction diagenesis (i.e., less than 15%). For most reservoirs, the porosity gradually decreases with increasing burial depth due to compaction and diagenesis. Theoretically, the porosity is less than 5% when the burial depth is greater than 4500 m (Sun et al. 2013). However, in the case of deep structural changes and fluid transformation, secondary pores and fractures can form, maintaining high porosity (Gordon et al., 2022; Yu et al., 2014; Cao et al., 2014; Zhou et al.,

Table 8.1 Division table of the density layer and interface in Northern Jiangsu Basin [according to Xu et al. (2016)]

Eon	Era	Period	Density/(g.cm⁻¹)		Density layer	High density layer	I	II
Phanerozoic	Cenozoic	Neogene	2.10	2.20	A₁	A		
		Paleogene	2.29		A₂			D₁
							D₂	
	Mesozoic	Cretaceous	2.53	2.56	B₁	B		D₃
		Jurassic	2.59		B₂			
		Triassic					D₄	
	Upper Paleozoic	Permian						
		Carboniferous	2.69		C₁			
		Devonian						
	Lower Paleozoic	Silurian		2.72	C	C		D₅
		Ordovician						
		Cambrian						
Proterozoic	Neoproterozoic	Sinian	2.75		C₂			
	Mesoproterozoic							
Archaeozoic and Pelaoproterozoic								

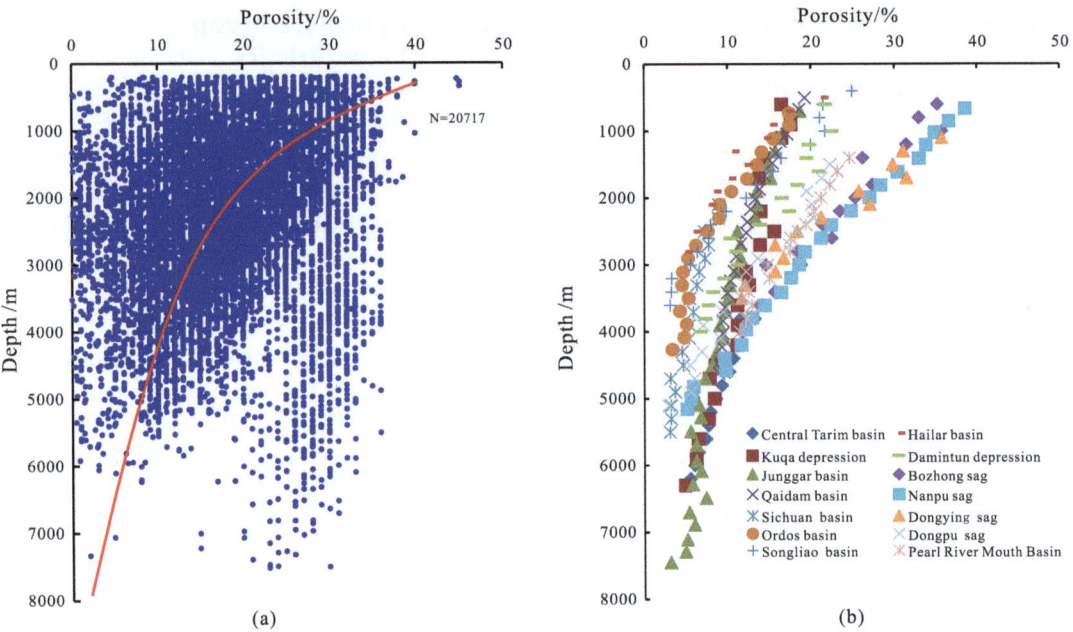

Fig. 8.2 Characteristics of the reservoir porosity variation with burial depth in a petroliferous basin: **a** major petroliferous basins worldwide and **b** major petroliferous basins in China

2013; Zhang et al., 2014). Overall, the porosities of deep reservoirs are generally low. The porosity data for 20,717 global oil and gas reservoirs (Fig. 8.2a) reveal that the reservoir porosity gradually decreases with increasing depth. When the depth exceeds 4500 m, the reservoir porosity distribution ranges from 5 to 15%, with the lowest being less than 2%. The decrease in reservoir porosity in different petroliferous basins differs with increasing burial depth. The porosity in some petroliferous basins decreases rapidly with increasing depth, whereas that in some basins decreases slowly and varies with different geological conditions. The trends of porosity evolution with increasing depth in several major petroliferous basins in China are similar. At burial depths exceeding 4500 m, the porosity is mostly less than 12%, with some basins even having porosities less than 5% (Fig. 8.2b). Moreover, porosity generally increases with increasing age, rapid burial, low geothermal gradient, and high quartz content (Maxwell 1964; Scherer 1987; Bei et al. 1985; Wilson 1994).

8.1.2.2 Permeability

Permeability is a key physical parameter in characterizing the fluid seepage capacity of reservoir rocks. Permeability, like porosity, gradually decreases with increasing burial depth. However, owing to structural changes and fluid transformation in deep layers, more secondary pores and fractures can form in reservoirs and promote high permeability. However, the permeability of deep reservoirs is still generally low. According to the statistics of 20,717 global oil and gas reservoirs, the permeability gradually decreases with increasing depth (Fig. 8.3a). At depths exceeding 4500 m, the permeability ranges mainly from 0.1 to 10,000 mD, with the lowest being less than 0.1 mD. However, the decrease in permeability in petroliferous basins varies with different geological conditions. Some exhibit a rapid decrease in permeability, whereas others show a slow decrease or an increase. However, a positive correlation generally exists between porosity and permeability; that is, the higher the reservoir porosity is, the greater the permeability

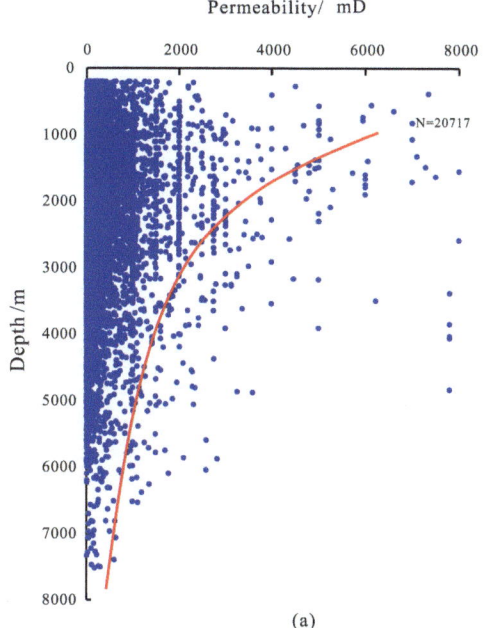

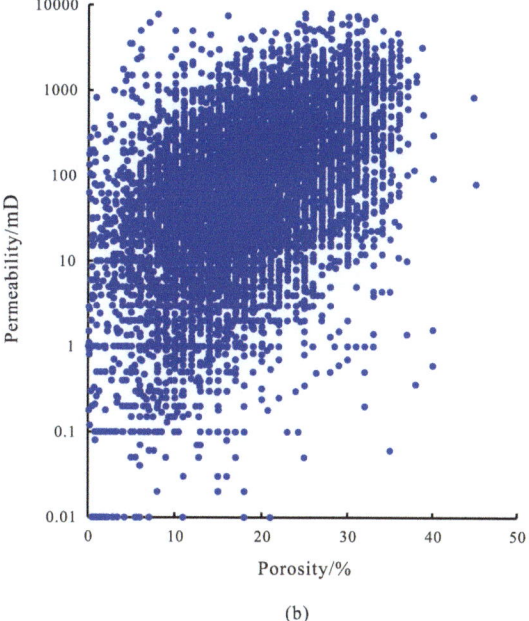

(a)   (b)

Fig. 8.3 Variation characteristics of reservoir permeability with burial depth and porosity in an oil-bearing basin: **a** variation characteristics of permeability with burial depth in the main oil–gas basins worldwide and **b** relationship between permeability and porosity of major petroliferous basins worldwide

(Fig. 8.3b). Many factors affect permeability, including fractures caused by tectonic stress and secondary pores caused by fluid dissolution. Fractures strongly affect increasing permeability, although they only slightly influence porosity. Secondary pores, which are caused mainly by fluid dissolution, generally increase the porosity, whereas the permeability remains low (Zhu et al., 2009; Biteau et al. 2006).

8.1.3 Hydrocarbon Enrichment Environment

8.1.3.1 Formation Temperature

Formation temperature is closely related to hydrocarbon generation, which is an important factor for determining the evolution of hydrocarbon generation (Tissot et al. 1974; Sun et al., 1995; Hao et al., 1996; Zhang et al., 2008). The geothermal gradient is an important geothermal parameter often used to describe the geothermal characteristics of sedimentary basins. Under the effects of the geothermal gradient, the formation temper-

ature gradually increases with increasing burial depth. The formation temperature of deep hydrocarbon reservoirs is also generally greater because of their greater depth. The temperatures of deep oil fields in Italy, the North Sea area, and the Williston Basin range from 153 °C, 165 °C–175 °C, and 180 °C, respectively. Moreover, the temperatures of the Washington and Barr Lake Oilfields in the United States and the Palaton and Bier Oilfields in the Gulf of Mexico Basin exceed 200 °C. The pay zone temperature of the Mullen Oilfield in the Persian Gulf exceeds 230 °C, whereas the deep reservoir temperature of the Brahai Reservoir in the Caspian Basin in Russia reaches 295 °C (Pusey 1973). According to the temperature variation characteristics of the 20,446 global oil and gas reservoirs with depth (Fig. 8.4a), the temperature distribution of the hydrocarbon reservoirs ranges from 50 °C to 370 °C at burial depths exceeding 4500 m, with the highest temperature reaching 370 °C. The increase in ground temperature differs with depth due to differences in geological conditions. Several major petroliferous basins in China have also shown increasing formation temper-

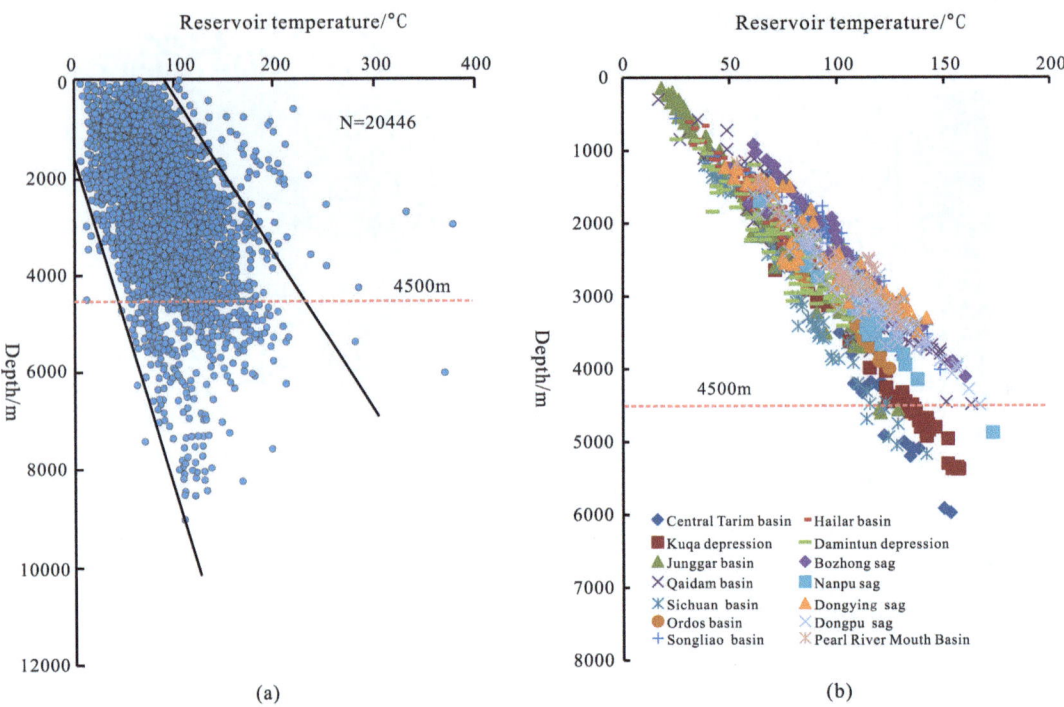

Fig. 8.4 Variation characteristics of the stratigraphic temperature with burial depth in a petroliferous basin: **a** major petroliferous basins worldwide and **b** major petroliferous basins in China

atures with increasing burial depth (Fig. 8.4b). However, the ground temperature in different basins varies with increasing burial depth, and its ground temperature gradient distribution ranges from 2.2 to 4.2 °C/hm. From east to west, the geothermal gradients of petroliferous basins gradually decrease, whereas deep hydrocarbon reservoirs greater than 4500 m in depth have higher temperatures and wider temperature distribution ranges.

8.1.3.2 Formation Pressure

Formation pressure is the direct expression of the underground hydrodynamic field (Liu et al. 2012), which significantly affects hydrocarbon formation and distribution in petroliferous basins. Exploration practices have shown that abnormal pressure is widely present in deep hydrocarbon reservoirs. The variation characteristics of the formation pressure with depth are reflected in the statistics of 19,772 global oil and gas reservoirs (Fig. 8.5a). The formation pressure distribution at depths exceeding 4500 m is between

30 and 150 MPa. However, the increase in pressure in different petroliferous basins differs with depth due to differences in geological conditions. The formation pressure in some basins rapidly increases with increasing depth, reflecting abnormally high pressure. In some basins, the layer pressures slowly increase with increasing depth— even less than the hydrostatic pressure—exhibiting abnormal low-pressure characteristics with a pressure coefficient far less than 1. In most cases, the pressure is equivalent to the hydrostatic pressure, with a pressure coefficient of approximately 1. Several major petroliferous basins face similar situations. A statistical analysis of the changes in the target formation pressure with depth in the Jiyang Depression and the Bohai Bay Basin revealed that the Dongying and Zhanhua depressions experienced lower pressures at shallow depths. An abnormally high pressure (Fig. 8.5b) is observed when the depth is greater than 2500 m. However, for specific petroliferous basins and target beds, abnormal pressure formation is influenced by many geological factors (Hunt 1990;

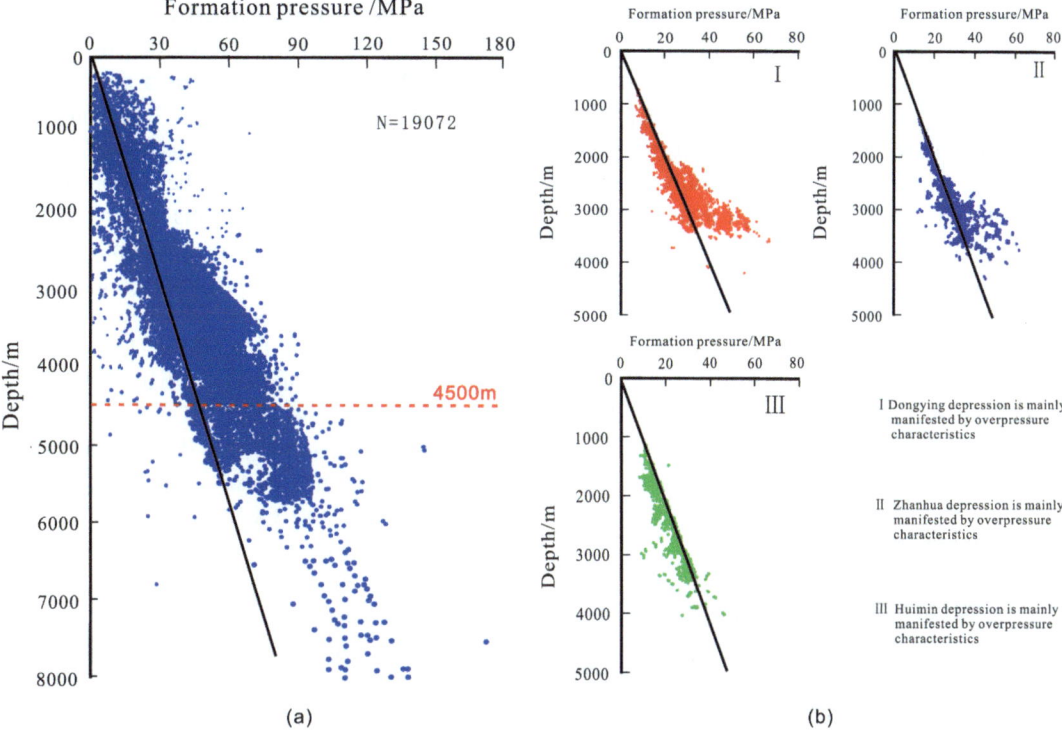

Fig. 8.5 Variation characteristics of the formation pressure with burial depth in a petroliferous basin: **a** major petroliferous basins worldwide and **b** major petroliferous basins in China

Osborne et al. 1997; Law et al. 1998). In general, there is more abnormal pressure in deep layers than in shallow layers.

8.1.3.3 Formation Water Salinity

Formation water is a major component of petroliferous basins. Its activities are closely related to petroleum circulation, generation, migration, accumulation, and dispersion. In the basin subsidence process, the flow–rock interaction is strengthened with increasing depth, and the formation water salinity gradually increases. Owing to the deep depth and high formation water salinity, the water medium conditions are mostly a reducing environment. According to the data of more than 10,000 oil and gas fields in China's petroliferous basins studied by Zhang Qin and Zhang Jinchuan, the formation water type in the eastern basin of China is mainly $NaHCO_3$, whereas that in western basins is mainly $CaCl_2$, indicating a deep closed reducing environment. The formation water salinity is high in the west and low in the

east and middle (Zhang et al. 2009). Man et al. (2009) reported that the salinity and the positive correlation coefficient of oilfield water increase with increasing depth, whereas the negative correlation coefficient decreases with increasing depth. Hydrocarbon preservation conditions are better in deep layers than in shallow layers. The variation characteristics of the salinity of different types of formation water with increasing depth in western Sichuan also indicate that the deep formation water is mainly of the $CaCl_2$ type. Moreover, the salinity of the same type of formation in the deep and middle–shallow layers greatly differs with increasing depth.

8.1.4 Thermal Maturation

The hydrocarbon generation model reveals that the kerogen begins to mature when $R_O > 0.5\%$ and when abundant liquid hydrocarbons are generated. When $R_o > 1.2\%$, kerogen enters the thermal cracking stage to generate moist gas. When $R_o >$ 2.0%, kerogen enters the thermal cracking stage to produce dry gas. Liquid hydrocarbons then begin to crack into gas. Owing to the deep depth and high temperature, the degree of thermal evolution of deep source rocks is generally high and essentially in the stages of high maturity ($R_o >$ 1.2%) and over maturity ($R_o >$ 2.0%). As shown in the analysis of the variation characteristics of the degree of thermal evolution of source rocks in major petroliferous basins in China with depth, the thermal maturity of source rocks in most basins is greater than 1.2%, with some exceeding 2.5% at depths exceeding 4500 m.

8.2 Basic Geological Characteristics of Deep Hydrocarbon Reservoirs

The compaction effect is continuously enhanced with increasing depth. The temperature and pressure conditions of the formation and physical characteristics of the reservoirs also change. Compared with middle and shallow hydrocarbon reservoirs, deep hydrocarbon reservoirs have significantly different basic geological characteristics: (1) multiphase accumulation characteristics in multirock masses; (2) multisource and stage accumulation characteristics; (3) multidynamic and category accumulation characteristics; and (4) wide distribution and variable production capacity.

8.2.1 Accumulation Characteristics of the Multiphase in Multirock Masses

Compared with those in middle and shallow hydrocarbon reservoirs, deep hydrocarbons develop in various rock types and present different fluid phases. The proportions of various rock reservoirs and the main fluid phases significantly differ due to the differences between the critical conditions of accumulation and the temperature and pressure conditions of the formation.

8.2.1.1 Deep Hydrocarbon Accumulation in Multiple Lithologic Reservoirs

Deep hydrocarbons develop in reservoirs with various lithologies in petroliferous basins. A series of hydrocarbon reservoirs have been found in deep sandstone beneath deep oceans (e.g., the Tiber clastic rock reservoir buried at a depth of 8740 m in the Gulf of Mexico). Similarly, a series of deep carbonate reservoirs have been found in deep layers in the mainland (e.g., Tahe, Tazhong, and Puguang in the Sichuan Basin) (Zhang 1999; Zhou et al. 2006; Ma et al. 2005). Large oil and gas fields have been discovered in deep metamorphic rock formations such as the Liaohe fault, Xinglongtai bedrock, and Qianshan Reservoir, indicating the presence of more than 100 million tons of petroleum geological reserves (Bi et al. 2014). Some large oil and gas fields were also found in deep volcanic rocks, such as the Yoshii–Higashi Kashiwazaki large gas reservoir in the deep rhyolite of the Niigata oil and gas basin in Japan (Schutte 2003). Overall, these discovered deep hydrocarbon reservoirs are dominated by clastic and carbonate rocks, with some volcanic and metamorphic rocks. The middle and shallow layers are dominated by clastic rocks, followed by carbonate and volcanic rocks and metamorphic rock reservoirs. By 2010, among the 1477 deep reservoirs discovered worldwide, 1035 (70.07%) were identified as clastic reservoirs, of which 429 (29.05%) and 13 (0.88%) were carbonate reservoirs and magmatic/metamorphic reservoirs, respectively (Pang et al. 2015a, b, c).

8.2.1.2 Mainly Gaseous Deep Hydrocarbons Coexist in Various Phases

Generally, the amount of gas is much greater than that of liquid hydrocarbons. With increasing burial depth, the cracking of crude oil and the gas of the source rock increase because of increasing temperature and pressure, and the ratio of gas to liquid hydrocarbons increases synchronously (Dyman et al. 2001a, b; Pang et al. 2015a, b, c; Tissot 2013; Price 1993). The types of deep hydrocarbon phases are complex and include gases,

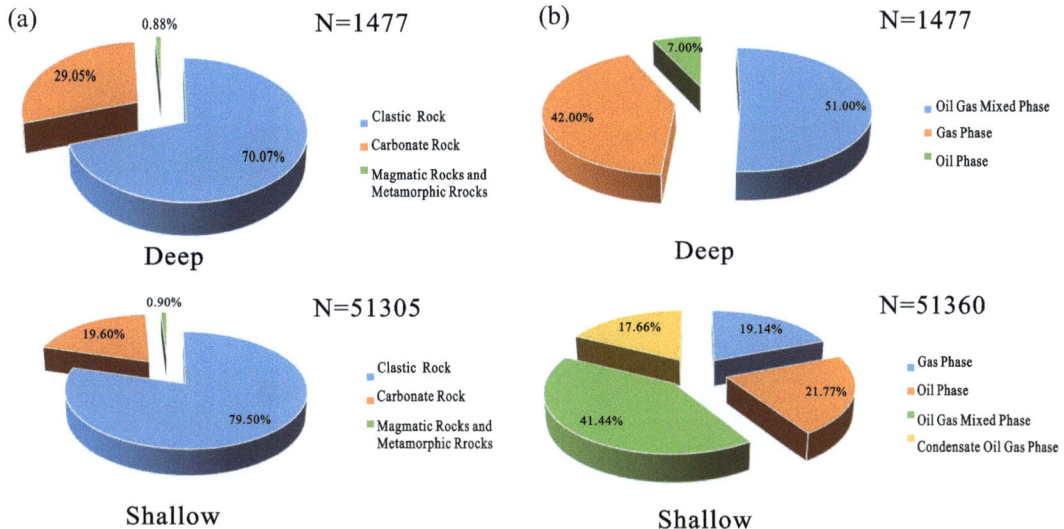

Fig. 8.6 Comparisons of the global geological characteristics of the discovered oil and gas reservoirs: **a** shallow–deep lithologic distributions in the discovered oil and gas reservoirs worldwide and **b** phase distributions between the shallow and deep oil and gas in the discovered oil and gas reservoirs worldwide

condensates, condensate gases, liquid hydrocarbons, and coexistent oil and gas (Thompson 1987; Pang 2015; Pang et al., 2015a, b, c; Shi et al. 2015). Through the phase statistics of 1477 deep oil and gas reservoirs worldwide, Pang et al. (2015a, b, c) reported that 51% of the deep hydrocarbon reservoirs comprises mixed oil and gas, 42% represents the gas phase, and 7% comprises the very small oil phase. The middle and shallow hydrocarbon phases are relatively simple and contain mainly liquid and gaseous hydrocarbons, mixed oil–gas, and a small amount of condensate gas. The statistics of 51,360 shallow and medium hydrocarbon reservoirs worldwide reveal that 41% represents mixed oil and gas, 22% represents the oil phase, and 19% comprises a gas phase and a considerable portion of condensate oil and gas (Fig. 8.6).

8.2.2 Characteristics of Multisource and Stage Accumulation

Compared with those in middle and shallow hydrocarbon reservoirs, the degree and frequency of tectonic influence are relatively large with increasing depth. Most petroliferous basins in

China are superimposed basins with multistage development and are widely affected by multistage tectonic movements (Pang et al., 2014a, 2014b, 2015a, b, c; Yang et al. 2013). Therefore, deep strata are often superimposed on each other, and faults, fractures and unconformities are generally interconnected. Source rock often undergoes multistage hydrocarbon generation and expulsion, and reservoirs undergo multiple adjustments and transformations. All these processes have resulted in multiple hydrocarbon sources and multistage hydrocarbon charges in deep layers (Jia et al. 2015; Pang et al. 2012a, b; Hu et al. 2022a; Wang et al. 2010, 2014; Zhang 2011).

Many source rock sets can be found in superimposed basins, which generally undergo different thermal evolution stages and multiple hydrocarbon generation and expulsion stages, leading to the formation of multistage oil and gas reservoirs and abundant hydrocarbon generation and expulsion hydrocarbons in the source control accumulation periods of oil and gas reservoirs. For example, the two sets of source rock formations in the Tarim Basin experienced four main hydrocarbon expulsion periods during their evolution. However, the hydrocarbon expulsion histories of the two sets of source rocks of the Middle and Lower Cambrian

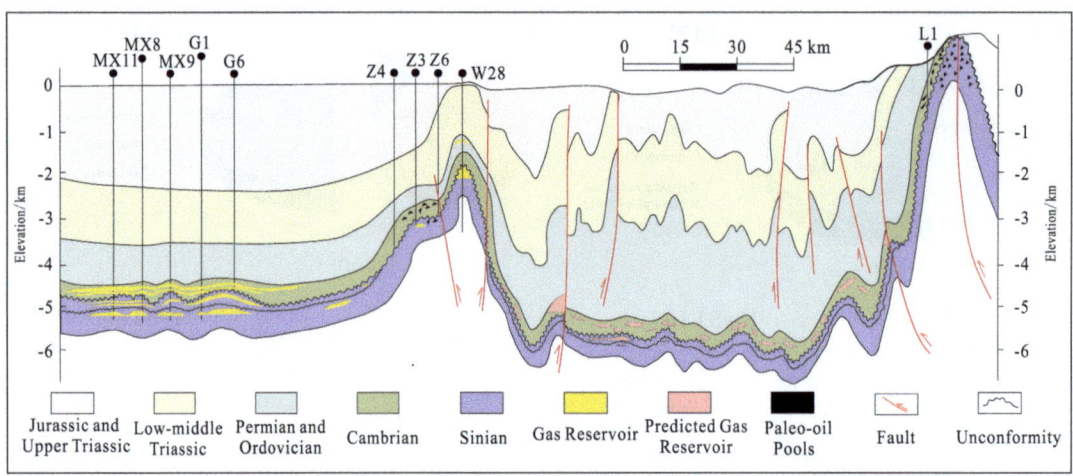

Fig. 8.7 Geological characteristics of the low-porosity and high-permeability dolomite reservoirs in Anyue Gas Field in the Sichuan Basin

and the Middle and Upper Ordovician are significantly different, as are the contributions in the four major hydrocarbon expulsion periods (Zhou et al. 2011; Li et al. 2007; Pang et al. 2010). The oil and gas of the Feixianguan Formation of Permian Changxing in northeast Sichuan are mainly deep gas with three accumulation stages and from the source rocks of the Longtan and Longmaxi formations, whereas the oil and gas of the Dengying Formation are from Cambrian and Sinian systems. They all underwent multiple hydrocarbon generation and expulsion processes. In the early stage, ancient oil reservoirs were generated and fractured into gas reservoirs due to the improvement in the degree of thermal evolution. Adjustment and transformation occurred during the Himalayan movement period, ultimately forming the present hydrocarbon distribution pattern (Fig. 8.7) (Zhang et al. 2022; Yan et al. 2011; Ma et al. 2007; LiKaika et al. 2014; Sweeney et al. 1990; Waples et al. 2000; Pang 2020).

8.2.3 Characteristics of Multidynamic and Category Accumulation

With respect to superimposed basins with multistage tectonic movements, many complex hydrocarbon reservoirs were formed by various combined dynamic actions (Pang et al. 2014c, a, b; Pang et al. 2018; Kang et al. 2019; Wang et al. 2019). Based on the correlation of various dynamic forces, Pang et al. (2021a) divided the discovered hydrocarbon reservoirs into three categories according to the types and mechanisms of dynamic forces and then subdivided them into 15 types of hydrocarbon reservoirs according to their characteristics (Fig. 8.8). They also expounded the relationship between reservoir-forming dynamics and reservoir types. The above buoyancy-driven hydrocarbon accumulation depth is dominated by buoyancy (Magoon and Dow 1994), resulting in the formation of five types of conventional hydrocarbon reservoirs. In contrast, the depths below buoyancy-driven hydrocarbon accumulation are dominated by nonbuoyancy (Jia et al. 2021), resulting in the formation of five types of tight unconventional hydrocarbon reservoirs. The five types of reformed oil and gas reservoirs were formed in some areas by external forces leading to hydrocarbon migration. All types of deep hydrocarbon reservoirs are well developed and contain nonbuoyancy-dominated unconventional oil and gas (Pang et al. 2014c, a, b). Hu et al. (2022b) illustrated the dynamic evolution of accumulation forces and periods of continuous hydrocarbons from shallow to deep strata and established a unified dynamic accumulation model.

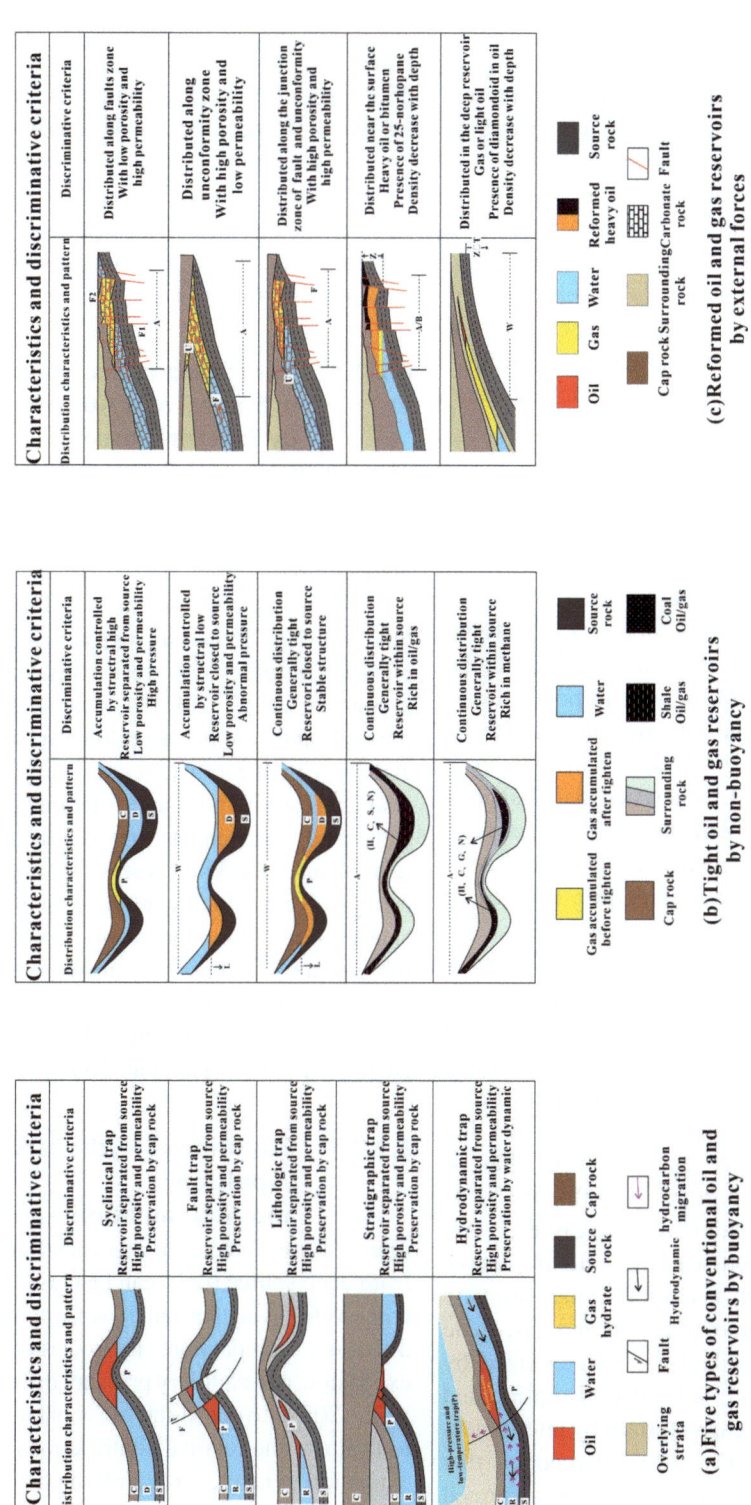

Fig. 8.8 Unified genetic classification of the hydrocarbon accumulation dynamic mechanism in the oil-bearing basin: **a** five types of conventional oil and gas reservoirs by buoyancy; **b** tight oil and gas reservoirs by nonbuoyancy; and **c** reformed oil and gas reservoirs by external forces

8.2.3.1 Conventional Hydrocarbon Accumulation Controlled by Buoyancy

The five types of conventional hydrocarbon reservoirs are dominated by buoyancy. Conventional hydrocarbons developed above the buoyancy-driven hydrocarbon accumulation depth, with the characteristics of high-point accumulation, high-level sealing, high-porosity enrichment, high-pressure reservoir formation, and source–reservoir separation, which include five types of hydrocarbon reservoirs, termed anticline, fault, lithologic, stratigraphic, and hydrodynamic seal reservoirs (Pang et al. 2021a).

8.2.3.2 Unconventional Hydrocarbon Accumulation Controlled by Nonbuoyancy

The five types of unconventional hydrocarbon reservoirs are dominated by nonbuoyancy. Relative permeability jails (Masters 1979), diagenetic traps (Cant 1986), critical pore throats (Berkenpas 1991), and lateral sealing of faults (Robert and Suzanne 2004) reduce the buoyancy effect. Hillis et al. (2001) suggested that fluid pressure differences control the hydrocarbon accumulation distribution, whereas Schenk and Pollastro (2001) suggested that medium–deep kerogen or gas from oil cracking might accumulate in unconventional reservoirs. Both porosity and permeability decrease with increasing depth (Bloch et al. 2002; Ehrenberg and Nadeau 2005), indicating that it is difficult for buoyancy to break through capillary resistance to become the main factor. A physical simulation experiment confirmed that the main driving force changes from shallow to nonbuoyancy (Guo et al. 2017a). Jia (2021) reported that unconventional oil and gas reservoirs can be formed by self-sealing, including heavy oil and bitumen with macromolecular viscosity and condensation forces, tight oil and gas, shale oil and gas, and coal bed gas characterized by capillary pressure and molecular adsorption and clathrate hydrate gas characterized by intermolecular clawing. Pang et al. (2021a) reported that the nonbuoyancy of tight oil and gas is formed from conventional source rocks characterized by low depression accumulation, low inversion, low-porosity accumulation, and low-pressure stability and is source reservoir-imminent. There are three types of reservoirs, namely, tight traps, tight deep basins, and tight composite reservoirs. The subtypes of unconventional intrasource tight oil and gas reservoirs generally have the characteristics of source–reservoir integration, extensive tightness, continuous distribution, low porosity, and low production, including coal bed oil and gas and shale oil and gas.

8.2.3.3 Hydrocarbon Accumulation in Some Areas Adjusted by External Forces

The five types of reformed oil and gas reservoirs are dominated by local external forces. They are widely distributed in superimposed basins with complex structures. There are three types of medium reformation: stress reformation fracture, fluid reformation vuggy, and joint reformation fracture–vuggy reservoirs. The physical properties of oil and gas reservoirs with high porosity and low permeability and low porosity and high permeability, respectively, are usually characterized by pore and fracture development. Reformed components are divided into high-temperature cracking reservoirs and microbial heavy oil asphalt oxidation, which are usually characterized by abnormal oil–gas viscosity and oil–gas ratios and are accompanied by abnormal oil–gas components (Pang et al. 2021a; Lorenz and Mroz 1999; Burchette 1996; Lu et al. 2017; Head et al. 2003; Hao et al. 2008).

8.2.3.4 Evolution of Hydrocarbon Expulsion and Accumulation

The driving forces for hydrocarbon expulsion from deep source rocks can be classified into five types as follows: thermal expansion, hydrocarbon diffusion, compaction, hydrocarbon-generating expansion, and capillary pressure difference. Pang et al. (2022a, b) used the mass balance principle to evaluate the relative contributions of different expulsion dynamics in Cambrian source rocks in the Tarim Basin. The results show that the relative contribution of the capillary force difference is the greatest, whereas that of diffusion is the smallest. The formation of deep oil and gas reservoirs is classified as follows: in the early stage, the amount of hydrocarbon expulsion from the middle and shallow layers is dominated by buoyancy, which results in the formation of conventional oil and gas reservoirs through migration and accumulation; in the middle and deep layers, various types of unconventional oil and gas reservoirs are dominated by a multidynamic combination; in the later stage, abnormal oil and gas reservoirs are formed mainly by the difference in the capillary force of the hydrocarbon expulsion power; and reformation-type oil and gas reservoirs are formed in areas with well-developed reformation (Pang et al. 2021a, 2022a, b).

8.2.4 Wide Distribution and Variable Production Capacity Characteristics

Most unconventional deep hydrocarbons are characterized by continuous distribution. They are widely distributed in petroliferous basins, continuously occur in sheets and zones and are not controlled by buoyancy. Hydrocarbon target intervals are generally compact and coexist with the source strata, which are more common in petroleum basins with cyclic sedimentary development and relatively stable structural evolution. Owing to its extensive and continuous distribution characteristics, the oil and gas productivities of a target zone in an exploration well also present local variation and complexity (Pang et al. 2014c, a, b, 2015a, b, c; Yang et al. 2013; Chen et al. 2016).

8.2.4.1 Continuous Development of Hydrocarbon Reservoirs

Superimposed continuous hydrocarbon reservoirs often exhibit extensive, continuous, and patchy belt distributions in oil- and gas-bearing basins because of the large distribution area and the large scale of resources and reserves. For example, the Shan 1, Shan 2, and He 8 members of the Upper Palaeozoic in the Ordos Basin formed a large area of tight gas reservoirs, with a gas-bearing range

of more than 10×10^4 km^2. The Xujiahe Formation in the Western Sichuan Depression consists of the Xinchang, Shifang, Majing, Xinduu, and Ludai oil and gas fields. The total gas-bearing area is thousands of square kilometres. However, the oil, gas, and water found in reservoirs are very complex. In addition, gas-rich areas coexist with water-cut areas. Deep continuous reservoirs are characterized by the coexistence of internal high- and low-pressure reservoirs and early- and late-formed reservoirs (Hu et al. 2022b; Pang et al. 2014c, a, b).

8.2.4.2 Deep Hydrocarbon Reservoirs Locally Enriched with Variable Productivity

Superimposed continuous reservoirs are rich in hydrocarbons in both the high and low parts of the structure. This reservoir is formed by the superimposition of a tight conventional reservoir, a tight deep basin reservoir, a tight composite reservoir, and so forth in space–time during the formation and evolution process of the limited fluid dynamic field formed in different stages and dynamic mechanisms. Vertically, it is close to the effective source rock series distribution (Pang et al. 2014c, a, b). For example, the Fuyang Reservoir in the Songliao Basin is generally a tight reservoir. In a macroenvironment, oil is enriched mainly in areas where source rocks are present (Wu et al. 2007). The target reservoirs in the 8th member of the Shihezi Formation, 1st member of the Shanxi Formation, and 2nd member of the Shanxi Formation in the Upper Palaeozoic in the Ordos Basin are closely adjacent to the effective source rocks and are distributed within the development range of the effective source rocks. These reservoirs are generally compact when the burial depth exceeds a certain critical limit. The gas-bearing layers in the high and low parts are continuous and not buoyancy-controlled (Yang et al. 2012). The tight gas reservoirs of the Xujiahe Formation in the Western Sichuan Depression of the Sichuan Basin are characterized by accumulation in both the high and low parts of structures within the gas-bearing range that are continuously distributed in the plane (Yang 2012). Studies have shown large differences in the degree of oil and gas enrichment within deep stacked and continuous oil and gas reservoirs and large variations in the productivity of the same target layer within the same oil- and gas-bearing region, with some producing up to 100,000 m^3 per day. Others produce only a few thousand cubic metres, whereas some depict only oil and gas. Others are dry layers controlled by the near-source-preferred facies-low potential energy (complex area of the same target zone). The deliverability of the target zone is related to the burial depth, structural location, reservoir porosity, fracture development, and other factors (e.g., oil and gas saturation) (Pang et al. 2014c, a, b; Yang et al., 2013).

8.3 Driving Forces and Relative Contributions to Hydrocarbon Expulsion from Deep Source Rock

8.3.1 Introduction

Tight oil and gas reservoirs are widely distributed in deep basin depressions and account for more than 85% of the total unconventional oil/gas resources (Jia 2017). High resource potential has attracted the attention of petroleum geologists and explorers for a long time (Zheng et al. 2019). Considering their different geological characteristics under different geological settings, tight reservoirs have been given different names by researchers. These names include deep basin oil/gas reservoirs, basin central oil/gas reservoirs (Law 2002), syncline oil/gas reservoirs (Wu et al. 2015), source-contacted oil/gas reservoirs, and continuous oil/gas reservoirs (Schmoker 2002), among many others. Given that these reservoirs are widely distributed in tight reservoirs and closely related to source rocks, most scholars believe that hydrocarbon migration and accumulation in tight reservoir layers are driven by nonbuoyant forces (Hu et al. 2022b; Pang et al. 2021b). However, the type of nonbuoyant force that is most important for driving hydrocarbons to migrate and accumulate in deep and tight reservoirs remains unclear. In this work, we focused on

the differences and correlations of these driving forces in oil/gas expulsion from source rocks to quantify their relative contributions to tight reservoir formation with increasing depth.

8.3.2 Classification of Driving Forces

Previous studies were performed on hydrocarbon expulsion by five types (T1–T5) and nine driving forces (F1–F9). Accordingly, sufficient evidence was proposed from each driving force. The differences in these driving forces in terms of the burial process of source rocks are shown in Fig. 8.9.

Hydrocarbon diffusion (T1) following the concentration gradient from source rocks to reservoirs is a type of driving force for hydrocarbon expulsion from source rocks (Leythaeuser 1987), expressed as F1.

The thermal expansion (T2) of minerals and fluids is considered another type of driving force for hydrocarbon expulsion from source rocks. This type of driving force can be subdivided into thermal expansion of the mineral skeleton (F2), water (F3), liquid oil (F4), and natural gas (F5) (Magara, 1975).

Source rock compaction (T3) by overlying strata is a type of driving force for hydrocarbon migration. As the porosity decreases due to compaction, the residual hydrocarbon saturation in the pores increases, and the free-phase hydrocarbons are largely discharged (Chen et al. 1989).

The product volume expanse (T4) of organic parent material pyrolysis transformation (Espitalie et al. 1980) and clay mineral dehydration (Lindgreen 1985) is regarded as the fourth type of driving force for hydrocarbon expulsion, expressed as clay dehydration (F7) and kerogen transformation to oil/gas (F8).

The difference in capillary pressure (T5) between the surrounding source rocks and adjacent reservoirs is regarded as the fifth type of driving force for hydrocarbon expulsion from the source rocks (Berg 1975) and is expressed as F9.

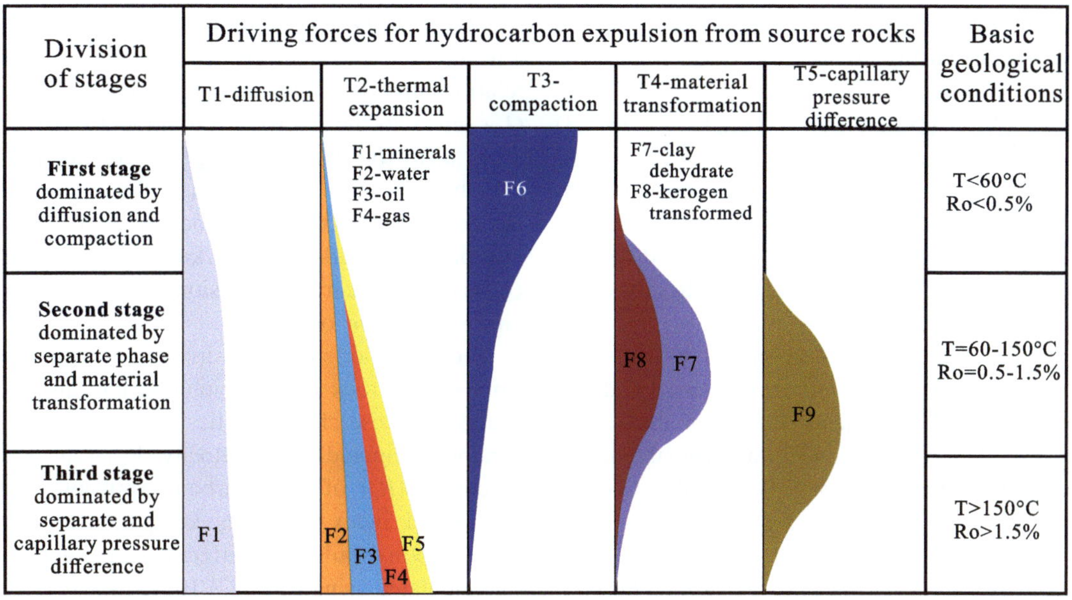

Division of stages	Driving forces for hydrocarbon expulsion from source rocks					Basic geological conditions
	T1-diffusion	T2-thermal expansion	T3-compaction	T4-material transformation	T5-capillary pressure difference	
First stage dominated by diffusion and compaction		F1-minerals F2-water F3-oil F4-gas	F6	F7-clay dehydrate F8-kerogen transformed		T<60°C Ro<0.5%
Second stage dominated by separate phase and material transformation				F8 F7	F9	T=60-150°C Ro=0.5-1.5%
Third stage dominated by separate and capillary pressure difference	F1	F2 F3 F5 F4				T>150°C Ro>1.5%

Fig. 8.9 Driving forces for hydrocarbon expulsion from source rocks confirmed by previous studies and the variation characteristics of the hydrocarbon expulsion amount under the action of each force type with increasing depth

8.3.3 Evaluation of the Contributions of Different Driving Forces for Oil/Gas Expulsion

8.3.3.1 Numerical Simulation of Hydrocarbon Expulsion in Different Phases

The numerical simulation was based on the recovery of the burial history and the thermal evolution of the source rocks. The methods used can be found in previous studies (Pang et al. 1993).

The hydrocarbon amount (Q_p) generated per cubic metre of the source rocks is related to the organic matter content (TOC), kerogen-type index (KTI), thermal evolution degree (Ro), or burial depth (Z) and density (D) of the source rocks, as expressed in Eq. (8.3.1):

$$Qp = Rp(Ro, KTI) \cdot D(z) \cdot TOC(Ro, KTI)/100 \tag{8.3.1}$$

where Q_p is the total hydrocarbon amount generated in kg/m^3 for oil or m^3/m^3 for gas, and $R_p(R_o, KTI)$ refers to the hydrocarbon amount generated in kg/t$_{toc}$ or m^3/t$_c$. The quantitative relationships among them have already been established (Peters et al. 1994).

The remaining hydrocarbon amount in the source rocks was calculated using Eq. (8.3.2):

$$Q_{ro} = \rho_o \cdot (\varphi_n + \Delta\varphi) \cdot (A_0 + A_1 \cdot (TOC) + A_2 \cdot (TOC)) \cdot \frac{1}{1 - B_k} \cdot e^{-\frac{\varphi_n}{D}(R_o - R')^2} \tag{8.3.2}$$

where Q_{ro} refers to the total hydrocarbon amount remaining, in kg/m^3 or m^3/km^3; Φn is the source rock porosity with normal compaction in %; $\Delta\Phi$ is the extra porosity of the source rock with abnormal compaction, %; R_o refers to the critical vitrinite reflectance corresponding to the maximum value of the residual hydrocarbon peak in %; ρ_o is the crude oil density in t/m^3; A_0, A_1, A_2, and D are constant values determined by a statistical analysis with the best fit between the actual data "S_1" or "A" and the calculation model in the study area; and B_k is the factor of light hydrocarbon compensation, which is the ratio of lost-to-residual light hydrocarbons quantitatively related to the major factors established previously (Pang et al. 1993).

The total residual gas amount in the source rock is calculated using Eq. (8.3.3):

$$Q_{rg} = (Q_{rgb} + Q_{rgo} + Q_{rgw}) \qquad (8.3.3)$$

where Q_{rg} is the total hydrocarbon amount remaining, in kg/m^3 or m^3/km^3, including the adsorbed gas amount of Q_{rgb}, the oil-dissolved gas amount of Q_{rgo}, and the water-dissolved gas amount of Q_{rgw}.

The retained hydrocarbon components are divided into the same three groups: liquid oil, heavy gas, and methane. Based on the above, the expelled hydrocarbon amount from the source rock in different phases is calculated using Eq. (8.3.4):

$$Q_e = R_e(Ro, KTI) \cdot D(z) \cdot TOC(Ro, KTI)/100 \qquad (8.3.4)$$

where Q_e is the total hydrocarbon amount expelled, in kg/m^3 for oil or m^3/m^3 for gas, and $R_e(R_o, KTI)$ is the hydrocarbon amount expelled, kg/T toc or m^3/Tc, which changes with KTI and R_o. Quantitative relationships among them have already been established (Pang et al. 1993). The hydrocarbon expulsion threshold (HET) of the source rock is determined in Eq. (8.3.5) to correspond to the critical conditions, where the generated hydrocarbon amount (Q_p) is equal to the residual hydrocarbon amount (Q_{rm}). The hydrocarbon amount expelled from the source rock as four different phases is calculated using Eqs. (8.3.6) to (8.3.9):

$$Q_p \geq Q_{rm} = Q_{ro} + Q_{rg} \qquad (8.3.5)$$

$$Q_{ew} = V_w \cdot \sum_1^4 q_w(i) \qquad (8.3.6)$$

$$Q_{ed} = \sum_1^4 \int_0^t D(i, T, \emptyset) \cdot \frac{d_c}{d_z} \cdot d_t \qquad (8.3.7)$$

$$Q_{eog} = Q_{eo} \cdot \sum_1^4 q_o(i) \qquad (8.3.8)$$

$$Q_{es} = Q_e - Q_{ew} - Q_{ed} - Q_{eog} \qquad (8.3.9)$$

where Q_{ew} is the hydrocarbon amount expelled as a water solution, in kg/km^2 or m^3/km^2, and the relationships are confirmed (Price et al. 1977). Q_{ed} refers to the hydrocarbon amount expelled as a diffusion phase, in kg/km^2 or m^3/km^2 (Leythaeuser et al. 1982). Q_{eog} is the amount of gas expelled from the source rocks as an oil–solution phase, in kg/km^2 or m^3/km^2 (Bruce 1984). Q_{es} is the hydrocarbon expulsion amount as a free phase, in kg/km^2 or m^3/km^2.

The characteristics of hydrocarbon expulsion from source rock during evolution can also be analysed. These characteristics are expressed as parameters, including the velocity (V_e), rate (S_e), efficiency (R_e) of hydrocarbon expulsion, and source rock index (SRI). The key parameters for hydrocarbon expulsion in different depth intervals are calculated using Eqs. (8.3.10) to (8.3.13).

$$V_e = \frac{\Delta Q_e}{\Delta Z} \times 100 \qquad (8.3.10)$$

$$S_e = \frac{\Delta Q_e}{H} \qquad (8.3.11)$$

$$R_e = \frac{Q_e}{Q_P} \qquad (8.3.12)$$

$$SRI = \frac{Q_{es}}{Q_e} \times 100 \qquad (8.3.13)$$

where V_e is the hydrocarbon expulsion velocity or the hydrocarbon amount expelled per square metre of the source rock and varies with $\triangle Q_e$ and depth. The SRI is the change in the source rock index with respect to the total amount of oil/gas expelled from the source rock as a separate phase (Q_{es}) and the total amount of oil/gas expelled in different phases (Q_e). The larger the SRI value is, the greater the source rock contribution to oil/gas reservoir formation.

8.3.3.2 Numerical Simulation of the Relative Contributions of the Nine Driving Forces

The hydrocarbon amount expelled from the source rock by the nine driving forces was calculated according to the corresponding relationships among these driving forces and the oil/gas amounts expelled in different phases.

The volume of liquid fluid expelled from the source rocks by the seven driving forces was calculated as follows: the volume of fluid expelled by each of the seven driving forces is denoted as $\triangle V_{fi}$, where i = 1, 2, ..., 8. The fluid volume ($\triangle V_{f2}$) expelled by F_2 from stages 1 to 2 is calculated using Eq. (8.3.14):

$$V_{f2} = (Ksr_2 - Ksr_1) \cdot (1 - \Phi_0) \qquad (8.3.14)$$

where Ksr is related to the expansion coefficient of the skeleton content $(1 - \Phi_0)$ at stages 1 and 2. The fluid volume ($\triangle V_{f3}$) related to F_3 from stages 1 to 2 was calculated using Eq. (8.3.15) (Barker 1984). The volume of fluids ($\triangle V_{f4}$) driven by F_4 was calculated using Eq. (8.3.16) (Magara 1976). $\triangle V_{f5}$ was calculated using Eq. (8.3.17) (Magara 1976). $\triangle V_{f6}$ was calculated using Eq. (8.3.18) and is related to the changes in porosity (Φ) and thickness (H) (Lee et al., 2018). $\triangle V_{f7}$ was calculated using Eq. (8.3.19) (Vernik et al. 1996). $\triangle V_{f8}$ related to F_8 was calculated using Eq. (8.3.20) (England et al. 1987).

$$V_{f3} = (Khw_2 - Khw_1) \cdot (\frac{\varphi_0 - \varphi_2}{1 - \varphi_2} - \frac{\varphi_0 - \varphi_1}{1 - \varphi_1})$$
$$(8.3.15)$$

$$V_{f4} = (Kho_2 - Kho_1) \cdot (\frac{\varphi_0 - \varphi_2}{1 - \varphi_2} - \frac{\varphi_0 - \varphi_1}{1 - \varphi_1}) \cdot S_o$$
$$(8.3.16)$$

$$V_{f5} = (Khog_2 - Khg_1) \cdot (\frac{\varphi_0 - \varphi_2}{1 - \varphi_2} - \frac{\varphi_0 - \varphi_1}{1 - \varphi_1}) \cdot S_G$$
$$(8.3.17)$$

$$V_{f6} = (\frac{\varphi_0 - \varphi_2}{1 - \varphi_2} - \frac{\varphi_0 - \varphi_1}{1 - \varphi_1}) \cdot H \qquad (8.3.18)$$

$$V_{f7} = (D_2 - D_1) \cdot (TOC_2 - TOC_1) \cdot K_v \qquad (8.3.19)$$

$$V_{f8} = 0.245 \cdot (C_{lay2} - C_{lay1}) \cdot (I_{m2} - I_{m1})$$
$$\cdot (D_2 - D_1) \qquad (8.3.20)$$

The inverse calculation of the driving forces for fluid expulsion from the source rock followed the improved Darcy's law (Germann 2018) using Eq. (8.3.21). The fluid viscosity (μi) was replaced by the water viscosity because the water volume

was the highest among the expelled fluid volumes, accounting for 70%.

$$F_i = \left[dV_i \cdot \mu w \cdot \frac{H}{dk_w \cdot d_t} \right] \cdot 10^{12} (i = 2, 3, \ldots, 8)$$
$$(8.3.21)$$

The ninth driving force (F9) was calculated as follows using Eqs. (8.3.22) and (8.3.23):

$$F_{9o} = Pc_{W/O} = 2 \cdot \delta_{W/O} \cdot \cos\theta \cdot \left(\frac{1}{r} - \frac{1}{R} \right)$$
$$(8.3.22)$$

$$F_{9g} = Pc_{W/G} = 2 \cdot \delta_{W/G} \cdot \cos\theta \cdot \left(\frac{1}{r} - \frac{1}{R} \right)$$
$$(8.3.23)$$

where Pc is the capillary pressure in the source rock; r is the throat radius of the surrounding rocks; R is the throat radius of the reservoir; and θ is the wetting angle of the hydrocarbon/water. The θ value was set to 0 in the calculation. The capillary pressures for oil (F_{9o}) and gas (F_{9g}) in the rocks were calculated separately and are denoted by Eqs. (8.3.22) and (8.3.23), respectively.

The amount of oil/gas expelled from the source rock by the nine driving forces was sequentially calculated. The hydrocarbon amount of Q_{ed} was contributed by F_1, denoted as Q_{e1}, and was calculated using Eq. (8.3.24) as a function of the diffusion coefficient (D), hydrocarbon concentration gradient (dc/dz), diffusion area (S), source rock porosity (φ),, and diffusion period (t). The amount of oil/gas (Q_{ei}) expelled by each of the other eight driving forces was calculated using Eq. (8.3.25). The related oil/gas amount (Q_{ew}) was expelled by the combination of $\sum Fi$, i = 2, 3,...8, and the oil/gas amount of Q_{eog} and Q_{es} contributed by the combination of $\sum Fi$, i = 2, 3,... 9.

$$Q_{e1} = D \cdot \frac{dc}{dz} \cdot \frac{1 - \varphi_0}{1 - \varphi_2} \cdot s \cdot \varphi_2 \cdot \Delta t \qquad (8.3.24)$$

$$Q_{ei} = (Q_{es} + Q_{eog})/F_i + Q_{ew}/F_j$$
$$(j = 2, 3, \ldots, 8; i = 2, 3, \ldots, 9)$$
$$(8.3.25)$$

Table 8.2 Geological parameters of the Cambrian source rocks in Tarim Basin

Geological parameters	Source rock thickness (m)	Total organic carbon TOC (%)	Kerogen type index (KTI)	Maturity Ro (%)	Gradient thermal (°C/100 m)
Maximum	450	0.2	50	0.8	2.7
Minimum	0	3.3	97	4.1	2.3
Mean	200	1.5	85	1.8	2.5
Source	Li et al. (2010, 2015), Shen et al. (2018)	Li et al. (2010, 2015), Shen et al. (2018)	Pang et al. (1992, 1993), Li et al. (2010, 2015)	Li et al. (2010, 2015), Shen et al. (2018)	Han et al. (2012), Hu et al. (2015), Liu et al. (2015)

Finally, the relative contributions of the nine driving forces were obtained by the ratio of the hydrocarbon amount by each driving force to the total hydrocarbon amount by the nine driving forces expressed using Eq. (8.3.26).

$$R_i = \frac{Q_{ei}}{\sum_{i=1}^{9} Q_{ei}} (i = 2, 3, \ldots, 8) \qquad (8.3.26)$$

8.3.3.3 Geological Settings

The Tarim Basin is rich in deep-seated oil and gas resources. Here, oil and gas are concentrated mainly in the Palaeozoic marine strata. The driving forces for hydrocarbon expulsion and their relative contributions to Cambrian source rocks were simulated in this work. The simulations were related to many geological parameters, five of which were essential. The parameter characteristics are listed in Table 8.2. Moreover, the other parameters used in the above equations were obtained from experiments, previous works, and actual data collected from the study area (Table 8.3).

8.3.4 Results

8.3.4.1 Hydrocarbon Expulsion in Four Phases from the Source Rocks

Figure 8.10 illustrates the case study results of the hydrocarbon expulsion from the source rocks of the Tarim Basin in four different phases. Figure 8.10a shows the variation characteristics of hydrocarbon generation, retention, and expul-

Table 8.3 Source of the driving force parameters

Driving force parameters		Source
Diffusion coefficient		Leythaeuser et al. (1982)
Thermal expansion coefficients	Rock	Magara et al. (1975b)
	Water	Barker (1984)
	Liquid oil	Magara (1975a)
	Natural gas	Magara (1975a)
Volume expansion coefficient	Kerogen products	Espitalie et al. (1980)
	Clay minerals dehydration	Dickinson et al. (1966)
Natural gas adsorption coefficient		Danial et al. (2007)
Hydrocarbon dissolution coefficient in water		Price (1976)
Gas dissolution coefficient in oil		Neglia et al. (1979)

sion with the increasing depth for methane (A1), heavy gas (A2), liquid hydrocarbons (A3), and *SRI* (A4), clarifying the relationships among the generated, retained, and expelled oil/gas amounts per cubic metre of source rock during its evolution. Figure 8.10b depicts the variation characteristics of the hydrocarbon expulsion along with the increasing depth. Oil/gas expulsion was characterized by four parameters, namely, the cumulative expelled hydrocarbon amount (B1), hydrocarbon expulsion velocity (B2), hydrocarbon expulsion rate (B3), and oil/gas expulsion efficiency (B4). Figure 8.10c shows the variation of the relative hydrocarbon amounts (%) from the source rocks expelled in four phases in the same depth interval, including methane gas, heavy hydrocarbon gas, and liquid hydrocarbon.

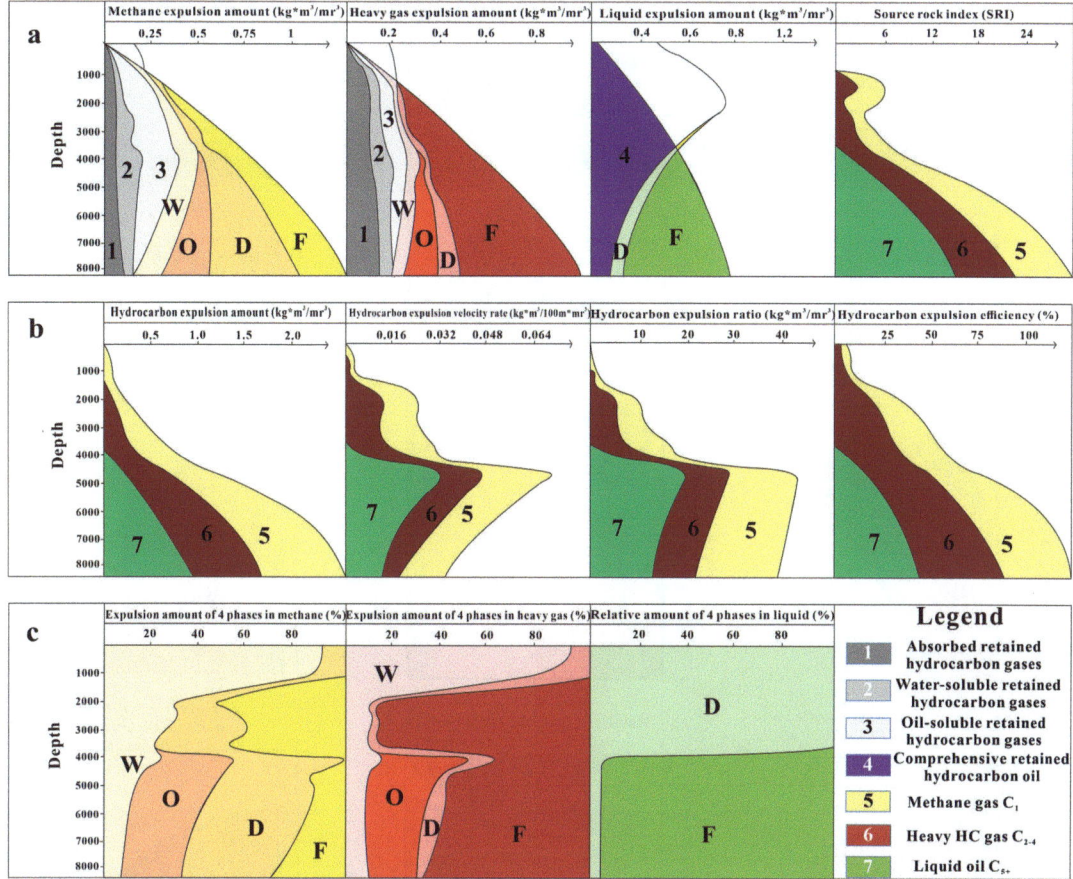

W, D, O, F refer to 4 phases of hydrocarbon expulsion in water-solution, diffusion, oil-solution and in free

Fig. 8.10 Case study of the numerical simulation results for hydrocarbon generation, retention, and expulsion in four different source rock phases in the Tarim Basin with essential parameters of TOC = 1.5%, KTI = 85, Ro = 1.36%, H = 200 m, and GT = 2.5 °C/100 m: **a** variation in hydrocarbon generation, retention, and expulsion by a cubic metre with increasing depth for methane (A1), heavy hydrocarbon gas (A2), liquid (A3), and source rock index (SRI); **b** variation in hydrocarbon expulsion characteristics, including the expulsion amount (B1), hydrocarbon expulsion velocity rate (B2), hydrocarbon expulsion rate (B3), and hydrocarbon expulsion efficiency (B4); and **c** variations in the relative oil/gas amounts expelled in four different phases from the source rocks, including methane (C1), heavy gas (C2), and liquid (C3). W, D, O, and F are the relative amounts of hydrocarbons expelled from the source rocks in the water solution, diffusion, oil solution, and free phase, respectively

8.3.4.2 Relative Contributions of the Nine Driving Forces to Oil/Gas Expulsion

Figure 8.11 illustrates the variation characteristics of the total water, natural gas (methane and heavy HC gas), and liquid oil expelled per cubic metre of source rock by the nine driving forces with increasing depth. Figure 8.11a shows the variation characteristics of the expelled water amount expressed as the instantaneous amount in the 100 m depth interval burial (A1), relative amount (A2), and cumulative amount (A3). Figure 8.11b displays the variation characteristics of the expelled gas amount expressed as an instantaneous amount in the 100 m depth interval burial (B1), relative amount (B2), and cumulative amount (B3). Figure 8.11c depicts the variation characteristics of the expelled liquid oil amount charac-

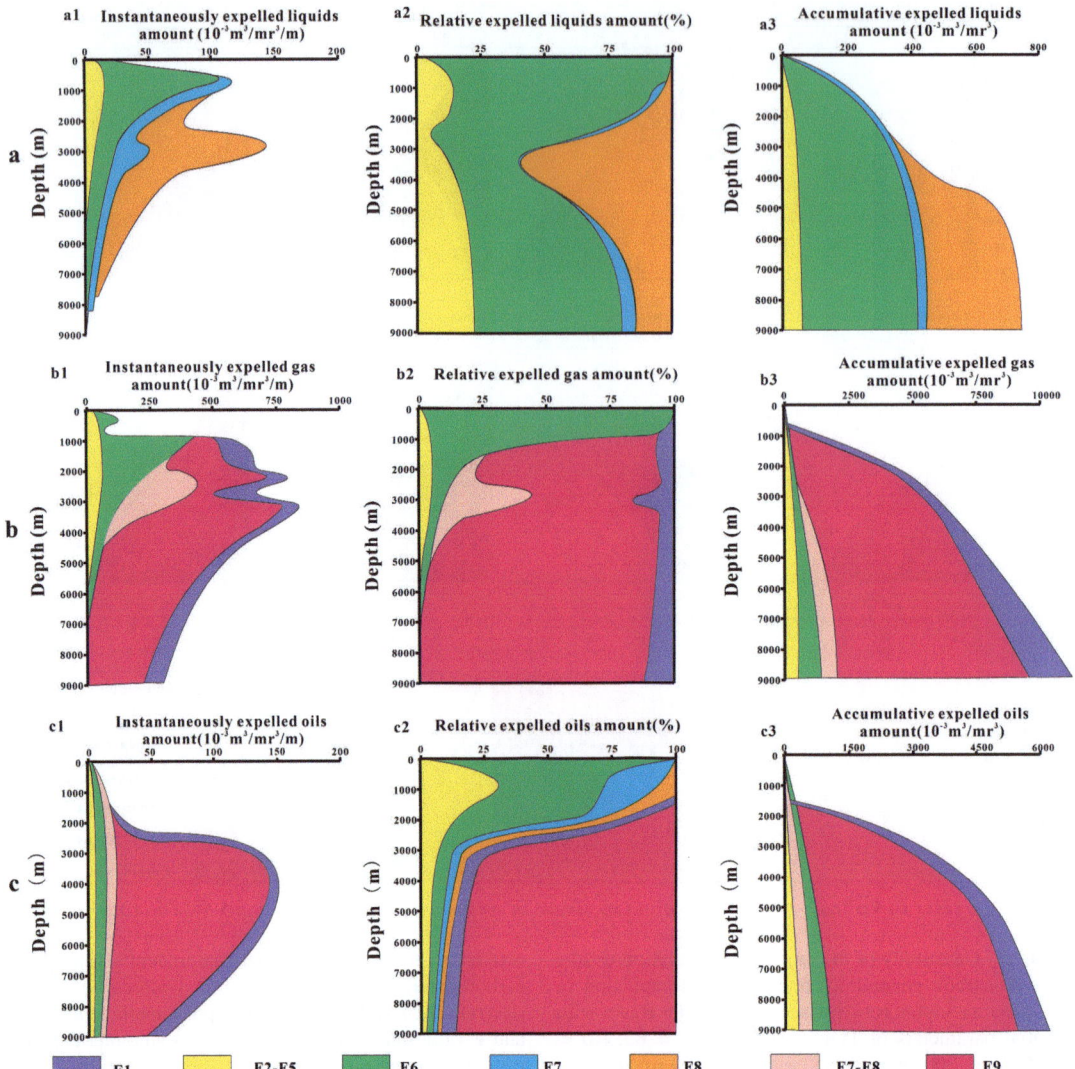

Fig. 8.11 Numerical simulation results for the fluid expulsion of water, oil, and gas by nine driving forces from the source rocks and their relative contributions with increasing burial depth: **a** variation characteristics of the expelled water amount from the source rock, including its instantaneous amount (A1), relative amount (A2), and cumulative amount (A3); **b** variation characteristics of the expelled gas amount, including its instantaneous amount (B1), relative amount (B2), and cumulative amount (B3); and **c** varia-tion characteristics of the expelled oil amount, including its instantaneous amount (C1), relative amount (C2), and cumulative amount (C3). F1 depicts the hydrocarbon diffusion. F2–F5 are the thermal volume expansions of the mineral skeleton, water, liquid oil, and natural gas, respectively. F6 represents compaction by the overlying strata. F7 and F8 are the product volume expansions induced by clay dehydration and kerogen transformation to oil/gas, respectively. F9 denotes the difference in capillary pressure

terized by instantaneous (C1), relative (C2), and cumulative (C3) amounts. The CPD contributed the most to the expelled gas amount (>80%). The contribution of diffusion was approximately 5%, and that of clay and kerogen transformation was 5%. The compaction (4%) and thermal expansion of the rock skeleton and fluids (<5%) contributed the least. The contributions of the nine driving forces to the expelled oil amount were similar to those to the expelled gas amount.

8.3.4.3 Dynamic Model for Oil/Gas Expulsion from Source Rocks with Increasing Depth

The dynamic process for hydrocarbon expulsion from the source rocks was divided into four stages with increasing depth (Fig. 8.12).

The first stage is from source rock deposition at the surface to the HET underground, which is dominated by the compaction of the overburden strata. Most of the oil/gas was expelled in the water–solution and diffusion phases. The relative contributions of compaction, diffusion, thermal expansion, and product volume expansion were approximately 40%, 25%, 20%, and 15%, respectively. This was unfavourable for oil/gas migration and accumulation in the reservoir layer.

The second stage is from HET to the LHED. Both oil and gas were expelled from the source rocks in the four different phase states. Most of the gas was expelled in the free phase, with some gas migrating in the oil solution dominated by multiple driving forces.

The third stage began with source rocks entering the LHED and lasted until the ASDL, indicating the end of hydrocarbon generation and expulsion. Natural gas with little liquid oil was expelled from the source rocks in the free phase.

More than 50% of the natural gas was expelled in this stage by the CPD.

8.3.5 Conclusions

Based on the study and discussion, the following two basic conclusions are drawn regarding the hydrocarbon expulsion dynamics of deep source rocks and the dynamic mechanism of unconventional tight oil and gas accumulation:

1. Continuous tight oil/gas reservoirs are widely developed in deep petroliferous basins. The distribution of the dynamic mechanism for hydrocarbon expulsion from source rocks was evaluated in this research. The results demonstrated that the effective CPD is the most important driving force for deep tight oil/gas reservoir formation.

2. The dynamic process and the variation characteristics of oil and gas expulsion from the source rocks were divided into four stages. The CPD did not contribute at the first stage before entering the HET, varied from 25 to 50% at the second stage, contributed more than 50% at the third stage in oil/gas expulsion, and contributed little at the fourth stage for the reformation of

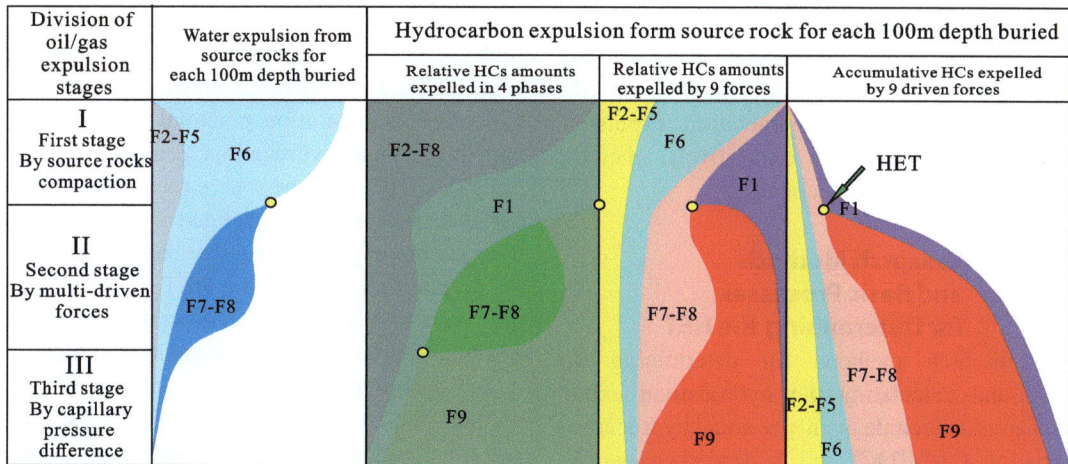

Fig. 8.12 Dynamic model of hydrocarbon expulsion in the four phase states from the source rock and stage division by the variation characteristics of the contributions of the driving forces with increasing depth. The Fs definitions are similar to those in Fig. 3.3

the reservoirs formed in the earlier stages due to tectonic movements.

8.4 Accumulation Dynamic Boundary and Geological Significance of Deep Hydrocarbon Reservoirs

8.4.1 HET of the Source Rock and the Earliest Time of Hydrocarbon Accumulation

8.4.1.1 Concept and Criterion of the HET

During the source rock burial and thermal evolution processes, the hydrocarbons generated from organic matter (OM) satisfy retention requirements, such as self-adsorption, water and oil dissolution, and capillary saturation, and are subsequently expelled from the source rocks in the free phase (Pepper 1992; Pang et al. 1995, 2005), as shown in Eq. (8.4.1). Pang et al. (2005) defined the geologic conditions at which hydrocarbon expulsion began in the free phase as the HET.

$$Q_e = Q_p - Q_r \begin{cases} = 0, Z \leq HET \\ > 0, Z > HET\# \end{cases} \quad (8.4.1)$$

where Z is the sample location at a given depth or measured by thermal maturity using vitrinite reflectance (Ro), and Q_p, Q_e, and Q_r are the amounts of hydrocarbon generation, expulsion, and residual hydrocarbon, respectively, when the source rock is buried at depth Z.

8.4.1.2 Research Methods and Basic Processes for Determining the HET

The research methods for determining the HET and calculating the hydrocarbon expulsion amount include mainly residual hydrocarbon variation ("A"/TOC) and hydrocarbon generation potential variation analyses and physical and numerical simulation experiments (Pang et al.,

2014c, a, b). Experimental measurements and basin modelling are two common methods. If the generated (Q_p) and residual (Q_r) hydrocarbon amounts in the source rocks are known, the expelled hydrocarbon (Q_e) amount is estimated using Eq. (8.4.1) based on the principle of mass balance. Whereas the residual hydrocarbon amount can be directly measured, the estimation of the generated hydrocarbon number is often problematic because the results vary with the different approaches used.

The Rock–Eval S_1 and S_2 peaks represent the free and pyrolyzed hydrocarbons in the rock, respectively. The hydrocarbon generation potential index is defined as ($S_1 + S_2$)/TOC × 100, reflecting the current amount of hydrocarbons generated per TOC unit in the source rock (Pang et al. 2005). The measured TOC value in a rock that experienced expulsion is generally lower than its original TOC value prior to hydrocarbon generation, depending on its maturity level. If we assume that the measured TOC represents its original TOC value, then a maximum value for the ($S_1 + S_2$)/TOC ratio is expected to occur at the early stage of thermal maturation. Herein, we considered that the maximum ($S_1 + S_2$)/TOC value represents the original total hydrocarbon potential. The hydrocarbon generation potential reaches its maximum value at the HET. Once a rock is buried beyond the depth of the expulsion threshold, the ($S_1 + S_2$)/TOC value in the rock decreases due to the hydrocarbon expulsion (Fig. 8.13). The difference between the original hydrocarbon generation potential index (HCI_o) and the residual hydrocarbon generation potential index (HCI_p) of the source rock is the hydrocarbon expulsion rate q_e, defined as the amount of hydrocarbon expelled per unit organic carbon after the source rock reaches the HET. Q_e represents the cumulative amount of hydrocarbons expelled from the source rock over the geological history.

In accordance with the abovementioned method, the basic flow of determining the HET and calculating the amount of hydrocarbon expulsion is as follows: (1) the Rock–Eval data of the source rocks in the study area are collected,

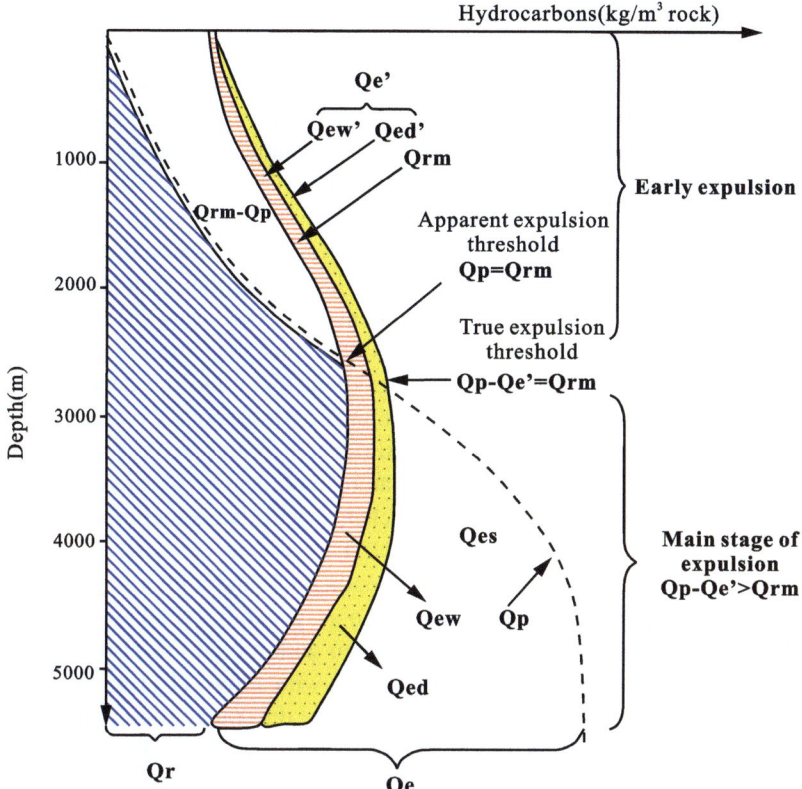

Fig. 8.13 Conceptual model of the quantities of generated and expelled hydrocarbons: Qp = amount of hydrocarbon generated; Qe = amount of hydrocarbon expelled; Qr = amount of residual hydrocarbon in the source rock; Qrm = minimal amount of hydrocarbon needed to saturate the source rock porosity; Qew = amount of hydrocarbon dissolved in water; Qed = amount of hydrocarbon diffused out of the source rock; Qes = amount of hydrocarbon expelled in a separate phase; Qe' = amount of hydrocarbon expelled early; Qew' = amount of hydrocarbon dissolved in water during early expulsion; and Qed' = amount of hydrocarbon diffused out of the source rock during early expulsion

and their hydrocarbon generation potential index profile is established; (2) the HET is determined, and the hydrocarbon expulsion rates of the source rocks are calculated as follows at different burial depths according to the variation characteristics and profile trends:

$$q_e = HCI_o - HCI_p(Z); \quad (8.4.2)$$

(3) the thickness, TOC content, and density of the source rocks are combined, and the hydrocarbon expulsion intensity is calculated as follows:

$$E_{hc} = \int_{Z_0}^{Z} 10^{-1} q_e(Z) H \rho(Z) TOCd(Z); \quad (8.4.3)$$

and (4) based on the calculation results of the hydrocarbon expulsion intensity, the amount of hydrocarbon expulsion is obtained as follows by area integration:

$$Q_e = \int_{1}^{n} \int_{Z_0}^{Z} 10^{-5} q_e(Z) HS(n) \rho(Z) TOCdZdn$$

$$(8.4.4)$$

where E_{hc} is the hydrocarbon expulsion intensity, t/km^2; $q_e(Z)$ is the hydrocarbon expulsion rate per unit TOC, mg/g; Q_e is the hydrocarbon expulsion amount, t; Z_0 is the HET, m; $\rho(Z)$ is the source rock density; H is the source rock thickness, m; and S(n) is the source rock area, m^2.

8.4.1.3 Main Controlling Factors and Variation Characteristics of the HET

The HET is mainly controlled by the hydrocarbon generation capacity, residual hydrocarbon capacity, and geothermal gradient of the basin (Pang et al. 2014c, a, b). Under the condition that the other source rock conditions remain unchanged, the HET changes with different types of OM. High-abundance source rocks with good OM type have earlier HET, whereas high-abundance source rocks with poor OM type have earlier gas expulsion than oil does. The HET of type II OM is between those of types I and III. With the other conditions remaining unchanged, the HET of the source rocks advances as the TOC increases. However, the effect of the TOC on the HET of the source rocks is not as great as that of the OM type because the increase in OM abundance leads to increases in both hydrocarbon generation and residual hydrocarbons. The HET of source rocks with good OM types is shallow. Furthermore, the variation range of the threshold with increasing OM abundance is small. However, the HET of source rocks with poor OM types is deep and greatly varies with increasing OM abundance. In addition, the higher the residual hydrocarbon capacity of the source rock is, the later the HET. The residual hydrocarbon capacity of the source rock is controlled mainly by the hydrocarbon solubility in the pore fluid, the capacity of the source rock to adsorb hydrocarbons, and capillary forces. Similarly, the hydrocarbon expulsion thresholds of the source rocks in different basins differ because the burial and thermal histories of compaction differ.

8.4.1.4 Variation Range and Prediction Method of the HET

Different types of OM have different hydrocarbon generation potentials and hydrocarbon expulsion thresholds. Taking the Songliao Basin as an example, the threshold depth of liquid hydrocarbon expulsion from source rocks with type III OM (TOC = 1%; KTI = 0) is 3000 m. At OM KTI values of 25%, 50%, 75%, and 100%, the hydrocarbon expulsion thresholds of the source

rocks are approximately 2300, 1300, 500, and 300 m, respectively. The fundamental reason for this phenomenon is that the OM type improves, consequently increasing the hydrocarbon generation amount of the source rock and causing it to reach the critical saturation amount of the residual hydrocarbon required before massive hydrocarbon expulsion in advance.

The geothermal heat flow of different sedimentary basins during different geological periods differs; thus, the variation rates of the geothermal gradient (GT) and the degree of thermal evolution (Ro) of the source rock with burial depth also differ across the basins. The formation temperature and degree of thermal evolution greatly change with burial depth at high heat flow. The HET of source rocks with high heat flow, high geothermal, and a high degree of thermal evolution is generally shallow. Taking the Songliao Basin with a high GT (1.6 HFU) and the Hailar Basin with a low GT (1.4 HFU) as examples, the HETs of the source rocks with similar conditions (TOC = 2%, KTI = 50, and Ro = 0.5%) in the Songliao Basin are shallower than those in the Hailar Basin (i.e., 1050 and 1400, respectively). Source rocks can easily expel hydrocarbons in sedimentary basins with high compaction rates. Under the same hydrocarbon generation conditions, a basin with a high compaction rate is more conducive to achieving a higher hydrocarbon saturation and reaching the critical saturation of the residual hydrocarbon required for hydrocarbon expulsion in advance to reach the HET.

8.4.1.5 Significance and Enlightenment of the HET

The concept of the HET not only indicates the objectivity of the critical geological conditions of hydrocarbon expulsion from source rocks but also reveals the relationships among hydrocarbon generation, residual hydrocarbon generation, hydrocarbon expulsion, and critical geological conditions of hydrocarbon expulsion from source rocks. The critical conditions of hydrocarbon expulsion mentioned by foreign scholars can be solved by applying the HET theory by simulating the lower limit standard of the minimum organic carbon content of hydrocarbon expulsion and the critical oil saturation of hydrocarbon expulsion. In addition, new oil and gas source rock concepts can be established by applying HET theory. The amount of oil discharged from the source rock can be calculated, the grade evaluation standard can be determined, the phase state of oil and gas discharged from source rock can be studied, and the mechanism of oil and gas migration and accumulation and the accumulation mode can be analysed. In addition, the stages of oil and gas discharge from the source rock can be divided, and oil and gas field exploration can be guided. After determining the HET of the source rocks, the distribution range of effective source rocks can then be delineated, and their hydrocarbon expulsion intensity and amount can be calculated to provide theoretical guidance for the optimization of favourable zones and the evaluation of resource potential. According to the hydrocarbon expulsion threshold, the spatial distribution of hydrocarbon source rocks can be effectively confirmed. Traps that may accumulate

oil and gas can also be further searched to improve exploration accuracy.

8.4.2 Buoyancy-Driven Hydrocarbon Accumulation Depth and Maximum Burial Depth of Conventional Oil/Gas Reservoirs

8.4.2.1 Concept and Criterion of Buoyancy-Driven Hydrocarbon Accumulation Depth

The buoyancy-driven hydrocarbon accumulation depth (BHAD) is defined as the critical condition corresponding to the change in the hydrocarbon driving force, where buoyancy-dominated hydrocarbon migration and accumulation change to nonbuoyancy-dominated migration and accumulation as the burial depth of the reservoir layer increases (Pang et al. 2021b; Fig. 8.14). Below the BHAD, the total effects of nonbuoyancy factors,

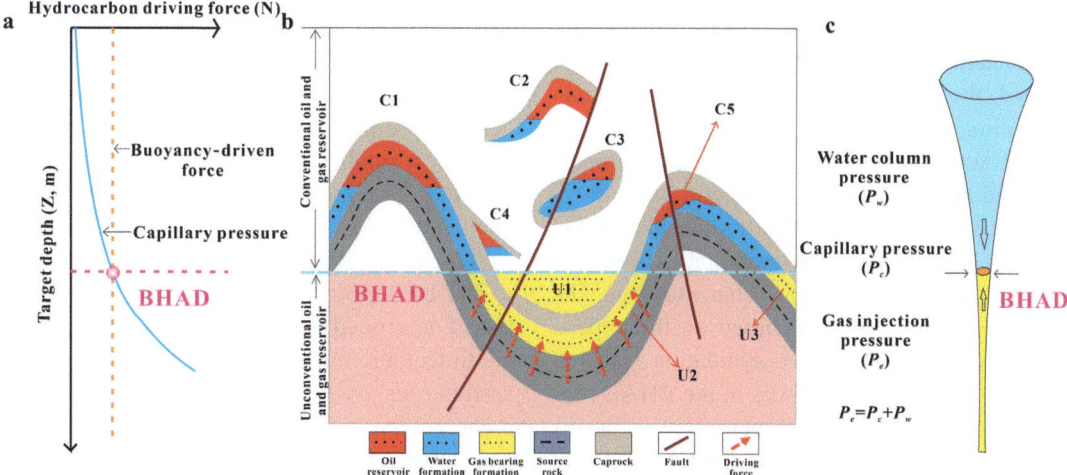

Fig. 8.14 Concept model of the buoyancy-driven hydrocarbon accumulation depth (BHAD) and how it controls the distribution of different hydrocarbon accumulations. **a** During the burial process, buoyancy is a dominant force for hydrocarbon accumulation in reservoir layers and does not change with depth. The capillary pressure increases with depth, resulting in a change in the hydrocarbon migration driving force from a buoyancy-dominated force above the BHAD to a capillary-dominated force (i.e., a

nonbuoyancy force) below it. **b** Changes in the driving force with depth lead to the formation and accumulation of conventional hydrocarbon reservoirs above the BHAD and the formation and accumulation of unconventional hydrocarbon reservoirs below it. **c** The pore throat radius in the reservoir layers decreases with increasing depth, resulting in the formation of hydrocarbon accumulation traps above the BHAD and a continuous hydrocarbon distribution in the reservoir layers below it

such as strata compaction, capillary forces (Wang et al. 2020), fluid and mineral thermal expansion, volume expansion due to clay and OM transformation, and hydrocarbon diffusion due to the concentration gradient, are more significant than the effects of buoyancy (Fig. 8.14a). The BHAD is identified by comparing the features of the conventional reservoirs formed by buoyancy within the reservoir layers above the BHAD and the features of the unconventional reservoirs formed by nonbuoyant forces below the BHAD (Fig. 8.14b). The maximum pore throat radius in the reservoir layers decreases with increasing burial depth, resulting in the disappearance of buoyancy-driven hydrocarbon accumulation at the BHAD. The driving force for hydrocarbon migration and accumulation is then changed to nonbuoyant forces below the BHAD (Fig. 8.14c).

8.4.2.2 Research Method and Basic Process of the BHAD

Three steps were taken to study the BHAD and predict its distribution in petroliferous basins. First, the hydrocarbon reservoirs discovered in petroliferous basins must be analysed through drilling results. Discriminant criteria were proposed to determine the BHADs in these basins. Second, the formation mechanism and the main controlling factors for the BHAD are revealed by two sets of physical simulation experiments designed to simulate the formation conditions, dynamic mechanics, and variations in the BHAD with geological factors. A quantitative dynamic force equilibrium equation characterizing the BHAD is then established based on the experimental results. Quantitative simulation research on the variation characteristics of the BHADs for the major reservoir layers in petroliferous basins is conducted based on the actual geological data and parameters. Finally, a model of the BHAD controlling hydrocarbon reservoir distribution is established and applied to explain the results obtained from geological analysis and physical and numerical simulations.

Some evidence from the following aspects has indicated that BHAD limits exist between conventional and unconventional hydrocarbon reservoirs in petroliferous basins: (1) differences in the distri-

bution of hydrocarbon reservoirs; (2) differences in accumulation in a petroleum system; (3) differences in accumulation in a petroliferous basin; (4) differences in pressure in hydrocarbon reservoirs; and (5) differences in the media of hydrocarbon reservoir layers, as discussed below.

8.4.2.3 Main Controlling Factors and Variation Characteristics of the BHAD

The BHADs in different basins are very different. BHAD variations are controlled by four geological factors: the lithology of reservoir layers, hydrocarbon components, tectonic movements, and heat flow in a petroliferous basin.

The BHAD increases with increasing grain size of the reservoir sandstone layers. The drilling results of the Upper Palaeozoic Carboniferous and Permian sandstone gas-bearing layers in the Ordos Basin indicate that the BHAD of a sandstone layer with a fine grain size is shallower than that of a sandstone layer with a coarse grain size. The BHAD corresponds to 1350 m depth with fine sandstone and 2700 m depth with coarse sandstone, and the BHAD tends to increase as the sandstone grain size increases.

The BHAD is presumably related to hydrocarbon components (Pang et al. 2014c, a, b) and increases with decreasing molecular size. Natural gas with a smaller molecular size is susceptible to greater buoyancy forces than liquid hydrocarbons with large molecular sizes under the same geological conditions are.

The BHAD decreases with increasing thickness of the eroded strata above the reservoir layers due to tectonic movement or uplift in the whole petroliferous basin. The original tight medium in the reservoir layer is damaged by fractures when a basin is reshaped by faulting and folding. Consequently, the nonbuoyancy-dominated status is changed to a buoyancy-dominated status, and the BHAD becomes deeper. A basin with a history of strong tectonic activity tends to have large variations in porosity and permeability in the same reservoir layer in different local regions, leading to large BHAD variations and a complicated distribution of conventional and unconventional hydrocarbon reservoirs throughout the basin.

The BHAD increases with decreasing heat flow in a petroliferous basin. When the geothermal gradient of a petroliferous basin changes from 1 °C to 5 °C/100 m, it changes from 9200 m to approximately 1500 m, and the entire hydrocarbon accumulation depth limits (both buoyancy- and nonbuoyancy-driven accumulation depths) change from 15,000 to 3000 m. The combined conventional and unconventional hydrocarbon resources in the "cold basin" with heat flow below 40 mW/m² are much deeper than those in the "hot basin" with heat flow above 60 mW/m².

8.4.2.4 Variation Range and Prediction Method of the BHAD

Through physical and numerical simulations, the formation mechanism of the BHAD is revealed. A prediction method for predicting the distribution of unconventional hydrocarbon accumulation and its resource potential in a petroliferous basin is also proposed.

The BHAD is a dynamic force equilibrium boundary in the reservoir layer outside of source rocks, where the driving force for hydrocarbon migration and accumulation is equal to the resistance. Guo et al. (2017a, b) performed two physical simulation experiments to characterize the BHAD. The first experiment with different pore throat radii employed a conical glass tube under variable fluid pressure conditions. In the second experiment with different sand particle sizes, a cylindrical sand column filled with increasing sizes of sand particles from bottom to top was used. The results of these two simulation experiments revealed that at the dynamic force equilibrium boundary (i.e., BHAD), the hydrocarbon expansion pressure (P_e) was equal to the sum of the overlying hydrostatic pressure (P_w) and the capillary pressure (P_c) in the reservoir layer (Eq. 8.4.5):

$$P_e = P_w + P_c \qquad (8.4.5)$$

All the parameters related to the expansion, hydrostatic, and capillary pressures in the force

equilibrium equation (Eq. 8.4.5) affect the BHAD, but it varies mainly across the four geological factors. The first factor is the reservoir layer lithology: the coarser the grain size is, the deeper the BHAD. The second factor is the hydrocarbon components and their physical properties. The BHAD becomes deeper when the pore throat radius in the reservoir layer decreases with increasing reservoir temperature while the other parameters remain constant. The BHAD depth becomes shallower with increasing oil or gas density, with increasing oil/gas–water interfacial tension, or with increasing wetting angle of the oil–gas. Although gas and oil have different physical states, their reservoir porosities at the BHAD are very close within the 10–12% range. The third factor is tectonic movement, where the BHAD decreases with uplift in the whole petroliferous basin. The fourth factor is underground heat flow, where the BHAD increases with decreasing heat flow in a petroliferous basin.

8.4.2.5 Significance and Enlightenment of the BHAD

Although the BHAD depths significantly differ from 1200 to 4200 m (altitude: −400 to −4000 m) in different basins, the critical parameters related to the BHAD (e.g., porosity and permeability) are almost the same, i.e., approximately 10% ± 2% for porosity and approximately 1 mD for permeability, which can be used as the discriminant criteria for BHAD determination. Therefore, the variation in and distribution of the BHAD in a basin can be predicted and confirmed by the distribution features of conventional and unconventional reservoirs. The accumulation of conventional and unconventional hydrocarbon reservoirs and their categories in petroliferous basins are also constrained by the BHAD. Above the BHAD, hydrocarbon reservoirs are formed by buoyancy and distributed in traps with unique geological features, such as hydrocarbons that accumulate at the high points of the trap structures, which are preserved by cap rocks at high positions, are enriched in layers of high porosity, form reservoirs with high fluid pressure, and are separated from their source rocks. Below the BHAD, hydrocarbon reservoirs are formed by nonbuoyancy forces and are distributed in tight reservoirs near the source rocks; these reservoirs feature unique features, such as hydrocarbons distributed in low depressions, are located in the inverted low position, accumulate in low-porosity layers, and are stable at low fluid pressures. The resource potential of hydrocarbon accumulations is also constrained by the BHAD. The original hydrocarbon for the conventional reservoirs above the BHAD is approximately 10% of the total amount generated in the basin. Moreover, the share of proven hydrocarbon reserves above the BHAD is disproportionally high at 82%, indicating the limitation of the remaining exploration potential of conventional hydrocarbons. The original hydrocarbons expelled for the unconventional reservoirs below the BHAD constitute approximately 40% of the total amount generated in the petroliferous basins, whereas the share of the proven reserves below the BHAD is only 18% of the total proven, indicating greater exploration potential (Pang et al. 2021b).

8.4.3 Hydrocarbon Accumulation Depth Limit and Maximum Burial Depth of Unconventional Oil/Gas Reservoirs

8.4.3.1 Concept and Criterion of the Hydrocarbon Accumulation Depth Limit

The hydrocarbon accumulation depth limit (HADL) refers to the maximum depth for oil and gas accumulation on Earth, beyond which hydrocarbons cannot accumulate to form commercial reservoirs in a petroliferous basin (Pang et al. 2021c). The HADL is a dynamic boundary characterized by a series of critical reservoir properties, such as porosity, permeability, pore throat radius, and degree of thermal evolution. Above the HADL, the geological conditions are favourable for oil and gas accumulation and for the formation of commercial reservoirs.

8.4.3.2 Research Method and Basic Process for Determining the HADL

The major challenge in determining the HADL is that no or very few exploration wells have reached the maximum depth of hydrocarbon accumulation in petroliferous basins. Two steps have been taken to solve this problem (Fig. 8.15). First, the drilling results of numerous exploratory wells in petroliferous basins are used to simulate the correlations between oil-bearing (i.e., dry layer ratio, oil and gas saturation, irreducible water saturation, and movable hydrocarbon ratio) and essential (i.e., porosity, permeability, pore throat radius, and thermal evolution degree) properties and determine the critical values of the essential properties corresponding to the HADL, where the dry layer ratios and the irreducible water saturation reach 100%, or the hydrocarbon saturation and movable hydrocarbon ratio tend to approach 0. Second, the vertical variation trends of these essential properties with depth are statistically analysed to confirm the HADL depth.

8.4.3.3 Main Controlling Factors and Variation Characteristics of the HADL

HADLs are affected by many geological factors, including hydrocarbon composition, reservoir lithology, reservoir age, thermal gradients, and tectonic movements, among which the thermal gradient is the most common and important factor. The maximum depth for the oil reservoir distribution is much shallower than that for the gas reservoirs. The size and the sorting of the fragment particles in the sandstone reservoir layers significantly influence the HADL. According to the drilling results, statistical analyses, and numerical simulations of the sandstone reservoirs, the larger the particle sizes are, the deeper the HADL. The HADLs of different reservoir layers within the same basin can differ. A younger reservoir layer generally has a deeper HADL. Compared with younger layers, older layers undergo a longer compaction process. The reservoirs also become tighter at the same depth, which is the key reason why the reservoir layers have the same lithology, but layers of different

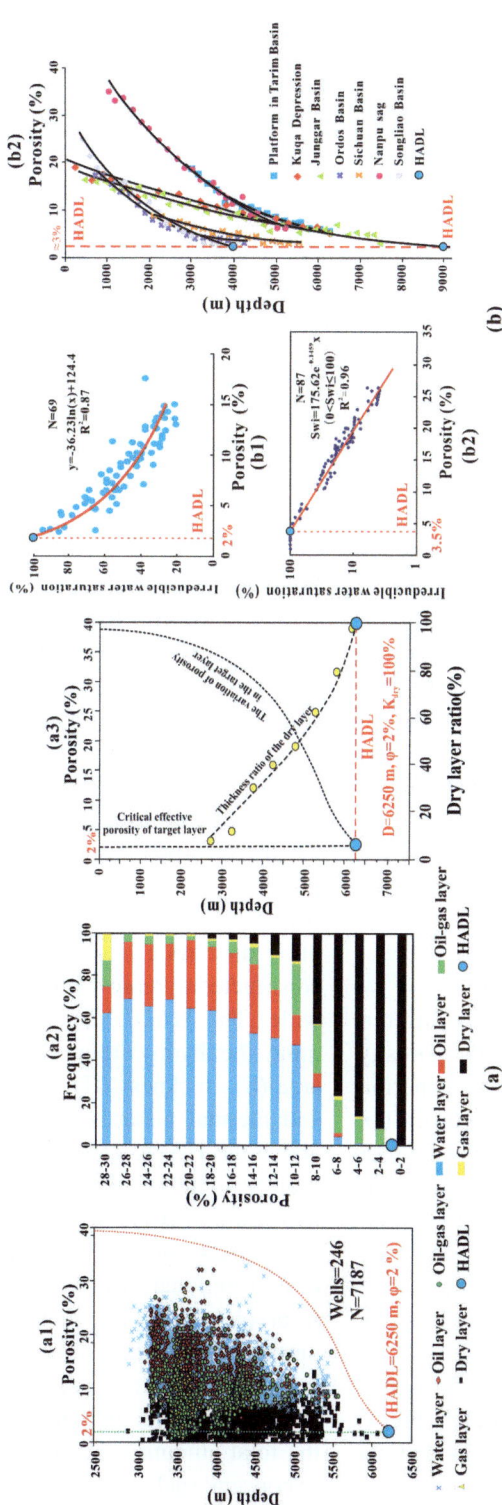

Fig. 8.15 Determination of the HADL by the vertical variation in the dry layer ratios and the bound water saturation in the reservoir layers with burial depth. (A) Case studies on the drilling results of Palaeozoic marine sandstone reservoirs in the Central Uplift in the Tarim Basin, from the (a) variation in the drilling results with burial depth; (b) variation in the drilling results; and (c) correlations among the critical parameters related to the HADLs, including the dry layer ratio, burial depth, and porosity. (B) Determination of the HADL by bound water saturation in the sandstone reservoir porosity with increasing burial depth: (a) a case study in the Kuqa Depression; (b) a case study in the Songliao Basin; and (c) vertical variation in the average porosity for the major reservoir layers with depth in the six basins and HADL determination

ages have different HADLs. The HADLs for petroliferous basins in China differ because of their different thermal evolution. Petroliferous basins with high heat fluxes or geothermal gradients have shallower HADLs. Compared with a lower heat flow or geothermal gradient, the higher heat flow and temperature in the reservoir layers lead to stronger compaction and lower porosities at the same burial depth, making hydrocarbon accumulation more difficult. In addition, geological factors, such as reservoir depressurization, temperature reduction, and pore fluid volume dilation caused by hydrocarbon leakage during uplift and erosion of overburden, complicate HADL determination (Wood and Sanei 2016).

Two key geological factors lead to HADL formation in petroliferous basins: (1) depletion of the hydrocarbon generation potential of the source rocks that cut off the oil or gas provision for reservoir formation in deep and tight areas; and (2) termination of differential compaction or capillary pressure differences between the inside and outside of reservoir layers with increasing depth, which ultimately results in the dominant driving force for hydrocarbon migration and accumulation in deep reservoir layers.

8.4.3.4 Variation Range and Prediction Method for the HADL

The determined HADLs differ across different basins. The HADL is determined by the porosity, permeability, pore throat radius, and thermal maturity of the reservoir layers. The hydrocarbon reservoir layers in different basins have similar critical values of essential properties, such as porosity ($3\% \pm 1\%$), permeability (0.01 md), pore throat radius (0.035 ± 15 μm), and vitrinite reflectance ($2.75\% \pm 0.25\%$) at the HADL, whereas the HADLs greatly vary from less than 4000 m to greater than 9000 m.

Theoretically, the maximum depth for hydrocarbon accumulation in a petroliferous basin can be deeper than the HADL predicted by a statistical analysis of the drilling results if the whole basin had been uplifted or if the overlying strata

had eroded. This problem can be solved in two ways (Pang et al. 2021c). One approach is to identify the thermal maturity (Ro) or porosity (Φ) of the reservoir layers with the HADL and calibrate the burial depth (D) by the established model of $D = f(Ro)$ or $D = f(\Phi)$ beforehand using the data from reservoirs without stratum erosion. This method is utilized to determine the HADLs for the Ordos, Songliao, and Sichuan Basins. The second method involves investigating basin evolution and restoring the eroded thickness of the overlying strata on the reservoir layers. The drilling data affected by stratum erosion are excluded from those used to establish a model for predicting the HADLs by statistical analysis. This method is utilized in the Tarim, Junggar, and Bohai Bay Basins.

8.4.3.5 Significance of and Information Gained from the HADL

The HADL generally controls the geological conditions of the maximum depth for hydrocarbon accumulation, a promising domain for exploration, and hydrocarbon resource potential. The HADL can be used to predict promising areas for hydrocarbon exploration. The hydrocarbon reservoir distribution in petroliferous basins is controlled by HADLs, and a complete pattern of hydrocarbon reservoir distribution can be constructed to represent many basin types worldwide. The HADL can also be used to determine the hydrocarbon resource potential. The area (S), thickness (H), OM (TOC), type (KTI), and thermal maturity (Ro) of the source rocks above the HADL in a petroliferous basin determine its total hydrocarbon generation amount and hydrocarbon resource potential. Combined with the BHAD concept, the hydrocarbon resources of conventional, unconventional tight, and shale reservoirs can be determined. Less than 10% of the proven hydrocarbon reserves come from unconventional resources worldwide, but their original hydrocarbon amount is greater than 90% of the total amount generated, implying that many unconventional resources remain in deep strata for future exploration.

8.4.4 ASDL and Distribution Limit of the Petroleum System

8.4.4.1 Concept and Criterion of the ASDL

Active source rocks are sedimentary rocks rich in OM. They are capable of generating hydrocarbons. The ASDL is the maximum burial depth of active source rocks, beyond which the source rocks no longer generate or expel hydrocarbons and become inactive (Pang et al. 2020a, b, c). In addition to the burial depth, the ASDL can be characterized by other physical parameters of the source rocks (e.g., thermal maturity).

Several parameters can be utilized to identify the ASDL (Fig. 8.16). In theory, the OM in source rocks eventually evolves into graphite with increasing thermal maturity. The atomic ratios of hydrogen to carbon (H/C) and oxygen to carbon

(O/C) of the remaining OM decrease to zero, indicating that the active source rocks no longer produce hydrocarbons and reach the ASDL. The Rock–Eval pyrolysis parameters [e.g., the hydrocarbon generation potential index ("$S_1 + S_2$"/ TOC)] can be utilized to identify the ASDL. When the index approaches zero, the source rocks can no longer expel hydrocarbons and reach the ASDL. Along with the evolution of the hydrocarbon generation potential index, the hydrocarbon expulsion ratio (Qe), hydrocarbon expulsion rate (Ve), and hydrocarbon expulsion efficiency (Ke) of the source rocks also evolve with thermal maturity. Qe represents the amount of hydrocarbons expelled from a unit weight of organic carbon. Ve represents the hydrocarbons expelled from a unit weight of organic carbon when the burial depth increases by 100 m. Ke represents the cumulative ratio of the hydrocarbons expelled from the source rocks

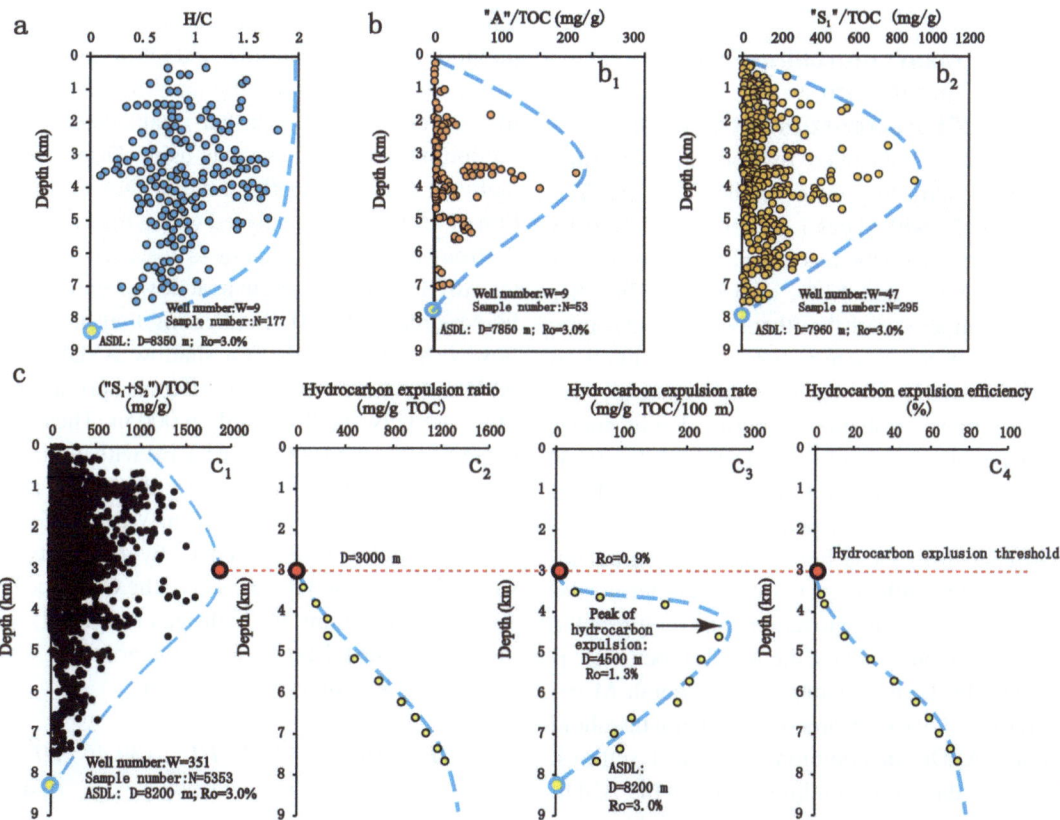

Fig. 8.16 Identification of the ASDL in the Junggar Basin using different indicators, including the H/C ratio variations (**a**), residual hydrocarbon amounts (**b**), ("S1 + S2"/TOC) (c1), Qe (c2), Ve (c3), and Ke (c4) versus depth

to the cumulative amounts of generated hydrocarbons. When the source rocks reach the ASDL, Qe and Ke approach constant values, whereas Ve approaches 0. Therefore, the residual hydrocarbon content index ("S_1"/TOC or "A"/TOC), which represents the quantity of hydrocarbons retained per unit weight of organic carbon, can be utilized to indicate the ASDL.

8.4.4.2 Research Method and Basic Process for Determining the ASDL

The theoretical threshold values of different geochemical parameters or indices for identifying the ASDL are provided. In practice, envelope lines enclosing all sample data points are utilized to show the overall trends of how these parameters change with increasing burial depth or thermal maturity. The interceptions of the envelope lines with these threshold values represent the source rocks that reach the ASDLs.

8.4.4.3 Main Controlling Factors and Variation Characteristics of the ASDL

An ASDL is controlled by four main factors. The first controlling factor is the OM type. The source rocks with types I (oil-prone), II, and III (gas-prone) kerogens reach ASDLs at Ro values of 3.0%, 3.5%, and 4.0%, respectively. In other words, compared with the other two types of source rocks, oil-prone source rocks are more likely to reach the ASDL and stop generating and expelling hydrocarbons at shallower burial depths under similar geological conditions. The second controlling factor is heat flow and the geothermal gradient. ASDLs are shallow in petroliferous basins with high heat flow and a high geothermal gradient. They are also influenced by the two other important factors: tectonic movement and stratigraphic age. As previously stated, the ASDL is better characterized by thermal maturity than by depth. Moreover, Ro = 3.5% is regarded as a general threshold for the ASDL in common geological settings. However, the corresponding depth of the ASDL for different source rock layers is highly variable. Owing to the irreversible nature of the vitrinite reflectance (Hayes 1991; Peters et al. 2018),

the ASDL depth for the source rocks historically uplifted after reaching the original ASDL is shallower than that of the younger source rocks that were not uplifted.

In addition to these four main factors, deep thermal fluids and overpressure retardation may affect the ASDL (McTavish 1998; Hao et al. 2007; Fetter et al. 2019). Compared with unaffected source rocks, source rocks influenced by deep thermal fluids may have shallower ASDLs. The source rocks influenced by overpressure retardation also have deeper ASDLs.

8.4.4.4 Variation Range and Prediction Method for the ASDL

The thermal maturity corresponding to the ASDL remains the same, regardless of whether the ASDL becomes deeper or shallower. A source rock will reach the ASDL when its Ro increases to 3.5% ± 0.5%, and its hydrocarbon generation potential is depleted. According to the results of the previous analysis, heat flow and OM type are the two main factors controlling the ASDL. A quantitative relationship is further established in this section by statistical analysis using Origin 2019 software. First, the depths of the ASDLs as a function of heat flow are analysed with a linear model. A strong negative correlation is observed between the ASDLs and the present heat flows, with a coefficient greater than 0.9, indicating that high heat flow very likely leads to a shallow ASDL. The ASDLs for the basins with different current heat flows are between 3000 and 16,000 m. These are generally less than 6000 m in basins with high heat flow (>70 mW m^{-2}) and greater than 9000 m in basins with low heat flow (<40 mW m^{-2}). Given that the ASDL is also influenced by the OM type, we further analysed the effects of the OM type on the ASDL by adding the hydrogen index (HI), an indicator of the OM type, to the linear model. The following equation is then deduced:

$$ASDL = 16448 - 3.61 \cdot HI - 139.46 \cdot HF \tag{8.4.6}$$

where ASDL is the active source rock depth limit (in metres); HI is the hydrogen index value of the major source rocks in a basin (in mg HC/g TOC); and HF is the present average heat flow value of a basin (in mW m^{-2}).

The quantitative relationship in Eq. (8.4.6) provides preliminary insights into the geological basis and boundary conditions for predicting the fossil fuel distribution in basins and helps in the evaluation of the hydrocarbon potential.

8.4.4.5 Significance of and Information Gained from the ASDL

The ASDL is the critical condition or the dynamic boundary at which oil and gas expulsion ends. It controls the formation and distribution of all economical hydrocarbon reservoirs. Theoretically, the ASDL represents the maximum depth of fossil fuel formation and distribution. Approximately 97.7% of the coal resources in China and 97.3% of recoverable coal reserves worldwide are distributed above ASDLs, corresponding to 4.0% Ro (CCRR 1996; CNACG 2016; Conti et al. 2016). Therefore, the ASDL represents the maximum depth of the hydrocarbon reservoir distribution, which includes oil, gas, and coal. The ASDLs for natural gas (ASDLg) and oil (ASDLo) are different. Theoretically speaking, the burial depth and thermal maturity corresponding to the ASDLo should be shallower than those of the ASDLg. High-temperature (room temperature to 600 °C) and high-pressure (50 MPa) pyrolysis simulation experiments conducted on immature or low-maturity kerogens sampled from the Junggar Basin in a closed system revealed that the source rocks reach the ASDL at approximately 2.0% Ro. The same source rocks reached the ASDLg at Ro values of 3.0–4.0%. Basins can be vertically divided into three regions by combining the HET, ASDLo, and ASDLg concepts. The upper field is favourable for hydrocarbons migrating upwards to form conventional reservoirs in traps. The source rocks in this field are predominantly immature and/or less mature. The middle field is favourable for source rocks to generate, expel, and retain hydrocarbons to form various types of oil and gas reservoirs. The source rocks in this field supply hydrocarbons that may migrate into the upper area. The lower field is favourable for source rocks

to generate, expel, and retain natural gas to form mainly unconventional resources.

8.5 Dynamic Models and Predictive Evaluation Methods for Deep Hydrocarbon Reservoir Formation

8.5.1 Dynamic Field-Controlled Reservoir Distribution Model (Combined Conventional and Unconventional Reservoir Model)

The hydrocarbon-free, -restricted, and -bound dynamic fields can be classified according to the different dynamic fields enclosed by the three dynamic thresholds of the lower limit of the buoyancy reservoir formation, bottom limit of the hydrocarbon formation, and bottom limit of the hydrocarbon supply from the source rocks. Accordingly, a dynamic field-controlled reservoir distribution model can be established (Fig. 8.17).

8.5.1.1 Distribution of Conventional Reservoir Formation Controlled by the Free Dynamic Field

The free dynamic field is located between the surface and the shallow lower buoyant reservoir limit. It is delineated by determining the lower buoyant reservoir limit mainly by statistical and reservoir analyses and numerical simulations (Hu et al. 2022b; Pang et al. 2014c, a, b; Liu 2018; Wang 2020).

The hydrocarbons in the free dynamic field are dominated mainly by buoyancy, consequently forming conventional hydrocarbon reservoirs.

8.5.1.2 Distribution of the Tight Reservoir

The lower limit of the buoyancy and the bottom limit of hydrocarbon reservoir formation together determine the confined dynamic field distribution.

Fig. 8.17 Dynamic reservoir-controlled hydrocarbon accumulation model. *Note* C1: conventional back-slope reservoir; C2: conventional tectonic reservoir; C3: conventional lithological reservoir; C4: conventional stratigraphic reservoir; and C5: conventional composite reservoir

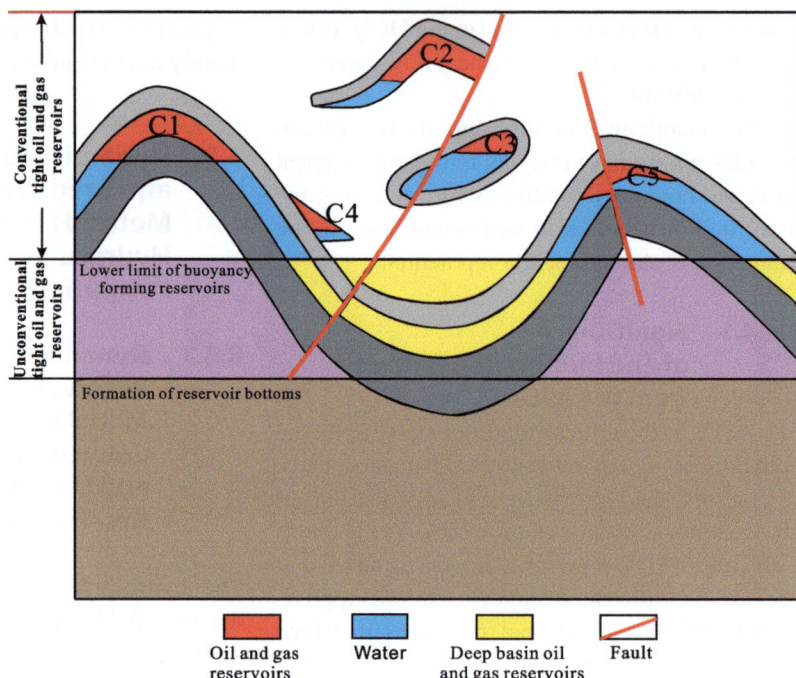

The restricted hydrodynamic field is controlled mainly by the lower limit of buoyancy formation and the bottom limit of hydrocarbon formation. The restricted hydrodynamic field mainly forms tight hydrocarbon reservoirs.

8.5.1.3 Distribution of Shale Reservoir Formation Controlled by the Bound Dynamic Field

The bound dynamic field is located below the bottom limit of the reservoir and above the bottom limit of source rock-generated hydrocarbons. Some hydrocarbons retained in the source rock layer can constitute shale oil and gas resources.

Together, the bottom limits of the reservoir and the source rock-generated hydrocarbons determine the bound dynamic field distribution (Pang et al. 2020a, b, c).

The formation and distribution of oil and gas reservoirs within the bound dynamic field of source rock formations exhibit the basic characteristics of source–storage integration; i.e., they are generally dense and have a wide distribution, low permeability, and low production. The burial depths of these reservoirs are highly variable, and

they usually consist of shale oil and shale gas (Pang et al. 2022a, b).

8.5.2 Combined Element-Controlled Reservoirs in the Free Dynamic Field and Conventional Reservoir Prediction and Evaluation Methods

8.5.2.1 Combining Element-Controlled Reservoirs to Predict Conventional Reservoir Evaluation Methods

Free Dynamic Field Boundary Recognition

The coupling relationship of each element under the free dynamic field by vertically stacking the elements can eventually lead to three types of favourable zones during a certain formation period. The first step is to study the geological conditions of hydrocarbons. The second step is to study the four functional elements. In the third step, the combination of functional elements is investigated to predict favourable reservoir formation periods, layers, and zones. In the fourth step, multiphase composite reservoir formation in the same target layer is studied to determine the most favourable reservoir formation zones (Yu et al. 2022).

8.5.2.2 Evaluation of Conventional Oil and Gas Reservoirs in the Permian Changxing Feixianguan Formation in Northeastern Sichuan

Functional Elements

Hydrocarbon source rocks. The hydrocarbon discharge centre of the Feixianguan Formation is east of the Puguang area. The whole area of Puguang is favourable for source control (Fig. 8.18a).

Sedimentary phases. The phase control favourable area is distributed mainly in sedimentary phases other than the leading edge slope (Fig. 8.18b). When the capping layer thickness is greater than 700 m (Yu et al. 2022), the relationship between the fracture distance interval of different single wells and the average hydrocarbon thickness is shown within the interval (Yu et al. 2022). On this basis, the fracture-controlled farthest distance of the reservoir is obtained, and a fracture-controlled favourable zone is constructed (Fig. 8.18d). The relationship between the distance interval from the top of the slope and the average hydrocarbon thickness within the interval is fitted for different single wells (Li et al. 2017; Yu et al. 2022). This is then used to construct the augmentation-controlled zone in conjunction with the palaeo-uplift plan (Fig. 8.18e). The method of controlling reservoirs with low interfacial energy (Pang et al. 2015a, b, c) is as follows. The potential index planar distribution and the potential favourable control zone are obtained (Fig. 8.18f).

Forecast of the Favourable Areas for Conventional Oil and Gas Reservoirs

Predicted Results of the Favourable Areas

The single-element reservoir control zones are superimposed to obtain three types of favourable zones (Fig. 8.19a).

8.5.2.3 Results Testing

The statistics of the drilled wells revealed that the greatest oil thickness occurred in Zone I (Fig. 8.19b). Verification of the lost wells revealed that there were no lost wells in Zone I, and the results were more reliable (Fig. 8.19c).

8.5.3 Method for Predicting and Evaluating Reservoir Formation and Tight Oil and Gas Reservoirs with a Combination of Restricted Dynamic Field Elements

8.5.3.1 Predictive Evaluation Method for Tight Reservoirs Controlled by a Confined Dynamic Field

The combination of functional elements in a confined dynamic field can be used to evaluate the coupling of reservoir-forming elements within a confined field. A planar superposition of the reservoir-forming elements was used to identify favourable zones. The main elements included hydrocarbon drainage intensity, reservoir depositional phase, confined field distribution, and fracture planes. This method was used to classify the confined dynamic fields in the study area into zones of reformed and tight reservoir development.

8.5.3.2 Evaluation of the Tight Oil and Gas Reservoirs in the Permian Changxing Formation in Northeastern Sichuan

Functional Elements

Hydrocarbon source rocks. The gas intensity of the Changxing Formation in the YB area of northeast Sichuan was less than 6×10^8 m^3/km^2. Four main types of sedimentary phases developed in the Yuanba area (Fig. 8.20b): open terrace phases, terrace margin phases, slope phases, and deepwater shelf phases. The dominant phase is the terrace margin (Hu et al. 2020). The lower limit of the buoyant reservoir formation was determined

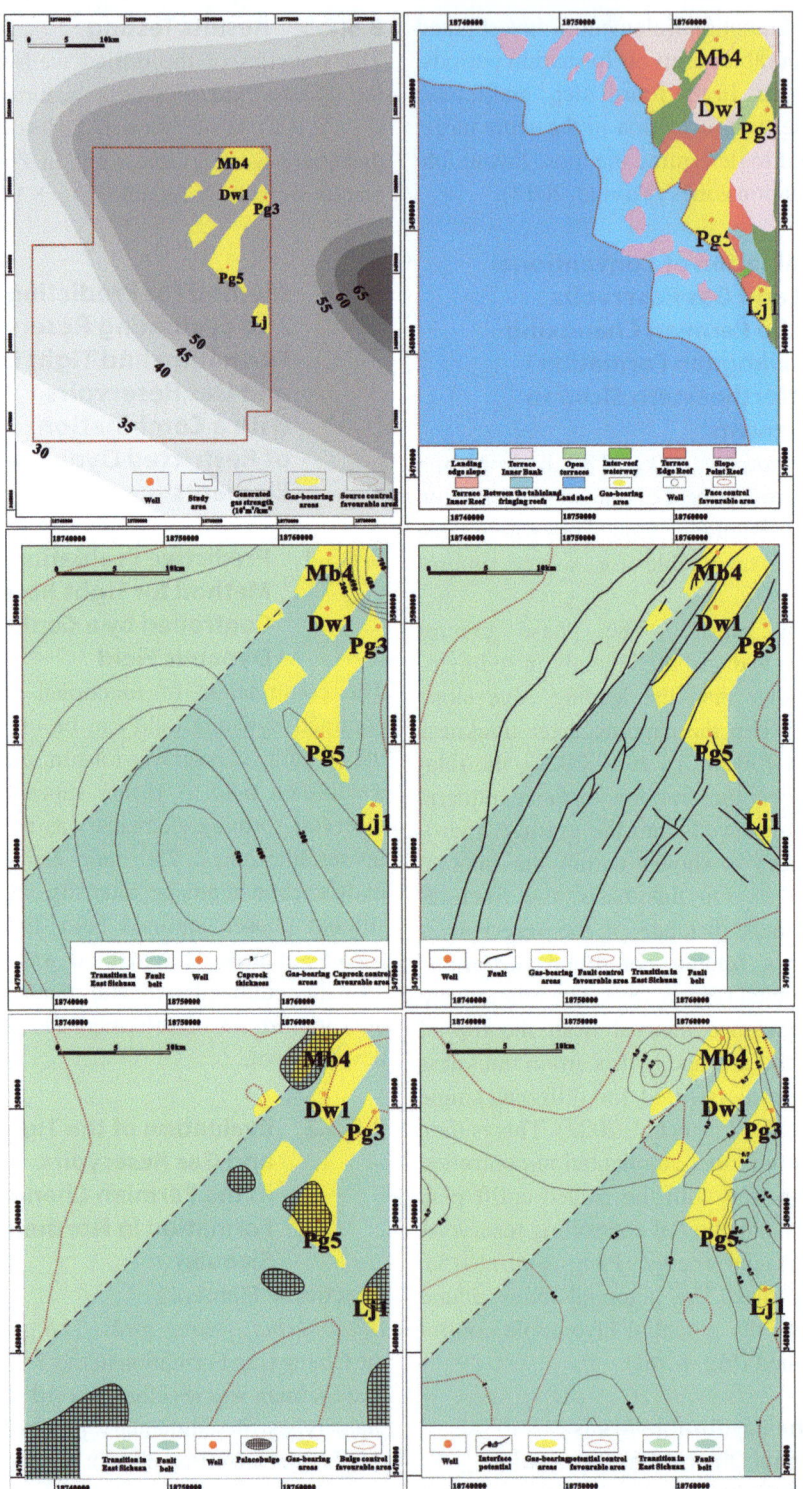

Fig. 8.18 Planar distribution of Changxing reservoir elements in the Puguang area: **a** gas generation intensity; **b** sedimentary phase; **c** caprock thickness; **d** fault distribu-tion; **e** palaeo-uplift distribution; and **f** potential difference distribution

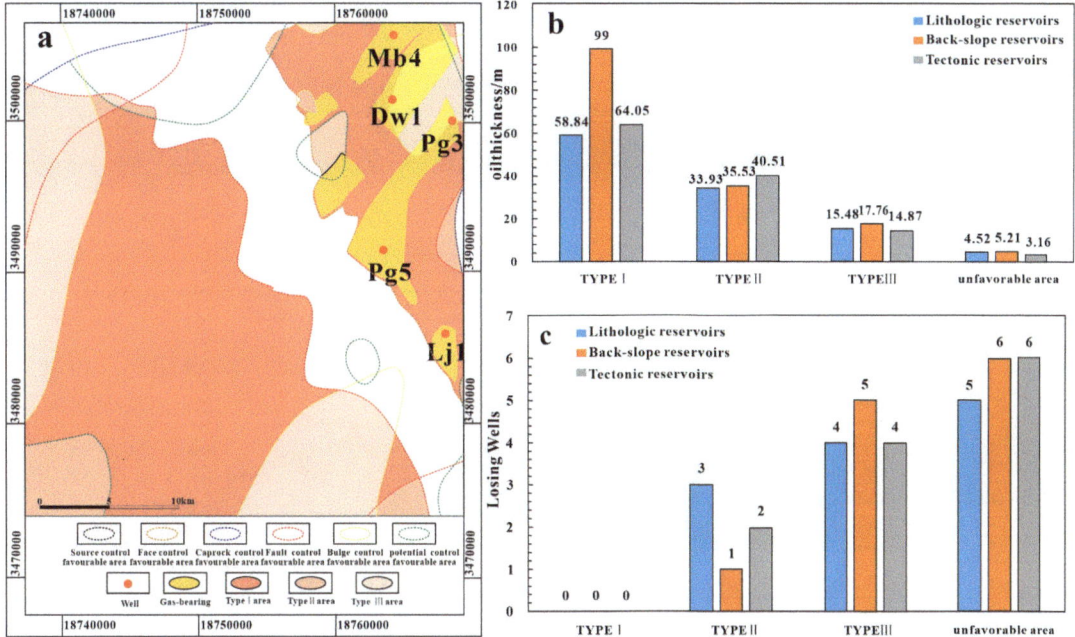

Fig. 8.19 Predicted results and tests of the favourable zones in the Changxing Formation in the Puguang area: **a** planar distribution of the favourable areas; **b** statistics of the oil layer thickness in different favourable zones; and **c** statistics of the number of failed wells in different favourable areas

to be 6850 m deep. The bottom limit of hydrocarbon formation is 10,370 m deep (Fig. 8.20e). The restricted dynamic field is distributed only in the shallow water shelf phase east of Yuanba and in the slope phase in the centre (Fig. 8.20c). The whole Yuanba area has good preservation conditions (Fig. 8.20c).

Prediction of Favourable Zones for Tight Oil and Gas Reservoirs

Predicted Results of Favourable Zones
The restricted power field, dominant sedimentary phase, hydrocarbon source gas, and tectonic stability zones in the Yuanba area were superimposed to obtain the development zones of the solution-altered and tight reservoirs within the restricted field (Fig. 8.20d).

Results Testing
The drilling results were compared with the predicted results. The evaluation results were reliable, with 80% accuracy (Fig. 8.20d).

8.5.4 Combination of Reservoir-Forming Factors in the Bound Dynamic Field and Prediction and Evaluation Methods for Shale Oil and Gas Reservoirs

Shale oil and gas reservoirs are distributed above the ASDL in the bound dynamic field (Hu et al. 2021a, 2021b; Pang et al. 2020a, b, c). The formation of the shale oil and gas reservoirs was dominated by nonbuoyancy. Shale oil and gas reservoirs do not have obvious trap boundaries and cannot use conventional oil and gas reservoir prediction and evaluation methods to guide their exploration (Hu et al. 2022c). In 2002, George and Jennie first proposed the term "sweet spot", where the "sweet spot" of shale oil and gas refers to the optimal area or horizon for shale oil and gas enrichment (Niu et al. 2021). The prediction and evaluation of shale

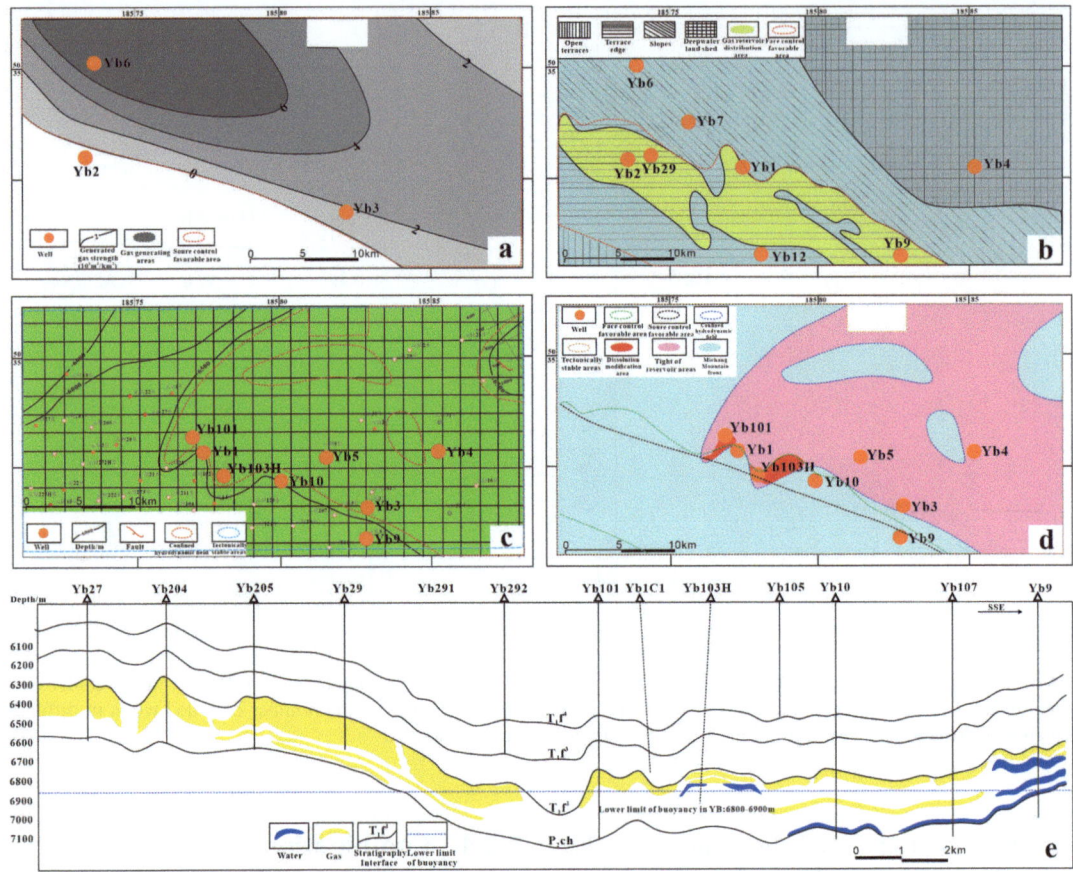

Fig. 8.20 Planar distribution and prediction of the Changxing reservoir elements in the Yuanba area: **a** gas generation intensity; **b** sedimentary phase; **c** tectonic planar distribution; **d** confined dynamic field prediction; and **e** hydrodynamic field discrimination

oil and gas reservoirs aim to predict and evaluate the "sweet spot" of shale oil and gas.

8.5.4.1 Prediction and Evaluation Methods for Shale Oil and Gas Reservoirs

The evaluation method for functional element combination refers to the prediction of shale oil and gas enrichment zones by orderly vertical superposition, effective horizontal superposition, and time matching of multiple functional elements. According to the evaluation method, the shale oil and gas reservoirs can be divided into four grades, namely, Class I–III sweet spot areas and nonsweet spot areas. The value ranges of various functional elements corresponding to different oil and gas enrichments were further determined. An evalua-

tion scheme of the functional elements of the shale gas reservoirs of the Permian Longtan Formation in the Sichuan Basin was established as an example. Table 8.4 and Fig. 8.21 present the value ranges of the functional elements of the Class I–III sweet spot areas and the nonsweet spot areas.

8.5.4.2 Prediction and Evaluation of the Longtan Formation Shale Gas Reservoir

Functional Elements

The material basis, reservoir properties, preservation conditions, and development conditions of the Longtan Formation in Northeast Sichuan were analysed. The distribution of the functional elements of the Longtan Formation in Northeast Sichuan is shown in Fig. 8.22.

Table 8.4 Evaluation scheme of the functional element combination for the shale gas reservoirs in the Longtan Formation in Sichuan Basin

Reservoir formation conditions	Functional elements	Class I sweet spot	Class II sweet spot	Class III sweet spot	Non-sweet spot
Material basis	TOC content (wt.%)	>3.0	2.0–3.0	1.0–2.0	<1.0
	Thickness of source rocks (m)	>40	30–40	20–30	<20
	Ro (%)	2.0–3.0	1.5–2.0 or 3.0–3.5	0.7–1.5	<0.7 or >3.5
Reservoir properties	Porosity (%)	>4.0	3.0–4.0	2.0–3.0	<2.0
	Gas content (m³/t)	>3.0	2.0–3.0	1.0–2.0	<1.0
Preservation conditions	Tectonic style	Gentle	Relatively gentle	Relatively closed	Closed
	Fault development	Barely developed	Poorly developed	Relatively developed	Developed
Development conditions	Pressure coefficient	>1.5	1.2–1.5	1.0–1.2	<1.0
	Fracture	Developed	Relatively developed	Poorly developed	Barely developed

Fig. 8.21 Sketch map of the functional element combination for the prediction and evaluation of shale oil and gas reservoirs

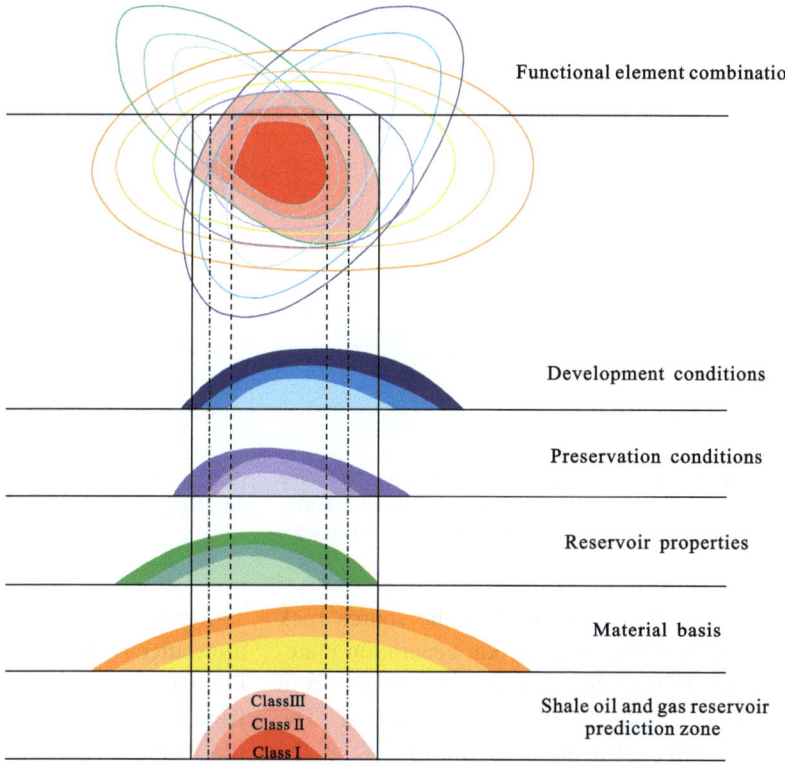

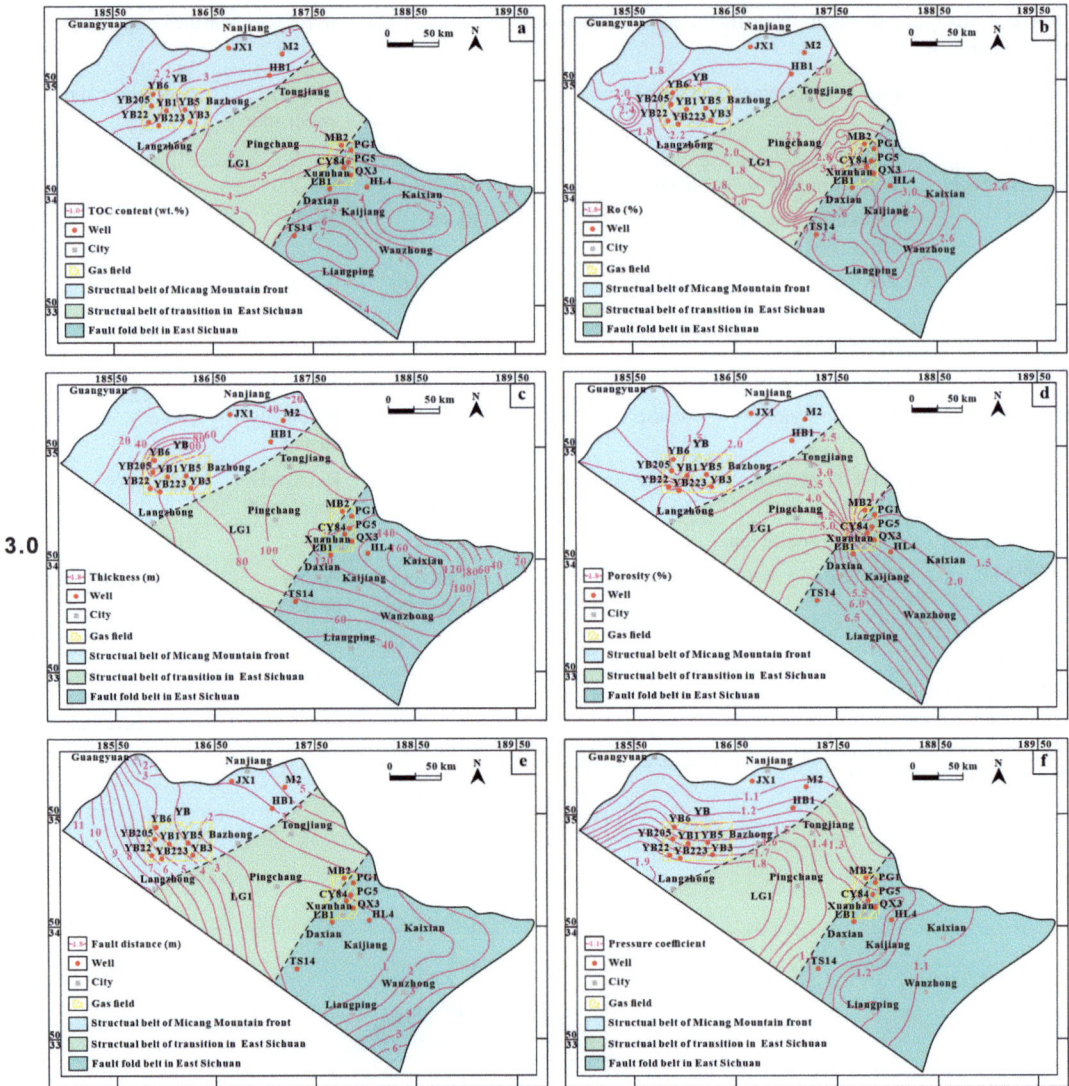

Fig. 8.22 Distribution of the functional elements of the shale gas reservoirs in the Longtan Formation in the northeast Sichuan Basin: **a** TOC content; **b** vitrinite reflectance (Ro); **c** source rock thickness; **d** reservoir porosity; **e** fault development; and **f** pressure coefficient

Material basis. The OM in the source rock is the basis of the hydrocarbon generation material. The TOC of the source rocks ranged from 1.4 to 8.5%, with a wide distribution range. The OM maturity (Ro) of the source rocks of the Longtan Formation ranged from 1.6 to 3.3%, which was high over the mature stage. The source rock thickness ranged from 15 to 170 m.

Reservoir properties. The physical properties of the source rocks of the Longtan Formation in the Puguang area were good, with porosities ranging from 4.5 to 7.3%. Moreover, the physical properties of the source rocks in the Yuanba area were poor.

Preservation conditions. Areas with less developed faults were conducive to shale oil and gas

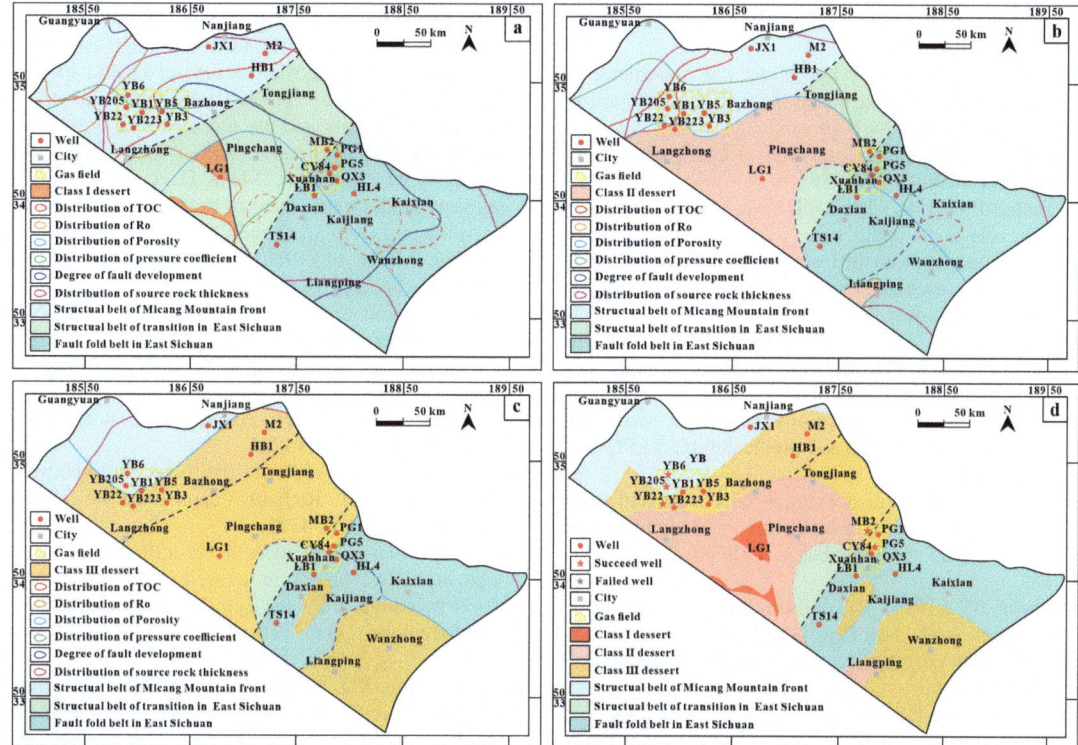

Fig. 8.23 Prediction and evaluation results of the shale gas reservoirs in the Longtan Formation in the northeast Sichuan: **a** prediction of Class I sweet spots by the evaluation method of functional element combination; **b** prediction of Class II sweet spots by the evaluation method of functional element combination; **c** prediction of Class III reservoir preservation. The faults in the Puguang area were more developed than those in the Yuanba area were.

sweet spots by the evaluation method of functional element combination; and **d** distribution of Class I–III sweet spots of the shale gas reservoirs. The area delineated by the solid line is the functional element distribution. The area delineated by the dotted line indicates that the functional elements are not distributed

Development conditions. Overpressure often indicates high-energy formation conditions conducive to improving the development effect of shale gas. The formation pressure coefficient of the Longtan Formation ranged mainly from 0.8 to 2.1.

Evaluation Results

The shale gas reservoirs of the Permian Longtan Formation in Northeast Sichuan were evaluated based on the evaluation method of the functional element combination (Fig. 8.23). The distribution of the Class I sweet spot area was small, mainly in the southern part of the study area. The Class II sweet spot area had a large distribution range and

was distributed in the middle of the study area. The Class III sweet spot areas were distributed in the east, north, and southwest of the study area.

The comprehensive logging interpretation data of the seven wells Verification of the seven wells proved that the evaluation results were reliable, yielding an 85% accuracy rate.

Acknowledgements The authors acknowledge the financial support of Natural Science Foundation of China (Grants U24B6001, U19B6003 and U20B6001).

References

Bai GP, Cao BF (2014) Global deep oil and gas reservoirs and their distribution rules. Oil Gas Geol 35(01):19–25

Bai XF, Liang JP, Zhang WJ et al (2018) Geological conditions, resource potential and exploration direction of deep gas in northern Songliao Basin. Nat Gas Geosci 29(10):1443–1454

Barker C (1984) Primary migration: the importance of water-mineral-organic matter interactions in the source rock. AAPG Bull 34:1–19

Berg RR (1975) Capillary pressures in stratigraphic traps. AAPG Bull 59(6):939–956

Berkenpas P G (1991) The Milk River shallow gas reservoir: role of the updip water trapand connate water in gas production from the reservoir. In: SPE annual technical conference and exhibition. Society of Petroleum Engineers October Dallas, Texas

Biteau JJ, Marrec AL, Michel LV et al (2009, 2006) The Aquitaine basin. Petroleum Geoscience 12(3):247–273

Bloch S, Lander RH, Bonnell L (2002) Anomalously high porosity and permeability in deeply buried sandstone reservoirs: origin and predictability. AAPG Bull 86(2):301–328

Bruce C (1984) Smectite dehydration; its relation to structural development and hydrocarbon accumulation in Northern Gulf of Mexico basin. AAPG Bull 68(6):673–683

Burchette TP (1996) Unconformities and porosity in carbonate strata. Mar Petrol Geol 13(5):596–597

Cant DJ (1986) Diagenetic traps in sandstones. AAPG Bull 70(2):155–160

Cao ZL, Yuan JY, Huang CG et al (2014) Effect of gypsum dissolution on calcite precipitation in high-temperature and high-pressure clastic reservoir. Acta Petrolei Sinica 35(3):450–454

CCRR (Editorial committee of China coal resources report) (1996) Comparison of coal resources between China and major coal-producing countries. China Coal 7:68–71

Chen DX, Pang XQ, Yang KM et al (2016) Genesis and formation process of deep laminated continuous tight sandstone gas reservoir in west Sichuan depression. J Jilin Univ (Earth Sci Edn) 46(06):1611–1623

Chen FJ, Tian SC (1989) Compaction and oil and gas migration. China University of Geoscience Press

CNACG (China National Administration of Coal Geology) (2016) China occurrence regularity of coal resources and resource evaluation. Science Press, Beijing

Conti, J, Holtberg, P, Diefenderfer, J, LaRose, A, Turnure, JT, and Westfall, L (2016) International energy outlook 2016 with projections to 2040 (No DOE/EIA-0484, 2016), USDOE Energy Information Administration (EIA), Office of Energy Analysis, Washington DC, US

Deng S, Li HL, Han J et al (2019) Activity characteristics and geological significance of No5 strike-slip fault in Shunbei, Tarim Basin. Oil Gas Geol 40(05):990–998+1073

Deng S, Li HL, Zhang ZP et al (2018) Characteristics of differential activity of main strike-slip fault zones along the north of Tarim Basin and its adjacent areas and its relationship with oil and gas enrichment. Oil Gas Geol 39(05):878–888

Dyman TS, Litinsky VA, Ulmishe GFK (2001) Geology and natural gas potential of deep sedimentary basins in the Former Soviet Union. AAPG Bull 67-C:1–24

Dyman T, Schmoker J, Root D et al (2001b) Deep gas resources have not been found in the United States-local evaluation by USGS. Nat Gas Geosci 12(4/5):40–47

Ehrenberg SN, Nadeau PH (2005) Evaluation of depth-dependent porosity and ulk modulus of a shear using permeability-depth trends. AAPG Bull 89(4):435–445

England W, Mackenzie A, Mann D et al (1987) The movement and entrapment of petroleum fluids in the subsurface. J Geol Soc Londn 144:327–347

Espitalie J, Madec M (1980) Tissot B Role of mineral matrix in kerogen pyrolysis; influence on petroleum generation and migration. AAPG Bull 64(1):59–66

Fetter, N, Blichert-Toft, J, Ludden, J, Lepland, A, Borque, JS, Greenhalgh, E, Garcia, B, Edwards, D, Télouk, P, Albarède F (2019) Lead isotopes as tracers of crude oil migration within deep crustal fluid systems. Earth Planet Sci Lett 525:115747

Germann P (2018) Viscosity-the weak link between Darcy's law and Richards' capillary flow. Hydrol Process 32(9):1166–1120

Gordon JB, Sanei H, Pedersen PK (2022) Secondary porosity development in incised valley sandstones from two wells from the flemish pass area, offshore newfoundland. Mar Pet Geol 140(105644):0264–8172

Guo YC, Pang XQ, Li ZX et al (2017a) The critical buoyancy threshold for tight sandstone gas entrapment: physical simulation, interpretation, and implications to the Upper Paleozoic Ordos Basin. J Pet Sci Eng 149:88–97

Guo YC, Pang XQ, Li ZX, Guo FT, Song LC (2017b) The critical buoyancy threshold for tight sandstone gas entrapment: physical simulation, interpretation, and implications to the Upper Paleozoic Ordos Basin. J Pet Sci Eng 149:88–97

Hao F, Guo T, Zhu Y et al (2008) Evidence for multiple stages of oil cracking and thermochemical sulfate reduction in the Puguang gas field Sichuan Basin, China. AAPG Bull 92(5):611–637

Hao F, Sun YC, Li ST et al (1996) Enhancement of thermal evolution of organic matter and hydrocarbon generation by active thermal fluid. J Earth Sci 21(1):68–72

Hao F, Zou H, Gong Z, Yang S, Zeng Z (2007) Hierarchies of overpressure retardation of organic matter maturation: case studies from petroleum basins in China. AAPG Bull 91:1467–1498

Hayes JB (1990) Porosity evolution of sandstones related to vitrinite reflectance. Org Geochem 17:117–129

He HQ, Wang ZY, Cheng YC (1999) Analysis of deep petroleum geological conditions in Bohai Bay Basin. Acta Sedimentol Sin 02:108–114

Head IM, Jones DM, Larter SR (2003) Biological activity in the deep subsurface and the origin of heavy oil. Nature 426(6964):344

Hillis RR, Morton JGG, Warner DS et al (2001) Deep basin gas: a new exploration paradigm in the Nappamerri Trough, Cooper Basin, South Australia. Appea J 4(1):185–200

Hu T, Pang XQ, Xu TW, Li CR (2022a) Identifying the key source rocks in heterogeneous saline lacustrine shales:

paleogene shales in the Dongpu depression, Bohai Bay Basin, eastern China. AAPG Bull 106(6):1325–1356

Hu T, Pang XQ, Jiang FJ, Zhang CX (2022b) Dynamic continuous hydrocarbon accumulation (DCHA): Existing theories and a new unified accumulation model. Earth Sci Rev 232:104109

Hu T, Wu GY, Xu Z, Pang XQ (2022c) Potential resources of conventional, tight, and shale oil and gas from Paleogene Wenchang formation source rocks in the Huizhou. Depression Adv Geo-Energy Res 6(5):402–414

Hu T, Pang XQ, Jiang FJ, Wang QF (2021a) Movable oil content evaluation of lacustrine organic-rich shales: methods and a novel quantitative evaluation model. Earth Sci Rev 214:103545

Hu T, Pang XQ, Jiang FJ, Wang QF (2021b) Key factors controlling shale oil enrichment in saline lacustrine rift basin: implications from two shale oil wells in Dongpu Depression, Bohai Bay Basin. Pet Sci 18:687–711

Hu XY, Zheng WB, You YC, Liu GP (2020) Probabilistic body-constrained geological modeling of reef-beach-phase gas reservoirs in the Yuanba Changxing formation Sichuan Basin. Petroleum Natural Gas Geol 41(01):157–163

Hunt JM (1990) Generation and migration of petroleum from abnormally pressured fluid compartments. AAPG Bull 74(1):1–12

Jia CZ, Pang XQ, Song Y (2021) The mechanism of unconventional hydrocarbon formation: hydrocarbon self-sealing and intermolecular forces. Petrol Explor Dev 48(3):507–526, 1876–3804

Jia CZ, Pang XQ (2015) Research progress and main development direction of deep oil and gas geology theory. Acta Petrolei Sinica 36(12):1457–1469

Jia CZ (2017) On the breakthrough of unconventional oil and gas to classical oil and gas geology theory and its significance. Petrol Explor Dev 44(1):1–11

Kang YZ, Xing SW, Li HJ et al (2019) Basin-controlling and oil-controlling distribution of structural systems in northern China. J Geomechanics 25(6):1013–1024

Law B, Spencer CW (1998) Abnormal pressures in hydrocarbon environments. AAPG Memoir 70:1–11

Law BE (2002) Basin-centered gas systems. AAPG Bull 86(11):1891–1919

Lee W, Kang P, Kim A et al (2018) Impact of surface porosity on water flux and structural parameter in forward osmosis. Desalination 439:46–57

Leythaeuser D, Schaefer RG, Yukler (1982) A role of diffusion in primary migration of hydrocarbons. AAPG Bull 66(4):408–429

Leythaeuser D (1987) Recognition and quantification of petroleum expulsion from shale source rocks: section iii geological environments and migration. AAPG Bull 81:341–344

Li HB, Pang XQ, Peng HJ, Wang ZC, Wang TB, Jiang DP, Wang WY (2017) Quantitative prediction of reservoir-controlling characteristics and favorable reservoir-forming zones in the Zhu Yi depression of the Zhujiangkou Basin. Mod Geol 31(04):802–813

Li J, She YQ, Gao Y et al (2019a) China's onshore deep-ultra deep gas exploration field and potential. China Petroleum Exploration 24(04):403–417

Li SB, Hu H, Song XB et al (2019) Main controlling factors of the formation of large-scale natural gas fields in Sichuan Basin and the next exploration direction. Natural Gas Ind 2019(S1):1–8[2020-03-2121]

Li YP, Chen LX, Wang Y et al (2007) The main geological factors controlling migration and accumulation of the Silurian reservoir in the Middle Tarim Basin of Tarim Basin Chinese. Sci Bull S1:185–191

Lindgreen H (1985) Diagenesis and primary migration in upper Jurassic claystone source rocks in North Sea. AAPG Bull 69(4):525–536

Liu Z, Zhu WQ, Sun Q et al (2012) Geothermal-ground pressure system in China's petroliferous basins. Acta Petrolei Sinica 33(1):1–17

Liu ZT (2018) Differences in the lower limit of buoyancy formation in different zones of the Nanbu depression and analysis of influencing factors. China University of Petroleum (Beijing)

Lorenz JC, Mroz TH (1999) Natural fracturing in horizontal core near a fault zone: the Rock Island Unit 4-H well, Green River basin, WY Consortium for Emerging Gas Resources in the Greater Green River Basin, Integrating Geoscience and Engineering Data to Characterize and Exploit Tight Gas Sand Sweet Spots, April Denver, Colorado

Lu X, Wang Y, Tian F, Li X et al (2017) New insights into the carbonate karstic fault system and reservoir formation in the Southern Tahe area of the Tarim Basin. Mar Petrol Geol 86:587–605

Luo B, Luo WJ, Wang WZ et al (2015) Formation mechanism of Sinian Gas Reservoir in Leshan-Longnusi ancient uplift in Sichuan Basin. Nat Gas Geosci 26(03):444–455

Ma H, Wang X, Li X et al (2012) Analysis on the exploration potential of deep tight sandstone gas in Bohai Bay Basin. J Shandong Univ Sci Technol (Natural Sci) 31(05):63–70

Ma YS, Cai XY, Guo TL (2007) The main controlling factors of oil and gas filling and accumulation in Puguang large gas field, Sichuan Basin China. Chin Sci Bull S1:149–155

Ma YS, Fu Q, Guo TL et al (2005) Accumulation model and process of the Changxing-feixianguan gas reservoir in Puguang gas field, northeastern Sichuan Basin. Pet Geol Exp 27(5):35–41

Magara K (1975) Reevaluation of montmorillonite dehydration as cause of abnormal pressure and hydrocarbon migration. AAPG Bull 59(2):292–302

Magara K (1976) Thickness of removed sediments paleopore pressure and paleotemperature south western part of western Canada Basin. AAPG Bull 60:554–565

Magoon LB, Dow WG (1994) The petroleum system: from source to trap. AAPG 1994:2–24

Man HM, Ye JR, Li LX et al (2009) Characteristics of oilfield water in the western slope belt of Dongpu depression and its oil and gas geological significance. Geol Sci Technol Inf 3:67–72

Masters JA (1979) Deep basin gas trap, Western Canada. AAPG Bull 63(2):152–181

Maxwell JC (1964) Influence of depth, temperature, and geologic age on porosity of quartzose sandstone. AAPG Bull 48(5):697–709

McTavish RA (1998) The role of overpressure in the retardation of organic matter maturation. Petrol Geol 21:153–186

Niu WT, Zhu DX, Jiang LW, Han B, Zhang WP, Zhou CJ (2021) Sweet spot "comprehensive evaluation technology of complex mountain shale gas reservoir: taking the Zhaotong National Shale Gas Demonstration Zone as an example. Nat Gas Geosci 32(10):1546–1558

Osborne MJ, Swarbrick RE (1997) Mechanisms for generating over-pressure in sedimentary basins: a reevaluation. AAPG Bull 81(6):1023–1041

Pang B, Chen JQ, Pang XQ et al (2022) Driving forces and their relative contributions to hydrocarbon expulsion from deep source rocks: a case of the Cambrian source rocks in the Tarim Basin. Petroleum Sci 1995–8226

Pang XQ, Jia CZ, Wang WY (2015) Petroleum geology features and research developments of hydrocarbon accumulation in deep petroliferous basins. Petroleum Sci 12(1):1–53

Pang XQ, Jiang ZX, Huang HD et al (2014a) Genetic mechanism, development model and distribution prediction of superimposed continuous reservoirs. Acta Petrolei Sinica 35(05):795–828

Pang XQ, Lin HX, Zheng DY et al (2020a) Basic characteristics, dynamic mechanism and development direction of formation and distribution of deep and ultra-deep carbonate reservoirs in China. J Geomechanics 26(05):673–695

Pang XQ, Shao XH, Li MW et al (2021a) Correlation and difference between conventional and unconventional reservoirs and their unified genetic classification. Gondwana Res 97:73–100

Pang XQ, Wang WY, Wang YX et al (2015b) Comparison of reservoir-forming conditions and characteristics between deep and middle-shallow layers in petroliferous basins. Acta Petrolei Sinica 36(10):1167–1187

Pang XQ, Zhou XY, Yan SH et al (2012a) Research progress and development direction of hydrocarbon accumulation in superimposed basins in China: a case study of Tarim Basin. Petrol Explor Dev 39(06):649–656

Pang XQ, Meng QY, Jiang ZX et al (2010) A hydrocarbon enrichment model and prediction of favorable accumulation areas in complicated superimposed basins in China. Pet Sci 7(1):10–19

Pang XQ (2014) Adjustment and reconstruction of oil and gas reservoirs and simulation of hydrocarbon amount of structural failure. Science Publishing, Beijing, pp 1–280

Pang X, Chen Z, Chen F (1993) Study on numerical simulation of geological history, thermal history, hydrocarbon generation-expulsion history and quantitative evaluation of hydrocarbon source rocks in petroliferous basins. Beijing Geological Press, Beijing

Pang XQ, Jia CZ, Song Y et al (2022b) Quantitative evaluation of whole oil and gas systems: methodological principles and practical applications. Acta Petrolei Sinica 43(06):727–759

Pang XQ, Jia CZ, Wang WY et al (2021b) Buoyance-driven hydrocarbon accumulation depth and its implication for unconventional resource prediction. Geosci Front 12(4):101133

Pang XQ, Jiang ZX, Huang HD et al (2014b) Genesis mechanism, development mode and distribution prediction of stacked continuous oil and gas reservoirs. Acta Petrolei Sinica 35(05):795–828

Pang XQ, Zhou XY, Jiang ZX et al (2012b) Formation, evolution and predictive evaluation of oil and gas reservoirs in the stacked basin. Acta Geol Sin 86(01):1–103

Pang XQ, Zhu WL, Lv XX et al (2015c) Study on oil and gas threshold-controlled reservoirs in the eastern Bohai Bay Basin and prediction and evaluation of favorable reservoir formation zones. J Petroleum 36(S2):1–18

Pang XQ, Jia CZ, Pang H et al (2018) Destruction of hydrocarbon reservoirs due to tectonic modifications: conceptual models and quantitative evaluation on the Tarim Basin China. Mar Petrol Geol 91:401–421

Pang X, Jia C, Chen J et al (2021c) A unified model for the formation and distribution of both conventional and unconventional hydrocarbon reservoirs. Geosci Front 12(2):395–711

Pang X, Jia C, Zhang K, Li M, Wang Y, Peng J, Li B, Chen J (2020b) The dead line for oil and gas and implication for fossil resource prediction. Earth Syst Sci Data 12(1):577–590

Pang XQ (1995) The theory and application of hydrocarbon expulsion threshold on hydrocarbon control. Beijing Petroleum Industry Press, Beijing

Pang XQ (2014b) The hydrocarbon migration and accumulation threshold and the resource potential evaluation. Beijing Science Press, Beijing

Pang XQ, Jia CZ, Zhang K et al (2020c) The dead line for oil and gas and implication for fossil resource prediction. Earth Syst Sci Data 12:577–590

Pang XQ, Jiang ZX, Huang HD et al (2014c) Formation mechanism, distribution model and prediction of superimposed, continuous hydrocarbon reservoir. Acta Petrol Sin 35(5):795–828

Pang XQ, Li MW, Li SM, Jin ZJ (2005) Geochemistry of petroleum systems in the Niuzhuang South Slope of Bohai Bay Basin: Part 3 estimating hydrocarbon expulsion from the Shahejie formation. Org Geochem 36:497–510

Pepper AS (1992) Estimating the petroleum expulsion behavior of source rocks: a novel quantitative approach In: England WA, Fleet AL (eds) Petroleum migration geological society (London). Special Publication, London, pp 9–31

Peters K, Cassa M (1994) Applied source rock geochemistry. In: Magoon LB, Dow WG (eds) The petroleum system from source to trap. American Association of Petroleum, Tulsa. Geologists Memoir, vol 60

Peters KE, Hackley PC, Thomas JJ, Pomerantz AE (2018) Suppression of vitrinite reflectance by bitumen generated from liptinite during hydrous pyrolysis of artificial source rock. Org Geochem 125:220–228

Price LC (1993) Thermal stability of hydrocarbons in nature: limits, evidence, characteristics, and possible controls. L Geochimica et Cos-Mochimica AcLa 57(14):3261–3280

Price L (1976) Aqueous solubility of petroleum as applied to its origin and primary migration. AAPG Bull 60(2):213–244

Pusey WC (1973) The ESR-kerogen method: how to evaluate potential gas and oil source rock. World Oil 176(5):71–75

Ran QC, Chen SM, Zhou X (2021) Geochemical characteristics and enrichment of deep natural gas in the Xingshan Sag in Songliao Basin. Nat Gas Geosci 32(05):727–737

Robert MC, Suzanne GC (2004) The origin of Jonah field, Northern Green River basin, Wyoming. In: Robinson JW, Shanley KW (eds) Jonah field: case study of a tight-gas fluvial reservoir: AAPG studies in geology, vol 52. The Discovery Group Inc, Denver, Colorado, USA

Schenk CJ, Pollastro RM (2001) Natural gas production in the United States: USGS Fact Sheet FS-113-01

Scherer M (1987) Parameters influencing porosity in sandstone: a mode for sandstone porosity prediction. AAPG Bull 71(5):485–491

Schmoker JW (2002) Resource-assessment perspectives for unconventional gas systems. AAPG Bull 86(11):1993–1999

Schutter SR (2003) Ocurrences of hydrocarbons in and around igneousrocks. Geol Soc Londn Special Publ 214(1):35–68

Shi X, Dai JX, Zhao WZ (2005) Analysis of exploration prospects for deep oil and gas reservoirs. China Petroleum Exploration 10(1):1–10

Sun YG, Fu JM, Liu DH et al (1995) The influence of volcanic activity on the evolution of sedimentary organic matter and its oil and gas geological significance-taking the eastern depression of Liaohe Basin as an example. Chin Sci Bull 40(11):1019–1022

Sweeney JJ, Burnham AK (1990) Evaluation of a sample method of vitrinite reflectance based on chemical kinetics. AAPG Bull 74:1559–1570

Thompson KFM (1987) Fractionated aromatic petroleums and the generation of gas condensates. Org Geochem 11:573–590

Tissot BP, Welte DH (2013) Petroleum formation and occurrence. Springer, Heidelberg, pp 215–228

Tissot B, Durand B (1974) Influence of natures and diagenesis of organic matter in formation of petroleum. AAPG Bull 58(3):438–459

Vernik L, Landis C (1996) Elastic anisotropy of source rocks: implications for hydrocarbon generation and primary migration1. AAPG Bull 80(4):531–544

Wang TG, Dai SF, Li MJ et al (2010) Thermal history of stratigraphic organic matter in the Tarim Basin Basin and its implications for regional geological evolution. Scientia Sinica (Terrae) 40(10):1331–1341

Wang TG, Song DF, Li MJ et al (2014) Gas sources and deep gas exploration prospects of the Ordovician Yingshan Formation in the Shunnan-Gucheng Tarim. Basin Oil Gas Geol 35(06):753–762

Wang WY (2020) Pore seepage evolution and lower limit of reservoir reservoir emplacement in deep carbonate reservoirs in the stacked basin. China University of Petroleum (Beijing)

Wang XH, Feng ZC, Liu ZH (2009) Study on distribution law of rock density. Chin J Rock Mech Eng 28(S2):3484–3489

Wang, WY, Pang, XQ, Chen, ZX et al (2020) Improved methods for determining effective sandstone reservoirs and evaluating hydrocarbon enrichment in petroliferous basins. Appl Energy 261(114457)

Waples DW (2000) The kinetics of in-reservoir oil destruction and gas formation: constrains from experimental and empirical data, and from thermodynamics. Org Geochem 31:553–576

Wilson MD (1994) Non-compositional controls on diagenetic processes reservoir quality assessment and prediction in clastic rocks. SPEM Short Course 30(1994):183–208

Wood JM, Sanei H (2016) Secondary migration and leakage of methane from a major tight-gas system. Nat Commun 7:13614

Wu HY, Liang XD, Xiang CF et al (2007) Characteristics and reservoir-forming mechanism of syncline reservoir in Songliao Basin. Scientia Sinica (Terrae) 37(2):185–191

Wu HY, Wang YW, Liang XD et al (2015) Theory of petroleum accumulation in syncline and its significance to petroleum geology. Earth Sci Frontier 22(1):181–188

Xu CC, Shen P, Yang YM et al (2014) Natural gas accumulation conditions and enrichment law of Sinian-Lower Cambrian Longwangmiao formation in Leshan-Longnusi ancient uplift. Nat Gas Ind 34(03):1–7

Yan L, Feng MG, Zhang CY (2011) Reservoir-forming model of Changxing formation in Yuanba area, northeastern Sichuan Basin. J Yangtze Univ (Natural Sci Edn) 8(10):19–21+275

Yang H, Fu JH, Liu XS et al (2012) Tight gas reservoir-forming conditions and exploration and development of Upper Paleozoic in Ordos Basin. Petrol Explor Dev 9(3):295–303

Yang KM, Zhu HQ (2013) Geological characteristics of the overlaid tight sandstone gas field in Kawanishi. Pet Geol Exp 35(1):1–8

Yu JW, Pang XQ, Zheng DY, Wang XJ, Xiao HY, Zhang PW, Zhuo SQ, Li HY (2022) Prediction of main control factors and favorable zones for deep carbonate gas reservoirs in Puguang area. Special Oil Gas Reservoirs 29(03):28–35

Yu WQ, Chen Y, Yang LG et al (2014) Corrosion of Shi Ying in tight sandstone reservoir in acid environment. Acta Petrolei Sinica 35(2):286–293

Zhang BM, Wang SJ, Xiang H (2008) Thermal evolution of magma intrusion in basin and its influence on oil generation window. Xinjiang Petroleum Geol 29(6):696–698

Zhang H (1999) The discovery of the Tahe oilfield and its geological significance. Oil Gas Geol 20(2):120–124

Zhang Q, Zhang JC, Tang X (2009) Characteristics of oilfield water in China's petroliferous basins. In: Wang T, Jia C, Pang X et al (eds) Proceedings of the 5th international symposium on hydrocarbon accumulation mechanism and evaluation of hydrocarbon resources. Beijing Petroleum Industry Press, Beijing

Zhang RH, Yang HJ, Wang JP et al (2014) Formation mechanism of ultra-deep low porosity tight sandstone reservoir in Kuqa depression and its significance for oil and gas exploration. Acta Petrolei Sinica 35(6):1057–1069

Zhang XG, Pang H, Pang XQ et al (2022) Characteristics of hydrocarbon generation and expulsion and resource potential of source rocks of the Longtan formation of the Upper Permian in Sichuan Basin. China Oil Gas Geol 43(03):621–632

Zhang YC, Zhang BM, Li BL et al (2011) The history of hydrocarbon accumulation in Chinese marine basins spanning major tectonic periods: a case study of Tarim Basin. Pet Explor Dev 38(01):1–15

Zhao R, Zhao T, Li HL et al (2019) Characteristics and main control factors of fault-controlled fractured-vuggy reservoirs in Shunbei oil and gas field in Tarim Basin. Special Oil Gas Reservoirs 26(05):8–13

Zhao WZ, Wang ZC, Jiang H et al (2020) The exploration position of deep Sinian system in Sichuan Basin from the formation conditions of ancient carbonate oil and gas fields. Nat Gas Ind 02:1–10

Zheng M, Li J, Wu X et al (2019) Potential of oil and natural gas resources of main hydrocarbon-bearing basins and key exploration fields in China. Earth Sci 44(03):833–847

Zhou SX, Song ZX, Wang BZ et al (2011) Characteristics of organic matter in Tarim Basin carbonate rocks. Petrol Explor Dev 38(03):287–293

Zhou XY, Wang ZM, Yang HJ et al (2006) The fifth example of marine oil and gas field exploration in China: exploration and discovery of large-scale Ordovician condensate gas field in Tazhong. Marine Origin Petroleum Geol 11(1):45–51

Zhou XY, Lv XX, Yang HJ et al (2013) Impact of strike-slip faults on the north slope of Tazhong on the differential enrichment of carbonate oil and gas. Acta Petrolei Sinica 34(4):628–637

Zhu G, Zhang Z, Zhou X et al (2019) The complexity, secondary geochemical process, genetic mechanism and distribution prediction of deep marine oil and gas in the Tarim Basin. China Earth-Sci Rev 102930

Zhu GY, Zhang SC, Wang HH et al (2009) Formation and distribution of deep weathered crust reservoirs in northern Tarim Basin. Acta Petrologica Sinica 25(10):2384–2398

Typical Examples and Exploration Potentials of Large-Scale Petroleum Accumulations in Deep and Ultradeep Carbonate Reservoirs in China

Xunyu Cai, Huili Li, Zicheng Cao, Bisong Li, Jun Han, Xiuxiang Zhu, Qingfeng Zhang, Zuxin Xu, and Chongyang Wu

Abstract

In deep and ultradeep marine carbonate fields, a number of large oil and gas fields with different types of reservoirs and different accumulation processes have been discovered in China. In this study, the basic conditions, accumulation processes and key reservoir-controlled elements of the Tahe Oil Field, Shunbei Oil and Gas Field, Puguang Gas Field and Yuanba Gas Field, which were discovered in the Tarim Basin and Sichuan Basin, were compared and analyzed. The Tahe Oil Field and Shunbei Oil Field have formed typical paleokarst fracture–cavity reservoirs and fault-controlled fracture–cavity reservoirs under the influence of paleo-uplifts development and strike-slip fault activity. After multistage accumulations and multiple transformations, the planar distribution of oil and gas accumulation in different phases is still regular with some order. Under the influence of the development of Kaijiang–Liangping continental shelf, the Puguang Gas Field and Yuanba Gas Field have formed typical platform-margin reef–shoal–type reservoirs with favorable near-source enrichment accumulation conditions, which have also experienced whole processes of oil and gas phase transformation. Large-scale source–reservoir efficient source and reservoir configurations of the source and reservoirs, and good preservation conditions are important conditions for the formation of deep and ultradeep large oil and gas fields. In the three superimposed basins of Tarim, Sichuan and Ordos, paleokarst reservoirs related to large-scale unconformities, fault-controlled reservoirs related to Multistage faults, platform edge mound-shoal reservoirs related to sedimentary differentiation, and widely distributed paleomicrobial rocks and dolomite reservoirs that developed in confined tidal flat environments, are all important exploration targets.

Keywords

Large oil and gas accumulation · Key elements of petroleum accumulation · Exploration potential · Deep carbonates

9.1 Introduction

Deep marine carbonate reservoirs have become a hot topic and the focus of hydrocarbon exploration

X. Cai (✉)
China Petroleum and Chemical Corporation (Sinopec), Beijing, China
e-mail: caixy@sinopec.com

H. Li · C. Wu
Petroleum Exploration and Production Research Institute, SINOPEC, Beijing, China

Z. Cao · J. Han · X. Zhu
Northwest Oilfield Company, SINOPEC, Urumqi, China

B. Li · Q. Zhang · Z. Xu
Exploration Company, SINOPEC, Chengdu, China

Y. Ma et al. (eds.), *Petroleum Geology and Exploration of Deep Marine Strata in China*, Advances in Oil and Gas Exploration & Production, https://doi.org/10.1007/978-3-032-02496-1_9

worldwide. With the continuous exploration and development of middle and shallow oil and gas reserves, the difficulty involved in oil and gas discovery is increasing. Increasing attention has been paid to the exploration and development of oil and gas reserves in deep and ultradeep carbonates, and new reserves of deep oil and gas in the world have shown a notable growth trend (Ma et al. 2020).

There are many different definitions of "deep" and "ultradeep" at home and abroad (Bai and Cao 2014). With respect to carbonate reservoirs, strata with burial depths of 4500–6000 m are defined as deep layers according to the definition of drilling engineering, and strata with burial depth greater than 6000 m are defined as ultradeep layers (Ma et al. 2020). Through statistical analysis of 349 basins around the world, Dyman et al. (2002) reported deep oil and gas accumulation in 87 basins. The oil and gas accumulations with burial depths of more than 6000 m abroad are concentrated in Mesozoic-Cenozoic clastic strata.

However, the deep and ultradeep marine carbonate rocks in China are distributed mainly in the Cambrian Ordovician system, Precambrian and Permian Triassic systems of the three major superimposed basins (namely, the Tarim Basin, Sichuan Basin and Ordos Basin) in central and western China. These sites are characterized by their deep burial depth and old age (Ma et al. 2020). Therefore, clarifying whether large amounts of oil and gas accumulation can occur in the deep-ultradeep environment, and further revealing the accumulation process and key elements involved in reservoir formation are key points for promoting exploration breakthrough.

After long-term exploration practices, Chinese oil companies have made great breakthroughs in deep and ultradeep marine carbonates in Sichuan and Tarim Basins (Ma et al. 2020; He et al. 2019). A number of large oil and gas fields with different reservoir types and distinct accumulation processes have been discovered, such as the Tahe Oil Field and Shunbei Oil and Gas Field in Tarim Basin (Liang 2008; Zhai 2011; Zhai and Yun 2008), the Puguang Gas Field and the Yuanba Gas Field, Anyue Gas Field and others in Sichuan Basin (Ma et al. 2007; Zhao et al. 2018; Zhu et al.

2020; Zou et al. 2014). All these oil and gas fields (reservoirs) exist in deep and ultradeep ancient marine carbonate strata with burial depths greater than 5000 m, with various types of reservoirs and different accumulation processes.

In this chapter, the basic conditions, accumulation processes and key petroleum accumulation elements of the Tahe Oil Field, Shunbei Oil and Gas Field, Puguang Gas Field and Yuanba Gas Field in the Tarim and Sichuan Basins, are compared and analyzed. The basic characteristics and enrichment laws of several typical large oil and gas accumulations are briefly introduced, and their exploration potential is discussed accordingly to provide a theoretical basis for greater breakthroughs in oil and gas exploration.

9.2 Tahe–Shunbei Deep Carbonate Rock Large-Scale Oil and Gas Accumulation Area in the Tarim Basin

As a typical superimposed basin in western China, the Tarim Basin is rich in oil and gas resources and is estimated to contain more than 25.8 billion tons of oil equivalent. After more than 70 years of arduous exploration in the basin, more than 70 oil and gas fields (reservoirs) have been discovered, which are distributed mainly in the two Cenozoic foreland basins of Kuqa and the southwest, and in the carbonate rocks of the Ordovician marine craton basin within the platform area. Among these reservoirs, the western margin of the Manjiaer Depression, which includes the Shaya Uplift, Shuntuoguole Low Uplift and Guchengxu Uplift, is the region with the greatest oil and gas enrichment in the platform basin (Fig. 9.1). Within this area, five oil and gas fields have been discovered with proven reserves of more than 2.0 billion tons of oil equivalent, in which Tahe Oilfield and Shunbei Oil and Gas Field are typical representatives.

In 1984, a new oil discovery in Ordovician marine carbonate rocks was made form the Shacan 2 well. After more than ten years of persistent exploration, in 1997, the Sha 46, Sha 47 and Sha 48 wells obtained industrial oil and gas flow

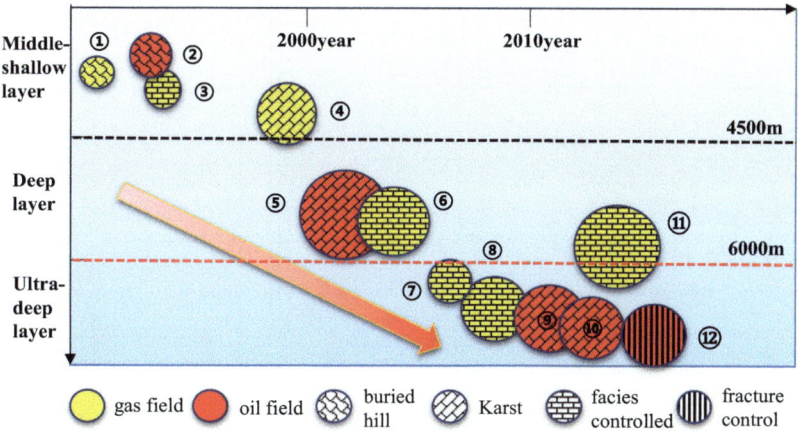

Fig. 9.1 Major carbonate oil and gas fields and reservoir types in China (modified from Ma et al. 2020). ① Weiyuan, Z ~ Є; ② Renqiu, Jx; ③ Goshiti, Z ~ Є; ④ Jingbian, O; ⑤ Tahe main body, O; ⑥ Puguang, P ~ T; ⑦ Longgang, P ~ T; ⑧ Yuanba, P ~ T; ⑨ Halrahatang, O; ⑩ Peripheral and deep layers of the Tahe, O; ⑪ Anyue, Z ~ Є; ⑫ Shunbei, O

in Ordovician carbonate rock buried more than 5000 m deep in the main part of the Shaya ancient uplift, marking the discovery of the Tahe Oilfield, a typical unconformity karst fracture–cave oilfield (Zhai 2011).To dateTo date, in the Tahe Oil Field, the proven reserves are more than 1.5 billion tons of oil equivalent, and the cumulative oil production is more than 110 million tons. Nearly 20 years after the discovery of the Tahe Oil Field, with breakthroughs in research on the strike-slip faults and related reservoirs, the Shunbei 1–1 Well and other wells, deployed in the strike slip fault zone of the Shuntuogole low uplift, successively drilled the fault-controlled fracture–cavern reservoir with a burial depth of more than 7000 m, and obtained high-yield industrial oil and gas flows with a daily output of more than 100 tons, which marked the discovery of the Shunbei Oil and Gas Field (Jiao 2018). To date, in the Shunbei Oil and Gas Field, the proven reserves are more than 200 million tons of oil equivalent, and the production capacity of oil and gas is up to one million tons.

9.2.1 Geological Background

The Tahe and Shunbei Oil and Gas Fields, located in the central and northern parts of the Tarim Basin (Fig. 9.2a), are located across multiple structural units such as the Shaya Uplift, Shuntuoguole Low Uplift, Awati Depression and Manjiaer Depres-

sion (Fig. 9.2b). The main body of the Tahe Oil Field is located on the Akekule Uplift of the Shaya Uplift, and the main body of the Shunbei Oil and Gas Field is located on the Shuntuoguole Low Uplift adjacent to the Shaya Uplift. The Shuntuoguole Low Uplift is adjacent to the Katake Uplift and the Shaya Uplift in the north and south, and adjacent to the Manjiaer Depression and the Awati Depression in the east and west, and is now a "saddle" geological tectonic structure.

Owing to the influence of Multistage activities of the surrounding oceanic basins and orogenic belts of the Tarim Basin, Both the Shaya Uplift and the Shuntuoguole Low Uplift have experienced three stages of tectonic evolution: stable tectonic subsidence with weak extension in the Early–Middle Caledonian craton weak extension, the formation of large paleo-uplifts and the development of low-uplifts with regional compression of Late Caledonian–Early Hercynian uplifts, and adjustment and transformation since the Middle and Late Hercynian and Indosinian (An et al. 2009; Deng et al. 2019, 2018; He et al. 2008). Multistage tectonic activity resulted in the development of strike-slip faults within the Shaya Uplift–Shuntuoguole Low Uplift, and activity characteristics of these faults differ across different tectonic locations (Deng et al. 2019, 2018) (Fig. 9.2b). In the Paleozoic, the Shaya Uplift and Shuntuoguole Low Uplift featured

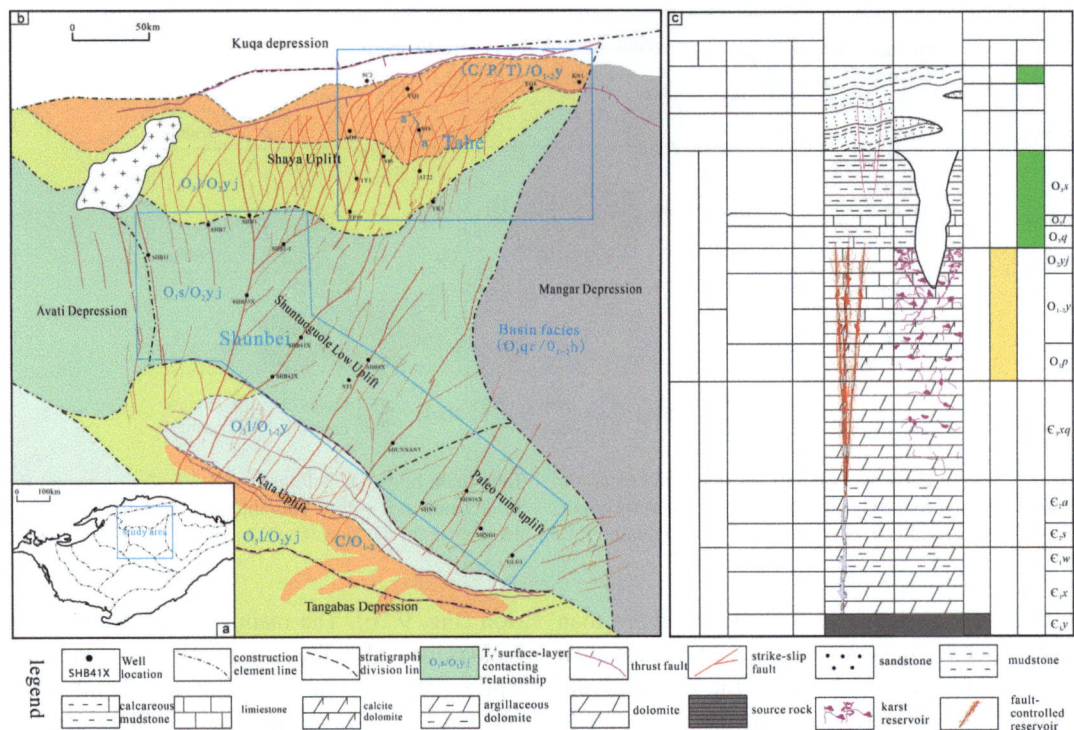

Fig. 9.2 Schematic diagram of the structural location and source–reservoir–cap combination within Tahe–Shunbei (a-location of the study area, b-structural division and fault distribution of the study area, c-schematic diagram of the source–reservoir–cap combination)

relatively consistent marine and marine–continental transitional sedimentary facies. During the Cambrian–Ordovician, they were composed mainly of platform facies carbonate rocks and shelf facies mudstone, with tidal flat and coastal clastic deposits generally in the Silurian–Devonian and deposits of marine–continental transitional facies rock and carbonate generally in the Carboniferous. During the Cambrian–Middle Ordovician, carbonate strata with a thickness of approximately 3000 m developed in areas of the Tahe and Shunbei and became the development material basis of large-scale unconformity karst fracture–cavity reservoirs in the Tahe Oil Field and fault-controlled fracture–cavity type large-scale reservoirs in the Shunbei Oil and Gas Field.

During the formation of the Shaya Paleo Uplift in the Late Caledonian–Early Hercynian, the Middle–Lower Ordovician carbonate rocks in the main part of the paleo uplift were denuded. As a result, denudation/hiatus was missing from the Upper Ordovician, Silurian and Devonian from, and the Carboniferous was covered and above the Middle and Lower Ordovician. Under the influence of these paleo uplifts and slopes, typical unconformity karst fracture–cave reservoirs, which developed in the Middle–Lower Ordovician carbonate rocks, formed a good reservoir–cap combination with the overlying Carboniferous mudstone and gypsum mudstone (Fig. 9.2c). Compared with the Shaya Uplift, the Shuntuoguole Low Uplift is always in the lower part of the paleostructure with a relatively complete strata development and preservation. Although the Middle and Lower Ordovician carbonate rocks do not have the conditions for unconformity karst development, they have been transformed by Multistage active strike-slip faults, formed distinctive fault-controlled fracture–cavity reservoirs, and further formed a superior reservoir-cap combination with the overlying Upper Ordovician mud rocks (Fig. 9.2c).

9.2.2 Development Characteristics of Karst Fracture–Cave Reservoirs in the Tahe Area and Fault-Controlled Karst Fracture–Cave Reservoirs in the Shunbei Area

Large amounts of drilling, logging, well testing, seismic and analysis data have revealed that the Tahe Oil Field and Shunbei Oil and Gas Field have developed two distinct types of large-scale reservoirs: unconformity karst fracture–cave reservoirs and fault-controlled fracture–cave reservoirs.

(1) Tahe unconformity karst fracture–cave reservoirs

Karst fracture–cave reservoirs within the Tahe Oil Field are mainly developed in the Middle–Lower Ordovician limestone with burial depths of 5300–7000 m. The carbonate matrix of this area is rela-tively dense and essentially no storage capacity. Its effective storage space is mainly caves, pores and fractures from multistage tectonic processes and paleokarstification in the Middle Caledo-nian–Late Hercynian (Fig. 9.3). The storage and seepage spaces have various geometric shapes with significant differences in size and uneven distribution (Zhai 2006; Zhai and Yun 2008). In accordance with the differences in the devel-opment degree of karst caves, dissolution pores and caves and dissolution fractures, unconformity karst fracture–cave reservoirs within the Tahe Oil Field can be divided into three types, namely fracture–karst reservoirs, fracture–cavity ones and fracture type ones. Within the high parts of the paleo-uplift and slope formed in the Late Caledo-nian–Early Hercynian, fracture–cavity-type reser-voirs are mainly developed and distributed along the unconformity. Within the lower part of the slope, fracture–cavity and fracture-type reservoirs are mainly developed and distributed along the fault (Fig. 9.4).

Fig. 9.3 Images of a typical Ordovician karst fracture–cavity core of in the Tahe Oil Field. **a** $O_{1-2}y$ from Well T615, karst breccia, with an interface of yellow and gray limestone breccias filled with light gray argillaceous silt-stone. **b** $O_{1-2}y$ of cavity chemical-filled giant crystal calcite from Well TS3, **c** $O_{1-2}y$ of yellow lime mud microcrys-talline limestone with crude oil-infiltrated dissolution frac-tures from Well S47. **d** $O_{1-2}y$ of yellow lime mud micro-crystalline limestone with crude oil-disseminated disso-lution fractures from Well S47. **e** $O_{1-2}y$ of yellow–gray microcrystalline limestone, and dissolved pores and cavi-ties inside and around the silicified nodules from Well AD19. **f** $O_{1-2}y$ of stalactites in karst caves from Well TS3. **g** $O_{2}yj$ of developed dissolved cavities, which are filled with asphalt, from Well T272

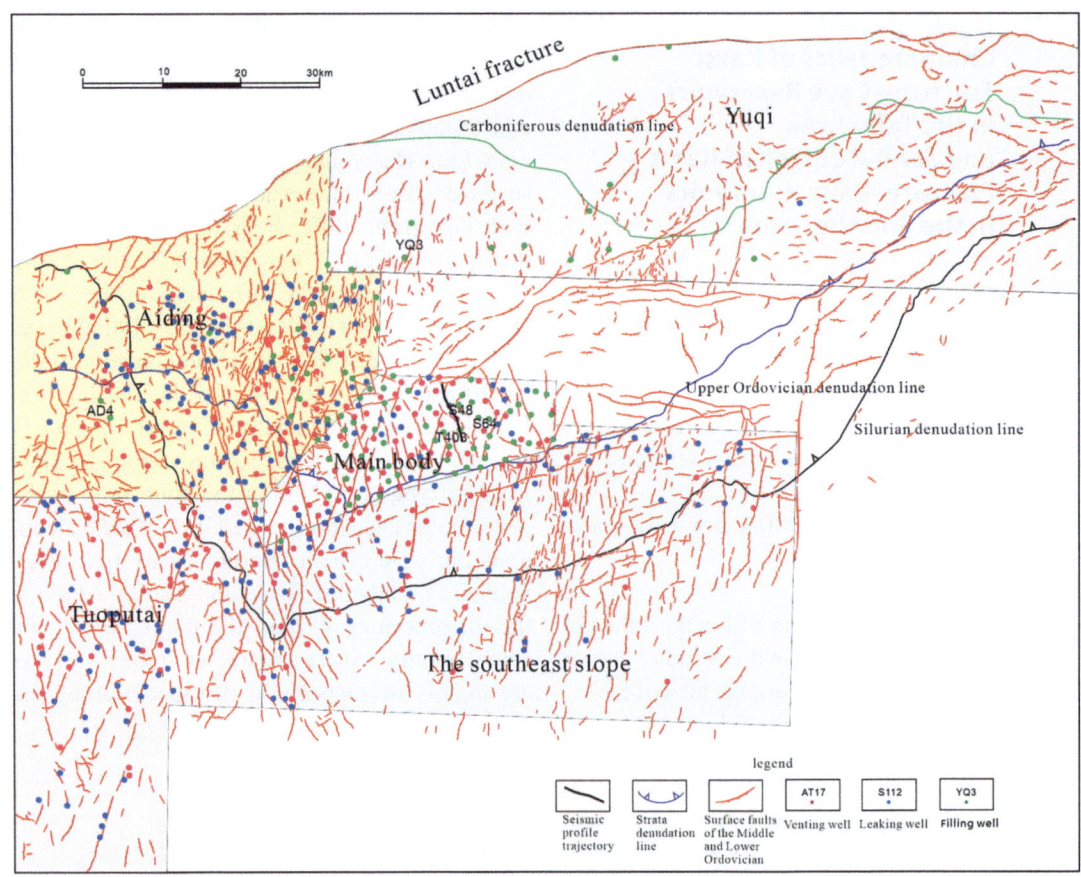

Fig. 9.4 Superimposition map of the venting well, leakageg well, filling well and fault in the Tahe Oil Field

Large fractures and caves are the most characteristic reservoir space types in the Tahe Oil Field, and are characterized mainly by venting and mud leakage during drilling. As shown in Table 9.1 and Fig. 9.4, large-scale karst fractures and caves are commonly encountered during drilling in different areas of the Tahe Oil Field, with several to tens of meters of empty spaces, and hundreds to thousands of cubic meters of drilling fluid leakage. Within the main area of the Tahe oil field, the superimposition of multistage paleokarstification had resulted in the development of large-scale fractures and caves which were filled and reformed to varying degrees. According to the statistics of 110 wells within the main area, both the venting and mud leakage rates can reach 38.2%. Because the karst caves in this area are mainly medium and large caves with heights of 5–100 m and are neither filled nor

half filled, they have very large storage spaces with very high productivity. Because the distribution of large-scale karst fractures and cavities is controlled by karst paleogeomorphology and paleowater systems (Qi and Yun 2010; Zhai 2011), these fractures and cavities can be densely distributed or developed in isolation spatially.

(2) Shunbei fault-controlled fracture–cavity reservoirs

Fault-controlled fracture–cavity reservoir in the Shunbei area have developed in the ultradeep layers with a burial depth greater than 7000 m and are a special type of reservoir group formed by the development of strike-slip faults (Jiao 2018; Lu et al. 2021; Ma et al. 2022; Zhai 2006). Their storage space is formed mainly by the fracture–cavity system from tectonic ruptures, including

Table 9.1 Statistics of caves encountered by typical drilling of Ordovician wells in the Tahe Oil Field

Study area	Well No	Layer position	Burial depth of Karst top (m)	Venting/leaking section (m)	Height of fracture/ cavity (m)	Leakage amount (m^3)
Tahe main body	Sha 48	O1-2y	5363.5	5364.26–5365.76 (venting)	1.5	2318.8
	Sha 61	O1-2y	5467.5	5665.9–5667.19 (venting)	1.29	673
	Sha 64	O1-2y	5470.5	5471.5–5498 (filling)	26.5	/
	T403	O1-2y	5405	5439–5447 (venting) 5488–5554 (venting)	8 66	/
	T60	O1-2y	5540	5583–5598.15 (venting)	15.15	829.4
	T61	O1-2y	5521	5534–5553/19 (venting)	19	/
	TK51	O1-2y	5445	5527–5573 (venting)	46	245
Southeast slope	Sha 106–1	O2yj	5796	5839.17–5840.3 (venting) 5851.89–5884.7 (venting)	1.13 32.81	380
Aiding	Aiding 4	O2yj	6495.5	6518–6543 (venting)	25	1704.1
Yuqi	Yuqi 3–1	O1-2y	5867	5930–5992 (venting)	62	/

fault cavities (caves), tectonic fractures, tectonic breccia fractures, and a small number of dissolved cavities and pores developed along the fault fractures (Table 9.2; Fig. 9.5).

Drilling has revealed that this reservoir space is distributed mainly along the strike-slip fault with a strong spatial heterogeneity as a "palette" (Table 9.2; Fig. 9.6). During the process of traversing the fault zone, laterally inclined wells and horizontal wells often encounter multiple reservoir sections with small-scale venting and mud leakage. According to drilling cores and logging interpretation, broken breccias developed inside the fault, which appear as "gravel-like bright spots" in imaging logging, and features high porosity and high permeability (Fig. 9.5e). Unlike from the drill well that horizontally traverses the fault, the drill well beyond 100 m from the fractured system rarely experiences any venting or leaking phenomena. The bedrock between faults is dense essentially and lacks any large-scale reservoir space. As shown in Well SHB41X, which traverses the Shunbei No. 4 fault (Fig. 9.6), the reservoir space has lateral and vertical heterogeneity within the strike-slip fault. Laterally, multiple sets of core-zone structures developed within the fault zone have their own system and formation characteristics (Ma et al. 2022, 2019), with a breccia zone width of 3.48–8.45 m, and a fracture width of 2.98 m (Fig. 9.6). Vertical differ-

ences in the development of fault-controlled scale reservoirs exist, and the types and scales of reservoir spaces within different drilling depths significantly differ (Aydin and Berryman 2010; Ferrill et al. 2014).

Manyfield and drilling studies have shown that the development scale and spatial distribution of fault-controlled fracture–cavity reservoirs within the Shunbei area are controlled mainly by the segmentation style, activity characteristics and activity scale of strike-slip faults (Ma et al. 2022; Yun and Deng 2022). Recent studies have also revealed that the activities of various types of fluids along the fault may have a certain effect on early formed fracture–cavity systems (Huang et al. 2022; Yu et al. 2022).

In addition to the alteration of tight Ordovician limestones into high-quality reservoirs through fault and fluid interactions, grain shoal facies limestones deposited in high-energy environments may also develop into higher quality reservoir units through the combined effects of sedimentary facies, faulting, and diagenetic fluids alteration. For example, intraplatform grain shoal limestone reservoirs in Wells Shunnan-7 and Shun-7 were enhanced by faulting and subsequent diagenetic alteration by fluids migrating through fractures, resulting in the formation of high-quality Ordovician reservoir. Well cores from the Yijianfang Formation at depths exceeding 6000 m in Well

Table 9.2 Statistics table of reservoir parameters of Yingshan Formation-Yijianfang Formation of typical high-yield wells in each fault within the Shunbei Oil and Gas Field

Fault	Well No	Full diameter physical properties of the core section		Well logging-based reservoir						Test results	
		Porosity (%)	Penetration (10^{-3} μm^2)	Fracture type		Cavity type		Pore type		Daily oil production/t	Daily gas production/ 10^4 m^3
				Thickness/ m	Number of layers	Thickness/ m	Number of layers	Thickness/ m	Number of layers		
No. 1	Shunbei 1–3	$\frac{0.09\sim1.60}{1.13(13)}$	$\frac{2.77\sim31.90}{10.4(12)}$	36.1	12	15.2	9			185.1	13.0
	Shunbei 1-7H	$\frac{1.80\sim4.70}{2.90(6)}$	$\frac{0.27\sim21.90}{5.23(6)}$							191.8	8.9
No. 4	Shunbei 41X	$\frac{0.40\sim0.60}{0.50(3)}$	$\frac{0.02\sim2.02}{0.77(3)}$	18.4	13	2.3	5	9.5	4	226.6	111.4
	Shunbei 42X	$\frac{0.60\sim1.60}{1.20(5)}$	$\frac{1.39\sim4.44}{3.58(5)}$	22.9	24	7.3	14	0.9	4	322.8	84.6
No. 5	Shunbei 5–1	$\frac{0.20\sim1.40}{0.80(21)}$	$\frac{0.02\sim0.32}{0.13(21)}$	27.7	4	0.9	1			121.5	2.7
	Shunbei 51X	$\frac{0.70\sim1.40}{0.96(16)}$	$\frac{0.06\sim14.40}{1.35(16)}$							154.9	4.5

Note The numerator represents the value range, the denominator represents the average value, and the number in parentheses is the total of samples

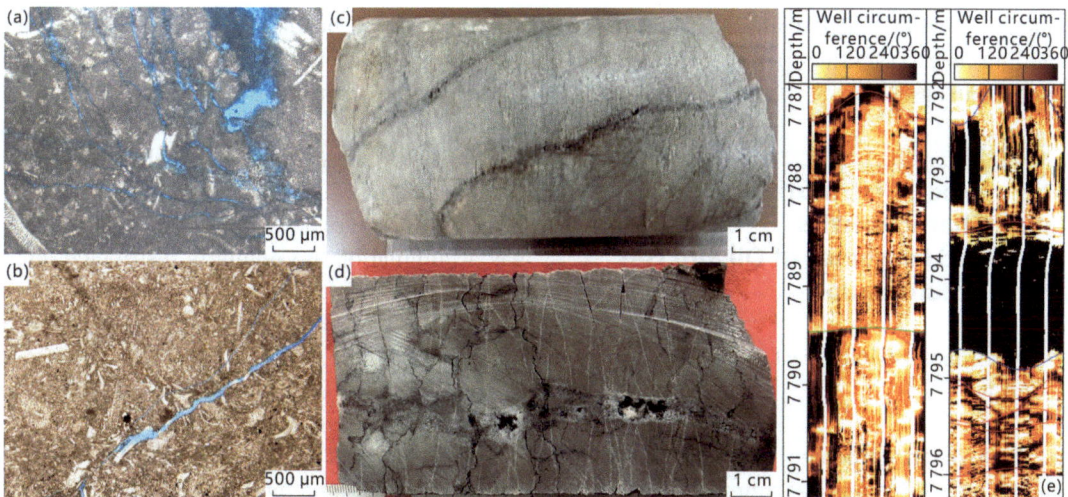

Fig. 9.5 Main reservoir space characteristics of fault-controlled fracture–cavity reservoirs in the Shunbei Oil and Gas Field (after Ma et al. 2022). **a** Well Shunbei 2 in the Yingshan Formation with bioclastic micrite, network microfractures and a depth of 7736.85 m, **b** Well Shunbei 1-7in the Yijianfang Formation with biogenic Detrital micrite, micro-fractures and, a depth of 7325.15 m, **c** Well Shunbei 5-1X in the Yijianfang Formation with micrite, the calcite-half-filled dissolution fractures and a depth of 7481.38–7481.61 m, **d** Shunbei Well 42X with sand-bearing micritic limestone, high-angle fractures and a depth of 7413.56–7413.62 m, and **e** drilling leaking location of Well Shunbei 7 in which imaging logging reveals fractures and semi-filled caves (black low-resistance part)

Fig. 9.6 Scheme of development of the strike-slip fault structure-controlled reservoirs within Well Shunbei 41X revealed through actual drilling (after Ma et al. 2022)

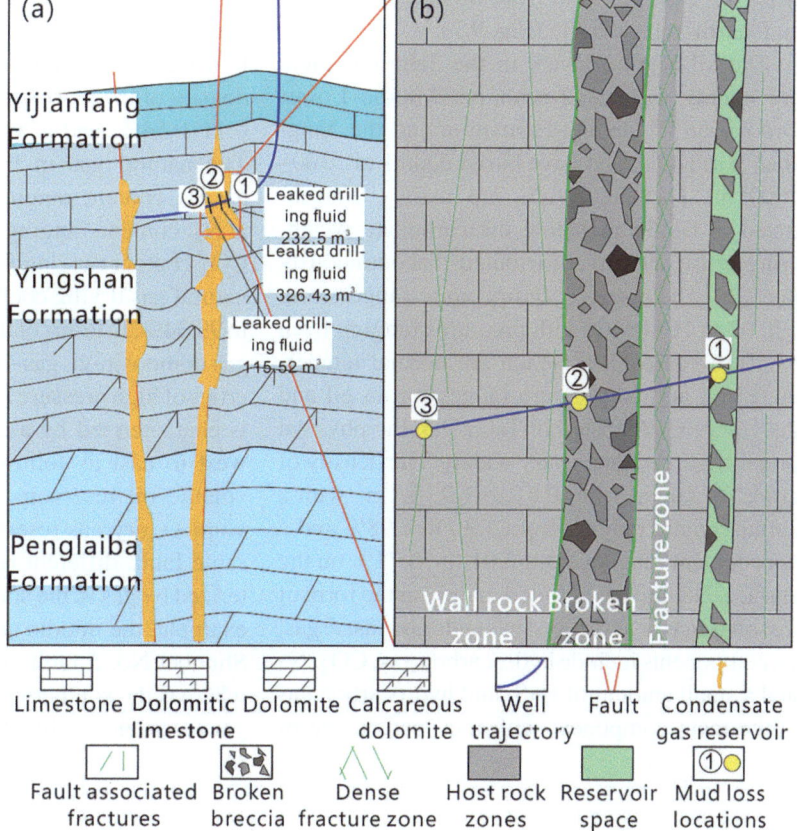

Shunnan-7 exhibit abundant reservoir spaces, including intercrystalline pores, fractures, and dissolution vugs.

9.2.3 Types, Distribution Characteristics and Processes of Oil and Gas Accumulation

(1) Types and distribution of oil and gas accumulation

The Ordovician carbonate rocks in the Tahe–Shunbei large oil and gas accumulation area have oil-bearing characteristics as a whole body. The phase states of oil and gas reservoirs in the area are diverse, and heavy oil reservoirs, medium oil reservoirs, light oil reservoirs, volatile oil reservoirs, condensate gas reservoirs and dry gas reservoirs have been drilled. Different oil and gas accumulations have the lateral distribution characteristics of "east–gas–and–west–oil and south–gas–and–north–oil" laterally (Fig. 9.7).

The oil-bearing layers in the Tahe Oil Field are mainly associated with the Middle–Lower Ordovician Yijianfang Formation and the Yingshan Formation, and have burial depths of 5300–7000 m. Their reservoirs are mainly concentrated in 0–150 m below the carbonate weathering crust, and are distributed discontinuously laterally. The reservoir temperature ranges from 120.02 to 163.01°C, with pressures ranging from 57.80 to 76.96 MPa, which are normal temperature and normal pressure ranges for an oil and gas reservoir. As listed in Table 9.3, the physical properties of crude oil vary widely. The density of crude oil ranges from 0.81 to 1.05 g/cm^3 with a sulfur content ranging from 1.30 to 2.98% and a wax content ranging from 2.01 to 10.71% on the surface. Natural gas is found mainly in the form of dissolved gas or associated gas and condensate gas. Its components include hydrocarbon gas, CO_2, N_2 and a small amount of H_2S, and hydrocarbon gas is the main component with total percentage of

89.17–98.57%. The distribution range of methane content is large, ranging from 49.8 to 93.64%, and its distribution mainly shows the characteristics of gradually decreases from east to west (Table 9.3; Fig. 9.7). In terms of the physical properties of surface crude oil, its production dynamic data and PVT high-pressure physical properties and oil and gas accumulation are regularly distributed laterally as a characteristic ring-shaped distribution, from south to north, of condensate gas reservoirs → volatile oil reservoirs → light oil reservoirs → medium oil reservoirs → heavy oil reservoirs → superheavy oil reservoirs (Fig. 9.7).

The oil and gas layers in the Shunbei Oil Field belong to the Yijianfang Formation and the Yingshan Formation of the Middle–Lower Ordovician with the oil and gas reservoir depth of 7300–8200 m, and a special "three-dimensional plate" spatial distribution along the fault. Its reservoir temperature is 150–180°C with a pressure of 81–95 MPa, characteristics of a high temperature and normal pressure oil and gas reservoir. As listed in Table 9.3, the crude oil density, natural gas composition and type of oil and gas accumulation are significantly different, with a crude oil density of 0.758–0.851 g/cm^3, a sulfur content of 0.11–0.54%, and a wax content of 0.46–9.89%. The natural gas in the Shunbei Oil Field has a CH_4 content ranging from 43.44 to 88.57%, a N_2 content ranging from 0.92 to 23.24%, a CO_2 content ranging from 0.20 to 16.72%, and a natural gas drying coefficient ranging from 51.69 to 99.34%. In terms of the crude oil density, natural gas composition, gas–oil ratio and physical properties of high pressure PVTs, the Shunbei Oil Field is characterized by a continuous transition from west to east as follows: light oil reservoirs → volatile oil reservoirs → condensate gas reservoirs → dry gas reservoirs (Fig. 9.7). Along the same fault, different reservoir facies are characterized by gas in the south and oil in the north. For example, the middle and northern sections of the Shunbei No. 5 fault are light oil reservoir areas, whereas the southern section features condensate gas reservoirs (Table 9.3).

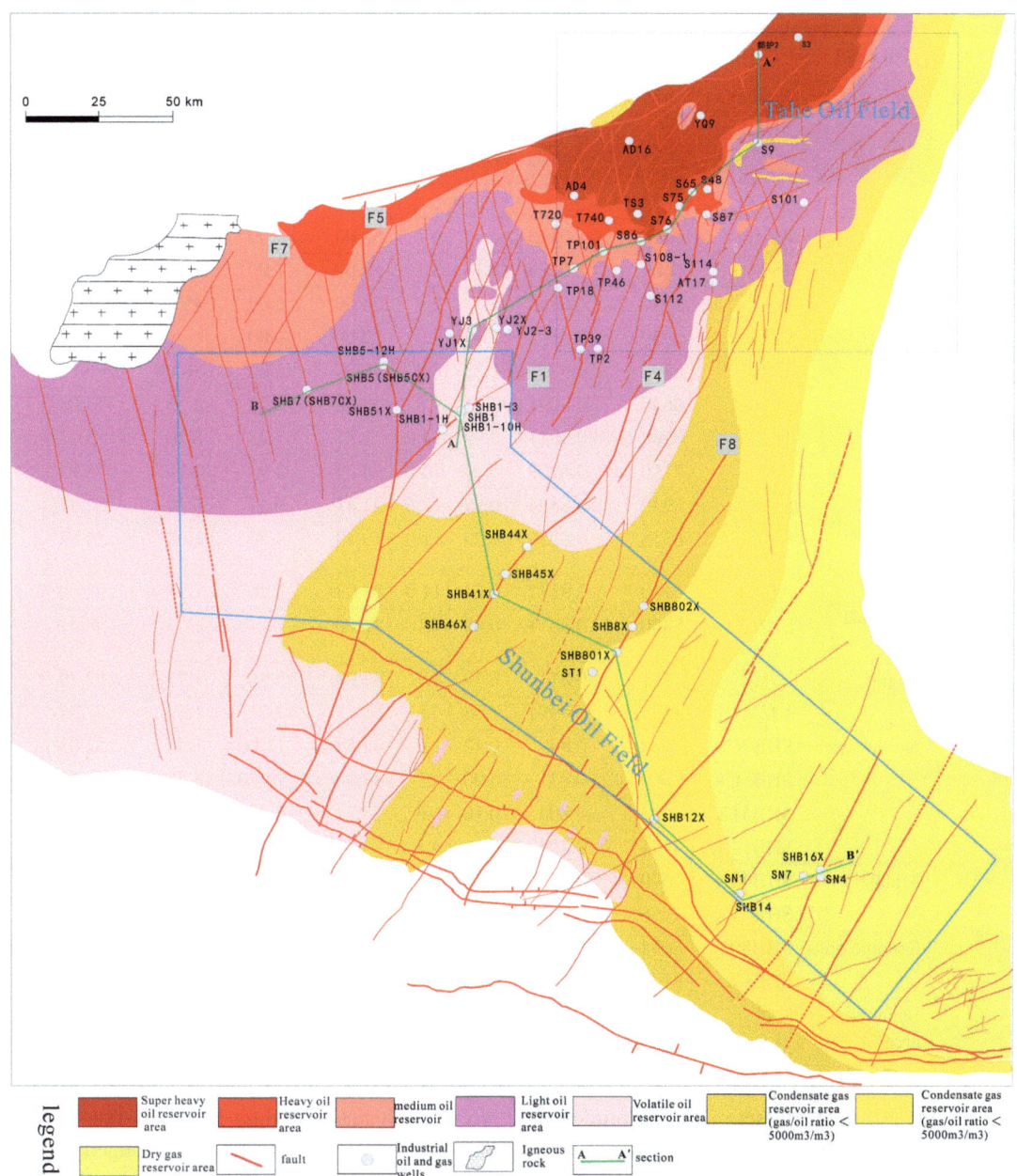

Fig. 9.7 Plane distribution of Ordovician oil and gas reservoir types in the Tahe–Shunbei area

(2) Oil and gas sources and their secondary transformation

Both oil and gas within the Tahe–Shunbei oil and gas accumulation area generally have similar geochemical characteristics because they are all from the Cambrian–Ordovician marine source rocks. However, there are regional differences in the maturity and secondary transformation of their crude oil, and their natural gas features kerogen cracking gas as well as crude oil cracking gas. Their oil and gas are of marine origin, but their lithofacies and layer locations may differ (Liang 2008; Ma et al. 2020; Zhai and Yun 2008).

Table 9.3 Statistics of temperature, pressure and other physical properties of oil and gas reservoirs in the Tahe Oil Field and Shunbei Oil Field

Oil and gas field	Area	Typical well	Oil and gas reservoir burial depth/m	Crude oil properties						
				Ground density (20 °C)	S (%)	Wax (%)	CH_4 (%)	N_2 (%)	CO_2 (%)	Drying coefficient (%)
Tahe	Main zone	AD16	6370	1.0513	2.47	6.1	58.00	8.00	13.00	77.33
		TS3	6608	1.0229	3.14	3.1	50.00	1.00	12.00	74.62
		YQ9	5760	1.0385	3.12	2.9	36.00	6.00	45.00	76.08
		S48	5369	0.9583	2.44	7.9	50.00	4.00	/	84.74
		T740	6290	0.9733	2.33	5.9	50.00	6.00	16.00	66.66
	Tofutai	TP20	6640	0.9091	1.11	9.6	68.66	3.33	2.53	75.03
		TP7	6588	0.8841	0.95	14.0	71.55	7.23	5.16	82.23
		TP39	7110	0.8209	0.23	10.0	74.39	3.9	1.76	78.98
	West of southeast slope	S108-1	5930	0.8894	0.99	12.0	74.81	9.57	0.20	82.91
		S112	6338	0.8259	0.29	11.0	79.49	4.56	1.68	84.84
	East of southeast slope	AT17	6440	0.8132	0.21	14.0	88.08	3.26	1.11	92.10
		S101	5800	0.8074	0.09	9.1	96.00	/	/	98.41
	Yuejin	YJ1X	7251	0.811	0.19	4.7	82.07	3.86	1.71	87.79
		YJ2X	7168	0.7994	2.68	4.6	83.15	2.21	1.29	87.35
Shunbei	Northwesten	SHB7CX	7864	0.849	0.12	3.6	42.54	11.00	5.48	52.03
		SHB5CX	7649	0.831	0.20	4.1	48.90	15.20	3.48	60.29
		SHB51X	7683	0.804	0.10	6.6	73.51	6.27	2.55	81.28
		SHB1-1H	7544	0.794	0.11	4.3	83.16	3.16	2.49	88.19
	Middle	SHB44X	7690	0.793	0.06	4.3	82.90	3.23	4.51	90.10
		SHB41X	7756	0.764	0.13	2.1	83.18	1.98	9.76	94.50
	Shuntuo area	SHB802X	8094	0.749	0.03	1.1	88.75	2.47	2.41	93.80
		SHB8X	8178	0.767	0.02	1.2	87.77	1.79	2.67	92.70
	Shunnan area	SN4	6647	/	/	/	88.86	2.56	8.05	99.40
		SN7	7147	/	/	/	94.65	0.09	5.18	99.83

The geochemical characteristics of crude oils from Tahe, Tuofutai, Yuejin and Shunbei Ordovician are similar (Figs. 9.8 and 9.9) with a Pr/Ph ratio of less than 1.5 (most of them are less than 1.2), a low content of C_{28} sterane (generally less than 25%), a lower C_{21} tricyclic terpene content than the C_{23} tricyclic terpene content, and relatively light carbon isotopes in the oil as a whole (less than $- 31.5‰$). Their heavy oil and normal crude oil products generally have aryl isoprenoid characteristics, indicating that the source rocks were deposited in a reducing anaerobic depositional environment. The crude oil in the Shunbei No. 7 Fault has a slightly heavy carbon isotope and relatively high content of rearranged sterane and rearranged hopane series, suggesting that their source rocks have relatively high contributions from clay source rocks (Fig. 9.9).

From the main body of Tahe to Toputai, to Yuejin, to Shunbei, to Shuntuo and then to Shunnan, the maturities of the crude oil and natural gas gradually increase, reflecting, on the one hand, the gas–oil ratio of the oil and gas accumulation, on the other hand, from gradually decreasing and even disappearing biomarkers in crude oil, the drying coefficient of

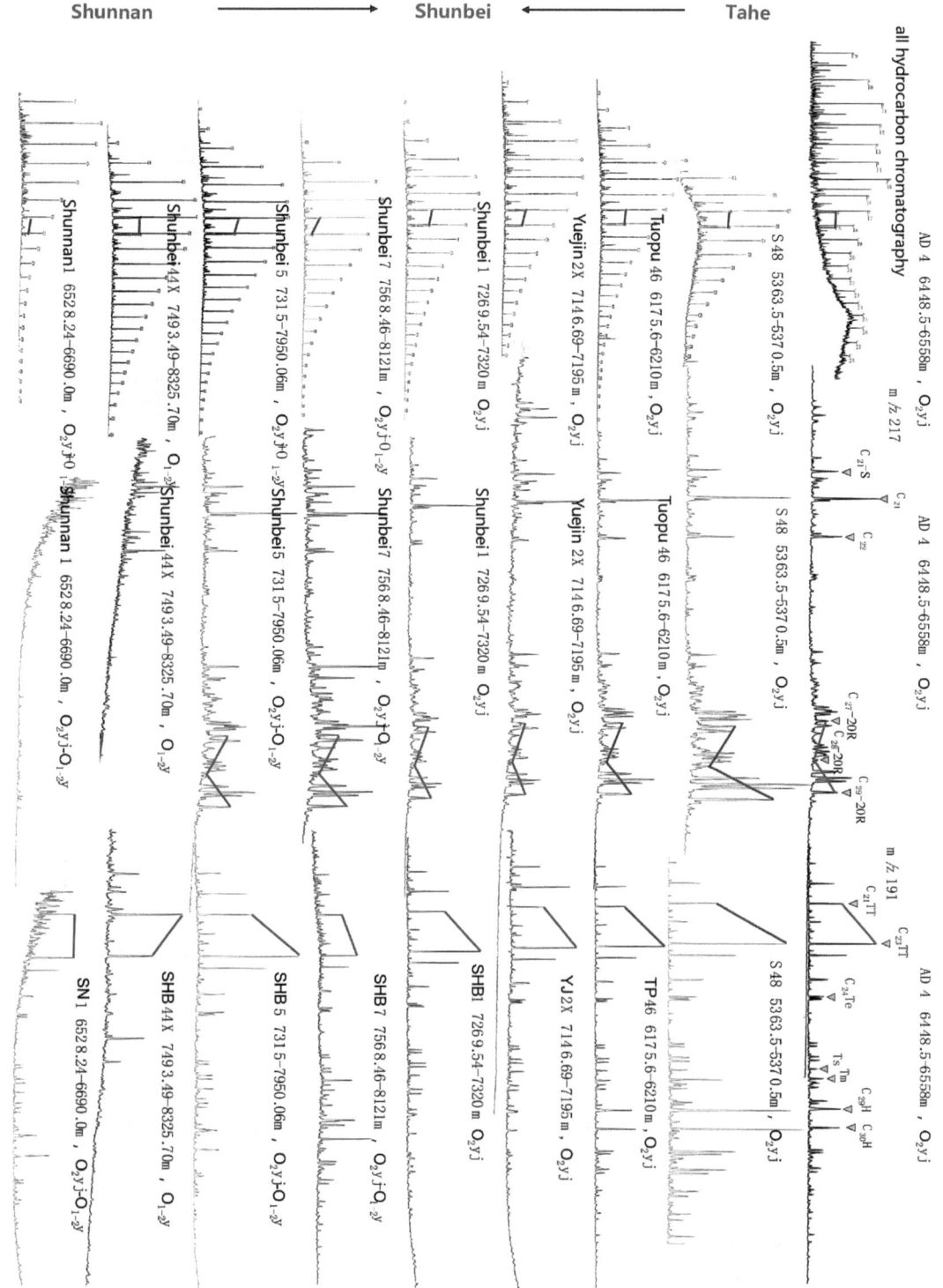

Fig. 9.8 Chromatogram of the distribution of saturated hydrocarbon, and sterane and terpane characteristics of crude oil from the Tahe Oil Field and Shunbei Oil Field

Fig. 9.9 Distribution of carbon isotopes in crude oil from the Tahe Oil Field and Shunbei Oil Field

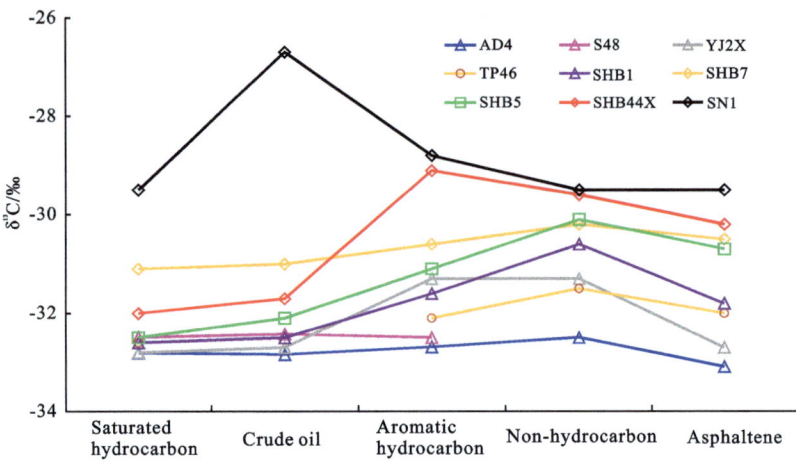

Fig. 9.10 Correlation diagram of thioadamantane-adamantane in crude oil from the Tahe Oil Field and Shunbei Oil Field, Tarim Basin

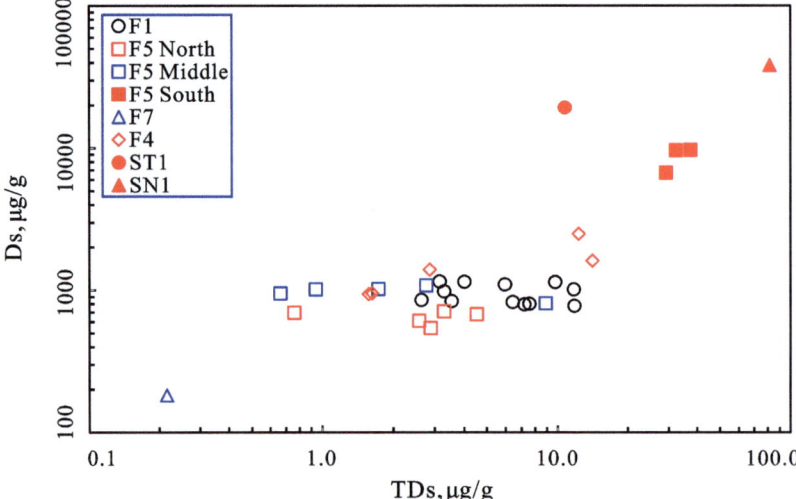

natural gas gradually increases, and the carbon isotope of natural gas methane gradually increases (Table 9.3; Fig. 9.7).

There are differences in the type and intensity of secondary hydrocarbon reformation in the main Tahe, Tuofutai, Yuejin, Shunbei, Shuntuo and Shunnan areas. Within the main area of the Tahe Oil Filed, biodegradation is widespread and manifests as a "UCM" bulge (Fig. 9.8). In Shunnan, the thermal cracking of crude oil is strong, which is also superimposed with the thermochemical reduction of sulfate, leading to a high H_2S content and relatively heavy crude oil isotopes (Fig. 9.10). In the eastern part of the Tahe and east of Yuqi, obvious gas invasion, occurs, indicating that the CH_4 content of natural gas in the eastern part of

the southeastern slope is significantly higher than that in the western part and that the carbon isotope is over heavy. However, the secondary reformation of oil and gas in the Tuofutai, Yuejin and Shunbei areas is relatively weak (Fig. 9.10).

(3) Oil and gas dynamic accumulation process

Based on comprehensive geological studies on the hydrocarbon generation history of source rocks, the differential activities of strike-slip faults and spatial–temporal configuration of source–reservoir, methods such as reservoir fluid inclusions and isotopic dating are used to study the processes of hydrocarbon accumulation and transformation. The results revealthat the Tahe-Shunbei oil and gas

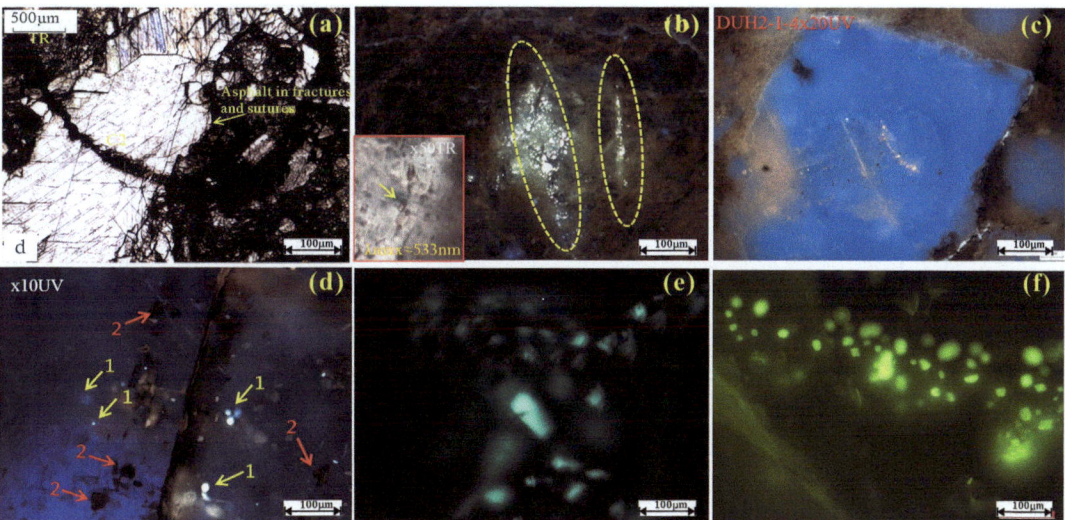

Fig. 9.11 Characteristics of fluid inclusions in Middle–Lower Ordovician carbonate reservoirs of the Tahe Oil Field and Shunbei Oil Field. **a** O_2yj of the Shunbei Well 5's asphalt fractures developed along the edge of the calcite vein with a depth of 7425.7 m; a sheet asphalt distribution observed under the electron microscope. **b** O_2yj of yellow–gray micrite sandy limestone with a large number of yellow–green fluorescent oil inclusions detected in fractured calcite at a depth of 7209.54 m from Well Yuejin 1X. **c** $O_{1-2}y$ of splendid grained limestone from Well Duhu 2 with a depth of 5786.89 m and yellow fluorescent oil inclu-sions in its dolomite cracks. **d** $O_{1-2}y$ of blue–green fluores-cent oil inclusions (1) and non-fluorescent black–brown oil inclusions (2) around fractures formed within giant crystal calcite–filled cavities and a depth of 6102.9 m from Well Tashen 3. **e** O_2yj of sandy micritic limestone with Ultravi-olet blue-green fluorescent oil inclusions in the core calcite veins at a depth of 7723.90 m from Well Shunbei 45X with. **f** O_2yj of sandy micritic limestone with blue-green fluores-cent oil inclusions in the blue core fluorite veins from Well Shunbei 45X

accumulation area is characterized by multistage accumulations and transformations. At different tectonic locations, owing to differences in the thermal evolution of source rocks, the characteris-tics of fault activity and the spatial configuration of source–reservoirs, the main accumulation periods and accumulation processes differ (Figs. 9.11 and 9.12).

(1) Tahe Oil Field

The oil and gas accumulation in the Tahe Oil Field occured mainly during the Late Caledo-nian–Early Hercynian period, and the accumula-tion adjusted during the Late Hercynian period and the Yanshan–Himalayan period (Figs. 9.11, 9.12 and 9.13).

In the Late Caledonian–Early Hercynian period, the karst fracture–cavity reservoirs in the Tahe Oil Field were developed on a large scale,

and the main body of the Middle–Lower Cambrian source rocks within the Shaya Uplift and its adja-cent areas experienced peak oil generation with strong fault activity and large-scale hydrocarbon accumulation within the Tahe Oil Field. More-over, owing to the shallow burial of the reser-voir, effective regional capping was lacking in some areas, the preservation conditions were poor, and the oil reservoir was generally subjected to biodegradation (Fig. 9.11). In addition to the previously-mentioned geochemical characteris-tics of crude oil that reflect the biodegradation, the heavy oil in the Aiding area, the bitumen prevalent in drilling cores of fractured-vuggy reservoirs, bitumen-bearing fluid inclusions and relatively low homogenization temperatures with the hydrocarbon-containing fluid inclusions with yellow fluorescence in the early filling cements of the fracture–cavity, indicate that the accumulation and reformation processes has occurred during

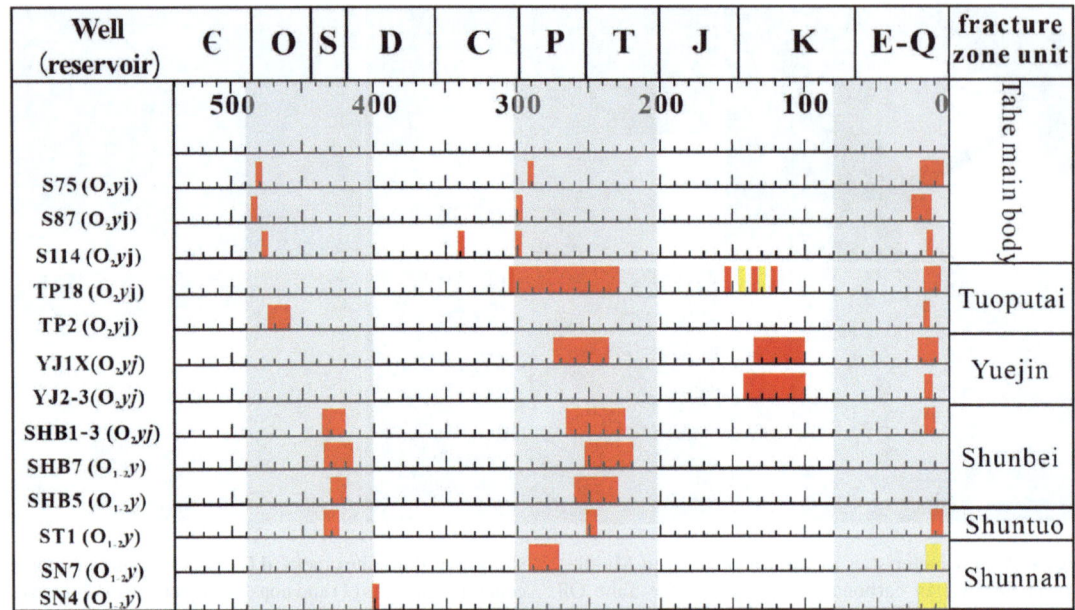

Fig. 9.12 Comparison of accumulation periods based on typical drill wells in different areas of the Tahe Oil Field and Shunbei Oil Field and the homogenization temperatures of reservoir fluid inclusions

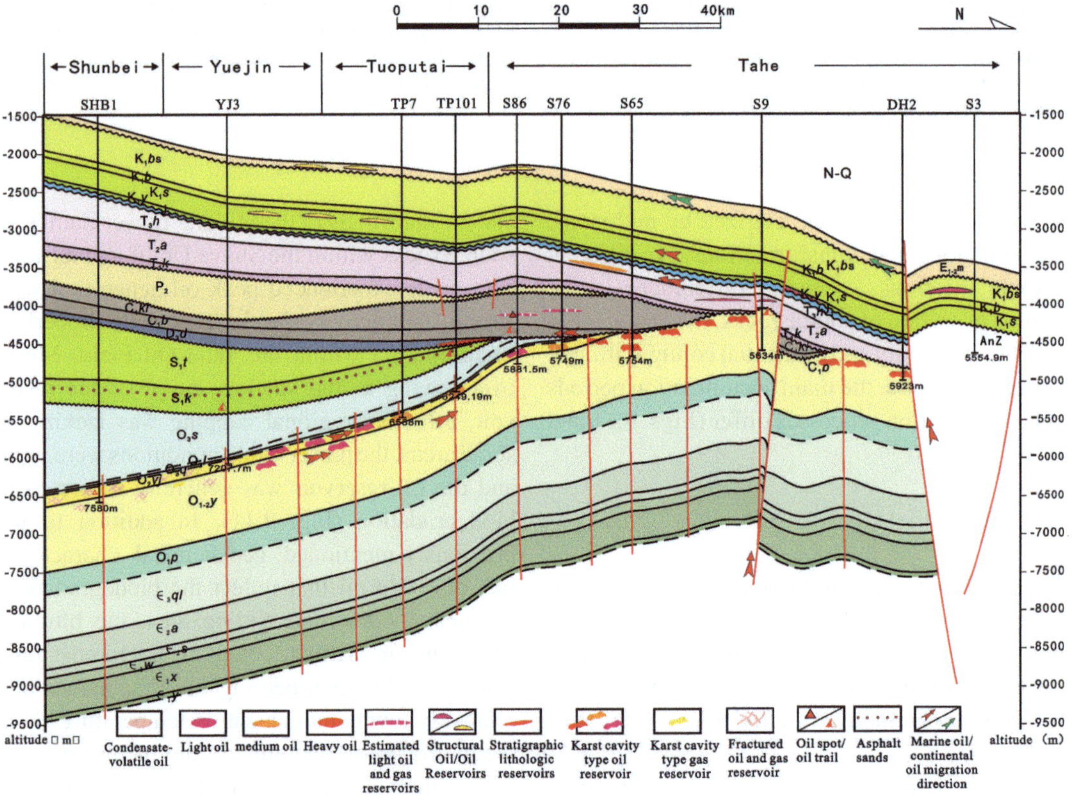

Fig. 9.13 Profile of oil and gas reservoirs in the Tahe Oil Field

this period (Fig. 9.11). Analysis of the organic matter Re–Os isotope dating of crude oil from Well AD19 in the Tahe Oil Field also indicated that early accumulation and reformation occurred at 419.3 ± 6.4 Ma.

During Late Hercynian period, the caprocks in the main area of the Tahe Oil Field were well developed, forming a good reservoir-cap combination with the karst fracture–cavity reservoirs. The Shaya paleo-uplift subsided and was buried during this period, the Cambrian-Middle–Lower Ordovician source rocks in the lower part of the slope generated and expelled hydrocarbons, and the faults were active. During the same period, on the one hand, the Ordovician karst fracture–cavity reservoirs could be recharged with oil and gas with relatively high maturity (Fig. 9.11). On the other hand, the oil and gas accumulation formed in the early stage may have been adjusted, and new oil and gas accumulation may have occured in the overlying traps. Yellow–green fluorescent hydrocarbon-bearing fluid inclusions were observed in the Tahe, Toputai, Yuejin and Shunbei areas, and the homogenization temperatures of the associated brine inclusions indicated that there was oil–gas charging occurred during the Late Hercynian period (Fig. 9.11).

During the Yanshan–Himalayan period, the Shaya paleo-uplift subsided and was buried rapidly. Its Lower Cambrian source rock in the slope area generated and expelled high-mature oil and gas again, and the Cambrian-Middle–Lower Cambrian source rock in the depression area generated and expelled natural gas. During this period, high-mature oil and gas migrated and accumulated along the fault from the south and east to the main part of the Tahe Oil Field. The eastern Tahe and eastern Yuqi areas currently have characteristics of late gas invasion (Fig. 9.11). In the Toputai and Yuejin areas, many blue-green fluorescent and liquid–containing hydrocarbon inclusions and gas-containing hydrocarbon inclusions are present, and the homogenization temperatures of their associated brine inclusionsindicate an accumulation and transformation processes during the Yanshan–Himalayan period (Fig. 9.12).

(2) Shunbei Oil and Gas Field

The Shunbei Oil and Gas Field has experienced a long-term accumulation process since the Late Caledonian–Early Hercynian. In its different structural locations, owing to the differences in the thermal evolution of source rocks and fault activity, the main accumulation periods are different (Figs. 9.11 and 9.12), although its distribution of reservoir types follows a certain rule.

The western part of the Shunbei Oil and Gas Field is dominated by oil reservoirs. Because this area is always in the lower part of the structure and has a continuous low geothermal field background (Ma et al. 2022), the Lower Cambrian source rocks have continued to generate hydrocarbons since the Late Caledonian–Early Hercynian period. Fault-controlled large-scale reservoirs are formed along with the development of strike-slip faults, and the fault activity promotes the accumulation of oil and gas in these reservoirs to form fault-controlled fracture–cave oil and gas accumulation. Because Late Caledonian–Early Hercynian oil and gas accumulation are shallow in burial with strong fault activity, some oil and gas accumulation may be damaged to form reservoir asphalt. Afterward, these reservoirs continue to be buried, their secondary reformations weaken with the multiperiod fault activity, and the oil and gas with different maturities can be charged multiple times. Studies on the differences between reservoir fluid inclusions and fault activities in the northern section of the Shunbei No. 1 fault and the Shunbei No. 5 fault have indicated that oil and gas charging occurred during the Late Caledonian–Early Hercynian and Late Hercynian periods. These fault activities are closely related to oil and gas charging, which further control the properties and distribution of oil and gas to a certain extent (Cao et al. 2020; Han et al. 2021; Wang et al. 2020).

The Shunnan area is dominated by gas reservoirs. The geothermal field background in this area is relatively high (Liu et al. 2020; Ma et al. 2021a, b; Ma et al. 2022) The Cambrian-Middle–

Lower Cambrian source rocks in this area and adjacent areas experienced earlier hydrocarbon generation and evolution processes than those in the Shunbei area did, with dry gas dominanting since the Himalayan period. In addition, the temperature of the Ordovician carbonate rock buried more than 8000 m deep has exceeded 160 °C, the cracking of crude oil has occurred, and the early-formed reservoir has undergone a phase state transformation. The relatively high content of adamantane in the crude oil of Well SN1, the discovery of bituminous inclusions, and the homogenization temperature of brine inclusions associated with methane-bearing inclusions which are essentially the same as the current geothermal temperature, indicate the accumulation of natural gas during the Himalayan period (Chen et al. 2020; Ma et al. 2021a, b). The homogenization temperatures of hydrocarbon-associated brine inclusions in Well SHB42X are distributed mainly between 90 and 110 °C and between 130 and140 °C, indicating that there are two phases of liquid hydrocarbon accumulation occurred during the Late Caledonian and the Late Hercynian periods in this area.

(3) Oil and gas enrichment conditions

A comprehensive analysis on the geological conditions and accumulation process of the Tahe and Shunbei Oil Fields reveals that superior source–reservoir conditions and space–time configurations are key elements for the formation of large oil and gas fields. The widely distributed high-quality source rocks in the oil and gas accumulation area are the basis for the formation of the Tahe–Shunbei large oil and gas accumulation area. Multistage tectonic movements have created two types of fracture–cavity reservoirs with their own characteristics and large-scale development (Fig. 9.14). The difference in the tectonic and geothermal field evolution within the area determines the regular changes in oil and gas maturity and secondary transformation in the oil and gas accumulation area. Currently, the "paleokarst geomorphology and faults" still control the local enrichment of oil and gas in these two large oil and gas fields. During the entire process of the formation and evolution of oil and gas accumulation, the tectonic action is particularly critical.

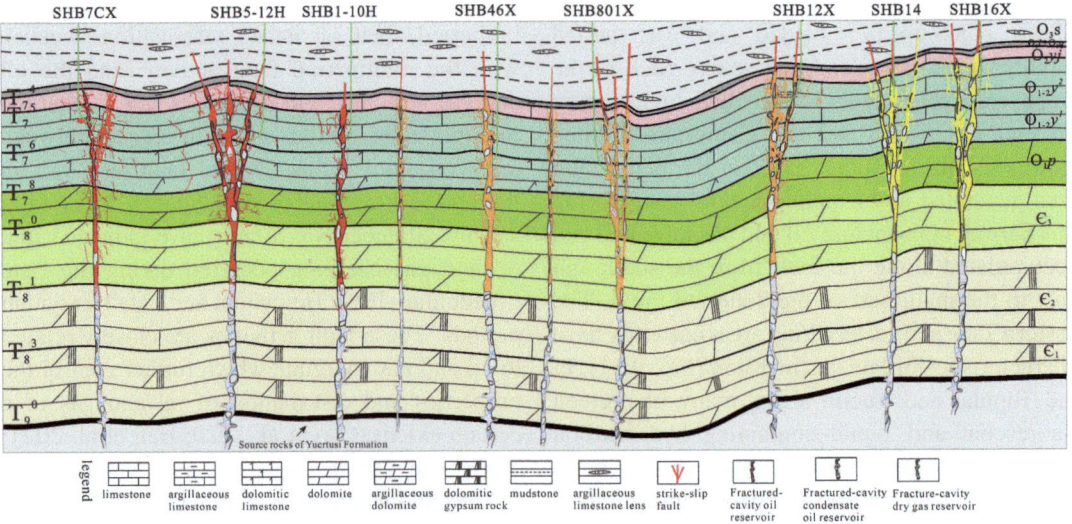

Fig. 9.14 Profile of oil and gas reservoirs in the Shunbei area

9.3 Puguang and Yuanba Gas Fields in the Sichuan Basin

The Sichuan Basin is a large superimposed petroliferous basin, that developed based on the Upper Yangtze Craton and has an area of approximately 18×10^4 km^2. Its resources in deep and ultradeep carbonate rock fields cover more than 12 trillion cubic meters with clustering and zonal distribution characteristics on the plane. To date, many large- and medium-scale gas fields have been identified in the Changxing–Feixianguanproven Formation around the Kaijiang–Liangping continental shelf, the Dengying Formation–Longwangmiao Formation around the Mianyang–Changning rift trough, the Carboniferous around the Kaijiang Paleo-Uplift, and the Permian along the periphery of the Luzhou Paleo-Uplift. Among them, the Changxing–Feixianguan Formation around the Kaijiang–Liangping Shelf has proven natural gas reserves of nearly one trillion cubic meters as a large natural gas accumulation area. Both the Puguang Gas Field and the Yuanba Gas Field are typical examples of these natural gas accumulation areas.

The Puguang 1 Well was drilled in November 2001 and completed at the top of the Upper Permian Changxing Formation in May 2003. The oolitic shoal dolomite reservoir on the edge of the platform in the lower part of the Lower Triassic Feixianguan Formation was drilled to 55 m, and the production of well completion reached 42.37×10^4 m^3/d, indicating the discovery of the Puguang Gas Field. The Yuanba Gas Field was discovered in 2007. To dateTo date, the proven total reserves have exceeded 200 billion cubic meters, and the cumulative gas production has exceeded 20 billion cubic meters. The Yuanba 1 Well, which is positioned south of the Kaijiang–Liangping paleocontinental shelf was drilled in May 2006. In November 2007, Well YB1 drilled in the second section of the Changxing Formation encountered a bioreef-covered dolomite reservoir on the edge of the platform with good oil and gas accumulation having an industrial gas flow of 50.30×10^4 m^3/d, which indicated that there was a breakthrough in the Changxing Formation gas reservoir within the Yuanba Gas Field. To date, Yuanba Gas Field has submitted a proven total reserves of more than 200 billion cubic meters with cumulative gas production of more than 20 billion cubic meters.

The discovery of the Puguang and Yuanba Gas Fields benefited from breakthroughs in the understanding of the development law of Permian–Triassic sedimentary facies and facies-controlled reef–shoal-scale reservoirs in the Sichuan Basin (Guo et al. 2014; Ma et al. 2007). Studies on the reservoir development characteristics, accumulation and enrichment laws of these two gas fields also provide a new reference for the exploration and development of deep carbonate rocks.

9.3.1 Geological Overview

Owing to the dragging effect of the subduction of the Paleo–Tethys Ocean on the western margin of the Yangtze plate, and effects of tectonic events during the "Emei Taphrogeny", the Middle and Late Permian experienced strong tectonic sedimentation differentiation on the early-unified carbonate rock platform and north of the Sichuan Basin, which formed the Kaijiang–Liangping continental shelf in the northwest direction. The third section of Maokou Formation in the Middle Permian formed a rift prototype, and Wujiaping Formation (Longtan Formation) in the Late Permian was further developed, the development of the Changxing Formation (Dalong Formation) rift reached its peak, and the Early Triassic Feixianguan Formation was filled. The seawater from the Late Jialingjiang Formation in the Early Triassic and the Leikoupo Formation in the Middle Triassic completely retreated from the platform, forming a limited evaporation environment. Because the development of the Kaijiang–Liangping Shelf controlled multiple sets of synchronous-heteropic sedimentary assemblages in the northeastern Sichuan Basin, it laid a solid foundation for the development of high-quality source rocks and reservoirs. The main source rock layers are three sets of deep-water sedimentary shelf facies, including the third section of the Maokou Formation, Wuji-

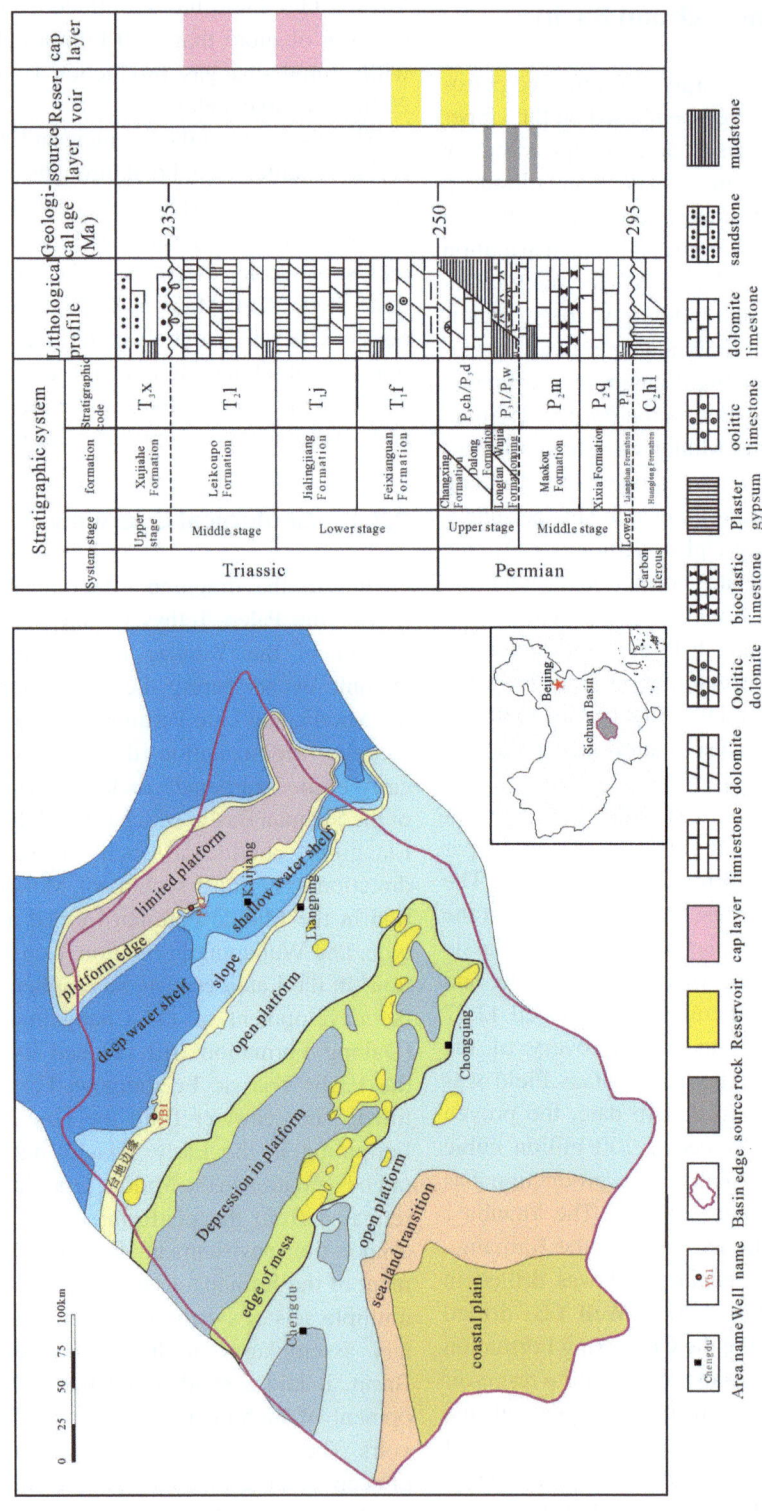

Fig. 9.15 Schematic diagram of the sedimentary facies of the Permian Changxing Formation, and the source–reservoir–cap assemblage of Puguang and Yuanba in the Sichuan Basin

aping Formation and Dalong Formation, and the main reservoir intervals have four sets of marginal reef–shoal facies sedimentary, including the third section of the Maokou Formation, Wujiaping Formation, Changxing Formation and Feixianguan Formation. In the Jialingjiang Formation and Leikoupo Formation, very large thick gypsum deposits have developed into regional caprocks. This superior source–reservoir–cap combination has laid a solid foundation for the formation of large natural gas fields around the Kaijiang–Liangping shelf (Huang 2014; Ma et al. 2010a, b). Puguang Gas Field and Yuanba Gas Field are located on the east and west sides of the Kaijiang–Liangping Shelf, and their source–reservoir–cap assemblages are shown in Fig. 9.15.

9.3.2 Development Characteristics and Main Controlling Factors of Facies-Controlled Reef–Shoal Reservoirs

(1) Reservoir Development Characteristics

According to a large amount of drilling data, the gas-bearing intervals of the Puguang and Yuanba Gas Fields are mainly reef–shoal facies reservoirs on the platform margins of the Permian–Triassic Changxing Formation and Feixianguan Formation, with residual reefs, bioclastic crystalline dolomite and residual oolitic crystalline dolomite as their most representative characteristics.

Residual reef and bioclastic crystalline dolomite reservoirs are present mainly within the Changxing Formation, which are subfacies sedimentary rocks such as platform edge reef–shoal facies barrier rocks, skeleton rocks, and exposed shoal facies bioclastic shoals at the platform edge. The content of reef-building organisms in the residual reef crystalline dolomite reservoir rocks varies from 35 to 50% and mainly includes sponges and small amounts of bryozoans and stratoporoids. Residual bioclastic crystalline dolomite reservoir rocks are rich in types of bioclasts, among which flies and fusulinids are the most common; large fossils such as brachiopods

and gills can also be found. Because the recrystallization and dissolution of the reservoir rock are strong, the original biological fabric has been mostly destroyed, and the dolomite has a medium–coarse crystal structure. The types of reservoir space are mainly dolomite intercrystalline pores, dissolution pores, and biological cavity pores, etc., which are often filled with asphalt (Guo et al. 2010, 2014) (Fig. 9.16 a–d). The physical properties of residual reefs and bioclastic crystalline dolomite reservoirs are good, and the logging data reveal that these reservoirs are mostly Class I and II reservoirs. For example, the drilling within the Yuanba 102 Well encountered reef facies on the edge of the Changxing Formation platform, and its logging interpretation thickness was 72.94 m. Its measured porosity ranged from 1.84 to 13.41% with an average of 4.62%. Its permeability ranged from 0.0233 to 46.4952 mD with an average of 0.7528 mD, and the test showed a productivity of 50.55×104 m^3/d. The drilling within Well Yuanba 12 encountered bioclastic shoal facies at the platform margin of the Changxing Formation. The logging interpretation reservoir thickness was 78.86 m, and testing showed a productivity of 53.14×104 m$_3$/d.

Residual oolitic crystalline dolomite reservoirs are present mainly within the Feixianguan Formation; they are distributed in Puguang, Longgang and other areas, and are deposited as oolitic shoal subfacies of exposed shoal facies at the platform edge. The oolitic content in the reservoir rock ranges from 70 to 80%. Owing to strong recrystallization, most oolites retain only their residual structure, and dolomite includes mainly medium-crystalline and fine-crystalline dolomite. The reservoir space is composed mainly of dolomite intercrystalline pores, dissolved pores, oolitic membrane pores, and intragranular dissolved pores, and is often filled with asphalt (Ma et al. 2007) (Fig. 9.16 e–f). Owing to strong recrystallization and dissolution, some rocks have coarse-grained structures and extremely developed dissolution intercrystalline pores. The rock has a light specific gravity, looks like "slag", and is called "sugar grain" dolomite. These reservoirs have good physical properties, and the logging interpretation reveals that most of them are type I and II reservoirs. For example, drilling activities through the lower part of Feixianguan Formation within the Puguang 2 Well encountered an oolitic dolomite reservoir. The measured porosity ranged from 0.94 to 28.86% with an average of 8.17%. The permeability ranged from 0.0112 to 3354.6965 mD with an average of 94.4234 mD, and testing revealed a productivity of 62.02 × 10^4 m³/d.

Additionally, silt-fine-crystalline dolomite reservoirs with good physical properties have developed in the Puguang and Yuanba areas.

These areas are composed sedimentary of platform edge reef–shoal facies and platform edge exposed shoal facies. However, reef limestone, bioclastic limestone and oosparite limestone have poor physical properties (Guo 2011a, b).

(2) Main factors Controlling Reservoir Development

According to many studies on sedimentary facies, diagenetic evolution and reservoir development from field and drilling practices, the develop-

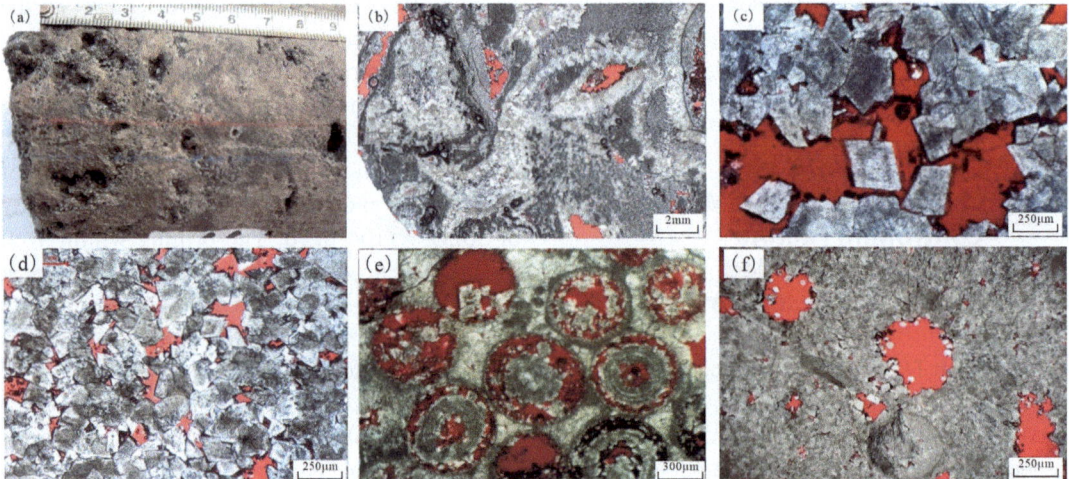

Fig. 9.16 Images of reservoir lithology and storage space of the Changxing Formation–Feixianguan Formation in the Puguang and Yuanba Gas Fields. **a** Changxing Formation dissolved porosity dolomite from Well Yuanba 224; **b** Changxing Formation reef dolomite from Well Yuanba 2, coral cavity pores and their casting thin sections; **c** Feixianguan Formation medium-coarse-grained dolomite from Well Puguang 2, intercrystalline pores and their thin cast-

ings; **d** Changxing Formation splendid bioclastic dolomite from Well Yuanba 27, intergranular dissolved pores and their casting thin sections; **e** Feixianguan Formation oolitic dolomite from Well Puguang 2, intragranular dissolved pores, intercrystalline pores and their casting thin sections; **f** Changxing Formation bioclastic dolomite from Well Yuanba 102, biological mold holes and their casting thin sections

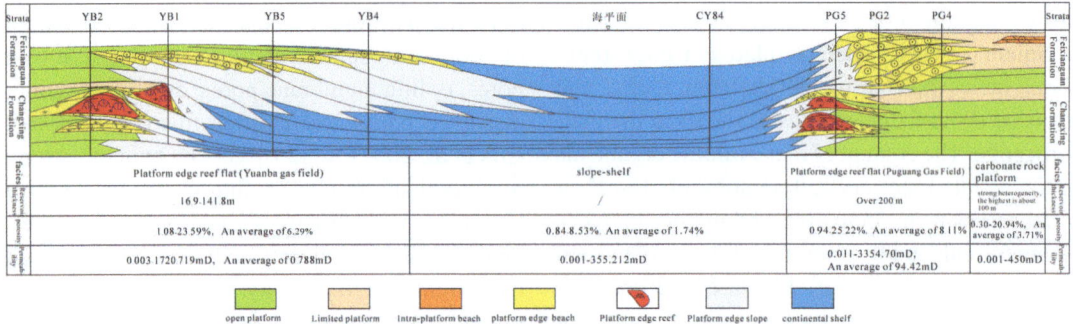

Fig. 9.17 Comparisons between the sedimentary facies models of the Changxing Formation and Feixianguan Formation from the Yuanba Gas Field and Puguang Gas Field (modified from Guo et al. 2011)

ment of high-quality reservoirs in the Puguang and Yuanba areas has distinct facies-controlled characteristics (Fan 2015; Ma et al. 2010a, b).

The platform edge facies belt is a positive geomorphological unit with strong hydrodynamic conditions and is more sensitive to the rise and fall of the sea level than the platform itself is. All these processes are generally favorable for the development of biological reefs, oolitic shoals and bioclastic shoals during the depositional period, as well as for dolomitization and selective dissolution of the fabric during the penecontemporaneous period and the early diagenetic stage. Therefore, high-quality reservoirs were more developed there. As shown in Fig. 9.17, the Puguang and Yuanba Gas Fields are in the platform edge facies zone on the east and west sides of the Kaijiang–Liangping continental shelf. Compared with the intraplatform reservoirs, these high-quality reservoirs are thicker and have better physical properties (Guo et al. 2017).

There are certain differences in the development characteristics of reef–shoal reservoirs in the Puguang Gas Field and Yuanba Gas Field. The Puguang Gas Field is dominated by Feixianguan Formation oolitic shoal facies reservoirs with large reservoir thicknesses and good physical properties, whereas the Changxing Formation reef–shoal facies belt is usually narrow. The Yuanba Gas Field is dominated by high-quality dolomite reservoirs of the Changxing Formation reefs and shoal facies behind the reefs. oolitic shoals of the Feixianguan Formation are widely distributed, but the reservoir thickness is small. Studies have shown that the difference in reservoir development characteristics between the Puguang and Yuanba Gas Fields is the result of the combined action of sedimentary paleogeomorphology, tectonic processes and degree of dolomitization (Hu 2011).

Within the platform edge facies zone, sedimentary microfacies further control the heterogeneity of the reservoir. The measured physical property data of the drilling cores from the Puguang and Yuanba Gas Fields revealed that the physical properties of oolitic shoal reservoirs were relatively good, followed by those of bioreefs and bioclastic shoals (Fig. 9.18). In the reefs, the reef–cap microfacies reservoirs have relatively good physical properties, with porosities concentrated at approximately 8.0%, followed by reef core microfacies, with porosities concentrated at 4.0–6.0%. The degree of dolomitization of the reef-base microfacies is relatively weak with a porosity of mainly 2.0–4.0%.

Beyond high-energy oolitic shoals and organic reefs related to sedimentary differentiation, recent exploration has indicated that localized faulting and multiphase fluid alterations along faults critically control reef–shoal reservoirs in the Puguang and Yuanba areas. Localized anomalously rapid water encroachment during the development of the Puguang Gas Field indicates the presence of local microfractures. A representative example, Well Puguang 104–1, encountered fracture-vuggy reservoir spaces at a depth of 5557 m in the Feixianguan Formation, formed by diagenetic fluid dissolution along fractures. These karst features, including dissolution fractures 2–3 cm wide and vugs 3–5 cm in size, constitute significant reservoir spaces in the well area. Many studies have further revealed that organic acid dissolution during the burial period, TSR and early oil and gas charging also play very important roles in the development and retention of reservoirs (Ding et al. 2020; Guo et al. 2017; Ma et al. 2010a, b). In some areas, spaces formed by various fractures play an important role in the processes of reservoir formation and retention too (Guo et al. 2017).

9.3.3 Oil and Gas Accumulation Characteristics and Their Main Controlling Factors

(1) Characteristics of gas reservoirs

The natural gas reservoirs discovered in the Puguang and Yuanba Gas Fields are mainly lithologic and structural–lithologic reservoirs, with the characteristics of "one reef, one shoal, one trap, and one gas reservoir".

The Puguang Gas Field is composed of three gas reservoirs, Puguang, Dawan and Maoba. The Yuanba Gas Field is composed of several biological reefs and shoals that are distributed in the north–west direction.

The Changxing–Feixianguan Formation gas reservoir in the Puguang Gas Field is buried at a depth of 5104.5 m (medium depth), and is a dry gas reservoir with normal pressure, low temperature, high H_2S content and medium CO_2 content. Its hydrocarbon component is mainly CH_4 with an average content of 74.74%. Its C_2H_6 average content is 0.09%, its H_2S average content is 14.64%, and its CO_2 average content is 9.95% (Table 9.4).

The Changxing Formation is the main producing layer of the Yuanba Gas Field, with a burial depth of 6682.9 m (medium depth). Its gas reservoir is a dry gas reservoir with normal pressure, high H_2S content and medium CO_2 content. The hydrocarbon components in the gas reservoir are mainly CH_4 with an average content of 86.41%. Its C_2H_6 average content is 0.05%, its H_2S average content is 6.47%, and its CO_2 average content is 6.85%. The Feixianguan Formation gas reservoir is distributed in the northwestern of Yuanba Gas Field, with a burial depth of 6317.0 m (medium depth), which is a dry gas reservoir with high pressure, a medium H_2S content and medium CO_2 content. Its hydrocarbon components in the gas reservoir are mainly CH_4 with an average content of 94.21%. Its C_2H_6 average content is 0.08%, its H_2S average content is 1.72%, and its CO_2 average content is 3.35% (Table 9.4).

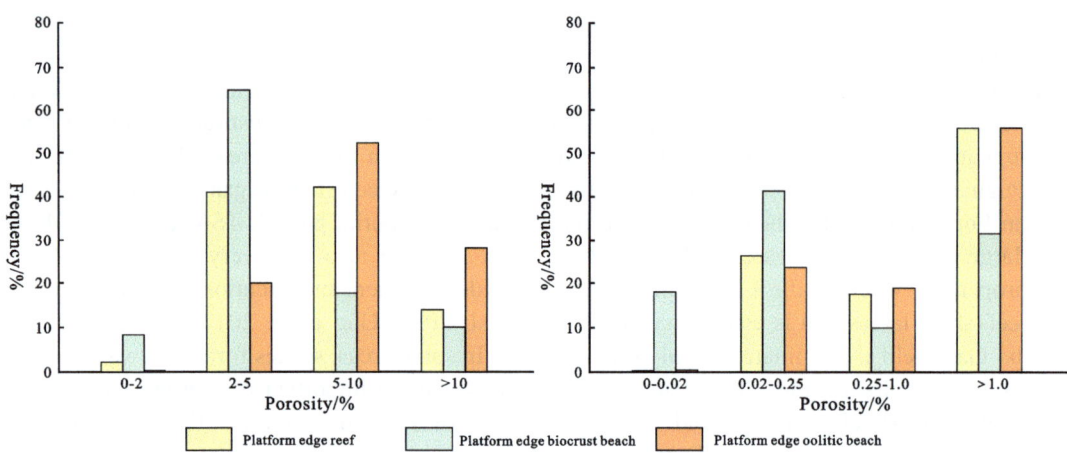

Fig. 9.18 Histograms of porosity and permeability in different facies of the Puguang and Yuanba Gas Fields

Table 9.4 Characteristics of gas reservoirs in Puguang and Yuanba Gas Fields

Gas filed	Location	Gas reservoir depth	Pressure coefficient	Geothermal gradient (°C/100 m)	Natural gas components (%)			
					CH_4	C_2H_6	H_2S	CO_2
Puguang Gas Filed	Changxing Formation—Feixianguang Formation	5104.5 m (medium)	1.07–1.1 with an average of 1.10	1.98–2.21	74.74	0.09	14.64	9.95
Puguang Gas Filed	Changxing Formation	6682.9 m (medium)	1.01–1.12 with an average of 1.03	1.90 (average)	86.41	0.05	6.47	6.85
	Feixianguang Formation	6317.0 m (medium)	1.95–1.96	2.11 (average)	94.21	0.08	1.72	3.35

(3) Dynamic Oil and Gas Accumulation Process

Based on studies evaluating natural gas geochemical characteristics, reservoir asphalt, reservoir fluid inclusions, regional source–reservoir–cap development conditions and tectonic evolution background, the dynamic accumulation processes of the Puguang and Yuanba Gas Fields have been determined.

a. Source of natural gas

Studies on natural gas geochemical characteristics, reservoir asphalt and regional temperature and pressure fields reveal that the natural gas of the Changxing Formation–Feixianguan Formation in the Puguang Gas Field and Yuanba Gas Field is mainly crude oil cracking gas, which originates from the source rock of the Upper Permian Wujiaping Formation, with some contributions from the source rocks of the Dalong Formation.

The evidence of crude oil cracking gas comes from two aspects. On the one hand, a large amount of solid asphalt (asphalt reflectance R_b = 2.5–3.0%) and asphalt inclusions in the reservoir indicate the ancient oil reservoir and the crude oil cracking process (Fig. 9.19). On the other hand, the hydrocarbon composition of natural gas is characterized by high $Ln(C_1/C_2)$ and low $Ln(C_2/C_3)$, indicating the characteristic secondary cracking of crude oil (Fig. 9.20).

Comparisons between the carbon isotope compositions of natural gas in the Puguang and Yuanba Gas Fields and those of source rocks in different members indicate that the natural gas in the Changxing Formation–Feixianguan Formation may have come mainly from source rocks of the Upper Permian Wujiaping Formation and the Dalong Formation, with partial contributions from the Dalong Formation source rocks (Fig. 9.21), which is consistent with the source rock distribution.

The source layers of the Wujiaping Formation (Longtan Formation) are distributed regionally in Sichuan Basin. The cumulative thickness of carbonaceous mudstone and calcareous mudstone in the Wujiaping Formation in the Puguang Gas Field is 100–120 m with an average total organic carbon (TOC) content of 4.44% and an estimated gas generation intensity of approximately (80–100) $\times 10^8$ m^3/km^2. The organic carbon content of the Dalong Formation (Changxing Formation) in this area is generally low (i.e., essentially less than 0.5%), and the hydrocarbon generation capacity is also low. The thicknesses of the dark marl and argillaceous lithofacies in the Wujiaping Formation in the Yuanba area range from 40 to 80 m with a total organic carbon (TOC) content of 0.54–7.20% and an average of 2.90%. Its estimated gas generation intensity is (20–40) $\times 10^8$ m^3/km^2. The organic carbon content of the Changxing Formation (Changxing Formation) is relatively high, with values between 0.5 and 2.0% in some samples. Therefore, the Dalong Formation (Changxing Formation) has a certain hydrocarbon generation potential (Duan 2016).

b. Oil and gas charging period

As shown in Fig. 9.21, large numbers of black solid bitumen inclusions, pure gaseous hydrocarbon inclusions, hydrocarbon-bearing brine inclusions and brine inclusions, and a small number of crude oil inclusions can find in the Changxing Formation–Feixianguan Formation reservoirs of the Puguang and Yuanba Gas Fields. Asphalt inclusions are formed from the high-temperature cracking of crude oil inclusions. Therefore, the duration of oil and gas charging can be determined by measuring the homogenization temperature of the brine inclusions associated with the bitumen inclusions. There are two main phases of the homogenization temperature of the brine inclusions in the Changxing Formation–Feixianguan Formation of Puguang Gas Field as: the first phase has a homogenization temperature of 106.1–112.6 °C, and the second phase has a homogenization temperature of 127.3–131.7 °C. According to their burial and thermal evolution histories, the charging times of the ancient crude oil are the Late Triassic and the Early Jurassic, respectively (Fig. 9.22). The homogenization temperatures of the brine inclusions in the Changxing Formation and Feixianguan Forma-

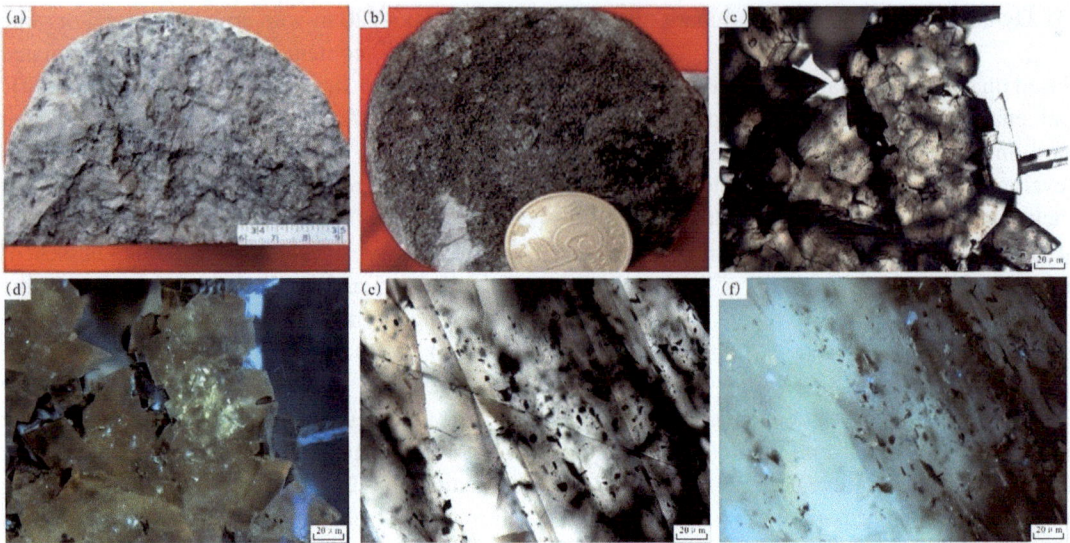

Fig. 9.19 Images of asphalt from the Changxing Formation–Feixianguan Formation reservoirs in the Puguang and Yuanba Gas Fields. **a** Feixianguan Formation dissolved pore fine-grained dolomite from Well Puguang 2, with asphalt filled in dissolved pores; **b** Changxing Formation residual oolitic dolomite from Yuanba 102, with a large amount of asphalt filled in the intergranular and dissolved pores; **c** Changxing Formation yellow and blue–white oil inclusions from Well Yuanba 2, and the transmitted light; **d** Changxing Formation yellow and blue–white oil inclusions from Well Yuanba 2, which have the same synoptic inclusion fluorescence as (**c**); (**e**) Changxing Formation light gray fine-crystalline dolomite and asphalt inclusions from Well Yuanba 2, and the transmitted light; **f** Changxing Formation light gray fine-crystalline dolomite and asphalt inclusion from Well Yuanba 2, which synoptic have the same synoptic inclusion fluorescence as (**e**)

tion of the Yuanba Gas Field concentrated mainly between at 100 and −130 °C. According to their burial and thermal evolution histories, it is determined that the crude oil charging period is from the Late Triassic to the Early Jurassic, which corresponds to the oil-generating peak period of the Upper Permian Wujiaping Formation–Dalong Formation source rocks.

Studies on the Re-Os isotopes of solid asphalt have shown that the Re-Os age of solid asphalt in the Feixianguan Formation of the Puguang Gas Field is 157 ± 57 Ma (Fig. 9.23). According to the burial and thermal evolution histories, the burial temperature of the reservoir is approximately 160 °C (Fig. 9.22), which is higher than the homogenization temperature of the brine inclusions associated with the asphalt inclusions but very close to the oil cracking temperature. Therefore, the Re-Os age of the solid asphalt from the Puguang Gas Field records the age of the solid asphalt produced by the cracking of crude oil, indicating that the solid asphalt may have been formed in the Middle and Late Jurassic when the crude oil in the Feixianguan Formation paleo-reservoir within the Puguang Gas Field had been significantly cracked into natural gas.

Therefore, paleo-oil charging and crude oil cracking occurred in the Changxing Formation–Feixianguan Formation of the Puguang and Yuanba Gas Fields. The paleo-oil charging period was the Late Triassic–Early Jurassic, and the crude oil cracking period was mainly the Middle–Late Jurassic.

c. Dynamic Accumulation Process

In terms of the natural gas source, oil and gas charging phase and regional geological evolution background, the accumulation of reef–shoal natural gas in the Puguang and Yuanba Gas Fields has experienced three general stages: early paleo-

Fig. 9.20 Natural gas Ln(C2/C3)–Ln(C1/C2) correlations for natural gas in the Changxing Formation-Feixianguan Formation in the Puguang and Yuanba Gas Fields (after Guo et al. 2014)

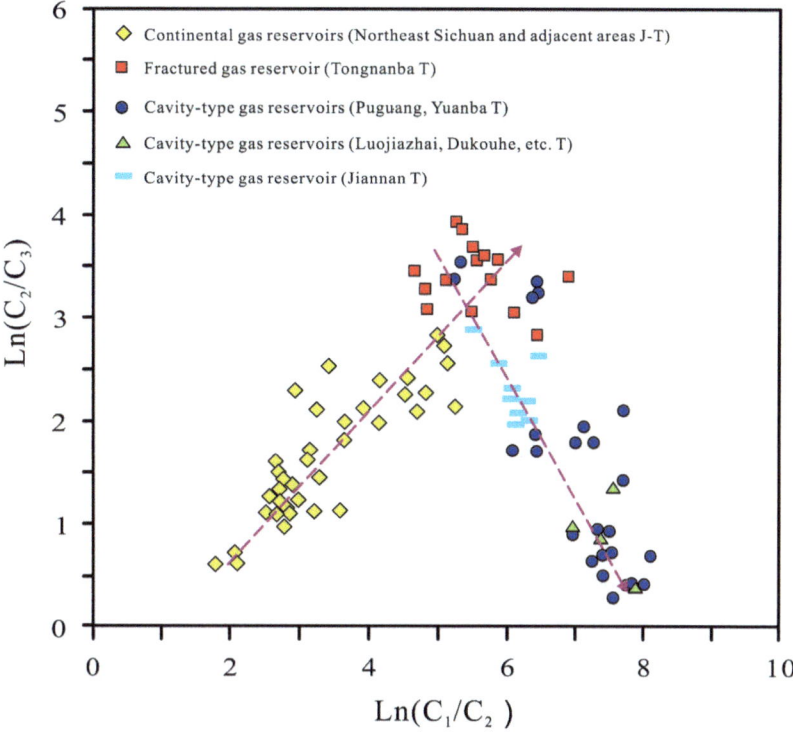

Fig. 9.21 Comparison of the δ^{13}C values of natural gas ethane and kerogen from various source rocks in the Sichuan Basin (after Guo et al. 2014)

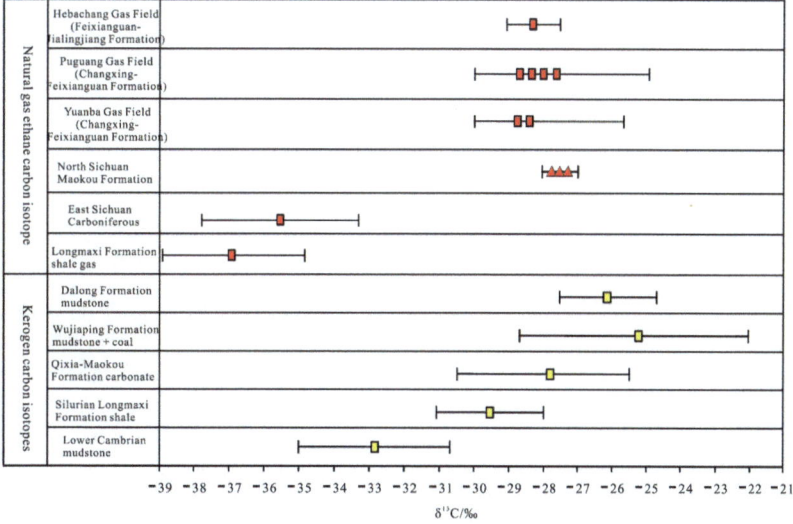

reservoir accumulation, crude oil cracking, and natural gas adjustment and reaccumulation (Duan et al. 2013; Ma and Cai 2006; Ma et al. 2010a, b). These two gas fields have different accumulation processes because of the differences in their tectonic evolutions (Duan 2016) (Fig. 9.24).

The Late Indosinian-Early Yanshan period was the formation period of the paleo-oil reservoir. The Changxing Formation–Feixianguan Formation reef–shoal bodies in the Puguang Gas Field were deposited on the slope northwest of the Kaijiang paleo-uplift. The inversion faults formed by the tectonic movement connected the Upper Permian source rocks and traps, when the source rocks entered the main hydrocarbon expulsion period. The paleo-oil migrated vertically along the faults and fractures to the Changxing Formation–Feixianguan Formation reservoirs, further forming structural–lithologic paleo-reservoirs. The tectonic background of the Yuanba Gas Field in the Late Indosinian was relatively stable, and no faults and fractures were developed. When the early Yanshan period began, the hydrocarbon generation peak of the Upper Permian source rocks occurred. Moreover, the Yuanba Gas Field was affected by the weak structural compression of the basin margin, and many interlayer fractures developed. These fractures connected the source rocks and reservoirsand formed the Changxing Formation–Feixianguan Formation lithologic paleo-reservoirs.

The Middle Yanshan is the cracking period of crude oil in paleo-oil reservoirs. With increasing of burial depth, the stratum temperature gradually exceeded 160 °C, and the paleoreservoirs in the two gas fields began to thermally crack and form gas reservoirs, which completed the phase transformation from oil to gas. Since the Late Yanshanian, natural gas has been adjusted and reaccumulated. In the Late Yanshanian, especially since the Himalayan period, the Puguang Gas Field has been strongly compressed and pushed over by the Daba Mountains, and its southwestern uplift has become a local high point, leading to the adjustment and reaccumulation of paleogas reservoirs, and the formation of present gas reservoirs under the effective transport of faults and reservoir transport layers. The gas–water interface has been adjusted accordingly, and the present gas reservoir is still controlled by structural–lithologic composite traps. Because the Yuanba Gas Field is affected by the anticline uplift of the Jiulongshan in the north, the strata in its northwest as a whole body continue to uplift, and the natural gas again flows through fractures to the northwest to migrate and accumulate with gas reservoirs. Compared with those in the Puguang Gas Field, the adjustment range of the trap heights in

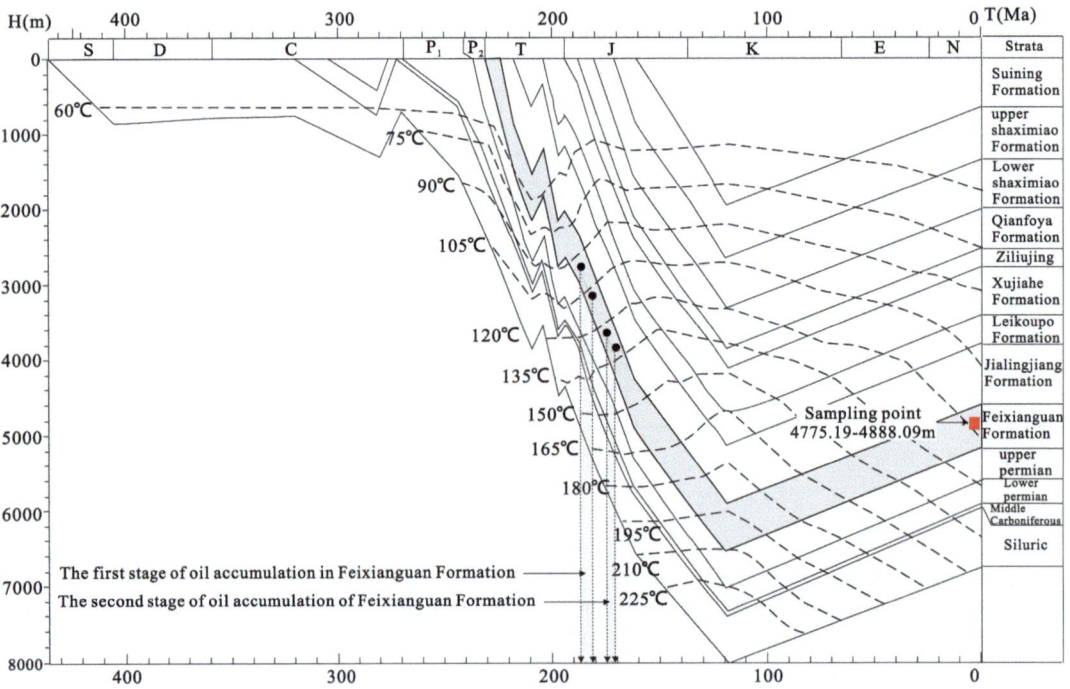

Fig. 9.22 Crude oil charging phases and times of the Changxing Formation–Feixianguan Formation in the Puguang Gas Field

Fig. 9.23 Re-Os isotope isochrones of solid asphalt from the Feixianguan Formation in the Puguang Gas Field

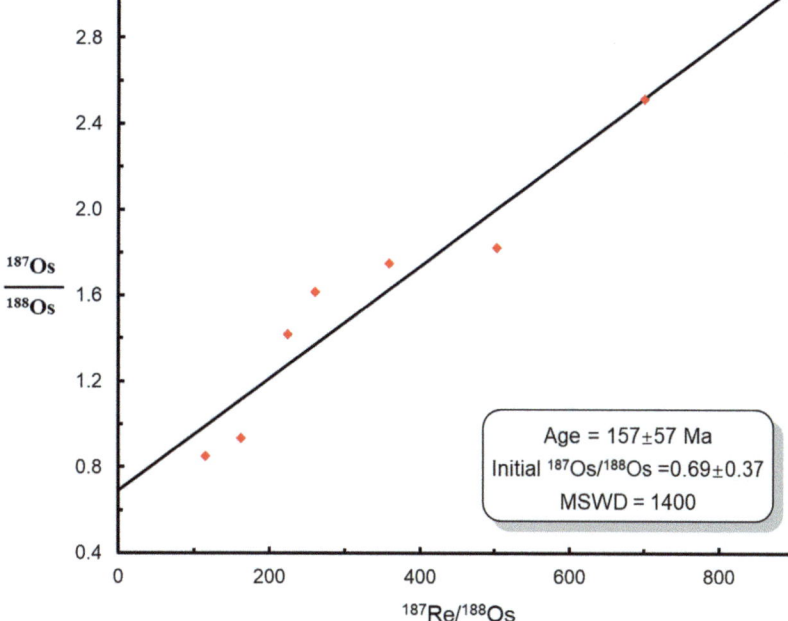

$$\frac{^{187}Os}{^{188}Os}$$

Age = 157±57 Ma
Initial $^{187}Os/^{188}Os$ =0.69±0.37
MSWD = 1400

$^{187}Re/^{188}Os$

the Yuanba Gas Field is smaller, and the adjustment and reaccumulation processes of natural gas is not significant, and are controlled mainly by a single reef–shoal lithologic trap.

d. Oil and Gas Accumulation and Enrichment Model

The formation and development of the Kaijiang–Liangping continental shelf and sedimentary differentiation under its control provide favorable accumulation conditions for the reservoir formation within the Puguang Gas Field and Yuanba Gas Field, which are typical representatives of "shelf-generation marginal storage and near-source enrichment" (Guo et al. 2014; Ma et al. 2010a, b). This accumulation process is characterized by "transformation of the hydrocarbon phase state, dynamic adjustment and continuous preservation" (Guo 2011a, b; Guo et al. 2020). During the formation and transformation of reservoirs and gas reservoirs, the original sedimentary facies of carbonate rocks played a key and controlling role.

9.4 Key Elements and Exploration Directions of Deep and Ultradeep Carbonate Oil and Gas Accumulation

As previously mentioned, various types of oil and gas fields (reservoirs) have been discovered in deep and ultradeep marine carbonate rocks in China, which arekey elements for further analysis of the formation of typical oil and gas fields, and can provide excellent information for identifying more potential exploration targets.

9.4.1 Key Elements of Oil and Gas Accumulation

(1) Scale Source–Reservoir Development and Efficient Allocation

According to the comparison and analysis of the basic elements of oil and gas accumulation in the Tahe Oil Field and Shunbei Oil and Gas Field in the Tarim Basin, Puguang Gas Field and Yuanba Gas Field in the Sichuan Basin, large-scale devel-

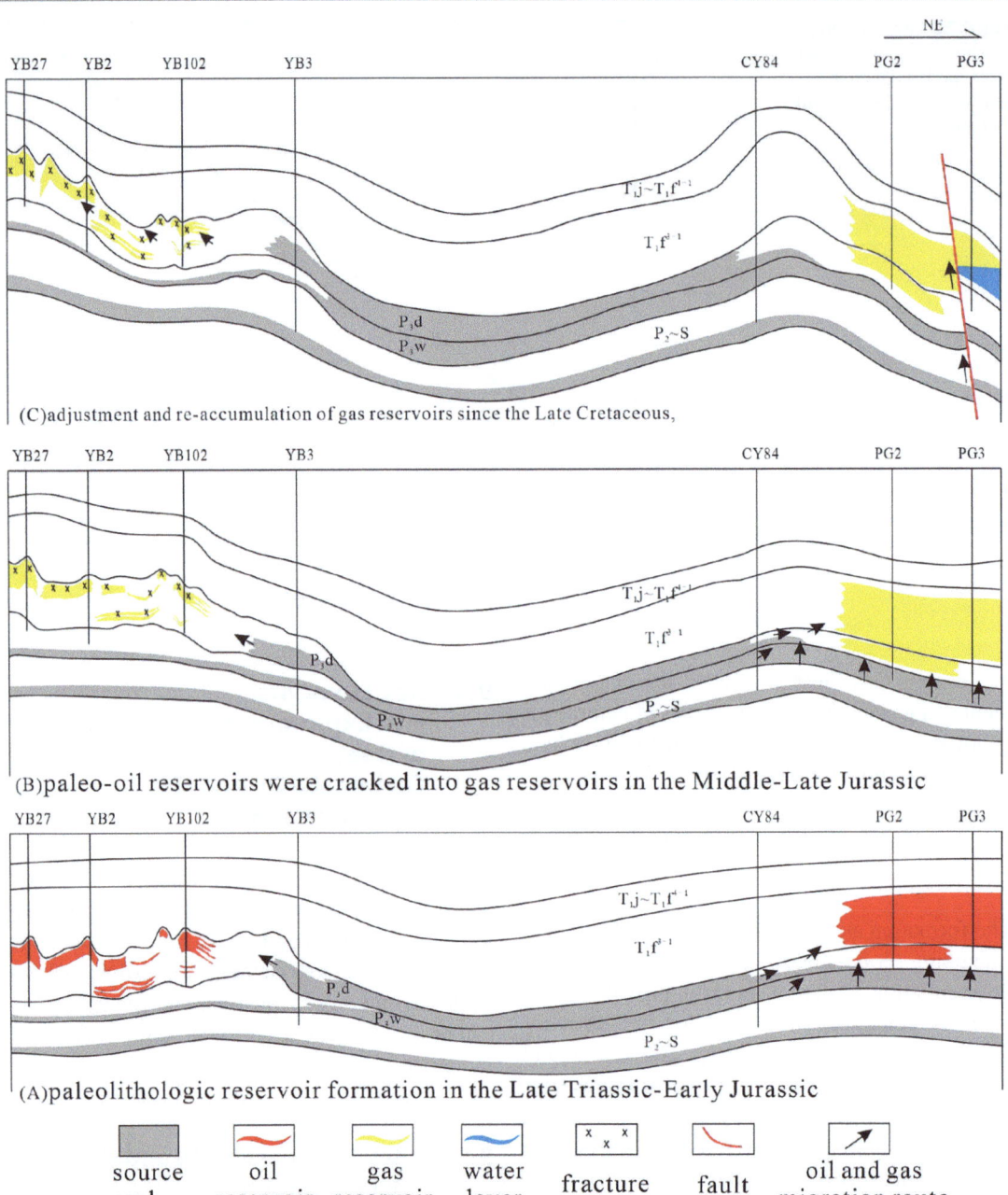

Fig. 9.24 Reservoir model of the Changxing Formation–Feixianguan Formation in the Puguang and Yuanba Gas Fields

oped reservoirs, widely-distributed high-quality source rocks and the efficient configuration of source–reservoirs during the process of temporal and spatial evolution are very important conditions for the formation of large oil and gas fields.

In the Tarim Basin, superior source–reservoir development and efficient allocation constitute the basis for the formation of the Tahe–Shunbei large oil and gas enrichment area. The Lower Cambrian Yuertusi Formation source rocks are widely distributed in the Tarim Basin, which is north of the Tazhong–Bachu paleouplift. Cambrian-Middle–Lower Ordovician facies source rocks are present in the Manjiaer Depression. Middle–Upper Ordovician source rocks may have developed in the Awati Depression. Owing to the influence of multicycle tectonic evolution and regional differences in the geothermal field, multiple sets of source rocks have the conditions for multi-source and Multistage continuous generation and expulsion of hydrocarbons from multiple sources and stages (Zhang et al. 2021, 2011, 2012), providing a sufficient material basis for the formation of large oil and gas fields. The development of the Tabei (Shaya) paleo-uplift and the activity of strike-slip faults provides important conditions for the formation of karst fracture–cavity reservoirs and fault-controlled fracture–cavity reservoirs. Strike-slip faults efficiently communicate sources and reservoirs vertically and laterally, and are active in multiple phases, providing important conditions for oil and gas migration and accumulation (Qi 2020, 2016; Yun 2021; Yun and Deng 2022).

In Sichuan Basin, the formation of the Puguang Gas Field and Yuanba Gas Field also had superior source–reservoir and configuration conditions. Controlled by the formation and development of the Kaijiang–Liangping continental shelf, multiple sets of source rocks of the Wujiaping Formation and Dalong Formation are superimposed, and form a source kitchen with sufficient hydrocarbon supply. High-quality reservoirs with different facies are developed at the carbonate platform edge during the same period with different facies, and the proximity of the source and reservoir is conducive to oil and gas transport, migration and accumulation. The Puguang Gas Field is in the steep structural belt in the eastern Sichuan Basin, where deep and large faults are present and can connect hydrocarbon sources and reservoirs to form an efficient transport system dominated by faults and reservoirs. The structure of the Yuanba Gas Field is relatively stable. In the Wujiaping Formation and Changxing Formation, many tiny faults, microfractures and interlayer fractures, which are filled with asphalt, are developed and form a "three-micro" transport system. This "three-micro" transport system results in

the oil and gas accumulation from shelf facies source rocks through the slope toward the platform margin reef belt (Guo 2019; Guo et al. 2014).

(2) Complex Accumulation–Transformation Process of Reservoirs and Their Effective Preservation

The formation and transformation of deep and ultradeep marine carbonate reservoirs involved long processes. With different tectonic evolution backgrounds, their dynamic accumulation processes also have their own characteristics. In the Tarim Basin, the Tahe Oil Field and Shunbei Oil and Gas Field continued to be buried after the Early Hercynian. Their dynamic oil and gas accumulation processes were occurred multiple phases with regional differences in their secondary reformation (Fig. 9.12). In the Sichuan Basin, after the source–reservoir formation of the Puguang Gas Field and Yuanba Gas Field, their reservoirs experienced rapid burial fromthe Indosinian to the Early Yanshanian, during which paleoreservoir formation and oil–gas transformation occurred. During the Late Yanshan, regional uplift and tectonic reformation resulted in the adjustment of gas reservoirs (Fig. 9.22).

Good preservation conditions during the later period are also important for the maintenance of deep and ultradeep large oil and gas fields. Although the Tahe Oil Field and Shunbei Oil Field in the Tarim Basin have gone through multiple stages of accumulation and transformation, the oil and gas resources have not been destroyed significantly, and the hydrocarbon phase state hasnot undergone large-scale transformation, mainly because the Tahe Oil Field is within a long-term stable ancient uplift and paleo-slope, whereas the Shunbei Oil and Gas Field is within a stable low uplift area. After the Late Caledonian–Early Hercynian period, the Shaya Uplift was generally in a stable subsidence period, the cap rock the within the Tahe Oil Field was not severely damaged, and the paleokarst fracture–cavity reservoirs were preserved (Fig. 9.13). Although the strike-slip faults in the Shunbei Oil Field were active during multiple periods, after the Late Hercynian, the fault activity was relatively weak, and its special vertical layered structure caused the petroleum reservoirs to have good conditions to connect with the source rock downward, with no strong damage to the caprocks upward (Fig. 9.14); thus, the Shunbei fault-controlled fracture–cavity reservoir has been preserved. Although the Puguang Gas Field and Yuanba Gas Field in the Sichuan Basin have undergone regional uplift and tectonic transformation, their main gas-bearing intervals are overlaid with gypsum-bearing formations with a thickness of approximately 500 m, which have become effective regional caprocks. In Puguang Gas Field, the later tectonic activity was strong, and faults and folds developed. However, these have not caused large-scale damage to the caprock in this area, and the natural gas reservoirs have been adjusted but preserved.

9.4.2 Key Exploration Directions

The sedimentary evolution of the deep and ultra-deep marine carbonate strata in the three superimposed basins of Tarim, Sichuan and Ordos are similar, although their tectonic transformation processes have their own characteristics. According to the sedimentary–tectonic evolution process of deep and ultradeep marine carbonate strata in these superimposed basins, the development conditions and temporal and spatial evolution characteristics of source–reservoir–cap under the control of macrogeological background, and the key accumulation elements of these discovered large oil and gas fields, it is clear that the paleokarst reservoirs related to the large-scale integration unconformities of paleo-uplifts, the fault-controlled reservoirs related to multiphase faults, the platform margin mound–shoal reservoirs related to sedimentary differentiation, and the widely distributed paleomicrobial rocks, and dolomite reservoirs developed under the influence of limited tidal flat environments, are important exploration targets for the future.

The paleokarst, related to the large-scale unconformity of the paleo-uplift, is the area where deep and ultradeep marine carbonate rocks have been explored earlier. In addition to the Tahe Oil Field and Jingbian Oil Field, the Sinian in the Tabei Paleo-Uplift of the Tarim Basin,

the Ordovician in the Tazhong Paleo-Uplift and Hetian Paleo-Uplift, the Cambrian–Ordovician in the Enan Paleo-Uplift of the Ordos Basin, and the Permian in the southern and southeastern Sichuan Basin, all have macrogeological conditions for the development of paleokarst reservoirs (He et al. 2019; Zou et al. 2014). However, owing to the differences in the original lithofacies, karst paleogeomorphology and duration, and the relationship with hydrocarbon source configuration, the development characteristics and scale of karst reservoirs in these areas, as well as the hydrocarbon enrichment rules, are not the same.

Platform–edge mound–shoal reservoirs related to sedimentary differences are also extremely important exploration targets. These reservoirs usually develop along the edges of carbonate rock platforms with steep slopes, and have superior accumulation conditions with lateral connections between the reservoir source and near-source. To date, in the Sichuan Basin, large-scale platform-edge reef–shoal/mound–shoal gas fields have been discovered successively in Puguang, Yuanba and Anyue. In addition to the Sinian paleo-rift trough and the Permian paleoshelf edges in the Sichuan Basin, the Middle–Upper Cambrian–Middle Ordovician in the Tarim Basin, the Upper Ordovician in the Tazhong paleo-uplift, and the Ordovician in the western and southern of Ordos Basin, similar reservoirs have developed (Wei et al. 2022; Yang et al. 2022; Zheng et al. 2022; Zhou et al. 2022). Certainly, differences in tectonic reformation in different basins and regions can affect their later adjustment, and the scale of oil and gas enrichment varies accordingly.

The discovery of the Shunbei Oil and Gas Field in the Tarim Basin has revealed a new type of deep and ultradeep-scale reservoir, whose development is controlled by multistage fault activity. In the Sichuan Basin and Ordos Basin, reservoirs, associated with faults and various fluid transformations, have also been discovered (Guo et al. 2010; He et al. 2021; Wei et al. 2019; Zhang et al. 2022; Zhong et al. 2021). The main controlling factors, fluid types and mechanisms of reservoir development, which are different from those of the Shunbei Oil and Gas Field in the Tarim Basin, should be further investigated.

Recent studies have also shown that paleomicrobial rocks, which are commonly developed in the Sinian of the Sichuan Basin and the Tarim Basin, can form widely distributed reservoirs after structural superimposition and diagenetic transformations (Liu et al. 2016; Yang et al. 2020; Zhu et al. 2020). Therefore, they are also excellent exploration targets. Additionally, during the Cambrian–Ordovician period, in the three basins of Tarim, Sichuan and Ordos, a set of reservoir–cap assemblages, composed of gentle slope-limited tidal-flat facies dolomite reservoirs and gypsum–salt caprocks, developed, such as the Middle and Lower Cambrian in the Tarim Basin, the Middle and Lower Cambrian in the east of the Sichuan Basin and the Ordovician in the east of the Ordos Basin. Although this set of pre-salt dolomite caprocks has superior sealing ability, the hydrocarbon source conditions and source–reservoir configurations are quite different in different basins and regions. Areas with better source–reservoir configurations have greater exploration potential.

9.5 Summary

(1) The Tahe Oil Field and Shunbei Oil and Gas Field have been discovered in the "Saddle" tectonic zone of the Tabei Paleo-Uplift and Tazhong-Tabei Paleo-Uplift in the Tarim Basin. The Tahe Oil Field is in the main body and slope of the Tabei Paleo-Uplift, in which the Ordovician carbonate rocks develop typical paleokarst fracture–cavity reservoirs. The Shunbei Oil and Gas Field is in the "saddle" of the two paleo-uplift structures, where typical fault-controlled fracture–cavity reservoirs are developed. These two types of large-scale reservoirs are strongly heterogeneous. The Tahe Oil Field is characterized mainly by large fractures and caves with very large storage spaces, whereas fault-controlled reservoirs in the Shunbei Oil and Gas Field have "grid" distribution characteristics, which is closely related to its faults. Within the Tahe Oil Field and Shunbei Oil and Gas Field, the Ordovician carbonate rocks are oil-bearing throughout the whole body with various phase states. Laterally, they are distributed as a pattern of "east–gas–and–west–oil and south–gas–and–north–oil". The geochemical characteristics of the oil and gas in the Tahe Oil Field and Shunbei Oil Field are generally similar, because both originate from Cambrian–Ordovician marine source rocks. Their crude oils have regional differences in maturity and secondary transformation, and natural gases are mainly kerogen cracking gases with some crude oil cracking gases. Both the Tahe Oil Field and the Shunbei Oil Field experience multistage accumulation and transformation. In different structural units, owing to the differences in thermal evolution, fault activity and source–reservoir space configuration of source rocks, the main accumulation periods and accumulation transformation processes differ. Today, "paleokarst landforms and faults" are still the key elements controlling the accumulation of oil and gas in these two large oil and gas fields. It is clear that throughout the entire formation and evolution process of these reservoirs, the tectonic action was particularly critical.

(2) The Puguang Gas Field and Yuanba Gas Field have been successively discovered on both sides of the Permian "Kaijiang–Liangping" continental shelf in the Sichuan Basin. In the Puguang Gas Field and the Yuanba Gas Field, platform edge reef–shoal facies reservoirs are present within Permian–Triassic Changxing Formation and Feixianguan Formation. These reservoirs have layer and strip-like distributions, and sedimentary microbes play key roles in controlling the development of these reservoirs. The natural gas in the Puguang Gas Field and Yuanba Gas Field is mainly cracked crude oil, and the process of oil and gas accumulation is characterized by "phase state transformation, dynamic adjustment and continuous preservation". During the formation and transformation of these reservoirs, the original sedimentary facies of carbonate rocks play a key role.

(3) China's deep and ultradeep marine carbonate rocks are distributed mainly in the Cambrian–Ordovician and Precambrian strata of three

superimposed basins in central and western China (the Tarim Basin, Sichuan Basin and Ordos Basin), and in the Permian–Triassic strata of the Sichuan Basin. Large-scale reservoirs, widely distributed high-quality source rocks, and efficient source–reservoir configurations during temporal and spatial evolution, are important conditions for the formation of large oil and gas fields. The formation and transformation of deep and ultradeep marine carbonate reservoirs are long processes, and good preservation conditions during the later period are important for their preservation. In addition to the discovered oil and gas fields, the paleomicrobial rocks of the Upper Sinian and Cambrian/Ordovician intrasalt dolomite also have great exploration potential.

Acknowledgements This study was financially supported by the National Natural Science Foundation of China (Grants U24B6001 and U19B6003).

References

An H, Li H, Wang J et al (2009) Tectonic evolution and its controlling on oil and gas accumulation in the northern Tarim Basin. Geotecton Metallog 33:142–147

Aydin A, Berryman JG (2010) Analysis of the growth of strike-slip faults using effective medium theory. J Struct Geol 32:1629–1642

Bai G, Cao B (2014) The global deep oil reservoirs and their distribution rules. Oil Gas Geol 35(1):19–25

Cao Z, Lu Q, Gu Y (2020) Characteristics of Ordovician reservoirs in Shunbei 1 and 5 fault zones, Tarim Basin. Oil Gas Geol 41:975–984

Chen Q, Xi B, Han J et al (2020) Preservation and influencing factors of ultradeep oil reservoirs in the Shuntuoguole area, Tarim Basin: evidence from fluid inclusions. China Petroleum Exploration 25:121

Deng S, Li H, Han J et al (2019) Characteristics of the central segment of Shunbei 5 strike-slip fault zone in Tarim Basin and its geological significance. Oil Gas Geol 40:990–998

Deng S, Li H, Zhang Z et al (2018) Characteristics of differential activities in major strike-slip fault zones and their control on hydrocarbon enrichment in the Shunbei area and its surroundings, Tarim Basin. Oil Gas Geol 39:878–888

Ding Q, He Z, Wang J et al (2020) Simulation experiment of carbonate reservoir modification by source rock-derived acidic fluids. Oil Gas Geol 41:223–234

Duan J (2016) The comparative study of natural gas accumulation characteristics# br# between Puguang and Yuanba Reef-bank Gas Field. J Southwest Pet Univ (Sci Technol Ed) 38:9

Duan J, Li P, Chen D et al (2013) Formation and evolution of the reef flat facies lithologic gas reservoir of Changxing Formation in Yuanba Gas Field, Sichuan Basin. Lithol Reserv 25:43–47

Dyman TS, Crovelli RA, Bartberger CE et al (2002) Worldwide estimates of deep natural gas resources based on the US Geological Survey World Petroleum Assessment 2000. Nat Resour Res 11(6):207–218

Fan X (2015) Control sedimentation characteristics of Upper Permian Changxing Formation in Yuanba Area and its control action for reservoir. J Southwest Pet Univ (Sci Technol Ed) 37:39

Ferrill DA, Mcginnis RN, Morris AP et al (2014) Control of mechanical stratigraphy on bed-restricted jointing and normal faulting: Eagle Ford Formation, south-central TexasMechanical Stratigraphy, Faults and Fractures in the Eagle Ford of Texas. AAPG Bull 98:2477–2506

Guo T (2011a) Basic characteristics of deep reef-bank reservoirs and major controlling factors of gas pools in the Yuanba Gas Field. Nat Gas Ind 31:12–16

Guo T (2011b) The characteristics of sedimentation and reservoirs of reef-beach gas fields in carbonate platform margins, Northeastern Sichuan Basin. Earth Sci Front 18:201–210

Guo T (2019) Gas accumulation conditions and key exploration & development technologies in Yuanba gas field. Acta Petrolei Sinica 40:748

Guo X, Guo T, Huang R et al (2010) Reservoir development characteristics and predication technologies of large Puguang-Yuanba Gas Field. Eng Sci 12:82–90

Guo X, Hu D et al (2011) Newest progress and key techniques in gas exploration of reef-bank reservoirs in the northeastern Sichuan Basin. Nat Gas Ind 31:6–11

Guo X, Huang R, Fu X et al (2014) Gas accumulation and exploration direction of the Permian and Triassic reservoirs of reef-bank facies in Sichuan Basin. Oil Gas Geol 35:295–302

Guo X, Hu D, Huang R et al (2017) Developing mechanism for high quality reef reservoir (Changxing Formation) buried in ultra-depth in the big Yuanba gas field. Acta Petrologica Sinica 33:1107–1114

Guo X, Hu D, Huang R et al (2020) Deep and ultradeep natural gas exploration in the Sichuan Basin: progress and prospect. Nat Gas Ind B 7:419–432

Han Q, Yun L, Jiang H (2021) Marine oil and gas filling and accumulation process in the north of Shuntuoguole area in northern Tarim Basin. J Jilin Univ (Earth Sci Ed) 51:645–658

He D, Ma Y, Liu B et al (2019) Main advances and key issues for deep-seated exploration in petroliferous basins in China. Earth Sci Front 26:1

He D, Zhou X, Yang H et al (2008) Formation mechanism and tectonic types of intracratonic paleo-uplifts in the Tarim Basin. Earth Sci Front 15:207–221

He Z, Ma Y, Zhu D et al (2021) Theoretical and technological progress and research direction of deep and ultradeep carbonate reservoirs. Oil Gas Geol 42:533–546

Hu D (2011) Differences in reef-bank reservoir features between Puguang and Yuanba gas fields and their reasons. Nat Gas Ind 31:17–21

Huang C, Yun L, Cao Z et al (2022) Division and formation mechanism of fault-controlled fracture-vug system of the Middle-to-Lower Ordovician, Shunbei area, Tarim Basin. Oil Gas Geol 43:54–68

Huang R (2014) Formation and evolution of Permian-Triassic Kaijiang-Liangping shelf and development of reef–shoal reservoirs in Sichuan Basin. China J Chengdu Univ Technol (Sci Technol Ed) 41:452–457

Jiao F (2018) Significance and prospect of ultradeep carbonate fault-karst reservoirs in Shunbei area, Tarim Basin. Oil Gas Geol 39:207–216

Liang D (2008) Review and expectation on the discovery of Ordovician Lunnan-Tahe oilfield, Tarim Basin. Acta Petrolei Sinica 29:153

Liu S, Song J, Luo P et al (2016) Characteristics of microbial carbonate reservoir and its hydrocarbon exploring outlook in the Sichuan Basin. China. J Chengdu Univ Technol (Sci & Technol Ed) 43:129–152

Liu Y, Qiu N, Chang J et al (2020) Application of clumped isotope thermometry to thermal evolution of sedimentary basins: a case study of Shuntuoguole area in Tarim Basin. Chin J Geophys 63:597–611

Lu H, Han J, Zhang J (2021) Development characteristics and formation mechanism of ultra-deep carbonate faultdis-solution body in Shunbei area, Tarim basin. Pet Geol Exp 43:14–22

Ma A, He Z, Yun L et al (2021a) The geochemical characteristics and origin of Ordovician ultradeep natural gas in the North Shuntuoguole area, Tarim Basin, NW China. J Nat Gas Geosci 6:289–300

Ma A, Lin H, Yun L et al (2021b) Diamondoids in oils from the ultradeep Ordovician in the North Shuntuoguole area in the Tarim Basin, NW China. J Nat Gas Geosci 6:89–99

Ma Y, Cai X (2006) Exploration achievements and prospects of the Permian-Triassic natural gas in northeastern Sichuan Basin. Oil Gas Geol 27:741–750

Ma Y, Cai X, Yun L et al (2022) Practice and theoretical and technical progress in exploration and development of Shunbei ultradeep carbonate oil and gas field, Tarim Basin, NW China. Pet Explor Dev 49:1–20

Ma Y, Cai X, Zhao P et al (2010a) Distribution and further exploration of the large-medium sized gas fields in Sichuan Basin. Acta Petrolei Sinica 31:347

Ma Y, Cai X, Zhao P et al (2010b) Formation mechanism of deep-buried carbonate reservoir and its model of three-element controlling reservoir: a case study from the Puguang Oilfield in Sichuan. Acta Geol Sin 84:1087–1094

Ma Y, Guo X, Guo T et al (2007) The Puguang gas field: new giant discovery in the mature Sichuan Basin, southwest China. AAPG Bull 91:627–643

Ma Y, He Z, Zhao P et al (2019) A new progress in formation mechanism of deep and ultradeep carbonate reservoir. Acta Petrolei Sinica 40:1415

Ma Y, Li M, Cai X et al (2020) Mechanisms and exploitation of deep marine petroleum accumulations in China: advances technological bottlenecks and basic scientific problems. Oil Gas Geol 41:655–672

Qi L (2020) Characteristics and inspiration of ultradeep fault-karst reservoir in the Shunbei area of the Tarim Basin. China Pet Explor 25:102

Qi L (2016) Oil and gas breakthrough in ultradeep Ordovician carbonate formations in Shuntuoguole uplift, Tarim Basin. China Pet Explor 21:38

Qi L, Yun L (2010) Development characteristics and main controlling factors of the Ordovician carbonate karst in Tahe oilfield. Oil Gas Geol 31:1–12

Wang B, Zhao Y, He S et al (2020) Hydrocarbon accumulation stages and their controlling factors in the northern Ordovician Shunbei 5 fault zone, Tarim Basin. Oil Gas Geol 41:965–974

Wei G, Yang W, Xie W et al (2022) Formation mechanisms, potentials and exploration practices of large lithologic gas reservoirs in and around an intracratonic rift: taking the Sinian—Cambrian of Sichuan Basin as an example. Pet Explor Dev 49:530–545

Wei G, Zhu Q, Yang W et al (2019) Cambrian faults and their control on the sedimentation and reservoirs in the Ordos Basin, NW China. Pet Explor Dev 46:883–895

Yang H, Shen A, Zheng J et al (2020) Sedimentary characteristics and reservoir significance of the microbial dolomite of Sinian Qigebrak Formation in the northwest margin of Tarim Basin. Mar Orig Pet Geol 25:1–10

Yang Y, Wen L, Chen C et al (2022) The oil and gas exploration potential of Permian multistage platform margin zone in the western Sichuan Basin. Earth Sci Front 1–12

Yu J, Shi K, Wang Q et al (2022) Structural diagenesis of deep carbonate rocks controlled by intra-cratonic strike-slip faulting: an example in the Shunbei area of the Tarim Basin, NW China. Basin Res 34:1601–1631

Yun L (2021) Hydrocarbon accumulation of ultradeep Ordovician fault-karst reservoirs in Shunbei area. Xinjiang Petroleum Geology 42:136

Yun L, Deng S (2022) Structural styles of deep strike-slip faults in Tarim Basin and the characteristics of their control on reservoir formation and hydrocarbon accumulation: a case study of Shunbei oil and gas field. Acta Petrolei Sinica 43:770

Zhai X (2011) Exploration practice and experience of Tahe giant oil-and-gas field, Tarim Basin. Pet Geol Exp 33:323–331

Zhai X (2006) Exploration practices in frontiers of Tahe oilfield. Oil Gas Geol 27:751–761

Zhai X, Yun L (2008) Geology of giant Tahe oilfield and a review of exploration thinking in the Tarim Basin. Oil Gas Geol 29:565–573

Zhang S, Su J, Zhang B et al (2021) Genetic mechanism and controlling factors of deep marine light oil and condensate oil in Tarim Basin. Acta Petrolei Sinica 42:1566

Zhang S, Zhang B, Li B et al (2011) History of hydrocarbon accumulations spanning important tectonic phases in marine sedimentary basins of China: taking the Tarim Basin as an example. Pet Explor Dev 38:1–15

Zhang S, Zhiyong G, Jianjun L et al (2012) Identification and distribution of marine hydrocarbon source rocks in the Ordovician and Cambrian of the Tarim Basin. Pet Explor Dev 39:305–314

Zhang Y, Cao Q, Luo K et al (2022) Reservoir exploration of the Permian Maokou Formation in the Sichuan Basin and enlightenment obtained. Oil Gas Geol 43:610–620

Zhao W, Hu S, Wang Z et al (2018) Petroleum geological conditions and exploration importance of Proterozoic to Cambrian in China. Pet Explor Dev 45:1–14

Zheng H, Tian J, Hu Z et al (2022) Lithofacies palaeo-geographic evolution and sedimentary model of the Ordovician in the Tarim Basin. Oil Gas Geol 43:733–745

Zhong Y, Yang Y, Wen L et al (2021) Reconstruction and petroleum geological significance of lithofacies paleogeography and paleokarst geomorphology of the Middle Permian Maokou Formation in northwestern Sichuan Basin, SW China. Pet Explor Dev 48:95–109

Zhou J, Yin C, Zeng L et al (2022) Development characteristics of grain shoals and favorable gas exploration areas in the 4th Member of Ordovician Majiagou Formation in the Ordos Basin. Nat Gas Ind 42:17–30

Zhu D, Liu Q, He Z et al (2020) Early development and late preservation of porosity linked to presence of hydrocarbons in Precambrian microbialite gas reservoirs within the Sichuan Basin, southern China. Precambr Res 342:105694

Zou C, Du J, Xu C et al (2014) Formation, distribution, resource potential, and discovery of Sinian-Cambrian giant gas field, Sichuan Basin, SW China. Pet Explor Dev 41:306–325

Geological Characteristics and Development Techniques of Ultra-deep Carbonate Sour Gas Reservoirs: A Case Study of the Puguang Gas Field

10

Daqian Zeng, Tong Li, Yuchun You, Rui Zhang, Qian Li, Zhaojie Song, Song Peng, and Cheng Zhang

Abstract

High-sulfur carbonate gas reservoirs, represented by the Puguang gas field, have large reserves and are characterized by high hydrogen sulfide contents, strong reservoir heterogeneity, and abundant edge and bottom water; thus, complex development techniques are needed. This chapter provides a detailed description of the reservoir characteristics of the Puguang gas field in terms of its stratigraphic, tectonic, and depositional characteristics; investigates the water intrusion law and water intrusion prediction methods during the development of the Puguang gas field; and proposes corresponding water control countermeasures and water draining and plugging processes. In accordance with the singlet sulfur dissolution–precipitation mechanism, the distribution of sulfur deposition in the wellbore and reservoir during the development of the Puguang gas field is studied, and a technique for the management of sulfur deposition in the wellbore and on the surface is developed. Finally, future directions for technological developments for gas field regression control and field recovery enhancement are outlined. The results of this study provide technical support for achieving high and stable production in the Puguang gas field.

Keywords

High-sulfur gas reservoirs · Reef-shoal facies reservoir · Water control technique · Sulfur deposition management technique · Puguang gas field

D. Zeng (✉) · T. Li · Y. You · R. Zhang · Q. Li
State Key Laboratory of Shale Oil and Gas Accumulation Mechanisms and Effective Development, Beijing, China
e-mail: zengdq.syky@sinopec.com

Sinopec Petroleum Exploration and Production Research Institute, Beijing, China

Z. Song
China University of Petroleum—Beijing, Beijing, China

S. Peng
Exploration and Development Research Institute of SINOPEC Zhongyuan Oilfield, Puyang, Henan, China

C. Zhang
Research Institute of Production Engineering, Zhongyuan Oilfield, Puyang, Henan, China

10.1 Introduction

Since the 1950s, the development of high-sulfur carbonate gas reservoirs has been rapidly progressing in Europe and the Americas; countries such as Canada, the United States, France, and Russia have successfully developed several high-sulfur carbonate gas reservoirs, established a relatively complete set of development techniques and management systems, and accumulated certain exploitation experience (Roberts 1997; Roberts 2017; Kuo 1972; Chesnoy and Pack 1997; Hyne and Derdall 1980; Bourrouilh et al. 1995; Espitalié and Drouet 1992; Paux and Zhou 1997; Vermersch 1968; Villien 1992; Coskuner 1994). China is rich in high-sulfur carbonate gas resources, with proven reserves of more than 1 trillion cubic meters; however, the development of gas reservoirs started late, and these gas fields generally have geological characteristics such as deep burial depths, strong reservoir heterogeneity, high H_2S contents, and complex edge and bottom water, making development difficult (Ma et al. 2008; Li et al. 2007; Guo et al. 2010).

The Puguang gas field is the largest high-sulfur gas field in China, with a depth of 4800–6000 m, a central reservoir temperature of 128 °C, and a molar content of 13–18% H_2S, with an average of 15%. The reservoir is characterized by a normal-pressure and low-temperature system with a formation pressure coefficient of 1.00–1.18 and a formation temperature gradient of 1.98–2.21 °C/100 m. It is located in the northeastern part of the Sichuan Basin in a high, steep tectonic zone (Luo et al. 2011; Liu et al. 2017). The main target sections are the Triassic Feixianguan Formation and the Permian Changxing Formation, which are strongly heterogeneous reservoirs with fracture development and complex gas–water relationships. These are ultradeep, tectonically and lithologically controlled, atmospheric carbonate, high-hydrogen-sulfide, and dry-gas reservoirs with edge water (Kong et al. 2011; Shi et al. 2014).

Exploration work in the Puguang area began in the 1950s. From the 1950s to the mid-1960s, mainly oil and gas geological investigations were conducted. From the late 1960s to the late 1970s, regional tectonic probing was conducted. The focus of this period was oil and gas discovery, and many exploratory wells were drilled to investigate local tectonics; however, the exploratory wells drilled were essentially shallow. From the 1980s to the 1990s, a tectonic zone census, a detailed local tectonic investigation, and deep exploration were conducted. Since 2000, Sinopec has reanalyzed the natural gas resource potential of the northeast Sichuan area based on a summary of previous exploration work, in addition to identifying the Changxing and Feixianguan Formations in the northeast Sichuan area as the preferred fields

for oil and gas exploration in the near future. On June 15, 2009, the last development well, Puguang 105-2, was completed; in July 2008, the Puguang 302-1 well was put into production; and in October 2009, it was officially put into production. The gas field is now in the stable production stage (Ma 2006; Huang 2015).

The edge water in the Puguang gas field has a large volume, strong energy, and fast intrusion speed (Wang et al. 2011a, b; Liu et al. 2015; Li et al. 2020). Owing to the presence of local fractures, the seepage characteristics and water intrusion during the development process are complex. As the reservoir and wellbore pressures decrease, elemental sulfur gradually precipitates and is deposited, affecting the reservoir's seepage capacity and the gas well's production ability, which in turn directly affects the stable production period and gas recovery (Guo et al. 2015; Mao et al. 2016; Li et al. 2022). Compared with other conventional gas reservoirs, more attention needs to be given to the research on and application of water intrusion patterns and water control techniques, as well as sulfur deposition patterns and sulfur control techniques. Therefore, on the basis of the geological characteristics of the development of the Puguang gas field, this chapter focuses on the water intrusion pattern and water control techniques, as well as the sulfur deposition pattern and management techniques of the gas reservoir.

10.2 Geological Characteristics of the Puguang Gas Field

10.2.1 Stratigraphic Features

Based on drilling and surface outcrops, the Lower Paleozoic stratigraphy of the Puguang gas field is relatively complete, with only the Upper Silurian missing. The Upper Paleozoic system is missing all of the Devonian and most of the Carboniferous, with only the Middle Carboniferous Huanglong Formation remaining. The Permian system is complete. The Triassic and Jurassic strata are well preserved, the Lower Cretaceous strata are well preserved, and the Upper Cretaceous strata are missing. There was essentially no sedimentary retention in the Eocene. Among the marine strata, the Silurian and Carboniferous, Carboniferous and Permian, and the Upper and Lower Permian are all in parallel unconformable contact. This is due to the movements of Garidon, Yunnan, and Dongwu, respectively. At the end of the Late Triassic, the Indo-Chinese movement lifted the Sichuan Basin as a whole above sea level, and the seawater of the ancient Tethys Sea completely withdrew from the Yangzi Plateau, leading to the deposition of lacustrine–deltaic–fluvial deposits dominated by terrestrial clastic rocks. The main target formations in the Puguang gas field are the Triassic Feixianguan Formation and the Permian Changxing Formation (Guan et al. 2010; Zhao et al. 2011; Wang et al. 2011a, b; Ma et al. 2014).

(1) Feixianguan Group (T_1f)

The stratigraphic thickness of the Feixianguan Formation in the Puguang gas field is 445–720 m, and it is in integrated contact with the underlying Changxing Formation and the overlying Jialingjiang Formation. The Feixianguan Formation is a terrace margin to terrace phase oolitic beach deposit, and the segmentation of sections Fei I to Fei III is not obvious, but this formation can be broadly divided into sections Fei I, II, III, and IV on the basis of regional stratigraphic comparisons, rock types, and logging curve characteristics.

Fei IV ($T_1 f_4$): The lithology is purple and gray dolomite, chert, gypsum dolomite, and gray–white gypsum interbedded at unequal thicknesses and is characterized by the development of gypsum. The stratigraphic thickness is generally approximately 30 m, with the overall thickness increasing from east to west and the gypsum layer tending to develop.

Fei III ($T_1 f_3$) is characterized by more developed tuff than sections Fei I and II and undeveloped pinacoidal dolomite. Overall, this section is thin in the east and thick in the west. The lithology in the Puguang area is gray and dark gray medium- to thick-bedded mud tuffs, microcrystalline tuffs, dolomite interbedded with oolitic tuffs, sandy debris tuffs, and lysergic dolomite, which formed during the shallow terrace phase. The thickness of the dolomite gradually decreases

and the tuff gradually increases along the line from Puguang 2 to Puguang 1, with obvious features to the west and south. To the east, the area around Yuezhai has become a tuff or marl sedimentary formation, which was formed mainly during the open terrace phase.

Fei I–II ($T_1 f_{1-2}$): The lithology in the Puguang area is gray and dark gray crystalline dolomite, lysergic dolomite, and oolitic dolomite, which are generally characterized by the development of lysergic dolomite. This section formed on the edge of the terrace during the exposed shallow phase, forming a dolomite development area in the Tieshanpo area in the north and the Dukouhe area in the east. It is in pseudointegrated contact with the underlying tuffs or biotite of the Changxing Formation.

(2) Changxing Group (P_2ch)

In this area, the Changxing Group is 92–290 m thick. Overall, it is thin in the west and thick in the east. The lithology is dominated by biogenic reef dolomite, crystalline dolomite, gray dolomite, and bioclastic tuff in the upper part and gray dolomite and tuff in the lower part, with a set of pure tuffs at the bottom. It is in integrated contact with the underlying impure tuffs and shale layers of the Longtan Formation.

At the end of the Permian, a massive decrease in sea level occurred on the Yangzi Plateau, and a set of dolomites was commonly developed at the top of the Changxing Formation in the shallow-water carbonate plateau area. Seawater erosion occurred at the beginning of the Triassic, and a set of thin-slabbed marl tuffs and marl-bearing tuffs was commonly deposited at the bottom of section Fei I (T_1f_1). The thin-slabbed marl tuffs and marl tuffs directly overlie the dolomites. Therefore, if dolomite developed between the Permian and Triassic, a boundary could be drawn on the contact surface between the dolomite and the tuff.

10.2.2 Constructive Features

The Puguang gas field is located in the northeast section of the Chuandong Fault Zone. This area underwent three stages of tectonic deformation during the Yanshan and Early and Late Xishan periods, primarily forming NNE- and NW-oriented structures. Three sets of brittle deformation layers and three sets of plastic deformation layers have developed. The upper part of the Jialingjiang Formation to the lower part of the Leikoupo Formation is the main slip layer, and the Silurian shale is the secondary slip layer, delineating the upper, middle, and lower deformation layers, which exhibit uncoordinated deformation characteristics. The middle tectonic deformation layer is the high-strain layer in this area, and it is also the main gas-bearing layer in the area (Ma et al. 2005; Ma 2007). The stratigraphy is Silurian–Middle Triassic, consisting of Middle and Lower Triassic plutonic rocks, tuffs, and dolomites; Permian tuffs; Carboniferous tuffs; and Silurian shales. The tectonic style is a series of smooth slip-overlay folding zones, with a series of reverse faults and related folds forming a stacked tile-like slip-overlay zone. The reverse faults are steep at the top and gentle at the bottom. They either disappear into the Silurian shale layers or cut downward through the Aurignacian–Cambrian and disappear into the slip surfaces between the cap and the basement. During the detachment and slip processes along different slip surfaces, several recoil reverse faults often formed to moderate motion. In this area, the middle deformation layer is the uplift and depression tectonic zone, which is controlled by the north–east fault, forming a tectonic pattern containing five uplifts and five depressions. The southeastern part of the study area has been modified by NW-oriented tectonic activity, resulting in the formation of a tectonic pattern involving two uplifts and two depressions. Tectonic deformation occurred mainly during the Late Yanshan period and was modified during the Early to Middle Xishan period.

The Puguang gas field is tectonically characterized by a nose-like structure in the Shuangshimiao–Puguang NE-trending tectonic zone in the northeast section of the Chuandong Fold Zone, and it is characterized by a large NNE-trending long-axis faulted backslope structure, with a high in the southwest and a low in the northeast. This structure is associated with retrograde faults. The Puguang fault is the main fault controlling the Puguang structure, and the Puguang 3 fault divides the Puguang structure into the Puguang 2 block and the Puguang 3 block (Fig. 10.1, left). The planar morphology is wide in the south and narrow in the north, with a steep western flank and a gentle eastern flank. A succession of formations developed, with the tectonic highs essentially overlapping (Fig. 10.1, right), mainly in the area of wells Puguang 2 to Puguang 6.

10.2.3   Depositional Characteristics

On the basis of the regional sedimentary background and analyses of various data from cores, well logging, mud logging, and seismic surveys, the sedimentary phase evolution of the Late Permian Changxing and Early Triassic Feixianguan Formations in the main body of the Puguang gas field shows strong inheritance, exhibiting a typical reef–beach sedimentary assemblage (Feng et al. 2008).

(1) Changxing period

During the Early Changxing period, the northeastern Sichuan area was in a gently sloping depositional mode, and the Puguang area was located in a medium-gently sloping environment. Under the influence of rifting, the differences in the northeastern Sichuan area increased and decreased. The western part sank and became a slope into a deep-water shelf depositional environment, while the eastern part sank into a shallow-water carbonate terrace depositional environment, and the Puguang area was located mainly on the carbonate terrace margin. The slope was located in the Liangping–Kaijiang area west of well P302-1. Platform margin bioreefs are distributed mainly in well areas Puguang 5 to Puguang 6 to Puguang 8 to Puguang 9. There are mostly bioreef tuffs in the lower part and bioreef dolomite in the upper part. The open terrace is located in the northeastern part of the platform margin bioreef phase zone, in which a set of biofractured tuff-dominated

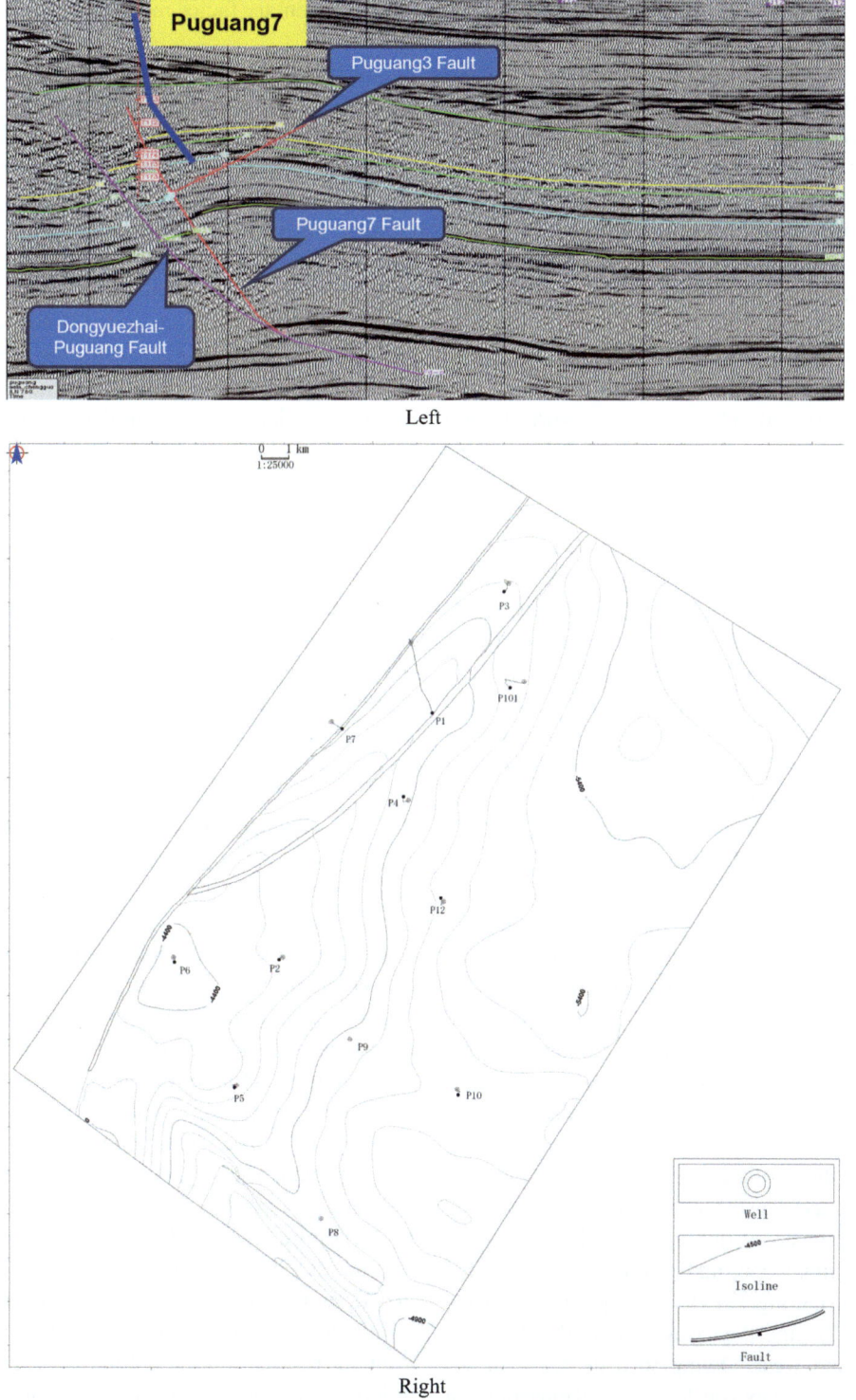

Left

Right

Fig. 10.1 Profile (left) and plane (right) diagram of the Puguang gas field structure

strata, which are generally nonreservoir rocks, was deposited. During the Late Changxing period, owing to the decrease in sea level, the Puguang area evolved into a confined terrace depositional environment, with the eastern part being mainly a confined terrace lagoon microphase and the western part of the former reef–beach phase zone evolving into a tidal ping microphase.

(2) Feixianguan period

During the early to mid Fei I–II period, the paleogeographic pattern was similar to that of the Changxing period. The phase zone spreads in a northwest–southeast direction, with the land shelf in the west and the carbonate plateau in the east. During the late Fei I–II period, the depositional environment changed because of a decrease in sea level and subsidence of the sedimentary basement. The eastern platform evolved from a confined platform to an evaporite platform, and the scale of the marginal shoals increased (Fig. 10.2). The most favorable reservoir phase zone in the field is located in well areas Puguang 6 to Puguang 5 and a relatively large area to the east, while the slope to the west of well Puguang 8 is located at the intersection of the western terrace and the land shelf in a narrow NW–SE-trending belt.

There was a brief sea intrusion in the early Fei III period, which left the Puguang area in an open terrace depositional environment. The subsequent sea retreat and the influences of various factors, such as sediment accumulation, led to significant changes in the paleogeographic pattern in the Puguang area. As a result of filling and replenishment, the Kaijiang–Liangping land shelf area in the west was gradually filled, and the northeastern Sichuan area became a unified terrace. The lithology is dominated by gray oolitic (graywacke) dolomite, sand (graywacke) dolomite, mudstone, and graywacke, with some reservoirs. During Fei IV, the topography of the Xuanhan–Daxian area was essentially shoals to the east and west. Owing to a decrease in sea level, a terrace evaporite depositional environment developed, and a set of purplish red mudstones and dolomites interspersed with gypsum, essentially nonreservoir rocks, was deposited.

(3) Sedimentary facies model

On the basis of the geological sedimentary reanalysis of the Changxing Formation–Feixianguan Formation, the key parameters for simulating the sedimentary process were obtained: parameters related to sea level changes, initial topography, tectonic subsidence, carbonate rock production, and sediment transport. On this basis and with an understanding of the stratigraphy and deposition in the Puguang gas field, stratigraphic sedimentary models were established for the Changxing Formation and sections Fei I–III through forward and inverse depositional simulations. Relying on our geological knowledge about reef–shoal facies reservoirs, high-precision geological, logging, and seismic data, as well as sedimentary simulation results (Fig. 10.3), were used to construct a training image of the reef–shoal facies reservoirs, and a 3-D model of the sedimentary microfacies of the Changxing Formation and Feixianguan Formation in the Puguang gas field was established using the multiple-point geostatistical method (Bracco et al. 2005, 2006; Huang et al. 2022).

10.2.4 Reservoir Characteristics

(1) Lithological characteristics

The Changxing Formation reservoir is characterized mainly by a set of reef–shoal dolomite assemblages, dominated by crystalline dolomite, bioclastic dolomite, arenaceous dolomite, gravelly dolomite, sponge reef dolomite, and sponge reef limestone, among which crystalline dolomite, gravelly dolomite, and sponge reef dolomite are important rock types. The lithology of the Feixianguan Formation reservoir is dominated by dolomite (Ma 2007; Zeng et al. 2011; Ren et al. 2008; Su et al. 2005; Chen et al. 2013), and six rock types are present, including oolitic dolomite, residual oolitic dolomite, sugar-granular residual oolitic dolomite, pebbly oolitic dolomite, arenaceous argillaceous dolomite, and crystalline dolomite, among which (residual) oolitic dolomite and crystalline dolomite are the most important.

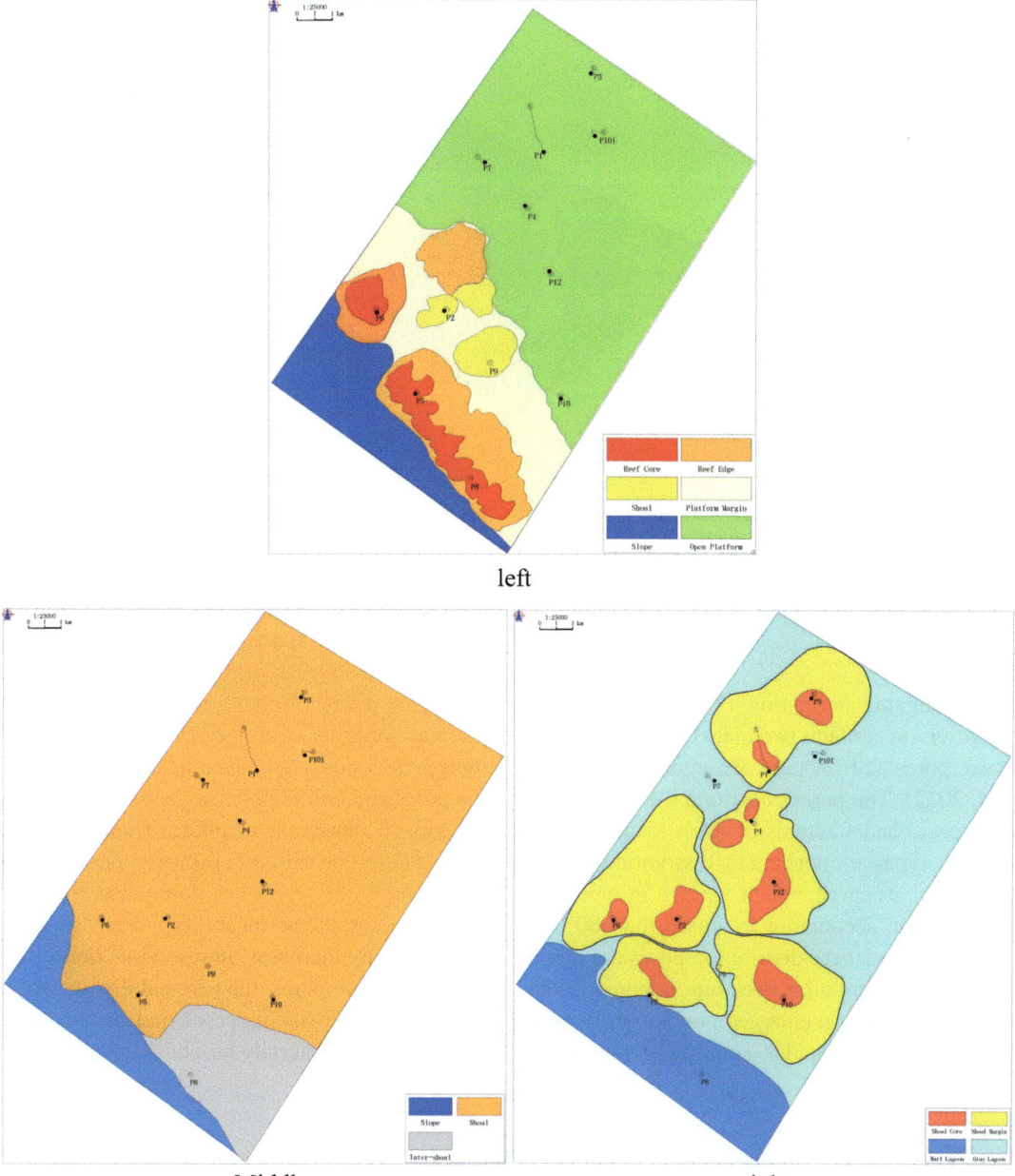

Fig. 10.2 Sedimentary phase diagrams of the Changxing Formation (left), Fei I and II (center) and Fei III (right) in the Puguang gas field

(2) Physical properties

The reservoir properties of the Changxing to Feixianguan Formations in the Puguang gas field are generally good. Among them, the Changxing Formation is dominated by medium-porosity and high-permeability reservoirs, with porosities of 0.79–23.05% (average of 6.12%) and permeabilities of 0.000509–9664.887×10^{-3} μm^2 (average of 121.92×10^{-3} μm^2). The Feixianguan Formation is dominated by medium-porosity and medium-permeability reservoirs and

Fig. 10.3 Sedimentary forward modeling of Changxing Formation in Puguang area

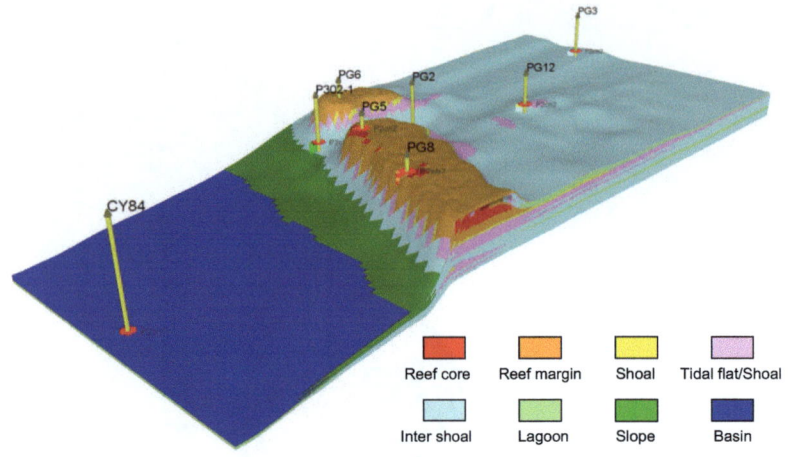

high-porosity and high-permeability reservoirs, with porosities of 0.26–28.86% and an average permeability of 56.86×10^{-3} μm^2 (Fig. 10.4).

(3) Storage space

The Changxing Formation–Feixianguan Formation reservoir contains two main types of reservoir space: pores and fractures (Li et al. 2014; Wang et al. 2022). The pores constitute the main reservoir space, and fractures develop locally. There are two types of pores: (1) dissolution-related pores, such as caves, which are rich in types and predominant, accounting for more than 80% of the pore space; these dissolution pores are dominated by intergranular dissolution pores, intergranular dissolution enlarged pores, oolitic mold pores, and intragranular solution pores, while the primary intergranular pores essentially disappear or are mostly difficult to identify after post-corrosion transformation; (2) intergranular pores unrelated to dissolution, accounting for only 10–15% of the pore space (Fig. 10.5). For convenience, the intergranular dissolution pores, intercrystalline pores, dissolution cavities, and intergranular pores are combined into the category of intergranular pores, and the intragranular dissolution pores and oolitic pores are combined into the category of intragranular pores. These two pore types lead to two different porosity–permeability relationships. In reservoirs dominated by intergranular pores, the permeability increases rapidly as the porosity increases. In reservoirs dominated by intragranular pores, the permeability increases slowly as the porosity increases, and the maximum permeability is generally around 1 mD.

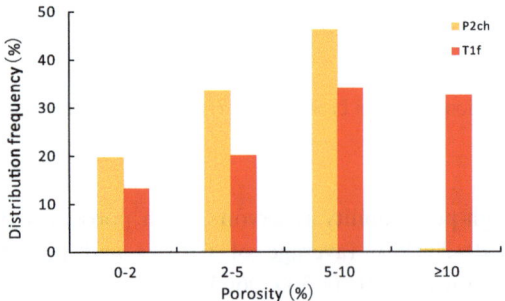
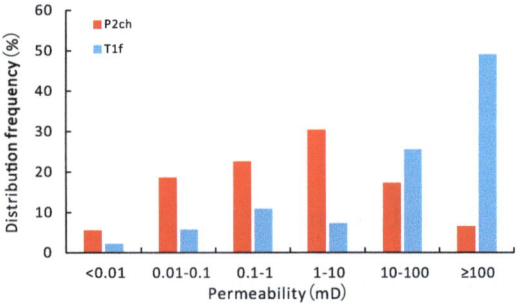

Fig. 10.4 Distribution histogram of porosity (left) and permeability (right) of Changxing Formation Feixianguan Formation reservoir in Puguang gas field

Fig. 10.5 Reservoir space types of Puguang gas field

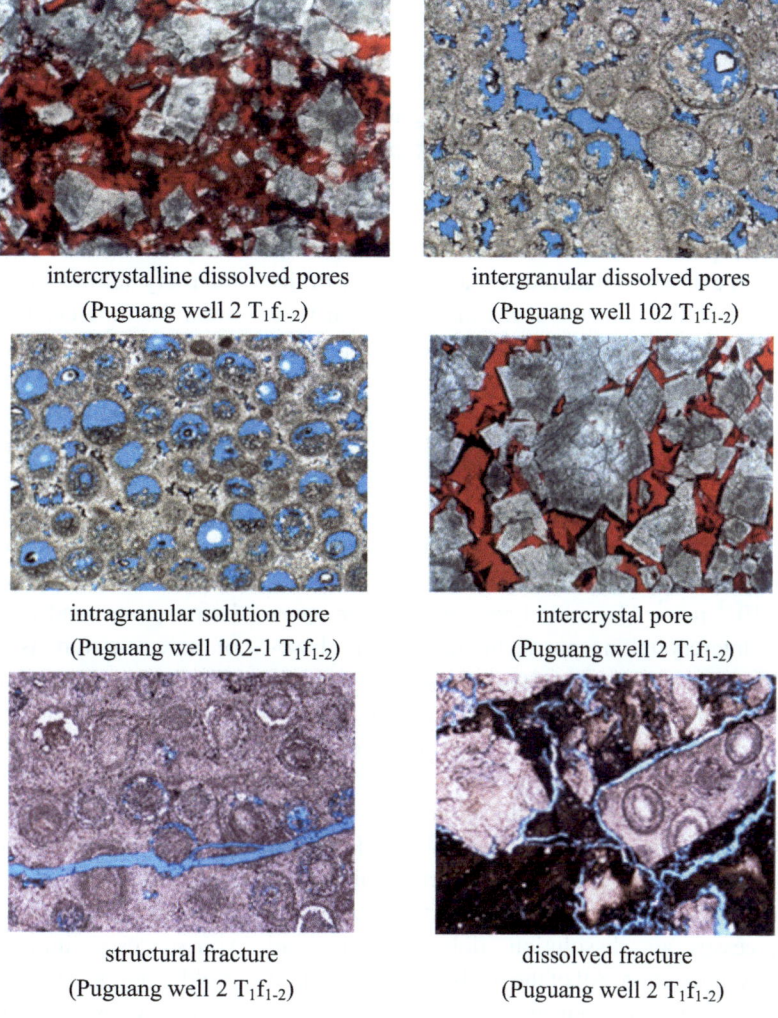

intercrystalline dissolved pores
(Puguang well 2 T_1f_{1-2})

intergranular dissolved pores
(Puguang well 102 T_1f_{1-2})

intragranular solution pore
(Puguang well 102-1 T_1f_{1-2})

intercrystal pore
(Puguang well 2 T_1f_{1-2})

structural fracture
(Puguang well 2 T_1f_{1-2})

dissolved fracture
(Puguang well 2 T_1f_{1-2})

The fractures in the Puguang gas field are developed only locally and are structural and dissolution fractures (Fig. 10.5). In general, fractures have little effect on productivity but have a greater effect on water intrusion during development. Core observations and EMI image logging revealed three stages of fractures in the Puguang area. Early fractures formed during the Late Yanshanian period and were affected mainly by NW-trending compressive stress. Second-stage fractures formed during the Early Himalayan period, and the principal stress direction was essentially the same as that during the Late Yanshanian period. Third-stage fractures formed during the Late Himalayan period and were affected mainly by NE-trending compressive stress.

(4) Reservoir classification

In the early stage of development of the Puguang gas field, the classification scheme of the reservoirs was based on porosity and was borrowed from that of the northeastern Sichuan region. However, owing to the variety of pore types and the extremely complex pore structures in the reservoirs in this area, there are various pore–permeability relationships. Layer classification cannot meet the needs of fine reservoir research in the development and adjustment stages (Anselmetti and Eberli 1997, 1999). To provide a basis for determining the distribution of the remaining gas and for tapping the potential in the later stage of gas field development, according to the reservoir physical properties, lithology, types of reservoir spaces, and pore structure characteristics of the Puguang gas field and based on the results of the production profile test, the permeability, which can best reflect the seepage capacity of the reservoir, was selected as the main reservoir type discrimination index. A new standard for reservoir classification and evaluation in the Puguang gas field that considers the production of the reservoir was established (Table 10.1, Fig. 10.6). The reservoirs were divided into two types and six categories.

Type I fully developed reservoirs have a porosity of $\geq$ 5% and a permeability of $\geq$ 0.3 $\times$ 10^{-3} μm^2. The main types of reservoir space are intercrystalline pores, intercrystalline dissolution pores, and intergranular dissolution pores, and the main types of pore structures are macropores with coarse–medium throats and mesopores with coarse–medium throats. The main reservoir rocks are coarse crystalline dolomite, medium crystalline dolomite, anisocrystalline dolomite, fine crystalline dolomite, and residual oolitic dolomite, and the main constructive diagenetic processes are dissolution, dolomitization, and recrystallization. All of the wells in the production profile test have produced.

Type I partially producing reservoirs have a porosity of 3.5–5% and a permeability of 0.1–0.3 $\times$ 10^{-3} μm^2. The main types of storage space are intercrystalline pores, intercrystalline dissolution pores, and intergranular dissolution pores. The main types of pore structures are mesopores with fine throats and small pores with fine throats, and the main reservoir rocks are powder crystal dolomite, powder fine crystal dolomite, micropowder crystal dolomite, microcrystalline dolomite, and residual oolitic dolomite. The main constructive diagenetic processes are dissolution, dolomitization, and recrystallization. Some of the wells in the production profile test have produced, but their gas production ratio is low.

Type I undeveloped reservoirs usually have a porosity of 2–3.5% and a permeability of 0.002–0.1 $\times$ 10^{-3} μm^2. The main types of reservoir spaces are intercrystalline pores, intercrystalline dissolution pores, and interparticle dissolution pores. The main types of pore structures are mesopores and micropores with thin throats. The main reservoir rocks are silty crystalline dolomite, silty microcrystalline dolomite, silty fine-crystalline dolomite, fine-crystalline dolomite, and residual oolitic dolomite. The main constructive diagenetic processes are dissolution, dolomitization, and recrystallization. The wells in the production profile testing have not produced.

Type II fully developed reservoirs usually have a porosity $\geq$ 16% and a permeability $\geq$ 0.3 $\times$ 10^{-3} μm^2. The main types of reservoir spaces are interparticle dissolution pores, intraparticle dissolution pores, and moldic pores. The main type of pore structure is macropores with medium–thin throats. The main reservoir rock is residual oolitic dolomite. The main constructive diagenetic processes are dissolution and dolomitization. All the wells in the production profile testing have produced.

Type II partially developed reservoirs usually have a porosity of 6–16% and a permeability of 0.1–0.3 $\times$ 10^{-3} μm^2. The main types of reservoir spaces are intraparticle dissolution pores and moldic pores. The main type of pore structure is mesopores with medium–thin throats. The main reservoir rocks are oolitic dolomite and residual oolitic dolomite. The main constructive diagenetic processes are dissolution and dolomitization. Some of the wells in the production profile testing have produced, but their gas production ratio is low.

Type II undeveloped reservoirs usually have a porosity of 2–6% and a permeability of 0.002–0.1 $\times$ 10^{-3} μm^2. The main types of reservoir spaces are moldic pores and intraparticle dissolution pores. The main type of pore structure is mesopores with thin throats. The main reservoir rocks are oolitic dolomite and residual oolitic dolomite. The main constructive diagenetic processes are dissolution and dolomitization. The wells in the production profile testing have not produced.

Table 10.1 Classification and evaluation criteria for carbonate reservoirs in Puguang gas field based on reservoir development

Reservoir type			Type I (interparticle pore)		
			Full developed	Partial developed	Undeveloped
Reservoir parameters	Quantify	Permeability $(10^{-3} \mu m^2)$	≥ 0.3	0.1–0.3	0.002–0.1
	Parameter	Porosity (%)	≥ 5	3.5–5	2–3.5
	Qualitative parameter	Main reservoir space type	Intercrystal pores, Intercrystal dissolved pores, Interparticle dissolved pores	Intercrystal pores, Intercrystal dissolved pores, Interparticle dissolved pores	Intercrystal pores, Intercrystal dissolved pores, Interparticle dissolved pores
		Main pore structure type	Macropore and mesopores middle-thick throats	Mesopores and micropore thin throats	Mesopores and micropore thin throats
		Main reservoir rock type	Coarse-crystalline dolomite, Medium-crystalline dolomite, Unequal-crystalline dolomite, Microcrystal dolomite, Residual oolite dolomite	Silt-crystalline dolomite, Silt-micro-crystalline dolomite, Silt-fine-crystalline dolomite, fine-crystalline dolomite, Residual oolite dolomite	Silt-crystalline dolomite, Silt-micro-crystalline dolomite, Silt-fine-crystalline dolomite, fine-crystalline dolomite, Residual oolite dolomite
		Main constructive diagenesis	Dissolution, dolomitization, recrystallization	Dissolution, dolomitization, recrystallization	Dissolution, dolomitization, recrystallization
		Production	Production profile testing in all Wells is produced	Production profile testing in partial Wells is produced, and it has low gas production ratio	Production profile testing have not produced
Reservoir parameters	Quantify	Permeability $(10^{-3} \mu m^2)$	≥ 0.3	0.1–0.3	0.002–0.3
	Parameter	Porosity (%)	≥ 16	6–16	2–6
	Qualitative parameter	Main reservoir space type	Moldic pores, Intraparticle dissolved pores	Moldic pores, Intraparticle dissolved pores	Moldic pores, Intraparticle dissolved pores
		Main pore structure type	Macropore middle-thin throat	Macropore middle-thin throat	Macropore thin throat
		Main reservoir rock type	Residual oolite dolomite	Residual oolite dolomite	Oolite dolomite, Residual oolite dolomite
		Main constructive diagenesis	Dissolution, Dolomitization	Dissolution, Dolomitization	Dissolution, Dolomitization
		Production	Production profile testing in all Wells is produced	Production profile testing in partial Wells is produced, and it has low gas production ratio	Production profile testing have not produced

Fig. 10.6 Porosity permeability relationship diagram of production profile section

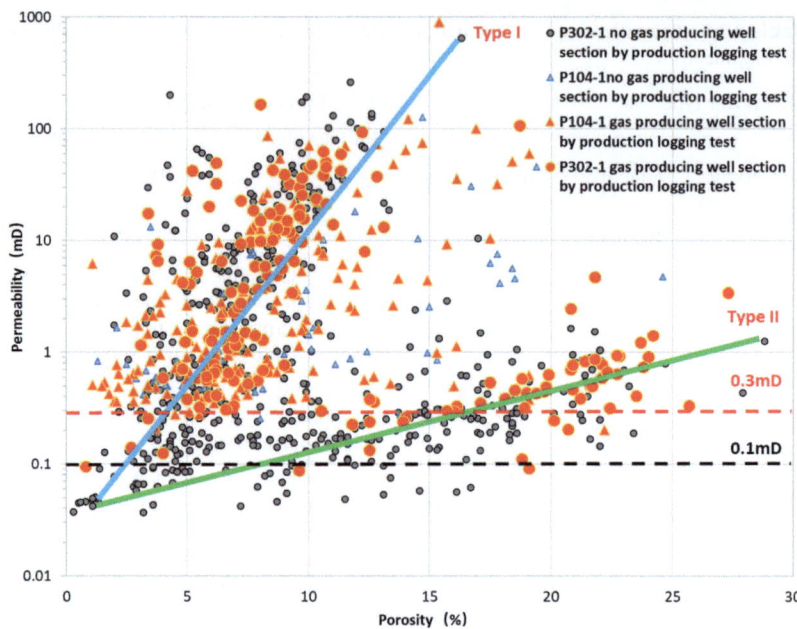

10.2.5 Gas–water Relationship and Gas Reservoir Type

(1) Gas–water relationship

The gas–water relationship in the Puguang gas field is complex, and many sets of gas–water systems exist. The gas–water systems in the Puguang-2 and Puguang-3 blocks are different, and the systems in the Feixianguan and Changxing Formations are also different (Li et al. 2008).

Puguang-2 block: There is a set of nonreservoir sections composed of micritic dolomite, micritic limestone, and argillaceous limestone at the bottom of the Feixianguan Formation. On the basis of reservoir comparison analysis of multiple wells, this set of sections was identified as the interlayer that divides the Feixianguan and Changxing Formations into two sets of gas–water systems. Furthermore, on the basis of the analysis of the data from electric measurements, gas tests, and water tests in four wells drilled in the water-contact layer in the Puguang-2 block, it was concluded that edge water is present in the Feixi-

anguan Formation in the Puguang-2 block, with a unified gas–water interface located at − 5125 m. There are also many sets of gas–water systems in the Changxing Formation, but there is no uniform gas–water interface.

Puguang-3 block: Only the Feixianguan Formation reservoir is developed in the Puguang-3 block. Three wells were drilled into the water layer in the lower part of the structure. From the perspective of drilling and reservoir prediction, sections 1 and 2 of the Feixianguan Formation contain a unified gas–water interface, which has been comprehensively determined to be located at − 4890.0 m.

(2) Gas reservoir type

According to the reservoir type, reservoir form, gas–water relationship, pressure system, and fluid properties, the Puguang gas field is an ultradeep, structure–lithology-controlled, normal-pressure, carbonate, high-hydrogen-sulfide, dry-gas reservoir with edge water.

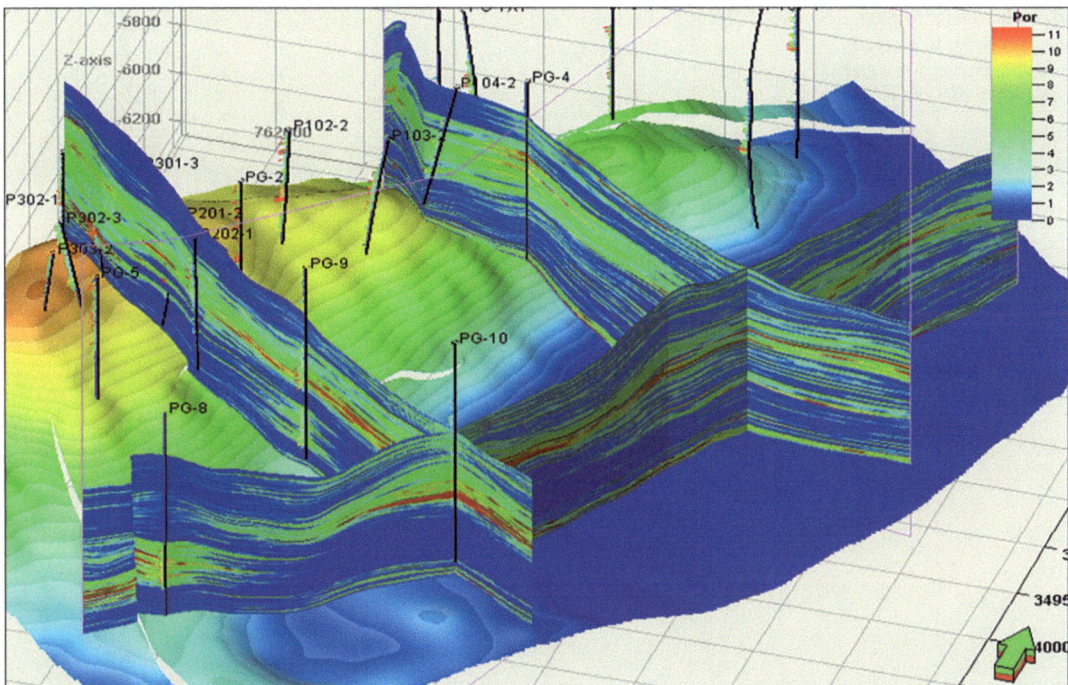

Fig. 10.7   Puguang gas field geological model

10.2.6   Three-Dimensional Geological Model of Gas Reservoirs

To determine the reservoir characteristics of the Puguang gas field, sedimentary simulations and a multipoint geostatistical modeling method were used to construct a reservoir model, and a detailed characterization of the spatial distribution of the heterogeneity of the reservoir matrix was performed. A multiscale fracture model was established using the grade-stage-group fracture modeling principle. This model was constrained by the genetic mechanism and main controlling factors, and it can accurately describe the distribution characteristics of the fracture space at different levels (Fig. 10.7).

10.3   Water Intrusion and Water Control Technique in the Puguang Gas Field

10.3.1   Analysis of Water Intrusion Characteristics in the Puguang Gas Field

(1) Type of water intrusion

The liquid-to-gas ratio and its derivative graphical representation are typically used to identify the type of water intrusion in gas wells. This method uses the actual dynamic production data for gas wells to obtain the relationship curve of the water-to-gas production ratio; its derivative, the production time, and the reason for the water production

Fig. 10.8 Water to gas ratio and water to gas ratio derivative curves for water producing wells

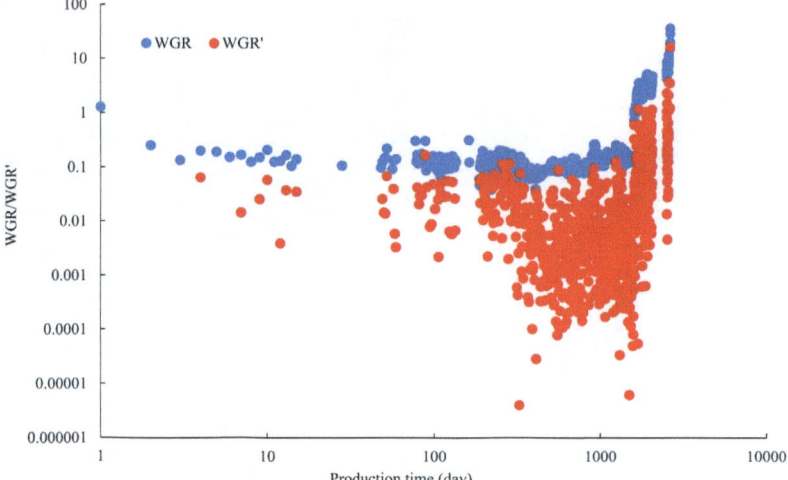

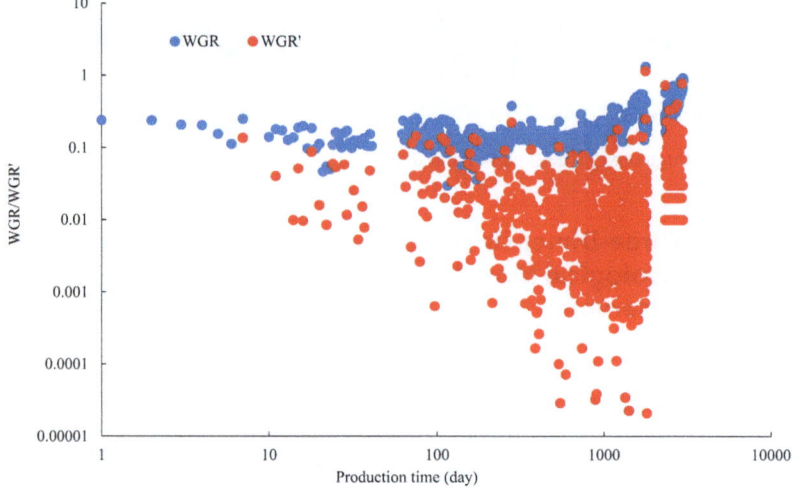

from the gas wells are determined on the basis of the shape of the curve. The water-to-gas ratio of each water-producing gas well and its derivative curve (Fig. 10.8) were plotted on the basis of actual production data, allowing for a preliminary determination of the type of water intrusion in the gas well. According to the map, the water-to-gas ratio of the first type of gas well gradually increases as the production time increases, and it increases faster in the later stage. The derivative of the water-to-gas ratio gradually increases with time and increases rapidly in the later stage, which clearly reflects the characteristics of water intrusion along fractures or the high-permeability layer. The second type of well is still charac-

terized by edge water intrusion along the high-permeability layer or fractures, but the edge water energy is relatively weak. Although the water-to-gas ratio and its derivative tend to increase, the rise is not as pronounced as in the first type of well. This is because the wells belong to the second line of water intrusion wells, and the first line of water intrusion wells is still in the drainage and gas extraction stage; thus, they have a more obstructive effect on edge water intrusion.

(2) Water volume and water intrusion

The degree of water intrusion in the gas reservoir can be calculated from the difference between the apparent geological reserves of the gas reservoir (GP) and the dynamic geological reserves of the gas reservoir (ΔG) as follows:

$$G_e = \Delta G \left\{ B_g - B_{gi} \left[1 - \left(\frac{C_p + S_{wc}C_w}{1 - S_{wc}} \right) \Delta P \right] \right\} \tag{10.1}$$

where B_g is the gas formation volume factor at the current reservoir pressure; B_{gi} is the initial gas formation volume factor; C_p is the compressibility coefficient of the reservoir rock, MPa^{-1}; C_w is the compressibility coefficient of the formation water, MPa^{-1}; S_{wc} is the initial water saturation in the reservoir, %; and ΔP is the reservoir pressure depletion, MPa.

The water volume can be calculated as follows:

$$W_i = W_e / \left[(C_w + C_p) f (p_i - p) \right] \tag{10.2}$$

where W_i is the volume of the natural aquifer, m^3; W_e is the water influx from the natural aquifer, m^3; P_i is the initial reservoir pressure, MPa; and P is the current reservoir pressure, MPa.

On the basis of the above method and the variations in reservoir formation pressure, the volume of edge water in the Puguang 2 gas reservoirs was calculated to be approximately 2037 million cubic meters. The water volume multiplier is 5.18 times

(Fig. 10.9). The calculation results are essentially consistent with the size of the water body calculated using the volumetric method.

10.3.2 Gas–Water Frontage Prediction in the Puguang Gas Field

(1) Core-scale gas–water frontage prediction method

To accurately simulate the dynamic changes in the gas–water front during actual gas reservoir development at the core scale, a physical simulation of water intrusion based on the gas state equation was established. A flow chart of the simulation method is presented in Fig. 10.10.

① Gas saturation: After evacuating the dry core for 12 h, the CH_4 in the intermediate containers was driven into the tandem core gripper using an ISCO pump in constant pressure mode, and the surrounding pressure of the gripper was always 2 MPa higher than the core pressure. When the core pore pressure reached the set pressure, the core was considered to be in saturated gas equilibrium, and the saturated gas volume was measured.

Fig. 10.9 Water intrusion analysis curve of Puguang main gas reservoir

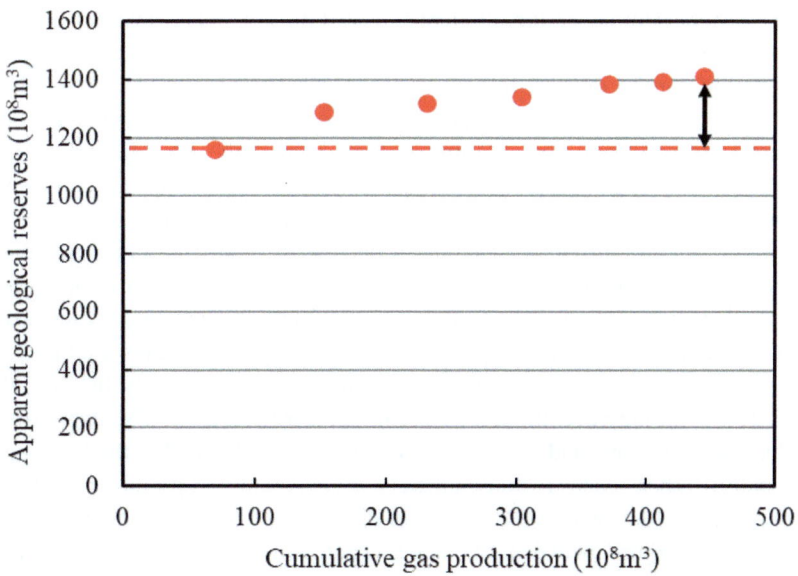

Cumulative gas production ($10^8 m^3$)

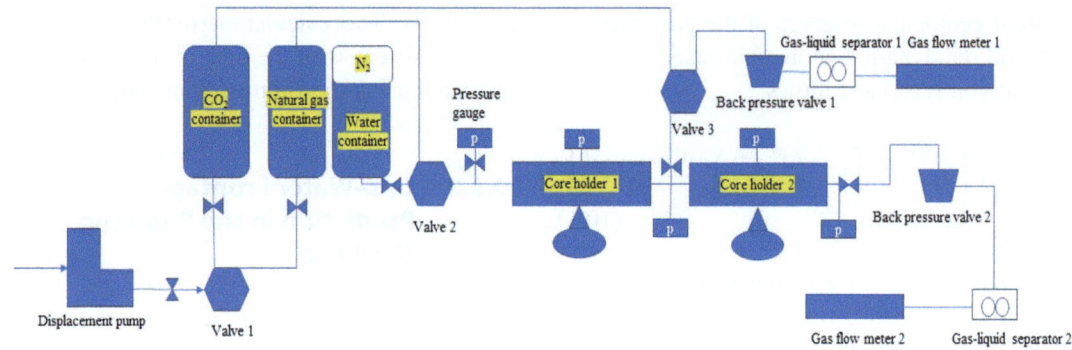

Fig. 10.10 Experimental setup drawing of gas–water frontage prediction

② High-pressure edge water preparation: N_2 was injected above the intermediate container until the pressure reached Pg0 in the equation. Formation water was injected below the intermediate container until the water pressure reached the core pore pressure.

③ Simulation of the water intrusion process in the core: The lower end of the intermediate container was connected to the front end of the core gripper under complete gas saturation, and the return valve pressure at the end of core 1 (front core) was set to the core pressure to begin production. The return pressure was decreased by 0.5 MPa per cycle until core 1 was completely flooded, and the gas production valve for core 1 was closed after flooding.

The pressure curve for the depleted development stage is shown in Fig. 10.11. Edge water intrusion can be divided into four stages. First, at the beginning of the core extraction, the gas energy is abundant, the difference in pressure between the edge water and the gas is small, and the edge waterfront at the end of the core is smooth. Second, as the core gas energy decreases, the difference in pressure between core 1 and the edge water increases, and the rate of water intrusion increases. Third, after core 1 reaches the water, gas, and edge water flow simultaneously within the core, the difference in pressure fluctuates sharply, and the edge water intrudes rapidly. When core 1 is completely flooded, only water flows within the core, and the difference in pressure between the edge water and

the end of the core decreases and then stabilizes.

Edge water intrusion strongly affects gas production dynamics. In the early stage of exploration, the edge water slowly percolates and slightly replenishes the reservoir energy, and gas production gradually decreases as the system pressure decreases. As development progresses, the water intrusion rate increases, the edge water replenishes the core energy, and gas production increases and remains stable. When the water reaches the end of core 1, the edge water occupies the high-permeability channel and reduces the gas flow capacity, and gas production decreases significantly. When core 1 is completely flooded, the core experiences only water phase flow, and gas is no longer produced.

The nuclear magnetic resonance (NMR) T2 map of the core during the depletion development process is shown in Fig. 10.12. During the depletion development process, there are two peaks on the T2 map, indicating the highly heterogeneous nature of the core sample. With further development, the right peak on the T2 map increases, and the corresponding relaxation time shifts to the right, indicating that the channel of the formation water is composed of large pores and that the pore radius gradually increases. When the core is completely flooded, the intrusion of edge water is 0.43 PV, approximately 90.41% of which is located in the large pores.

A pseudocolor image of the core during depletion development is shown in Fig. 10.13. The colored part is the signal area of the edge water,

Fig. 10.11 The pressure of multi-core series during natural depletion of the first core

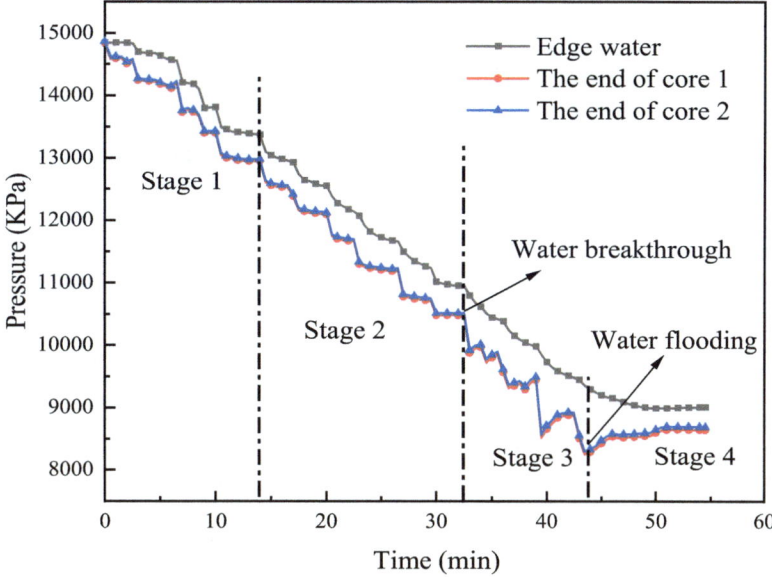

（a）Pressure curves of tandem cores

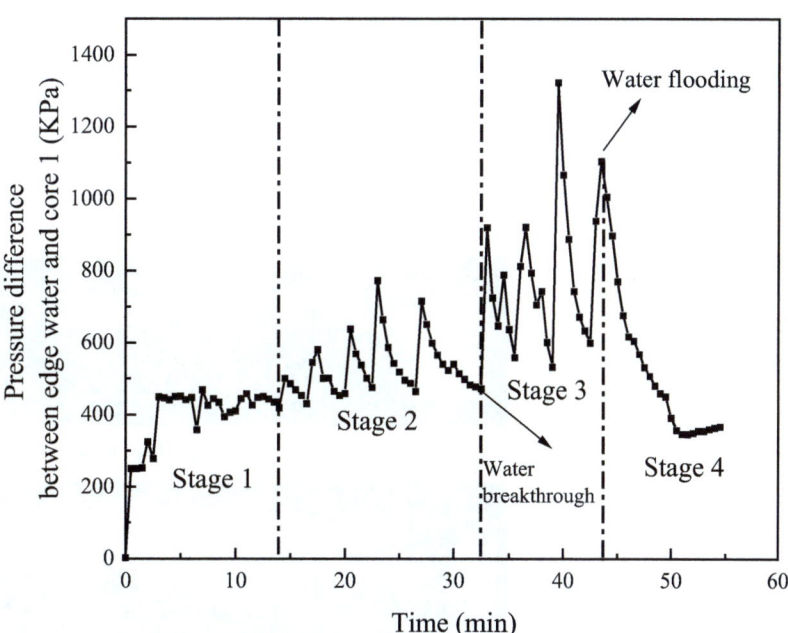

（b）Different pressure curve at both ends of core 1

and the black part is the signal area of the gas and core backbone. In the first and middle stages of depletion development, as shown in Fig. 10.13a, the edge waterfront is more uniform near the end of the core, and the water flows away from the end. When the water reaches the end of the core, as shown in Fig. 10.13b, the edge water exhibits fingering, and a water flow channel is essentially formed. When the core is completely flooded, edge water accumulates at the end of the core,

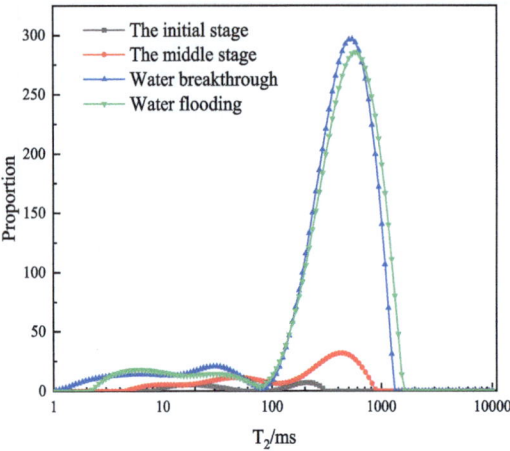

Fig. 10.12 T2 map during natural depletion development

and a large amount of gas is still blocked at this time. As shown in Fig. 10.13c, the distribution of the invading water is not uniform across the core sample, which means that the water invades through the high-permeability zone and that the remaining gas is heterogeneously blocked in the low-permeability zone.

(2) Method for predicting the gas–water front in gas reservoir engineering

The method of predicting the gas–water front in gas reservoir engineering mainly considers the heterogeneity of the reservoir by considering the differences in the permeability and physical properties of different parts, such as fractures, high-permeability layers, and medium- and low-permeability layers. According to the equivalence seepage resistance method, the seepage resistance of the water intrusion in the fractures, high-permeability layer, and medium- and low-permeability layer can be derived separately and substituted into the total water intrusion equation, and the stratified water front equation can be obtained. Using two example wells in which water has been observed—wells P103-1 and P104-3 in the first line—the time at which water appears in the gas wells can be predicted on the basis of the parameters of the dynamic history fitting analysis. The predicted time at which water appeared was 28–40 days later than observed (Table 10.2).

Fig. 10.13 MRI image during natural depletion development

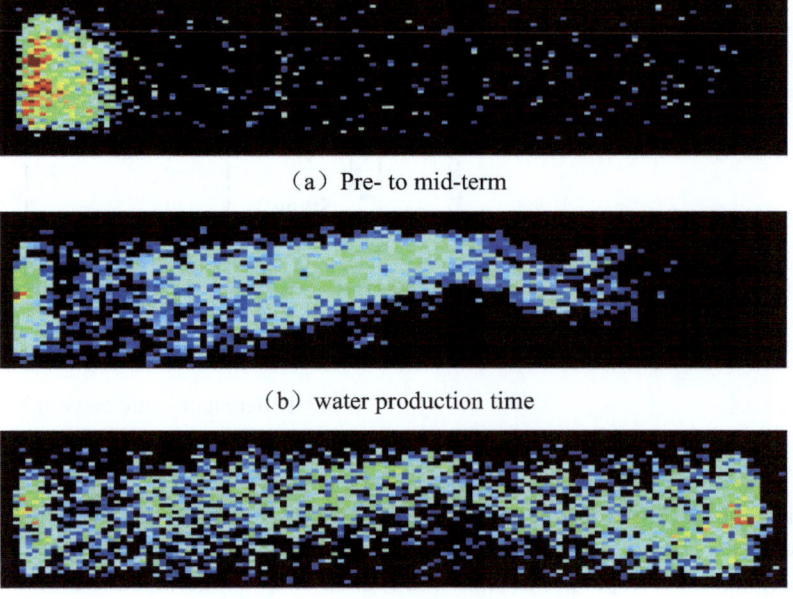

（a）Pre- to mid-term

（b）water production time

（c）water flooding time

Table 10.2 Prediction of water emergence time in well P103-1

Well number	Water production time, days	Forecast time, days
P103-1	830	870
P104-3	1588	1560

(3) Numerical simulation method for predicting the gas–water front

On the basis of the established three-dimensional geological model of the dual media in the Puguang gas field, a numerical simulation method was used to finely describe the distribution of the gas–water interface and the inrush of edge water. On the plane, according to the water intrusion distance and the distribution of gas saturation, the water intrusion areas in sections Fei-I and Fei-II in the Puguang 2 block can be roughly divided into four breakout directions. Direction 1 is along the line from P104-1 to P107-1H, direction 2 is along the line from P104-1 to P103-1, direction 3 is in the area of well P203-1, and direction 4 is in the area of well P304-1 (Fig. 10.14). The usage degree of section Fei-III in the Puguang gas field is poor, and the overall degree of water invasion of the gas reservoir is low. In addition, the upper part of sections Fei-I and Fei-II in the Puguang 3 block has essentially been flooded, while the middle and lower parts are relatively weakly flooded. The gas reservoir in the Changxing Group is not strongly affected by the bottom water, and the degree of flooding of each reef body is low.

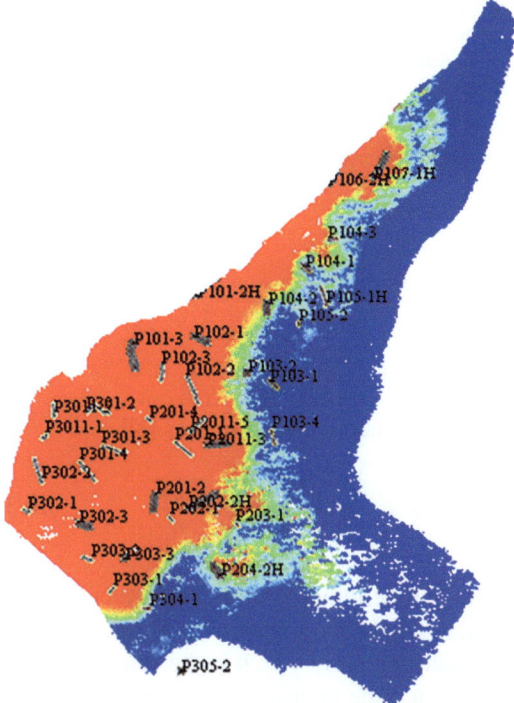

Fig. 10.14 Direction of water intrusion in fracture model Fei-I and Fei-II sections of Puguang gas field

10.3.3 Countermeasures for Water Control in the Puguang Gas Field

(1) Optimization of the production allocation method

The reservoir plane in the Puguang gas field is strongly heterogeneous, and the edge water is channeled along the local high-permeability zone. To balance the gas–water front, the domi-

nant water intrusion channel and nondominant production channel prediction models were used to conduct differential production allocation. The differential pressure of the gas wells in the high-permeability zone was decreased to slow the front, while the differential pressure of the gas wells in the non-high permeability zone was increased appropriately to balance the seepage rate and make the gas–water interface front more uniform. On the basis of these two production models, a plot of the relationship between the production and the water emergence time was constructed (Fig. 10.15), which can be quickly used to conduct differential production allocation of the gas wells.

(2) Drainage capacity prediction of gas extraction wells

On the basis of the stable seepage theory and the output profile test data for water-producing gas wells, a water production prediction model was established. The maximum fluid production

Fig. 10.15 Plot of gas well production versus time to water at different permeability

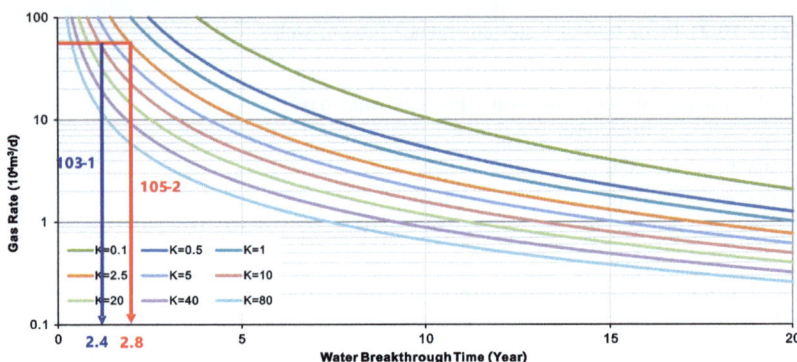

capacity prediction equation is as follows:

$$Q = \frac{543\pi \, Kh(P_e - P_w)}{\mu B \ln \frac{R_e}{R_w}} \qquad (10.3)$$

where Q is the volume of formation water injected, m^3/d; K is the effective permeability of the water-producing formation, D; h is the reservoir thickness, m; u is the viscosity of the formation water, mPa s; R_e and R_w are the control radius and well radius, respectively, m; and P_e and P_w are the formation pressure and wellbore injection pressure, respectively, MPa.

The calculation results reveal that in the drainage gas extraction well, as the formation pressure decreases, the drainage volume and gas production gradually decrease, and when the pressure decreases to a certain level, the drainage gas extraction cannot lift the mixture to the wellhead, at which time the drainage gas extraction well can only be converted to a strong drainage well.

(3) Reasonable drainage volume of forced drainage wells

As the formation pressure decreases, water production increases, and drainage gas extraction is unable to lift the fluid to the wellhead. At this time, some of the drainage gas extraction wells are converted to forced drainage wells, which can slow down the front of the edge water and cause it to continue to channel into the high-permeability zone after forced drainage. Taking the area around well P103 as an example, a simulation is used to predict the edge water-front characteristics, as well as the development index situation, according to drainage/intrusion values of 1, 1/2, 1/3, 1/4, 1/5, and 0. When the discharge/intrusion ratio is higher, the flooded area is smaller. When the drainage ratio is 0.2–1, the flooded area is equivalent to 0.95–0.68 of the undrained area.

According to the predicted development index, when the drainage/intrusion ratio is < 1/3, the recovery rate changes less; and when the drainage/intrusion ratio is > 1/3, the recovery rate of the gas reservoir increases faster. Thus, the drainage/intrusion ratio of the gas reservoir should be controlled above 1/3.

10.3.4 Water Drainage and Plugging Technique in the Puguang Gas Field

(1) Foam drainage technology for high-sulfur gas wells

In response to the high-temperature and high-pressure characteristics of the Puguang gas field, which has a high hydrogen sulfide content, we developed a high-efficiency foam drainage agent with high sulfur resistance and a device for evaluating the performance of the high-temperature, high-pressure, hydrogen-sulfide-resistant foam drainage agent. Then, we evaluated the performance of the new foam drainage agent in terms of the temperature resistance, salt resistance, acid resistance, and environmental protection level. A safe filling process for high-sulfur gas well foam drainage was designed, and a new mixed steam–water compound foam drainage process was developed.

For flooded wells, only the bubble discharge technique is used. In this technique, the bubble discharge agent is directly injected through the tubing, and then, the well is opened to spray. Owing to the lack of airflow disturbance, it is difficult to fully bubble and resume production. Therefore, it is necessary to develop new drainage and gas extraction techniques [10]. Liquid nitrogen-induced injection is a drainage process that is typically used after acidification in the early stage of production of the Yuanba marine gas wells. In this process, liquid nitrogen is injected into the wellbore using a liquid nitrogen pump, and the lower-density nitrogen replaces the accumulated higher-density liquid in the wellbore, reducing the density of the accumulated liquid and establishing a production pressure difference to realize the resumption of production of the gas well. However, this process has been used in new wells with high formation pressures, and when used in flooded wells with depleted formation pressures, production may fail to resume because of insufficient pressure and apparent slippage.

To further improve the success rate of production resumption in water-flooded wells, a new method involving compound bubble drainage with mixed soda is proposed. The gas–water mixture is used to reduce the density of the fluid in the wellbore and to withstand a certain formation back pressure to prevent the packer from unsealing. In addition, the foam discharge agent can reduce gas–liquid slippage and improve drainage ability.

(2) Water plugging technique for high-sulfur gas wells

Aiming at the fractured reservoir in the Puguang gas field, a wellbore bottom water plugging technique—the mechanical double plug-tubing bridge plug and cement plug technique—was developed. Using the newly developed high expansion rate bridge plug with a constant pressure closing protection mechanism and supporting tools, the problems of the bursting of the bridge plug rubber tube and a low success rate in water plugging construction of low-pressure gas wells were solved; water plugging through tubing in inclined vertical wells with a water outlet in fractured-porous reservoirs was also achieved.

10.4 Sulfur Deposition Characteristics and Treatment Technology in the Puguang Gas Field

To address the problem of sulfur deposition during the development of the Puguang gas field, dissolution, deposition, and seepage models of elemental sulfur were established on the basis of the results of solubility and relative permeability experiments conducted on elemental sulfur. Sulfur deposition and seepage in reservoirs and wellbores were investigated, and corresponding sulfur deposition prevention and control techniques for high-sulfur gas reservoirs were developed.

10.4.1 Mechanism of Sulfur Dissolution and Precipitation

(1) Determination of the elemental sulfur content

The elemental sulfur content of natural gas with a high H_2S content is a key parameter for quantitatively analyzing sulfur precipitation and deposition (Brunner and Woll 1980; Brunner et al. 1988; Bian et al. 2010; Roberts 2017). The initial elemental sulfur content is quite different in gas reservoirs with unsaturated dissolved elemental sulfur in the original state, and the only way to obtain this parameter is to obtain and analyze underground gas samples. At present, there are two major methods for determining the sulfur content in acidic gas: the gravimetric method and chromatography. The gravimetric method requires many gas samples, and the lower the sulfur content is, the more gas samples are needed. Owing to the high cost of underground sampling, the gravimetric method is an economically poor

Table 10.3 Test results of condensation point of elemental sulfur

Pressure	(MPa)	65	55	45	39	33	23
Condensation point	(°C)	118.65	116.99	115.23	114.25	113.37	112.1
Pressure	(MPa)	17	11	5	3	0.5	
Condensation point	(°C)	111.52	111.42	111.91	112.4	114.25	

choice. Gas chromatography separates the various components of a complex mixture and performs quantitative analysis on the basis of the proportional relationship between the response signal (peak area or peak height) of each component and the amount of each component. In recent years, chromatography has been used to determine the elemental sulfur content of acidic gas (Peng et al. 2018).

On the basis of underground pressure–volume–temperature (PVT) gas samples such as those from wells Puguang 104-1 and 303-2, the sulfur content of the main body in Puguang was measured using improved chromatography. The sulfur content was determined to range from 0.35 to 0.39 g/m^3. Further testing revealed that the saturated elemental sulfur content of the gas sample was greater than 1.23 g/m^3 under the sampling conditions, which indicates that the Puguang gas field was unsaturated and contained dissolved elemental sulfur in its original state.

(2) Phase state change of elemental sulfur precipitation

The phase state of elemental sulfur precipitated in the main gas reservoir formation in Puguang was analyzed using a high-temperature and high-pressure melting point testing device. The test results revealed that the solidification temperature of elemental sulfur in natural gas with a high H$_2$S content initially decreased but then increased with increasing pressure. When the pressure was less than 11 MPa, the solidification temperature of elemental sulfur decreased with increasing pressure. When the pressure was greater than 11 MPa, it increased with increasing pressure. The solidification temperature of elemental sulfur decreased from 120.40 °C to 111.42 °C as the pressure decreased from 75 to 11 MPa. The

solidification temperature of elemental sulfur increased from 111.42 °C to 114.25 °C as the pressure decreased from 11 MPa to 0.5 MPa (Table 10.3).

The temperature of the main formation in the Puguang gas field is greater than 120 °C, the elemental sulfur precipitated in the formation is liquid, and solid sulfur is precipitated in the upper part of the wellbore and on the ground surface. The elemental sulfur precipitated in the stratum and wellbore, and the ground surface is solid.

(3) Solubility characteristics of elemental sulfur under different temperature and pressure conditions

As the temperature and pressure decreased, the thermodynamic saturation point of elemental sulfur in acidic natural gas decreased, which led to a reduction in the sulfur dissolved in the natural gas. The solubility of sulfur in acidic gas at different temperatures (70 °C, 100 °C, 118 °C, and 130 °C) and pressures (20–55 MPa) was tested. The results revealed that at temperatures ranging from 70 to 130 °C and pressures ranging from 20 to 55 MPa, the solubility of sulfur in natural gas with a high H$_2$S content ranged from 0.0369 to 3.1208 g/m^3. The solubility of sulfur increased with increasing pressure at a constant temperature, and the solubility of sulfur increased with increasing temperature at a constant pressure.

In light of the above, establishing a simple, applicable, and accurate prediction model for the solubility of elemental sulfur on the basis of existing experimental results is one method for determining the solubility of elemental sulfur under different temperature and pressure conditions. At present, prediction models for the solu-

Fig. 10.16
Micro-deposition
experiment of liquid Sulfur
in fractured rock core

| a Core without gas driving | b Core after gas driving for 20 hours |

| c Core after gas driving for 60 hours | d Core after gas driving for 100 hours |

bility of elemental sulfur mainly include a thermodynamic model based on the equation of state, a prediction model based on a neural network, and a simple calculation model based on density (Chrastil's semiempirical model) (Chrastil 1982; Karan et al. 1998; Heidemann et al. 2001; Bian et al. 2018; Fu et al. 2019). Comparatively speaking, Chrastil's semiempirical prediction model has a simpler form and fewer calculations and is convenient to use, making it the most suitable for engineering applications. It is currently the major model used to predict the solubility of elemental sulfur. As Chrastil's prediction model is a semiempirical formula, it is necessary to regress the model parameters according to the actual situation of the gas reservoir and the experimental data to improve the prediction accuracy. A prediction model for the solubility of sulfur suitable for the Puguang gas field was constructed by improving the empirical parameter fitting method on the basis of the test results of sulfur solubility at different pressures and temperatures in the Puguang gas field. This model can be used to quickly evaluate the change in saturated sulfur solubility with the pressure drop in a gas reservoir or wellbore at a certain temperature. It subsequently calculates the amount of sulfur precipitation at any time, which can accurately predict the amount of sulfur deposition in the reservoir/wellbore, facilitating the implementation of sulfur removal countermeasures.

$$c_r = \left(\frac{28.96\gamma_g p}{ZRT}\right)^{-0.0141*T+9.8917}$$
$$\times \exp\left(\frac{-813.69}{T} - 12.71\right) \quad (10.4)$$

10.4.2 Regular Distribution Pattern of Sulfur Deposition in High-Sulfur Gas Reservoirs

(1) Characteristics of sulfur deposition in high-sulfur gas reservoirs

During the development of high-sulfur gas fields, the energy in the formation gradually decreases, and the sulfur in the formation fluid precipitates and is deposited, thus affecting the porosity and permeability of the reservoir. An experiment on the microscopic distribution of sulfur deposition in a reservoir was carried out using a microscopic seepage visualization device; the microscopic migration, deposition morphology, and distribution characteristics of the gas–liquid sulfur in the porous core were determined.

Under a temperature of 130 °C and pressures decreasing from 55 to 10 MPa, a microdeposition experiment was conducted on liquid sulfur in a fractured rock core. The results revealed that as the amount of injected gas increased, a certain portion of the liquid sulfur that had precipitated in the fractures migrated into the pores. When the precipitated liquid sulfur reached saturation, the liquid sulfur distributed in the fractures and pores was relatively stable (Fig. 10.16).

The experiments revealed that the liquid sulfur was distributed mainly in relatively loose fractures and large pores in the core and that most of the micropores did not contain liquid sulfur deposits.

① In the areas where fractures and matrix coexisted, liquid sulfur was preferentially deposited in the fractures. Liquid sulfur deposition occurred mainly in areas where the gas flowed easily, i.e., where the pressure drop occurred faster and deposition occurred more easily.

② The heterogeneity of the reservoir strongly influenced sulfur deposition. Under fracture and matrix coexistence conditions, sulfur was preferentially deposited in the fractures, with an overall heterogeneous distribution.

③ Once the liquid sulfur was deposited, it could flow when the deposition reached a certain saturation level.

(2) Evaluation of reservoir permeability damage due to sulfur deposition

As the development of this high H_2S-bearing gas field progresses, when the formation pressure decreases to a critical pressure, the singlet sulfur in the formation fluid precipitates and is deposited, thus affecting the porosity and permeability of the reservoir. Cores with different permeabilities obtained from the gas field were used to test the permeability under different sulfur deposition saturation conditions to quantitatively evaluate the pattern and extent of the reservoir permeability damage caused by sulfur deposition. The experimental results revealed the following:

① The reservoir permeability decreased rapidly with increasing sulfur deposition (saturation), and the degree of damage was as high as 80%.

② A comparison of the extent of sulfur deposition-induced damage to cores with different permeabilities revealed that the lower the permeability of the reservoir core was, the more pronounced the damage caused by sulfur deposition was.

An empirical equation for liquid sulfur damage to reservoirs was established via regression analysis of the experimental data.

$$K_s = K_0 \frac{\exp(0.8633 \times \ln K_0 \times S_s)}{\exp(7.3852 \times S_s)} \quad (10.5)$$

where K_s is the permeability under sulfur saturation S_S (mD), K_0 is the initial permeability of the reservoir (mD), and S_S is the saturation of sulfur deposition in the reservoir.

(3) Distribution pattern of sulfur deposition in different types of reservoirs

In terms of gas percolation theory, the near-well zone is characterized by a high pressure drop and a high flow rate. A high flow rate results in a negative non-Darcy effect, but it can also cause a positive effect, i.e., the capillary number effect, which together affect the distribution of sulfur saturation. In addition, the compressibility of the gas and the Joule–Thomson cooling effect in the near-well zone, where the gas expands because of the lower pressure, can lead to temperature changes that affect the solubility of sulfur. To this end, a predictive model for the distribution of sulfur deposition saturation was developed on the basis of the Forcheimer non-Darcy flow equation, considering the Joule–Thomson effect and the gross tube number effect in the near-well zone, to characterize the distribution of sulfur deposition saturation in the reservoir.

1. Temperature and pressure distribution characteristics

The gas well drainage area was divided into 50 radial grids, and the pore-type reservoir sulfur saturation deposition prediction model was applied to the unfractured gas wells. For the acid fracture-modified gas wells, the fractured reservoir sulfur deposition saturation prediction model was used to calculate the area within the fracture half-length as the boundary, and the pore-type model was used for the area outside of this boundary. The results indicate the following:

① The temperature and pressure decrease in the near-well area. In the nearly 1000 m drainage area, the temperature and pressure drops occur mainly in the area close to the wellbore, and the temperature and pressure drops are greater closer to the wellbore.

② Significant temperature and pressure decreases occur in both fractured gas wells within half the length of the fracture. In the unfractured wells, the temperature and pressure drops occurred mainly within 3 m of the wellbore, while in the two fractured wells, the temperature and pressure drops were more pronounced starting approximately 50 m from the wellbore, which is comparable to the half-length of the artificially created fractures.

2. Sulfur deposition saturation distribution characteristics

The prediction of sulfur deposition saturation in gas wells requires a combination of seepage effects in the near-well area, that is, the non-Darcy effect increases sulfur saturation and the capillary number effect caused by high-velocity flow decreases saturation. The results of the predictions reveal that without considering the capillary effect, the sulfur deposition saturation is greater closer to the wellbore, reaching more than 40% near the bottom of the well. This is because the gas velocity and capillary number are greater closer to the wellbore, i.e., the capillary number effect is more pronounced and the sulfur deposition saturation is lower. Further comparative analysis was conducted considering the effects of temperature variations in the near-well area on the distribution characteristics of sulfur deposition saturation. The prediction results reveal that owing to the compressibility of the gas, the high-velocity flow and the Joule–Thomson cooling effect in the near-well region result in significant temperature changes, with the temperature decreasing

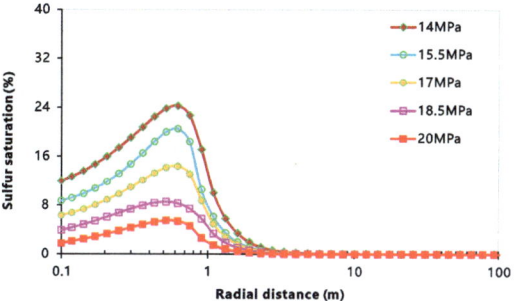

Fig. 10.17 Predicted sulfur deposition saturation distribution at different formation pressures (P201-2)

closer to the wellbore. Because the solubility of sulfur is quite sensitive to temperature, the predictions are too small if the temperature drop is not considered.

The predicted radial distribution of the sulfur deposition saturation during the decrease in the formation pressure from 26 to 14 MPa reveals that the sulfur deposition saturation increases as the formation pressure decreases, but the sulfur is deposited mainly within a certain range in the near-well zone, with very little monomeric sulfur deposited in more distant formations even though the formation pressure decreases significantly (Fig. 10.17).

Taking a formation pressure of 18 MPa as an example, the distributions of sulfur deposition saturation under different permeabilities and different fracture widths were compared. When the permeability of the reservoir is lower and the fracture width is smaller, the sulfur deposition saturation in the near-well area is higher, and the location of the maximum sulfur deposition saturation is farther from the bottom of the well. In addition, based on a comparison of the distribution characteristics of sulfur deposition saturation in different types of reservoirs, the range of sulfur deposition saturation in the near-well area in pore-type reservoirs is smaller than that in fractured reservoirs. The range of sulfur deposition in pore-type reservoirs is within approximately 3 m of the near-well area, and the highest sulfur saturation occurs at approximately 0.5–1 m. Meanwhile, the peak deposition occurs somewhat further away (3–10 m), and the area where sulfur deposition occurs in the formation may extend to 15–25 m from the well.

Based on our understanding of the mechanism of sulfur dissolution and precipitation and the distribution pattern of sulfur deposition, sulfur deposition is expected to occur gradually in the Puguang gas field formation by 2023.

10.4.3 Distribution of Sulfur Deposition in Gas Wells in a High-Sulfur Gas Reservoir

As the Puguang gas field entered the middle and late stages of development, the gas reservoir pressure decreased. The elemental sulfur in the sulfur-containing natural gas fluid was constantly precipitated, gradually deposited along the surface process–Christmas tree–wellbore direction, and attached to the surfaces of the gathering and transportation equipment, pipelines, wellheads, and inner walls of the tubing, resulting in blockage and severely affecting the normal production of the gas wells.

(1) Due to the distribution of the temperature and pressure of the wellbore flow, the sulfur blockage in the wellbore gradually moved downward from the wellhead.

Several coiled tubing operations and caliper logs from wells P102-3, P102-2, and D402-3 revealed that the wellbore blockage point gradually moved downward from the wellhead over time. According to field statistics, wellbore plugging occurred after an average of 15.5 months.

By adopting a three-phase gas–water–sulfur wellbore sulfur deposition model, which considers diffusion adsorption, a one-well–one-model principle was established to predict that the movement of the sulfur blockage position (the maximum deposition thickness) accelerates downward as the pressure decreases. Under the current formation pressure and production conditions, the sulfur blockage position should be 366–1095 m away from the wellhead, with an average of 739 m. It was estimated that 13 sulfur-blocked gas wells are added in 2022 (Fig. 10.18, Table 10.4).

(2) Influenced by the melting point of elemental sulfur, the maximum depth of solid sulfur deposition in the wellbore is approximately 4650 m, and liquid sulfur accumulation at the bottom of the wellbore may occur in the later stage.

The formation temperature of the Puguang gas field was 128 °C, which is higher than the melting point of elemental sulfur (111.4–120.4 °C). In the flow process of high-sulfur natural gas from the bottom to the wellhead, the maximum depth of the transition from liquid sulfur to solid sulfur is approximately 4650 m. If the depth is shallower than this value, the wellbore absorbs and deposits sulfur. If the depth is deeper than this value, the wellbore does not absorb liquid sulfur stably, and liquid sulfur might accumulate at the bottom of the well later (Figs. 10.19 and 10.20).

10.4.4 Treatment of Sulfur Deposition in the Wellbore and on the Ground Surface for a High-Sulfur Gas Reservoir

Owing to the influence of external additives such as surface corrosion inhibitors, the composition and morphology of sulfur deposition in the wellbore and on the ground surface station are quite different. By solving key problems, a sulfur deposition prediction method was innovatively established, and a sulfur dissolution agent system and treatment process with different functions were developed.

(1) Treatment of sulfur deposition in a wellbore

Since the Puguang gas field was put into production, dimethyl disulfide (DMDS), polyamine, and binary amine surface sulfur-dissolving agents have been developed, and the sulfur-dissolving effect and safety performance have gradually improved, reaching the domestic standard. This initial surface sulfur dissolving agent system did not consider the corrosion of rubber seals in the wellbore, and there were serious safety risks when this system was applied to high-sulfur gas wells. A fast-dispersing sulfur-dissolving agent with a rubber protection function was developed, which can quickly peel off honeycomb sulfur deposits, with a dispersion rate of more than 85% and a rubber damage rate of less than 2%. It was combined with a multiphase fusion jet flushing process, and the average gas recovery volume was 138,300 m²/day; thus, it achieved a good blocking removal effect.

① Low-damage, high-efficiency sulfur-dissolving and plug-removing formula system

Based on the principle of sulfur amine solution, a system for sulfur dissolution and plug removal with amine groups and long short-chain polyamines as the main agent and a dispersant and permeant as the auxiliary agent was designed. The amount of dissolved sulfur was 12.3 g/100 ml (50 °C, 1 h). On the basis of the vulcanization principle, a rubber protective agent was developed for the first time to promote the vulcanization, crosslinking, and repolymerization of damaged rubber molecules, and the swelling rate of fluorine rubber was controlled to within 2%, which achieved the goals of dissolving the sulfur, removing the plugs, and protecting the rubber equipment components. This also ensured the safety of the downhole pipe string during the process of sulfur dissolution.

The sulfur-soluble agent has good anticorrosion performance at 90 °C. The seven-day corrosion rate of a G3/TP110SS steel sheet was ≤ 0.0499 mm/a. At high temperatures (50 °C and 90 °C), the sulfur solubilizer and its reaction products have good compatibility with the output solution, without the occurrence of delamination, precipitation, or crystallization. It had good compatibility with the corrosion inhibitor used in the Puguang gas field and had little effect on sulfur solubility.

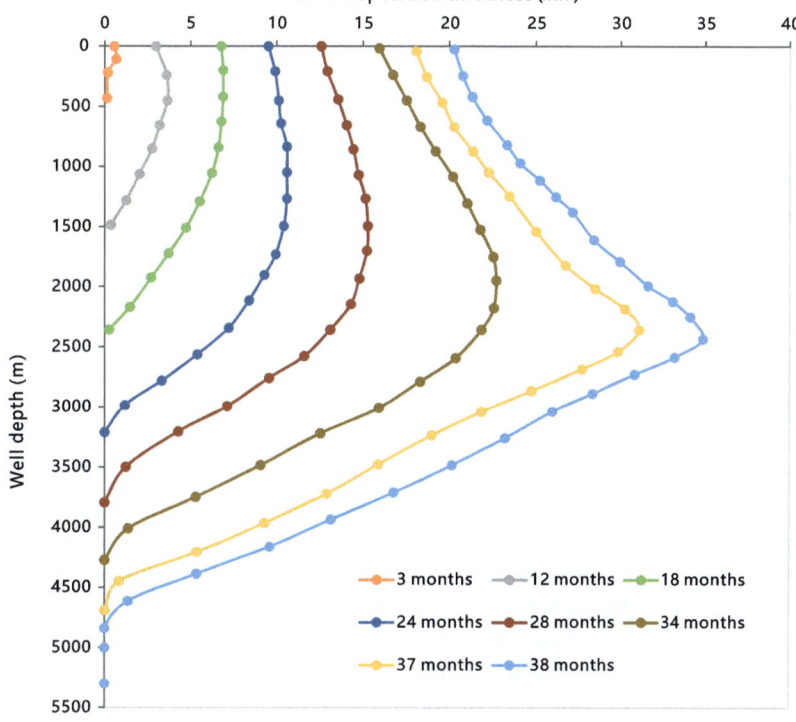

Fig. 10.18 Prediction of Puguang 301-3 wellbore sulfur deposition thickness

Table 10.4 Prediction of sulfur plugging position and speed by wellbore sulfur deposition

Production time after ground sulfur plugging (months)	Deposition point downward velocity (m/month)	Sulfur deposition rate (mm/month)	Sulfur plugging point (m)	Deposition thickness of sulfur plugging points (mm)	Production decrease degree (%)
3	15	0.22	45	0.67	0
12	23	0.36	249	3.58	1.3
18	66	0.45	643	6.62	8.0
24	87	0.67	1166	10.62	21.3
28	133	1.16	1698	15.24	43.8
34	79	1.23	2175	22.59	63.3
37	60	2.83	2356	31.07	84.5
38	77	3.81	2432	34.88	96.2

Fig. 10.19 Melting point of elemental sulfur under different pressures in Puguang gas field

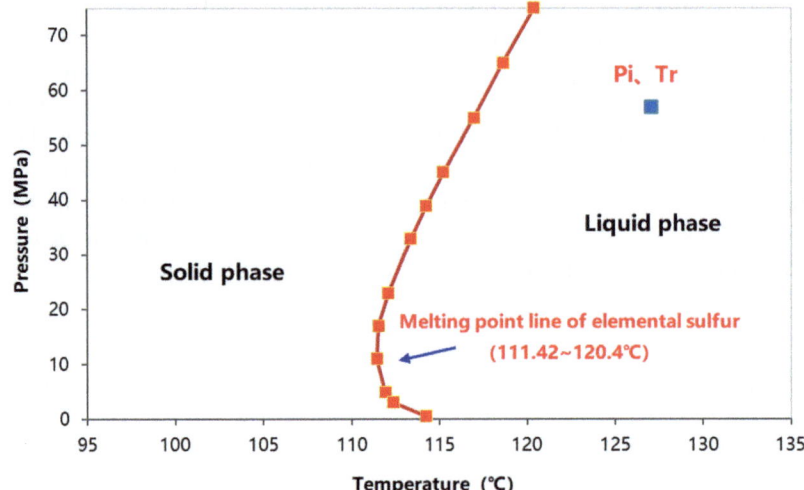

② Supporting sulfur-dissolution and blockage-removal technology

A water bath heating device–cement truck–multifunctional pumping process was established, and the key parameters of the construction dosage, pumping, and shutdown of the soaking well were determined. A technique for controlling the dissolution of sulfur plugs in different well types was established, which provided a safe technological guarantee for dissolving sulfur plugs in the wellbore.

In addition, from the perspective of temperature control, wellbore hot washing as a measurement of wellbore sulfur removal has also been applied in the field. Although the cost of the method is low per well, the validity period of the measure is short (5 days), and the recovery gas rate of a single well is low (50,000 m³/day). Therefore, this method is only suitable for the initial stage of wellbore sulfur plugging when the well is unblocked and production is minimally affected.

(2) Treatment of sulfur deposition at ground stations

The sulfur deposits in the ground station pipe network were compact and difficult to dissolve. A spontaneous thermal polyamine sulfur dissolution system was developed, with a sulfur dissolution capacity of 32 g/100 ml and a composite sediment dissolution rate greater than 90%. In a field application, the restored gas volume was 7.3997 million square meters per day. The sulfur

Fig. 10.20 Wellbore flow temperature profile at different production rates

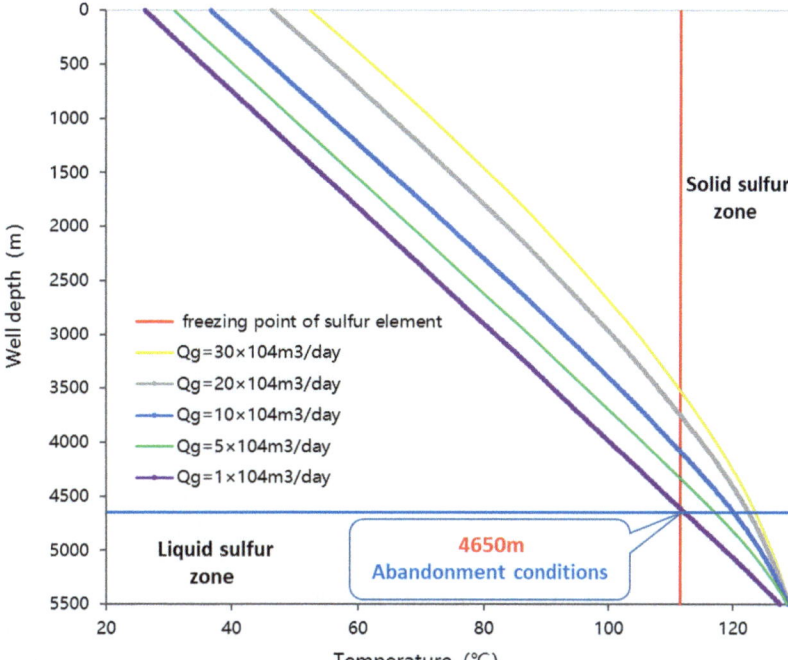

deposits in ground station containers had complicated compositions and were sticky and difficult to clean. An inorganic alkali-catalyzed sulfur solubilizing agent was developed for sand washing at stations, with a sulfur dissolution capacity of 36 g/100 ml. A continuous hot flow cycle cleaning process was used, reducing the cost of the agent by more than 80% and the cleaning time by more than 60%.

① Formulation system of an efficient plug-removing agent for organic composite sulfur deposition

Aiming at the combination of elemental sulfur in bulk or paste form and organic matter (mainly corrosion inhibitor) in the sediment in the gathering and transportation system, a large amount of polyamino nanometer organic hydrocarbon dissolution promoter was introduced, which effectively dissolved the organic matter and increased the blocking efficiency by 50%. A formulation system for a highly efficient composite plug remover, with polyamine as the main agent, auxiliary agent, and accelerator, was developed. The amount of dissolved sulfur reached 40.6 g/100 ml, and

the dissolution rate of the compound sediment reached more than 90%.

② Blockage removal via circular soaking and subsection cleaning

A density and viscosity online two-parameter saturated sulfur solubility evaluation method was established, and the utilization rate of the plug remover agent was increased by two times. The process of circular soaking and subsection cleaning was designed according to the degree of plugging of the gathering and transportation pipeline. This process was applied to a total of 11 wells, and the sulfur blockages were effectively removed.

10.5 Challenges in Gas Field Development and the Direction of Enhanced Recovery

At present, the development of the Puguang gas field has entered a declining period, and the formation pressure has decreased to less than 50%

Table 10.5 Pressurization effect prediction of the Puguang gas field

Wellhead pressure MPa	Abandonment pressure MPa	EGR %	Improvement of the gas reservoir recovery %	Increased recoverable reserves bcm
9	15.5	62.52		
7	14.0	65.11	2.59	43.11
6	13.3	66.31	3.78	62.81
5	12.7	67.38	4.86	80.71
4	12.1	68.31	5.79	96.18
3	11.7	69.03	6.50	108.05
2	11.5	69.38	6.86	113.94
1	11.3	69.68	7.16	118.95

of the original value. The wellhead pressure is close to the export pressure, and later, the sulfur deposition gradually spreads from the wellbore to the formation. In addition, the distribution of residual gas in the reservoir is very complex because of edge water flooding; thus, to further enhance gas field development recovery, several key techniques, such as pressurized production, fine descriptions of the residual gas, sulfur deposition treatment, water plugging, CO_2 injection techniques, and mechanisms for water control, need to be investigated further.

the application of pressurized production based on domestic research on pressurized production, evaluation of the pressurization potential, and analysis of a pressurization pilot test. Therefore, the application of pressurized production in the Puguang gas field needs to be considered, and a reasonable, optimized pressurization scheme for the gas reservoir should be developed according to a comparison and analysis of development indicators under different pressurization ranges and modes to further improve the development efficiency of the Puguang gas field (Fig. 10.5, Table 10.5).

10.5.1 Pressurized Production

The high supply requirements of the Puguang gas field have restricted its current situation of continuous stable production and enhanced recovery. Pressurized production is a countermeasure for achieving stable production and enhanced recovery in the gas field. In the middle and late stages of gas field development, pressurized production prolongs the stable production period and enhances gas recovery, such as in the West Sichuan gas field, Wolonghe gas field, and Daniudi gas field. The Wolonghe gas field is an elastic gas-driven reservoir with heterogeneous carbonate, medium–high sulfur content, and limited edge water energy. Pressurization effectively delayed the decrease in gas production and improved gas recovery by 15%. The Puguang gas field is considered to have great potential for

10.5.2 Fine Description and Potential Tapping of the Gas Field

The low-permeability reservoir in the Feixianguan Formation has relatively poor production because of the influence of the development well and gas release radius of the classified reservoir. In the nonwater invasion area, improving the production degree of the class II gas reservoir and the residual gas reserves in the grain bore in sections Fei I–II was necessary. In the relatively centralized and large-scale class II reserve area, large-scale fracturing of horizontal wells was implemented to communicate with the joint production of high-quality reservoirs, and measures were implemented to tap the potential of the class II reserves distributed in small-scale areas and scattered around the well. For the remaining gas-rich areas in sections Fei-I and Fei-II in the main gas area and the potential areas of edge water intrusion, adjustment wells and measurement wells were arranged to implement the remaining gas adjustment and potential tapping.

10.5.3 Sulfur Deposition Treatment

By the end of 2021, nearly 70% of the gas wells in the Puguang gas field had experienced sulfur deposition damage, and the impact of surface flow and wellbore sulfur deposition had become one of the main challenges in gas field development. The formation pressure of the Puguang gas field is expected to gradually decrease to the critical pressure for elemental sulfur precipitation in 2023. In addition, sulfur deposition is expanding deeper into the wellbore and reservoir, which presents new challenges for the stable production of gas wells and recovery from gas reservoirs. In view of the rule and prevention of sulfur deposition in deep wellbores and reservoirs, the following aspects should be strengthened. First, the experimental data on sulfur solubility should be expanded and improved, the changes in the solubility of sulfur under different compositions and temperature and pressure conditions should be clarified, and the rule of water–liquid sulfur and gas–liquid sulfur two-phase flow and the damage to the reservoir caused by sulfur precipitation should be investigated further. Second, molecular simulations and experiments should be combined to predict sulfur solubility and obtain three-phase seepage interaction parameters. Third, on the basis of the experimental and molecular simulation results, a new sulfur solubility prediction model, a three-phase equilibrium model considering chemical reactions, and a three-phase relative permeability model of gas–water–liquid sulfur should be established. Fourth, numerical simulations of sulfur deposition in deep wellbores and reservoirs should be carried out to predict the distribution of sulfur deposition and its impact on gas production. Fifth, integrated prevention and control measures and techniques for sulfur deposition in reservoirs and wellbores should be developed to address key problems and provide new technical means for long-term stable production and enhanced gas recovery in the Puguang gas field.

10.5.4 Water Plugging Technique

The water production from a fracture-type outlet well is large, and mechanical water plugging in the early stage mainly achieves general water plugging below the water breakthrough section. Four measures should be taken to address this problem in the future. First, the mechanical water plugging technique should be further developed to achieve accurate water plugging in any water-producing layer. Second, a chemical water plugging technique should be developed, and a large slug polymer solution should be added to pretreat the formation, reduce the deep-water phase permeability of the gas well, and inhibit the flow of formation water around the area. High-temperature-resistant and strong-adsorption gel should be used to plug the main water-producing layer of the formation and improve the precipitation effect. Third, the two-stage injection method should be adopted, that is, the polymer solution/tubing head pressure (THP) gel system should be alternately injected, and the second-stage plugging agent should be injected after the first-stage plugging agent solution has been gelled to enhance the transformation effect of the plugging agent system on the formation heterogeneity. Fourth, a temperature-resistant and sulfur-resistant surfactant solution developed in the laboratory should be used as the last-stage replacement liquid instead of clean water, which could generate a large amount of foam, reducing the liquid column pressure and improving the liquid discharge efficiency during a subsequent blowout.

10.5.5 CO₂ Injection Techniques and Mechanisms for Collaborative Water Control Tests

Core experiments and NMR results revealed that, driven by edge water energy, the formation water quickly occupied the high-permeability pore channel, which significantly reduced the gas production capacity of the core. CO_2 injection can supplement the core energy and inhibit edge water channeling. Therefore, CO_2 injection has good water control potential. However, this work is only in the experimental exploration stage at present. The main research goal in the next step is to apply the simple one-dimensional core CO_2 injection water control method to a complex three-dimensional geological reservoir.

10.6 Summary

The Puguang gas field is an ultradeep, high-sulfur, carbonate gas reservoir. Reservoir heterogeneity, sulfur deposition, and edge-bottom water channeling have important effects on gas field development. In the past two decades, through in-depth research on reservoir refinement prediction, gas reservoir water intrusion prediction, and sulfur deposition distribution, we have developed a deep carbonate gas reservoir water intrusion prediction technique based on reservoir refinement descriptions, formulated gas reservoir water control countermeasures, effectively mitigated the influence of formation water, and reduced the rate of water intrusion. The phase and percolation rules of sulfur-containing gas have been clarified, the reservoir sulfur deposition distribution has been revealed, a reservoir sulfur deposition distribution model for high-sulfur gas reservoirs has been established, and a preliminary sulfur deposition management system has been developed. The above results have enriched and improved the development technology for high-sulfur gas reservoirs and have been directly applied to the development of the Puguang gas field, making

important contributions to maximizing the utilization of geological reserves, formulating scientific and effective measures and countermeasures for water and sulfur control and for adjusting the tapping potential, and extending the stable production period and improving the recovery rate of the gas field.

By further deepening the research on techniques such as the fine description of the remaining gas, water control and plugging, prevention and control of sulfur deposition, and gas reservoir booster extraction, we aim to increase the recovery rate of the reserves to more than 80% and further support the efficient development of the Puguang gas field.

Acknowledgements The authors acknowledge the financial support of Natural Science Foundation of China (Grants U24B6001 and U19B6003), and Sinopec management for permission to publish this work.

References

Anselmetti FS, Eberli GP (1997) Sonic velocity in carbonate sediments and rock. Geophys Dev 6:53–74

Anselmetti FS, Eberli GP (1999) The velocity-deviation log: a tool to predict pore type and permeability trends in carbonate drill holes from sonic and porosity or density logs. AAPG Bull 83:450–466

Bian XQ, Du ZM, Guo X et al (2010) Measurement of the solubility of sulfur in natural gas with a high H_2S content. Nat Gas Ind 30(12):57–58

Bian XQ, Zhang L, Du ZM et al (2018) Prediction of sulfur solubility in supercritical sour gases using grey wolf optimizer-based support vector machine. J Mol Liq 262:431–438

Bourrouilh R, Richert JP, Zolnaï G (1995) The North Pyrenean Aquitaine Basin, France: evolution and hydrocarbons. AAPG Bull 79(6):831–853

Bracco G, Wagner PD, Baechle GT et al (2005) Obtaining permeability from seismic data: A new breakthrough in carbonate reservoir modeling. In: Proceedings of the international petroleum technology conference, p 10577

Bracco G, Weger GR, Nasser M et al (2006) Permeability from seismic in carbonate reservoirs, Part II: The field example: Presented at the Annual Meeting, AAPG

Brunner E, Woll W (1980) Solubility of sulfur in hydrogen sulfide and sour gases. SPE J 20(05):377–384

Brunner E, Place JRMC, Woll WH (1988) Sulfur solubility in sour gas. J Petrol Technol 40(12):1587–1592

Chrastil J (1982) Solubility of solids and liquids in supercritical gases. J Phys Chem 86(15):3016–3021

Chen PY, Tan XC, Yang HT (2013) Quantitative characterization of interlayer heterogeneity of reef flat reservoir. Lithol Reserv 25(4):27–32

Chesnoy AB, Pack DJ (1997) S8 threatens natural gas operations. Oil Gas J 74–78

Coskuner G (1994) Flow of sour gas, liquid sulfur, and water in porous media. J Colloid Interface Sci 165(2):526–531

Espitalié J, Drouet S (1992) Petroleum generation and accumulation in the Aquitaine Basin (France). Generation and accumulation of Europe's hydrocarbons II. EAGE Special Publication, vol 2, pp 127–149

Feng RW, Wang XZ, Zhang F et al (2008) Sedimentary facies of isolated carbonate platform of the first to third members of the Lower Triassic Feixianguan Formation in the northeastern part of the Sichuan basin and other related aspects. Geol China 35(1):54–66

Fu L, Hu JH, Zhang Y et al (2019) Investigation on sulfur solubility in sour gas at elevated temperatures and pressures with an artificial neural network algorithm. Fuel 262

Guan HL, Wang SY, Jiang XQ (2010) Characteristics and identification marks of burial dissolution pores in the reservoirs of the Changxing and Feixianguan formations in the Puguang gas field. Nat Gas Ind 30(3):31–34

Guo X, Guo T, Huang R et al (2010) Reservoir development characteristics and predication technologies of large Puguang-Yuanba Gas Field. Eng Sci

Guo X, Zhou XF, Zhou BH (2015) Prediction model of sulfur saturation considering the effects of non-Darcy flow and reservoir compaction. J Nat Gas Sci Eng 22:371–376

Heidemann RA, Phoenix AV, Karan K et al (2001) A chemical equilibrium equation of state model for elemental sulfur-containing fluids. Ind Eng Chem Res 40(9):2160–2167

Huang LM (2015) New progresses in safe, clean and efficient development technologies for high-sulfur gas reservoirs. Nat Gas Ind 2(4):360–367

Huang Y et al (2022) Understanding depositional evolution of ancient carbonate systems using 3D stratigraphic forward modelling: a case study in Puguang area (Late Permian, northeast Sichuan Basin, China). In: The 21st international sedimentological congress (21st ISC), China University of Geosciences, Beijing, China, 22–26 Aug

Hyne JB, Derdall GD (1980) How to handle sulfur deposited by sour gas. World Oil 11:111–120

Karan K, Heidemann RA, Behie LA (1998) Sulfur solubility in sour gas: predictions with an equation of state model. Ind Eng Chem Res 37(5):1679–1684

Kong FQ, Wang SP, Zeng DQ (2011). Key techniques for the development of the Puguang Gas Field with a high content of H2S. Nat Gas Ind 2011

Kuo CH (1972) On the production of hydrogen sulfide-sulfur mixtures from deep formations. J Petrol Technol 24(9):1142–1146

Li J, Yang S, Qi Z et al (2020) A calculation model for water breakthrough time of gas wells in gas reservoirs with edge water considering interlayer heterogeneity: a case study of the Lower Triassic Feixianguan gas reservoirs in the Puguang Gas Field. Nat Gas Ind B 7(6):631–638

Li PP, Zou HY, Zhang YC et al (2008) Paleo-oil-water contact and present-day gas-water contact: implication for evolution history of Puguang Gas Field, Sichuan Basin, China. J China Univ Geosci 19(6):715–725

Li SL, Du JF, G P et al (2007) Suggestions on high-sulfur gas field development. Nat Gas Ind 27(2):137–140

Li T, Ma YS, Zeng DQ et al (2022) Research progress and prospect of formation sulfur deposition in high sulfur gas reservoirs. Fault Block Oil Gas Field 29(4):433–440

Li XY, Wang Q, Han YB et al (2014) Main control factors and analysis of reservoir space evolution of reef-beach facies in Changxing-Feixianguan Formation of northeastern Sichuan Area. Nat Gas Geosci 25(10):1594–1602

Liu AH, Han YK, Liang HJ et al (2015) Water invasion identification and water production mode in gas well in Puguang Gas Field. Spec Oil Gas Reserv 22(3):125–127

Liu QY, Jin ZJ, Bing Z et al (2017) Main factors for large accumulations of natural gas in the marine carbonate strata of the Eastern Sichuan Basin, China. J Nat Gas Geosci 2(2):81–97

Luo KP, Huang ZG, Jiang XQ et al (2011) Reformation mechanism of high-quality carbonate reservoirs in northeastern Sichuan Basin. Pet Geol Exp 33(6):559–563

Ma YS, Cai Y, Li GX (2005) Basic Characteristics and Concentration of the Puguang gas field in the Sichuan Basin. Acta Geol Sin 79(6):858–865

Ma YS, Cai XY, Zhao PR (2014) Characteristics and formation mechanisms of reef-shoal carbonate reservoirs of Changxing-Feixianguan formations, Yuanba gas field. Acta Petrolei Sinica 35(6):1001–1011

Ma YS (2006) Cases of discovery and exploration of marine fields in China (Part 6): Puguang Gas Field in Sichuan Basin. Mar Origin Petrol Geol 11(002):35–40

Ma YS (2007) Generation mechanism of Puguang Gas Field in Sichuan Basin. Acta Petrolei Sinica 28(2):9–21

Ma YS, Zhang SC, Guo TL et al (2008) Petroleum geology of the Puguang sour gas field in the Sichuan Basin, SW China. Mar Petrol Geol 25(4):357–370

Mao JC, Yang XJ, Wang DL et al (2016) Optimization of effective sulfur solvents for sour gas reservoir. J Nat Gas Sci Eng 36:463–471

Paux F, Zhou H (1997) Field case-match of a 600 bars depletion and 40 years of history in a fractured carbonate sour gas field. In: SPE annual technical conference exhibition, San Antonio, SPE 38909, pp 693–701

Peng S, Jiang YW, Su YX et al (2018) Content determination and critical precipitation pressure of elemental sulfur in sour gas rich in H_2S in Puguang Gas Field. Petrol Geol Exp 40(4):573–576, 582

Ren LJ, Zhang CS, Liu DZ et al (2008) Reservoir diagenesis and pore evolution in Puguang gas field. Nat Gas Expl Dev 31(3):10–41

Roberts B (1997) The effect of sulfur deposition on gaswell inflow performance. SPE Reserv Eng 12(02):118–123

Roberts B (2017) Flow impairment by deposited sulfur—a review of 50 years of research. J Nat Gas Eng 2(1):84–105

Shi XC, Zeng DQ, Zhang SQ (2014) Development technology and practice of Puguang high sour gas field. Sinopec Press, Beijing

Su LP, Luo P, Luo Z et al (2005) Characteristics of Feixianguan formation oolitic shoal reservoirs in northeast Sichuan basin. Nat Gas Ind 25(6):14–17

Vermersch F (1968) Problems and techniques in producing gas wells in South-East France. J Inst Petrol 54(53):251–258

Villien A (1992) 3D seismic and deep gas exploration in the Aquitaine Basin. In: EAPG annual conference, Paris, p 26

Wang WH, Liu CX, Mu L et al (2011a) Technical policy optimization for the development of carbonate sour gas reservoirs. Oil Gas Geol 32(2):302–310

Wang XP, Chen SP, Feng GM et al (2022) Delaminated fracturing and its controls on hydrocarbon accumulation in carbonate reservoirs of weak deformation regions: a case study of the Yuanba ultra-deep gas field in Sichuan Basin, China. Front Earth Sci 10

Wang ZH, Guo TL, Tan QY et al (2011b) Reservoir physical properties of Changxing-Feixianguan formations in Xuanhan district. J Oil Gas Technol 33(5):36–41

Zhao WZ, Xu CC, Wang TS et al (2011) Comparative study of gas accumulations in the Permian Changxing reefs and Triassic Feixianguan oolitic reservoirs between Longgang and Luojiazhai-Puguang in the Sichuan Basin. Chin Sci Bull 56(31):3310–3320

Fluid Flow Mechanism and Numerical Simulation Method in Deep Fracture-Cavity Reservoirs, Tarim Basin

11

Zhijiang Kang, Yun Zhang, Hongkai Li, Dawei Wu, Ke Sun, and Ziyan Deng

Abstract

China's deep carbonate fracture-cavity oil and gas resources are abundant, and the cumulative proven oil in place is 40.66×10^8 tons; exploring these resources has become an important area for oil and gas exploration and development in China, as well as increasing reserves and production. Owing to the low description accuracy, diverse flow patterns, and difficulty in simulating and predicting carbonate fracture-cave bodies that are buried at depths of 6500–8000 m, water and gas injection channelling is likely to cause low oil displacement efficiency. Efficient development of such reservoirs is a worldwide problem. After the continuous and efficient development of theoretical and technological research and innovation, a detailed description and geological modelling technology for deep fracture-cavity reservoirs was developed, and on the basis of physical experiments and theoretical derivation, the stress-sensitive properties of rock permeability during the production process were determined, and the fluid flow law of multiscale composite media was established. On the basis of multiphase fluid dynamics, rock solid mechanics and thermodynamics, a fluid–solid–thermal joint mathematical model of composite media was established, and a numerical simulation solution and parallel computing software for the joint consideration of cavity, fracture, and hole weight changes were developed, which improved reservoir simulation accuracy and speed. Through research on an efficient development mechanism for numerical simulation, efficient water injection technology that is based on the fracture-cave structure and nitrogen injection-induced cave top flooding-enhanced oil recovery technology was developed. Since the popularization and initial application of the technology, it has supported the large-scale development and sustained high and stable production of the Tahe oilfield, which is the world's first deep superlarge fracture-cavity reservoir. Moreover, it has supported the major discovery and large-scale production of the Shunbei oilfield (also known as the "Shendi No. 1" oilfield), which is the world's deepest (8000 m) large-scale fault-controlled fractured-cavity oil and gas reservoir, and has achieved a significant transition from continental to marine facies in China's oilfield development. This study provides a technical reference and demonstrations for the development of deep and ultradeep carbonate reservoirs.

Z. Kang · Y. Zhang (✉) · H. Li · D. Wu
Sinopec Petroleum Exploration and Production Research Institute, Beijing, China
e-mail: zhangyun.syky@sinopec.com

K. Sun
China University of Petroleum, Beijing, China

Z. Deng
China University of Geosciences, Beijing, China

© The Author(s) 2026
Y. Ma et al. (eds.), *Petroleum Geology and Exploration of Deep Marine Strata in China*, Advances in Oil and Gas Exploration & Production, https://doi.org/10.1007/978-3-032-02496-1_11

Keywords

Deep fracture-cavity reservoirs · Multifield coupling · Numerical simulation · Fluid flow mechanism

11.1 Introduction

Carbonate reservoirs provide extremely important global oil and gas resources. According to the statistics of the IHS, the proven recoverable reserves of carbonate reservoirs account for approximately 50% of the global recoverable oil reserves, and their production accounts for approximately 60%. Carbonate oil and gas reservoirs are widely distributed in North America, the Middle East, Central Asia, and other regions. The main types are reefs, grain beaches, dolomite, and weathering crusts. They are usually large in scale and less than 3000 m deep. They are concentrated mainly in the Jurassic, Cretaceous, and Neogene and are dominated by porous media. The development and design of carbonate reservoirs outside of China have been mostly based on the continuum theory. In terms of drilling, completion and engineering technology, a series of technologies, such as complex structural wells, have been developed to improve production and recovery.

China's marine carbonate rocks are rich in oil and gas resources. The carbonate deposits are distributed in an area of 450×10^4 km^2, and the oil and gas resources are 358×10^8 tons of oil equivalent. By the end of 2018, the cumulative proven carbonate rock oil in the Tarim Basin, Ordos Basin, and Bohai Bay Basin was 40.66×10^8 t. In September 1984, the Shashen 2 well in the northern Tarim Basin obtained high-yield oil and gas flow, thus achieving the first major breakthrough in Palaeozoic marine carbonate reservoirs in China, which became an important milestone in the history of oil and gas exploration in China. In 1990, the Sha-23 well discovered China's first Palaeozoic ultradeep marine superlarge oil field, the Tahe oilfield. In 1997, the Tahe oilfield was put into development, and the world's largest oil and gas production base for deep carbonate fracture-cave reservoirs was built, with an annual produc-

tion capacity of 600×10^4 t oil equivalent. With the development of the Tahe and Shunbei deep and ultradeep carbonate oil and gas fields, crude oil production has increased rapidly, and carbonate oil and gas reservoirs have become important areas for China's oil and gas exploration and development, along with oil and gas reserves and production.

Compared with carbonate reservoirs in other countries, China's carbonate reservoirs are of old geological age, are deeply buried, and have undergone a multistage tectonic movement transformation.

Reservoir development faces four major main challenges: (a) fracture-cavity reservoirs are highly discrete, which makes identification, description, and characterization difficult; (b) multiphase fluid flow in multiscale fracture cavities is complex, which makes predicting development and production difficult; (c) injecting water flows along fractures is difficult, along with designing injection-production well patterns and high-efficiency displacement; and (d) the recovery factor is low, and the calibration recovery factor is only 15.0%. After more than 20 years of research and exploration by developers, new progress has been made in terms of the genetic mechanism, fluid complex flow mechanism, and characterization of various types of carbonate reservoirs. Moreover, fine description and geological modelling technology for deep fracture-cavity reservoirs has been innovatively developed. On the basis of multiphase fluid flow dynamics, rock mechanics and thermodynamics, a fluid–solid–thermal coupled mathematical model of composite media has been established, and numerical simulation solutions and parallel computing software for the integration of cavity, fracture, and hole weight changes have been developed, which improve the accuracy and speed of reservoir simulation. On the basis of research on high-efficiency development mechanisms via numerical simulation, high-efficiency water injection technology that is based on fracture–cavity structure and nitrogen injection cave top flooding-enhanced oil recovery technology has been innovatively developed. It supports the large-scale development and sustained high and stable production of the

Tahe oilfield, which is the world's first deep, very large fracture-cavity oil reservoir, and the major discovery and large-scale production of the Shunbei oilfield (also known as the "Shendi No. 1" oilfield), which is the world's deepest (8000 m) large-scale fault-controlled fractured-cavity oil and gas reservoir, and it has achieved a significant transition from continental to marine facies in China's oilfield development. This study provides a technical reference and demonstrations for the development of deep and ultradeep carbonate reservoirs.

(1) Geological modelling technology for deep fault-controlled reservoirs

A geological reservoir model is the final result of a comprehensive study of reservoir characterization, which is a digital representation of reservoir type, geometry, scale, parameters, and pore structure. The basic strategy is to use logging data as control data; seismic data, geological data, logging data as constraints; and dynamic production data as inspection conditions and to perform 3D digitization and quantification of oil reservoirs through deterministic or random mathematical algorithms. With respect to sedimentary reservoirs, the laws of geostatistics are relatively stable, and geostatistics has become the key to the realization of geological models. With the development of geostatistics, modelling technology that is based on geostatistics has developed from the original simple kriging method to the classic method. Multipoint geostatistics has become a research and development hotspot. In 1951, D. Krige proposed the "weighted moving average", which was named the kriging method by G. Matheron; this method predicts the distribution of reservoirs between wells through interpolation on the basis of well point data, but it is insufficient for characterizing the heterogeneity of the distribution of reservoirs between wells. In 1984, Dr. H. Haldorsen and Professor L. Lake proposed the concept of stochastic simulation, which was developed into classical two-point statistics on the basis of the kriging method. The heterogeneity of reservoir development between wells is characterized by the introduction of random

functions and variogram functions. Because it is based on the statistics of two wells, the characterization of the structural characteristics of reservoirs is weak. In 2000, Strebelle proposed a multipoint geostatistics method, which effectively improved the computational efficiency of the multipoint statistical algorithm through the search tree algorithm. Multipoint geostatistics solves for the spatial structure and configuration relationships of reservoirs by constructing training images and effectively characterizing the geometry of channel-type reservoirs.

The above commonly used geological modelling methods are not suitable for deep fractured-cavity carbonate reservoirs, and a multiscale classification and karst facies-controlled geological modelling method has been developed. The development of fracture-cavity carbonate reservoirs is controlled by multiple factors, such as palaeogeomorphology, water systems, and faults. Reservoirs can develop along layers or faults and phreatic surfaces. Reservoirs have diverse shapes and large-scale differences; thus, a single geostatistical modelling method is not suitable for all reservoirs. In 2002, Wang Genjiu established a porosity model for fractured-vuggy reservoirs on the basis of the relationship between porosity and wave impedance (Wang et al. 2002) but did not establish a pore-vuggy reservoir model; thus, their approach was insufficient for reservoir characterization. In 2003, Kang Zhihong established a logging interpretation model for fractured-cavity reservoirs by using the logging curves for 104 wells in the Tahe oilfield (Kang et al. 2003). In combination with seismic and drilling data, he classified and established a three-dimensional model of reservoir types and property parameters of matrix pores, large karst caves, and fractures. In 2004, Yang Huiting initially explored the method of modelling different reservoir spaces divided into three layers vertically (Yang et al. 2004). In 2008, Zhao Min et al. strengthened the seismic identification and prediction of various types of reservoirs and proposed a seismic resampling modelling method for 3D seismic waveforms. On the basis of the statistics of the relationships between different types of reservoirs and earthquakes, various types of reservoir models are established through cut-off values (Zhao et al. 2008). In 2006, relying on the national "973" key basic research and development program, Li Yang, Lu Xinbian, et al. proposed a "multiclass and multiscale modelling" method that is based on large-scale caves, dissolved pores, large-scale fractures, and small-scale fractures. In 2011, on the basis of the 12th Five-Year National Major Special Project of Fracture-Cavity Reservoir Geological Modelling, Hu Xiangyang, HouJiagen, et al. proposed the method of "classified karst facies control and multitype fusion" and established a fusion principle on the basis of karst origin. The fusion of different types of fracture-cavity reservoirs was realized. In 2016, on the basis of the 13th Five-Year National Major Project, we systematically analysed the randomness, complexity, ambiguity, and probability of understanding fractured-cavity reservoirs and proposed the establishment of multiple uncertainty models that meet specified conditions. Dynamic reserves, dynamic connectivity, and production history matching methods are used to reduce model inaccuracies. In 2021, He Zhiliang proposed a method for extracting reservoir development characteristics and laws on the basis of a knowledge base, constructing training images, and simulating underground river reservoirs via multipoint geostatistical methods to characterize the structure and morphology of ultradeep underground river reservoirs (He et al. 2021).

An ultradeep fault-controlled reservoir is a special fracture-cave type reservoir, and the burial depth is generally greater than 6000 m. Unlike that of a fracture-cave oil reservoir, the development of an ultradeep fault-controlled reservoir is mainly controlled by the associated fault and is affected by the fault grade, stage, trend, and other factors. Influenced by the combination mode, the reservoirs are distributed in the fault core and fracture zone controlled by the fault and are characterized by a planar distribution along the fault and a large vertical thickness (generally greater than 500 m). Geological modelling technology for fracture-cavity reservoirs lacks the shape constraints of fault-controlled bodies and the prediction of vertical fracture-cavity bodies. For this reason, a "multiple-constraint, five-step" 3D

geological modelling method for fault-controlled reservoirs was developed (Kang et al. 2020a, b), which involves the following steps: Step 1: The fracture pattern and fracture system model are established, and the structural pattern and combination relationship of the fracture system are clarified. Step 2: A fault-controlled reservoir profile characterization method with static and dynamic constraints is developed, and a fault-controlled reservoir profile model (fault core and fracture zone) is established via deterministic and artificial combination. Step 3: An internal structure characterization method that is based on fault-control model guidance, a target-based method for karst cave modelling, and sequential indication fracture–vuggy reservoir modelling is developed to clarify the internal structural characteristics of fault-controlled reservoirs. Step 4: A multivariate constrained fracture–vuggy porosity model that integrates logging interpretation, production performance, wave impedance inversion and facies control is created; a fracture–cavity permeability model that is based on well testing and porosity constraints is built; an equivalent model of fracture properties that is constrained by imaging logging and well testing is established; and the heterogeneity of the internal physical properties of fault-controlled reservoirs is clarified. Step 5: A reservoir and attribute model that characterizes the planar and vertical distribution characteristics of fault-controlled reservoirs, as well as the internal structure and heterogeneity, is formed by fusion. With respect to the geological model of the No. 1 fault zone in the Shunbei oilfield, in Tarim Basin, the drilling coincidence rate of 5 new wells reached 80%.

(2) Flow mechanism of the fracture–cavity composite medium

The fluid flow mechanism of carbonate reservoirs, especially fracture-pore and karst fracture-cavity reservoirs, has always been a research focus at home and abroad. In the 1950s, Bruce G. H. et al. established a mathematical model of a single continuum, which solved the problem of characterizing the seepage mechanism of a porous continuum. In the 1970s, Warren et al. established a dual-continuum mathematical model for carbonate reservoirs with network fractures. The bedrock is assumed to be separated into groups by fractures, and the groups are assumed to form a relatively independent hydrodynamic system with fluid exchange between them. This kind of reservoir bedrock has high porosity and low permeability, whereas fractures have low porosity and high permeability. In the 1990s, owing to the rapid development of computing speed, Zhou Dehua et al. proposed a discrete fracture network model to solve the problem of fine characterization of flow in fractures.

For deep fractured-cavity carbonate reservoirs, in 2010, on the basis of the 12th Five-Year National Major Project, we proposed a method that involves coupled fluid flow in composite media, which solved the problem of fluid flow characterization of complex multiscale fractured-cavity media. First, on the basis of the theoretical method for the seepage characterization unit, the fluid flow in rock blocks with large karst caves and large fractures does not have the characteristics of continuous medium flow (Kang 2010). Large-scale karst caves and fractures have become the main body of flow in rock blocks, which does not align with the basic characteristics of single-signature units. Large-scale karst caves and fractures can only flow as embedded discrete media. That is, the fluid flow in multiscale fracture–vuggy reservoirs is a kind of composite medium flow that includes not only the continuous medium flow of matrix karst pores but also the coupled flow of discrete media of large karst caves and fractures. Large-scale karst caves and fractures correspond to free flow, small-scale dissolved pores are collectives of seepage flow, and the flow form is compound flow. The flow characteristics are as follows: Free flow has an inertial force, which is nonlinear flow. There is a sudden change in flow velocity at the interface between free flow and seepage; the multiphase flow of oil, gas, and water in a cave has rapid gravity differentiation, and there are obvious interfaces between different phases (Kang et al. 2015).

Owing to the extreme heterogeneity and complex flow environment of this kind of carbonate formation, efficiently developing fractured–vuggy reservoirs is highly challenging. Thus, the effective stress law for the permeability of carbonate fractured-vuggy rock (CFVR), which can further be used to analyse the stress sensitivity performance during the process of liquid production, urgently needs to be studied.

Terzaghi (Terzaghi 1923) first introduced the concept of effective stress ($\sigma_{\mathrm{eff}}^{\mathrm{T}}$) in experimental studies of geotechnics and proposed the classical effective stress model:

$$\sigma_{\mathrm{eff}}^{\mathrm{T}} = \sigma - p \qquad (11.1)$$

where σ represents the external stress (pressure) in MPa, and p represents the internal stress (pressure) in MPa.

The physical properties of underground rock, which is another typical kind of porous material, are very different from those of soils, and scholars have reported that the Terzaghi classical model cannot be used to calculate the effective stress in subsurface rock (Li and Xiao 2008). That is, the Terzaghi model is applicable only to extremely loose porous media but not tight rock media. Thus, Biot (1941) proposed a widely used correction model for the effective stress ($\sigma_{\mathrm{eff}}^{\mathrm{B}}$):

$$\sigma_{\mathrm{eff}}^{\mathrm{B}} = \sigma - \alpha p \qquad (11.2)$$

In Eq. (11.2), α represents the Biot coefficient, where $\alpha = 1 - c_{\mathrm{s}}/c_{\mathrm{b}}$.

Robin (1973) reported that certain physical properties of a porous solid, such as density, length, seismic velocity, porosity and permeability, are functions of the confining pressure and pore pressure. For the scenario in which only the rock permeability is studied, Bernabe (1986) expressed the relationship among the permeability (K), effective stress ($\sigma_{\mathrm{eff}}^{\mathrm{K}}$), confining pressure ($p_{\mathrm{c}}$), and pore pressure ($p_{\mathrm{p}}$) as follows:

$$\sigma_{\mathrm{eff}}^{\mathrm{K}} = p_{\mathrm{c}} - \kappa p_{\mathrm{p}} \qquad (11.3)$$

In Eq. (11.3), σ_{eff} represents the effective stress coefficient for permeability (ESCP), which was expressed by Bernabe (1986) as follows:

$$\kappa = -\left(\partial K / \partial p_{\mathrm{p}}\right)_{p_{\mathrm{c}}} \Big/ \left(\partial K / \partial p_{\mathrm{c}}\right)_{p_{\mathrm{p}}} \quad (11.4)$$

Equation (11.4) clearly indicates that if the rock permeability is more sensitive to changes in confining pressure, then $\kappa < 1$, whereas if the rock permeability is more sensitive to changes in pore pressure, then $\kappa > 1$. In particular, for the classical Terzaghi model, according to Eq. (11.1), the following equation is obtained:

$$\sigma_{\mathrm{eff}}^{\mathrm{T}} = p_{\mathrm{c}} - p_{\mathrm{p}} = \sigma_{\mathrm{net}} \quad (11.5)$$

As shown in Eq. (11.5), for convenience of analysis, many engineers have directly applied the Terzaghi model to treat the net stress (σ_{net}, the difference between the confining pressure and the pore pressure) as the effective stress in the field; i.e., the value of the ESCP is generally considered to be constant at 1. Many theoretical and experimental explorations of the magnitude and variation range of the ESCP have been reported in the literature (Al-Wardy and Zimmerman 2004; Warplnski and Teufel 1992). All the results revealed that different rock samples have different ESCP values under different confining and pore pressures, and both the magnitude and range of the ESCP that were obtained in different studies varied significantly. Clearly, the true ESCP values for most rocks are not always equal to 1 and may even deviate substantially from 1 (Xiao et al. 2012; Kang et al. 2017). Therefore, in most cases, net stress cannot be used in place of effective stress.

To date, no related studies on the ESCP features of CFVR with different structural characteristics have been reported. Thus, whether the ESCP is a constant value and can be approximated as 1 for this kind of rock is unknown. To reveal the differences in effective stress and stress sensitivity for various types of carbonate fractured-vuggy rocks, four experimental core models with different structural characteristics were designed and constructed, namely, a one-fracture core model, a two-horizontal-fracture core model, a fracture–vug core model, and a through-hole core model.

(3) Deep Reservoir Complex Continuum Coupling Numerical Simulation Technology

Reservoir simulation technology is important for oilfield production index prediction, remaining oil characterization, and the efficient exploitation of reservoirs. Research on reservoir simulation began in the 1930s, and the results were applied in the petroleum industry in the 1950s. In the 1980s, reservoir simulation technology was developed into complete, large-scale, and multifunctional comprehensive software. In the 1990s, the dual-continuum numerical approach was introduced, and integrated software for compositional modelling, chemical flooding, and wellbore ground integration was developed. In the twenty-first century, reservoir simulation technology is developing towards faster computation speeds and more accurate results. Multicontinuum models are being further developed, and parallel computing technology has been applied to numerical simulations of reservoirs.

Research on deep reservoir complex continuum coupling numerical simulation technology started in 2005. Relying on the National Program on the Key Basic Research Project of China (973 Program), numerical simulation software (KarstSim) for fracture-cavity reservoirs has been developed for research on and the development of large-cavity multiphase free flow, embedded discrete fracture-cavity models, coupled free flow and seepage, variable continuum methods, multicomponent phase behaviour simulations, complex continuum numerical well testing, CPU + GPU parallel computing, and other functions. This software systematically solves the problems that are associated with simulating deep multiscale reservoirs. Through numerical applications, water flooding technology and nitrogen cave top flooding-enhanced oil recovery technology that is based on the fracture-cavity structure have been proposed. These technologies have supported the large-scale development, as well as high and stable production, of the world's first deep, very large fracture-cavity reservoir, the Tahe oilfield (Kang et al. 2017, 2020a, b).

For the numerical simulation of ultradeep (> 6000 m) carbonate reservoirs, owing to the high-temperature (greater than 135 °C) and high-pressure (greater than 60 MPa) environments, the reservoir fluid, rock skeleton, and temperature experience complex physical and chemical interactions and affect each other during production. Therefore, the numerical simulation of such reservoirs is a thermal–hydro–mechanical multifield coupling problem. In 1943, Terzaghi first proposed the concept of effective stress and developed a one-dimensional consolidation model (Terzaghi 1943). Biot later established a complete three-dimensional consolidation and wave propagation theory with the stress balance equation of the rock skeleton and the continuity equation of the pore fluid as the main control equations (Biot 1941, 1956). Zienkiewicz and Shiomi proposed the generalized Biot equation on the basis of Biot's three-dimensional consolidation equation in 1984 and numerically solved it via the finite element method (Zienkiewicz and Shiomi 1984). In 1986, Mctigue extended Biot's three-dimensional consolidation theory to nonisothermal conditions and established a mathematical model for thermal–hydromechanical coupling problems that involve single-phase flow (Mctigue 1986). Coussy summarized the thermal–hydromechanical coupling model and named it thermoporoelasticity theory (Coussy 2004). In 2004, Kong and Wang proposed a preliminary theoretical model of the thermal–hydro–mechanical coupling system (Kong et al. 2006; Wang et al. 2004). In 2018, Yao and Huang summarized the key mechanical problems to be solved in deep oil and gas reser-

voirs and noted the development direction of thermal–hydro–mechanical multifield coupling simulations in deep reservoirs (Yao et al. 2018). In 2020, on the basis of the Joint Funds of the National Natural Science Foundation of China project on addressing ultradeep thermal–hydromechanical multifield coupling numerical simulation technology, a multifield coupling model that is based on a complex continuum was proposed. The governing equations for multifield coupling problems were established. The mechanical relationship and mathematical characterization method were determined, and multifield coupled numerical simulation software for ultradeep fractured reservoirs was developed (Li et al. 2018).

11.2 Geological Models of Ultradeep Fracture-Cavity Reservoir Systems

Ultradeep fault-controlled fracture-cavity reservoirs are a new type of oil reservoir, with burial depths generally greater than 6000 m. Due to the lack of development of pores, the reservoirs are developed mainly near the extensional, compressive, pure strike-slip and intersecting faults that are controlled by strike-slip faults. The internal structure is divided into a fracture-cavity core and fracture fragmentation, and the vertical thickness is generally greater than 500 m (Fig. 11.1). On the basis of the multiscale complex development characteristics of fault-controlled reservoirs, a "multiple-constraint, five-step" 3D geological modelling method was established, and the technical route is shown in Fig. 11.2.

Fig. 11.1 Reservoir distribution pattern of ultra-deep fault-controlled reservoirs

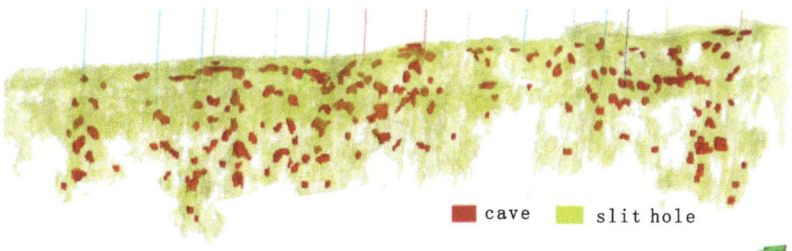

■ cave ■ slit hole

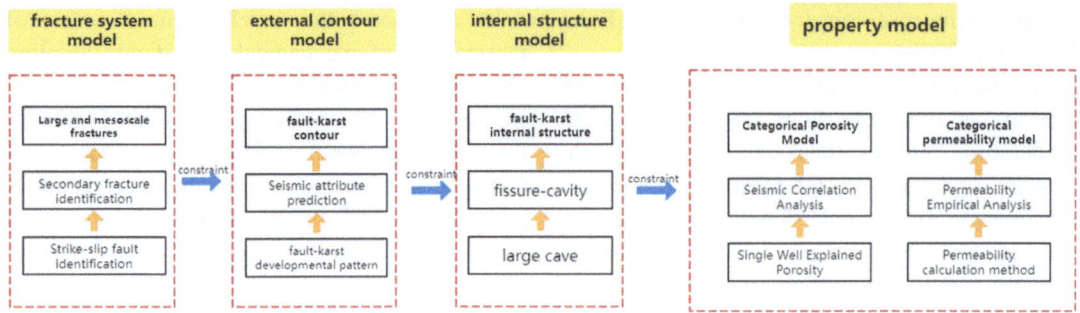

Fig. 11.2 Technical roadmap for geological modeling of fault-controlled reservoirs. Technical roadmap for a "multiple constraints, five-step method" 3D geological modeling method

(1) Modelling of fault systems at various scales

On the basis of the results of drilling leakage and logging fault interpretation, spatial distribution models of faults at various scales, stages, and directions were constructed. The scales included large-scale faults and mesoscale faults. Large-scale faults refer to faults that are directly interpretable by earthquakes. An analytical and modelling method for small-slip strike-slip faults that is based on stage, strength, properties, and combination relationships was established. The primary and secondary relationships and spatial distribution model of large-scale fractures. Mesoscale faults refer to faults that can

be predicted by seismic attributes. The curvature ant volume method and the coherent ant volume method were established to describe the azimuth, dip, inclination, and position, and a 200-m-long fault was identified. On the basis of the prediction volume of mesoscale faults, faults with the same azimuth, dip angle, and inclination as the large-scale faults were excluded, the mesoscale fault model was deterministically established, and the tension–torsion, compression–torsion and pure strike–slip segments were divided (Fig. 11.3).

(2) Profile modelling of fault-controlled reservoirs

Fig. 11.3 Causes and division of compressive-torsion and compressive segmentation of strike-slip faults

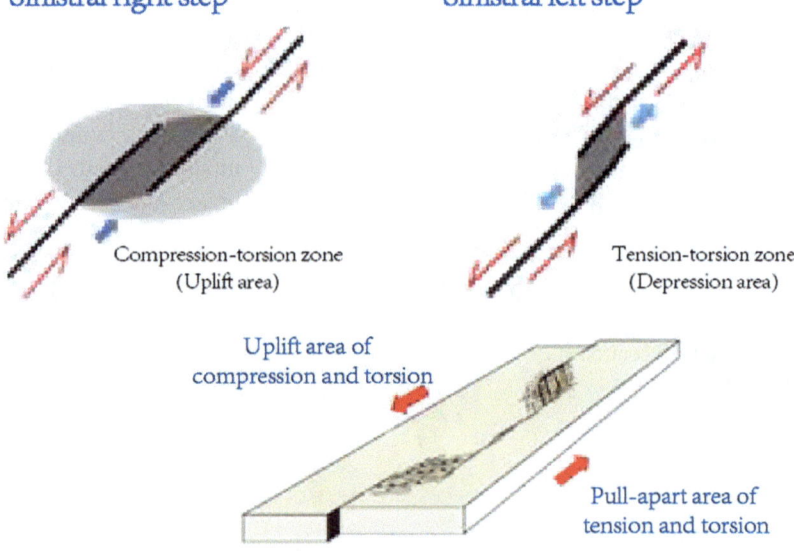

The fault-controlled reservoir profile is the boundary and extent of the effective reservoir. A static identification method for reservoir contours with sensitive seismic attributes that is based on well logging, drilling mud logging, drilling time, and structural tensors was established. A method for determining the dynamic boundary of a reservoir on the basis of well testing and a fault-controlled reservoir development model was developed, and the comprehensive static profile and dynamic boundary were deterministically and artificially corrected.

(3) Internal structure modelling of fault-controlled reservoirs

The internal structure refers to the karst-vug model and the fracture–vug model within the outline of the fault-controlled reservoir. Discrete distribution model of large-scale karst caves: A single-well karst cave model that is based on single-well venting, lost circulation, logging curves, and dynamics was established, and a karst-vug probability body construction method that is based on a single-well karst cave model and the sensitivity of seismic attributes such as wave impedance difference and seismic frequency division energy was developed. Constrained by the geological model and probability, a karst cave reservoir model was established via the combination of target-based and artificial methods. Fracture–vug reservoir distribution model: Seismic attributes such as fracture–vug and amplitude spectrum gradients were compared to establish a probability development body, the relationship between crack-holes and fractures was analysed, and a fault control probability body was established. The fracture–vug reservoir at the well point was taken as the hard data; the fault-controlled reservoir contour, seismic probability volume, and fault probability volume were used as the multiconstraints; and a fracture–vug reservoir model was established by the sequential indication simulation method (Fig. 11.3).

(4) Fault-controlled reservoir property modelling

The porosity and permeability models of three types of reservoirs, namely, karst cave, fracture–

vug and fracture (fracture system) reservoirs, were established. Attribute modelling of caves, fractures and holes was conducted. A chart was constructed on the basis of the porosity and cumulative production of the caves, and interpretations were comprehensively logged to establish a single-well porosity model. A dynamic reserve calculation method was developed on the basis of the production performance and geological characteristics of a single well, a porosity development trend surface was formed, and the development characteristics of the porosity plane were clarified. The nonlinear relationship between single well porosity and wave impedance was analysed, and a porosity probability body was established. With the well point porosity taken as the hard data, the cave and fracture–vug reservoir models regarded as the "phase control" constraints, and the porosity plane trend and the porosity probability volume taken as synergistic constraints, the sequential Gaussian simulation method was used to establish the porosity attribute model. The permeability model was constrained by the permeability explained by the well test and was calculated on the basis of the empirical formulas for porosity and permeability.

Modelling crack properties: A single-well fracture porosity and fracture opening model was established, and an equivalent method was used to construct a fracture porosity model on the basis of the fracture opening; the relationships between the equivalent permeability of fractures on the basis of the opening and the effective permeability of fractures on the basis of well testing were established, the influence of fracture filling on fluid flow was clarified, and a fracture permeability model was established.

(5) Fault-controlled reservoir fusion model

Karst caves, fracture vugs, and fracture reservoirs are established separately from each other via the above classification method. However, for real reservoirs, only one type of reservoir exists per grid. According to the genesis and characteristics of fault-controlled reservoir development, the "isotopic assignment" reservoir fusion method is formed, which involves the fusion of karst

Fig. 11.4 Reservoir distribution model in the SHB1 fault zone. A three-dimensional geological model of the fault-controlled reservoir in the SHB1 fault zone

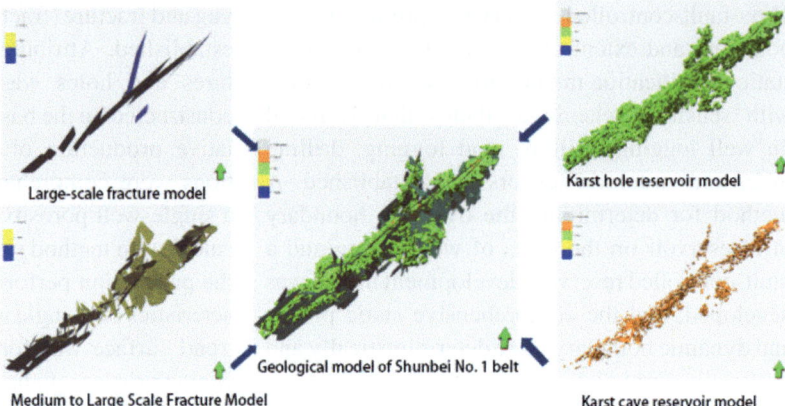

caves first, faults second, and fracture–vug last. A unique reservoir type is assigned to each grid to form a fault-controlled reservoir model. For reservoir properties, a method of summing the porosity and taking the maximum value of the permeability is established, which is fused to form an attribute fusion model.

On the basis of this modelling method, a three-dimensional geological model of the fault-controlled reservoir in the SHB1 fault zone was established, and five new wells were used to verify the coincidence rate of the geological model. All five wells encountered karst caves or fractures and lost leakage, which was consistent with the model prediction results, but one well had a deviation in the cave position of 12 m, and the model conformity rate was 80% (Fig. 11.4).

11.3 Effective Stress Law for Evaluating the Permeability and Stress Sensitivity of Fractured-Vuggy Carbonate Rocks

First, two typical natural core samples (a porous core and a fractured-vuggy core) were selected to investigate the basic fluid flow laws under various pressure and temperature conditions for deep reservoirs. For the single-phase flow, Fig. 11.5a and b shows the relationship curves between the pressure gradient and displacing velocity for the two types of cores. Oil follows Darcy's seepage law in the fractured-vuggy core. Under high-pressure conditions, the higher the temperature is, the faster the

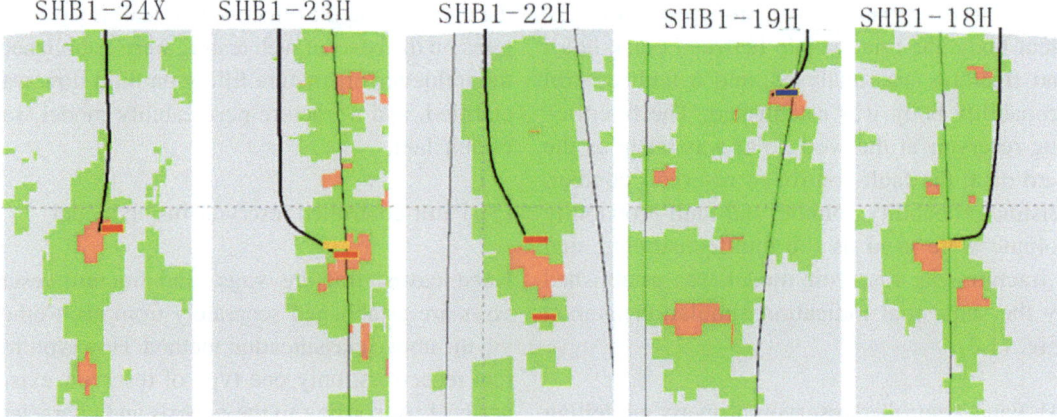

Fig. 11.5 Cross-sectional view of new well and model comparison. Cross-sectional view of 5 new wells and comparison with that prediction of the model established by the method

corresponding displacement velocity. In contrast, oil shows a low-velocity non-Darcy seepage law in the porous core, and the threshold pressure gradient decreases with increasing temperature. For the oil–water two-phase flow, Fig. 11.5c and d shows the tested relative permeability curves for the two types of cores. Under high-pressure conditions, as the temperature increases, the hydrophilicity of the rock increases, and the connate water saturation also increases. The residual oil saturation is not significantly changed by the combined effect of the oil polar components and the water–rock interactions.

Additionally, four experimental core models with different structural characteristics were designed to determine the effective stress for fractured-vuggy carbonate rocks, and the absolute permeability under alternating stress conditions was tested by the loading method of constant confining pressure and cyclic pore pressure. On the basis of the experimental data, the ESCP for the core sample was obtained via several calculation methods [differential methods (Bernabe 1986), translation methods (Al-Wardy and Zimmerman 2004), cross-plotting methods [14], and response surface methods (Warplnski and Teufel 1992)], which were ultimately used to calculate the corresponding effective stress for permeability under various confining and pore pressure conditions.

When the response surface method (RSM) is applied, the intrinsic physical properties of the experimental object need not be considered, and the empirical model is built only by fitting the approximate response surface of the data. This approach is particularly useful for tight rock samples with relaxation microcracks because such a microstructure can invalidate any mechanistic model that is based on linear elasticity (Warplnski and Teufel 1992). Thus, the RSM, which is a completely empirical approach, can be used to avoid biasing the real results. Owing to the significant advantages of the RSM, this approach was used to calculate the ESCP values for the core models.

In general, the RSM consists of three main steps: (1) transforming the permeability data to a simpler form and weighting the variance; (2) fitting

the converted data with linear or quadratic surfaces; and (3) using standard statistical and graphical techniques to determine the value of the ESCP.

The ESCP surfaces for the four representative core models that were calculated via the RSM are shown in Figs. 11.1 and 11.2. Overall, the calculation results demonstrate that (1) the effective stress for the permeability of the CFVR clearly has nonlinear characteristics since the ESCP values are not constant and vary with changes in the confining pressure and pore pressure; (2) not all of the obtained ESCP values are equal to 1, but the variation range is concentrated between 0 and 1, which indicates that the permeability of the CFVR is more sensitive to changes in the confining pressure than to changes in the pore pressure; and (3) the CFVR with different structural characteristics exhibits different variation features of the ESCP. Xiao et al. (2012) summarized previous research results on the relationship between ESCP and rock deformation response and reported that single-component fractured rocks without clay minerals have significant features in which the ESCP value is less than 1 and varies with changes in stress. The results of this study are in agreement with this conclusion, thus indicating that the fracture deformation effect of the rock models is very significant.

As shown in Fig. 11.6a and b, the ESCP values for the one-fracture core model and the two-horizontal-fracture core model range from 0.002–0.534 and 0.254–0.672, respectively. The ESCP decreases with increasing confining pressure and increases with increasing pore pressure. Obviously, the ESCP values for both models vary within a range much smaller than 1. Theoretically, when the initial pore pressure is relatively high, the fracture is in an open state, so the corresponding ESCP value of the ideal infinite fracture should be equal to 1 (Bernabe 1986). Since infinite fractures do not exist in real rocks, the initial value of the ESCP is often less than 1. With the gradual decrease in pore pressure, the rock skeleton is deformed and compressed, the contact points on the two fracture surfaces gradually increase, the fracture becomes difficult to close, and the ESCP decreases continuously. As the pore pressure continues to decrease, the fracture reaches the ultimate closure status, and the matrix pore space also gradually reaches

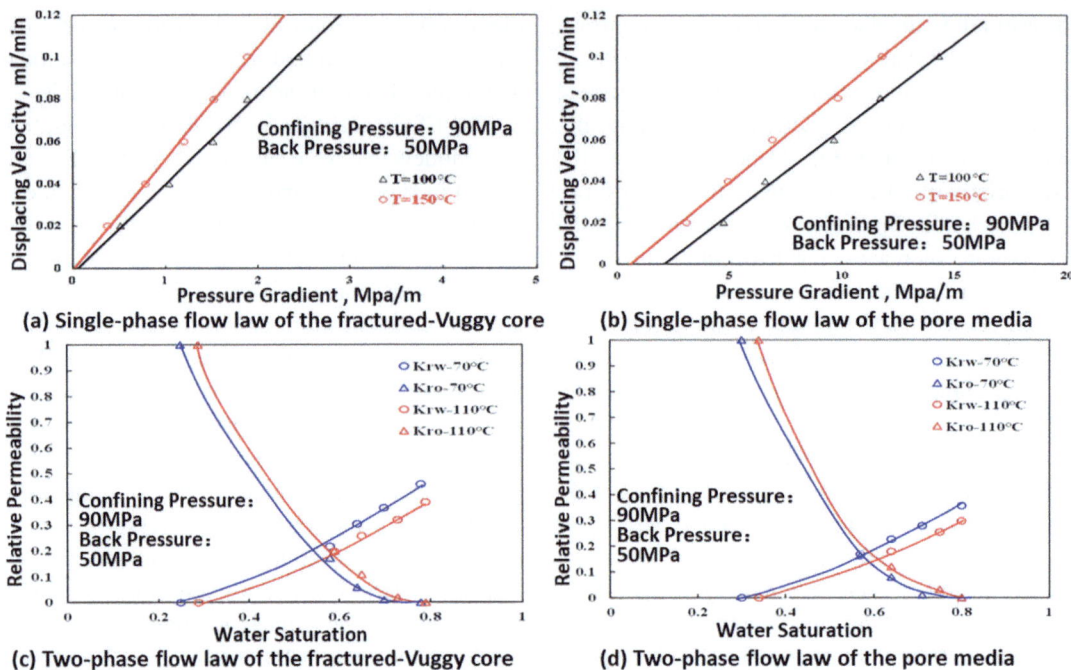

Fig. 11.6 The basic fluid flow laws of the porous core and the fractured-vuggy core under different pressure and temperature conditions

the compression limit status. At this point, the total porosity and matrix porosity are lower than those of the initial state, and accordingly, the ESCP decreases to the minimum value. Thus, the lower the initial porosity is, the lower the minimum ESCP value. Since the carbonate fractured core models used have very low porosity, the obtained lower limit value of the ESCP is also relatively small. For the one-fracture core model, the ESCP is much smaller than 1 and close to 0 at $p_p = 10$ MPa. In comparison, the lower limit value of the ESCP for the two-horizontal-fracture core model is 0.254, and the value surface is shifted upwards overall. As mentioned above, an increase in the number of fractures increases the initial porosity of the core model. Accordingly, the porosity in the ultimate compression state is also greater than that in the one-fracture core model.

The calculated ESCP surfaces for the fracture–vug core model and the through-hole core model are shown in Fig. 11.7a and b, respectively. Compared with those of fractured rocks, the ESCP distribution characteristics of vuggy rocks are more complicated. The ESCP distribution interval

for the fracture–vug core model is more concentrated (0.768–0.997), and the ESCP value gradually increases with increasing pore pressure, which proves that the value of the ESCP for the fracture–vug rocks can be approximated as 1. When the pore pressure is relatively high, both the fracture and vug spaces are fully open, which is much closer to the ideal infinite fracture model that was assumed by Bernabe (1986); thus, the upper limit of the ESCP is close to 1. Unlike only fracture-developed carbonate rocks, the compressibility of the vug structure is extremely limited regardless of the variations in the confining pressure and pore pressure, and the contactable area for the fracture surfaces is significantly reduced. This leads to the fracture–vug rocks quickly reaching the limit state of contactable fracture closure under the combined action of internal and external pressure; thus, the lower limit value of the ESCP is relatively high, and the range of ESCP variation is quite narrow, which indicates that the deformation response characteristics of porous rocks are strong. In addition, the ESCP distribution interval for the through-hole core model is more dispersed (0.100–0.982),

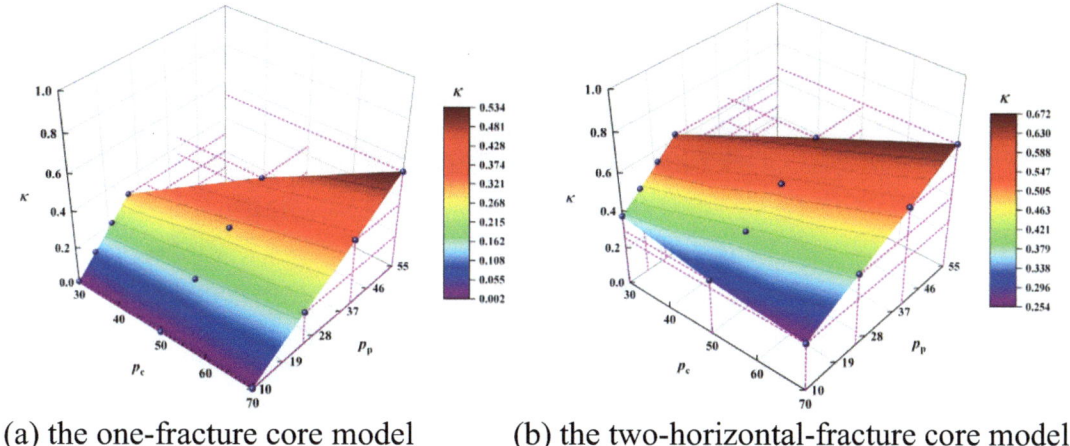

(a) the one-fracture core model   (b) the two-horizontal-fracture core model

Fig. 11.7 The ESCP surface of carbonate fractured core models calculated by RSM: **a** one-fracture core model; **b** two-horizontal-fracture core model

and the value of the ESCP decreases continuously with increasing confining pressure. As the flow space for the through-hole rock is a through-hole channel, the compressibility of the whole medium is quite high. The larger the applied confining pressure is, the more the through-hole and matrix pore space is compressed; thus, the porosity decreases, and the value of the ESCP also decreases gradually. When the confining pressure is equal to 70 MPa, the ESCP is distributed around 0.1, which is much smaller than the corresponding value for the fracture–vug core model.

Because the power model has the best adaptability and simplicity, it was selected to further analyse and evaluate the stress sensitivity of the CFVR. On the basis of the regression curves of the power model, the relationships between permeability and effective stress for the four representative core models were plotted, as shown in Fig. 11.8. Although the rock permeability values that correspond to various effective stresses can be directly obtained from the figure to guide engineering practices, the original permeability is distinct among different rocks, and the ranges and trends of permeability variation also vary considerably from rock to rock.

To quantitatively evaluate the stress sensitivity for various types of CFVR on the basis of a unified standard, the novel concept of the "permeability decline rate" was introduced to charac-

terize the relative change in rock permeability with increasing effective stress, i.e., the permeability decrease fraction within unit effective stress (D_K, %/MPa), which is expressed as follows:

$$D_K = -\frac{1}{K}\frac{dK}{d\sigma_{eff}} \times 100\% \qquad (11.6)$$

Substituting the power model expression Eq. (11.7) into Eq. (11.6) yields Eq. (11.8):

$$K = \alpha \cdot \sigma_{eff}^{-\beta} \qquad (11.7)$$

$$D_K = \frac{\beta}{\sigma_{eff}} \times 100\% \qquad (11.8)$$

Equation (11.8) shows that the rate of decrease in permeability D_K is related only to the coefficient β in the power model. The greater the value of the coefficient β is, the greater the rate of decrease in permeability D_K under the same effective stress conditions. The two values are proportional to each other. Therefore, the value of β can be directly used to evaluate the strength of the stress sensitivity of CFVR. The larger the value of β is, the stronger the stress sensitivity of the tested rock; the smaller the value of β is, the weaker the stress sensitivity of the tested rock.

According to the calculated results, the values of β for the one-fracture core model, the two-horizontal-fracture core model, the fracture–vug

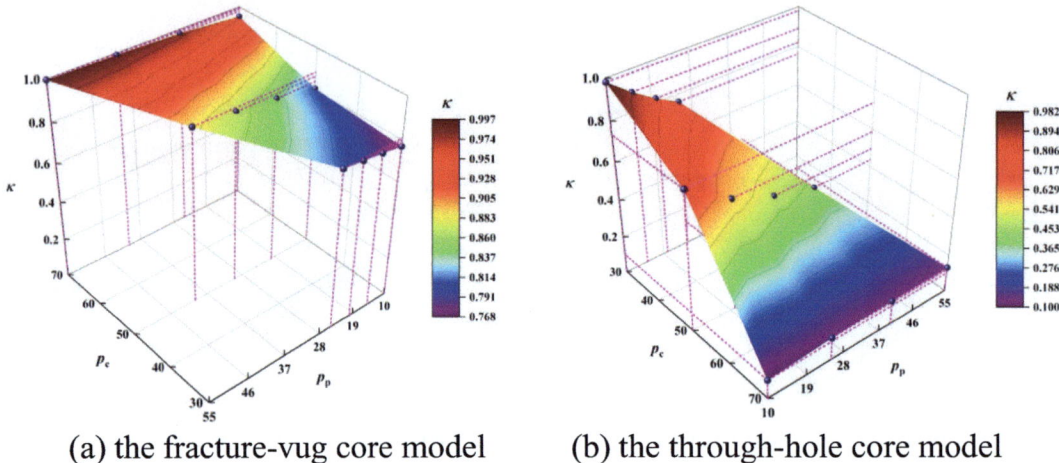

(a) the fracture-vug core model   (b) the through-hole core model

Fig. 11.8 The ESCP surface of carbonate vuggy core models calculated by RSM: **a** fracture-vug core model; **b** through-hole core model

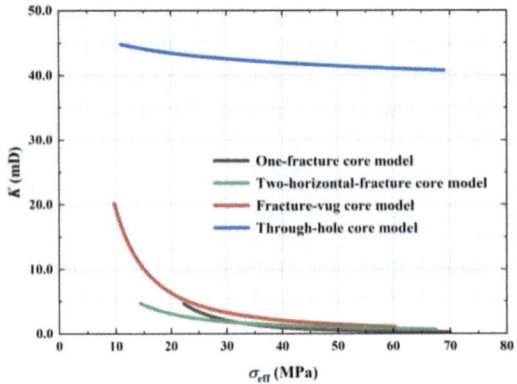

Fig. 11.9 Variation of permeability with effective stress for the four representative carbonate fractured-vuggy core models

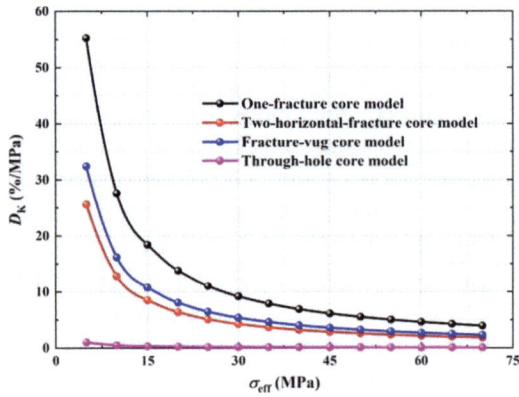

Fig. 11.10 Variation of permeability decline rate with effective stress for the four representative

core model, and the through-hole core model are 2.76, 1.28, 1.62, and 0.052, respectively. Accordingly, the stress sensitivity of the one-fracture core model is the strongest, followed by those of the fracture–vug core model and the two-horizontal-fracture core model, and the stress sensitivity of the through-hole core model is the weakest. The relationships between the permeability decline rate and effective stress for the four representative core models are plotted in Fig. 11.9.

In Fig. 11.10, the main decreasing stage of D_K is concentrated when the effective stress is between 0 and 15 MPa, and when the effective stress is greater than 30 MPa, the value of D_K is

generally less than 10%/MPa. When the effective stress is relatively low, the stress sensitivity of the CFVR is strong, and as the effective stress gradually increases, the stress sensitivity gradually weakens. In this process, D_K decreases rapidly at low values of σ_{eff} and levels off at high values of σ_{eff}. When the effective stress increases to 70 MPa, the D_K values for the four representative core models are less than 5%/MPa, and the stress sensitivity is minimal.

Therefore, to minimize the negative effect of the stress sensitivity effect on the production performance, the drawdown pressure and production rate should be reasonably controlled during the early stage of the increase in effective stress during the development of fractured-vuggy carbonate reservoirs.

11.4 Multifield Coupled Numerical Simulation Method for Fracture-Cavity Reservoirs

To address the problem of fluid flow in fracture-cavity reservoirs with various sizes of cavities, dissolved pores, faults, and fractures, a complex continuum numerical simulation method was established. Fluid flow in large-scale cavities and fractures is considered free flow. In this method, these cavities and fractures are treated explicitly and embedded into the reservoir. Seepage in vugs and fracture networks is simulated via the variable continuum model. In terms of the actual reservoir, some of the grids use the dual continuum model, and the remainder use the triple continuum model. Dissolved pores and small fractures are simulated via an effective single-continuum model.

Large fractures are embedded into the reservoir. Fractures are embedded via an integrally embedded discrete fracture model. This model has a new gridding method that can arbitrarily grid the fractures according to the requirements rather than finely subdividing the fracture elements. A more precise pressure distribution assumption inside the matrix block is adopted, and a more accurate semi-analytic calculation method for matrix–fracture transmissibility is obtained. Since cavities and fractures have similar high-permeability characteristics, large cavities can be regarded as fractures with large volumes and special shapes. Afterwards, the cavities can be treated similarly to the fractures and embedded into the reservoir, and the transmissibility of the matrix–cavity connection can be calculated.

The in situ stress, temperature, and pore fluid pressure of deep reservoirs are very high. The deformation characteristics of reservoir rocks are complex, and strong mechanical, thermal, and physicochemical interactions occur between the reservoir fluid and the rock skeleton. To this end, a numerical simulation method for the thermal–hydromechanical coupling problem in ultradeep fracture–cavity carbonate reservoirs is established. In this method, the reservoir space includes cavities, fractures, and dissolved pores; reservoir bodies include cavities + fractures + dissolved pores, fractures + dissolved pores, fractures, and cavities; the fluid flow field includes pressure, fluid flux, phase saturation, and flow velocity; the stress field includes stress and strain; the temperature field includes heat flux and temperature; the coupling between the temperature field and flow field includes density, viscosity, and the thermal conductivity change; the coupling between the flow field and stress field includes porosity, permeability, and the effective stress change; and the coupling between the stress field and temperature field includes the thermal strain, thermal effect and thermal conductivity change (Fig. 11.11).

11.4.1 Mathematical Model for Multifield Coupling Problems

In a multifield coupling system, the comprehensive consideration of the interactions among the seepage field, stress field, and temperature field is necessary. The system of governing equations of the multifield coupling problem consists of three parts: a mass conservation equation, an energy conservation equation, and a force balance equation. In addition, a series of auxiliary equations need to be introduced to consider the coupling between multiple fields.

(1) Mass and Energy Conservation Equations

The basic mass and energy conservation equations can be written in the following general form:

$$\frac{\partial M}{\partial t} = \nabla \cdot \vec{F} + q \tag{11.9}$$

where M represents the mass or heat accumulation term, $\vec{F}$ represents the mass or heat flux term, and q represents the source term.

Fig. 11.11 Coupling between the multi-fields. Coupling among temperature field, seepage field and stress field

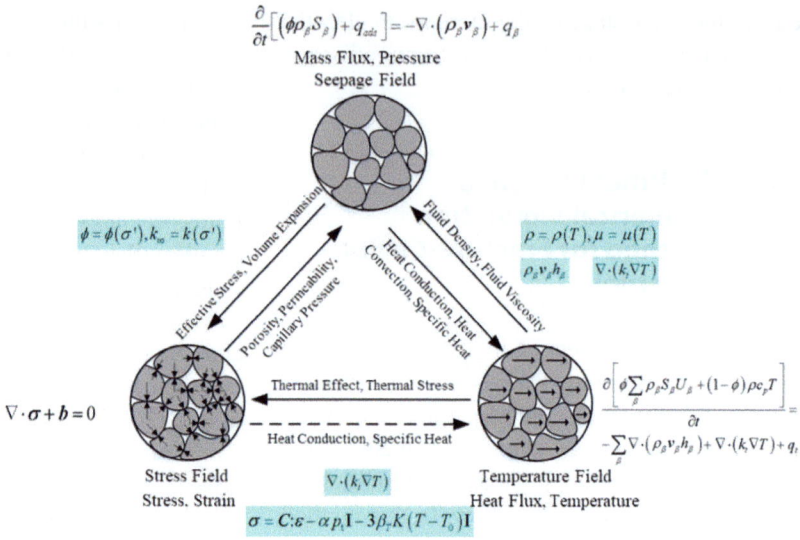

A mass conservation equation needs to be established for each component. For component β, the mass accumulation term is given by

$$M_\beta = \phi S_\beta \rho_\beta \qquad (11.10)$$

where ϕ represents the porosity, ρ_β represents the density of β, and S_β represents the saturation of β.

The mass flux term of phase β is as follows:

$$\vec{F}_\beta = -k \frac{k_{r\beta} \rho_\beta}{\mu_\beta} \left(\nabla P_\beta - \rho_\beta g \nabla D \right) \qquad (11.11)$$

where μ_β represents the viscosity of phase β, k represents the absolute permeability, $k_{r\beta}$ represents the relative permeability to phase β, and P_β represents the fluid pressure in phase β.

The heat accumulation term in the energy conservation equation is given by

$$M_T = (1 - \phi)\rho_R C_R T + \phi \sum_\beta S_\beta \rho_\beta u_\beta \qquad (11.12)$$

where ρ_R represents the density of the rock grains, C_R represents the specific heat of the rock, T represents the temperature, and u_β represents the specific internal energy in phase β.

Ignoring radiative heat transfer in the reservoir, the heat flux term includes conduction and convection components:

$$\vec{F}_T = -K\nabla T + \sum_\beta h_\beta \vec{F}_\beta \qquad (11.13)$$

where K represents the thermal conductivity and h_β represents the specific enthalpy in phase β.

(2) Force Balance Equation

Under a temperature change, the solid skeleton of reservoir rock undergoes thermal expansion or contraction. Therefore, in addition to the mechanical strain that is caused by effective stress, the deformation of the rock skeleton also includes thermal strain that is caused by temperature changes. The additional load that is caused by thermal strain is called thermal stress, which is expressed as follows (where the vectors and tensors are expressed in components):

$$\sigma_{Tij} = 3K_b\beta_T(T - T_0)\delta_{ij} \qquad (11.14)$$

where T represents the current temperature, T_0 represents the initial temperature, β_T represents the coefficient of thermal expansion, and K_b represents the bulk modulus.

When the thermal stress is considered, the relationship between the total stress tensor and strain tensor becomes

$$\sigma_{ij} = D_{ijkl}\varepsilon_{kl} - bP\delta_{ij} - 3K_b\beta_T(T - T_0)\delta_{ij} \tag{11.15}$$

The force balance equation is expressed in terms of the total stress as

$$\rho\ddot{u}_i^s = \frac{\partial\sigma_{ij}}{\partial x_j} + \rho f_i \tag{11.16}$$

By substituting Eq. (11.15) into Eq. (11.16), the expression of the force balance equation in the multifield problem can be obtained.

(3) Auxiliary Equations

Owing to the coupling between multiple physical fields, updating the physical parameters in the process of solving the multifield coupling problem is necessary. During production, a pressure decrease in the reservoir leads to mechanical deformation of the rock skeleton and causes porosity changes. The change in temperature causes additional deformation of the rock skeleton, which further affects the porosity change:

$$\phi = \phi_0 + \left(\frac{1}{K_b} - \frac{1+\phi_0}{K_s}\right)(p - p_0)$$
$$- \left(\frac{1}{K_b} - \frac{1}{K_s}\right)\left(\sigma_m - \sigma_m^0\right)$$
$$- 3(1 - \phi_0)\beta_T(T - T_0) \tag{11.17}$$

In addition, the change in temperature causes a change in fluid viscosity, which can be described by the following empirical formula:

$$\ln\mu = A + \frac{B}{T} + CT + DT^2 \tag{11.18}$$

where A, B, C and D represent fluid parameters that need to be measured experimentally.

(4) Stress–Strain Constitutive Relations of Fractured Rock

In the calculation of the force balance equation, the stress–strain constitutive relation of the fractured rock needs to be determined. The presence of fractures leads to a reduction in rock stiffness; hence, the stress–strain constitutive relation of fractured rock is different from that of porous rock.

A fractured rock consists of two separate continua, namely, the rock matrix and the fracture, which behave differently under stress. The concept of the equivalent continuum is introduced to establish the stress–strain constitutive relations for fractured rock. In the equivalent continuum model, an equivalent material is introduced to represent the overall deformation of the fractured rock. The deformation behaviour of the introduced equivalent material is a combination of rock matrix deformation and the displacement of fracture planes. Therefore, the compliance matrix of the fractured rock can be expressed as the weighted average of the rock matrix and fracture compliance matrix. For fractured rock with multiple fractures, the contribution of each fracture to the stress–strain constitutive relation needs to be considered, which yields the expression for the overall compliance matrix:

$$[C_{overall}] = [C_M] + \sum_{i=1}^{N_f}\alpha_{Fi}[T_i]^T[C_{Fi}][T_i] \tag{11.19}$$

where N_f represents the number of fractures in the considered element, α_{Fi} represents the volume fraction of fracture i, $[C_{Fi}]$ represents the compliance matrix of fracture i, and $[T_i]$ represents the coordinate transformation matrix of fracture i.

11.4.2 Numerical Computation Method for the Multifield Coupling Problem

For multifield coupling problems, the fully implicit solution needs to simultaneously establish the governing equations of each physical field and solve them uniformly through numerical discretization. The calculation results of the fully implicit solution are accurate, and the convergence is good, but coupled multiphysics problems have many governing equations and many unknowns to be solved. Therefore, the fully implicit solution has the problem of low computational efficiency and is not appropriate for field applications. To improve the computational efficiency for multi-field coupling problems, a semi-implicit solution method can be adopted. In each time step, some of the governing equations are first solved implicitly, and then the remaining governing equations are solved according to the calculation results of the implicit solution. The force balance equation has the most unknowns and the most equations and causes the greatest computational burden. Therefore, the computational efficiency of multifield coupling problems can be improved by implicitly solving the mass conservation equation and heat conservation equation while explicitly solving the force balance equation. To ensure calculation accuracy and stability, the fixed-stress split iterative coupling algorithm is used to perform the semi-implicit numerical calculation of the governing equations.

11.4.3 Testing and Application of Multifield Coupling Simulation

(1) Case of a Deep Fracture-Cavity Reservoir with Bottom Water

A case in which a deep fracture-cavity reservoir is produced was investigated. The configuration and mesh of the reservoir are displayed in Fig. 11.12. There are 5 clusters of fractures in the reservoir, as shown in the figure. Under the initial stress, the porosity of the reservoir is 0.195, and the permeability of the matrix is 2.62 mD. The thermal conductivity of the reservoir rock is 2.0 W/(m^3 °C). The temperature at the reservoir bottom is 110 °C, and the geothermal gradient is 2.5 °C/100 m. The fluid pressure at the reservoir bottom is 65 MPa. The production well is produced at a fixed bottom hole pressure of 50 MPa, and the bottom water is continuously replenished into the reservoir as the production proceeds.

The thermal–hydromechanical coupled model, the hydromechanical model, the thermal–hydro model, and the hydraulic model were used for calculation. The production rates for oil and water, which are shown in Fig. 11.13, reveal that oil and water production decrease when the influence of the stress field is considered but increase when the influence of the temperature field is considered.

The temperature distribution at 30,000 days is shown in Fig. 11.14a. In a fractured reservoir, fractures are the main flow channels, which not only affect the oil–water displacement process but also accelerate the heat convection process. The temperature increases significantly near the fracture cluster in the centre of the reservoir, which is directly connected to the production well. The distribution of thermal strain at 30,000 days (positive for compression) is shown in Fig. 11.14b. In this case, the solid skeleton of the rock expands with increasing temperature, so the thermal strain is mainly tensile.

(2) Numerical Simulation of the Shunbei Oilfield Reservoir

The top and bottom depths of the fracture-cavity geological model are 6373.4 m and 8164.7 m, respectively (Fig. 11.15), the thickness

Fig. 11.12
a Configuration of the reservoir; **b** grids in the reservoir. **a** Configuration and **b** grids of a reservoir containing five sets of fractures

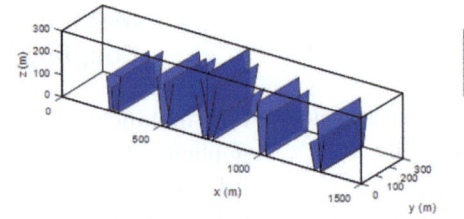

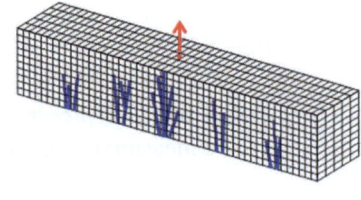

(a) Configuration   (b) Grids

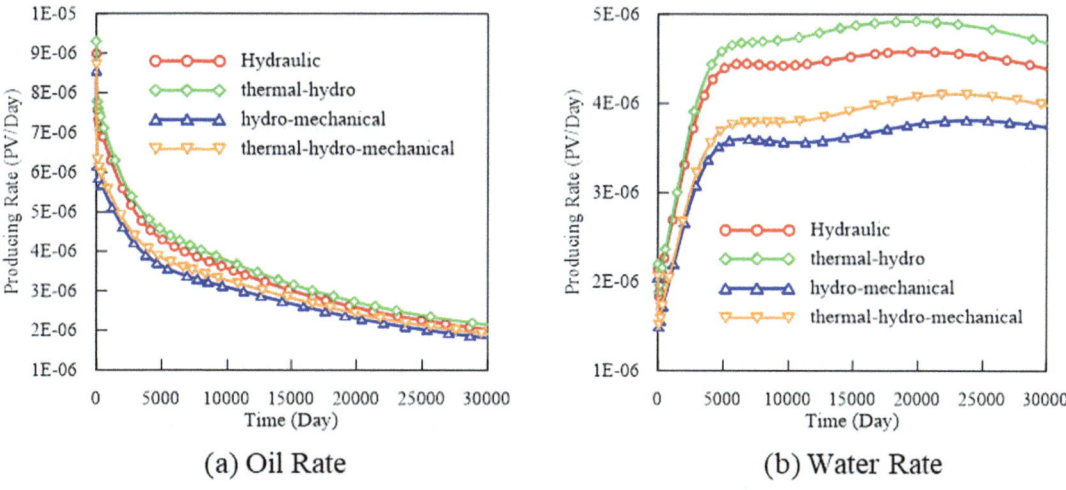

Fig. 11.13 Histories of producing rate for **a** oil and **b** water. Histories of producing rate for **a** oil and **b** water Considering the influence of different fields

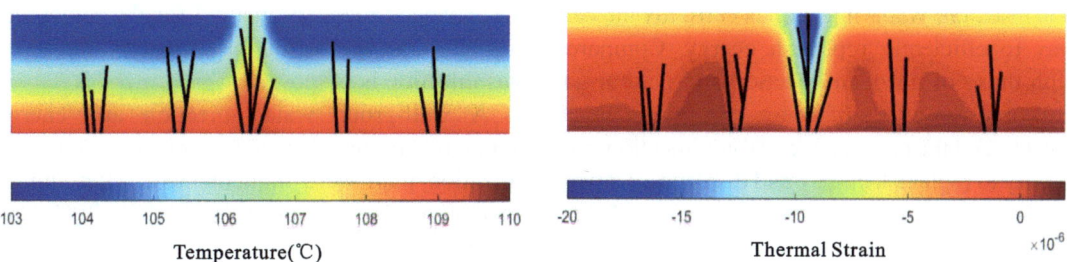

Fig. 11.14 Distribution of **a** temperature and **b** thermal strain. Distribution of **a** temperature and **b** thermal stress in production wells during 30,000 day

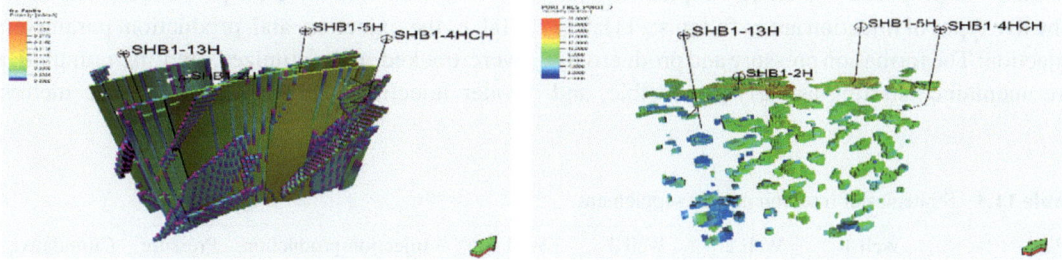

Fig. 11.15 Geological model of fractures and cavities in Shunbei Oilfield Reservoir

is 1791.3 m, and the oil column height is 615 m. The initial formation pressure is 87.1 MPa, the saturation pressure is 35 MPa, the viscosity of the oil is 0.29 cp, the oil density is 0.55×10^3 kg/m^3, and the oil volume coefficient is 2.17. The reservoir is produced by four wells, with a cumulative oil production of 211,000 tons and a recovery rate of 6.1%. After the physical parameters are

corrected, the fitting rate of single-well production history matching is 87.6%.

The mechanism of the rapid decline in production of fault-controlled reservoirs is revealed, and the rapid decline in single-well production is discussed. Because the fault-controlled reservoir has a plate-like structure, the volume of bottom water is smaller than that of layered reservoirs, which results in insufficient reservoir energy. In addition, the fracture-cavity medium deforms because of the changing pressure, which results in rapid reductions in permeability and accelerates the decline in productivity. The formation energy drops by 37%, the single-well productivity loss has reaches 50%, and the annual decline rate is as high as 32–48%. Therefore, accelerating research on methods for supplying formation energy is suggested.

Efficient strategies for supplying formation energy have been developed. The key factors to be determined are what should be injected, how it should be injected, and how much of it should be injected into this new type of thick and plate-like reservoir. Energy supplementation strategies have been investigated on the basis of reservoir simulation research. Water injection is currently the best way to efficiently replenish energy. Compared with those of nitrogen injection and natural gas injection, the effect of water injection is much greater, and the cost is lower. Water injection uses gravity to replenish energy and displace oil downwards. Owing to the large thickness of the oil layer in a fault-controlled reservoir, if the difference between the injection pressure and production pressure is well controlled, lateral water channelling does not occur easily. A "five injection method" is proposed for energy replenishment. The five types of injection are as follows: (1) Early injection: The formation pressure and productivity are maintained starting as early as possible, and water injection is most needed when the pressure drops by 20%. (2) Slow injection: Fast injection, which results in early water channelling, is avoided, and the water injection speed should be controlled at 200–500 cubic meters per day. (3) Low injection: Gravity is used to reduce channelling and increase the sweep volume, and the vertical spacing between injection and production wells should be greater than 100 m. (4) Spacing injection. The influence range of water injection is expanded, and pairing one injection well with multiple production wells, which results in a large injection volume and early water channelling near the production wells, is avoided. (5) Trial injection. The trial injection period is set to 1–6 months, and dynamic detection is performed during the trial injection period to reduce the risk. In addition, an enhanced recovery method for the remaining top oil in the late period that can be produced by gas injection is proposed.

A well-group water injection and energy supplementation plan and a tracking and optimization plan were implemented. After implementation, the pressure recovered by 30%, and the rate of decrease was controlled at 16%. Well-group water injection and energy supplementation strategies were developed (Table 11.1 and Fig. 11.14). The injection–production ratio was 1.16:1, the cumulative oil production was 65 × 10⁴ t, the injection and production parameters were tracked and optimized, and the cumulative water injection was 1.86 million cubic metres.

Table 11.1 Strategies of reservoir energy supplement

Plan	Well 1 (m³/day)	Well 2 (m³/day)	Well 3 (m³/day)	Well 4 (m³/day)	Injection-production ratio	Pressure recovery (MPa)	Cumulative oil production (thousand tons)
1	Water injecting 400	Producing 70	Producing 60	Producing 50	1:1	52	640
2 (recommended)	Water injecting 220	Producing 110	Producing 70	Water Injecting 240	1.16:1	63	650
3	Producing 70	Nitrogen injecting 45,000	Producing 70	Water injecting 350	1.19:1	64	420

The average formation pressure recovered from 31.2 MPa to 40.2 MPa after water injection. The wellbore risks of 11 wells were eliminated, and the annual decline rate of the converted oil production decreased from 48.6% to 16% after water injection; thus, the energy supplementation effect was obvious.

11.5 Development Method for Deep Fractured-Cavity Reservoirs

The Tahe oilfield was the first deep carbonate fracture-cavity reservoir to be successfully developed on a large scale worldwide, with an annual crude oil output of more than 550×10^4 t and a cumulative production of 100 million tons of crude oil. "Spatiotemporal difference" water injection technology and nitrogen injection cave top flooding-enhanced oil recovery technology that is based on the fracture-cavity structure have been developed, which have supported the stable production of the Tahe oilfield for 15 years.

11.5.1 Water Flooding Technology that is Based on the Fracture-Cavity Structure

A water injection development model for fracture-cavity reservoirs was established, and a water injection method of "spatiotemporal difference" was proposed. Spatially, fracture-injection-cavity production, low-injection high-production, and same-layer injection-production occur; in time, early test injection, followed by moderate injection, periodic injection, and later water injection pressure cone, reverses oil displacement, which solves the problem of easy water channelling in large fractures. realize high-efficiency water injection in the early stage of fractured-cavity reservoirs. With a large amount of water injection and a gradual deterioration in the water injection utilization rate, the remaining oil after water flooding is mainly residual oil in the roof of the cave (attic oil), runner shield, corners, and the pore filling. Various water flooding methods have been proposed, such as variable-intensity water injection, periodic water injection, pulse water injection, reversing oil flooding (also known as injection-production reversal), surfactant water injection, and conversion of water injection to gas injection, which results in the formation of a spatial structure well pattern design, along with injection–production parameter optimization technology.

(1) Spatial Structure Well Pattern Design Technology

Well pattern deployment is the key to improving reserve production, productivity, and recovery. The conventional reservoir well patterns are mostly five-point, nine-point, diamond-shaped, and row-shaped. For deep discrete fracture-cavity reservoirs, the use of conventional well pattern deployment results in many nonproductive wells or low-productivity wells, and the degree of reserve control is low. Therefore, changing the strategy and method for well pattern design, adapting to the efficient displacement of water injection of discrete fracture-cavity reservoirs, and establishing a design method for injection-production well patterns with spatial structure are necessary. The specific method involves "determining the locations of oil wells in karst caves, the locations of connected water wells, and the numbers of wells for injection and production of reserves". Wells that are located on large karst caves are designed as oil production wells, and wells on fractured and porous reservoirs are designed as water injection wells. In accordance with the water content of the wells, water injection wells are gradually deployed to achieve fracture-injection-cavity production, low injection and high production, and injection and production in the same layer. The relationship between the injection and production of water injection wells and oil production wells is optimized to improve the control of water flooding in the remaining reserves. A reversing oil well group is designed between the karst caves (exchange between oil production wells and water injection wells). In the specific design, the karst residual mounds and caves, the main fracture-controlled fracture-cave bodies, and the main underground river (the ancient underground river is divided into the main underground river and the branch underground river). Considering the scale of reserves and economic factors, the number of infill wells and the numbers of injection and production wells are determined according to the connection situation, and the relationship between injection and production is established; in the process of designing the well pattern and determining the well position, in consideration of the scale of the reserves and economic factors, the number of infill wells and the numbers of injection and production wells are determined according to the scale of the unit reserves. The demonstration unit of the Tahe oilfield has implemented spatial structure well pattern design technology, and the water flooding control reserve level has reached 89%.

(2) Injection–production parameter optimization technology

Owing to the strong heterogeneity of fractured-cavity reservoirs and the unbalanced flow of injected water, optimizing the amounts of water injection and oil production between injection and production wells to guide the injected water to displace more crude oil is necessary. The design of injection and production volumes is an optimization problem. To maximize the economic benefits, the injection and production parameters of oil and water wells are automatically optimized and solved by a computer so that the water injection volume of the well group or unit is the smallest and the oil production volume is the highest, thereby regulating the production of oil and water wells and improving the development effect; compared with the artificial design method, this method is more accurate, efficient and fast and has better applicability to fracture-cavity reservoirs with strong heterogeneity and complex interwell connectivity.

Three injection–production parameter optimization methods have been developed: The first is an optimization method that is based on an established geological model. On the basis of the established geological model and reservoir numerical model, the water injection or oil produc-

tion of each well is simulated and calculated to optimize the oil production of the well group or unit. The optimization calculation method transforms the control variables to the logarithmic domain of the same order of magnitude to solve for the stochastic disturbance gradient. The second method is a robust optimization method that is based on an uncertain geological model, which is based on multiple conditions that conform to the geological model; it takes production dynamics as the constraint condition and eliminates the geological model if it does not meet the production requirements. Combined with the numerical simulation prediction method, the design ranges of the injection and recovery of various oil and water wells are input into all the models, and the optimal expected value of the economic recoverable volume among all the models is used to establish the optimal injection and recovery plan. The third method is an optimization method that is based on the interwell connectivity model, which is based on the controlled reserves of a single well and the degree of interwell connectivity. In accordance with the theories of injection–production balance and oil–water two-phase front advancement, a new oil–water flow and production prediction calculation model for reservoirs is established to simulate the production process of fractured-cavity reservoirs. On this basis, the optimal injection and recovery scheme is simulated through the objective function of optimal injection and recovery and economically recoverable expected values. Among these methods, the robust optimization method has the strongest anti-risk performance, and the optimization method that is based on the interwell connectivity model is the fastest.

After the implementation of the optimization plan, the cumulative oil production at the well level significantly increased, and the corresponding cumulative water production and cumulative water injection significantly decreased, thus achieving the optimal control objectives of less water injection and more oil production. After the S80 unit was applied, the effects of precipitation and oil increase were obvious. After optimization, the cumulative oil increase was 23.1 $\times 10^4$ t, the comprehensive water content was reduced by 6.8%, the cumulative water production was reduced by 4.5×10^4 m^3, and the water consumption rate was reduced by 41.6%.

11.5.2 Nitrogen Injection and Cave Top Flooding Enhanced Oil Recovery Technology

Through the numerical simulation of nitrogen injection into cave reservoirs, the mechanism of gravity differentiation of nitrogen injection into caves and the displacement of cave top oil was revealed for the first time, and nitrogen injection "cave top flooding" was proposed as an effective method for enhancing oil recovery in fractured-cavity reservoirs. The Tahe oilfield has been developed by nitrogen injection on a large scale, which has become an important enhanced oil recovery method for fractured-cavity reservoirs.

(1) Intelligent Well Selection Technology for Nitrogen Injection in Fractured-Cavity Reservoirs

A nitrogen injection effect evaluation and prediction method, a gas channelling risk assessment calculation, and a gas channelling prediction method were established to realize the organic combination of qualitative analysis and quantitative evaluation and prediction. Taking three karst oil reservoirs, namely, a weathering crust, a palaeo-underground river, and a fault-karst body, as geological targets, the principles and standards of well selection for nitrogen injection in fractured-cavity reservoirs were determined, and a standardized process for quick and intelligent well selection for nitrogen injection in fractured-cavity reservoirs was established. A software platform for intelligent well selection for nitrogen injection in fractured-cavity reservoirs was developed, which solves the problems of long cycles and slow ageing of well selection for nitrogen injection in fractured-cavity reservoirs. The effective rate of gas injection in the oilfield increased from 81 to 90%, and the reserves that were covered by nitrogen injection increased by 32%.

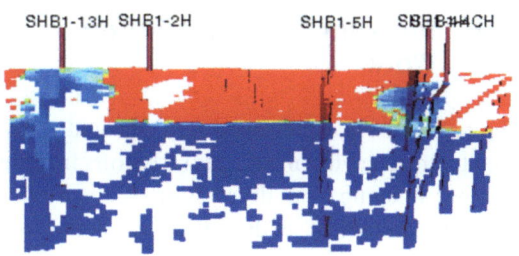

Fig. 11.16 Saturation distribution after water flooding

(2) Nitrogen Injection Rate Optimization Technology for Single Wells and Well Groups

On the basis of the residual oil occurrence pattern after water flooding, a differential injection–production method for nitrogen cave top flooding was established (Fig. 11.16). Weathered crust oil reservoirs are dominated by small karst caves, dissolved vugs, and small-grade fractures, with good plane continuity, and the remaining oil is controlled mainly by the interwell structure. High injection and low recovery can be used to better tap the remaining oil. The fractured-cavity reservoirs in underground river reservoirs develop along the extension direction of the channel, and the remaining oil is mainly closed oil on both sides of the high-conductivity channel. Branch river injection and main channel mining are adopted, and the double-layer channel adopts an injection–production well pattern with high injection and low recovery. The karst caves and karst pores in fault–karst oil reservoirs are distributed around the fault zone, and they are distributed in a band. The development degree of the karst caves from the core to the wing is weakened. The remaining oil is mainly separated oil that was formed by the poor connection of the wings. It is suitable for the injection–production well pattern of wing injection and core production. A gas–water synergistic development method is proposed, and the synergistic effect is best when the downwards displacement of injected gas and the upwards displacement of water injection + bottom water reach a balance.

After the implementation of the gas injection scheme, the amount of oil increased by 176.1×10^4 t, the recoverable reserves increased by 248.4×10^4 t, and the recovery rate increased from 14.91% to 17.07%.

11.6 Problem Areas and Future Work

The problems that are faced by deep and ultra-deep fracture-cavity oil and gas reservoirs include the low imaging and description accuracy for complex geological bodies that are controlled by faults in deep reservoirs, the difficulty in characterizing complex phase states of ultradeep condensate gas reservoirs, and the problem of intelligent and efficient development of such oil and gas reservoirs. Therefore, efficient development technology for deep and ultradeep fracture-cavity reservoirs still needs continuous research, especially in the following main areas:

(1) Fine seismic prediction technology for ultra-deep small fracture-cave bodies;
(2) Fault-controlled reservoir development models and geological knowledge bases;
(3) Pressure-preserving development methods and enhanced oil recovery technologies for ultradeep condensate gas reservoirs;
(4) Real-time optimization of deep oil and gas reservoirs and intelligent oilfield research.

11.7 Summary

China's deep carbonate fracture-cave-type oil and gas resources are abundant and have become important for China's oil and gas exploration and development and for its oil and gas reserves and production. A series of innovative high-efficiency development technologies for deep and ultra-deep fracture-cavity reservoirs were established in this study to support the large-scale development and continuous stable production of Tahe superdeep fracture-cavity oilfields and the major discovery of ultradeep fault-controlled fracture-cave oil and gas reservoirs in Shunbei. This work will be put into production on a large scale. It provides a technical reference and demonstration for the development of deep and ultradeep carbonate reservoirs.

(1) A method for geophysical identification and detailed description of deep fault-controlled fractured-cavity reservoirs was developed, and a "multiple constraint, four-step" geological modelling method was established. The prediction coincidence rate for new wells was 80%.
(2) On the basis of high-temperature and high-pressure strain flow experiments in fracture–cavity media with various structural characteristics, the stress-sensitive characteristics of deep composite media and the coupled flow mechanism of media deformation were revealed.
(3) A mathematical model of multiscale composite media and multifield coupling was established, a fluid flow–stress–temperature coupling numerical simulation method for multiple fractures was deduced, the numerical solution method was optimized, and a numerical simulation method for fault-controlled reservoirs was developed, which resulted in a single-well history matching rate of 87.6%.
(4) The formation of deep karst fracture-cavity reservoirs via "spatiotemporal difference" water injection technology and nitrogen injection cave top flooding enhanced oil recovery technology on the basis of the fracture-cavity structure was achieved, and a "five-injection method" for water injection supplementary energy in ultradeep fault-controlled reservoirs was established to support the stable production of the Tahe oilfield and Shunbei oilfield.

Moreover, the established geological model can help people better understand the control of faults on reservoirs, optimize reservoir exploration and development strategies, and improve reservoir production efficiency and economic benefits.

carbonate fractured-vuggy core models.

Acknowledgements This work was supported by the JointFund of NSFC for Enterprise Innovation and Development (Grants Nos. U24B6001 and U19B6003). The authors would like to sincerely acknowledge the fund for their financial support.

References

Al-Wardy W, Zimmerman RW (2004) Effective stress law for the permeability of clay-rich sandstones. J Geophys Res Solid Earth 109(B4)

Bernabe Y (1986) The effective pressure law for permeability in Chelmsford granite and Barre granite. Int J Rock Mech Min Sci Geomech Abs 23(3):267–275

Biot MA (1941) General theory of three-dimensional consolidation. J Appl Phys 12(2):155–164

Biot MA (1956) Theory of propagation of elastic waves in a fluid-saturated porous solid. I. Low-frequency range. J Acoust Soc Am 28(2):168–178

Coussy O (2004) Poromechanics. Wiley, New York

He ZL, Sun JF, Guo PH et al (2021) Construction method of carbonate reservoir knowledge base and its application in fracture-cavity reservoir geological modeling. Petrol Sci Technol 48(4):710–719

Kang ZJ (2010) A new method for coupled numerical simulation of fracture-cavity carbonate reservoirs. Xinjiang Petrol Geol 31(5):514–516

Kang ZJ, Di Y, Cui SY (2017) Technology and application of numerical simulation of carbonate fracture-cave reservoirs. China University of Petroleum Press, Beijing

Kang ZJ, Lu XB, Zhang Y (2020a) The basic theory of enhanced oil recovery in carbonate fracture-cave reservoirs. China University of Petroleum Press, Beijing

Kang ZJ, Shang GH, Li HK et al (2020b) The inaccuracy of understanding and production optimization theory of carbonate fracture-cave reservoirs in China. China University of Petroleum Press, Beijing

Kang ZH, Lu XB, Yang M et al (2003) Overall reservoir description of ordovician proven reserve areas in Tahe oilfield. Sinopec North-West

Kang ZJ, Zhao YY, Zhang DL (2015) Theory and method of numerical simulation of fracture-cavity carbonate reservoirs. Geo Press

Kong XG, Li DL, Xu XZ et al (2006) Study on the mathematical models of coupled thermal-hydrological-mechanical (THM) processes. J Hydrodyn 20:269–275

Li M, Xiao W (2008) Experimental study on permeability-effective-stress law in low-permeability sandstone reservoir. Chin J Rock Mech Eng 27(2):3535–3540

Li Y, Kang ZJ, Xue ZJ et al (2018) Theories and practices of carbonate reservoirs development in China. Petrol Explor Dev 4:669–678

Mctigue DF (1986) Thermoelastic response of fluid-saturated porous rock. J Geophys Res Solid Earth 91(B9):9533–9542

Robin PYF (1973) Note on effective pressure. J Geoph Res 78(14):2434–3437

Terzaghi K (1923) Die berechnung der durchladdikesitsziffer des tones aus dem verlauf der haydrodynamischen spannungserscheinungen. Sber Akad Wiss Wien 132(2):105

Terzaghi K (1943) Theoretical soil mechanics. Wiley, New York

Walsh JB (1981) Effect of pore pressure and confining pressure on fracture permeability. Int J Rock Mech Min Sci Geomech Abs 18(5):429–435

Wang GJ, Wang GH, Yu GY et al (2002) Geological model of carbonate reservoirs in Tahe oil field. Petrol Sci Technol 29(1):109–111

Wang ZM, Du ZM, Song WJ et al (2004) A reservoir fluid-solid-heat coupling model and its application. J Univ Sci Technol China 34:508–514

Warplnski NR, Teufel LW (1992) Determination of the effective stress law for permeability and deformation in low-permeability rocks. SPE Form Eval 7(02):123–131

Xiao W, Zhao J, Li M et al (2012) Deformation response characteristics of low-permeability and rich-clay sandstones. Rock Soil Mech 33(8):2444–2450

Yang HT, Jiang TW, Yan QB et al (2004) Preliminary study on 3D geological modeling method of fractured-cavity carbonate reservoirs. Pet Geol Oilf Dev Daqing 04:11–17

Yao J, Huang CQ, Liu WZ et al (2018) Key mechanical problems in the development of deep oil and gas reservoirs. Sci China 48(04):5–31

Zhao M, Kang ZH, Liu J (2008) Modeling and application of Fracture-Vuggy carbonate reservoirs. Xinjiang Pet Geol 29(3):318–320

Zienkiewicz OC, Shiomi T (1984) Dynamic behaviour of saturated porous media; the generalized Biot formulation and its numerical solution. Int J Numer Anal Method Geomech 8(1):71–96

Deep Shale Gas Accumulation, Flow Mechanisms and Development Methods in the Sichuan Basin

12

Dongfeng Hu, Xianming Xiao, Weihong Wang, Ruobing Liu, Baojian Shen, Jing Wang, and Tao Yuan

Abstract

Deep shale gas has become a strategic alternative resource in China's oil and gas energy portfolio. The core scientific challenges in the exploration and development of marine deep shale gas in the Sichuan Basin primarily include the formation mechanisms of shale reservoirs, the enrichment patterns of gas reservoirs, and multiphase/multiphysics-coupled seepage mechanisms. This study integrates experimental analysis with theoretical modeling, achieving the following breakthroughs. A CO_2–N_2—high-pressure mercury intrusion joint characterization technique based on density functional theory (DFT) was developed, enabling accurate quantitative characterization of full-aperture pores in deep shale. Nuclear magnetic resonance (NMR) T_2 spectrum distribution technology was used to quantitatively analyze organic pores, inorganic pores, and microfractures in shale. Quantitative pore–fracture characterization revealed that the coupled effects of quartz compression-resistant pore preservation and reservoir fluid overpressure are key to the high porosity of marine deep shale in the Sichuan Basin. A novel paleotherm/pressure evaluation method was innovatively developed using laser Raman spectroscopy, fluid inclusion homogenization temperature measurements, and Sm–Nd dating, enhancing the precision of paleothermobaric reconstruction during different uplift stages. Through laboratory experiments on shale adsorption–desorption and molecular dynamics simulations, a new method for evaluating gas content was established. Integrating geological characteristics, the adsorption–desorption behavior and main controlling factors of enrichment and accumulation in deep shale were clarified, confirming that free gas is dominant in deep shale gas reservoirs. By comprehensively considering adsorption/desorption, stress sensitivity, microscale flow, surface diffusion, and fracturing fluid effects, a multiphase/multimechanism seepage model was established, enabling dynamic predictions of gas–water two-phase transient productivity in horizontal wells, characterization of complex

D. Hu (✉) · W. Wang · R. Liu · B. Shen · T. Yuan
State Key Laboratory of Shale Oil and Gas Enrichment Mechanisms and Efficient Development, SINOPEC, Beijing, China
e-mail: hudf.ktnf@sinopec.com

D. Hu · R. Liu · T. Yuan
Sinopec Exploration Company, Chengdu, Sichuan, China

X. Xiao
School of Energy Resources, China University of Geosciences (Beijing), Beijing, China

W. Wang · B. Shen
Exploration and Production Reservoir Institute, Key Laboratory for Marine Oil and Gas Exploitation, Sinopec, Beijing, China

J. Wang
State Key Laboratory of Petroleum Resources and Prospecting, China University of Petroleum (Beijing), Beijing, China

© The Author(s) 2026
Y. Ma et al. (eds.), *Petroleum Geology and Exploration of Deep Marine Strata in China*, Advances in Oil and Gas Exploration & Production, https://doi.org/10.1007/978-3-032-02496-1_12

fracture network distributions, and description of fracture parameter evolution during production. These technologies have played a crucial role in facilitating breakthroughs in the exploration of deep marine shale gas.

Keywords

Marine · Deep shale gas · Accumulation · Development mechanism

12.1 Formation Mechanism of Deep Effective Shale Reservoirs and the Preservation of Shale Gas

12.1.1 Quantitative Characterization of Pores in Deep Shale Reservoirs

There are more than ten methods for characterizing the pore structure of shale reservoirs, among which high-pressure mercury intrusion, gas absorption and nuclear magnetic resonance are commonly used for the quantitative characterization of pore structure. However, these methods are suitable for different test ranges and applicable pore scales; it is difficult for a single method to characterize the pore structure of shale reservoirs fully and effectively. Therefore, multiple methods should be used to characterize the distribution of all pores in shale reservoirs in a comprehensive study. Importantly, different methods are based on different principles and methods, and accurate connections between the characterization results of different methods have become a critical issue.

Studies have shown that organic-rich shales have multiple types of pores, such as micropores, mesopores and macropores (> 50 nm), and are typical composite materials ranging from micropores to macropores, with various pore shapes and complex media. Gas adsorption is a preferable method for the comprehensive characterization of porous materials and can provide important information about specific surfaces, pore distributions and porosities. The most common

characterization methods are nitrogen adsorption and carbon dioxide adsorption. Nitrogen adsorption is a conventional method for characterizing medium pores and macropores and is mainly used to calculate the pore size distribution by the BJH model established in the study of cylindrical pores. However, when used for the characterization of shales with multilevel pores and medium-sized pores with uneven distributions, great errors occur. The density functional theory method uses the mean field method to approximately calculate the residual free energy attraction, which can accurately describe the fluid structure limited by simple geometric shapes such as narrow, cylindrical and spherical pores. This method provides not only a microscopic model of adsorption but also more realistically reflects the thermodynamic properties of fluids in pores and has been developed as an effective method to describe the adsorption and phase behaviors of heterogeneous fluids confined by multiple pore materials. In particular, the quenching solid density functional theory (QSDFT), which has been continuously improved in recent years, can accurately describe the pore size distribution of disordered micromedium porous materials by further considering the lattice distribution and surface roughness of the material. Its greatest advantage is that it can be used for the analysis of micropores and mesopores (Landers et al. 2013; Robert et al. 2000; Ustinov and Do 2004; Ustinov et al. 2006).

Using the density functional theory method to process the results of nitrogen adsorption and carbon dioxide adsorption tests, the hybrid model for calculating porous type can be used to characterize pores over a wider range of diameters; furthermore, two gas adsorption techniques can be used within the framework of the same theory to process the results, and the connection among the results obtained using different model methods can be easily solved.

Mercury injection can be used to measure the macropores of deep shales, nitrogen adsorption for medium-pore determination and carbon dioxide adsorption for micropores; the nitrogen adsorption and carbon dioxide adsorption results can then be processed according to density functional theory to precisely characterize pores below

Fig. 12.1 Distribution characteristics of pores of all sizes in deep shales of the Pushun 1 Well

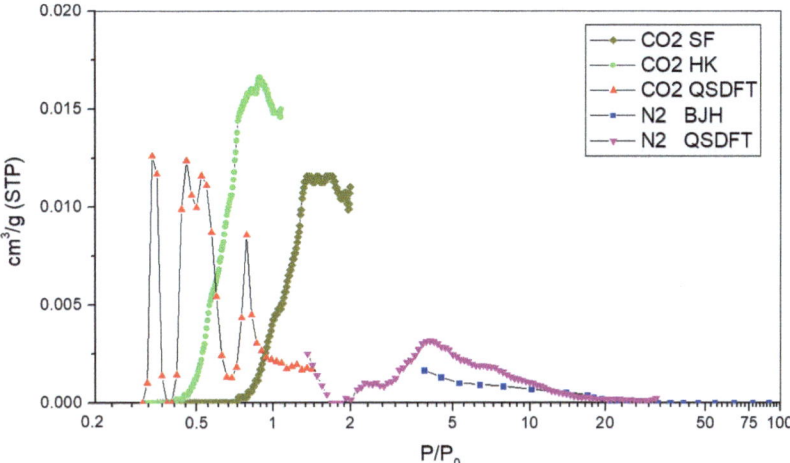

2 nm in size and within the range of 2–50 and to achieve an effective connection of the two results. Moreover, by adopting the connection points determined by the first-order differential point of the pore volume obtained by nitrogen adsorption and high-pressure mercury intrusion, a comprehensive quantitative characterization and analysis technology suitable for the full-pores of deep shales is developed (Fig. 12.1).

Studies on the deep shales in Pushun Well 1, Dongyeshen Well 1, Dingye Well 8 and Leba Well 1 (in the Dalong Formation) in the Sichuan Basin revealed that the numbers of micropores in the deep shales characterized using this newly established method are nearly greater than those characterized by BJH and nitrogen adsorption. These results, combined with electron microscopy results, further prove that deep shales have more micropores and that micropores cover a greater proportion of the total pore area.

12.1.2 Pore Development and Preservation Mechanism

(1) Planktons in deep-water shelves and their solid bitumen are the main sources for the development of organic pores.

Organic pores develop mainly in type I and type II kerogen, whereas organic pores generally did not develop in type III kerogen (Curtis et al. 2012; Jarvie et al. 2007; Loucks et al. 2012; Löhr et al. 2015). However, in the high evolution stage, most of the kerogen will undergo strong compaction and physicochemical changes and will be mixed with the residual oil/bitumen, making it difficult to distinguish different kerogen types and bitumen. With the help of advanced experimental analysis techniques such as high-resolution scanning electron microscopy and atomic force microscopy and the theory of hydrocarbon generation from organic matter, for some organic matter particles with typical characteristics, these pores can be identified as asphalt organic matter pores, acritarch organic matter pores, algae–bacterial organic matter pores and graptolite organic matter pores according to the organic matter shape, internal structure and distribution characteristics.

1. Asphalt organic matter pores

Pores commonly developed in asphalt organic matter can be divided into two categories according to pore characteristics. One type is an irregularly shaped pore, which is uniformly distributed throughout the organic matter, with a high pore density; pore diameters are substantially smaller than 50 nm in diameter and have irregular shapes. Conversely, bubble-like pores are close to circular or elliptical in shape. The pore sizes can vary from 10 to 100 nm, whereas pores of different sizes often coexist. In individual asphalts, very large pores can be found in bitumen, with diameters ranging from hundreds of nanometers or even up to 1 μm.

2. Acritarch organic matter pores

The micropores in this type of organic matter develop unevenly, and the levels of development of pores in the cores and peripheral organic matter differ markedly. Most of the pores do not develop or develop poorly in the core portion, while the pore density in the peripheral organic matter is relatively greater and the pore size is essentially larger than 20 nm, indicating that there is a marked difference in the chemical properties between the peripheral organic matter and core organic matter. Spherical organic matter can be seen in multiple well samples, which are spherical, ellipsoid or hemispherical, with diameters ranging from 2 μm to several times larger, with clear edge contours, mostly double-layered structures, and scattered phosphate minerals in the inner part of the organic matter.

3. Organic pores of algae fragments

There are two main types of pores associated with algae fragments. One type takes the shape of irregular and amorphous pieces; the diameter of the pores is greater than 20 μm and can reach 100 μm, and the periphery of the pores is often surrounded by minerals such as quartz. Fiber-like clay minerals can be found in the inner of the organic matter. Most of the pores in this type of organic matter are relatively developed; the pores are irregularly angular, the pore size is relatively uniform, and the surface porosity is usually high. The other type of pore has a relatively clean interior without clays, and the overall outline is clear. The organic matter generating these types of pores is amorphous and has polygonal or arc shapes, with pore sizes less than 20 μm. The pores in the organic matter are abundant, and the surface porosity is high. The pore shapes are irregular and angular, with a uniform size and good connectivity.

4. Organic pores of bacterium-like aggregation

Pores in this type of organic matter are all interspherulite pores, with irregular and angular shapes and sizes ranging from 10 to 50 nm. Owing to compaction of the aggregations of spherulites, some pores are retained. In this case, the original shapes of the spherulites are difficult to identify.

(2) The mechanism of anti compression by quartz and pore preservation is the basis for the development of pores in deep high-quality shale reservoirs.

Biogenic siliceous minerals exist in the form of amorphous opal, a type of disordered hydrous amorphous siliceous substance with a nearly amorphous structure. Biosilica (Opal A/CT) is continuously dehydrated through diagenesis and transformed into rigid crystalline quartz (Aplin and Macquaker 2011; Mastalerz et al. 2013; Milliken et al. 2013, 2016). Dehydration and recrystallization of biological opal A occur in the early stage; at the early stage of diagenesis, the transformation from quasicrystalline opal CT to crystalline quartz is completed; thus, siliceous shales developed, and the later diagenetic transformation was weak (Fig. 12.2). Siliceous organisms produce planktonic forms with abundant pores, which settle into the sediments after death, resulting in loose and porous sediments with high primary porosity.

The hardness of the siliceous minerals increases as the opals continue to transform into quartz crystals via recrystallization, and the hardness of the siliceous shales, which are composed of a large number of quartz grains, also continues to increase, as well as their ability to resist compaction and the formation of a rigid framework. The compaction resistance of biogenic siliceous shales is greatly improved, and many primary pores are well preserved. As the evolution continues, the liquid hydrocarbons begin to split, and organic pores develop. At this point, the biogenic siliceous material has completed its transformation into quartz crystals, and the resis-

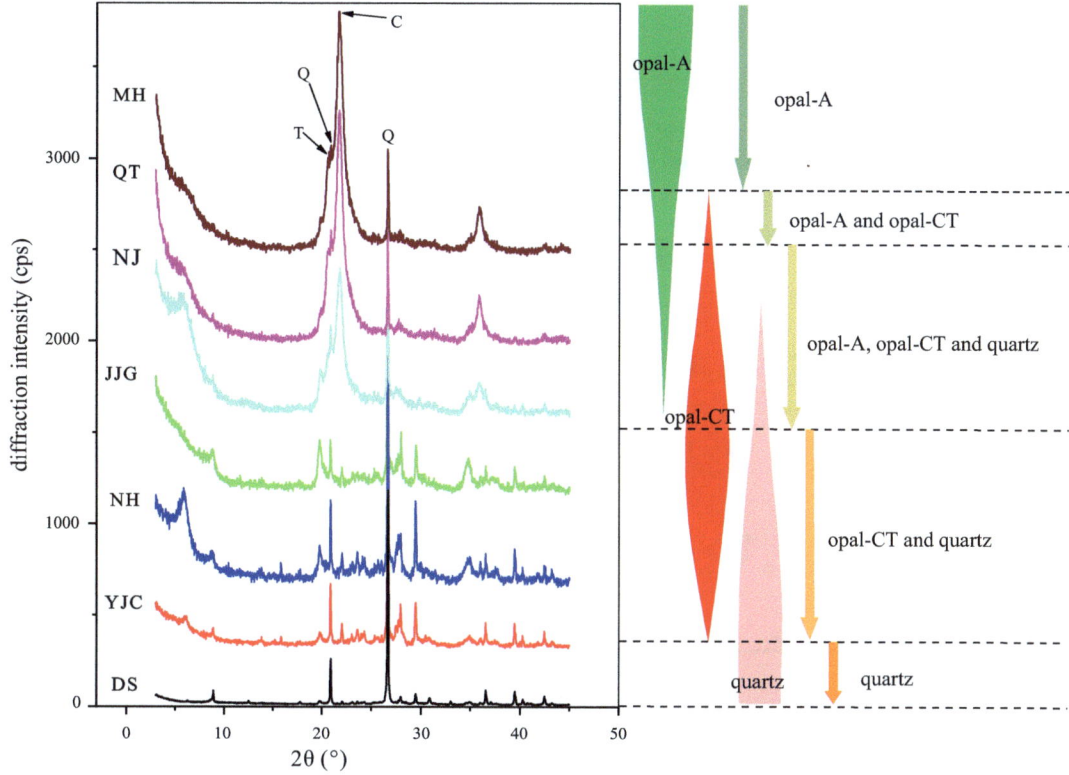

Fig. 12.2 X-ray diffraction pattern and diagenetic model of the biogenic opal shale

tance to compaction of the shales is increased, which is beneficial for the preservation of plastic organic pores. Thus, the pores are effectively preserved under the circumstances of strong structural movements in the late stage (Fig. 12.2).

In the early diagenetic stage of biogenic siliceous shales, mechanical compaction and chemical compaction (solution by compression) occurred almost simultaneously and strongly affected the physical properties of the shale. Although a considerable number of shale pores are lost and reduced, the hardness of the shales increases, and the ability to resist compaction increased; thus, the loss rates of primary pores in the middle and later stages of diagenesis decrease, and effective protection for secondary organic pores is provided (Fig. 12.2). During the early diagenetic process, biogenic siliceous shales are characterized by rapid diagenesis, which is important for maintaining high porosity in middle and late diagenesis.

(3) Overpressure of deep shale reservoirs is preferable for the preservation of organic pores.

The deep shale structures of the Wufeng Formation–Longmaxi Formation in the southern Sichuan Basin exhibit weak deformation and are generally characterized by overpressure. Overpressure clearly protected the development and maintenance of shale pores and could offset the mechanical compaction of the shale reservoirs caused by the overlying strata, preserving the plastic organic pores that formed. The higher fluid pressure in the pores could slow the compaction of plastic pores. When the preservation conditions become poor, the escape of shale gas leads to a in fluid pressure in shale pores, and the preservation conditions of organic pores become poor. Under the conditions of similar TOCs in deep-quality shale layers, the porosity of the shale gas intervals with overpressure coefficients significantly

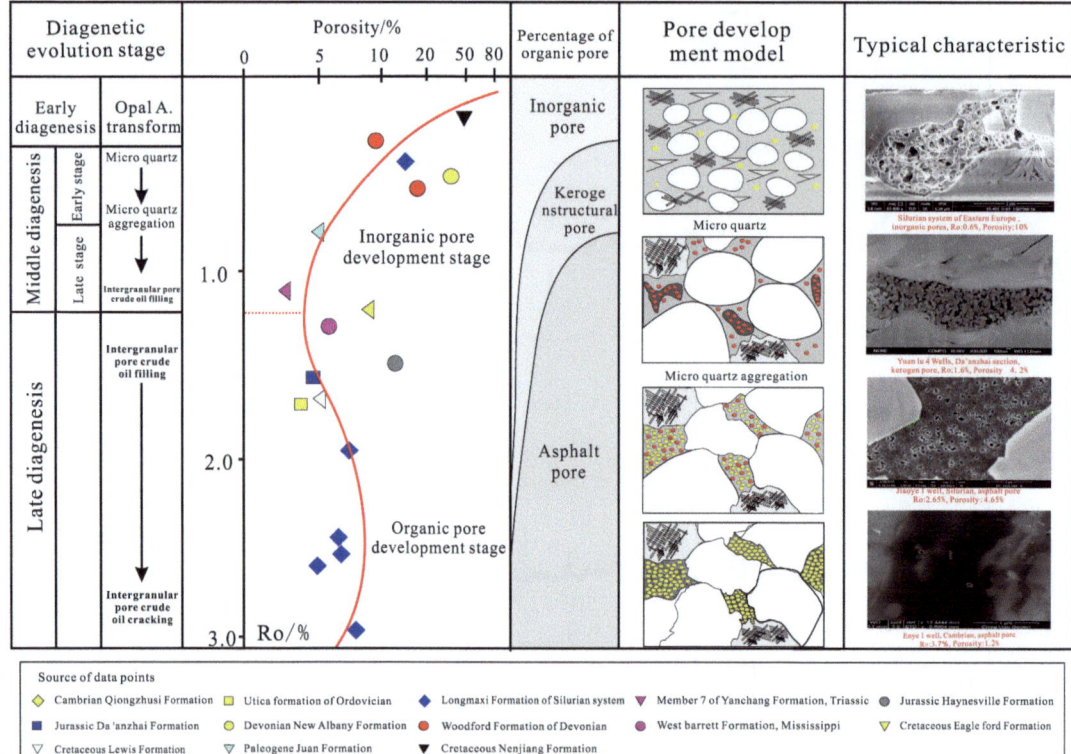

Fig. 12.3 Evolution model diagram of pore development and retention in marine deep shale gas

increased (Fig. 12.3). Wells with poor preservation conditions also have lower pressure coefficients and porosity. Scanning electron microscopy reveals that wells with small coefficients have worse pore development and significantly smaller pore diameters than wells with high overpressure coefficients do. This further confirms that the organic pores cannot be well preserved without overpressure. Therefore, the overpressure fluid conditions of shale reservoirs are crucial to the preservation of pores in deep shale reservoirs.

12.2 Storage Mechanism and Variation Laws of Deep Shale Gas

The gas present in deep marine shales exists mainly in adsorbed and free states; however, the phase state varies widely (Curtis 2002; Kong et al. 2014). The development of shale gas is a dynamic process from the stage of free gas release to adsorbed gas desorption to free gas release. The content of adsorbed gas has an important impact on the development of shale gas wells and their long-term and stable production (Jia et al. 2012; Zou et al. 2012). Therefore, studies on the adsorption, adsorbed gas content, phase state transition processes and laws of variation for different types of shale play important roles in the analysis of shale gas storage mechanisms.

12.2.1 Storage Characteristics and Phase State Variation Laws of Deep Shale Gas

(1) Storage characteristics of deep shale gas

High-quality shale gas layers on deep shelves are generally characterized by ultrahigh pressure. Under conditions of good preservation and high fluid pressure, the adsorption capacity of shale gas layers decreases, the proportion of free gas increases, and the amount of free gas becomes more enriched as the depth and temperature increase (Hu et al. 2022).

1. Gas-bearing characteristics of shales in different structural parts

In the later stage of the sedimentary basin, the effects of uplift are weak, the amount of denudation is low, and the preservation conditions are good with overpressure. Research on the genesis and the evolution of overpressure in shale gas reservoirs reveals that the quality of shale gas layers in deep shelves is generally characterized by high gas contents and abnormally high pressure at the moment of maximum burial depth before the beginning of continuous uplift. In the early and deep stages of the Wufeng–Longmaxi Formation, favorable sealing conditions in the roof and floor plates, combined with high total organic carbon (TOC), mixed sapropel and para-sapropel kerogens and moderate thermal evolution, promote the retention and enrichment of hydrocarbons generated in the shale reservoirs, which is favorable for the formation of overpressure and high gas contents within the reservoirs. The tectonic uplift in the southern Sichuan Basin was weak. Although the shale gas that formed during the early stage was lost during the process of uplift and denudation, the characteristics of high gas contents and overpressure were maintained. However, shale gas can easily escape, and the resulting formation pressure is released in the strongly deformed area on the edge or outside of a basin. As discussed, the pressure coefficient of Well YZ1 is less than 1.0, and the average gas contents of the quality shale intervals are less than 0.5 m^3/t, which indicates that the deep shale has extremely low gas contents in areas with poor preservation conditions.

2. Storage characteristics of deep shale gas

In cases of high pressure, deep shale gas is dominated mainly by free gas, which is beneficial for production. The variations in adsorbed gas and free gas at different depths and under different pressure coefficients are simulated by using samples from the Longmaxi Formation in the JY1 well. The results show that the adsorbed gas contents of the Wufeng–Longmaxi Formation shales first increase with increasing depth but clearly decrease when the depth exceeds 1000 m; however, the free gas contents increase with increasing depth and pressure coefficient. These findings indicate that the deeper the depth is, the greater the degree of gas enrichment, and the greater the pressure coefficient is, the greater the free gas content becomes.

On the basis of the understanding above, the quantitative characterizations of adsorbed gas and free gas are carried out in overpressure areas, such as Jiaoshiba, Dingshan and Dongxi. The results show that free gas accounts for 68% and adsorbed gas accounts for 32% in Jiaoshiba; free gas accounts for 81% and adsorbed gas accounts for 19% in Dingshan; and free gas accounts for 84% and adsorbed gas accounts for 16% in Dongxi. Although shale gas is dominated by free gas in these areas, the contents of free gas in the Jiaoshiba overpressure area are generally lower than those in Dingshan and Dongxi due to the shallow depth and low formation temperature.

(2) Phase state variation of adsorbed gas and free gas

On the basis of the understanding of the characteristics of gas-bearing shales with different mineral compositions and different pore types, a theoretical prediction model based on molecular simulation is established to study the transformation of gas from the adsorbed state to the free state under geological temperature and pressure conditions (the model is composed of an accurate van der Waals molecular force field and a shale pore atomic-scale model).

1. Laws of transformation of shale gas in organic matter pores

A comparison reveals that with increasing depth, the proportion of methane adsorbed in all pore sizes gradually decreases, whereas the proportion of free methane gradually increases. When the pore size is less than 2 nm, the proportion of methane adsorbed is significantly greater than

Fig. 12.4 Modes of free and adsorbed methane in pores of different sizes in kerogen I

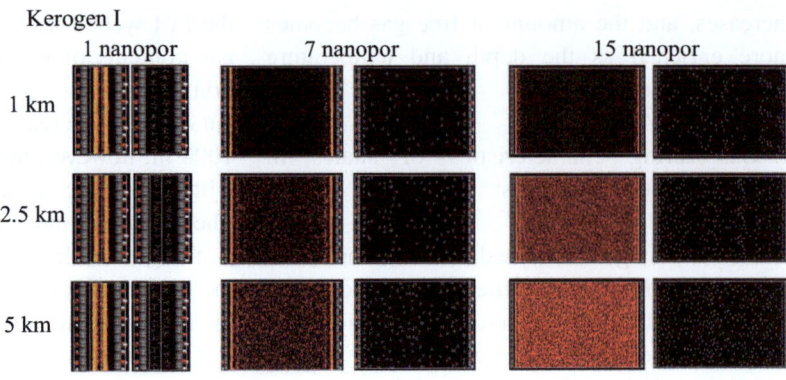

that of free methane, and the overall proportion of the adsorbed state is greater than 80%. When the pore size is greater than 2 nm, the large pore space no longer affects the limiting factor for the transformation from free methane to enter the pore, and the free methane starts to increase significantly. The increase in temperature accelerates the transformation from adsorbed gas to free gas. The proportion of adsorbed gas begins to decrease rapidly. The van der Waals force decreases faster with increasing intermolecular distance, and the temperature increases with increasing depth, resulting in very weak interactions between the structure of the second adsorption layer and the pore wall. In addition to the first adsorption layer, the second adsorption layer can also be observed. Therefore, the methane molecules adsorbed in the second layer cannot be considered to be in the adsorbed state. Moreover, with increasing depth, the amount of free methane in the pores significantly increased (Fig. 12.4).

2. Laws of transformation of shale gas in inorganic pores

Studies on the transformation of shale gas from the adsorption state to the free state in montmorillonites, illites, kaolinites and quartz with increasing depth under geological temperature and pressure reveal that the proportion gas in the free state and gas in the adsorption state in the three kinds of clay minerals is essentially consistent with that in organic matter. The conversion between free gas and adsorbed gas is affected by the pore diameter and the geological temper-

ature and pressure.; When the pore size is less than 2 nm, the amount of adsorbed gas is greater than that of free gas; as the pore size increases, the proportion of gas adsorbed begins to decrease rapidly with increasing depth, and the proportion of free gas increases rapidly. At a certain depth, the proportion of adsorbed gas begins to increase slowly, and the proportion of free gas begins to decrease; however, the free gas content is still higher than that of adsorbed gas. In addition, at depths almost all the methane is free in the pores of quartz minerals, which are predominantly in large pores. When the depth increases to 3 km, the adsorbed gas begins to gradually increase because of the high pressure of the formation. The interaction between quartz and methane is weak, and the adsorbed gas is released rapidly with decreasing pressure during the mining process (Fig. 12.5).

12.2.2 Dynamic Simulation of Gas Contents in Deep Shales

On the basis of methane adsorption isotherm experiments and adsorption potential theory, a mathematical geological model of gas contents in deep shale is established to simulate the dynamic evolution of gas contents in deep shale. The dynamic simulation evolution process includes two main parts. First, nonlinear fitting of the experimental data is performed to obtain important parameters for the conversion of the excess adsorption capacity, the data into absolute adsorption capacity; second, in combination with the Gibbs equation and the supercritical SDR model,

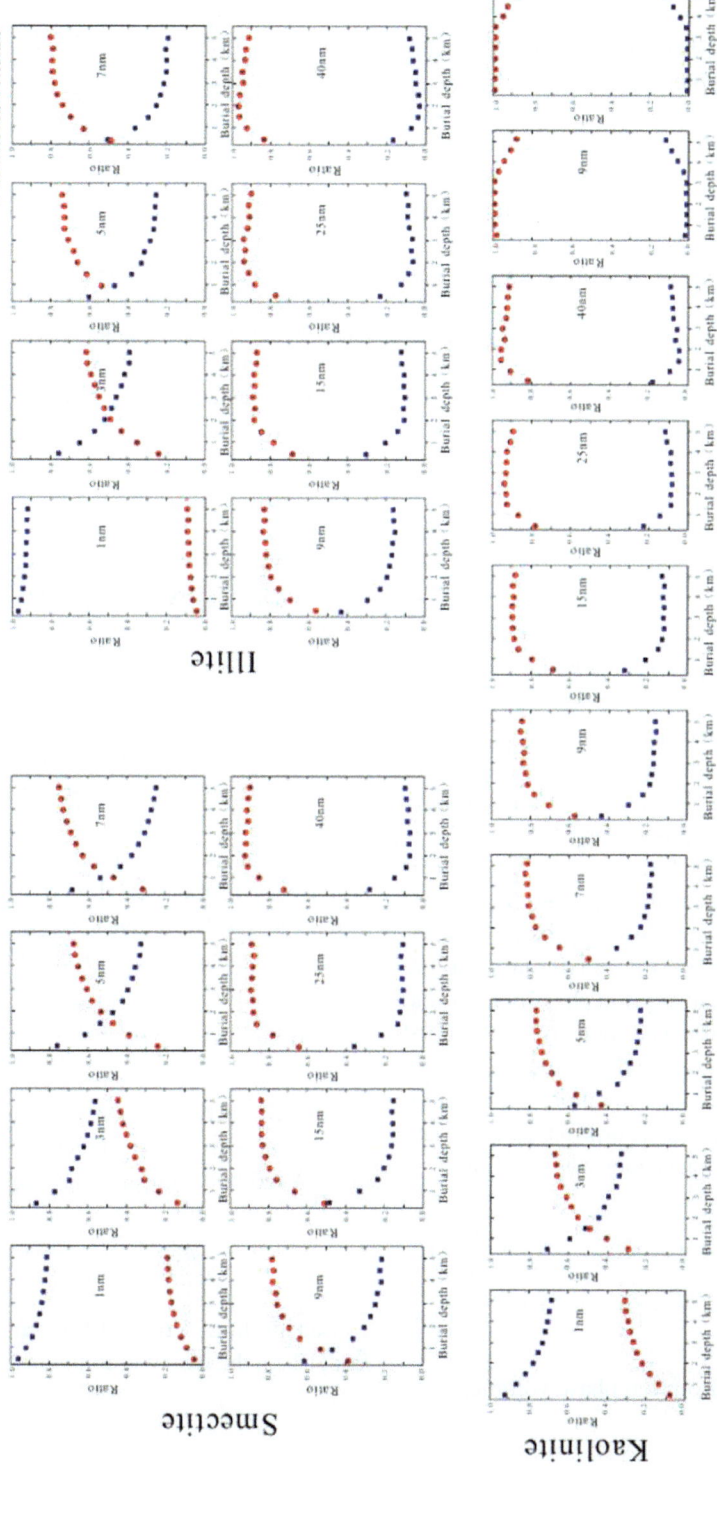

Fig. 12.5 Laws of transformation from adsorbed gas to free gas in montmorillonite, illite, kaolinite and quartz with increasing depth

mathematical calculation models for the adsorbed gas, free gas and total gas contents are derived.

(1) Calculation models of free gas, adsorbed gas and total gas contents

When converting the excess adsorption capacity into absolute adsorption capacity, the adsorption phase density and the maximum absolute adsorption capacity are the key parameters to be determined. In this study, the SDR adsorption potential theory and the Gibbs equation are combined to obtain the formula for calculating the excess SDR adsorption capacity:

$$SC_{excess, adsorbed\ gas} = SC_{\infty, adsorbed\ gas}$$
$$\times \exp\left\{-D \times \left[\ln\left(\frac{\rho_{abs}}{\rho_{abs}}\right) \times R \times T\right]^2\right\}$$
$$\times \left(1 - \frac{\rho_{free}}{\rho_{ads}}\right) \tag{12.1}$$

where $SC_{excess, adsorbed\ gas}$ is the excess adsorption capacity, m^3/t; $SC_{\infty, adsorbed\ gas}$ is the maximum absolute adsorption capacity, m^3/t; ρ_{free} and ρ_{ads} are the densities of free and adsorbed gas, respectively, kg/m^3; R is the universal gas constant, 8.314×10^{-3} kJ/(mol K); T is the temperature, K; and D is a constant, mol^2/kJ.

In accordance with formula (12.1), nonlinear fitting is performed using the experimental data for methane isotherm adsorption (i.e., excess adsorption capacity) to obtain the maximum absolute adsorption capacity, adsorption phase density and constant D; the free phase density is obtained according to temperature and pressure.

According to the Gibbs equation, the relationship between the excess adsorption capacity and the absolute adsorption capacity is as follows:

$$SC_{excess, adsorbed\ gas} = SC_{adsorbed\ gas} \times \left(1 - \frac{\rho_{free}}{\rho_{ads}}\right)$$
$$= SC_{adsorbed\ gas} - \rho_{free} \times V_{ads} \tag{12.2}$$

where V_{ads} is the adsorption phase volume.

Both formula (12.1) and formula (12.2) can be used to convert the excess adsorption capacity into the absolute adsorption capacity. The resulting calculation formula is as follows:

$$SC_{adsorbed\ gas} = SC_{\infty, adsorbed\ gas}$$
$$\times \exp\left\{-D.\left[\ln\left(\frac{\rho_{abs}}{\rho_{abs}}\right) \times R \times T\right]^2\right\} \tag{12.3}$$

Shale gas occurs mainly in the free and adsorbed phases of shale reservoirs, so the total gas content ($SC_{total\ gas}$) is the sum of the adsorbed gas content ($SC_{adsorbed\ gas}$) and the free gas content ($SC_{free\ gas}$), namely:

$$SC_{total\ gas} = SC_{adsorbed\ gas} + SC_{free\ gas} \tag{12.4}$$

Combining the above formulas, the total gas content can be expressed as follows:

$$SC_{total\ gas} = SC_{excess, adsorbed\ gas} + \rho_{free} \times V \tag{12.5}$$

where V is the sum of the volume of the adsorbed phase and the volume of the free phase per unit mass of rock, that is, the total pore volume per unit mass of rock. The total pore volume of rock is not easy to measure directly, but it is related to the rock skeleton density ($\rho_{skeletal}$, kg/m^3) and bulk density (ρ_{bulk}, kg/m^3) as follows:

$$V = \frac{1}{\rho_{bulk}} - \frac{1}{\rho_{skeletal}} \tag{12.6}$$

In addition, rock porosity (Φ) has the following relationships with rock skeleton density ($\rho_{skeletal}$, kg/m^3) and bulk density (ρ_{bulk}, kg/m^3):

$$\varphi = 1 - \frac{\rho_{bulk}}{\rho_{skeletal}} \tag{12.7}$$

Substituting Eqs. (12.6) and (12.7) into Eq. (12.5), the total gas contents expressed by the rock skeleton density ($\rho_{skeletal}$, kg/m^3) and rock porosity (Φ) can be obtained as follows:

$$SC_{total\ gas} = SC_{excess,\ adsorbed\ gas}$$
$$+ \rho_{free} \times \frac{\Phi}{\rho_{skeletal} \times (1 - \Phi)} \quad (12.8)$$

The above mathematical functions can be used to quickly and accurately fit the experimental methane isotherm adsorption data and create a necessary premise for converting the excess adsorption capacity into absolute adsorption capacity. However, this approach is invalid for the simulation of gas content in an actual geological environment. Therefore, no water is present under the experimental conditions, and the water saturation and the water balance/dry adsorption capacity ratio need to be considered in the mathematical model to more accurately simulate the gas contents under geological conditions.

The percentage of shale adsorption capacity reduction (k) and water saturation (Sw) are introduced, and the supercritical SDR model is corrected to obtain the excess adsorption, absolute adsorption, and free gas and total gas contents in the shale reservoirs. The mathematical function expressions are as follows:

Excess adsorption capacity:

$$SC_{excess,\ adsorbed\ gas} = SC_{\infty,\ adsorbed\ gas}^T$$
$$\times \exp\left\{-D \times \left[\ln\left(\frac{\rho_{abs}^T}{\rho_{free}^{P,T}}\right) \times R \times T\right]^2\right\}$$
$$\times \left(1 - \frac{\rho_{free}^{P,T}}{\rho_{abs}^T}\right) \times k \quad (12.9)$$

Absolute adsorption capacity:

$$SC_{adsorbed\ gas} = SC_{\infty,\ adsorbed\ gas}^T$$
$$\times \exp\left\{-D \times \left[\ln\left(\frac{\rho_{abs}^T}{\rho_{free}^{P,T}}\right) \times R \times T\right]^2\right\} \times k \quad (12.10)$$

Free gas:

$$SC_{free\ gas} = SC_{total\ gas} - SC_{adsorbed\ gas} \quad (12.11)$$

Total gas:

$$SC_{total\ gas} = SC_{excess,\ adsorbed\ gas}$$
$$+ \rho_{free}^{P,T} \times \frac{\Phi \times (1 - S_W)}{\rho_{skeletal} \times [1 - \Phi]} \quad (12.12)$$

The mathematical functions established above for the excess adsorption capacity (Eq. 12.9), absolute adsorption capacity (Eq. 12.10), free gas amount (Eq. 12.11) and total gas content (Eq. 12.12) are important for revealing the transformation of free gas into adsorbed gas and obtaining shale gas contents with variations in paleotemperature and pressure;. Later studies are based on the results above.

(2) Variation in shale gas content with paleotemperature and pressure during the uplifting stage

With respect to geological history, temperature and pressure change significantly, which affects the gas content and gas storage state. According to Eqs. (12.9), (12.10) and (12.12), temperature and pressure are important parameters for recovering the shale gas content during the uplifting stage. The relationships among temperature, pressure, maximum theoretical adsorption capacity and temperature as well as the relationship between the adsorption phase density and temperature in paleo-buried shales are key to the recovery of gas contents in shales during the uplifting stage. Information on the paleotemperature and paleopressure of shales can be obtained from studies on fluid inclusions captured by fractured shale veins by thermodynamics. On the basis of the relationships between the maximum theoretical adsorption capacity, adsorption phase density and temperature, methane adsorption experiments under different temperature conditions are performed to obtain the maximum theoretical adsorption capacity and adsorption phase density, and an equation is obtained to describe the linear relationship equation.

Dynamic simulation of the evolution of shale gas contents is conducted using samples selected from the First Member of the Wufeng–Longmaxi Formation in the Dingshan area (DY5 well) and Dongxi area (DYS1 well) in the south-

Table 12.1 Comparisons of key fitting parameters of SDR formula for isothermal adsorption line of high pressure methane from shales in Wufeng-Longmaxi Formation

Well	Fitting parameters	Temperature (°C)				Linear relationship with temperature
		30	50	70	90	
Well DY5	m_{max} (m³/t)	3.53	3.15	2.90	2.84	$-0.0116*T + 3.8026$ ($R^2 = 0.92$)
	D (mol²/kJ²)	0.01	0.01	0.01	0.01	0.0097125 (average)
	ρ_{ads} (kg/m³)	0.41	0.41	0.38	0.37	$-0.0008*T + 0.4389$ ($R^2 = 0.89$)
Well DYS1	m_{max} (m³/t)	2.534	2.195	3.049	2.175	$-0.0011 \times T + 2.86$ ($R^2 = 0.92$)
	D (mol²/kJ²)	0.0112	0.0102	0.0151	0.0110	0.0119 (average)
	ρ_{ads} (kg/m³)	421	421	421	421	421

eastern Sichuan Basin. On the basis of the experimental data for methane isothermal adsorption and the mathematical equation for excess adsorption derived using the SDR model (Eq. (12.2)), the parameters (mmax, D, ρads) are obtained by fitting, and then a linear equation for the relationships between temperature and pressure is established (Table 12.1).

On the basis of the previous results and the logging and experimental analysis data from the area studied, parameters such as the surface temperature, formation pressure coefficient, water balance/adsorbed gas reduction ratio under dry conditions and water saturation (Table 12.2) are substituted into formulas (12.9), (12.10) and (12.12).

After the linear relationships between the maximum theoretical adsorption capacity and the density of the adsorbed phase and temperature (Table 12.1) and the information on each param-

Table 12.2 Parameters required for gas contents calculation of Well DY5 and Well DYS1

Parameter	Well DY5	Well DYS1
Formation pressure coefficient (PC)	1.47	1.85
Porosity (%)	4.78	6.05
Grain density (g/cm³)	2.48	2.46
Adsorption gas reduction ratio under water balance/drying conditions (K) (%)	50	50
Water saturation (S_w) (%)	35.8	37.4

eter (Table 12.2) are obtained, these values are substituted into mathematical models of absolute adsorption capacity, free gas content and total gas content to simulate the evolution of shale gas content in the first member of the Wufeng–Longmaxi Formations in wells DY5 and DYS1 in the southeastern Sichuan Basin. The simulation results are shown in Fig. 12.6.

During the course of uplift, the total gas and free gas contents of the first member of the Wufeng–Longmaxi Formations in Well DY5 gradually decrease, and the adsorbed gas contents increase slightly compared with those at the maximum depth (Fig. 12.6). The total gas content and free gas content decrease from the values of 8.58 m³/t and 7.99 m³/t to 6.14 m³/t and 4.84 m³/t, respectively, and the adsorbed gas content increases from 0.58 m³/t to 1.30 m³/t. Well DYS1 and Well DY5 have similar characteristics in terms of the overall evolution trend of gas contents, with the total gas and free gas contents gradually decreasing and the adsorbed gas contents increasing (Fig. 12.6). The total gas content and free gas content decrease from 8.16 m³/t to 7.23 m³/t, respectively, at the maximum depth to 6.04 m³/t and 4.21 m³/t, respectively, at present, and the adsorbed gas content increases from 0.93 m³/t to 1.83 m³/t. In addition, there is a certain correlation between the dynamic evolution of gas contents and the tectonic cycle, with three stages evident in the gas dissipation rate. The total gas content and free gas loss rates are fast in the middle–late Yanshan period, slow in the late

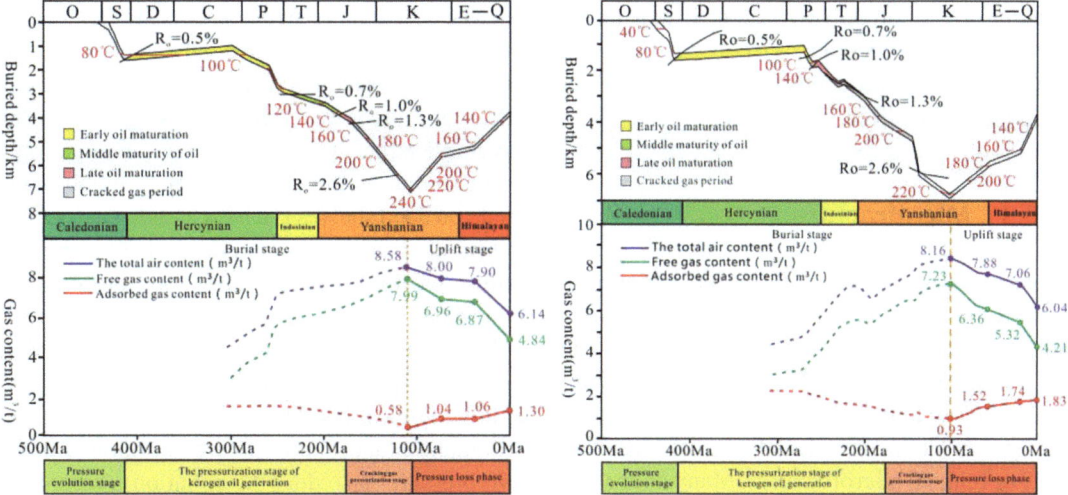

Fig. 12.6 Simulation of the dynamic evolution of shale gas content in the 1st member of the Wufeng–Longmaxi formation in Wells DY5 (left) and DYS1 (right)

Yanshan period to the middle Himalayan period, and accelerate in the middle–late Himalayan period.

12.3 Main Factors Controlling Deep Shale Gas Enrichment and High Production

Deep shale gas is generally characterized by great depth, high temperature and pressure, and great difficulties in construction (Chang et al. 2022; Guo 2021; Liu et al. 2021). Therefore, the "sweet spot" from which high yields can be obtained is controlled by many factors.

12.3.1 Quality Shales in Deep Shelves

High-quality shales in deep shelves are the basic conditions for evaluating sweet spots for deep shale gas. The quality shale on the deep shelf of the Wufeng–Formation–Longmaxi Formation has a strong hydrocarbon-generating capacity, adaptable thermal evolution and high-quality reservoirs. In the complex structural belt of the southeastern part of the Sichuan Basin, the thickness of the quality shales on the deep shelf ranges from approximately 30 m to 40 m, with a well-laminated structure and high organic carbon and silica contents. The average TOC content is 3.5%, the silica content is generally above 40%, and the clay mineral content is low. In general, quality shales on deep shelves with thicknesses greater than 30 m, high TOCs, high brittleness, and well-laminated structures are the basic conditions for evaluating the sweet spot of deep shale gas.

12.3.2 Fluid Pressure

High fluid pressure is favorable for the enrichment of deep shale gas. On the one hand, shale reservoirs with high fluid pressure exhibit good preservation conditions and low levels of escape so that the shale gas in reservoirs can be preserved. Shale reservoirs with high fluid pressure are generally characterized by high gas contents. On the other hand, triaxial experiments under different confining pressures reveal that confining pressure plays a leading role in the brittle–ductile transformation of shales; with increasing confining pressure, rock mechanical parameters such as peak strength, elastic modulus and residual strength increase, and the degree of fragmentation gradually decreases. However, owing to the high fluid pressure, the effective stress acting on the rock skeleton can be effectively reduced; that is, the actual confining pressure is reduced, the shale brittleness is subsequently improved, and the compressible quality is increased.

12.3.3 Microfracture

Microfractures in shale gas reservoirs under overpressure are favorable for high production. Exploration practices reveal that three exploration wells (DYS1, L203 and Z202-H1) with test production levels of more than 300,000 cubic meters of deep shale gas in China are located in areas with relatively developed microfractures. Many microfractures and microbedding structures develop in shales, which are combined with a large number of pores to form a fracture–matrix pore network system. Small faults and microfractures cause stress release, which can reduce the stress to a certain extent. In the case of overpressure, microfractures may function as weak planes; therefore, it is easier to reduce the initiation pressure of shales to produce fractures.

12.3.4 Stress

Low stress and low differences in horizontal bidirectional stress are favorable geological conditions for effective reservoir stimulations. Deep shale gas is generally characterized by high stress. Finding targets with low stress and reducing the difficulty of construction is important for evaluating sweet spots. The aim of finding a smaller difference in the value of the bidirectional stress is favorable for realizing uniform fracture initiation, forming a complex fracture network, improving the fracture volume effectively and utilizing shale gas layers effectively. Studies have shown that the present stress is influenced by many factors, such as depth, regional stress, paleostress and fractures. With increasing depth, the overall stress increases, and the differences in the values of the bidirectional stress and the stress gradient become markedly differ across regions. Affected by the current regional stress, the stress gradient near the basin-controlling fault is greater, whereas the

stress gradient far from the large basin-controlling fault is lower. At the same depth, areas with wide and gentle structures with small differences in stress and areas with small stress gradients are favorable targets for deep shale gas exploration.

In summary, the development of high-quality shale on deep shelves is a basic condition for evaluating deep shale gas, and deep shale reservoirs with high fluid pressure have good preservation conditions and high gas contents, which are favorable for enrichment Low stress and small differences in horizontal bidirectional stress are favorable geological conditions for effective reservoir stimulation, and microfractures developed under overpressure can reduce the initiation pressure and contribute to high production.

12.4 Multiphase and Multimechanistic Flow in High-Temperature and High-Pressure Deep Shale Gas

12.4.1 Fluid Distribution and Adsorption Under Reservoir Conditions

On the basis of analyses of the geochemical and pore characteristics of high-pressure and high-temperature shale rocks, the distribution and storage space of connate water in shale gas samples can be investigated using low-temperature nitrogen adsorption, CO_2 adsorption and water vapor adsorption/desorption experiments (Qi et al. 2017; Xu et al., 2022; Zhang et al. 2020; Zolfaghari et al. 2015). The results reveal that the content of connate water in deep and ultradeep shale gas is significantly influenced by the content of clay minerals. Connate water exists mainly in nanopores with diameters less than 4 nm, and the connate water undergoes secondary adsorption. Adsorption and desorption isotherm measurements were conducted under reservoir conditions using weight and volume measurement methods (Yang et al. 2016). The results that there is a peak in the excess adsorption isotherm curves and that the excess adsorption tends stabi-

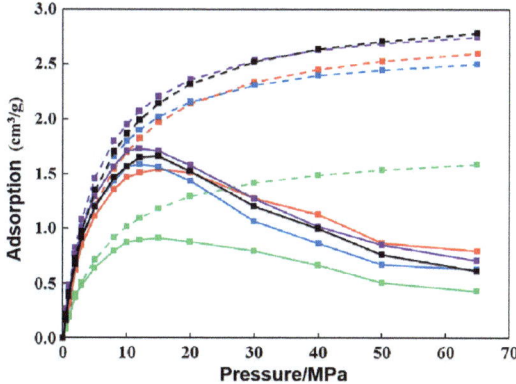

Fig. 12.7 Excess adsorption and adsorption isotherm curves of different shale samples (Thesolid and dotted lines represent excess adsorption and adsorption, respectively)

lize at 50 MPa (Fig. 12.7). In addition, there is a lag between the adsorption and desorption results. Adsorption is positively correlated with the organic content of shale, and the adsorption capacity decreases as the reservoir temperature increases.

The measured adsorption/desorption isotherm results have been fitted using 3 classical models: Langmuir, DR and BET (Brunauer et al. 1940; Hatch et al. 2012; Sawa et al. 2015; Yang et al. 2016; Yu et al. 2016). Under pressures less than 40 MPa, the Langmuir model has a good correlation. However, the adsorption isotherms of high-pressure and high-temperature (> 40 MPa) shale gases, which are close to or greater than the maximum density, do not obey the assumptions of the Langmuir model (Yu et al. 2014). Methane molecules adsorbed onto kerogen nanopores at 0 ~ 80 MPa and 430 K were simulated using molecular dynamics methods to investigate the adsorption/desorption mechanism and calculate the adsorption/desorption isotherms. On the basis of the simulation results, an analytical model for calculating the methane adsorption isotherm was developed, which was verified by our experimental results. The results reveal that when the pressure increases to 20 MPa, the excess adsorption in the nanopores of shale approaches the maximum density. With increasing pressure, the excess adsorption gradually decreases. For instance, when the pressure reaches ~ 40 MPa, the adsorption layer of methane approaches the

maximum density. When the pressure increases to > 40 MPa, the methane in the adsorption layer is compressed, and the density slightly increases.

12.4.2　Gas/Water Two-Phase Flow Under High Pressure

An experimental method for measuring the unsteady gas–water relative permeability in a shale matrix fractures was developed to test the gas–water relative permeability in a 10^{-3} mD matrix. Experimental studies focused on the influence of temperature, displacement pressure difference, effective stress and other factors on the gas water relative permeability of shale were conducted. The results revealed that the two-phase coprecipitation zone of the matrix is narrow, as shown in Fig. 12.8. When a gas channel forms, the sweep efficiency of the gas is constant. In combination with the various nanopores in the shale matrix, most of the water cannot be displaced by gas because of the strong interactions between water and pores. The relative permeability of water is slightly influenced by the increase in the pressure drop, whereas the relative gas permeability increases significantly. Increasing the temperature can increase the gas–water two-phase flow capacity in shale, reduce the residual water saturation and broaden the gas–water co-percolation zone. By increasing the effective stress of shale gas during the gas production process, the

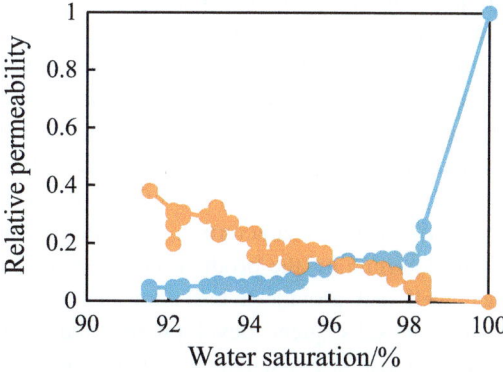

Fig. 12.8　Relative permeability curves of gas/water (with pressure drops of 5 and 10 MPa)

relative gas permeability decreases exponentially, the water saturation decreases, and the influence on the flow capacity increases.

On the basis of the experimental results, the gas/water flow in kerogen and quartz nanopores—at 0 ~ 80 MPa and 430 K is simulated using a molecular dynamics. The results show that the water saturation is ~ 0.6 in small nanopores and ~ 0.8 in midsized nanopores after the invasion of water. Owing to the stronger water/wall interactions and shorter equilibrium distance, water can displace methane in small nanopores, creating an interface between methane and water. Water molecules in the free gas zone are continuous and form small clusters. With respect to the midsized nanopores, water can adsorb to the organic walls, and the distribution of water molecules is relatively continuous. In the free gas zone of midsized nanopores, there is no clear interface between the methane and water. The MSD results reveal that invading water decreases the mobility of methane in the nanopores and that methane molecules are bound in small nanopores. In contrast, methane molecules in the free gas zone of midsized nanopores can move through the discontinuous water molecules and flow out of the nanopores.

12.4.3 Variations in Petrophysical Properties and Fluid Flow During Depletion

The current production method for deep shale gas is primarily depletion. During the production process, the formation pressure gradually decreases, and the effective stress increases, allowing the natural fractures and artificial fractures close to a certain extent, resulting in a sharp decrease in the initial permeability, after which the gas reserves decrease. Therefore, revealing the stress-sensitive characteristics of shale gas can aid in determining the production scenarios for each well and provide a basis for the optimal design of subsequent development plans. At present, most stress-sensitive experiments use experimental methods for constant flow pressure and variable confining pressure (Choi et al. 2017; Tian 2014; Yu et al. 2014). In these experimental

methods, the stress on the rock skeleton is quite different from that in the original reservoir conditions, especially for deep shale oil and gas reservoirs. The effective stress of deep shale reservoirs is greater, and the amount of deformation of the rock skeleton or particle contact is more complex (Li et al. 2021; Wang et al. 2021). To overcome the insufficiencies of traditional test methods, a flow-overburden pressure synergistic in situ stress sensitivity measurement method was proposed, and an ultrahigh-pressure system with a pressure resistance of 180 MPa and a temperature resistance of 200 °C was developed, which was applied to the stress sensitivity measurement of deep shale. A core test of 3 deep shale gas wells in southern Sichuan revealed that it is almost impossible for gas to flow through without penetrating fractures; the permeability of deep shale decreases exponentially with increasing effective stress, and the stress sensitivity index is greater than that of shallow shale and tight gas reservoirs. The permeability loss from production is greater than 80%; and the angle between the effective stress and the fracture is greater. When the effective stress is perpendicular to the fracture direction, the permeability loss exceeds 90%, as shown in Fig. 12.9.

During the production period, shale gas reservoirs are always accompanied by a gas–water two-phase flow process, and the flow capacity of the water phase controls the flow conditions of the shale gas. By applying focused ion beam scanning electron microscopy (FIBSEM), an experimental study on the microscopic characterization of marine shale gas reservoir cores was carried out, and three-dimensional FIBSEM images of marine shale gas cores were obtained. On the basis of the lattice Boltzmann method (Chen et al. 1998; Ning et al. 2015), a pore-scale multiphase flow simulation method was established, and a simulation study of gas–water two-phase flow in deep shale gas formations was carried out. The simulated relative permeability curves of deep marine shale gas reservoirs are shown in Fig. 12.10. As illustrated in the figure, shale gas reservoirs are different from conventional oil and gas reservoirs; the two-phase span is narrow, and the gas phase permeability decreases rapidly after Sw > 0.3,

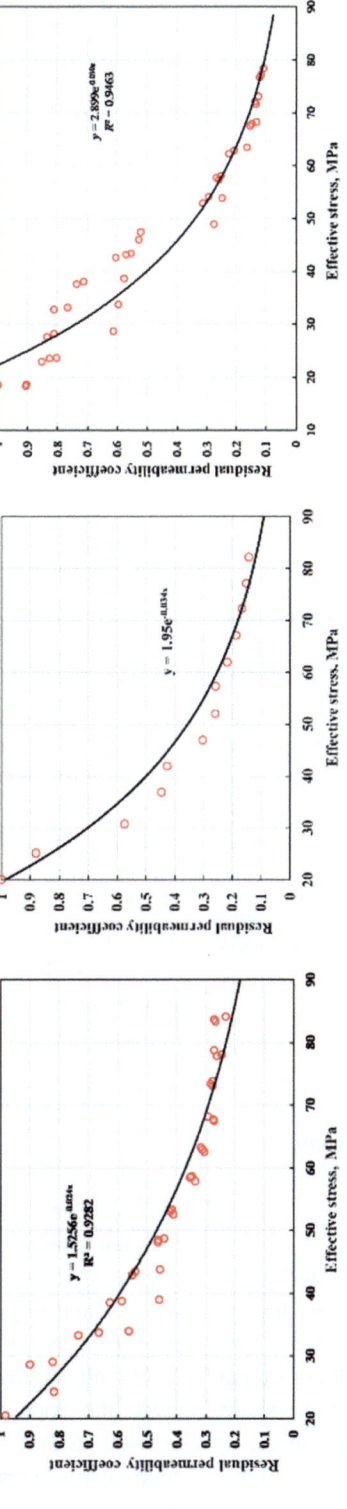

Fig. 12.9 In situ stress sensitivity of deep shale in southern Sichuan

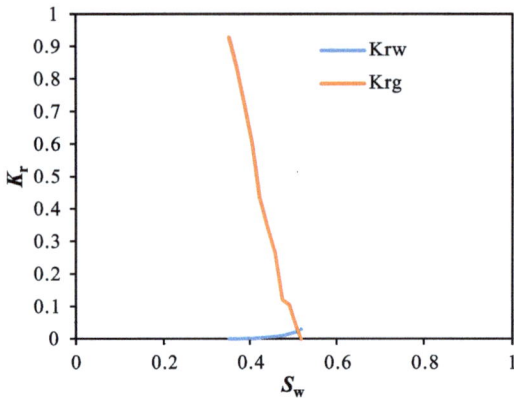

Fig. 12.10 Gas and water relative permeabilities of deep marine shale obtained by pore-scale simulation

whereas the water phase permeability increases slowly, and the gas phase permeability is close to zero after Sw > 0.5.

Because spontaneous imbibition always occurs during the development process of shale gas reservoirs (Roychaudhuri et al. 2013; Xu et al. 2022; Zolfaghari et al. 2015), a dynamic simulation of the hydrocarbon production process coupled with spontaneous imbibition was carried out. This study revealed that the spontaneous invasion rate of the water phase gradually slowed, and the decrease in the gas phase permeability was limited; thus, a relatively stable hydrocarbon flow channel was ultimately formed (the Krg was ultimately reduced by 23% according to this pore-scale simulation), which would not lead to severe water blockage.

12.4.4 Multiphase and Multimechanism Flow Rules and Their Mathematical Models

As shown in previous research, the production process of shale gas is accompanied by desorption of adsorbed gas, pore deformation under stress sensitivity, and microscale effects of gas flow in micro- and nanopores; water in shale, and the surface diffusion effect under the action of the concentration potential difference of the adsorbed layer strongly affects shale gas transmis-

sion. Considering the various effects, a multifield and multiple-coupling seepage dynamic model of shale gas was established to clarify the percolation characteristics during shale gas development. On the basis of the shale pore model, the expression of apparent permeability under the effects of gas dynamic adsorption and desorption, stress sensitivity, non-Darcy effect and surface diffusion can be written as follows:

$$K_{apparent} = \frac{\phi_e R_e^2}{8\tau_m}\left[1 + \frac{\alpha(4-b)Kn^2 + (4+\alpha)Kn}{1-b\cdot Kn}\right] + K_s$$

In the formula above:

$$K_s = \frac{\mu D_S \alpha_\rho \rho_r V_L}{p_0 Z}$$

$$\left\{\frac{C}{(1-x)^2}\frac{1-(n+1)x^n + nx^{n+1}}{1+(C-1)x-Cx^{n+1}} - \frac{Cn(n+1)x^n}{1+(C-1)x-Cx^{n+1}}\right.$$
$$\left.-\frac{Cx[C-1-C(n+1)x^n][1-(n+1)x^n+nx^{n+1}]}{(1-x)[1+(C-1)x-Cx^{n+1}]^2}\right\}$$
$$\left(1 - \frac{R_e^2}{R_0^2}\right)$$

Water is ubiquitous in shale reservoirs, present mostly in the shale matrix and in fractures as free water (Hu et al. 2019). The influence of water on shale gas is twofold: one affects gas flow and permeability, and the other reduces gas absorption by blocking pore throats. On the basis of the influence of water on adsorption, desorption and gas phase flow, the expression of gas apparent permeability considering the water cut is obtained as follows:

$$K_{apparent} = \begin{cases} \frac{\phi_e^* R_e^{*2}}{8\tau_m}\left[1 + \frac{\alpha(4-b)Kn^{*2}+(4+\alpha)Kn^*}{1-b^*\cdot Kn^*}\right] + K_s^* & 0 \le S_W \le 0.5 \\ \frac{\phi_e^* R_e^{*2}}{8\tau_m}\left[1 + \alpha(4-b)Kn^{*2} + (4+\alpha)Kn^*\right] + K_s^* & 0.5 < S_W \le 1 \end{cases}$$

Taking a deep shale gas well in southern Sichuan as an example, on the basis of the experimental results of adsorption and desorption and stress sensitivity, the permeability variation with pressure was calculated (Fig. 12.11).

Owing to the stress-sensitive and adsorbed layer, the apparent permeability is significantly lower than the absolute permeability. The difference becomes more pronounced as the reservoir pressure decreases. If the abandonment pressure

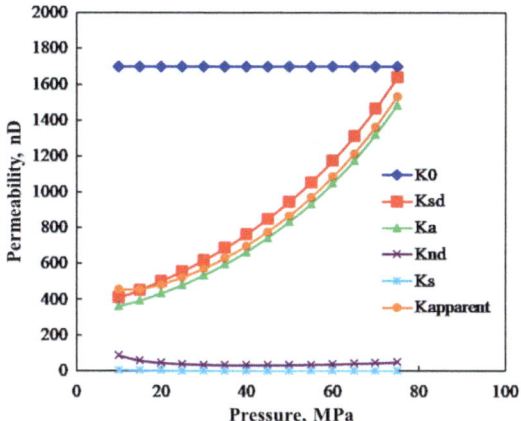

Fig. 12.11 Variations in permeability with pressure under different mechanisms

is 10 MPa, when the abandonment pressure is reached, the effective permeability is only approximately 20% of the absolute permeability, and the effect of stress sensitivity is the most significant. Considering the contributions of various factors to the effective permeability, nonlinear seepage flow and surface diffusion have little effect.

The influence of moisture content on apparent permeability under different pressures, is shown in Fig. 12.12. The higher the moisture content is, the lower the permeability; the higher the pres-

sure is, the more sensitive the permeability is to water.

12.5 Production Evaluation and Analysis Method for Deep Shale Gas Development

12.5.1 Dynamic Prediction Method for Gas–water Productivity

A large amount of fracturing fluid remains in the reservoir after fracturing in deep shale gas wells, and the characteristics of gas–water two-phase seepage are marked. One portion of the fracturing fluid is retained in the complex fracture network, and the other part invades the matrix around the fracture, which can be characterized by the concept of a fracturing fluid invasion layer (Wu et al. 2021).

In this paper, by considering fractures near propped wellbores and fractures far from incompletely propped wellbores in the fluid-invasion layer and the fracturing fluids existing in the main fracture and the water-invasion layer at the same time, a gas–water two-phase productivity

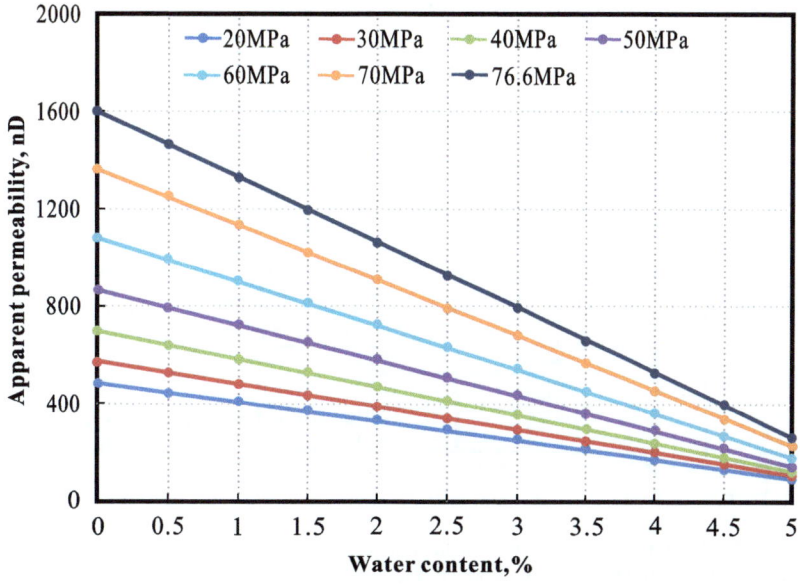

Fig. 12.12 Effect of moisture content on apparent permeability under different pressures

prediction method for deep shale gas wells is established.

(1) Physical model

In accordance with the distribution of proppant in hydraulic fractures and the transformation effect perpendicular to the horizontal wellbore, the SRV area can be divided into 4 areas (Wu et al. 2021) (Fig. 12.13): the near-well propped fracture area, the inner SRV area, the far-well unpropped fracture area and the outer SRV area. Owing to the extremely low permeability of the original shale reservoir and its lower contribution to gas production, the influence of the unreconstructed area is not considered.

On the basis of the four-zone model, the SRV zone is divided into a single-phase flow zone and a gas–water two-phase flow zone of the fracturing fluid intrusion layer according to the distribution of the fracturing fluid. The flow of fluids in each area is shown in Eq. (12.5.1), and the following assumptions are proposed:

① Adsorbed gas and free gas exist in the matrix.
② Considering the mechanisms of gas desorption, diffusion, slippage and seepage, free gas and water exist in the main fractures.
③ For fractures and the matrix, the stress sensitivity of the fracture should be considered.
④ The flow of fluids in each area is linear.

(2) Mathematical mode

A mathematical seepage model is deduced and established (the matrix flow model in the outer SRV area, the gas–water flow model in the unpropped fracture area, the matrix model in the inner SRV area and the gas–water flow model in the propped fracture area) by introducing dimensionless variables and pseudotime and pseudopressure functions. The gas–water flow equation in the fracturing fluid invasion layer is as follows:

$$\begin{cases} \frac{\partial^2 \psi_c}{\partial x^2} = \frac{1}{0.0864} \frac{\mu_{gi}(\phi c_t)_{ci}}{k_c k_{crg}} \frac{\partial \psi_c}{\partial t_a} \\ \psi_c(x, t_a)|_{t_a=0} = \psi_i \\ \psi_f(x, t_a)|x = 0 = \psi_c(x, t_a)|x = 0 \\ \frac{\partial \psi_c(x,t_a)}{\partial x}|x = h_c = \frac{k_{ma}}{k_{ca}k_{crg}} \frac{\partial \psi_m(x,t_a)}{\partial x}|x = h_c \end{cases}$$

$$\begin{cases} \frac{\partial^2 p_w}{\partial x^2} = \frac{1}{0.0864} \frac{\mu_w \phi_c c_{tc}}{k_c k_{crw}} \frac{\partial p_w}{\partial t} \\ p_w(x, t)|_{t=0} = p_i \\ \frac{\partial p_w(x,t)}{\partial x}|x = h_c = 0 \\ p_f(x, t)|x = 0 = p_w(x, t)|x = 0 \end{cases} \quad (12.5.1)$$

In Eq. (13.1), ψ_c is the gas-phase pseudopressure in the fracturing fluid invasion layer, and p_w is the water-phase pressure in the fracturing fluid invasion layer.

To facilitate the solution, the model was transformed using a perturbation transformation equation and Laplace transform, and the gas and water production levels were obtained in the Laplace domain.

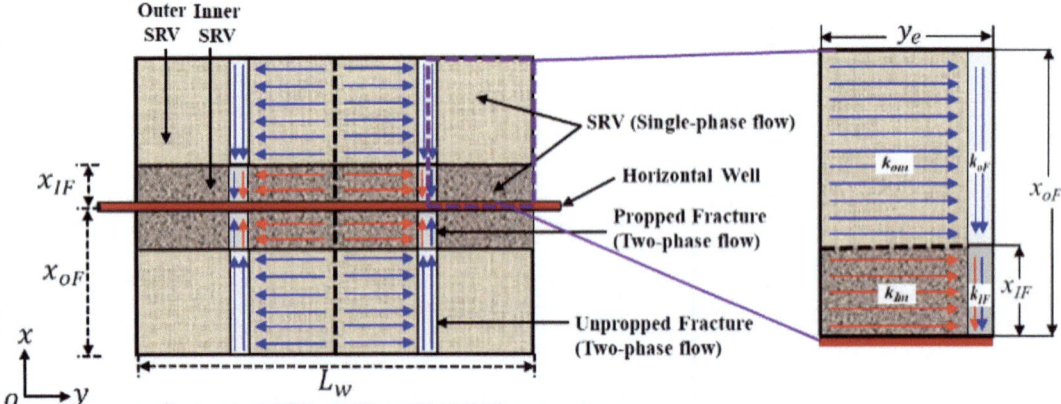

Fig. 12.13 Schematic illustration of the conceptual model for the two phases gas–water flow in fractured horizontal wells considering incompletely propped fractures and the fracturing fluid invasion layer

$$\bar{q}_{gD} = \frac{k_{frg}k_{fD}w_{fD}}{\pi} \frac{\sqrt{\alpha_f}}{s} \tan h\left(\sqrt{\alpha_f}y_{fD}\right) \quad (12.5.2)$$

$$\bar{q}_{wD} = \frac{k_{frw}k_{fD}w_{fD}}{\pi} \frac{\sqrt{\beta_f}}{s} \tan h\left(\sqrt{\beta_f}\cdot y_{fD}\right) \quad (12.5.3)$$

$$\alpha_f = \frac{s}{k_{frg}\eta_{fgD}} - \frac{2k_{crg}}{C_{fD}k_{frg}}\sqrt{\frac{s}{k_{crg}\eta_{cgD}}}$$
$$\frac{\tan h\left(\sqrt{\frac{s}{\eta_{mgD}}}\cdot(h_{cD}-x_{mD})\right) - \frac{k_{cD}\sqrt{k_{crg}}}{k_{mD}}\sqrt{\frac{\eta_{mgD}}{\eta_{cgD}}}\tan h\left(\sqrt{\frac{s}{k_{crg}\eta_{cgD}}}\cdot h_{cD}\right)}{\frac{k_{cD}\sqrt{k_{crg}}}{k_{mD}}\sqrt{\frac{\eta_{mgD}}{\eta_{cgD}}} - \tan h\left(\sqrt{\frac{s}{\eta_{mgD}}}\cdot(h_{cD}-x_{mD})\right)\tan h\left(\sqrt{\frac{s}{k_{crg}\eta_{cgD}}}\cdot h_{cD}\right)} \quad (12.5.4)$$

$$\beta_f = \frac{s}{k_{frw}\eta_{fwD}} + \frac{2k_{crw}}{C_{fD}k_{frw}}\sqrt{\frac{s}{k_{crw}\eta_{cwD}}}\tan h\left(\sqrt{\frac{s}{\eta_{cwD}k_{crw}}}\cdot h_{cD}\right) \quad (12.5.5)$$

The gas and water production solutions can be inverted to the time domain using the Stehfest method, and after which the time steps can be divided. At each time step, the average pressure and average water saturation obtained by the material balance method and the Newton–Ralph method are used to update the parameters in the model to calculate the production in the next time step (Huang et al. 2021).

For a single fracturing stage, the material balance equation for water in the fracture area can be written as follows:

$$W_p = W_{cp} + W_{fp} = V_{cp}\left(\frac{S_{cwi}}{B_{wi}} - \frac{\hat{S}_{cw}}{B_w}\right)$$
$$+ V_{fp}\left(\frac{S_{fwi}}{B_{wi}} - \frac{\hat{S}_{fw}}{B_w}\right) \quad (12.5.6)$$

where V_p is the total pore volume of the fracture area and is calculated as follows:

$$V_p = V_{cp} = +V_{fp} = 4y_f h_c H\phi_c + 2y_f \omega_f H\phi_f.$$

(3) Practical example for calculating deep shale gas productivity

A deep shale gas well in the Weirong gas field is used to test the application of the proposed gas–water two-phase flow model. The length of the horizontal well is 1500 m, with 63 clusters and 23 fracturing stages.

By matching the field data with the proposed model, as shown in Fig. 12.14, we can determine that the half-lengths of the fracture and the propped fracture are 107 m and 68.2 m, respectively, and that the permeabilities of the outer and inner SRVs are 1.06×10^{-44} and 3.12×10^{-33} mD, respectively. Furthermore, the cumulative gas production over 20 years is $8.7 \times 10^7 \text{m}^3$ (Fig. 12.14).

12.5.2 Numerical Well Testing Analysis Method for Fracturing Horizontal Wells

At present, well test models reflecting the flow characteristics of shale gas in complex fracture networks and the well test interpretation methods are not perfect. Existing analytical and numerical well test technologies are aimed mainly at vertical wells, horizontal wells with single-phase flow and horizontal wells; with uniform fracturing, these technologies are not capable of evaluating the distribution of complex fractures after gas well fracturing and the dynamic changes in fractures

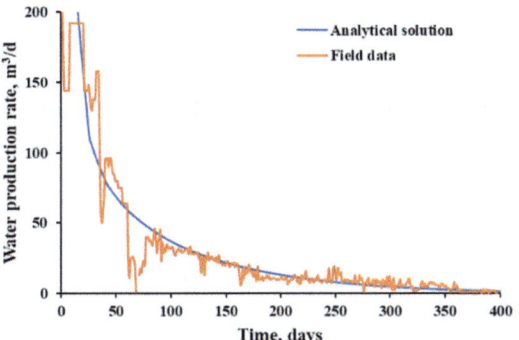
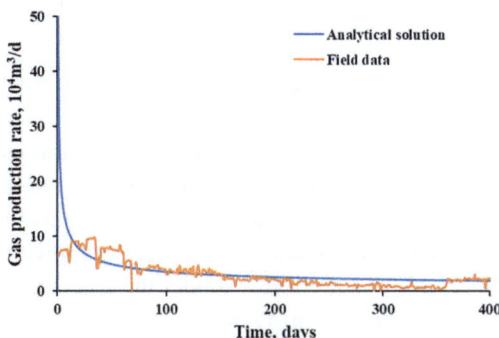

Fig. 12.14 History matching results for gas and water production data with the proposed model

during production cannot be performed (Jia et al. 2016).

On the basis of studies on the flow mechanism of deep shale gas reservoirs, a coupled model of multiphase complex pore–fracture medium seepage in shale gas-fractured horizontal wells was established to develop an effective numerical solution method to establish a numerical well testing analysis method on the basis of pressure recovery data by shutting in wells and dynamic fracture network parameters with production data for evaluating reservoirs.

(1) Mathematical model for the two-phase gas–water flow and seepage

To address complex fractures that occur after fracturing and comprehensively consider the dynamic changes in fracture permeability, matrix adsorption and desorption, matrix nonlinear seepage and other mechanisms, a mathematical model for the two phases of gas–water flow and seepage in horizontal wells during fracturing was established.

The equation for the gas phase:

$$\nabla \cdot \left[\frac{k(p_g)k_{rg}}{B_g \mu_g} \nabla p_g \right] + q_g = \frac{\partial}{\partial t} \left(\frac{\phi s_g}{B_g} + V_L \frac{p_g}{p_g + p_L} \right) \tag{12.5.7}$$

The equation for the water phase:

$$\nabla \cdot \left[\frac{k(p_w)k_{rw}}{B_w \mu_w} \nabla p_w \right] + q_w = \frac{\partial}{\partial t} \left(\frac{\phi s_w}{B_w} \right) \tag{12.5.8}$$

The constraint equations are:

$$s_w + s_g = 1 \tag{12.5.9}$$

$$p_{cgw} = p_g - p_w \tag{12.5.10}$$

In this paper, the outer boundary conditions are all closed boundaries (Neumann boundary conditions), and the inner boundary conditions involve flow conditions toward the wellbore, as shown in Fig. 12.15. The well flow model the fractures is shown as follows Fig. 12.16:

$$q = \frac{2\pi k_f w_f}{\mu B} \frac{p_j - p_{wf}}{\ln(r_e/r_w)} \tag{12.5.11}$$

In the equation,

$$r_e = 0.14(h^2 + \Delta y^2)^{1/2} \tag{12.5.12}$$

(2) Unstructured grid meshing and finite element solution

On the basis of the Delaunay triangulation meshing algorithm, combined with the geometric characteristics of discrete fractures (Hyman et al. 2015), unstructured grid meshing is established for fractures vertical to the wellbore or at a certain angle to the wellbore and complex fracture networks.

The discrete solution of the model is divided into two stages (Fang and Cheng 2021). The calculation of early flow is based on the linear flow element method, and the mixed boundary element method is used in the middle and later stages. The discrete element relationship based on the Delaunay triangular mesh is shown in Fig. 12.15. Two numerical methods are used to construct a complex fracture network hybrid model (Laplace space) for fractured horizontal wells, and multithread parallel technology is used to solve the model efficiently. The dimensionless pressure distribution of the complex fracture network is shown in Fig. 12.16.

The mixed boundary element method discrete equation is as follows:

$$R_{ij}h_j + L_{ij}(\frac{q}{K})_j - U_{imj}\psi_m h_j$$
$$+ W_{imj}\left[\sigma_m \frac{dh_j}{dt} + \upsilon_m f_j \right] = 0 \tag{12.5.13}$$

In the equation, $R_{ij} = \int_{\Gamma_e} \Omega_j \nabla G(r, r_i) \cdot \boldsymbol{n} ds - \delta_{ij}\lambda$, $L_{ij} = \int_{\Gamma_e} \Omega_j G(r, r_i) ds$.

$$U_{imj} = \iint_{\Lambda_e} G(r, r_i) \frac{\partial \Omega_m}{\partial x} \frac{\partial \Omega_j}{\partial x} dA$$
$$+ G(r, r_i) \frac{\partial \Omega_m}{\partial y} \frac{\partial \Omega_j}{\partial y} dA$$

Fig. 12.15 Schematic diagram of the numerical discrete relationship between adjacent grids

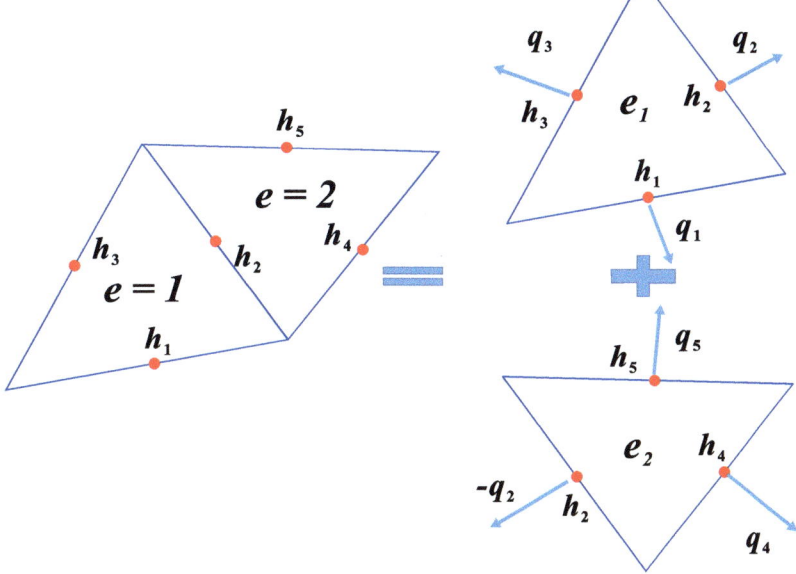

Fig. 12.16 Dimensionless pressure distribution of the complex fracture network

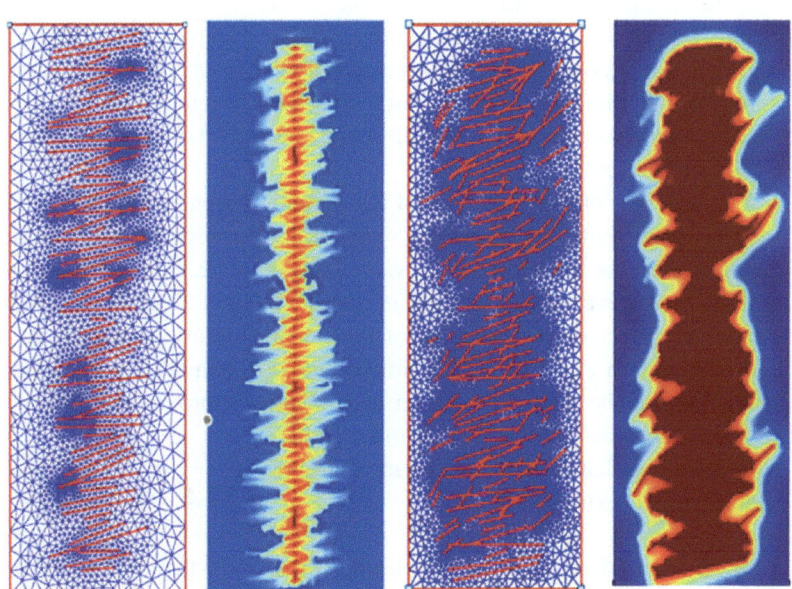

$$W_{imj} = \iint_{\Lambda_e} G(r, r_i)\Omega_m\Omega_j dA$$

This method overcomes the large error of the finite element numerical well test model in the early pressure stage and the pressure derivative solutions. Moreover, it improves the speed of the simulation calculation.

(3) Analysis of the well test curve characteristics of fracturing horizontal wells

On the basis of the numerical well test model, a well test curve the gas well pressure and the pressure derivative is established. The flow characteristics can be divided into multiple stages: the

Fig. 12.17 Influence of the main fracture conductivity on the well testing curves

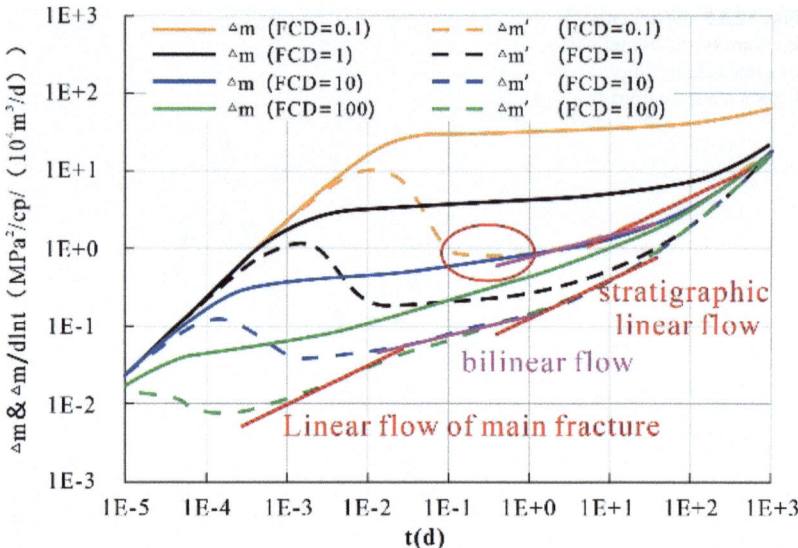

wellbore storage effect, linear flow in the main fracture, bilinear flow, formation linear flow, outer area linear flow, transition flow and boundary flow. Moreover, the dimensionless pressure distribution of the planes corresponding to different flow stages can be calculated.

A comparison of the influences of the dimensionless conductivity (FCD) on the well testing curves reveals that the higher FCD values result in lower derivative values, an earlier onset of linear flow, and a shorter duration of the linear flow, as shown in Fig. 12.17. The dimensionless conductivity of the main fractures in deep shale gas wells is characterized by an infinite conductivity, reflecting that the effective permeability of the formation is extremely low.

(4) Inversion of fracture network parameters in deep shale gas wells after fracturing

The numerical well test model of the horizontal fracturing well established in this paper is used to invert and analyze the pressure recovery data of a shale gas well in Yongchuan. Combined with the multiparameter automatic history matching algorithm (Cheng et al. 2014), the gas well production history is fitted, and the parameters of the fracture network can be ultimately obtained. They are then compared with the results of well testing interpre-

tation in the main area of Fuling to understand the seepage characteristics of deep shale gas wells.

According to the interpretations, the effective fracture half-length in Well Yongye X is 34–109 m, the secondary fracture permeability is 1.8 $\times 10^{-4}$ mD, and the matrix permeability is 10 nD, suggesting that the effective permeability of the reservoirs after fracturing is extremely low. The effective permeability of fractures in deep shale gas wells is 10^{-4} mD, which is approximately 2 orders of magnitude lower than that in the main area of Fuling (the permeability after fracturing is approximately 0.01 mD) (Fig. 12.18).

12.6 Summary

Deep shale gas has emerged as a critical alternative domain within China's oil and gas resource strategy. This research focuses on the formation and preservation mechanisms of effective reservoirs in deep marine shale, the occurrence mechanisms and evolutionary patterns of shale gas, key controlling factors for enrichment and high yield, and multiphase/multimechanism flow dynamics, leading to the innovative development of a series of technical methods that have demonstrated successful application in the exploration of deep shale gas in complex structural zones and

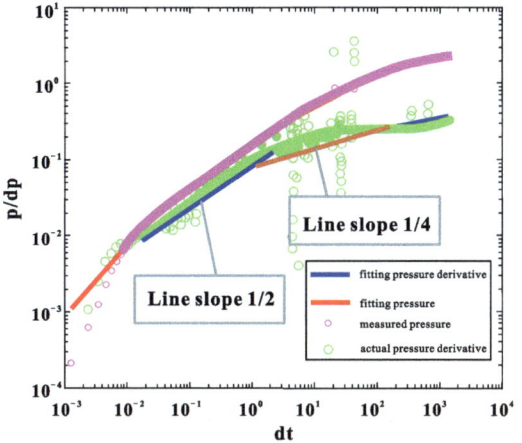

Fig. 12.18 Numerical well testing interpretation for the recovered well pressure recovery for the Yongye X Well

the new Permian Dalong Formation in northern Sichuan.

(1) Full-scale quantitative analysis of pores and fractures revealed the pore size distribution characteristics and pore types in deep shale reservoirs, confirming that the coupled effects of "quartz compression resistance preserving pores" and "reservoir fluid overpressure" are key to maintaining high porosity in deep marine shale in the Sichuan Basin.

(2) Laboratory experiments and molecular simulations indicate that free gas content increases with burial depth, with deep shale being predominantly free gas; on the basis of methane isothermal adsorption experiments and adsorption potential theory, a mathematical geological model for gas content in deep shale was established, enabling the dynamic simulation of gas content evolution.

(3) Comprehensive analysis demonstrates that deep-water shelf facies form the foundation of shale "sweet spots," where formation overpressure facilitates gas accumulation, whereas well-developed microfractures, low geostress, and minimal horizontal stress differentials constitute optimal engineering conditions for effective stimulation.

(4) This study clarified fluid occurrence states under reservoir conditions, methane desorption patterns, and gas–liquid two-phase flow characteristics under high pressure, ultimately establishing a multiphase/multimechanism seepage model for deep shale gas that incorporates adsorption–desorption, stress sensitivity, microscale seepage, surface diffusion, and fracturing fluid effects.

(5) To construct a deep shale gas seepage model, a two-phase (gas–liquid) production prediction method was developed, with the simulation results showing high consistency with actual well production data; additionally, a coupled seepage well-test analysis model for multiphase complex pore-fracture media was created, enabling the quantitative evaluation of postfracture complex fracture distribution and dynamic changes in fracture parameters during production.

Acknowledgements The authors acknowledge the financial support of Natural Science Foundation of China (Grants U24B6001 and U19B6003-03), and Sinopec management for permission to publish this work.

References

Aplin AC, Macquaker HS (2011) Mudstone diversity: origin and implications for source, seal, and reservoir properties in petroleum systems. AAPG 95(12):2031–2059

Brunauer S, Deming LS, Deming WE, Teller E (1940) On a theory of the van der Waals adsorption of gases. J Am Chem soc 62(7):1723–1732

Chang C, Zhang J, Hu H et al (2022) Molecular simulation of adsorption in deep marine shale gas reservoirs[J]. Energies 15(3):944

Chen S et al (1998) Lattice boltzmann method for fluid flows. Annu Rev fluid mechanics, 30(1):329–364

Cheng M (2014) Mechanisms analysis of shale gas supercritical adsorption and modeling of isorption adsorption. J china coal soc 39(1):179–183

Choi C S et al (2017) Effect of pore and confining pressure on the supercritical CO2 permeability of sandstone: implications for the effective pressure law. J Geophys Res: Solid Earth 122(8):6231–6246

Curtis JB (2002) Fractured shale-gas systems. AAPG Bull 86(11):1921–1938

Curtis ME, Cardott BJ, Sondergeld CH et al (2012) Development of organic porosity in the Woodford Shale with increasing thermal maturity[J]. Int J Coal Geol 103(23):26–31

Dai C, Li H, Zhang D (2014) Efficient and accurate global sensitivity analysis for reservoir simulations by use of the probabilistic collocation method. SPE J 19(04):621–635

Fang S, Cheng L (2021) An improved green element method and its application in seepage problems[J]. Chin J Comput Mech 38(06):787–795

Guo T (2021) Progress and research direction of deep shale gas exploration and development[J]. Reserv Eval Dev 11(1):1–6

Hatch CD, Wiese JS, Crane CC, Harris KJ, Kloss HG, Baltrusaitis J (2012) Water adsorption on clay minerals as a function of relative humidity: application of BET and Freundlich adsorption models. Langmuir, 28(3): 1790–1803

Hu D, Wei Z, Li Y et al (2022) Deep shale gas exploration in complex structure belt of the southeastern Sichuan Basin: progress and breakthrough[J]. Nat Gas Ind 42(08):35–44

Hu S et al. (2019). The influence of oil shale in situ mining on groundwater environment: A water-rock interaction study. Chemosphere 228:384–389

Huang S, Zhang J, Fang S, Wang X (2021) An analytical method for parameter interpretation of fracture networks in shale gas reservoirs considering uneven support of fractures. Geofluids

Hyman JD et al (2015) DfnWorks: a discrete fracture network framework for modelling subsurface flow and transport. Comput Geosci 84:10–19

Jarvie DM, Hill RJ, Ruble TE et al (2007) Unconventional shale-gas systems: the Mississippian Barnett Shale of north-central Texas as one model for thermogenic shale-gas assessment[J]. AAPG Bull 91(4):475–499

Jia C, Zheng M, Zhang Y (2012) Unconventional hydrocarbon resources in China and the prospect of exploration and development[J]. Petrol Exp Dev 39(02):129–136

Jia P et al (2016) A semi-analytical model for the flow behavior of naturally fractured formations with multiscale fracture networks. J Hydrol 537:208–220

Kong D, Ning Z, Yang F et al (2014) Study of methane adsorption on shales[J] 14(06):108–111+117

Landers J, Gor GY, Neimarkav A (2013) Density functional theory methods for characterization of porous materials. Colloids and Surf A 437:3–32

Liu Y, Lai F, Zhang H et al (2021) A novel mineral composition inversion method of deep shale gas reservoir in Western Chongqing[J]. J Petrol Sci Eng 202:108528

Löhr SC, Baruch ET, Hall PA et al (2015) Is organic pore development in gas shales influenced by the primary porosity and structure of thermally immature organic matter [J]. Org Geochem 87:119–132

Li J et al (2021) Geochemical characteristics and resource potential of shale gas in Sichuan Basin, China. J Nat Gas Geosci 6(6):313–327

Loucks RG, Reed RM, Ruppel SC et al (2012) Spectrum of pore types and networks in mudrocks and a descriptive classification for matrix-related mudrock pores [J]. AAPG Bull 96:1071–1098

Mastalerz M, Schimmelmann A, Drobniak A et al (2013) Porosity of devonian and mississippian new albany shale across a maturation gradient: Insights from organic petrology, gas adsorption, and mercury intrusion. AAPG Bull 97(10):1621–1643

Milliken KL, Esch WL, Reed RM, Zhang T (2013) Grain assemblages and strong diagenetic overprinting in siliceous mudrocks, Barnett Shale (Mississippian), Fort Worth Basin, Texas. AAPG Bull 96(8):1553–1578

Milliken KL, Ergene SM, Ozkan A (2016) Quartz types, authigenic and detrital, in the upper cretaceous eagle ford formation, South Texas, USA. Sed Geol 339:273–288

Ning Y et al. (2015). Numerical modeling of slippage and adsorption effects on gas transport in shale formations using the lattice boltzmann method. J Nat Gas Sci Eng 26:345–355

Qi Y et al (2017) Nanoporous structure and gas occurrence of organic-rich shales. J Nanosci Nanotechnol, 17(9), 6942–6950

Robert JD, Daniel RH, Christian ML (2000) Pore size analysis of activated carbons from argon and nitrogen porosimetry using density functional theory. Langmuir 16:5041–5050

Roychaudhuri B et al. (2013). An experimental investigation of spontaneous imbibition in gas shales. J Pet Sci Eng 111:87–97

Sawa Y, Liang Y, Murata S, Matsuoka T, Akai T, Takagi S (2015) Pore-filling nature of CH4 adsorption behavior in kerogen nanopores: a molecular view based on full-atom kerogen models. In: SPE Asia Pacific unconventional resources conference and exhibition (pp. SPE– 176999). SPE

Tian K Q (2014) Research on the policy of advanced water injection in ma 2 abnormally high pressure and extremely low permeability reservoirs. China University of petroleum (East China)

Ustinov EA, Do DD (2004) Application of density functional theory to analysis of energetic heterogeneity and pore size distribution of activated carbons. Langmuir 20:3791–3797

Ustinov EA, Stinov EA, Do DD, Fenelonov VB (2006) Pore size distribution analysis of activated carbons: application of density functional theory using nongraphitized carbon black as a reference system. Carbon 44(4):653–663

Wang Y et al. (2021). Gas adsorption characterization of pore structure of organic-rich shale: Insights into contribution of organic matter to shale pore network. Nat Resour Res 30(3):2377–2395

Wu Y et al (2021) A transient two-phase flow model for production prediction of tight gas wells with fracturing fluid-induced formation damage. J Petrol Sci Eng 199:108351

Xu Y, Lun Z, Zhou X, Zhang G, Wang H, Zhao C, Zhang H, Zhang D (2022) Influences of microwave irradiation on pore, fracture and methane adsorption of deep shale. J Nat Gas Sci Eng 101:104489

Yang B et al (2016) Measurement of the surface diffusion coefficient for adsorbed gas in the fine mesopores and micropores of shale organic matter. Fuel 181: 793–804

Yu W (2014) Sensitivity analysis of hydraulic fracture geometry in shale gas reservoirs. J Pet Sci Eng 113:1–7

Yu W et al (2016) Modeling gas adsorption in Marcellus shale with Langmuir and bet isotherms. Spe J 21(02): 589–600

Zhang B et al (2020) Review of formation and gas characteristics in shale gas reservoirs. Energies, 13(20), 5427

Zolfaghari A et al (2015) A comparative study of pore size distribution in gas shales. In: SPE Asia Pacific unconventional resources conference and exhibition (pp. SPE–176908). SPE

Zou C, Tao S, Yang Z et al (2012) New advance in unconverntional petroleum exploration and research in China[J]. Bulletion Mineral, Petrol Geochem 31(04):312–322

Advances in Drilling and Completion for Deep Marine Reservoirs

13

Yijin Zeng, Gensheng Li, Bing Hou, Huaizhong Shi, Mao Sheng, Daqi Li, Junhai Chen, Rengguang Liu, Yang Xia, Shiming Wei, Wenhao He, and Yayun Zhang

Abstract

Deep marine shale gas (3500–4500 m) and ultradeep marine carbonate oil and gas ($\geq$ 6000 m) are among the most important development areas for China's future oil and gas resources. Some key problems related to the safety and efficient drilling and completion of marine oil and gas wells are addressed in this paper: (1) a prediction method for pore pressure in deep marine reservoirs; (2) a method to solve the problems of low penetration rates and high drilling costs; (3) a method to maintain wellbore sealing integrity under cyclic loading to simulate staged fracturing; (4) a wellbore stability control method; and (5) how to carry out efficient hydraulic fracturing. To focus on the efficient drilling and completion of oil and gas wells in ultradeep shale and carbonate formations, we carried out research to solve the aforementioned problems and share our results in this paper.

Y. Zeng · B. Hou · D. Li · J. Chen · R. Liu · Y. Zhang
Sinopec, Beijing, China

G. Li (✉) · B. Hou (✉) · H. Shi · M. Sheng · Y. Xia ·
S. Wei · W. He
China University of Petroleum-Beijing, Beijing, China
e-mail: ligs@cup.edu.cn

B. Hou
e-mail: binghou@cup.edu.cn

B. Hou
China University of Petroleum-Beijing at Karamay,
Karamay, China

Keywords

Deep shale reservoirs · Deep carbonate reservoirs · Pore pressure prediction · Wellbore integrity · Hydraulic fracturing

13.1 Introduction

As the difficulty of developing oil reservoirs increases, efficient development techniques are gradually gaining attention. The demand for unconventional reservoir development is growing with the increasing global economy and population. Acid fracturing and hydraulic fracturing are important techniques for unconventional reservoirs ($\geq$ 3500 m).

The production of gas and oil in unconventional reservoirs is related mainly to whether hydraulic fracturing results in the formation of sufficient and effective fractures. In theory, hydraulic fractures propagate along the maximum horizontal principal stress direction. However, the propagation direction of fractures can change due to geological and engineering factors. The fracturing results are controlled by natural fractures and pores, fracturing fluid viscosity, pump rate and other factors. Engineers often monitor fracture propagation using a microseism technique.

Various studies have investigated fracture propagation. The most effective hydraulic fractures should cross and connect to discontinuities;

Y. Ma et al. (eds.), *Petroleum Geology and Exploration of Deep Marine Strata in China*, Advances in Oil and Gas Exploration & Production, https://doi.org/10.1007/978-3-032-02496-1_13

however, fracture arrest, diversion, or offset can be observed when discontinuities are encountered. The first two behaviours could limit the fracture length, and offset could reduce the fracture width, resulting in proppant bridging and premature screen-out (Beugelsdijk et al. 2000). Furthermore, numerous geological factors affect these interactions. Daneshy (1974) reported that the cement strength and orientation of weak planes affect the activation of discontinuities by fractures. Blanton (1982) reported that the stress state and approach angle between hydraulic and natural fractures also influence this interaction. Teufel and Clark (1984) reported that an interface with a low friction coefficient or low cohesion may alter the extension path of a hydraulic fracture. Zhou et al. examined the effect of the shear strength of preexisting fractures on induced fracture propagation by inserting paper sheets with different shear strengths in a cement block. Olson et al. (2012) reported that the fracture propagation path is hindered and that its direction is deflected owing to barriers. Zou et al. reported that a hydraulic fracture tends to cross over a natural fracture at a high horizontal differential stress. Moreover, Rickman et al. (2008) reported that a high clay mineral content was counterproductive in the formation of a complex fracture network, but brittle mineral-rich shale had positive effects. Olsen et al. (2009) reported that complex fractures were more likely to be generated in shale formations that had a low Poisson's ratio and extremely low permeability. Bing et al. (2014) also reported from numerous laboratory experiments that a complex fracture network was easily produced when the content of brittle minerals was high. Furthermore, a successful complex fracture network not only activates and connects the discontinuities but also ensures that a long fracture length is essential for a large SRV.

The difficulty of acid fracturing treatment is the leakage of acid, which hinders the ability of acid-etched fractures to propagate and reduces the effective acid penetration distance (Economides and Nolte 1989). Therefore, it is necessary to design specific experiments that help solve this problem. For example, Wang showed that self-generating acid is highly compatible with gelled acid and self-diverting acid, which suggests that self-generating acid can be used as a future fluid in experiments. Hou reported the successful utilization of alternating acid fracturing, in which self-generating acid was used as the preceding fluid to penetrate and cool reservoirs and gelled acid was used to deeply penetrate carbonate reservoirs. These studies provide a reference for experimental design. Hou introduced an innovative method to perform acid fracturing experiments based on true triaxial experiments for the first time (Hou et al. 2019b).

To address the problems of high fracture initiation pressure, poor reconstruction results, and low fracturing efficiency in current development, the following key scientific issues are investigated in this chapter. The characterization of rock at high temperature and high stress is described in Sect. 13.6.1. The effects of strong stress on the fracture network in deep shale formations are discussed in Sect. 13.6.2. The initiation and propagation of acid fracturing fractures in ultradeep carbonate reservoirs are discussed in Sect. 13.6.3. The study of the fracture conductivity of acid fracturing in carbonate is discussed in Sect. 13.6.4. The field test of hydraulic/acid fracturing in tight oil reservoirs is discussed in Sect. 13.6.5.

13.2 Prediction Method for Pore Pressure in Deep Marine Reservoirs

Marine deep reservoirs are characterized by strong heterogeneity, long geological evolution times, and complex diagenesis and reservoir formation mechanisms. Thus, the distribution of pore pressure is difficult to predict, which severely restricts the safe and efficient drilling and completion of marine oil and gas wells. In this section, quantitative characterization and prediction methods for determining pore pressure in marine deep reservoirs are explored to improve the application accuracy in oil and gas fields.

13.2.1 Prediction Method for Determining Pore Pressure in Deep Shale Reservoirs

13.2.1.1 Reservoir and Mechanical Characteristics of Deep Shale Reservoirs

Commercial exploitation of shale reservoirs has already been successful in the middle–shallow layers of China. However, more than 65% of shale reservoir resources are buried in deep formations, which have not yet been effectively developed. Deep burial depth is the source of a series of challenges, including the prediction of pore pressure (Jin and Chen 2019).

The marine shale in deep formations mainly contains quartz, albite, calcite, pyrite, and clay minerals, with the contents of quartz and the dominant minerals decreasing gradually from bottom to top (Wang et al. 2021). Confining pressure strongly affects mechanical parameters, especially the peak deviatoric stress and Young's modulus. Confining pressure restrains lateral expansion and microrupture, which enhances the stiffness and ultimate strength of the rock. The higher the confining pressure is, the stronger the effect on improving the strength and Young's modulus. Because confining pressure can induce large differences in the mechanical parameters of shale reservoirs, triaxial tests could be more representative than uniaxial tests for characterizing the mechanical behaviour of rock in deep formations.

There are many sets of Cambrian and Silurian shale formations in the Sichuan Basin, in which the shale is a set of highly efficient self-generating and self-storing source rocks with high organic carbon content and stable distribution. Owing to good reservoir sealing conditions and high tectonic stress, predicting pore pressure in deep shale reservoirs is highly challenging.

13.2.1.2 Prediction Model and Method for Determining Pore Pressure in Deep Shale Reservoirs

Because pore pressure responds differently to different pore elastic parameters, studying the effects of pore elastic parameters on pore pressure is very useful for predicting pore pressure (Song et al. 2019a, b). The relationship between pore pressure and porosity is influenced by porous elastic parameters, such as overburden pressure, fluid volume modulus, Young's modulus, Passion's ratio, solid volume modulus, and solid shear modulus. The pore pressure increases with increasing overburden pressure, and the overburden pressure determines the maximum pore pressure. The maximum pore pressure is equal to the overburden pressure. The pore pressure reaches a maximum when the porosity is equal to one.

In accordance with the pore elasticity theory (Biot 1939), the constitutive equation is expressed as follows:

$$\sigma_y = \lambda \varepsilon_{vol} + 2G\varepsilon_y - C\xi \tag{13.1}$$

$$\sigma_z = \lambda \varepsilon_{vol} + 2G\varepsilon_z - C\xi \tag{13.2}$$

$$\tau_{yz} = 2G\Gamma_{yz} \tag{13.3}$$

$$\tau_{xz} = 2G\Gamma_{xz} \tag{13.4}$$

$$\tau_{xy} = 2G\Gamma_{xy} \tag{13.5}$$

$$\varepsilon_{vol} = \varepsilon_x + \varepsilon_y + \varepsilon_z \tag{13.6}$$

$$\xi = -\phi \nabla \cdot \left(\vec{u_s} - \vec{u_f} \right) \tag{13.7}$$

Where $\sigma_x, \sigma_y, \sigma_z,$ $\tau_{yz}, \tau_{xz}, \tau_{xy},$ $\varepsilon_x, \varepsilon_y, \varepsilon_z,$ $\Gamma_{yz}, \Gamma_{xz}, \Gamma_{xy}$ represent the stress and strain in different directions; ε_{vol} is the volumetric strain; ξ is the volumetric strain of the fluid relative to the solid; $\xi = 0$ represents that the object of study is undrained; $\overrightarrow{u_s}$ and $\overrightarrow{u_f}$ are the displacements of the solid and fluid in porous media, respectively; λ, G is the Ramet constant of porous media; C represents the compressibility of rock under the change in pore pressure; and ϕ is porosity.

It is known that:

$$E = G\frac{3\lambda + 2G}{\lambda + G} = 3K(1 - 2\upsilon) \quad (13.8)$$

$$\upsilon = \frac{\lambda}{2(\lambda + G)} \quad (13.9)$$

Then, it follows that:

$$P_{p\prime} = \frac{(2a + b - c)}{(1 - 2\phi)(a + 2b + 3c)}\sigma_v \quad (13.10)$$

$$a = (1 - \phi)/E$$
$$b = (1 - \upsilon)/2E$$
$$c = \phi/2(1 - \phi)E \quad (13.11)$$

where $P_{p\prime}$ is the pore pressure in deep shale reservoirs; E is the modulus of elasticity; and υ is the Poisson's ratio. The prediction formula is based on the constitutive equation of pore elasticity and considers the deformation of undrained rock, and the quantitative relationships between pore pressure and elastic modulus, Poisson's ratio and porosity are clarified. The model reflects the relationships among the overburden stress, shale mechanical parameters, porosity and pore pressure in deep shale reservoirs.

In particular, the effect of natural fractures on pore pressure should be considered when natural fractures are present in deep shale reservoirs. The prediction formula is as follows:

$$P_p'' = P_{p\prime} + \rho g h \quad (13.12)$$

where h corresponds to the height of the natural fracture.

13.2.2 Pressure Reconstruction of Deep Shale Reservoirs After Stimulation

13.2.2.1 Modelling Pressure Conduction in Shale

The horizontal section of a shale gas well is usually approximately 1000 m. Multistage fracturing can produce a fracture network that includes main fractures and different scales of subfractures (Fig. 13.1a). To investigate the pore pressure disturbance at the horizontal well scale, the modelled size is Lx = 1000 m in length and Ly = 500 m in width, which represents half of the horizontal plane along the horizontal wellbore section (Fig. 13.1b). Here, we use an ellipse (semimajor

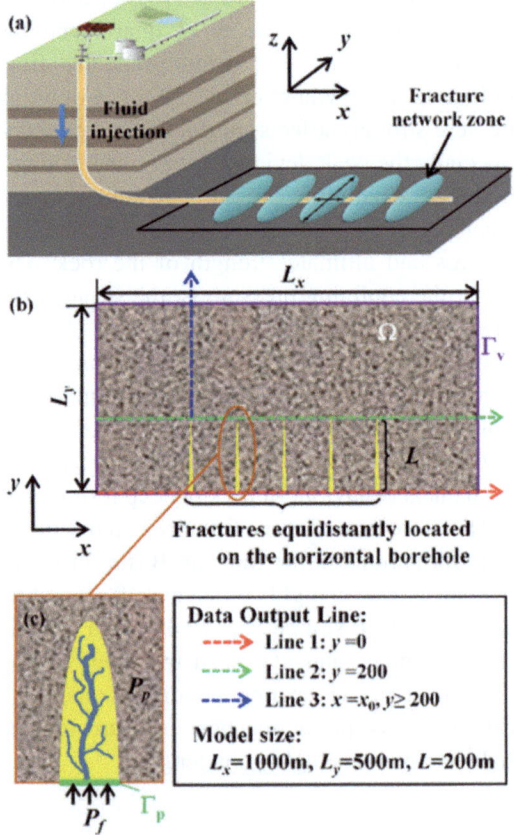

Fig. 13.1 Numerical model of pore pressure disturbance by hydraulic fracturing. The modelled domain size is 1000 × 500 m. The fracture network of each stage is represented by an elliptical zone (major axis = 200 m, short axis = 2 m) with high permeability

axis L $= 200$ m, short axis d $= 2$ m) with a very high permeability of $k_f = 10^8$ mD to represent the fracture network (as shown in Fig. 13.1c) compared with that of the low permeability of the shale formation ($k_m < 1$ mD).

The transportation of any single phase (fluid or gas) within a porous formation is controlled by the continuity equation according to the law of conservation of mass:

$$\frac{\partial(\rho\phi)}{\partial t} + \nabla \cdot (\rho \mathbf{V}) = 0 \qquad (13.13)$$

where ρ is the density of the fluid; t is time; ϕ is the porosity, and $\mathbf{V}$ is the percolation velocity. If fluid percolation follows Darcy's law, the motion equation should be

$$\mathbf{V} = -\frac{k}{\mu}\nabla P \qquad (13.14)$$

where μ is the viscosity of the fluid, k is the permeability of the formation, and ∇P is the pressure gradient. Combining Eqs. (13.13) and (13.14) yields the following:

$$\nabla \cdot \left[\rho\frac{k}{\mu}\nabla P\right] = \frac{\partial(\rho\phi)}{\partial t} \qquad (13.15)$$

Because the compressibility of the shale rock is much lower than that of fluid, the decrease in porosity with pressure is very small; thus, the right side of Eq. (13.15) can be further derived as follows:

$$\frac{\partial(\rho\phi)}{\partial t} = \phi\frac{\partial\rho}{\partial t} = \phi\rho_0\beta_p\frac{\partial(P - P_0)}{\partial t} \qquad (13.16)$$

where β_p is the fluid compressibility and P_0 is the reference pressure.

Therefore, Eq. (13.15) can be written as follows:

$$\nabla \cdot \left(\rho\frac{k}{\mu}\nabla P\right) = \rho_0\phi\beta_p\frac{\partial(P - P_0)}{\partial t} \qquad (13.17)$$

The boundary conditions in Fig. 13.1 can be expressed as follows:

$$P|_{t=0,\Omega} = P_0 \qquad (13.18)$$

$$P|_{t,\Gamma_p} = P_f \qquad (13.19)$$

$$\frac{k}{\mu}\nabla P \cdot \vec{n}\bigg|_{t,\Gamma_v} = 0 \qquad (13.20)$$

13.2.2.2 Pressure Reconstruction After Reservoir Stimulation

The modelling results of pressure reconstruction after reservoir stimulation for a group of parameters are shown in Fig. 13.2. The zone affected by pore pressure disturbance increases with increasing duration of differential pressure.

13.2.3 Prediction Method for Determining Pore Pressure in Deep Carbonate Reservoirs

13.2.3.1 Current Studies of Pore Pressure Prediction Methods in Carbonate Reservoirs

Carbonates represent a key exploration area for deep and ultradeep oil and gas resources, and monitoring pore pressure is key for ensuring the safety of well control; however, owing to complex rock formations and the large differences in the properties of tuffs and dolomites, monitoring the pore pressure of carbonates has become a considerable problem. Long-term research on pore pressure prediction and monitoring methods for deep and ultradeep carbonate rocks has been carried out worldwide; the pore pressure-related carbonate rock properties, such as the physical properties, fluid characteristics, acoustic properties and mechanical properties of carbonate rocks, have been analysed, and the pore pressure genesis mechanism of carbonate rocks has been explored, on the basis of which four types of pore pressure prediction and monitoring methods for carbonate rocks have been developed.

First, in the effective stress method, the effective stress is the stress acting on the rock skeleton particles, and the overburden at any point in

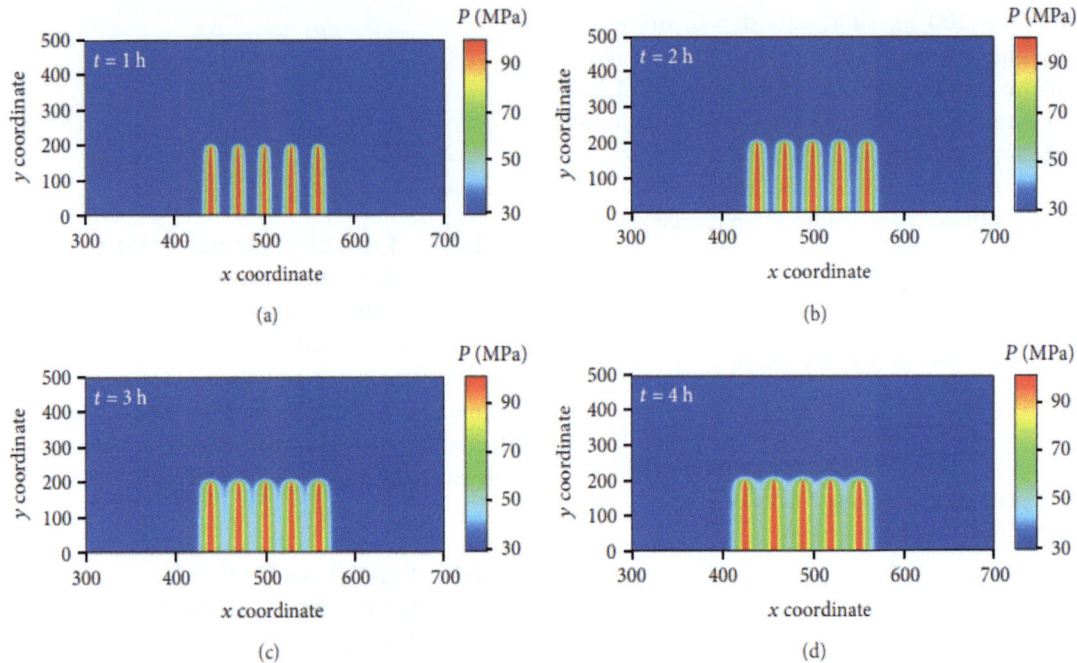

Fig. 13.2 Pore pressure distributions at different times from the modelling ($k = 0.2$ mD, $\mu = 1$ mPa $\cdot$ s, $P_0 = 30$ MPa, $\Delta P = 70$ MPa)

the formation is borne by the skeleton particles and the fluid in the pores. Second, the volume variation measurement method directly establishes a comprehensive model of pore pressure and stress (effective stress, mean principal stress, and stress difference) and strain (porosity, volumetric compression coefficient, pore compression coefficient, skeleton volumetric modulus, matrix volumetric modulus, and pore fluid volumetric modulus)-related parameters, is used. Third, the sound velocity separation method is based on the theory that small variations in the longitudinal wave velocity of carbonate rocks at different pore pressures are caused by fluid sound velocity. Fourth, the causal contribution method fully considers the three major pressure causal contributions from rock pore volume change, pore fluid volume change, and fluid movement.

13.2.3.2 Acoustic and Mechanical Response of Carbonate Rocks

The acoustic velocity of stratigraphic rocks is governed by the rock skeleton, pore structure and pore fluid together, and changes in the rock skeleton deformation and pore fluid pressure cause changes in the rock acoustic velocity. By conducting experiments on the acoustic response characteristics of carbonate rocks, the relationships among the variations in the acoustic velocity, rock skeleton deformation and pore fluid pressure were established.

Accordingly, experiments on the acoustic response characteristics of carbonate rocks were carried out. Two common marine carbonate formations, fractured and porous, were used for the experiments (Ø25 mm × 50 mm), as shown in Fig. 13.3). The porosity, permeability, and longitudinal velocity of the samples were measured in the dry state.

The samples were then saturated with water, and the longitudinal and transverse velocities of the samples were measured under different pore pressures, circumferential pressures, and axial pressures using a rock mechanics experimental device and an ultrasonic instrument.

The experimental results reveal that the change in pore pressure of carbonate rock has no effect

Fig. 13.3 Carbonate rock samples used in experiments

(1) Porosity carbonate rocks   **(2) Fractured carbonate rocks**

on the transverse wave velocity, whereas the longitudinal wave velocity decreases slightly with increasing pore pressure (the maximum decrease is only 3.5%). The longitudinal wave velocity of the skeleton of carbonate rocks is essentially unchanged because the pore pressure of the formation has less influence on the structure of the skeleton of carbonate rocks. The small variation in the longitudinal wave velocity of carbonate rock is caused by the variation in pore fluid velocity, and the pore pressure of carbonate rock formation can be determined based on the longitudinal wave velocity of the pore fluid.

13.2.3.3 Prediction Model and Method for Determining Pore Pressure in Carbonate Reservoirs

The longitudinal wave velocity of carbonate rock obtained by experiments and logging is the longitudinal wave velocity under the combined effect of the rock skeleton and pore fluid. In accordance with the signal characteristics of the logging acoustic velocity, the db3 wavelet basis function of the Daubechies wavelet family is selected to decompose the logging longitudinal velocity signal by combining the definition of the wavelet function and the more mature and commonly used wavelet basis functions to obtain the low-frequency and high-frequency components to separate the longitudinal velocities of the pore fluid and rock skeleton.

The adaptive method based on the white noise test is combined with the logging data signal containing white noise, which is a finite random sequence. According to time series analysis theory, autocorrelation function estimation can be used to check whether this sequence is consistent with the white noise characteristics to determine the number of wavelet decomposition layers. The results of the two methods can be mutually verified to obtain a reasonable number of decomposition layers.

Taking a well as an example, using the adaptive method based on the white noise test, the number of decomposition layers of the longitudinal time difference of a carbonate formation in the well is 5, and the longitudinal time difference of the formation is decomposed using the db3 wavelet base; the decomposition result is shown in Fig. 13.4, which shows that the change trend of ca6 obtained after the decomposition of 6 layers is completely different from that of the original longitudinal time difference. This finding shows that the trend value obtained after decomposition is distorted and cannot reflect the original longitudinal time difference trend. A comparison of the trend of ca5 with the original longitudinal time difference trend reveals that the trend changes are essentially the same. Thus, choosing the 5-layer decomposition is reasonable, and the trend of cd5 is used to identify the anomalous high-pressure zone.

For the high- and low-frequency parts decomposed from the longitudinal wave velocity of carbonate rock, the pore pressure prediction method of carbonate rock formation suitable for

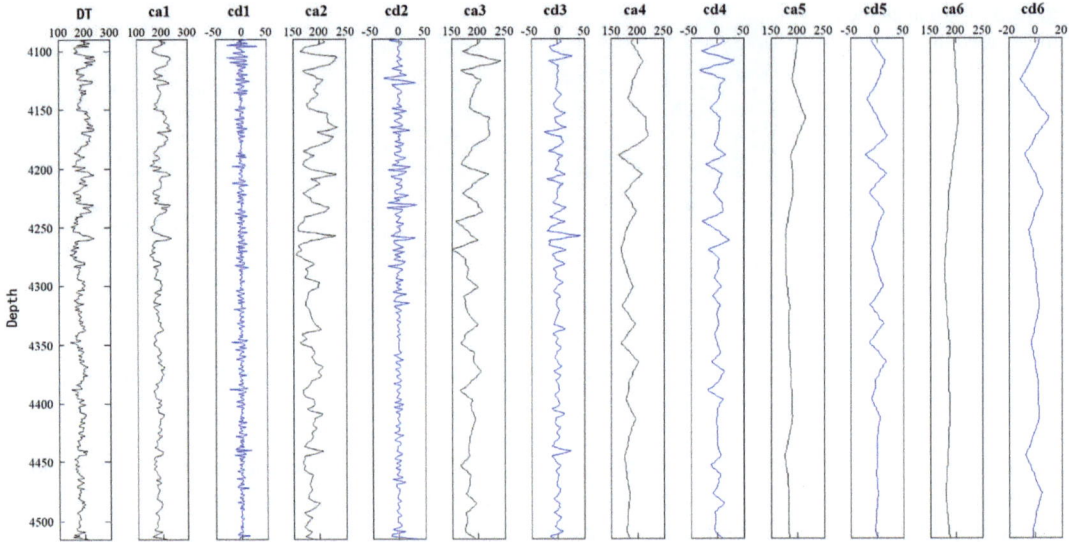

Fig. 13.4 Determination of the number of layers of longitudinal time difference decomposition of carbonate formations in a well

the region can be obtained by comparing the field-measured pore pressure value of the formation with the acoustic wave velocity of the high-frequency part.

13.2.4 Field Application

13.2.4.1 Application of the Pore Pressure Prediction Method in Shale Reservoirs

The predicted pore pressures of several wells in the Sichuan Basin, China, are in good agreement with the measured values in Fig. 13.5, indicating that this method has good applicability. The selected wells, in the range of 3500–4200 m, are all deep and contain deep shale formations that have been explored.

13.2.4.2 Application of the Pore Pressure Prediction Method in Carbonate Reservoirs

The initial application of the fluid velocity-based pore pressure prediction method for carbonate

formations has been carried out in the northeastern Sichuan and Yada regions of the Sichuan Basin and Iran, and longitudinal wave velocity decomposition and high-pressure formation identification processes for carbonate formations in the corresponding regions have been established, with a compliance rate of more than 85% for high-pressure formation identification, which scientifically guides drilling design and risk prevention and control.

In a well from 4250 m for carbonate formations shown in Fig. 13.6, combined with the longitudinal velocity decomposition of the low-frequency part (ca5) and the measured formation pressure value comparison, the 4250–4400 m low-frequency curve markedly deviates from the centerline; 4400 m below the stratum low-frequency curve is essentially near the centerline, and the 4250–4400 m stratum pore pressure is higher than the stratum below 4400 m. Therefore, the high-frequency part of the longitudinal velocity curve strongly fluctuates in the high-pressure strata, according to which a certain area can be used to identify the high-pressure strata well using the change law of the high-frequency part.

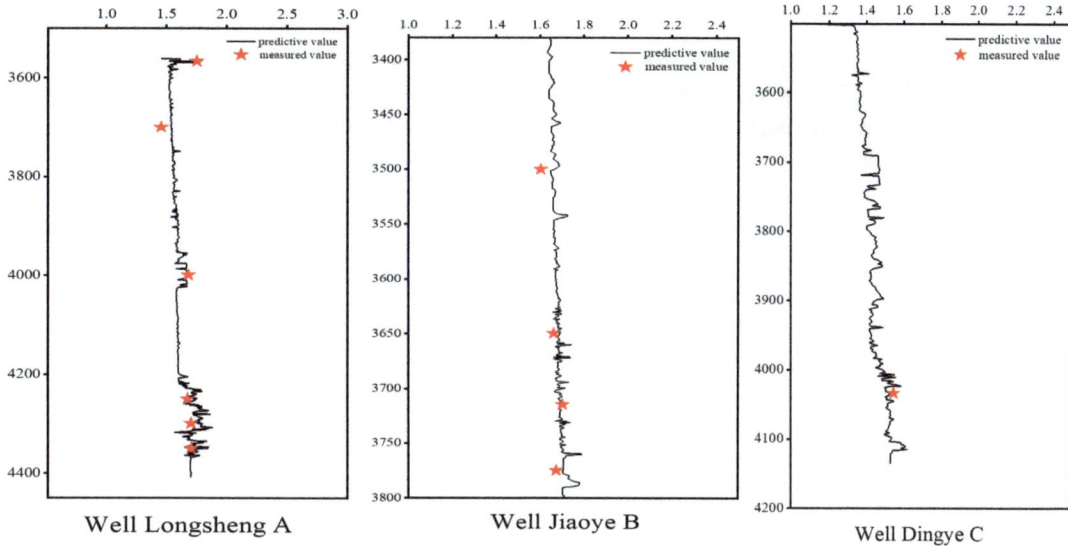

Fig. 13.5 Prediction results of pore pressure in the well

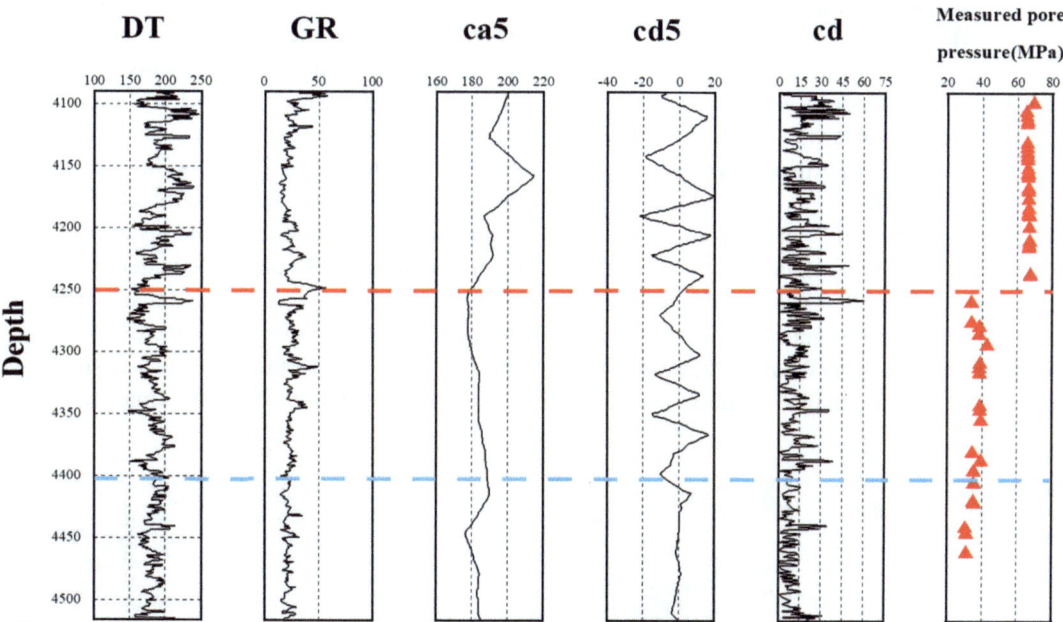

Fig. 13.6 Longitudinal wave velocity decomposition of carbonate formation in a well compared with the measured formation pore pressure

13.3 Rock Breakage Mechanisms and Methods Using PDC Bits Assisted by Axial–torsional Coupled Percussion

Hard rocks are often encountered in deep well drilling. The excessive wear of drill bits and abnormal vibrations in drilling tools lead to low penetration rates and high drilling costs. Percussive drilling can significantly increase the breaking efficiency of hard brittle rock. Generally, percussive drilling includes axial percussive drilling, torsional percussive drilling, and axial–torsional coupled percussive drilling.

13.3.1 Mechanisms of a PDC Cutter Breaking Hard Rocks

13.3.1.1 Rock Breakage Mechanisms of a Conventional PDC Cutter

PDC bits were first put into use in oil drilling in the 1970s, and more than 90% of oil and gas wells have been drilled using them to this day due to their longevity and high rate of penetration (ROP). To elucidate the rock breakage mechanism of conventional PDC cutters, a series of cutting experiments were conducted on a single cutter test rig (Fig. 13.7). The rock sample is fixed on a horizontal sliding table, which can move at a constant velocity (1.6–230 mm/s) on two horizontal slide rails. When the experiment begins, the PDC cutter remains still, and the rock moves in a straight line. The PDC cutter is mounted on the cutter holder and locked by three locking screws once the cutting depth and cutting angle are set. Four load sensors are mounted in front of the rock sample along the cutting direction, which can measure cutting forces ranging from 0 to 10 kN with a precision of 1 N. The control/measurement system can control the cutting speed and the cutting length through user programming and record the cutting force.

During the cutting tests, the formation of major cracks was captured in situ with a high-speed camera. The subsurface cracks were then studied by reproducing the cutting groove after some rock blocks were cut and further observed by means of thin-section optical microscopy. Finally, the development sequence of the cracks and crushed zone was determined by recording the cutting process on a rock-like silica glass block. The major cracks before the cutter were mainly curved shear cracks, and the cracks under the cutter were linear tensile cracks. The rock at the contact position of the cutter was compressed to powder to form a crushed zone. The formation of the crushed zone occurred more than 95% of the time during the

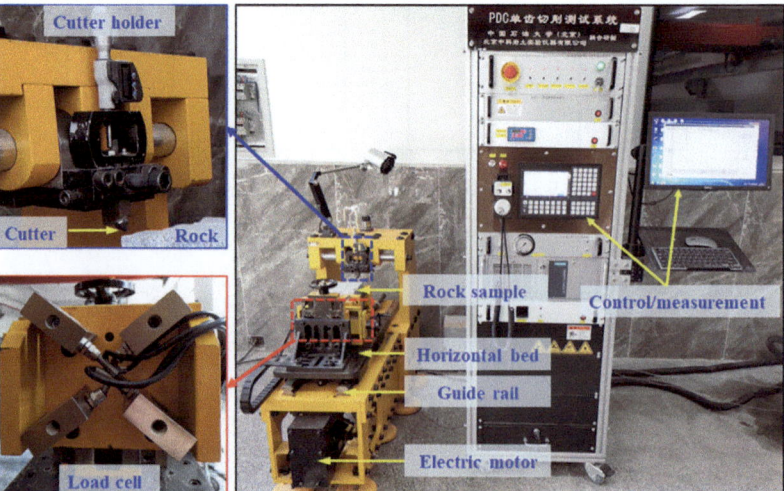

Fig. 13.7 Experimental setup for the single cutter tests

rock breaking process, and an arc-shaped shear crack in front of the cutter initiated in the crushed zone. When the major crack propagated to the rock-free surface, large cuttings formed and flew away. The PDC cutter accelerated suddenly and generated an impact on the rock, resulting in the formation of a tensile crack under the cutter. Thus, the rock breaking mechanisms of the conventional PDC cutter were compaction crushing on the cutter surface, shear crushing on the cutter front, and tensile crushing on the cutter bottom.

13.3.1.2 Rock Breakage Mechanism of a Special-Shaped PDC Cutter

An innovative Stinger PDC cutter has been proposed for high rock-breaking performance in hard formations. The Stinger PDC cutter is conical with a three-dimensional cutting surface, and the conventional PDC cutter is cylindrical with a flat cutting edge. Different geometric shapes cause completely different rock-breaking mechanisms (Xiong et al. 2020). To investigate the cutting mechanism of Stinger PDC cutters, a series of cutting tests were conducted on a single cutter test rig, and the rock stress response and damage evolution characteristics of the Stinger PDC cutter breaking rock were studied by numerical simulation.

The Stinger PDC cutter could create concentrated point loading at the tip of the cutter, which compressed the rock and formed a crushed zone, and numerous microcracks were initiated at the periphery of the crushed zone. As the cutter moved forwards, the microcracks at the front of the cutter propagated to the rock surface and formed major arc-shaped cracks under tensile stress. Volumetric breakage then occurred, and large cuttings were produced. Furthermore, owing to the microcracks, a damaged area at the bottom of the cutting groove formed, which could help to improve the rock-breaking efficiency of the next cutting. Thus, the Stinger cutter first applied high concentrated point loading to induce microcracks in the rock, and the tensile stress significantly promoted the propagation of microcracks and produced large debris, leading to high rock-breaking efficiency.

13.3.1.3 Rock Breakage Mechanism of a Multicutter Combination

Compared with conventional PDC cutters, Stinger PDC cutters exhibit higher rock breaking efficiency and excellent impact resistance, wear resistance, and life duration. However, the Stinger PDC cutter has a low bottom hole coverage ability, which presents challenges for cutter arrangement in PDC bits. At present, the Stinger PDC cutter is usually arranged on the PDC bit together with conventional PDC cutters in the hybrid PDC bit. To elucidate the rock breaking mechanism of the combination of the Stinger PDC cutter and conventional PDC cutter, an experiment involving mixed tool cutting of carbonate rock with the Stinger PDC cutter and conventional PDC cutter was carried out.

Under the action of the Stinger PDC cutter, lateral cracks and bottom cracks were generated around the cutting groove, causing damage to the rock. The conventional PDC cutter gave full play to the advantages of a large rock breaking volume and strong bottom hole coverage and stripped the damaged rock around the cutting groove of the Stinger PDC cutter so that the mixed tool cutting had both high rock breaking efficiency and high bottom hole coverage ability.

13.3.2 Hard Rock Breakage Mechanism Under Axial–Torsional Coupled Percussion

13.3.2.1 Axial–Torsional Coupled Percussive System Model

To study the rock breakage mechanism of axial–torsional coupled percussive drilling, a three-dimensional percussive system model in ABAQUS was established, as shown in Fig. 13.8a. Rocks were meshed with the structured grid. The total number of elements of the rock model is 317,269. The hemispherical cutter and the multiple-cutter bit were selected. The size of the cutter is $\Phi 12$ mm. In the percussive process, the impact stress wave is transmitted to the bit along

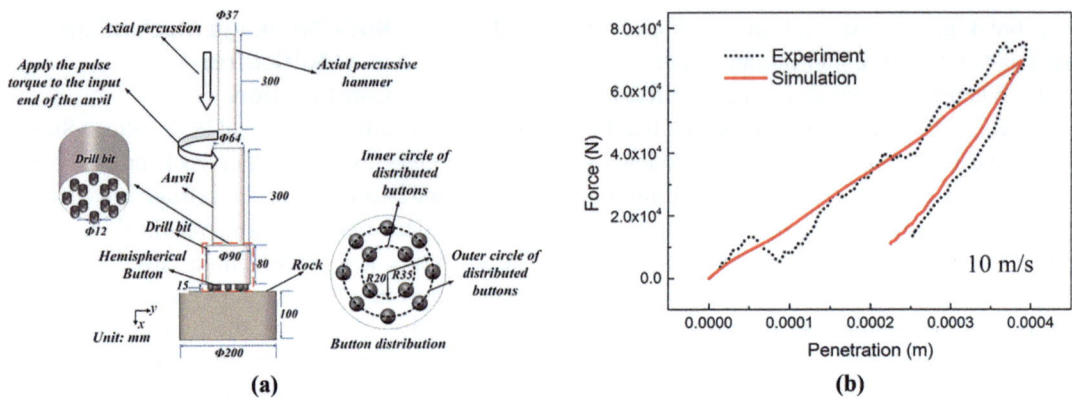

Fig. 13.8 Model development and verification: **a** geometric model of the axial–torsional coupled percussive system; **b** force–penetration curves

Table 13.1 Model parameters for rock

Parameter	Value
Elastic modulus (GPa)	33.7
Poisson's ratio	0.26
Dilation angle (°)	5
Shape factor (Kc)	0.667
Flow stress ratio	1.16
Eccentricity	0.1
Viscous factor	0.0005
Tensile strength (MPa)	12.8
Compressive strength (MPa)	91.1
Density (kg/m^3)	2630
Compressive stiffness recovery	1
Tensile stiffness recovery	0

the anvil and then exerted through the cutter on the rock (Song et al. 2019a, b).

The plasticity model of damaged concrete is selected as the rock damage model. The model parameters for the rock are listed in Table 13.1. We regard the percussive hammer, anvil and drill bit as elastic deformation bodies and regard the cutters as rigid bodies. The elastic modulus of the drill bit, the anvil and the axial percussive hammer is set to 210 GPa. The Poisson's ratio of the drill bit, the anvil and the axial percussive hammer is set to 0.28. The density of the drill bit, the anvil and the axial percussive hammer is set to 7800 kg/m^3. The density of the bit buttons is set to 15,630 kg/m^3.

At present, few studies have considered the percussive process of torsional percussion and axial–torsional coupled percussion. However, many studies have investigated the percussive process of axial percussion. Based on the rock damage model constructed above, we constructed the same percussive system model in ABAQUS as the axial percussive test system in previous research (Saksala et al. 2014). From the simulation results, we can obtain the force versus bit penetration responses, as shown in Fig. 13.8b. There is good agreement between the simulated and experimental bit force–penetration curves at a velocity of 10 m/s.

13.3.2.2 Characteristics of Stress Response and Damage Evolution

Under axial–torsional coupled percussion, the bit impacts the rock along the axial direction and simultaneously shears the rock along the circumferential direction (Song et al. 2020). The dynamic stress distribution of the rock under axial–torsional coupled percussion is shown in Fig. 13.13a. The tensile, compressive, and shear stresses on the rock samples were analysed. The maximum principal stress distribution could distinguish between the tensile and compressive stress zones of the rock. The compressive stress was distributed mainly in the area directly under the cutter, which easily formed crushed craters. The tensile stress was distributed mainly between adjacent cutters, which were prone to the generation of tensile cracks, as well as the connection of existing tensile zones. The shear stress was distributed mainly in the front area of the cutter, which could cause shear fragmentation of the rock. These three forms of fragmentation represent the main rock-breaking mechanisms of axial–torsional coupled percussive drilling. Coupled percussive drilling combines the advantages of axial percussion and torsional percussion to form multimodal rock fragmentation. Compared with single axial percussion, the damaged areas between the cutter were larger under axial–torsional coupled percussion, which was more conducive to improving rock breaking efficiency.

Next, we analyse the percussive process and rock damage distribution in axial–torsional coupled percussive drilling. The percussive process and the bit penetration process are shown in Fig. 13.9b. Under axial–torsional coupled percussion, the bit impacts the rock along the axial direction and simultaneously shears the rock along the circumferential direction. In the percussive process, the damaged area of rock rapidly expands. The damaged areas caused by the cutters gradually become interconnected. Compared with single axial percussion (Song et al. 2019a, b), the damaged areas between cutters are larger under axial–torsional coupled percussion, which is more conducive to efficient rock breaking.

13.3.2.3 Influence of Various Impact Parameters

In this section, the effects of the impact velocity of the axial percussive hammer and pulse torque amplitude on the percussive process in axial–torsional coupled percussive drilling are analysed. The simulation parameter settings are as follows. The impact velocities of the axial percussive hammer are set to 4, 6, 8, 10, 12, 14, and 16 m/s. The pulse torque amplitude is set to 0, 4000, 8000, 12,000, 16,000, 20,000, 24,000, and 28,000 N m.

To clearly show the characteristics of the above relationship curves, we present only the output force–penetration curves and the output torque–angle curves at impact velocities of 4, 8, 12, and 16 m/s. Figure 13.10a presents the relationships between the output force and the penetration depth. Both the maximum output force and the maximum penetration depth increase with increasing impact velocity. A decrease in the output force to zero means that the percussive process is over. The penetration depth at the end of the percussive process also increases as the impact velocity increases.

The output force versus penetration depth responses at different pulse torque amplitudes are shown in Fig. 13.13b. The pulse torque amplitude has a small effect on the maximum output force and the maximum bit penetration. With increasing amplitude, the maximum output force slightly decreases, whereas the maximum penetration slightly increases.

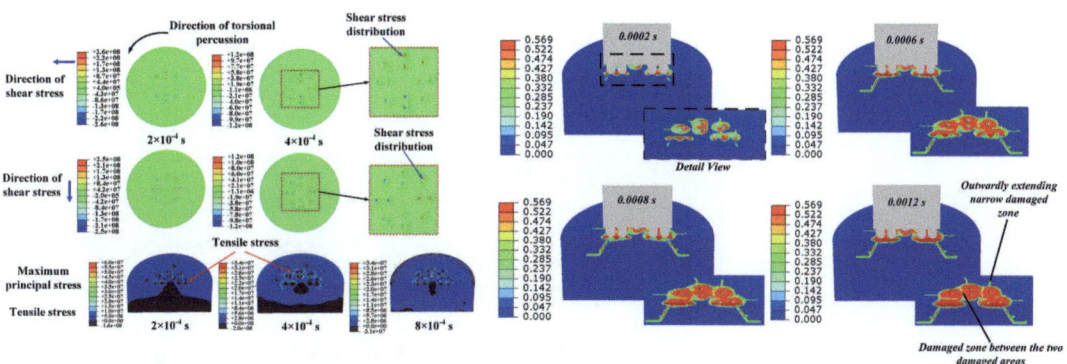

Fig. 13.9 Rock breakage characteristics: **a** dynamic stress distribution of the rock under axial–torsional coupled percussion; **b** percussive process and rock damage distribution

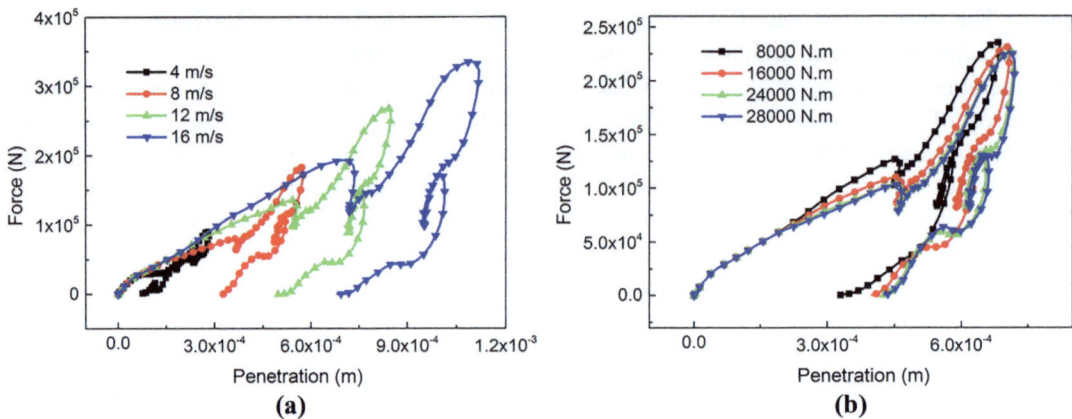

Fig. 13.10 **a** Relationships between the output force and penetration depth at different impact velocities; **b** relationships between the output force and penetration depth at different pulse torque amplitudes

13.3.3 Speed-Up Technology for Axial–Torsional Coupled Percussive Drilling

Axial–torsional coupled impact drilling technology significantly enhances drilling speed and efficiency by converting stable weights on bits into high-frequency torsional impacts, thereby directly reducing operational costs, prolonging tool longevity, and improving drilling safety.

13.3.3.1 Axial–Torsional Coupled Impact Tool and PDC Bit Design

As shown in Fig. 13.11, the axial–torsional coupling impact tool is composed mainly of an axial impact unit and a torsional impact unit. During an axial impact process, the drilling fluid drives the disc valve to rotate. When the arc of the disc valve rotates to the lower liquid flow channel of the axial hammer seat, the drilling fluid pressure drives the axial impact hammer to strike upwards,

Fig. 13.11 Structural diagram of the axial–torsional coupling impact tool

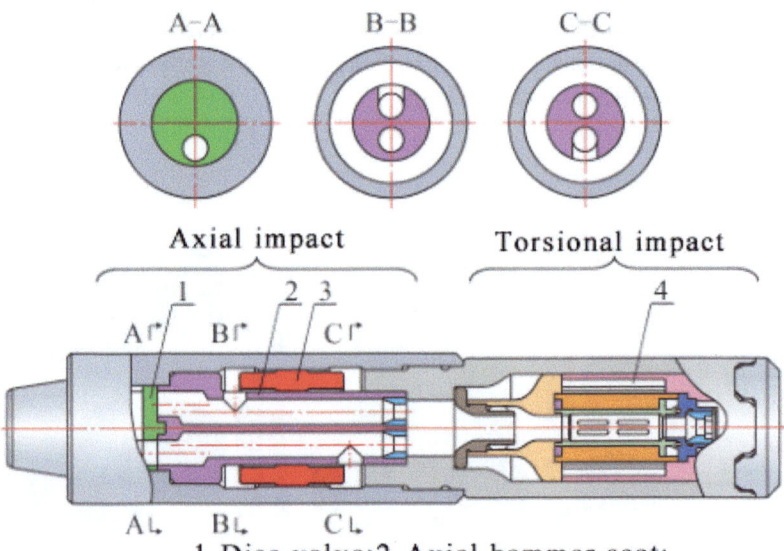

and vice versa. Thus, the periodical rotation of the disc valve can result in periodic changes in the axial impact force. During a torsional impact process, a high pressure can be generated on one side of the torsional impact hammer, and a low pressure is generated on the other side. Thus, the difference in pressure drives the impact hammer to the origin, leading to periodic changes in the torsional impact force. Consequently, the periodic changes in both the axial impact force and the torsional impact force can help the bit break the rock.

To better adapt to the characteristics of the axial–torsional coupling tool, in deep hard formations, a 5-blade PDC bit with 16-mm-diameter cutters is suggested. After the preliminary design is completed, the cutting structure parameters and hydraulic structure parameters of the PDC bit are optimized using the finite element analysis method. This approach is expected to further improve the bit footage and ROP and other performance indicators.

13.3.3.2 Field Application

A field application has been conducted to verify the rock-breaking efficiency of PDC bits assisted by axial–torsional coupled percussive drilling in Well Y*** in Chongqing City, China. The drilling history revealed that the section beyond the drilling depth of 3000 m was common, with severe bit failure and low ROP. Then, the axial–torsional coupling impact tool and supporting PDC bit were applied in the 3102–4230 m section. It operated for 212 h in the wellbore downhole and achieved a drilling footage of 1128 m. Meanwhile, the average ROP reached 5.32 m/h, which was an increase of 43.59% compared with that of the adjacent wells. After the tool and the bit were pulled out, there was no significant deformation or damage to the outside wall of the tool and no vibration damage to the internal structure of the tool or to the PDC cutters.

13.4 Failure Mechanism and Control Method of Wellbore Integrity in a Shale Gas Well

The phenomenon of annulus pressure in shale gas wells appears after fracturing in China, which affects shale gas exploitation and highlights safety and environmental protection issues. This annular pressure buildup frequently stems from formation–fluid migration through compromised cement sheaths or microannuli, creating significant drilling challenges such as wellbore instability in laminated shales and casing deformation under cyclic stress. With respect to completion strategies, sustained casing pressure impedes multistage fracturing efficiency by limiting diverter effectiveness and increasing risks during plug-and-perf operations. Additionally, uncontrolled methane migration increases groundwater contamination risks and carbon emission penalties under China's "dual carbon" policy. Therefore, it is necessary to study the wellbore sealing integrity under cyclic loading to simulate staged fracturing and provide a control method.

13.4.1 Mechanical Characteristics of Cement Under Complex Conditions

13.4.1.1 Mechanical Properties Under High Temperature and Pressure

The mechanical properties of cement were tested at room temperature, 95 °C and 130 °C under 15 MPa of confining pressure. At room temperature, the initial stress–strain curve increased linearly, the elastic modulus was approximately 8.5 GPa, and the Poisson's ratio was 0.20. After the deviatoric stress exceeded approximately 41 MPa, the loading curve gradually deviated from the original linear state, and the cement entered the nonlinear deformation stage. The peak stress was 56.8 MPa. After the test, oblique macroscopic fracture was caused by shear failure of the sample. At 95 °C, the stress–strain curve also increased linearly during the initial stage; the elastic modulus was 7.41 GPa, and the Poisson's ratio was 0.105. The critical value was 28.9 MPa, which was lower than that at room temperature. The subsequent stress–strain curves revealed a significant strain hardening phenomenon; that is, as the axial strain increased, the bearing capacity of the cement continued to increase. The axial strain exceeded 4.0%, the deviatoric stress reached approximately 66.4 MPa, there were still no macroscopic cracks on the sample surface, and the sample height was significantly shortened. At 130 °C, the same characteristics as those at 95 °C were observed, and the plastic deformation ability was greater. The results of triaxial compression tests at room temperature and high temperature suggested that the mechanical properties of the cement greatly differed under different temperature conditions: at room temperature, the elastic modulus of the cement was relatively high; at high temperature, the elastic modulus decreased to some extent, and its deformation properties greatly improved.

13.4.1.2 Deformation Characteristics Under Cyclic Loading

To clarify the deformation characteristics of cement under staged fracturing of shale gas wells, a cyclic loading test was carried out. In the initial loading stage of the cyclic loading test, the stress–strain curve essentially coincided with the triaxial test curve at room temperature; however, the curve in the unloading phase no longer coincided with the curve in the loading phase, and there was significant hysteresis, indicating that unrecoverable residual deformation was generated inside the sample. As the number of cycles increased, the "hysteresis loop" formed by the loading curve and the unloading curve continuously moved to the right; accumulative additional strain was formed for each

cycle, and the residual deformation continuously increased.

The cyclic loading/unloading stress–strain curve at high temperature presented a different morphology, which illustrated that the loading and unloading curve could be divided into four stages: the initial stage of loading, with a small sample deformation and a rapid increase in stress; the later stage of loading, with relatively slow stress growth and relatively rapid sample deformation; the initial stage of unloading, in which the stress decreased rapidly, the deformation recovery of the sample showed an obvious lag, and no obvious rebound was observed; and the later stage of unloading, in which the axial strain of the sample decreased rapidly with decreasing stress, and the rebound was significant. Moreover, as the number of cycles increased, the hysteresis loop became denser, indicating that the growth of plastic deformation accumulation slowed. At high temperature, the unloading curve exhibited a significant "rebound lag", and the unloading secant modulus also increased.

13.4.1.3 Deterioration Mechanism Under Complex Conditions

Under confining pressure, the failure mode of the cement is no longer brittle and results in a large plastic deformation, the elastic modulus decreases, and the strength greatly improves, especially at high temperatures. The main reason is that microcracks easily occur in cement and expand rapidly under uniaxial compression, resulting in brittle failure; however, it is not easy to produce microcracks and expand when there is a confining pressure constraint. Therefore, the deformation capacity and strength are improved. Under cyclic loading, residual strain is produced and gradually accumulates with increasing loading time. Deterioration occurs inside the cement, and the deterioration mechanism is that the cement contains a large number of pores of different sizes and a small number of microcracks before being subjected to external loading. Under the action of loading, stress concentration occurs at the defects and crack tips. The internal part of the pore structure collapses, the microcracks expand, and viscous flow of the gel occurs, which results in plastic deformation and unrecoverable residual strain after unloading. Under repeated cyclic loading, the damage at the crack tip increases, the crack tends to grow, the plastic deformation increases, and the cumulative residual strain increases gradually after unloading.

13.4.2 Failure Mechanism of the Sealing Integrity of the Cement Sheath Under Dynamic Loading

13.4.2.1 Full-Size Evaluation Device for a Cement Sheath

A full-size evaluation device for a cement sheath was developed, which consists of a wellbore system, a temperature control system, a pressure system, a gas injection and gas channelling detection system, a control and acquisition system, an ejection system and other components. The wellbore system includes a casing, a cement sheath and an outer cylinder. A casing with an OD of 139.7 mm and a wall thickness of 7.72 mm is preferred with a steel grade of P110, and metallic cylinders of different thicknesses with a steel grade of N80 are applied. On the one hand, the temperature control system can be used for high-temperature curing of the cement sheath by heating the water inside the casing; on the other hand, the system can be matched with cooling by the continuous circulation cooling water. The pressure system mainly applies pressure to and releases pressure from the casing to test the sealing of the cement sheath under cyclic pressure with a pressure no greater than 120 MPa. The gas injection system injects nitrogen under a certain pressure into the annulus and determines the gas injection flow rate and the total flow rate. A dedicated control and acquisition system was developed to provide automatic experimental control and enable data acquisition. To reuse the borehole

system and reduce test costs, a demoulding system was developed.

13.4.2.2 Failure Mechanism of Cement Sheath Sealing

With the full-size cement sheath evaluation device described above, the sealing failure rules of the cement sheath under different internal pressures were determined. In the test, when the internal pressure load reaches its maximum value, no sealing failure is observed, which indicates that there is no microcracking in the cement sheath body caused by tensile or shear strength failure. After several cycles of alternating stress, gas channelling begins to occur when the pressure is reduced to a lower value, and a lower pressure may cause a larger gas vent volume. After that, the integrity of the cement sheath is restored by pressurizing the sheath, which indicates that a microannulus is formed on the sheath to cause cement sheath sealing failure. Under the condition of lower internal pressure, after several cycles of alternating internal pressures, the integrity of the cement sheath is maintained. Under higher internal pressures, the sealing of the cement sheath fails after repeated internal pressure cycles. As the internal pressure increases, the number of stress cycles for sealing failure decreases. When the internal pressure is low, the entire cement sheath interface is in the elastic force stage. Along with the increase in internal pressure, the region near the 1st interface experiences greater stress and first enters the plastic force stage, whereas the outside region is still in the elastic force stage. When the internal pressure exceeds a certain value, the cement sheath may completely enter the plastic stress state, with no elastic zone. After several cycles of alternating stress, gas channelling begins to occur when the pressure is reduced to a lower value, and a lower pressure may cause a larger gas vent volume. After that, the integrity of the cement sheath is restored by pressurizing the sheath, which indicates that a microannulus is formed on the sheath to cause cement sheath sealing failure.

13.4.2.3 Determination Criteria and Control Method for Cement Sheath Sealing Integrity

According to the research findings, three forms of failure of the cement sheath seal are considered here: compression failure, circumferential tensile failure and interface microannulus. The determination criteria are as follows.

(1) Compressive plastic failure

The criterion of compressive damage is as follows:

$$\sigma_e \geq \sigma_s \qquad (13.21)$$

$$\sigma_e = \sqrt{\frac{1}{2}\left[(\sigma_r - \sigma_\theta)^2 + (\sigma_\theta - \sigma_z)^2 + (\sigma_z - \sigma_r)^2\right]} \qquad (13.22)$$

where σ_e is the equivalent stress, in MPa, and σ_s is the yield stress, which is the stress at which the stress–strain curve deviates from the initial straight line segment. σ_r, σ_θ, and σ_z are the radial, circumferential and axial principal stresses, respectively, in MPa.

(2) Circumferential tensile failure

The tensile failure criterion is as follows:

$$\sigma_\theta \geq \sigma_t \qquad (13.23)$$

where σ_θ is the circumferential principal stress, in MPa. σ_t is the axial tensile strength of the cement, in MPa.

(3) Interface cementation failure

The criterion for cementation failure is as follows:

$$\sigma_N \geq \sigma_B \qquad (13.24)$$

where σ_N is the interface normal stress, in MPa, and σ_B is the interface tensile bonding strength, in MPa.

According to the failure mechanism and judgement criteria, the failure control method can be determined. First, the elastic modulus of the cement can be reduced, which can reduce the stress in the cement sheath under loading, such as during fracturing. This purpose can be achieved by adding elastic particles and latex into the cement. Second, the strength of the cement, including its compressive strength and tensile strength, should be maintained at high levels. This can be achieved by adding nanomaterials and fibre materials. Strength failure does not easily occur with high-strength cement; the upper limit of stress under cyclic loading is low, the residual strain produced is low, and cementation failure does not easily occur. Finally, the interfacial cementation performance is improved; thus, the cement sheath is not prone to interface cementation failure. Interface cementing enhancers and improving cementing quality are beneficial to this process.

13.4.3 Casing Damage Mechanism and Control Method for Shale Gas Wells

13.4.3.1 Geomechanical Environment for Casing Damage

The Weiyuan anticline is a basement-involved anticline that varies from gentle to steep. The fault is present only in the basement and does not extend across the shale formation. However, tectonic movement does not occur only in the basement. Hence, the accumulation of large tectonic movements can induce high geostress in shale formations, and the dominant natural fractures can be easily reactivated. Ultimately, reactivated natural fractures may cause shear deformation of the casing. Natural reactivation is the main reason for the deformation of the casing in this area.

13.4.3.2 Natural Fracture Reactivation for Casing Damage

The slip of natural fractures is reactivated by the hydraulic fracturing fluid invading the dominant natural fractures, and ultimately, shear deformation of the casing occurs in wells because of the slip of natural fractures.

The method for calculating the trend factor of shear activity (Tong et al. 2015) can be expressed as follows:

$$f_s = \frac{\sqrt{\begin{array}{l}(\sigma_1^2 \sin^2\theta + \sigma_2^2 \cos^2\theta \cos^2\alpha + \sigma_3^2 \cos^2\theta \sin^2\alpha) \\ -(\sigma_1 \sin^2\theta + \sigma_2 \cos^2\theta \cos^2\alpha + \sigma_3 \cos^2\theta \sin^2\alpha)^2\end{array}}}{C_p + \mu_p(\sigma_1 \cos^2\theta + \sigma_2 \cos^2\theta \cos^2\alpha + \sigma_3 \cos^2\theta \sin^2\alpha)} \quad (13.25)$$

σ_1, σ_2 and σ_3 are the maximum in situ stress, vertical stress and minimum in situ stress of the formation rock, respectively; θ is the angle between the interface and σ_1; α is the angle between σ_3 and the intersection line of the interface of the σ_2–σ_3 plane; C_p is the cohesion of the interface; and μ_p is the internal friction coefficient.

When fs = 1, the interface of the fracture is in a critical shear-activated state. When fs > 1, the fracture is in a shear-activated state. When fs < 1, the fracture is in a stable state.

13.4.3.3 Casing Damage Control Technology

(1) Temporary plugging technology

According to previous studies, when a channel for fracturing fluid flows into a natural fracture, the fracturing fluid of the whole fracturing section preferentially converges together into a natural fracture, and a "funnel" is generated in the natural fracture. This type of "funnel" will be useful for temporary plugging. When the "funnel" is formed in a natural fracture, temporary plugging material can be added to the fracturing fluid, and the fracturing fluid with temporary plugging material will preferentially converge into the "funnel" to plug the "funnel". If the "funnel" is plugged, the normal hydraulic fracturing treatment can be restarted.

After the hydraulic fracturing is complete, the temporary plugging materials are dissolved by the formation fluid and do not affect subsequent production. According to the above analysis, it is not only necessary but also feasible to plug the high-speed leakage channel (large fracture) of the fracturing fluid using temporary plugging technology.

(2) Fracture identification and analysis

Three-dimensional earthquake interpretation is used to identify and interpret faults, including the fault features, fault horizon, extension length and fault distance. However, no fault can be directly identified by seismic data in any of the platform data. Therefore, no fault is considered, and three-dimensional earthquake interpretation is used to identify and interpret natural fractures. Seismic attribute analysis techniques such as curvature, coherence, ant body tracking and trace integration are used to identify and interpret large fractures (including seismic subresolution faults). The extension length, depth and occurrence (dip and dip angle) of large fractures are interpreted. The drilling and logging data are compared with the detected parameters to evaluate their reliability. The purpose of fault and natural fracture identification and interpretation is to construct a fault–fracture network system (high-speed fluid flow channel) for the platform and provide a reference for predicting and evaluating the risk of casing deformation.

(3) Changes in the casing size

When the shear load acts on the casing, the casing easily reaches its yield limit, and yield deformation occurs in the casing (Liu et al. 2016). The scale of the shear deformation of the casing increases rapidly when yield deformation occurs. The casing with a diameter–thickness ratio of 10.7 can bear a shear load of 40 MPa. In the geological environment of the Weirong shale gas reservoir, the diameter–thickness ratio of the Q125 steel casing should be less than 9.8.

13.5 Wellbore Stability Control for High-Stress Fractured Formations

The exploration and development of (ultra)deep fractured formations are fraught with a synergistic convergence of extreme geological and engineering challenges. These conditions not only amplify drilling risks but also fundamentally threaten the economic viability of such projects.

(1) Drilling and Wellbore Stability Challenges

Drilling in this environment is complicated primarily by a perilously narrow mud weight window. The high overburden stress and complex tectonic history create formations in which the pore pressure required to prevent wellbore collapse is only slightly lower than the fracture pressure, resulting in catastrophic loss of circulation. This delicate balance is exacerbated by the following:

High Temperature and High Pressure (HTHP): Temperatures exceeding 150 °C and pressures greater than 10,000 psi degrade conventional drilling fluids, making precise density and rheology control difficult. Furthermore, these conditions cause rock mechanics to behave in less predictable, often more ductile ways, complicating stress modelling around the wellbore.

Geomechanical Instability: The presence of multiscale discontinuous structures, such as natural fractures, faults, and bedding planes, in broken formations creates inherent weaknesses. When subjected to stress concentrations from drilling, these preexisting discontinuities are prone to reactivation, leading to frequent sloughing, caving, and borehole collapse. Attempts to suppress this collapse by increasing the mud weight often fracture weak formations, leading to a vicious cycle of wellbore instability and fluid loss.

(2) Completion and well cementation

The challenges extend beyond the drilling phase into well completion, where long-term integrity is paramount.

Cementing in Hostile Conditions: Achieving effective zonal isolation is a major hurdle. An HTHP environment can cause cement slurries to flash-set prematurely or suffer from strength retrogression over time. In zones of lost circulation, the cement slurry can escape into the formation instead of forming a hydraulic seal in the annulus, compromising well integrity and potentially allowing for dangerous cross-flow between zones.

Equipment and Hardware Installation: Running casings, liners, and other completion hardware through long, unstable sections of a wellbore is operationally complex and risky. The potential for becoming stuck is high, which can lead to costly fishing operations or even complete loss of the well.

(3) Economic and Safety Implications

Ultimately, these technical issues translate directly into significant economic and safety risks.

Economic Feasibility: Frequent drilling problems—lost circulation, stuck pipe, and wellbore collapse—result in substantial nonproductive time (NPT). With deep-water rig rates often exceeding $500,000 per day, even minor delays can cripple project economics. The need for specialized HTHP-rated equipment, advanced drilling fluids, and complex cementing systems further inflates the already high capital expenditure.

Safety Risks: A failure to manage the narrow mud weight window can have catastrophic consequences. An uncontrolled influx of formation fluids (a "kick") can escalate into a blowout, posing grave danger to personnel, the environment, and the asset itself. Similarly, poor zonal isolation can lead to long-term sustained casing pressure issues, creating a significant safety risk throughout the life of the well.

This study directly addresses these multi-faceted challenges by focusing on the fundamental geomechanical drivers of instability. By investigating the dynamic evolution of stresses around the wellbore and the mechanisms of rock deterioration at both the macro- and microscales and developing advanced methods for controlling wellbore stability, we aim to provide the engineering foundation needed to unlock these resources safely and economically.

13.5.1 Wellbore Instability Mechanism in Fractured Shale Formations

13.5.1.1 Physical and Chemical Characteristics of Fractured Shale Formation

A mineral composition analysis of 87 Longmaxi Formation shale samples from the Fuling shale gas field revealed that the Longmaxi shale is mainly composed of quartz and clay minerals and contains small amounts of feldspar, calcite, dolomite and pyrite (Table 13.2). The high contents of quartz and clay minerals indicate that the Longmaxi Formation shale is strongly heterogeneous. Additionally, its clay minerals mainly include an illite and montmorillonite mixed layer (I/S) and illite, which indicates a weak hydration capacity and poor expansion performance.

(1) Microstructure characteristics

SEM revealed a significant difference in the microstructure distribution between the parallel bedding direction and the vertical bedding plane direction of the Longmaxi shale. In the direction parallel to the bedding plane, minerals present a flat directional arrangement. Different types of nanopores, such as organic matter pores, skeleton mineral pores, and clay intergranular pores, are densely distributed on the bedding plane. Many microcracks can be seen in the direction perpendicular to the bedding plane, and the width of the cracks is 0.1–40 μm and is regularly distributed in parallel layers.

(2) Hydration characteristics

The rolling recovery rate of the rock cuttings is between 94 and 98%, indicating that the rock in this formation has poor dispersion and is not easily dispersed. Similarly, in the linear expansion rate test, the final expansion rate of the Longmaxi outcrop artificial core is 10.5% under the condition of clean water for 23 h, 8.4% under the condition of 5% KCl, 6.0% under the condition of the potassium-based anti-sloughing drilling fluid, and 2.5% under the condition of the oil-based drilling fluid, indicating that the hydration expansion performance of the Longmaxi outcrop cuttings is weak and that the oil-based drilling fluid system has a strong inhibition ability. The CEC of the Longmaxi shale is between 3.5 and 6.8 mmol/100 g, and the water sensitivity is weak.

Table 13.2 Mineral composition characteristics of Longmaxi shale

Number of samples	Mineral component content (%)								
	Quart	Feldspar	Calcite	Quartz + Feldspar + Calcite + Dolomite	Clay minerals				
					Total	Kaolinite	Illite	I/S	Chlorite
87	18~70.6/ 37.3	3.2~15.0/ 9.3	0~11.8/ 3.8	33.9~80.3/ 56.5	16.6~62.8/ 40.9	0~13/0.1	12~68/39.4	25~85/54.4	0~20/6

13.5.1.2 Micro Instability Mechanism

The water–rock damage to shale has significant anisotropic distribution characteristics (Fig. 13.12). Clay minerals are key factors in fracture propagation in fractured shale. Through a soaking experiment of tuff without clay minerals and shale rich in clay minerals in clean water, it was found that the structure of tuff remains unchanged after soaking in clean water for 60 min, whereas the fracture expansion of shale is obvious after soaking in clean water for 30 min. Owing to the poor expansibility of illite, the minerals on the fracture surface of the cracks that develop in illite tend to peel off, and the width of the cracks is small. The I/S mixed layer has strong expansibility, in which the fracture surface minerals with cracks are completely stripped, and the fracture width changes greatly.

Using a molecular dynamics model of water–clay interactions, the montmorillonite and I/S mixed layer show the characteristics of longitudinal expansion and lateral migration, and the lateral migration of I/S mixed layer is more significant. Moreover, water content is the main factor controlling clay expansion, and temperature and pressure have little effect on its expansion performance. Therefore, water control is the key to controlling the deterioration of water–rock interactions.

To elucidate the chemical inhibition mechanisms of different inhibitory cations on the hydration of clay minerals in shale, a micromolecular dynamics evaluation model of the cation inhibition of shale hydration was constructed. The predominant mechanisms through which inhibitory cations (K^+/NH_4^+/Cs^+/Ca^{2+}) inhibit hydration include controlling the charge repul-

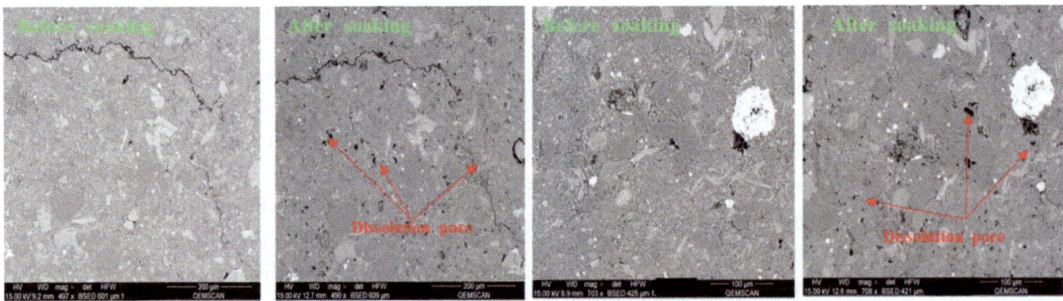

(a) Parallel to bedding plane: development of dissolution pores

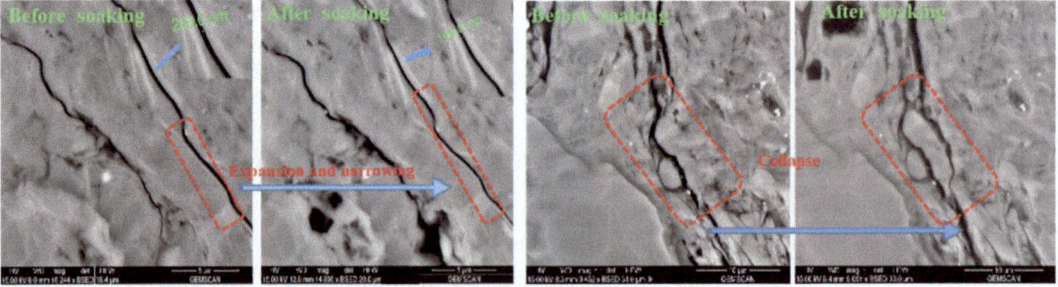

(b) Perpendicular to the bedding plane: crack initiation, narrowing and collapse

Fig. 13.12 Anisotropic distribution characteristics of water-induced rock damage: **a** parallel to the bedding plane: development of dissolution pores; **b** perpendicular to the bedding plane: crack initiation, narrowing and collapse

sion between layers, promoting decreased layer spacing, and reducing particle intrusion ability.

13.5.1.3 Macroscopic Instability Mechanism

The shale formation of the Longmaxi Formation has weak clay hydration and expansion capacity, a high content of brittle minerals and developed microcracks; thus, it is a typical hard brittle shale. The mechanism of wellbore instability caused by the action of fractured hard and brittle shale formations and drilling fluid is significantly different from that of traditional hydratable expansive shale formations. In general, the local high formation pressure and stress concentration around the wellbore caused by spontaneous imbibition of the shale gas reservoir and the uneven fracture of the formation and rock strength degradation are key to the macro wellbore instability.

13.5.2 Wellbore Instability Mechanism in Deep Dolomite Formations

13.5.2.1 Physical and Chemical Characteristics

A series of physical and chemical experimental analyses are carried out on realistic dolomite samples (Fig. 13.13). Notably, some of the samples are comprehensive and tight; some are characterized by dissolved pores, and the pores are filled with quartz and asphalts; in some samples, both dissolved pores and fissures are well developed, and the void space is filled with quartz and asphalts. The results of the XRD tests reveal that the dolomite content is above 91% and that the clay content is between 0.7% and 1.9%, indicating a prevailing dolomite content. The results show that in the tight dolomite samples, there is a powder and mud crystal structure, a laminar structure, and thin interbedding of powder and mud crystals. The results also present a small amount of algal traces, which are distributed in flower spots. Microcracks have developed in the rock samples, and some of them are filled with mud crystals and fine particles of protolith.

13.5.2.2 Mechanical Characteristics

In this analysis, the changes in rock mechanical properties before and after water–rock interaction are demonstrated. The confining stresses of the compression tests can reach 90 MPa, and the soaking time mimicking the water–rock interaction is 7 days in this analysis. Standard rock cores with a diameter of 2.5 cm and a length of 5 cm are used in the study. Both uniaxial and triaxial compression tests are involved. Based on the results, the compression test has a total of six stages: compaction, elasticity, yield, failure, soft-

Fig. 13.13 Various dolomite samples obtained from the field

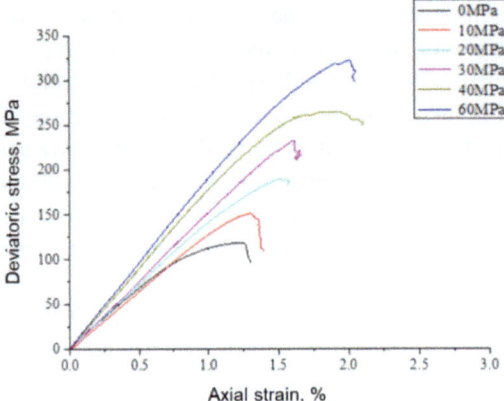

Fig. 13.14 Stress–strain curves associated with soaking

ening, and residual (Fig. 13.14). For this specific test, the peak strength during yield and failure can exceed 100 MPa. The longer the soaking time is, the lower the peak strength. After 7 days of soaking in clean water, the peak strength is the lowest.

13.5.2.3 Microscopic Deformation and Damage Mechanism

During the compression tests, pictures are taken at specific time steps to visualize the displacement changes during the increase in confining stress, which can be used to reveal the initiation and propagation of hydraulic fractures at the microscale. The displacement contours are strongly affected by changes in axial stress and mineral content. This is especially noticeable as large strains and large displacements occur primarily near the cementing of minerals. Cementing is weaker than the rock matrix and is more easily affected by external loading.

13.5.2.4 Instability Mechanisms

Using the experimental setup and an 8-mm artificial wall cylinder, increasing the confining pressure can result in changes in the wellbore shape and wellbore instability. After the wellbore in the dolomite becomes unstable, the wellbore shape changes from circular to elliptical, and the effective area of the wellbore increases after fluid immersion. Under high stress, the short axis of the wellbore instability ellipse changes little, with

the main changes concentrated on the long axis of the ellipse.

According to the analysis in this section, the three types of dolomite have high strength, reflecting obvious hard brittleness; the microcracks of the three types of dolomite contain argillaceous filling, and the dolomite with dissolved pores and fissures is the most obvious; dense dolomite has the fewest pores and the highest compressive strength; the stress–strain threshold of dolomite fracture with dissolved pores and cracks is the lowest; the compressive strength and microfracture conditions of dolomite with dissolved pores are between those of the other two dolomites; and under high stress, the formation of cemented interface fractures in dolomite is the key to instability failure.

13.5.3 Kinetics Modelling of Wellbore Instability

13.5.3.1 Stress Characteristics in the Borehole Wall in Fractured Formations

Because the drilling operation in the reservoir disturbs the original equilibrium state, the stress is redistributed around the wellbore, and obvious stress concentration occurs around the wellbore. The von Mises stress at the wellbore is the greatest, and the greater the radial distance from the wellbore is, the lower the von Mises stress, and the instability phenomenon is most likely to occur at and near the wellbore.

Under the pressure of the drilling fluid column, the borehole wall is subjected to tensile damage along the direction of the maximum horizontal principal stress. The variation diagram of the minimum principal stress in different directions around the well reveals obvious periodic characteristics. The maximum tensile stress occurs in the direction of the maximum horizontal principal stress. When the density of the drilling fluid is too high, the tensile stress in this direction first exceeds the tensile strength of the rock, and tensile failure occurs. Thus, selecting the appropriate density of drilling fluid is very important.

13.5.3.2 Kinetics of Wellbore Instability Modelling in Fractured Formations

To further explore the effects of high temperatures, stress concentrations, and physical water–rock reactions, a fluid–solid–thermal multifield coupling wellbore stability modelling method is proposed. In accordance with the law of conservation of mass, the first law of thermodynamics and the conservation of momentum, theoretical models representing the temporal and spatial evolution of the seepage field, temperature field and stress field are established. The governing equation of the coupled system can be given as follows:

$$\frac{\partial(\phi\rho_f)}{\partial t} + \nabla \cdot (\phi\rho_f u_{fs}) = \rho_f q_f \quad (13.26)$$

$$\rho c_r \frac{\partial(T)}{\partial t} - \nabla \cdot (\lambda \nabla T) = q_h \quad (13.27)$$

$$\nabla \cdot \sigma = 0 \quad (13.28)$$

where ϕ is the porosity of rock, ρ_f is the water density, t is time, q_f is the sink or source term, u_{fs} is the velocity of fluid molecules, ρ is the density of rock, c_r is the specific heat of the rock, T is temperature, λ is the rock thermal conductivity, q_h is the volumetric heat sink/source, and σ is the stress tensor. These equations describe the mass transfer, heat transfer, and momentum balance in a hydrocarbon medium that is porous and fully saturated.

Because the reservoir fluids are slightly compressible, an isothermal compressibility is used as follows:

$$c_f = \frac{1}{\rho_f} \frac{d\rho_f}{dp} \quad (13.29)$$

where p is the pressure. Thus, the relationship between the density and pressure is formulated. It depicts the compressibility of the reservoir fluids in deeply buried formations. It is important to couple the flow with the mechanical problem and to couple the flow with heat. Based on poromechanics and heat advection:

$$\rho c_r \frac{\partial(T)}{\partial t} - \nabla \cdot (\lambda \nabla T) + \rho_f c_f u_{fs} \nabla T = q_h \quad (13.30)$$

$$\frac{\partial(\phi\rho_f)}{\partial t} + \nabla \cdot [\phi\rho_f(u_s + u_{fs})] = \rho_f q_f \quad (13.31)$$

Thus, the thermal field, fluid flow field, and mechanical field are sequentially coupled within a system for the multiphysical simulation.

13.5.3.3 Prediction of Collapse Pressure in Fractured Formations

At a certain depth of the formation, the formation fracture pressure indicates the hydraulic pressure generated by the drilling fluid column to fracture the formation, opening and extending its original fractures and forming new fractures. The fracture pressure of the formation is closely related to the in situ stress. At present, there are two essentially different views on the cause of formation fracture. One is that the underground rock stratum is full of bedding, joints and fractures, and the fluid pressure in the well just invades along these weak surfaces and makes them open. Therefore, the fluid pressure that causes the fracture to open only needs to overcome the in situ stress perpendicular to the fracture surface. Second, the formation fracture depends on the stress concentration in the wellbore. Increasing the fluid pressure in the well will change the stress state on the borehole wall. When the tangential stress on the wall exceeds the tensile strength of the rock, the formation will fracture.

The fracture pressure of the formation and the direction of the fracture are affected and controlled by the stress on the borehole wall, in situ stress, formation characteristics, borehole trajectory and other related factors. In addition, if one needs to know the orientation of fracture extension, one should also know the main direction of horizontal in situ stress. The factors influencing fracture pressure can be divided into two categories. First are the characteristics of the formation itself, including in situ stress, the mechanical characteristics of the formation rock, formation bedding, and natural fracture occurrence, which are uncontrollable factors. Second, artificial factors, including well trajectory, well pressure, and drilling fluid performance, are considered. When different prediction models are used, the prediction results can be quite distinct because of various key assumptions.

13.5.4 Wellbore Stability Control for High-Stress Fractured Formations

13.5.4.1 Instability Control Method for Fractured Shale Formations

With respect to the wellbore instability mechanism of the Longmaxi Formation in the Fuling area, a fractured shale wellbore stability control method can be summarized as external prevention and internal treatment, which can also be described as preventing water phase invasion and internal treatment of hydration and expansion.

Preventing water phase invasion involves three main technical measures: (a) Reasonable drilling fluid density provides efficient stress support and maintains the wellbore stress balance. (b) Effective plugging and excellent mud cake performance can prevent water from invading the formation. (c) Efficient wetting reversal chemical agents should be used to modify the wettability of various pores and fractures to prevent water from invading the formation and to reduce spontaneous infiltration at the same time.

Internal treatment of hydration and expansion is performed mainly to control the hydration reaction after the fluid invades the shale, including two aspects of technical measures: (a) improving the ion concentration of the drilling fluid and effectively inhibiting osmotic hydration; and (b) using an efficient clay hydration inhibitor can effectively inhibit surface hydration.

Moreover, maintaining good lubricity of the drilling fluid is one of the key measures for maintaining wellbore stability by potentially avoiding the effects of the drilling tool.

13.5.4.2 Instability Control Method for High-Stress Deep Dolomite

To prevent instability control in high-stress deep dolomite formations, stress support through reasonable drilling fluid density should be provided to maintain the stress balance and weaken the rockburst tendency of the rock mass, preventing wellbore collapse. Moreover, the strength deterioration of rock caused by water–rock interactions, such as alkali solution corrosion, can be effectively controlled. Chemical means can effectively inhibit the damage to the composition and structure of dolomite formation after fluid intrusion and reduce the degree of deterioration of rock strength.

13.6 Hydraulic/Acid Fracturing for Tight Shale/Carbonate Reservoirs

Deep marine shale gas (3500–4500 m) and ultra-deep marine carbonate oil and gas ($\geq$ 6000 m) are among the most important development areas for China's future oil and gas resources. To address the problems of high fracture initiation pressure, poor reconstruction results, and low fracturing efficiency in current development, studies on the following key scientific issues, such as the interaction between artificial fractures in high-temperature and high-stress reservoirs and multiscale natural fractures/fractures and caves, have been conducted. These studies evaluated the following ① mechanism and characterization of rock fracture at high temperature and high stress; ② mechanism of the fracture network in deep shale formation under strong stress interference; and ③ mechanism of the initiation and propagation of acid fracturing fractures in ultra-deep carbonate fractured reservoirs. The results of these studies revealed the elastoplastic fracture mechanical properties of rocks under temperature and pressure and the characteristics of the local in situ stress evolution of artificial fracture propagation under fluid–solid–thermal coupling and the acid fracturing fractures and crack–karst cave–fracture interaction criteria in carbonate rocks in three dimensions. These findings can innovate the basic theory of safe and efficient fracturing in deep marine formations and provide scientific theories and key technical solutions for the efficient development of deep marine oil and gas reservoirs.

13.6.1 Rock Micromechanics of Laminated Shale

13.6.1.1 Micromechanical Properties by Grid Nanoindentation Tests

Mechanical heterogeneity is a major characteristic of organic-rich shale. The relationship between mechanical heterogeneity and formation in situ stress has seldom been addressed but is important for understanding hydraulic fracture propagation, wellbore stability, and hydrocarbon flow. Thus, the grid nanoindentation technique was used to characterize the heterogeneity of the mechanical properties of Longmaxi organic-rich shales from various burial depths and under various in situ stresses. Nanoindentation is a measurement that is conducted at the micrometer scale. A rigid indenter with a known hardness and geometry is pressed to a certain depth in the testing material. The Young's modulus and hardness of the material can be calculated from the loading and unloading curve within an indentation cycle. Grid nanoindentation testing was able to characterize the elastoplasticity of rock samples. Furthermore, a deconvolution method was applied to distinguish three mechanical phases with distinct hardnesses and Young's moduli. The indentation topography was imaged by SEM with a high resolution of 10 μm. According to the deconvolution results, the influence of confining stress on individual mechanical phases was determined. Our results confirm that both confining stress and minerology control the heterogeneity of organic-rich shale.

The grid nanoindentation was run through commercial Keysight Nano Indenter G200 equipment. In particular, the continuous stiffness measurement (CSM) mode was selected to continuously measure the mechanical properties with increasing indentation depth. The CSM is a mature technique mode in nanoindentation that applies a harmonic oscillator to an indenter. A sinusoidal load was applied with a constant frequency. The highest load in each load cycle is higher than that in the previous cycle. Therefore, the load and penetration depth increase. During each load cycle, the elastic modulus and hardness at that depth are obtained using the same calculation method as follows (Phani et al. 2020). Owing to the complex composition and strong heterogeneity of shale, the mechanical properties of distinct minerals vary greatly. A Berkovich diamond indenter was selected because of its high spring constant range and suitability for complex materials with a wide range of elastic moduli. An indenter was used with a maximum loading of 30 mN and a maximum effective penetration depth of 25 μm. Indenter loading was conducted at a constant strain rate of 0.05 s-1 and a frequency of 45 Hz. The ultimate indent depth was set to 3000 nm to enable multiscale measurements of minerals and mineral phases. The room temperature was maintained at 25 °C ($\pm$0.5 °C).

The indenter tip penetrates the sample surface until the indentation depth reaches a prescribed magnitude. With increasing indentation depth, the material deformation transforms from elastic deformation to plastic deformation. During indenter withdrawal, only elastic deformation recovers. The area enclosed by the loading, holding and unloading curves in the load–displacement curve represents the irreversible energy. Notably, the energy consists of not only the plastic deformation energy of the indentation point but also the fracture energy of the generated or extended cracks (Cheng et al. 2002). A grid nanoindentation scheme was proposed to obtain a statistically significant evaluation of the mechanical properties. The individual points were separated by 90 μm, which was adequate to eliminate interactions among individual points and reach the scale to obtain the average response of the composite (Ulm and Abousleiman 2006).

Three distinct shale samples were collected from the downhole cores of the lower Silurian Longmaxi Formation, Sichuan Basin, China. The vertical burial depths of N-213, Z-202 and N-

222 are 2502.0 m, 3858.3 m and 4278.7 m, respectively. Their mineral compositions indicate that the target samples are classical black shale containing many quartz and clay minerals. Samples N-213 and N-222 are composed mainly of clay minerals, whereas sample Z-202 is composed mainly of quartz minerals (Table 13.3).

13.6.1.2 Mechanical Heterogeneity of Shale Under in Situ Stress

Strong heterogeneity in the mechanical properties of organic-rich shale at the scale of the constituent minerals was evaluated and confirmed in this study. The heterogeneity of shale properties has been confirmed by previous work. The local mechanical behaviour is strongly heterogeneous at the microscale. Moreover, the fracture and deformation characteristics around indentation imprints verify the diverse mineral compositions and strong heterogeneity of organic-rich shales.

In addition, a nonuniform shift in the Young's modulus and the corresponding frequency of the individual phases are observed in N-213 and N-222. The compositions of the two samples are similar, but the burial depths are different. The deconvolution results reveal that the frequency of the first phase of N-222 is less than that of N-213, which indicates that high in situ stresses result in the hardening of soft minerals. The red area represents soft minerals, organic matter and clay, and the blue area represents other minerals. The yellow area shows the hardening effect of in situ stresses on soft minerals, which demonstrates that high in situ stress weakens the heterogeneity of the mechanical properties of the downhole shale.

At the mineral scale, local heterogeneity is controlled by both confining stress and mineralogy. There are obvious shifts in the Young's

modulus and hardness distribution at different confining stresses. The mineral types and components strongly affect the deformation behaviour. For example, the Young's modulus decreases with increasing amounts of ductile materials such as clay and kerogen. However, the increase in brittle minerals such as quartz will lead to an increase in the Young's modulus.

13.6.2 Experimental Mechanics of Hydraulic Fracturing in Shale

13.6.2.1 Experimental Equipment

The true triaxial pressure machine used for the experiments includes a true triaxial test frame, a triaxial hydraulic voltage source, an oil–water separator, an MTS pressurization and controller, and a data acquisition and processing system (Hou et al. 2019a, b).

13.6.2.2 Main Factors Controlling Hydraulic Fracture Morphology

Researchers have demonstrated that the hydraulic fracture geometry can be categorized into four types in shallow shale formations: simple fractures, fishbone-like fractures, fishbone-like fractures with fissure openings, and multilateral fishbone-like fracture networks (Hou et al. 2022a). However, in this study, only deflection and crossing occurred when the fractures encountered discontinuities, and the fracture geometry could be classified broadly into four types in the deep shale formations in our experiments: transverse fracture, natural fracture with bedding planes, transverse fracture with bedding planes, and transverse fracture with bedding planes and

Table 13.3 Mineral composition of three distinct core samples

Sample	Depth (m)	Quartz (wt%)	Clay (wt%)	Calcite (wt%)	Dolomite (wt%)	Feldspar (wt%)	Pyrite (wt%)	Others (wt%)
N-213	2502.0	31.2	39.3	8.5	7.5	10.4	3.1	/
Z-202	3858.3	48.1	31.7	2.3	0.2	9.3	1.2	7.1
N-222	4278.7	34.6	39.5	10.6	/	13.2	2.1	/

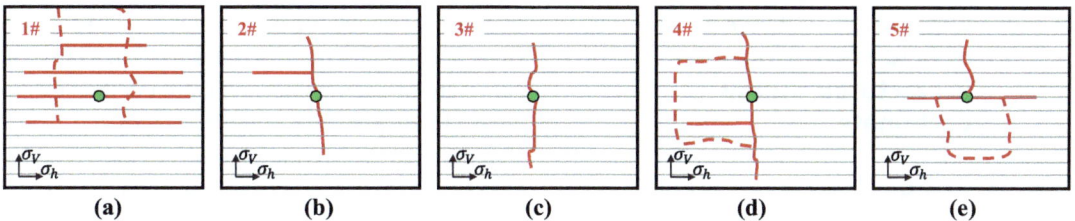

Fig. 13.15 Fracture morphology of each shale sample under different experimental conditions. The green points represent the initiation points, and the dashed area represents the fracture that is vertical to the direction of the maximum horizontal stress. (For interpretation of the references to colour in this figure legend, the reader is referred to the Web version of this article)

natural fractures. A transverse fracture is likely to form when the discontinuities are undeveloped or strongly bonded. Natural fractures with bedding planes are typically induced when discontinuities with weak bonding are located at or near the initiation point. Natural fractures are activated by fracturing fluid and propagate along their own direction. Some weakly cemented beddings are opened along the natural fracture propagation path. However, transverse fracture with bedding planes can be generated when hydraulic fractures encounter developed bedding planes. A high horizontal stress contrast facilitates a transverse fracture to extend but within a short distance and deflects into the developed beddings, propagating along the beddings. Under the two factors of a high horizontal stress contrast and developed beddings, the fracture geometry will ultimately have a sidestep shape. Transverse fractures with bedding planes and natural fractures occur under conditions in which the developed discontinuities interact with a crossing and deflecting hydraulic fracture, resulting in a complex fracture geometry. The above four kinds of fracture geometries have been found more frequently in deep shale formations than in shallow shale formations. Because of the deep burial, the presence of a few microfractures and the surrounding high horizontal stress present contrast. Therefore, approaches to facilitate fracture propagation and increase the SRV in deep shale formations should be emphasized in the future.

A complex fracture network is easily generated in a low horizontal stress contrast, similar to that in test 1 compared with test 2, as depicted in Fig. 13.15a, b. However, horizontal stress is a natural factor that cannot be changed or optimized. Only engineering factors such as the fracturing fluid viscosity and pump rate can be effectively designed to obtain a complex fracture network and increase the SRV in a deep shale formation, as illustrated in Fig. 13.15d, e. By applying a periodic variable pump rate of a low-viscosity fluid for fracturing, the low pump rate can activate and connect the bedding planes and natural fractures, whereas the high pump rate enhances the fracture length and produces new fractures, resulting in the formation of a large SRV.

AE testing is a valid and reliable method for determining where fractures initiate and how fractures propagate. However, the results of the AE tests in this study were not ideal; in some cases, no AE signal points were present, and in others, a few signals with positional deviation were observed. Adjusting the acoustic velocity and receiving frequency for good monitoring was ineffective. However, the existence of discontinuities and the properties of the shale could account for this. Numerous well-developed discontinuities intercept and stop the AE signals: discontinuities prevent signals from crossing beddings and natural fractures, and thus, it is better to attach the AE probe parallel to the beddings for better monitoring.

13.6.3 Experimental Mechanics of Acid Fracturing in Fractured-Vuggy Carbonate

13.6.3.1 Large-Scale True Triaxial Hydraulic Fracturing Experimental Equipment

Acid fracturing is also simulated by a true triaxial simulation experimental system. The experimental system includes the following components: a high-pressure alternating injection system, a triaxial pressure loading system, and a data collection system.

13.6.3.2 Analysis of Experimental Results

In carbonate reservoirs, vugs and fractures are common (Hou et al. 2021, 2022b; Dai et al. 2021). The interaction between artificial fractures and natural fractures has been widely studied. In the process of acid fracturing carbonate formations, the distribution of natural fractures and vugs is related to the final fracture complexity. The results of the comparative experiment reveal that natural fractures and vugs affect the propagation direction of fractures during acid fracturing (Fig. 13.16).

13.6.3.3 Posttreatment Method

During acid fracturing, the acid will corrode the carbonate reservoir, which is macroscopically evidenced by the roughness of the fracture surface. 3D scanning technology is a new quantitative

method to describe fracture surfaces. We select some acid-etched fracture surfaces. The Gaussian curvature of each point on the area is then calculated. To display the relative size of the Gaussian curvature, different colours are used: red for high Gaussian curvature and rougher surfaces and blue for low Gaussian curvature and smoother surfaces. The results are shown in Fig. 13.17. The etching of the fracture surface by the acid solution is uneven such that the red spot areas where the Gaussian curvature suddenly increases are distributed unevenly on the image. Analysis of the image revealed that the roughness of the continuous vug region on the right of and below the wellbore is significantly greater than that in the left region.

13.6.4 Fracture Conductivity of Acid Fracturing in Carbonate

13.6.4.1 Experimental Evaluation of the Conductivity of Acid-Etched Fracture

Fracture conductivity is the ability of fractures to pass through fluid. Too low of a conductivity means that the flow capacity in the fracture is less than the formation fluid supply capacity, and the production decreases; if the conductivity is too high, it means that although the fracture has sufficient flow capacity, but the formation fluid supply cannot keep up, resulting in unnecessary waste.

Fig. 13.16 Continuous vugs in the acid fracturing turning step

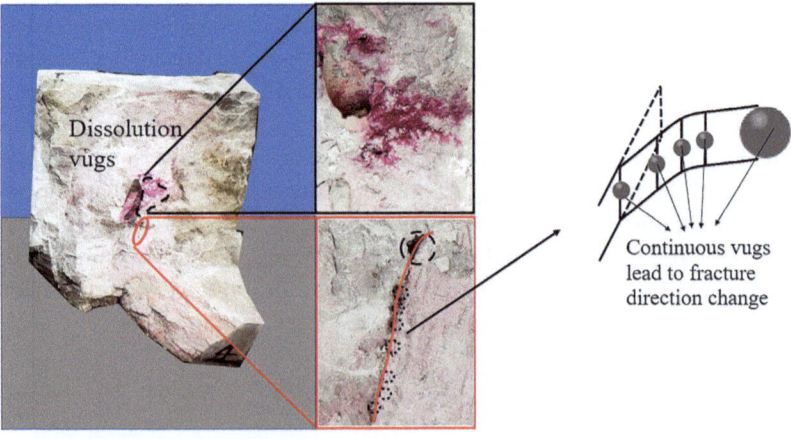

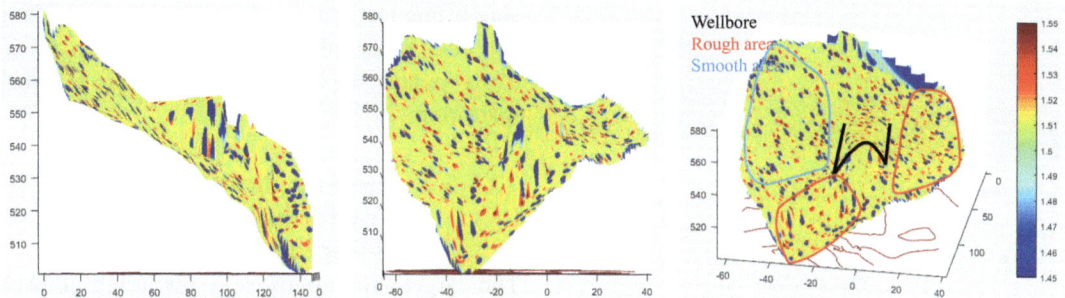

Fig. 13.17 Gaussian curvature of the selected area: **a** Left view; **b** front view; **c** stereogram. There are more red spots in the red area, indicating that the area has higher Gaussian curvature and a rougher surface. There are more blue dots in the blue area, indicating that the area has a lower Gaussian curvature and a smoother surface

An appropriate conductivity is very important for evaluating the economic benefits of fracturing. After fracturing, the stimulation effect and validation period are strongly related to the fracture conductivity.

13.6.4.2 Experimental Method

Fracture conductivity is tested mainly using laboratory short- or long-term experimental methods. In short-term conductivity experiments, the proppant sample is pressed step by step from small to large, and the flow rate and width of the propped fracture through the fracture support zone at each pressure level are measured to obtain the permeability and fracture conductivity of the support zone. Long-term conductivity experiments involve placing the proppant sample under a certain constant pressure and specified experimental conditions to investigate the degree of decrease in the conductivity with time.

13.6.4.3 Experimental Scheme and Results

The formula for calculating the permeability and conductivity of the proppant pack is as follows:

$$kW_f = \frac{5.555\mu \cdot Q}{\Delta P} \tag{13.32}$$

where kW_f is the propped fracture conductivity; Q is the flow rate, in cm^3/min; μ is the viscosity of the experimental fluid, in mPa s; and ΔP is pressure difference, in KPa.

The same ceramic particles with different particle sizes, 20/40 mesh and 30/60 mesh, are shown in Fig. 13.18. When the sand-laying concentration is 10 kg/m^2, the conductivity of ordinary ceramsite (20/40 mesh) is greater than that of CP ceramsite (30/60 mesh). The same is true when the sand-laying concentration is 5 kg/m^2. Even the long-term conductivity of proppant fractures is similar when the sand concentration of ordinary ceramsite (20/40 mesh) is 5 kg/m^2 and that of CP ceramsite (30/60 mesh) is 10 kg/m^2. Although the CP ceramsite (30/60 mesh) was of good quality, the larger proppant was more effective.

13.6.5 Field Test of Hydraulic/Acid Fracturing in Tight Oil Reservoirs

13.6.5.1 Field Application of Hydraulic Fracturing in Shale Gas Reservoirs

The vertical depth of deep shale gas well A in southeastern Sichuan is 3846.9–4055.31 m, the TOC content is 4.01%, the logging porosity is 4.4%, the Young's modulus is 39.2 GPa, the Poisson's ratio is 0.3, the reservoir temperature is 118 °C, and the formation pressure coefficient is 1.1–1.2. In this well, the formation dip angle is large, the well inclination angle is approximately 70°, and the interfracture interference is severe. Therefore, this well is moderately enlarged based on cluster spacing. According to the characteris-

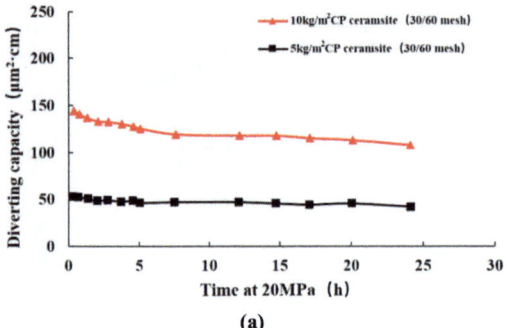

(a)

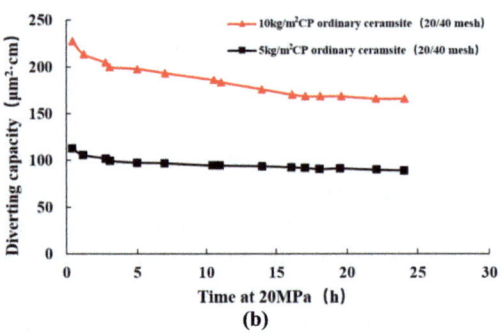

(b)

Fig. 13.18 Fracture conductivity test results: **a** 30/60 mesh ceramic particles; **b** 20/40 mesh ceramic particles

tics of fracture development, microstructure and in situ stress, the construction parameters are optimized. In the well section with a deep burial depth near target B, a preglue solution is adopted to ensure the perforation effect. During the entire process, high-viscosity drag reduction water is used to carry 40/70 mesh- and 30/50 mesh-coated sand. In the extrusion and loss section, a preglue solution is used to reduce the influence of filtration loss on fracturing, and "rapid extraction + large displacement (16–20 m³/min)" is used to reduce the influence of formation loss on fracturing.

The total amounts of fluid and sand in the actual fracture construction are 42,570 m³ and 2227 m³, respectively, and the coincidence rate of liquid and sand addition is 100%, which verifies the rationality of the description of the fracture geological environment and technological countermeasures of this well. Through the construction pressure curve, it is found that the pressure curve is mainly the "climbing type", which indicates that the net pressure of this well can be continuously increased by various net pres-

sure lifting methods, such as pumping rate lifting and temporary plugging technology, to effectively reduce the difficulty of fracture network formation.

13.6.5.2 Field Application of Acid Fracturing in Carbonate Reservoirs

The target stratum of acid fracturing in well TH121146 of the Tahe oilfield is the 6125.00–6194.00 m open hole section, and the lithology is light yellow–grey micritic limestone. The well is located at the intersection of faults, and the seismic time migration profile shows a typical "beaded" reflection. The formation pressure coefficient is 1.10, which is a normal pressure reservoir. The stratum temperature gradient is 2.26 °C/100 m. The reservoir oil in the well area is extremely heavy crude oil with high viscosity and high wax and sulfur contents.

Considering the relationship between well and storage, the complex acid pressing process of slickwater + high viscosity retarded acid (without thickening agent) + slickwater (carrying fibre) + fracturing fluid + high viscosity retarded acid was adopted. First, the formation pressure and reduced formation temperature of the slickwater are tested. Then, an uncrosslinked and low-viscosity acid solution is injected to pretreat the transformation section to form an acid etching channel and guide the subsequent liquid to create fractures in the direction of the favourable reservoir. The ability of acid slickwater to open natural fractures was also tested. After steering, gels are injected to create the main cracks along the guide channel, and then high-viscosity slow acid is injected to etch the cracks at depth to improve the fracture conductivity. After acid fracturing construction, the average production reached 28 T/d and remained stable for 3 months.

13.7 Summary

The pressure reconstruction of deep shale reservoirs after stimulation is presented in this chapter. Extensive experimental work has been performed to study the physical, chemical, and mechanical properties of fractured hard brittle shale and high-stress deep dolomite. This chapter also presents advances in the stimulated hydraulic/acid fracturing method for tight shale and carbonate reservoirs involving fundamental rock mechanics, triaxial fracturing tests, and field applications from the microscale to the macroscale.

Field applications have been conducted to verify the rock-breaking efficiency of PDC bits assisted by an axial–torsional-coupling percussive drilling tool, and a numerical model and experimental results are presented in this chapter.

Acknowledgements This work was funded by the National Natural Science Foundation of China (Grants U24B6001 and U19B6003).

References

Beugelsdijk LJL, De Pater CJ, Sato K (2000) Experimental hydraulic fracture propagation in a multi-fractured medium. In: SPE 59419 presented at the Asia Pacific conference on integrated modelling for asset management, Yokohama, Japan, 25–26 April 2000

Bing HOU, Mian CHEN, Zhimeng LI, Yonghui WANG, Ce DIAO (2014) Propagation area evaluation of hydraulic fracture networks in shale gas reservoirs. Pet Explor Dev 41(6):833–838. https://doi.org/10.1016/S1876-3804(14)60101-4

Biot MA (1939) Non-linear theory of elasticity and the linearized case for a body under initial stress. Phil Mag 27(183):468–489

Blanton TL (1982) An experimental study of interaction between hydraulically induced and pre-existing fracturing. In: Paper SPE/DOE 10847 presented at the SPE/DOE unconventional gas recovery symposium of the society of petroleum engineers, Pittsburgh, PA, 16–18 May 1982

Cheng YT, Li Z, Cheng CM (2002) Scaling relationships for indentation measurements. Philos Mag A 82(10):1821–1829

Dai Y, Hou B, Zhou C et al (2021) Interaction law between natural fractures-Vugs and acid-etched fracture during steering acid fracturing in carbonate reservoirs. Geofluids 2021(5):1–16

Daneshy AA (1974) Hydraulic fracture propagation in the presence of planes of weakness. Paper presented at the SPE European Spring Meeting, Amsterdam, Netherlands. https://doi.org/10.2118/4852-MS

Economides MJ, Nolte KG (1989) Reservoir stimulation. Old Tappan, NJ; Prentice Hall Inc (6):1–14

Hou B, Zhang R, Chen M et al (2019a) Investigation on acid fracturing treatment in limestone reservoir based on true tri-axial experiment. Fuel 235:473–484

Hou B, Chang Z, Fu W, Muhadasi Y, Chen M (2019b) Fracture initiation and propagation in a deep shale gas reservoir subject to an alternating-fluid-injection hydraulic-fracturing treatment. SPE J 24:1839–1855. https://doi.org/10.2118/195571-PA

Hou B, Dai Y, Zhou C et al (2021) Mechanism study on steering acid fracture initiation and propagation under different engineering geological conditions. Geomech Geophys Geo-Energy Geo-Resour 7(3):1–14

Hou B, Cui Z, Ding JH et al (2022a) Perforation optimization of layer-penetration fracturing for commingling gas production in coal measure strata. Petrol Sci

Hou B, Dai YF, Fan M et al (2022b) Numerical simulation of pores connection by acid fracturing based on phase-field method. Acta Petrolei Sinica 43(6):849–859

Jin Y, Chen KP (2019) Fundamental equations for primary fluid recovery from porous media. J Fluid Mech 860:300–317

Liu K, Gao DL, Wang YB et al (2016) Effects of local load on shale gas well casing deformation. Nat Gas Ind 36(11):76–82

Olson JE, Bahorich B, Holder J (2012) Examining hydraulic fracture—natural fracture interaction in hydrostone block experiments. Paper presented at the SPE Hydraulic Fracturing Technology Conference, The Woodlands, Texas, USA. https://doi.org/10.2118/152618-MS

Olsen N, Mandea M, Sabaka TJ, Tøffner-Clausen L (2009) CHAOS-2—a geomagnetic field model derived from one decade of continuous satellite data. Geophys J Int 179(3):1477–1487. https://doi.org/10.1111/j.1365-246X.2009.04386.x

Phani PS, Oliver WC, Pharr GM (2020) Understanding and modelling plasticity error during nanoindentation with continuous stiffness measurement. Mater Design 194:108923

Rickman R, Mullen M, Erik P, Bill G, Donald K (2008) A practical use of shale petrophysics for stimulation design optimization: all shale plays are not clones of the Barnett Shale. Paper presented at the SPE Annual Technical Conference and Exhibition, Denver, Colorado, USA. https://doi.org/10.2118/115258-MS

Saksala T, Gomon D, Hokka M et al (2014) Numerical and experimental study of percussive drilling with a triple-button bit on Kuru granite. Int J Impact Eng 72(4):56–66

Song H, Shi H, Ji Z et al (2019b) The percussive process and energy transfer efficiency of percussive drilling with consideration of rock damage. Int J Rock Mech Min 119:1–12

Song H, Shi H, Li G et al (2020) Numerical simulation of the energy transfer efficiency and rock damage in axial-torsional coupled percussive drilling. J Petrol Sci Eng 196:107675

Song B, Chen S, Zhong Q et al (2019) Pore pressure prediction based on elastic parameters inversion in shale oil reservoir. 81st EAGE conference and exhibition

Teufel LW, Clark JA (1984) Hydraulic fracture propagation in layered rock: experimental studies of fracture containment. SPE J 24:19–32. https://doi.org/10.2118/9878-PA

Tong HM, Chen ZL, Liu RX (2015) Generalized shear activation criterion. Chin J Nat 37(6):441–447

Ulm FJ, Abousleiman Y (2006) The nanogranular nature of shale. Acta Geotech 1(2):77–88

Wang L, Guo Y, Zhou J et al (2021) Rock mechanical characteristics of deep marine shale in southern China, a case study in Dingshan block. J Petrol Sci Eng 204:108699

Xiong C, Huang Z, Yang R et al (2020) Comparative analysis cutting characteristics of stinger PDC cutter and conventional PDC cutter. J Petrol Sci Eng 189:106792